General data and fundamental constants

Quantity	Symbol	Value	Power of ten	Units
Speed of light	c	2.997 924 58*	10^{8}	$m\ s^{-1}$
Elementary charge	e	1.602 176	10^{-19}	C
Faraday constant	$F = N_A e$	9.648 53	10^{4}	$C\ mol^{-1}$
Boltzmann constant	k	1.380 65	10^{-23}	$J\ K^{-1}$
Gas constant	$R = N_A k$	8.314 47		$J\ K^{-1}\ mol^{-1}$
		8.314 47	10^{-2}	$L\ bar\ K^{-1}\ mol^{-1}$
		8.205 74	10^{-2}	$L\ atm\ K^{-1}\ mol^{-1}$
		6.236 37	10	$L\ Torr\ K^{-1}\ mol^{-1}$
Planck constant	h	6.626 08	10^{-34}	J s
	$\hbar = h/2\pi$	1.054 57	10^{-34}	J s
Avogadro constant	N_A	6.022 14	10^{23}	mol^{-1}
Atomic mass unit	u	1.660 54	10^{-27}	kg
Mass				
electron	m_e	9.109 38	10^{-31}	kg
proton	m_p	1.672 62	10^{-27}	kg
neutron	m_n	1.674 93	10^{-27}	kg
Vacuum permittivity	$\varepsilon_0 = 1/c^2 \mu_0$	8.854 19	10^{-12}	$J^{-1}\ C^2\ m^{-1}$
	$4\pi\varepsilon_0$	1.112 65	10^{-10}	$J^{-1}\ C^2\ m^{-1}$
Vacuum permeability	μ_0	4π*	10^{-7}	$J\ s^2\ C^{-2}\ m^{-1}$ $(= T^2\ J^{-1}\ m^3)$
Magneton				
Bohr	$\mu_B = e\hbar/2m_e$	9.274 01	10^{-24}	$J\ T^{-1}$
nuclear	$\mu_N = e\hbar/2m_p$	5.050 78	10^{-27}	$J\ T^{-1}$
g value	g_e	2.002 32		
Bohr radius	$a_0 = 4\pi\varepsilon_0\hbar^2/m_e e^2$	5.291 77	10^{-11}	m
Fine-structure constant	$\alpha = \mu_0 e^2 c/2h$	7.297 35	10^{-3}	
	α^{-1}	1.370 36	10^{2}	
Second radiation constant	$c_2 = hc/k$	1.438 78	10^{-2}	m K
Stefan-Boltzmann constant	$\sigma = 2\pi^5 k^4/15h^3c^2$	5.670 51	10^{-8}	$W\ m^{-2}\ K^{-4}$
Rydberg constant	$R = m_e e^4/8h^3 c\varepsilon_0^2$	1.097 37	10^{5}	cm^{-1}
Standard acceleration of free fall	g	9.806 65*		$m\ s^{-2}$
Gravitational constant	G	6.673	10^{-11}	$N\ m^2\ kg^{-2}$

* Exact value

The Greek alphabet

Α, α	alpha	Ν, ν	nu
Β, β	beta	Ξ, ξ	xi
Γ, γ	gamma	Ο, ο	omicron
Δ, δ	delta	Π, π	pi
Ε, ε	epsilon	Ρ, ρ	rho
Ζ, ζ	zeta	Σ, σ	sigma
Η, η	eta	Τ, τ	tau
Θ, θ	theta	Υ, υ	upsilon
Ι, ι	iota	Φ, ϕ	phi
Κ, κ	kappa	Χ, χ	chi
Λ, λ	lambda	Ψ, ψ	psi
Μ, μ	mu	Ω, ω	omega

Physical Chemistry

Physical Chemistry

Seventh Edition

Peter Atkins
University of Oxford

Julio de Paula
Haverford College

W. H. Freeman and Company
New York

Cover design: Cambraia Fornseca Fernandes
Cover image: DNA: Scanning tunneling microscopic view (Science VU/IBMRL/Visuals Unlimited)

Library of Congress Cataloging-in-Publication Data

Atkins, P. W. (Peter William), 1940–
Physical chemistry / Peter Atkins and Julio DePaula, - 7th ed.
p. cm.
Includes bibliographical references and index.

1. Chemistry, Physical and theoretical. I. DePaula, Julio. II. Title.
QD453.3 .A74 2001
541.3–dc21 2001051228

ISBN 0-7167-6407-5 (EAN 9780716764076)
First printing

Preface for the student

Welcome to your study of physical chemistry, the subject that underlies all chemistry and increasingly biology, engineering, and medicine. This text will give you an introduction to the wide range of topics that constitute physical chemistry. It is unlikely that you will be expected to cover everything in the text, but in your later careers you may find it an invaluable reference to topics you did not cover as well as those you did.

Mathematical support

We have to be honest: physical chemistry lies at the mathematical end of the spectrum that constitutes modern chemistry. We have done our best, though, to make the mathematical content as accessible and interesting as possible.

- The text is heavily illustrated, to help you grasp the qualitative aspects of a concept pictorially.
- We have taken great care to explain why a particular mathematical step is being taken and, wherever appropriate, to interpret equations physically.
- You will find that many of the derivations are in the sections called *Justifications.*

These *Justifications* enable you to read through a topic without getting bogged down in too much detail initially, and then return to the derivation once you have grasped the big picture. Because the *Justifications* are set off from the main text, they are in a convenient form for revision for examinations. Some of you might be advised to skip them altogether: take your instructor's advice. They are there to make the text authoritative and complete as a reference, and to be used according to the level of the course you are taking.

- There are short marginal notes designed to remind you of the mathematics being invoked.

The background to these commentaries will be found in the *Further information* section at the back of the book.

Problem solving support

We have tried to provide a lot of problem-solving practice material.

- There are numerous worked *Examples* throughout the text.
- Each one has a *Method* section which is designed to advise you how to approach the problem and collect the information you need.
- They all end with a *Self-test* so you can test yourself immediately after working through the example.

You will also find many *Illustrations*. These are short worked examples, showing in a succinct way how a concept or equation is used. They too have *Self-tests*.

The list of *Key ideas* at the end of each chapter will be particularly useful for review. We have collected there all the newly defined terms from the chapter, so the section acts as a glossary of important concepts. We have included the most important equations too. However, we do not recommend that you learn off these equations by rote: it is far better in almost every case to know how to derive an equation than simply to memorize it.

Media support

We draw your attention to the web site for this text at **www.whfreeman.com/pchem7**. You will find a lot of additional information there, such as pointers to sources of data and further information about selected topics. One of the most important components of the site is the section we call *Living graphs*. These are dynamic versions of the graphs marked with the icon in the text, and will enable you to explore what happens as you change the parameters in an equation. Richer versions, including three-dimensional displays of some concepts and equations, will be found in the Mathcad® worksheets that we have provided on the CD-ROM *Explorations in physical chemistry*. Another feature of the site is that it makes available chapters of the *Student's solutions manual* for on-line sale, so if you don't have the printed version and are desperate for help, you can purchase the relevant chapter.

Physical chemistry is a demanding but richly rewarding subject. It underlies all chemistry, and is becoming increasingly important as the investigative and computational aspects of chemistry become more sophisticated. We wish you well with your studies, and hope that you find this text rewarding.

Oxford P. W. A.
Haverford J. de P.

Preface for the instructor

As has been the habit of the text, this edition has been substantially revised from its predecessor. We have aimed to make the text more flexible and to render it more accessible while retaining its authority and scope. At the same time we have sought to modernize the content and take into account the electronic aids that have become an essential part of teaching. The most significant new feature of this edition is:

- A new co-author. This edition welcomes Julio de Paula, Haverford College, to the book.

His research interests are in the general field of biophysical chemistry, and you will see his influence on raising the profile of that aspect of physical chemistry throughout the text. Together we have developed a modern text that we believe is well suited to the 21st century. One outcome is that:

- We have included many new topics, relating not only to biophysical chemistry but also to materials science, the environment, spectroscopy, engineering, and astrophysics.

We have also taken the opportunity to check all the exposition and equations thoroughly (see the *Acknowledgements*).

Overall structure of the text

The text is still divided into three parts, but material has been reallocated between chapters and the chapters have been reorganized. There is a continuing argument in the physical chemistry community about the choice of a 'quantum first' or a 'thermodynamics first' approach. In response:

- A strategic aim of this revision is to make it possible to work through the text in a variety of orders.

To help the instructor navigate, we have suggested two sequences at the end of this Preface.

- Another central aim is to enhance the treatments of modern applications of physical chemistry by providing numerous *Boxes*.

We have noted the attention that biophysical chemistry is now receiving, and many of the Boxes deal with these topics. However, physical chemistry is not just a subject underpinning biology: it is essential to chemical engineering, materials science, environmental science, and to the rest of chemistry. Therefore, we have also paid attention to applications in these areas. The use of *Boxes* to hold these applications away from the main exposition in the text itself is a part of our strategy to make the book flexible. However, to facilitate the incorporation of this material in courses:

- Each box has two *Discussion questions*, so the material can be assigned and tested, and suggestions for *Further reading*.

Mathematical support

The level of mathematical competence of students is of perennial but increasing concern. There is no denying that physical chemistry is a mathematical subject, but the manner in which the mathematics is handled is a crucial concern. We have adopted a number of strategies to help you and your students.

- The mathematics is motivated and interpreted.

The text has always been attractive to users for the way in which mathematics is motivated and interpreted physically.

- As in the 6th edition, we have continued the practice of putting key derivations into *Justifications*.

The idea here is that a student can gain an overall picture of the subject before grappling with the derivations; the more recondite derivations can be ignored but are present for reference; the *Justifications* are readily available for review or assignment in lectures. In short, they are a component of the flexibility of the text, for their judicious use is a way of tailoring the level of the text to the needs of the course.

- We use marginal notes to remind the student of a key mathematical topic that is being invoked in a derivation.

The background to these mathematical comments will be found in the revised and extended *Further information* section at the end of the text.

Order of topics

Users of earlier editions will see that there have been changes in the order of topics: these changes have been made in response to suggestions from users and to achieve a better logical and pedagogical flow. The kinetic theory of gases is now treated in a single place, where molecular motion is introduced (Chapter 24). A new chapter on the solid state (Chapter 23) collects a lot of material into one place which previously was dispersed and has added a discussion of mechanical and electrical properties, to act as a bridge to courses on material science. The old and rarely used chapter on electric and magnetic properties has been moved and repackaged as a chapter on molecular interactions (Chapter 21). The most notable additions to other parts of the text are the strengthening of quantum theory, the introduction of time-dependent perturbation theory as a way of unifying the presentation of spectroscopy, and the inclusion of a discussion of the mechanism of electron transfer reactions (Marcus theory). We have streamlined some of the more classical aspects of thermodynamics and its applications.

Problem solving support

We have paid special attention to providing help and practice with problem solving.

- Each chapter now ends with a glossary of *Key ideas* and key equations.

We hope this innovation will be regarded as a study aid rather than as an aid to rote memorization.

- Each chapter has a large number of structured worked *Examples*, each accompanied by a *Self-test*.
- There are also numerous *Illustrations*, which are short, unstructured examples of how a concept or equation is used.

All these elements—the *Examples*, *Illustrations*, and *Self-tests*—are numbered, so they can be assigned easily.

- As always, there are numerous end-of-chapter *Exercises* and more demanding *Problems*.

The innovations in this edition are that:

- The *Exercise* section begins with a selection of *Discussion Questions*, designed to promote discussion and deeper understanding of the material.
- The *Problems* have also been extended by the inclusion of a new category: *Applications*.

To help with making assignments, solutions to the 'a' *Exercises* and the *Problems* with blue numbers are given in full in the *Student's solutions manual*; solutions to the 'b' *Exercises* and the *Problems* with black numbers are given in the *Instructor's solutions manual*. Both manuals have been prepared with great care by Charles Trapp with help from Carmen Giunta and Marshall Cady. Chapters from the former can be purchased on-line from the book's web site (see below).

Media support

Electronic media are now an intrinsic part of pedagogy. This revision has a closely integrated *web site* at **www.whfreeman.com/pchem7**. On this site you will find:

- All the line-art drawings from the text, for use in lectures.
- *Living graphs*: dynamic versions of all the graphs in the text that are marked with a 'Living graph' icon ( ).
- Web addresses for sources of data and other links.
- Chapters of the *Student's solutions manual* for purchase on-line.

Marginal notes throughout the text alert the student to useful links to be found at the text's web site. In short, the web site is intended to be a closely integrated utility to enrich the text and to provide a dynamic resource.

We have also prepared a CD-ROM called *Explorations in physical chemistry*, which contains about 75 Mathcad® worksheets which extend the concept of Living Graphs and provide a great deal of additional information and practice.

Oxford P. W. A.
Haverford J. de P.

Traditional approach

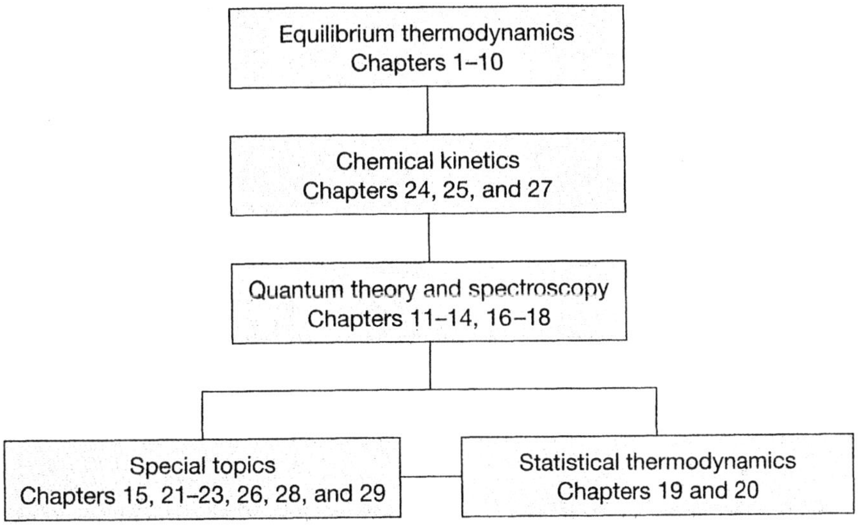

Molecular approach

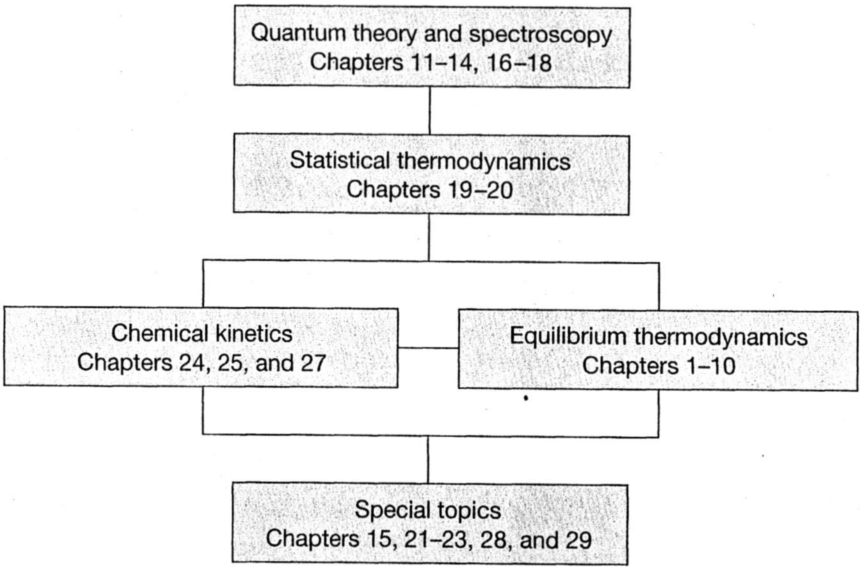

Acknowledgements

A book as extensive as this could not have been written without significant input from many individuals. We would like to reiterate our thanks to the hundreds of people who contributed to the first six editions. In addition, we have many new and old friends to thank for their contribution to this edition. First among them is Valerie Walters, Haverford College, who read the whole set of proofs through in great detail, and made many valuable suggestions. Our warm thanks go Charles Trapp, who has produced many editions of the *Solutions manuals* over the years, and has made numerous useful remarks. Carmen Giunta, one of his collaborators on the solutions manuals, also provided us with thoughtful input. Hideaki Chihara, Osaka, who has translated many editions into Japanese, has been particularly assiduous in sending us comments.

Many people gave their advice based on the sixth edition, and others reviewed the draft chapters as they emerged. We therefore wish to thank the following colleagues most warmly:

Warren Beck, Michigan State University
Dor Ben-Amotz, Purdue University
Isaac Bersuker, The University of Texas at Austin
Michael Blandamer, University of Leicester
Roland Böhmer, University of Mainz
Eric Borguet, University of Pittsburgh
Peter Budd, University of Manchester
David Cadenhead, University of Buffalo
Mark Dadmun, The University of Tennessee
Robert Dunbar, Case Western Reserve University
Thomas Engel, University of Washington
Paul Fletcher, University of Hull
Deborah Glover-Fischer, North Carolina Wesleyan College
James Goll, West Virginia University
Edward Grant, Purdue University
Susan Green, St. Olaf College
Arthur Halpern, Indiana State University
Judith Howard, University of Durham
Thomas J. Murphy, University of Maryland
Graham Jackson, University of Cape Town
Evguenii Kozliak, University of North Dakota
Matthew Kubasik, Fairfield University
Barbara Lawrence, Eastern Illinois University
Lawrence Lohr, University of Michigan
Peter Lykos, Illinois Institute of Technology
David Metcalf, University of Virginia
Gerald Morine, Bemidji State University
Gareth Morris, University of Manchester
Richard Moss, University of Southampton
Thomas Murphy, University of Maryland
Dwight Myers, East Central University
Breandán Ò Cochláin, National University of Ireland, Galway
John Parker, Heriot-Watt University
Dvora Perahia, Clemson University
Sanford Safron, Florida State University
Hans Sillescu, University of Mainz
Kathy Singfield, Saint Mary's University
Durwin Striplin, Davidson College
Göran Sundholm, Helsinki University of Technology
Peter Thistlethwaite, University of Melbourne
Alexander van Hook, University of Tennessee, Knoxville
Michael Wagner, George Washington University
Robert Wyatt, University of Texas at Austin
Nancy Xu, Old Dominion University
Lucy Ziurys, University of Arizona

We would also like to thank our two publishers, Oxford University Press and W.H. Freeman & Co., for their constant encouragement, advice, and assistance. Authors could not wish for a more congenial publishing environment.

Copyright acknowledgements

The permission of the respective copyright holders to reproduce tables of data and extracts from them is gratefully acknowledged as follows: Butterworths (28.1, 28.2 [© 1964]), Professor J. G. Calvert (17.1 [© 1966]), Dr J. Emsley (13.4 [© 1989]), Chapman and Hall (16.3 [© 1975], 2.5, 9.1 [© 1986]), CRC Press (1.4, 9.1, 10.2, 10.7, 14.3, 22.1 [© 1979]), Dover Publications Inc. (12.2 [© 1965]), Elsevier Scientific Publishing Company (22.1 [© 1978]), McGraw Hill Book Company (1.4, 1.5, 2.3, 3.1, 3.7, 4.1, 13.4, 16.2, 22.2, 24.3, 24.7 [© 1975], 5.2 [© 1961], 25.1, 25.2 [© 1965], 3.7 [© 1968]), National Bureau of Standards (2.5, 2.6 [© 1982]), Dr J. Nicholas (25.1, 25.2, 25.4 [© 1976]), Oxford University Press (25.1, 25.2, 25.4 [© 1975], 28.2, 28.3, 28.4 [© 1974]), Pergamon Press Ltd (22.3 [© 1961]), Prentice Hall Inc. (23.1, 23.3 [© 1971]), John Wiley and Sons Inc. (7.1 [© 1975], 21.6 [© 1954], 22.25 [© 1961]. The sources of the material are quoted at the foot of each table.

The permission of the following individuals is gratefully acknowledged: NMR spectroscopy, John Wiley and Sons Ltd, Professor H. Günther (18.7), Professor G. Ehrlich (28.29), Dr J. Evans (21.17), Dr A. J. Forty (28.4), Dr J. Foster (illustration in Box 12.1), Professor H. Ibach (28.9), Professor M. Karplus (27.23, 27.24), Professor D. A. King (28.26), Professor G. Morris (18.32, 18.46), Dr A. H. Narten (21.27), Oxford University Press (28.4, 28.5), Professor C. F. Quaite (illustration in Box 12.1), Professor M. W. Roberts (28.8, 28.11), Dr H. M. Rosenberg (28.5), Professor G. A. Somorjai (28.14, 28.17), Dr A. Stevens (23.17), Professor A. H. Zewail (illustration in Box 27.1).

Summary of contents

Contents

List of boxes

Conventions

Names of quantities

A **substance** is a distinct, pure form of matter.

The **amount of substance**, n (more colloquially 'number of moles' or 'chemical amount'), in a sample is reported in terms of the **mole** (mol): 1 mol is the amount of substance that contains as many objects (atoms, molecules, ions, or other specified entities) as there are atoms in exactly 12 g of carbon-12. This number is found experimentally to be approximately 6.02×10^{23} (see the endpapers for more precise values). If a sample contains N entities, the amount of substance it contains is $n = N/N_A$, where N_A is the Avogadro constant: $N_A = 6.02 \times 10^{23}\ \text{mol}^{-1}$. Note that N_A is a quantity with units, not a pure number.

An **extensive property** is a property that depends on the amount of substance in the sample. Two examples are mass and volume. An **intensive property** is a property that is independent of the amount of substance in the sample. Examples are temperature, mass density (mass divided by volume), and pressure.

A **molar property**, X_m, is the value of an extensive property, X, of the sample divided by the amount of substance present in the sample: $X_m = X/n$. A molar property is intensive. An example is the **molar volume**, V_m, the volume of a sample divided by the amount of substance in the sample (the volume per mole). The one exception to the notation X_m is the **molar mass**, which is denoted M. The molar mass of an element is the mass per mole of its atoms. The molar mass of a molecular compound is the mass per mole of molecules, and the molar mass of an ionic compound is the mass per mole of formula units. A **formula unit** of an ionic compound is an assembly of ions corresponding to the chemical formula of the compound; so the formula unit NaCl consists of one Na^+ ion and one Cl^- ion. The names *atomic weight* and *molecular weight* are still widely used in place of molar mass (often with the units omitted), but we shall not use them in this text.

The **molar concentration** ('molarity') of a solute in a solution is the amount of substance of the solute divided by the volume of the solution. Molar concentration is usually expressed in moles per litre (mol L^{-1} or mol dm^{-3}; 1 L is identical to 1 dm^3). A solution in which the molar concentration of the solute is 1 mol L^{-1} is prepared by dissolving 1 mol of the solute in sufficient solvent to prepare 1 L of solution. Such a solution is widely called a '1 molar' solution and denoted 1 M. The term **molality** refers to the amount of substance of solute divided by the mass of solvent used to prepare the solution. Its units are typically moles of solute per kilogram of solvent (mol kg^{-1}).

Notational conventions

We use SI units and IUPAC conventions throughout, except in a small number of cases. These conventions are specified in the 'green book', *Quantities, units, and symbols in physical chemistry*, Blackwell Scientific, Oxford (1993).

The default numbering of equations is ($C.n$), where C is the chapter; however, [$C.n$] is used to denote a definition and {$C.n$} is used to indicate that a variable x should be interpreted as $x/x^{\ominus}$, where $x^{\ominus}$ is a standard value. Cross-references within a chapter are denoted simply eqn n. A subscript r attached to an equation number indicates that the equation is valid only for a reversible change. A superscript ° indicates that the equation is valid for an ideal system, such as a perfect gas or an ideal solution.

The symbol $p^{\ominus}$ denotes 1 bar (10^5 Pa) exactly, and $b^{\ominus}$ denotes 1 mol kg^{-1} exactly.

When referring to temperature, T denotes a thermodynamic temperature (for instance, on the Kelvin scale) and θ a temperature on the Celsius scale.

For numerical calculations, unless otherwise specified, assume that zeros in data like 10, 100, 1000, etc. are significant (that is, interpret such data as 10., 100., 100., etc.).

Part 1
Equilibrium

Part 1 of the text develops the concepts that are needed for the discussion of equilibria in chemistry. Equilibria include physical change, such as fusion and vaporization, and chemical change, including electrochemistry. The discussion is in terms of thermodynamics, and particularly in terms of enthalpy and entropy. We see that we can obtain a unified view of equilibrium and the direction of spontaneous change in terms of the chemical potentials of substances. The chapters in Part 1 deal with the bulk properties of matter; those of Part 2 will show how these properties stem from the behaviour of individual atoms.

1

The properties of gases

This chapter establishes the properties of gases that will be used throughout the text. It begins with an account of an idealized version of a gas, a perfect gas, and shows how its equation of state may be assembled experimentally. We then see how the properties of real gases differ from those of a perfect gas, and construct an equation of state that describes their properties.

The simplest state of matter is a **gas**, a form of matter that fills any container it occupies. Initially we consider only pure gases, but later in the chapter we see that the same ideas and equations apply to mixtures of gases too.

The perfect gas

We shall find it helpful to picture a gas as a collection of molecules (or atoms) in continuous random motion, with speeds that increase as the temperature is raised. A gas differs from a liquid in that, except during collisions, the molecules of a gas are widely separated from one another and move in paths that are largely unaffected by intermolecular forces.

1.1 The states of gases

The **physical state** of a sample of a substance, its physical condition, is defined by its physical properties. Two samples of a substance that have the same physical properties are in the same state. The state of a pure gas, for example, is specified by giving its volume, V, amount of substance (number of moles), n, pressure, p, and temperature, T. However, it has been established experimentally that it is sufficient to specify only three of these variables, for then the fourth variable is fixed. That is, it is an experimental fact that each substance is described by an **equation of state**, an equation that interrelates these four variables.

The general form of an equation of state is

$$p = f(T, V, n) \tag{1.1}$$

This equation tells us that, if we know the values of n, T, and V for a particular substance, then the pressure has a fixed value. Each substance is described by its own equation of state, but we know the explicit form of the equation in only a few special cases. One very important example is the equation of state of a 'perfect gas', which has the form

$$p = \frac{nRT}{V} \tag{1.2}$$

where R is a constant. Much of the rest of this chapter will examine the origin of this equation of state and its applications.

(a) Pressure

Pressure is defined as force divided by the area to which the force is applied. The greater the force acting on a given area, the greater the pressure. The origin of the force exerted by a gas is the incessant battering of the molecules on the walls of its container. The collisions are so numerous that they exert an effectively steady force, which is experienced as a steady pressure.

The SI unit of pressure, the *pascal* (Pa), is defined as 1 newton per metre-squared:

$$1\ \text{Pa} = 1\ \text{N m}^{-2} \qquad [1.3a]$$

In terms of base units,

$$1\ \text{Pa} = 1\ \text{kg m}^{-1}\ \text{s}^{-2} \qquad [1.3b]$$

Several other units are still widely used (Table 1.1). A pressure of 10^5 Pa (1 bar) is the **standard pressure** for reporting data; we denote it $p^{\ominus}$.

Example 1.1 Calculating pressure

Suppose Isaac Newton weighed 65 kg. Calculate the pressure he exerted on the ground when wearing (a) boots with soles of total area 250 cm^2 in contact with the ground, (b) ice skates, of total area 2.0 cm^2.

Method Pressure is force divided by area ($p = F/A$), so the calculation depends on being able to calculate the force, F, that Newton exerts on the ground, and then to divide the force by the area, A, over which the force is exerted. To calculate the force, we need to know (from elementary physics) that the downward force an object of mass m exerts at the surface of the Earth is $F = mg$, where g is the acceleration of free fall, 9.81 m s^{-2}. Note that $1\ \text{N} = 1\ \text{kg m s}^{-2}$ and $1\ \text{Pa} = 1\ \text{N m}^{-2}$. Convert the areas to square metres by using $1\ \text{cm}^2 = 10^{-4}\ \text{m}^2$. To reduce rounding errors, leave the evaluation of numerical expressions to the end of the calculation.

Table 1.1 Pressure units

Name	Symbol	Value
pascal	1 Pa	$1\ \text{N m}^{-2}$, $1\ \text{kg m}^{-1}\ \text{s}^{-2}$
bar	1 bar	10^5 Pa
atmosphere	1 atm	101 325 Pa
torr	1 Torr	(101 325/760) Pa = 133.32 ... Pa
millimetre of mercury	1 mmHg	133.322 ... Pa
pound per square inch	1 psi	6.894 757 ... kPa

Answer The force exerted by Newton is

$$F = (65\ \text{kg}) \times (9.81\ \text{m s}^{-2}) = 65 \times 9.81\ \text{kg m s}^{-2} = 65 \times 9.81\ \text{N}$$

(This force works out to 6.4×10^2 N.) The force is the same whatever Newton's footwear. The areas 250 cm^2 and 2.0 cm^2 correspond to $2.50 \times 10^{-2}\ \text{m}^2$ and $2.0 \times 10^{-4}\ \text{m}^2$, respectively. The pressure he exerts in each case is:

$$\text{(a)}\ p = \frac{65 \times 9.81\ \text{N}}{2.50 \times 10^{-2}\ \text{m}^2} = 2.6 \times 10^4\ \text{N m}^{-2} = 2.6 \times 10^4\ \text{Pa (26 kPa)}$$

$$\text{(b)}\ p = \frac{65 \times 9.81\ \text{N}}{2.0 \times 10^{-4}\ \text{m}^2} = 3.2 \times 10^6\ \text{Pa (3.2 MPa)}$$

Comment One Newton exerts a force of much more than 1 newton. A pressure of 26 kPa corresponds to 0.26 atm and a pressure of 3.2 MPa corresponds to 32 atm.

Self-test 1.1 Calculate the pressure exerted by a mass of 1.0 kg pressing through the point of a pin of area $1.0 \times 10^{-2}\ \text{mm}^2$ at the surface of the Earth.

[0.98 GPa, 9.7×10^3 atm]

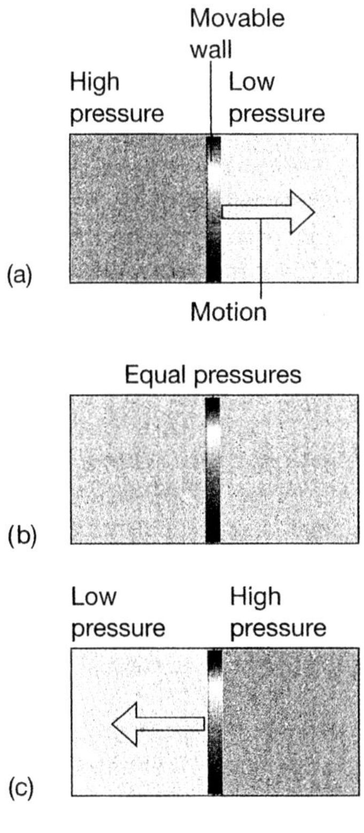

1.1 When a region of high pressure is separated from a region of low pressure by a movable wall, the wall will be pushed into one region or the other, as in (a) and (c). However, if the two pressures are identical, the wall will not move (b). The latter condition is one of mechanical equilibrium between the two regions.

If two gases are in separate containers that share a common movable wall (Fig. 1.1), the gas that has the higher pressure will tend to compress (reduce the volume of) the gas that has lower pressure. The pressure of the high-pressure gas will fall as it expands and that of the low-pressure gas will rise as it is compressed. There will come a stage when the two pressures are equal and the wall has no further tendency to move. This condition of equality of pressure on either side of a movable wall (a 'piston') is a state of **mechanical equilibrium** between the two gases. The pressure of a gas is therefore an indication of whether a container that contains the gas will be in mechanical equilibrium with another gas with which it shares a movable wall.

(b) The measurement of pressure

The pressure exerted by the atmosphere is measured with a **barometer**. The original version of a barometer (which was invented by Torricelli, a student of Galileo) was an inverted tube of mercury sealed at the upper end. When the column of mercury is in mechanical equilibrium with the atmosphere, the pressure at its base is equal to that exerted by the atmosphere. It follows that the height of the mercury column is proportional to the external pressure.

Example 1.2 Calculating the pressure exerted by a column of liquid

Derive an equation for the pressure at the base of a column of liquid of mass density ρ (rho) and height h at the surface of the Earth.

Method As in Example 1.1, $p = F/A$ and $F = mg$, so we need to know the mass of a column of liquid. The mass of a column of liquid is its mass density, ρ, multiplied by its volume, V. So, the first step is to calculate the volume of a cylindrical column of liquid.

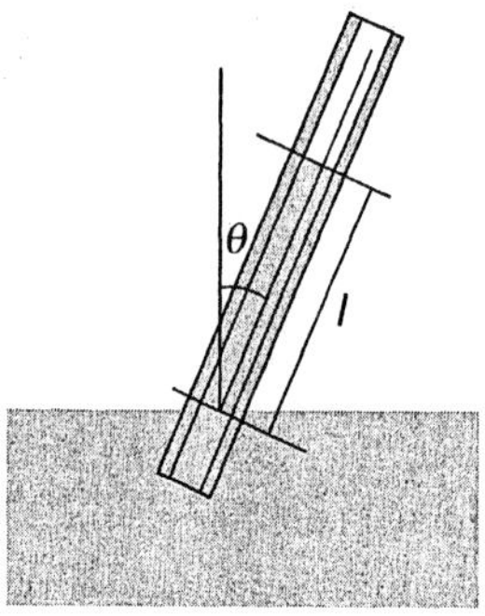

1

Answer Let the column have cross-sectional area A; then its volume is Ah and its mass is $m = \rho Ah$. The force the column of this mass exerts at its base is

$$F = mg = \rho Ahg$$

The pressure at the base of the column is therefore

$$p = \frac{F}{A} = \frac{\rho Ahg}{A} = \rho gh$$

Comment Note that the pressure is independent of the shape and cross-sectional area of the column. The mass of the column increases as the area, but so does the area on which the force acts, so the two cancel.

Self-test 1.2 Derive an expression for the pressure at the base of a column of liquid of length l held at an angle θ (theta) to the vertical (1). [$p = \rho gl \cos \theta$]

The pressure of a sample of gas inside a container is measured with a **manometer** (Fig. 1.2). In its simplest form, a manometer is a U-tube filled with a liquid of low volatility (such as silicone oil). The pressure, p, of the gaseous sample balances the pressure exerted by the column of liquid, which is equal to ρgh if the column is of height h (see Example 1.2), plus the external pressure, p_{ex}, if one tube is open to the atmosphere:

$$p = p_{ex} + \rho gh \tag{1.4}$$

It follows that the pressure of the sample can be obtained by measuring the height of the column and noting the external pressure. More sophisticated techniques are used at lower pressures. Methods that avoid the complication of having to account for the vapour from the manometer fluid are also available. These superior methods include monitoring the deflection of a diaphragm, either mechanically or electrically, or monitoring the change in a pressure-sensitive electrical property. These techniques have displaced the use of manometers in laboratory experiments.

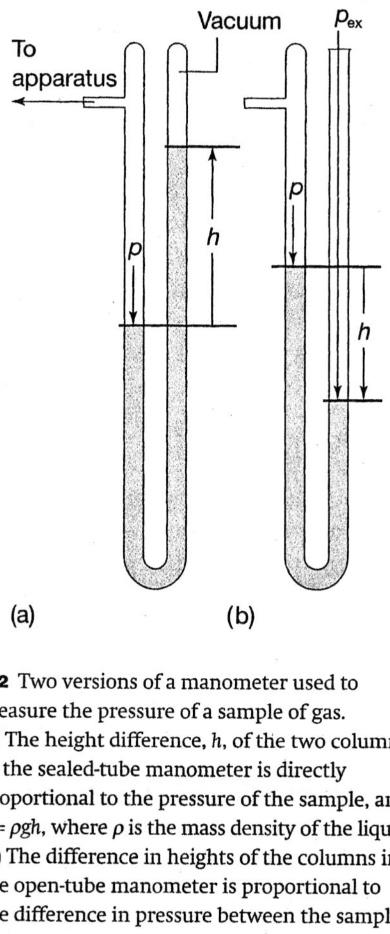

1.2 Two versions of a manometer used to measure the pressure of a sample of gas. (a) The height difference, h, of the two columns in the sealed-tube manometer is directly proportional to the pressure of the sample, and $p = \rho gh$, where ρ is the mass density of the liquid. (b) The difference in heights of the columns in the open-tube manometer is proportional to the difference in pressure between the sample and the atmosphere. In the example shown, the pressure of the sample is lower than that of the atmosphere.

(c) Temperature

The concept of temperature springs from the observation that a change in physical state (for example, a change of volume) can occur when two objects are in contact with one another, as when a red-hot metal is plunged into water. Later (Section 2.1) we shall see that the change in state can be interpreted as arising from a flow of energy as heat from one object to another. The **temperature**, T, is the property that indicates the direction of the flow of energy through a thermally conducting, rigid wall. If energy flows from A to B when they are in contact, then we say that A has a higher temperature than B (Fig. 1.3).

It will prove useful to distinguish between two types of boundary that can separate the objects. A boundary is **diathermic** if a change of state is observed when two objects at different temperatures are brought into contact.[1] A metal container has diathermic walls. A boundary is **adiabatic** if no change occurs even though the two objects have different temperatures.

The temperature is a property that indicates whether two objects would be in 'thermal equilibrium' if they were in contact through a diathermic boundary. **Thermal equilibrium** is established if no change of state occurs when two objects A to B are in contact through a diathermic boundary. Suppose an object A (which we

1. The word dia is from the Greek for 'through'.

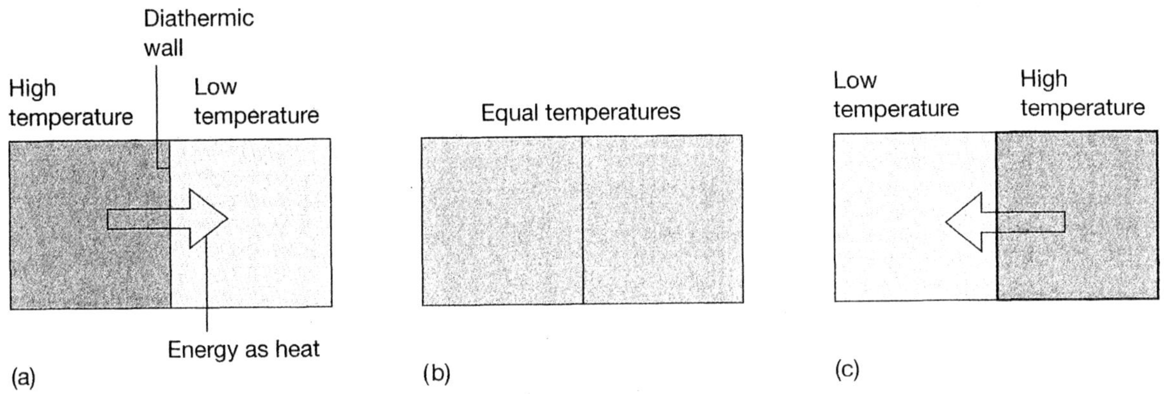

1.3 Energy flows as heat from a region at a higher temperature to one at a lower temperature if the two are in contact through a diathermic wall, as in (a) and (c). However, if the two regions have identical temperatures, there is no net transfer of energy as heat even though the two regions are separated by a diathermic wall (b). The latter condition corresponds to the two regions being at thermal equilibrium.

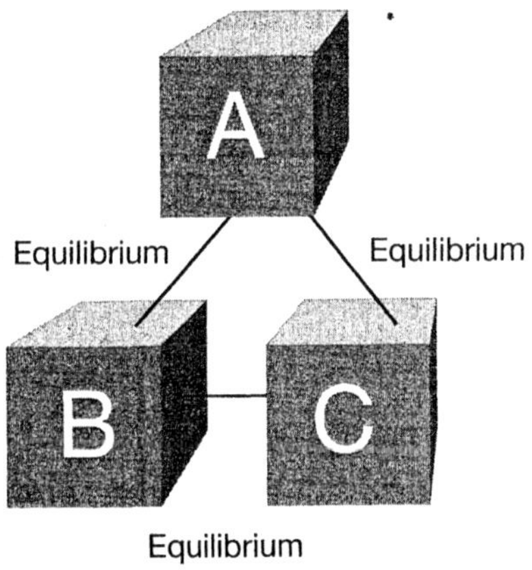

1.4 The experience summarized by the Zeroth Law of thermodynamics is that an object A is in thermal equilibrium with B and B is in thermal equilibrium with C, then C is in thermal equilibrium with A.

can think of as a block of iron) is in thermal equilibrium with an object B (a block of copper), and that B is also in thermal equilibrium with another object C (a flask of water). Then it has been found experimentally that A and C will also be in thermal equilibrium when they are put in contact (Fig. 1.4). This observation is summarized by the **Zeroth Law of thermodynamics**:

If A is in thermal equilibrium with B, and B is in thermal equilibrium with C, then C is also in thermal equilibrium with A.

The Zeroth Law justifies the concept of temperature and the use of a **thermometer**, a device for measuring the temperature. Thus, suppose that B is a glass capillary containing a liquid, such as mercury, that expands significantly as the temperature increases. Then, when A is in contact with B, the mercury column in the latter has a certain length. According to the Zeroth Law, if the mercury column in B has the same length when it is placed in contact with another object C, then we can predict that no change of state of A and C will occur when they are in contact. Moreover, we can use the length of the mercury column as a measure of the temperatures of A and C.

In the early days of thermometry (and still in laboratory practice today), temperatures were related to the length of a column of liquid, and the difference in lengths shown when the thermometer was first in contact with melting ice and then with boiling water was divided into 100 steps called 'degrees', the lower point being labelled 0. This procedure led to the **Celsius scale** of temperature. In this text, temperatures on the Celsius scale are denoted θ and expressed in *degrees Celsius* (°C). However, because different liquids expand to different extents, and do not always expand uniformly over a given range, thermometers constructed from different materials showed different numerical values of the temperature between their fixed points. The pressure of a gas, however, can be used to construct a **perfect-gas temperature scale** that is independent of the identity of the gas. The perfect-gas scale turns out to be identical to the **thermodynamic temperature scale** to be introduced in Section 4.2c, so we shall use the latter term from now on to avoid a proliferation of names. On the thermodynamic temperature scale, temperatures are denoted T and are normally reported in kelvins, K. Thermodynamic and Celsius temperatures are related by the exact expression

$$T/\mathrm{K} = \theta/{}^\circ\mathrm{C} + 273.15 \tag{1.5}$$

This relation is the current definition of the Celsius scale in terms of the more fundamental Kelvin scale. It implies that a difference in temperature of 1°C is equivalent to a difference of 1 K.

Illustration 1.1

To express 25.00°C as a temperature in kelvins, we write

$$T/\text{K} = (25.00°\text{C})/°\text{C} + 273.15 = 25.00 + 273.15 = 298.15$$

Note how the units are cancelled like numbers. This is the procedure called 'quantity calculus' in which a physical quantity (such as the temperature) is the product of a numerical value (25.00) and a unit (1°C). Multiplication of both sides by the unit K then gives $T = 298.15$ K.

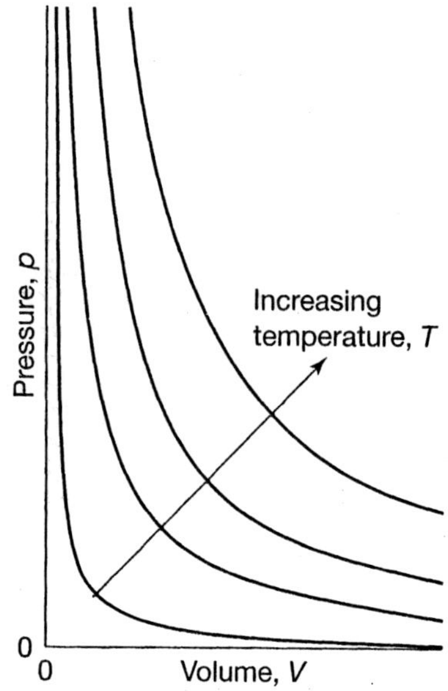

1.5 The pressure–volume dependence of a fixed amount of perfect gas at different temperatures. Each curve is a hyperbola (pV = constant) and is called an *isotherm*.

A hyperbola is a curve obtained by plotting y against x with xy = constant.

1.2 The gas laws

The equation of state of a gas at low pressure was established by combining a series of empirical laws. We introduce these laws below, and then show how they can be combined into the single equation of state $pV = nRT$.

(a) The individual gas laws

Robert Boyle, acting on the suggestion of a correspondent John Townley, showed in 1661 that, to a good approximation, the pressure and volume of a fixed amount of gas at constant temperature are related by

$$pV = \text{constant} \qquad (1.6)°$$

This relation is known as **Boyle's law**. That is, at constant temperature, the pressure of a sample of gas is inversely proportional to its volume, and the volume it occupies is inversely proportional to its pressure:

$$p \propto \frac{1}{V} \qquad V \propto \frac{1}{p} \qquad (1.7)°$$

The variation of the pressure of a sample of gas as the volume is changed is depicted in Fig. 1.5. Each of the curves in the graph corresponds to a single temperature and hence is called an **isotherm**. According to Boyle's law, the isotherms of gases are hyperbolas. An alternative depiction, a plot of pressure against 1/volume, is shown in Fig. 1.6.

More recent experiments have shown that Boyle's law is valid only at low pressures, and that real gases obey the law only in the limit of the pressure approaching zero (which we write $p \to 0$). Boyle's law is therefore an example of a **limiting law**, a law that is strictly true only in a certain limit, in this case $p \to 0$. Equations valid in this limiting sense will be signalled by a ° on the equation number, as in eqn 6.

The molecular explanation of Boyle's law is that, if a sample of gas is compressed to half its volume, then twice as many molecules strike the walls in a given period of time than before it was compressed. As a result, the average force exerted on the walls is doubled. Hence, when the volume is halved the pressure of the gas is doubled, and $p \times V$ is a constant. The reason why Boyle's law applies to all gases regardless of their chemical identity (provided the pressure is low) is that at low pressures the average separation of molecules is so great that they exert no influence on one another and hence travel independently.

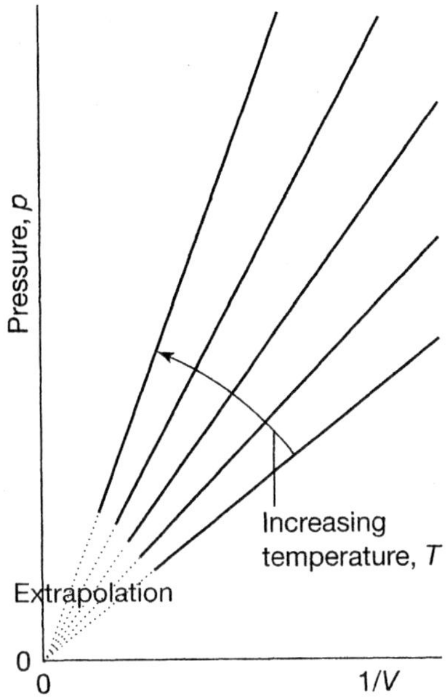

1.6 Straight lines are obtained when the pressure is plotted against $1/V$ at constant temperature.

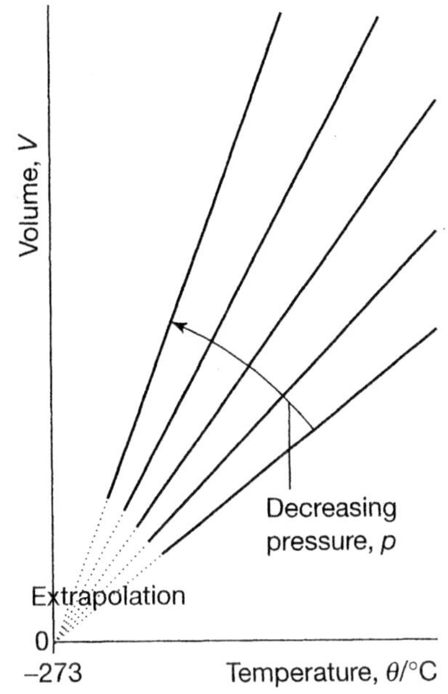

1.7 The variation of the volume of a fixed amount of gas with the temperature at constant pressure. Note that in each case the isobars extrapolate to zero volume at −273°C.

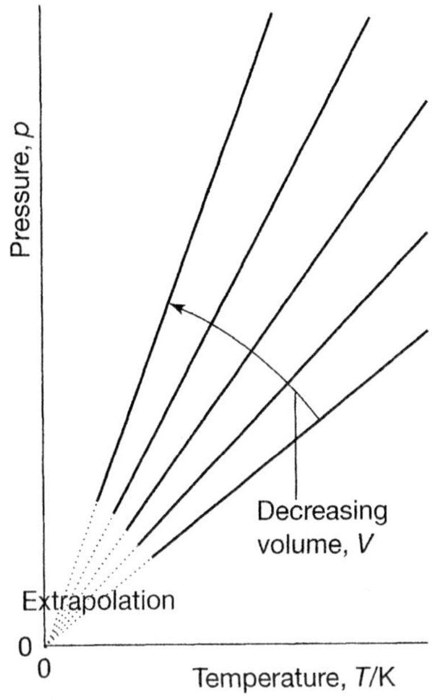

1.8 The pressure also varies linearly with the temperature at constant volume, and extrapolates to zero at $T = 0$ (−273°C).

The French scientist Guillaume Amontons established the next important property of gases, but his results were ignored for almost a century. Then Jacques Charles and Joseph Louis Gay-Lussac studied the effect of temperature on the volume of a sample of gas subjected to constant pressure. They found independently that the volume increased linearly with the temperature, whatever the identity of the gas, provided it was at low pressure. Specifically, they established that[2]

$$V = \text{constant} \times (\theta + 273.15\,°\text{C}) \qquad \text{(at constant pressure)} \qquad (1.8)°$$

The linear variation of volume with temperature summarized by this expression is illustrated in Fig. 1.7. The lines are examples of **isobars**, or lines showing the variation of properties at constant pressure.

Equation 1.8 suggests that the volume of any gas should extrapolate to zero at $\theta = -273.15$°C and therefore that −273.15°C is a natural zero of a fundamental temperature scale. As we have already indicated, a scale with 0 set at −273.15°C is equivalent to the thermodynamic temperature scale devised by Kelvin, so **Charles's law** (which is also known as **Gay-Lussac's law**) can be written

$$V = \text{constant} \times T \qquad \text{(at constant pressure)} \qquad (1.9)°$$

An alternative version of Charles's law, in which the pressure of a sample of gas is monitored under conditions of constant volume, is

$$p = \text{constant} \times T \qquad \text{(at constant volume)} \qquad (1.10)°$$

This version of the law indicates that the pressure of a gas falls to zero as the temperature is reduced to zero (Fig. 1.8).

2. Recall that we use θ to denote temperatures on the Celsius scale.

The molecular explanation of Charles's law lies in the fact that raising the temperature of a gas increases the average speed of its molecules. The molecules collide with the walls more frequently and with greater impact. Therefore they exert a greater pressure on the walls of the container.

The final piece of experimental information we need is that, at a given pressure and temperature, the molar volume, $V_m = V/n$, the volume per mole of molecules, of a gas is approximately the same regardless of the identity of the gas. This observation implies that the volume of a sample of gas is proportional to the amount of molecules (in moles) present, and that the constant of proportionality is independent of the identity of the gas:

$$V = \text{constant} \times n \qquad \text{(at constant pressure and temperature)} \qquad (1.11)^{\circ}$$

This conclusion is a modern form of **Avogadro's principle**, which states that equal volumes of gases at the same pressure and temperature contain the same numbers of molecules.[3]

Table 1.2 The gas constant in various units

R
8.314 47 J K^{-1} mol^{-1}
8.205 74 × 10^{-2} L atm K^{-1} mol^{-1}
8.314 47 × 10^{-2} L bar K^{-1} mol^{-1}
8.314 47 Pa m^3 K^{-1} mol^{-1}
62.364 L Torr K^{-1} mol^{-1}
1.987 21 cal K^{-1} mol^{-1}

(b) The perfect gas law

The empirical observations summarized by eqns 6–11 can be combined into a single expression:

$$pV = \text{constant} \times nT$$

This expression is consistent with Boyle's law (pV = constant) when n and T are constant, with both forms of Charles's law ($p \propto T$, $V \propto T$) when n and either V or p are held constant, and with Avogadro's principle ($V \propto n$) when p and T are constant. The constant of proportionality, which is found experimentally to be the same for all gases, is denoted R and called the **gas constant**. The resulting expression

$$pV = nRT \qquad (1.12)^{\circ}$$

is the **perfect gas equation**. It is the approximate equation of state of any gas, and becomes increasingly exact as the pressure of the gas approaches zero. A gas that obeys eqn 12 exactly under all conditions is called a **perfect gas** (or *ideal gas*). A **real gas**, an actual gas, behaves more like a perfect gas the lower the pressure, and is described exactly by eqn 12 in the limit of $p \to 0$. The gas constant R can be determined by evaluating $R = pV/nT$ for a gas in the limit of zero pressure (to guarantee that it is behaving perfectly). However, a more accurate value can be obtained by measuring the speed of sound in a low-pressure gas and extrapolating its value to zero pressure. Table 1.2 lists the values of R in a variety of units.

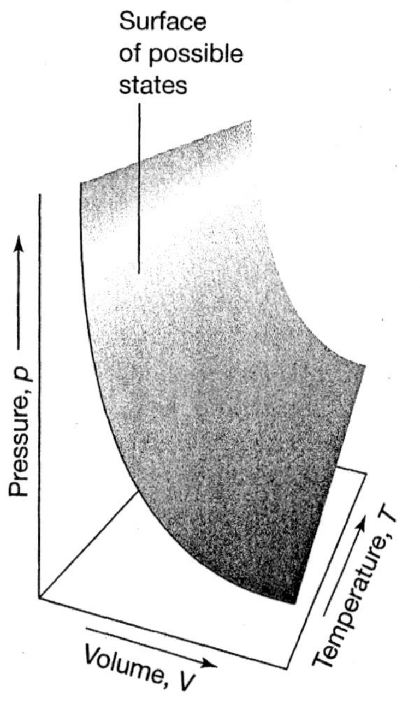

1.9 A region of the p, V, T surface of a fixed amount of perfect gas. The points forming the surface represent the only states of the gas that can exist.

The surface in Fig. 1.9 is a plot of the pressure of a fixed amount of perfect gas against its volume and thermodynamic temperature as given by eqn 12. The surface depicts the only possible states of a perfect gas: the gas cannot exist in states that do not correspond to points on the surface. The graphs in Figs. 1.5 and 1.7 correspond to sections through the surface (Fig. 1.10).

The perfect gas equation is of the greatest importance in physical chemistry because it is used to derive a wide range of relations that are used throughout thermodynamics. However, it is also of considerable practical utility for calculating the properties of a gas under a variety of conditions. For instance, the molar volume of a perfect gas under the conditions called **standard ambient temperature and pressure** (SATP), which means 298.15 K and 1 bar (that is, exactly 10^5 Pa), is easily

3. Avogadro's principle is a principle rather than a law (a summary of experience) because it depends on the validity of a model, in this case the existence of molecules.

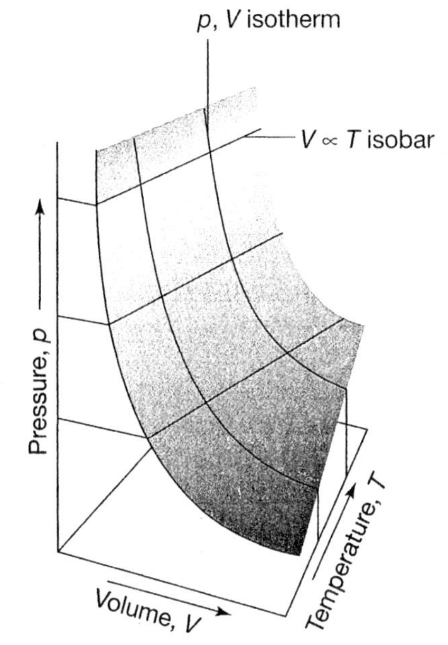

1.10 Sections through the surface shown in Fig. 1.9 at constant temperature give the isotherms shown in Fig. 1.5 and the isobars shown in Fig. 1.7.

calculated from $V_m = RT/p$ to be 24.789 L mol^{-1}. An earlier definition, **standard temperature and pressure** (STP), was 0°C and 1 atm; at STP, the molar volume of a perfect gas is 22.414 L mol^{-1}.

The perfect gas equation can be developed as follows to calculate the change in conditions when a constant amount of gas is subjected to different temperatures and pressures and allowed to occupy a different volume. First, write eqn 12 for the initial and final sets of conditions:

$$\frac{p_1V_1}{T_1} = nR \qquad \frac{p_2V_2}{T_2} = nR$$

Because n (in this problem) and R are both constant, it follows that

$$\frac{p_1V_1}{T_1} = \frac{p_2V_2}{T_2} \qquad (1.13)°$$

This relation is sometimes called the **combined gas equation**. Among other applications, eqn 13 can be used to discuss processes in the atmosphere that give rise to the weather (Box 1.1).

Example 1.3 Using the combined gas equation

In an industrial process, nitrogen is heated to 500 K in a vessel of constant volume. If it enters the vessel at 100 atm and 300 K, what pressure would it exert at the working temperature if it behaved as a perfect gas?

	n	p	V	T
Initial	Same	100	Same	300
Final	Same	?	Same	500

2

Method The known and unknown data are summarized in (**2**). We expect the pressure to be greater on account of the increase in temperature. Moreover, because pressure is proportional to temperature, we can anticipate that the pressure will increase by a factor of T_2/T_1, where T_1 is the initial temperature and T_2 is the final temperature.

Answer Cancellation of the volumes (because $V_1 = V_2$) on each side of eqn 13 results in

$$\frac{p_1}{T_1} = \frac{p_2}{T_2}$$

which can be rearranged into

$$p_2 = \frac{T_2}{T_1} \times p_1$$

Substitution of the data then gives

$$p_2 = \frac{500\text{ K}}{300\text{ K}} \times (100\text{ atm}) = 167\text{ atm}$$

Comment Experiment shows that the pressure is actually 183 atm under these conditions, so the assumption that the gas is perfect leads to a 10 per cent error.

Self-test 1.3 What temperature would result in the same sample exerting a pressure of 300 atm? [900 K]

Box 1.1 The gas laws and the weather

The biggest sample of gas readily accessible to us is the atmosphere, a mixture of gases with the composition summarized in the table below. The composition is maintained moderately constant by diffusion and convection (winds, particularly the local turbulence called *eddies*) but the pressure and temperature vary with altitude and with the local conditions, particularly in the troposphere (the 'sphere of change'), the layer extending up to about 11 km.

The pressure and temperature vary with altitude. In the troposphere the average temperature is 15°C at sea level, falling to −57°C at the bottom of the tropopause at 11 km. This variation is much less pronounced when expressed on the Kelvin scale, ranging from 288 K to 216 K, an average of 268 K. If we suppose that the temperature has its average value all the way up to the tropopause, then the pressure varies with altitude, h, according to the *barometric formula*:

$$p = p_0 e^{-h/H}$$

where p_0 is the pressure at sea level and H is a constant approximately equal to 8 km. More specifically, $H = RT/Mg$, where M is the average molar mass of air and T is the temperature. The barometric formula fits the observed pressure distribution quite well even for regions well above the troposphere (see the first illustration). It implies that the pressure of the air and its density fall to half their sea-level value at $h = H \ln 2$, or 6 km.

Local variations of pressure, temperature, and composition in the troposphere are manifest as 'weather'. A small region of air is termed a *parcel*. First, we note that a parcel of warm air is less dense than the same parcel of cool air. As a parcel rises, it expands adiabatically (that is, without transfer of heat from its surroundings), so it cools. Cool air can absorb lower concentrations of water vapour than warm air, so the moisture forms clouds. Cloudy skies can therefore be associated with rising air and clear skies are often associated with descending air.

The motion of air in the upper altitudes may lead to an accumulation in some regions and a loss of molecules from other regions. The former result in the formation of regions of high pressure ('highs' or anticyclones) and the latter result in regions of low pressure ('lows', depressions, or cyclones). These regions are shown as H and L on the weather map opposite. The lines of constant pressure—differing by 4 mbar (400 Pa, about 3 Torr)—marked on it are called *isobars*. The elongated regions of high and low pressure are known, respectively, as *ridges* and *troughs*.

In meteorology, large-scale vertical movement is called *convection*. Horizontal pressure differentials result in the flow of air that we call *wind*. Because the Earth is rotating from west to east, winds are deflected towards the right in the Northern hemisphere and towards the left in the Southern hemisphere. Winds travel nearly parallel to the isobars, with low pressure to their left in the Northern hemisphere and to

The composition of dry air at sea level

Component	Percentage By volume	Percentage By mass
Nitrogen, N_2	78.08	75.53
Oxygen, O_2	20.95	23.14
Argon, Ar	0.93	1.28
Carbon dioxide, CO_2	0.031	0.047
Hydrogen, H_2	5.0×10^{-3}	2.0×10^{-4}
Neon, Ne	1.8×10^{-3}	1.3×10^{-3}
Helium, He	5.2×10^{-4}	7.2×10^{-5}
Methane, CH_4	2.0×10^{-4}	1.1×10^{-4}
Krypton, Kr	1.1×10^{-4}	3.2×10^{-4}
Nitric oxide, NO	5.0×10^{-5}	1.7×10^{-6}
Xenon, Xe	8.7×10^{-6}	1.2×10^{-5}
Ozone, O_3; summer	7.0×10^{-6}	1.2×10^{-5}
winter	2.0×10^{-6}	3.3×10^{-6}

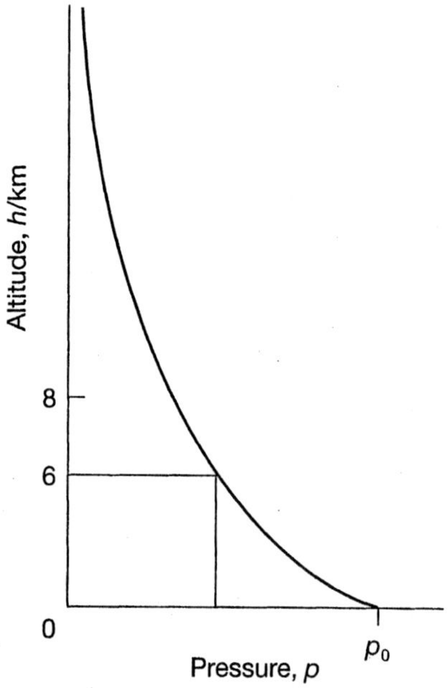

The variation of atmospheric pressure with altitude, as predicted by the barometric formula and as suggested by the 'US Standard Atmosphere', which takes into account the variation of temperature with altitude.

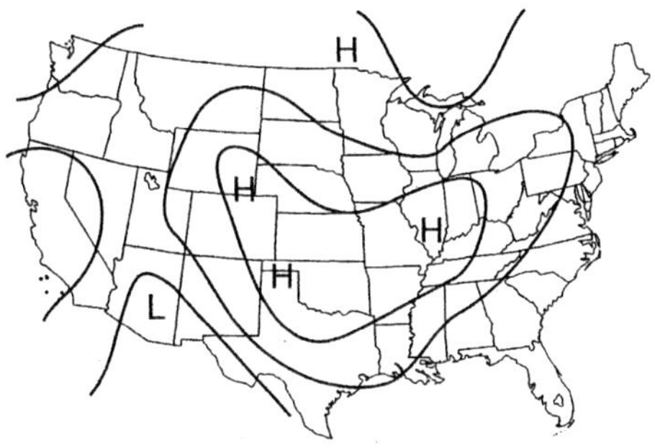

A typical weather map; in this case, for the United States on 1 January 2000.

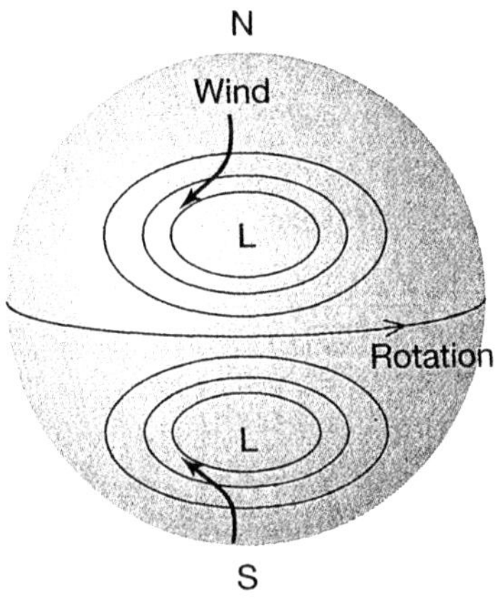

The flow of air ('wind') around regions of high and low pressure in the Northern and Southern hemispheres.

the right in the Southern hemisphere. At the surface, where wind speeds are lower, the winds tend to travel perpendicular to the isobars from high to low pressure. This differential motion results in a spiral outward flow of air clockwise in the Northern hemisphere around a high and an inward counterclockwise flow around a low (third illustration).

The air lost from regions of high pressure is restored as an influx of air converges into the region and descends. As we have seen, descending air is associated with clear skies. It also becomes warmer by compression as it descends, so regions of high pressure are associated with high surface temperatures. In winter, the cold surface air may prevent the complete fall of air, and result in a temperature *inversion*, with a layer of warm air over a layer of cold air. Geographical conditions may also trap cool air, as in Los Angeles, and the photochemical pollutants we know as *smog* may be trapped under the warm layer.

Discussion questions

1. The barometric formula relates the pressure of a gas of molar mass M at an altitude h to its pressure p_0 at sea level. Derive this relation by showing that the change in pressure dp for an infinitesimal change in altitude dh where the density is ρ is $dp = -\rho g dh$. Remember that ρ depends on the pressure. Evaluate the pressure difference between the top and bottom of (a) a laboratory vessel of height 15 cm, and (b) 1350 ft. Ignore temperature variations.

2. Ozone, O_3, is a trace atmospheric gas that plays an important role in screening the earth from harmful ultraviolet light, and the abundance of ozone is commonly reported in *Dobson units*. One Dobson unit is the thickness, in thousandths of a centimetre, of a column of gas if it were collected as a pure gas at 1.00 atm and 0°C. What amount of O_3 (in moles) is found in a column of atmosphere with a cross-sectional area of 1.00 dm^2 if the abundance is 250 Dobson units (a typical mid-latitude value)? In the seasonal Antarctic ozone hole, the column abundance drops below 100 Dobson units; how many moles of O_3 molecules are found in such a column of air above a 1.00 dm^2 area? Most atmospheric ozone is found between 10 and 50 km above the surface of the earth. If that ozone is spread uniformly through this portion of the atmosphere, what is the average molar concentration corresponding to (a) 250 Dobson units, (b) 100 Dobson units?

Further reading

S.K. Lower, Thermal physics (and some chemistry) of the atmosphere. *J. Chem. Educ.* **75**, 837 (1998).

R.P. Wayne, *Chemistry of atmospheres, an introduction to the chemistry of atmospheres of earth, the planets, and their satellites.* Oxford University Press (2000).

(c) Mixtures of gases

When dealing with gaseous mixtures, we often need to know the contribution that each component makes to the total pressure of the sample. In the nineteenth century John Dalton made observations that provide the answer and summarized them in **Dalton's law**:

> **The pressure exerted by a mixture of gases is the sum of the partial pressures of the gases.**

The **partial pressure** of a perfect gas is the pressure that it would exert if it occupied the container alone at the same temperature. That is, if a certain amount of H_2 exerts 25 kPa when present alone in a container, and an amount of N_2 exerts 80 kPa when present alone in the same container at the same temperature, then the total pressure when both are present is the sum of these two partial pressures, or 105 kPa (provided the individual gases and the mixture behave perfectly). More generally, if the partial pressure of a perfect gas A is p_A, that of a perfect gas B is p_B, and so on, then the total pressure when all the gases occupy the same container at the same temperature is

$$p = p_A + p_B + \cdots \qquad (1.14)^\circ$$

where, for each gas J,

$$p_J = \frac{n_J RT}{V} \qquad (1.15)^\circ$$

Example 1.4 Using Dalton's law

A container of volume 10.0 L holds 1.00 mol N_2 and 3.00 mol H_2 at 298 K. What is the total pressure in atmospheres if each component behaves as a perfect gas?

Method It follows from eqns 14 and 15 that the total pressure when two gases, A and B, occupy a container is

$$p = p_A + p_B = (n_A + n_B)\frac{RT}{V}$$

To obtain the answer expressed in atmospheres, use $R = 8.206 \times 10^{-2}$ L atm K^{-1} mol^{-1}.

Answer Under the stated conditions

$$p = (1.00 \text{ mol} + 3.00 \text{ mol}) \times \frac{(8.206 \times 10^{-2} \text{ L atm K}^{-1} \text{ mol}^{-1}) \times (298 \text{ K})}{10.0 \text{ L}}$$

$$= 9.78 \text{ atm}$$

Self-test 1.4 Calculate the total pressure when 1.00 mol N_2 and 2.00 mol O_2 are added to the same container (with the nitrogen and hydrogen still inside) at 298 K. [17.1 atm]

(d) Mole fractions and partial pressures

We can take a step closer to the discussion of mixtures of real gases by introducing the mole fraction of each component J. The **mole fraction**, x_J, is the amount of J expressed as a fraction of the total amount of molecules, n, in the sample:

$$x_J = \frac{n_J}{n} \qquad n = n_A + n_B + \cdots \qquad [1.16]$$

When no J molecules are present, $x_J = 0$; when only J molecules are present, $x_J = 1$. A mixture of 1.0 mol N_2 and 3.0 mol H_2, and therefore of 4.0 mol molecules in all, consists of mole fractions 0.25 of N_2 and 0.75 of H_2. It follows from the definition of x_J that, whatever the composition of the mixture,

$$x_A + x_B + \cdots = 1 \qquad (1.17)$$

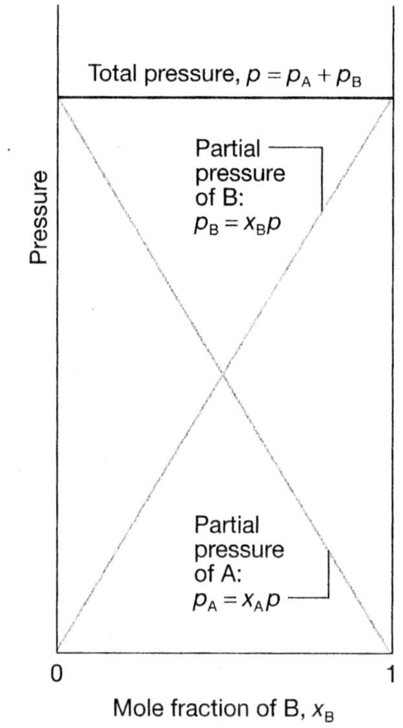

1.11 The partial pressures p_A and p_B of a binary mixture of (real or perfect) gases of total pressure p as the composition changes from pure A to pure B. The sum of the partial pressures is equal to the total pressure. If the gases are perfect, then the partial pressure is also the pressure that each gas would exert if it were present alone in the container.

Next, we define the **partial pressure**, p_J, of a gas J in a mixture (any gas, not just a perfect gas) as

$$p_J = x_J p \qquad [1.18]$$

where p is the total pressure of the mixture (Fig. 1.11). Only for a mixture of perfect gases can we identify p_J with the pressure that the gas J would exert if it were alone in the container and calculate its value from eqn 15. For real gases, the interactions between molecules result in the total pressure being slightly different from the sum of their partial pressures as defined in eqn 15.

It follows from eqns 17 and 18 that the sum of the partial pressures is equal to the total pressure:

$$p_A + p_B + \cdots = (x_A + x_B + \cdots)p = p \qquad (1.19)$$

This relation is true for both real and perfect gases.

Example 1.5 Calculating partial pressures

The mass percentage composition of dry air at sea level is approximately N_2: 75.5; O_2: 23.2; Ar: 1.3. What is the partial pressure of each component when the total pressure is 1.00 atm?

Method We expect species with a high mole fraction to have a proportionally high partial pressure. Partial pressures are defined by eqn 18. To use the equation, we need the mole fractions of the components. To calculate mole fractions, which are defined by eqn 16, we use the fact that the amount of molecules J of molar mass M_J in a sample of mass m_J is $n_J = m_J/M_J$. The mole fractions are independent of the total mass of the sample, so we can choose the latter to be 100 g (which makes the conversion from mass percentages very easy).

Answer The amounts of each type of molecule present in 100 g of air are

$$n(N_2) = 0.755 \times \frac{100\ \text{g}}{28.02\ \text{g mol}^{-1}} = \frac{0.755 \times 100}{28.02}\ \text{mol}$$

$$n(O_2) = 0.232 \times \frac{100\ \text{g}}{32.00\ \text{g mol}^{-1}} = \frac{0.232 \times 100}{32.00}\ \text{mol}$$

$$n(\text{Ar}) = 0.013 \times \frac{100\ \text{g}}{39.95\ \text{g mol}^{-1}} = \frac{0.013 \times 100}{39.95}\ \text{mol}$$

These three amounts work out as 2.69 mol, 0.725 mol, and 0.033 mol, respectively, for a total of 3.45 mol. The mole fractions and partial pressures (obtained by multiplying the mole fraction by the total pressure, 1.00 atm) are as follows:

	N_2	O_2	Ar
Mole fraction	0.780	0.210	0.0096
Partial pressure/atm	0.780	0.210	0.0096

Comment We have not had to assume that the gases are perfect: partial pressures are defined as $p_J = x_J p$ for any kind of gas.

Self-test 1.5 When carbon dioxide is taken into account, the mass percentages are 75.52 (N_2), 23.15 (O_2), 1.28 (Ar), and 0.046 (CO_2). What are the partial pressures when the total pressure is 0.900 atm?

[0.703, 0.189, 0.0084, 0.000 27 atm]

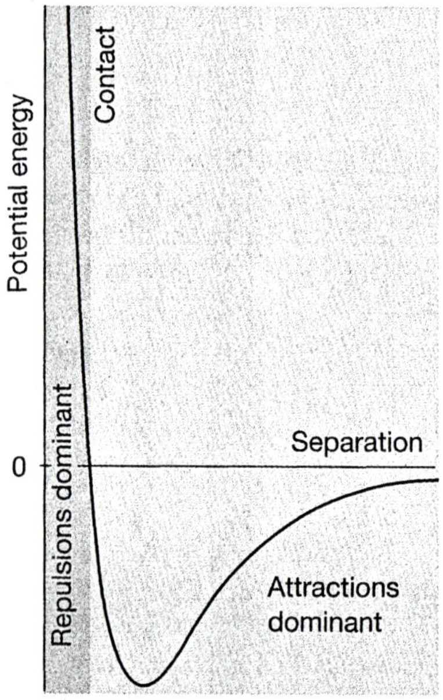

1.12 The variation of the potential energy of two molecules on their separation. High positive potential energy (at very small separations) indicates that the interactions between them are strongly repulsive at these distances. At intermediate separations, where the potential energy is negative, the attractive interactions dominate. At large separations (on the right) the potential energy is zero and there is no interaction between the molecules.

Real gases

Real gases do not obey the perfect gas law exactly. Deviations from the law are particularly important at high pressures and low temperatures, especially when a gas is on the point of condensing to liquid.

1.3 Molecular interactions

Real gases show deviations from the perfect gas law because molecules interact with each other. Repulsive forces between molecules assist expansion and attractive forces assist compression.

Repulsive forces are significant only when molecules are almost in contact: they are short-range interactions, even on a scale measured in molecular diameters (Fig. 1.12). Because they are short-range interactions, repulsions can be expected to be important only when the average separation of the molecules is small. This is the case at high pressure, when many molecules occupy a small volume. On the other hand, attractive intermolecular forces have a relatively long range and are effective over several molecular diameters. They are important when the molecules are fairly close together but not necessarily touching (at the intermediate separations in Fig. 1.12). Attractive forces are ineffective when the molecules are far apart (well to the right in Fig. 1.12). Intermolecular forces are also important when the temperature is so low that the molecules travel with such low mean speeds that they can be captured by one another.

At low pressure, when the sample occupies a large volume, the molecules are so far apart for most of the time that the intermolecular forces play no significant role, and the gas behaves virtually perfectly. At moderate pressure, when the average separation of the molecules is only a few molecular diameters, the attractive forces dominate the repulsive forces. In this case, the gas can be expected to be more compressible than a perfect gas because the forces help to draw the molecules together. At high pressure, when the average separation of the molecules is small, the repulsive forces dominate and the gas can be expected to be less compressible because now the forces help to drive the molecules apart.

(a) The compression factor

The **compression factor**, Z, of a gas is the ratio of its molar volume, V_m, to the molar volume of a perfect gas, V_m°, at the same pressure and temperature:

$$Z = \frac{V_m}{V_m^\circ} \qquad [1.20a]$$

Because the molar volume of a perfect gas is equal to RT/p, an equivalent expression is

$$pV_m = RTZ \qquad (1.20b)$$

We can explore the distance-dependence of intermolecular forces by plotting Z against pressure. Because for a perfect gas $Z = 1$ under all conditions, deviation of Z from 1 is a measure of departure from perfect behaviour.

Some experimental values of Z are plotted in Fig. 1.13. At very low pressures, all the gases shown have $Z \approx 1$ and behave nearly perfectly. At high pressures, all the gases have $Z > 1$, signifying that they have a larger molar volume than a perfect gas. Repulsive forces are now dominant. At intermediate pressures, most gases have

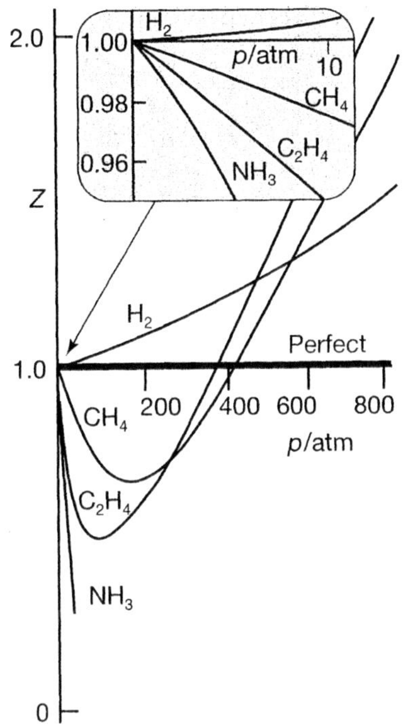

1.13 The variation of the compression factor, Z, with pressure for several gases at 0°C. A perfect gas has $Z = 1$ at all pressures. Notice that although the curves approach 1 as $p \to 0$, they do so with different slopes.

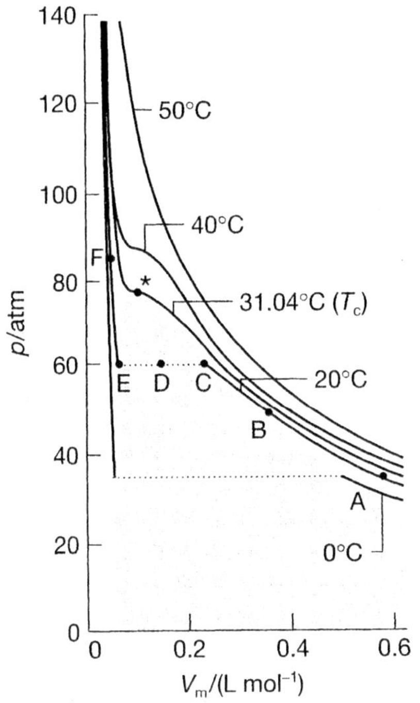

1.14 Experimental isotherms of carbon dioxide at several temperatures. The 'critical isotherm', the isotherm at the critical temperature, is at 31.04°C. The critical point is marked with a star.

$Z < 1$, indicating that the attractive forces are reducing the molar volume relative to that of a perfect gas.

(b) Virial coefficients

Figure 1.14 shows the experimental isotherms for carbon dioxide. At large molar volumes and high temperatures the real-gas isotherms do not differ greatly from perfect-gas isotherms. The small differences suggest that the perfect gas law is in fact the first term in an expression of the form

$$pV_m = RT(1 + B'p + C'p^2 + \cdots) \tag{1.21}$$

This expression is an example of a common procedure in physical chemistry, in which a simple law (in this case $pV = nRT$) is treated as the first term in a series in powers of a variable (in this case p). A more convenient expansion for many applications is

$$pV_m = RT\left(1 + \frac{B}{V_m} + \frac{C}{V_m^2} + \cdots\right) \tag{1.22}$$

These two expressions are two versions of the **virial equation of state**.[4] In each case, the term in parentheses can be identified with the compression factor, Z.

The coefficients B, C, ... , which depend on the temperature, are the second, third, ... **virial coefficients** (Table 1.3); the first virial coefficient is 1. The third virial coefficient, C, is usually less important than the second coefficient, B, in the sense that at typical molar volumes $C/V_m^2 \ll B/V_m$.

Table 1.3* Second virial coefficients, $B/(\text{cm}^3\ \text{mol}^{-1})$

	Temperature	
	273 K	600 K
Ar	−21.7	11.9
CO_2	−149.7	−12.4
N_2	−10.5	21.7
Xe	−153.7	−19.6

* More values are given in the *Data section*.

4. The name comes from the Latin word for force. The coefficients are sometimes denoted B_2, B_3, ...

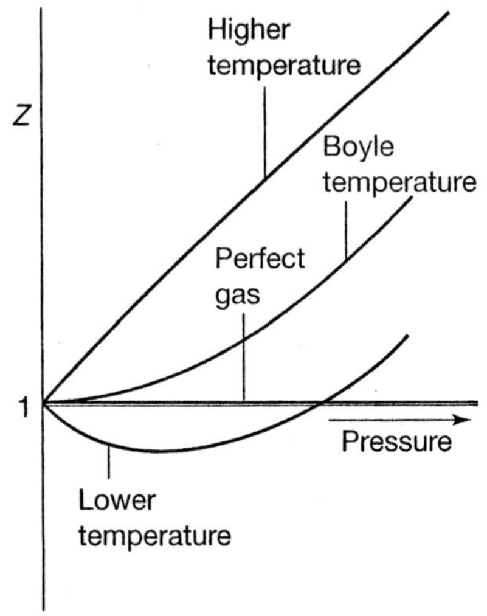

1.15 The compression factor, Z, approaches 1 at low pressures, but does so with different slopes. For a perfect gas, the slope is zero, but real gases may have either positive or negative slopes, and the slope may vary with temperature. At the Boyle temperature, the slope is zero and the gas behaves perfectly over a wider range of conditions than at other temperatures.

We can use the virial equation to demonstrate the important point that, although the equation of state of a real gas may coincide with the perfect gas law as $p \to 0$, not all of its properties necessarily coincide with those of a perfect gas in that limit. Consider, for example, the value of dZ/dp, the slope of the graph of compression factor against pressure. For a perfect gas $dZ/dp = 0$ (because $Z = 1$ at all pressures), but for a real gas

$$\frac{dZ}{dp} = B' + 2pC' + \cdots \to B' \text{ as } p \to 0 \tag{1.23}$$

However, B' is not necessarily zero, so the slope of Z with respect to p does not necessarily approach 0 (the perfect gas value). Because several properties depend on derivatives (as we shall see), the properties of real gases do not always coincide with the perfect gas values at low pressures. By a similar argument,

$$\frac{dZ}{d(1/V_m)} \to B \text{ as } V_m \to \infty, \text{ corresponding to } p \to 0 \tag{1.24}$$

Because the virial coefficients depend on the temperature, there may be a temperature at which $Z \to 1$ with zero slope at low pressure or high molar volume (Fig. 1.15). At this temperature, which is called the **Boyle temperature,** T_B, the properties of the real gas do coincide with those of a perfect gas as $p \to 0$. According to eqn 1.24, Z has zero slope as $p \to 0$ if $B = 0$, so we can conclude that $B = 0$ at the Boyle temperature. It then follows from eqn 1.22 that $pV_m \approx RT_B$ over a more extended range of pressures than at other temperatures because the first term after 1 (that is, B/V_m) in the virial equation is zero and C/V_m^2 and higher terms are negligibly small. For helium $T_B = 22.64$ K; for air $T_B = 346.8$ K; more values are given in Table 1.4.

(c) Condensation

Now consider what happens when we compress a sample of gas initially in the state marked A in Fig. 1.14 at constant temperature (by pushing in a piston). Near A, the pressure of the gas rises in approximate agreement with Boyle's law. Serious deviations from that law begin to appear when the volume has been reduced to B.

At C (which corresponds to about 60 atm for carbon dioxide) all similarity to perfect behaviour is lost, for suddenly the piston slides in without any further rise in pressure: this stage is represented by the horizontal line CDE. Examination of the contents of the vessel shows that just to the left of C a liquid appears, and there are two phases separated by a sharply defined surface. As the volume is decreased from C through D to E, the amount of liquid increases. There is no additional resistance to

Table 1.4* Critical constants of gases

	p_c/atm	V_c/(cm^3 mol^{-1})	T_c/K	Z_c	T_B/K
Ar	48.0	75.3	150.7	0.292	411.5
CO_2	72.9	94.0	304.2	0.274	714.8
He	2.26	57.8	5.2	0.305	22.64
O_2	50.14	78.0	154.8	0.308	405.9

* More values are given in the *Data section.*

the piston because the gas can respond by condensing. The pressure corresponding to the line CDE, when both liquid and vapour are present in equilibrium, is called the **vapour pressure** of the liquid at the temperature of the experiment.

At E, the sample is entirely liquid and the piston rests on its surface. Any further reduction of volume requires the exertion of considerable pressure, as is indicated by the sharply rising line to the left of E. Even a small reduction of volume from E to F requires a great increase in pressure.

(d) Critical constants

The isotherm at the temperature T_c (304.19 K, or 31.04°C for CO_2) plays a special role in the theory of the states of matter. An isotherm slightly below T_c behaves as we have already described: at a certain pressure, a liquid condenses from the gas and is distinguishable from it by the presence of a visible surface. If, however, the compression takes place at T_c itself, then a surface separating two phases does not appear and the volumes at each end of the horizontal part of the isotherm have merged to a single point, the **critical point** of the gas. The temperature, pressure, and molar volume at the critical point are called the **critical temperature**, T_c, **critical pressure**, p_c, and **critical molar volume**, V_c, of the substance. Collectively, p_c, V_c, and T_c are the **critical constants** of a substance (Table 1.4).

The web site contains links to online databases of properties of gases.

At and above T_c, the sample has a single phase that occupies the entire volume of the container. Such a phase is, by definition, a gas. Hence, the liquid phase of a substance does not form above the critical temperature. The critical temperature of oxygen, for instance, signifies that it is impossible to produce liquid oxygen by compression alone if its temperature is greater than 155 K: to liquefy oxygen—to obtain a fluid phase that does not occupy the entire volume—the temperature must first be lowered to below 155 K, and then the gas compressed isothermally. The single phase that fills the entire volume when $T > T_c$ may be much denser than we normally consider typical of gases, and the name **supercritical fluid** is preferred.

1.4 The van der Waals equation

We can draw conclusions from the virial equations of state only by inserting specific values of the coefficients. It is often useful to have a broader, if less precise, view of all gases. Therefore, we introduce the approximate equation of state suggested by J.H. van der Waals in 1873. This equation is an excellent example of an expression that can be obtained by thinking scientifically about a mathematically complicated but physically simple problem; that is, it is a good example of 'model building'. Van der Waals himself proposed his equation on the basis of experimental evidence available to him in conjunction with rigorous thermodynamic arguments.

The **van der Waals equation** is

$$p = \frac{nRT}{V - nb} - a\left(\frac{n}{V}\right)^2 \tag{1.25a}$$

and a derivation is given in the *Justification* below. The equation is often written in terms of the molar volume $V_m = V/n$ as

$$p = \frac{RT}{V_m - b} - \frac{a}{V_m^2} \tag{1.25b}$$

The constants a and b are called the **van der Waals coefficients**. They are characteristic of each gas but independent of the temperature (Table 1.5).

Table 1.5* Van der Waals coefficients

	$a/(\text{atm L}^2\text{ mol}^{-2})$	$b/(10^{-2}\text{ L mol}^{-1})$
Ar	1.337	3.20
CO_2	3.610	4.29
He	0.0341	2.38
Xe	4.137	5.16

* More values are given in the *Data section*.

Justification 1.1

The repulsive interactions between molecules are taken into account by supposing that they cause the molecules to behave as small but impenetrable spheres. The nonzero volume of the molecules implies that instead of moving in a volume V they are restricted to a smaller volume $V-nb$, where nb is approximately the total volume taken up by the molecules themselves. This argument suggests that the perfect gas law $p=nRT/V$ should be replaced by

$$p=\frac{nRT}{V-nb}$$

when repulsions are significant.

The pressure depends on both the frequency of collisions with the walls and the force of each collision. Both the frequency of the collisions and their force are reduced by the attractive forces, which act with a strength proportional to the molar concentration, n/V, of molecules in the sample. Therefore, because both the frequency and the force of the collisions are reduced by the attractive forces, the pressure is reduced in proportion to the square of this concentration. If the reduction of pressure is written as $-a(n/V)^2$, where a is a positive constant characteristic of each gas, the combined effect of the repulsive and attractive forces is the van der Waals equation of state as expressed in eqn 25.

In this *Justification* we have built the van der Waals equation using vague arguments about the volumes of molecules and the effects of forces. The equation can be derived in other ways, but the present method has the advantage that it shows how to derive the form of an equation out of general ideas. The derivation also has the advantage of keeping imprecise the significance of the coefficients a and b: they are much better regarded as empirical parameters than as precisely defined molecular properties.

Example 1.6 Using the van der Waals equation to estimate a molar volume

Estimate the molar volume of CO_2 at 500 K and 100 atm by treating it as a van der Waals gas.

Method To express eqn 1.25b as an equation for the molar volume, we need to rearrange it into

$$V_m^3-\left(b+\frac{RT}{p}\right)V_m^2+\left(\frac{a}{p}\right)V_m-\frac{ab}{p}=0$$

Although closed expressions for the roots of a cubic equation can be given, they are very complicated. Unless analytical solutions are essential, it is usually more expedient to solve such equations with commercial software.

Answer According to Table 1.5, $a=3.610\ \mathrm{L^2\,atm\,mol^{-2}}$ and $b=4.29\times10^{-2}\ \mathrm{L\,mol^{-1}}$. Under the stated conditions, $RT/p=0.410\ \mathrm{L\,mol^{-1}}$. The coefficients in the equation for V_m are therefore

$$b+RT/p=0.453\ \mathrm{L\,mol^{-1}}$$

$$a/p=3.61\times10^{-2}\ (\mathrm{L\,mol^{-1}})^2$$

$$ab/p=1.55\times10^{-3}\ (\mathrm{L\,mol^{-1}})^3$$

Therefore, on writing $x = V_m/(\text{L mol}^{-1})$, the equation to solve is

$$x^3 - 0.453x^2 + (3.61 \times 10^{-2})x - (1.55 \times 10^{-3}) = 0$$

The acceptable root is $x = 0.366$, which implies that $V_m = 0.366\ \text{L mol}^{-1}$.

Comment For a perfect gas under these conditions, the molar volume is $0.410\ \text{L mol}^{-1}$.

Self-test 1.6 Calculate the molar volume of argon at 100°C and 100 atm on the assumption that it is a van der Waals gas. [$0.298\ \text{L mol}^{-1}$]

(a) The reliability of the equation

We now examine to what extent the van der Waals equation predicts the behaviour of real gases. It is too optimistic to expect a single, simple expression to be the true equation of state of all substances, and accurate work on gases must resort to the virial equation, use tabulated values of the coefficients at various temperatures, and analyse the systems numerically. The advantage of the van der Waals equation, however, is that it is analytical (that is, expressed symbolically) and allows us to draw some general conclusions about real gases. When the equation fails we must use one of the other equations of state that have been proposed (some are listed in Table 1.6), invent a new one, or go back to the virial equation.

That having been said, we can begin to judge the reliability of the equation by comparing the isotherms it predicts with the experimental isotherms for carbon dioxide in Fig. 1.14. Some calculated isotherms are shown in Figs 1.16 and 1.17. Apart from the oscillations below the critical temperature, they do resemble experimental isotherms quite well. The oscillations, the **van der Waals loops**, are unrealistic because they suggest that under some conditions an increase of pressure results in an increase of volume. Therefore, they are replaced by horizontal lines drawn so the loops define equal areas above and below the lines: this procedure is called the **Maxwell construction (3)**. The van der Waals coefficients, such as those in Table 1.5, are found by fitting the calculated curves to the experimental curves.

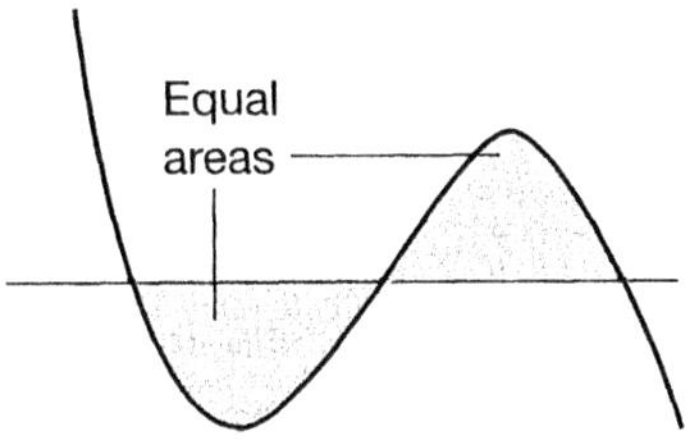

3

Table 1.6 Selected equations of state

	Equation	Reduced form*	Critical constants		
			p_c	V_c	T_c
Perfect gas	$p = \dfrac{RT}{V_m}$				
Van der Waals	$p = \dfrac{RT}{V_m - b} - \dfrac{a}{V_m^2}$	$p_r = \dfrac{8T_r}{3V_r - 1} - \dfrac{3}{V_r^2}$	$\dfrac{a}{27b^2}$	$3b$	$\dfrac{8a}{27bR}$
Berthelot	$p = \dfrac{RT}{V_m - b} - \dfrac{a}{TV_m^2}$	$p_r = \dfrac{8T_r}{3V_r - 1} - \dfrac{3}{T_r V_r^2}$	$\dfrac{1}{12}\left(\dfrac{2aR}{3b^3}\right)^{1/2}$	$3b$	$\dfrac{2}{3}\left(\dfrac{2a}{3bR}\right)^{1/2}$
Dieterici	$p = \dfrac{RTe^{-a/RTV_m}}{V_m - b}$	$p_r = \dfrac{e^2 T_r e^{-2/T_r V_r}}{2V_r - 1}$	$\dfrac{a}{4e^2b^2}$	$2b$	$\dfrac{a}{4Rb}$
Virial (Kammerlingh Onnes)	$p = \dfrac{RT}{V_m}\left\{1 + \dfrac{B(T)}{V_m} + \dfrac{C(T)}{V_m^2} + \cdots\right\}$				

* Reduced variables are defined in Section 1.5.

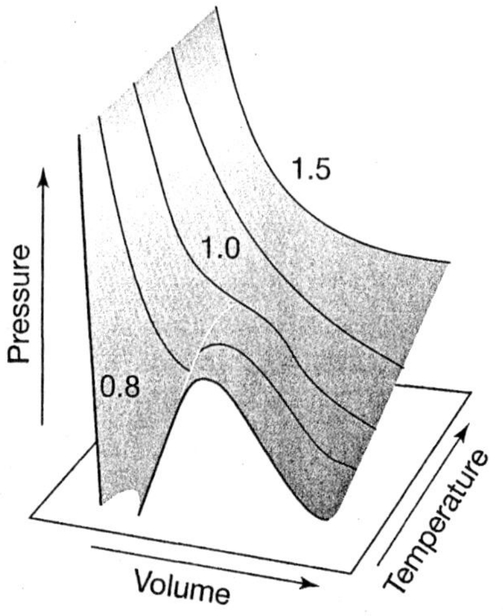

1.16 The surface of possible states allowed by the van der Waals equation. Compare this surface with that shown in Fig. 1.9.

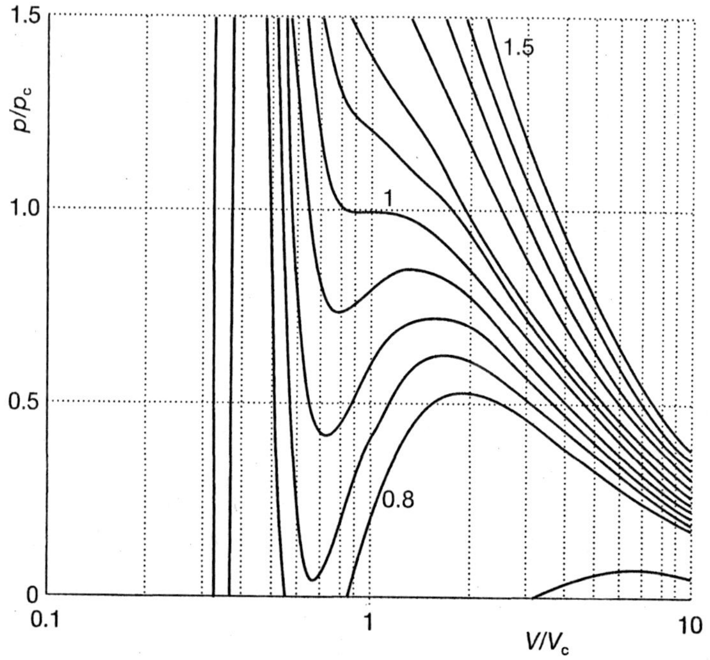

1.17 Van der Waals isotherms at several values of T/T_c. Compare these curves with those in Fig.1.14. The van der Waals loops are normally replaced by horizontal straight lines. The critical isotherm is the isotherm for $T/T_c = 1$.

(b) The features of the equation

The principal features of the van der Waals equation can be summarized as follows.

(1) Perfect gas isotherms are obtained at high temperatures and large molar volumes.

When the temperature is high, RT may be so large that the first term in eqn 1.25b greatly exceeds the second. Furthermore, if the molar volume is large in the sense $V_m \gg b$, then the denominator $V_m - b \approx V_m$. Under these conditions, the equation reduces to $p = RT/V_m$, the perfect gas equation.

(2) Liquids and gases coexist when cohesive and dispersing effects are in balance.

The van der Waals loops occur when both terms in eqn 1.25b have similar magnitudes. The first term arises from the kinetic energy of the molecules and their repulsive interactions; the second represents the effect of the attractive interactions.

(3) The critical constants are related to the van der Waals coefficients.

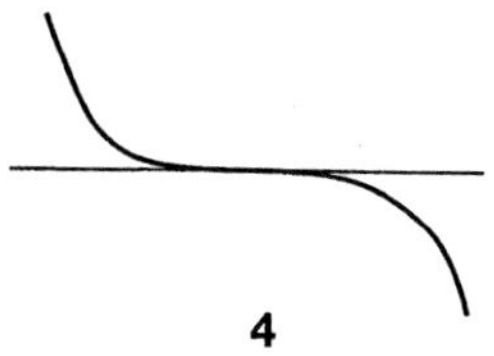

4

For $T < T_c$, the calculated isotherms oscillate, and each one passes through a minimum followed by a maximum. These extrema converge as $T \to T_c$ and coincide at $T = T_c$; at the critical point the curve has a flat inflexion (4). From the properties of curves, we know that an inflexion of this type occurs when both the first and second derivatives are zero. Hence, we can find the critical constants by calculating these derivatives and setting them equal to zero:

$$\frac{dp}{dV_m} = -\frac{RT}{(V_m - b)^2} + \frac{2a}{V_m^3} = 0$$

$$\frac{d^2p}{dV_m^2} = \frac{2RT}{(V_m - b)^3} - \frac{6a}{V_m^4} = 0$$

at the critical point. The solutions of these two equations are

$$V_c = 3b \qquad p_c = \frac{a}{27b^2} \qquad T_c = \frac{8a}{27Rb} \tag{1.26}$$

These relations can be tested by noting that the **critical compression factor**, Z_c, is predicted to be equal to

$$Z_c = \frac{p_c V_c}{RT_c} = \frac{3}{8} \tag{1.27}$$

for all gases. We see from Table 1.4 that although $Z_c < \frac{3}{8}$ (or 0.375), it is approximately constant (at 0.3) and the discrepancy is reasonably small.

1.5 The principle of corresponding states

An important general technique in science for comparing the properties of objects is to choose a related fundamental property of the same kind and to set up a relative scale on that basis. We have seen that the critical constants are characteristic properties of gases, so it may be that a scale can be set up by using them as yardsticks. We therefore introduce the **reduced variables** of a gas by dividing the actual variable by the corresponding critical constant:

$$p_r = \frac{p}{p_c} \qquad V_r = \frac{V_m}{V_c} \qquad T_r = \frac{T}{T_c} \qquad [1.28]$$

If the reduced pressure of a gas is given, we can easily calculate its actual pressure by using $p = p_r p_c$, and likewise for the volume and temperature. Van der Waals, who first tried this procedure, hoped that gases confined to the same reduced volume, V_r, at the same reduced temperature, T_r, would exert the same reduced pressure, p_r. The hope was largely fulfilled (Fig. 1.18). The illustration shows the dependence of the compression factor on the reduced pressure for a variety of gases at various reduced temperatures. The success of the procedure is strikingly clear: compare this graph with Fig. 1.13, where similar data are plotted without using reduced variables. The observation that real gases at the same reduced volume and reduced temperature

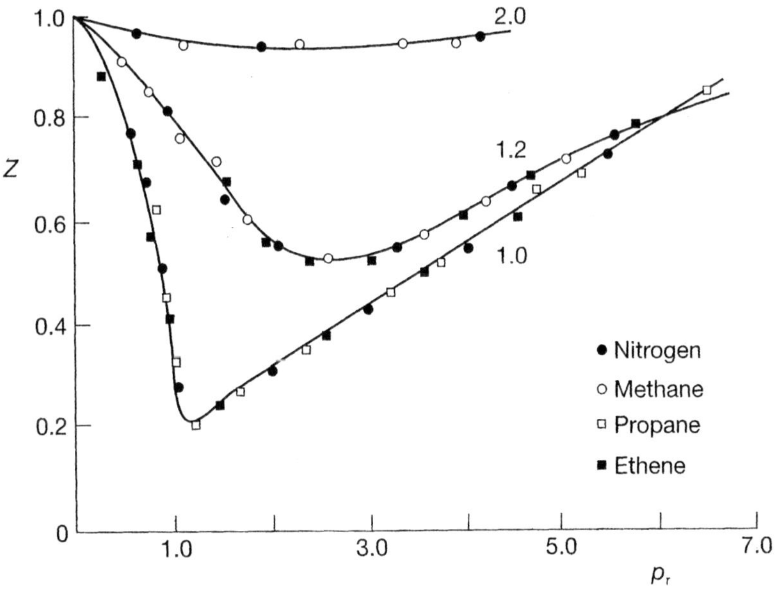

1.18 The compression factors of four gases, including two from Fig. 1.13, plotted using reduced variables. The use of reduced variables organizes the data on to single curves.

exert the same reduced pressure is called the **principle of corresponding states**. The principle is only an approximation. It works best for gases composed of spherical molecules; it fails, sometimes badly, when the molecules are non-spherical or polar.

The van der Waals equation sheds some light on the principle. First, we express eqn 25b in terms of the reduced variables, which gives

$$p_r p_c = \frac{RT_r T_c}{V_r V_c - b} - \frac{a}{V_r^2 V_c^2}$$

Then we express the critical constants in terms of a and b by using eqn 26:

$$\frac{ap_r}{27b^2} = \frac{8aT_r}{27b(3bV_r - b)} - \frac{a}{9b^2 V_r^2}$$

which can be reorganized into

$$p_r = \frac{8T_r}{3V_r - 1} - \frac{3}{V_r^2} \qquad (1.29)$$

This equation has the same form as the original, but the coefficients a and b, which differ from gas to gas, have disappeared. It follows that, if the isotherms are plotted in terms of the reduced variables (as we did in fact in Fig. 1.17 without drawing attention to the fact), then the same curves are obtained whatever the gas. This is precisely the content of the principle of corresponding states, so the van der Waals equation is compatible with it.

Looking for too much significance in this apparent triumph is mistaken, because other equations of state also accommodate the principle (Table 1.6). In fact, all we need are two parameters playing the roles of a and b, for then the equation can always be manipulated into reduced form. The observation that real gases obey the principle approximately amounts to saying that the effects of the attractive and repulsive interactions can each be approximated in terms of a single parameter. The importance of the principle is then not so much its theoretical interpretation but the way that it enables the properties of a range of gases to be coordinated on to a single diagram (for example, Fig. 1.18 instead of Fig. 1.13).

Checklist of key ideas

- ☐ gas, a form of matter that fills any container it occupies

The perfect gas

1.1 The states of gases

- ☐ physical state, the condition of a substance
- ☐ equation of state, an equation that interrelates pressure, volume, temperature, and amount of substance: $p = f(T, V, n)$
- ☐ pressure, force divided by the area to which the force is applied
- ☐ standard pressure, $p^\ominus = 1$ bar (10^5 Pa)
- ☐ mechanical equilibrium, the condition of equality of pressure on either side of a movable wall
- ☐ barometer, a device for measuring atmospheric pressure
- ☐ manometer, a device for measuring the pressure of gas in a system
- ☐ temperature, the property that indicates the direction of the flow of energy through a thermally conducting, rigid wall
- ☐ diathermic boundary, a boundary that permits the passage of energy as heat
- ☐ adiabatic boundary, a boundary that prevents the passage of energy as heat
- ☐ thermal equilibrium, a condition in which no change of state occurs when two objects A and B are in contact through a diathermic boundary
- ☐ Zeroth Law of thermodynamics: If A is in thermal equilibrium with B, and B is in thermal equilibrium with C, then C is also in thermal equilibrium with A
- ☐ thermometer, a device for measuring temperature
- ☐ Celsius scale, a temperature scale on which $\theta/°C = T/K - 273.15$
- ☐ perfect-gas temperature scale, a temperature scale based on the properties of a perfect gas

- ☐ thermodynamic temperature scale, a temperature scale defined in terms of thermodynamic quantities

1.2 The gas laws

- ☐ Boyle's law: At constant temperature, the pressure of a sample of gas is inversely proportional to its volume, $p \propto 1/V$ at constant temperature
- ☐ isotherm, curves in a graph corresponding to a single temperature
- ☐ limiting law, a law that is strictly true only in a certain limit
- ☐ isobar, curves in a graph corresponding to a single pressure
- ☐ Charles's law: At constant pressure, the volume of a gas is proportional to the absolute temperature, $V \propto T$ at constant pressure
- ☐ Avogadro's principle: Equal volumes of gases at the same pressure and temperature contain the same numbers of molecules
- ☐ gas constant, R
- ☐ perfect gas equation, $pV = nRT$
- ☐ perfect gas, a gas that obeys $pV = nRT$ exactly under all conditions
- ☐ real gas, an actual gas
- ☐ standard ambient temperature and pressure (SATP), 298.15 K and 1 bar
- ☐ standard temperature and pressure (STP), 0°C and 1 atm
- ☐ combined gas equation, $p_1V_1/T_1 = p_2V_2/T_2$
- ☐ Dalton's law: The pressure exerted by a mixture of gases is the sum of the partial pressures of the gases
- ☐ mole fraction, the amount of J, n_J, expressed as a fraction of the total amount of molecules, n, in the sample, $x_J = n_J/n$
- ☐ partial pressure of a perfect gas, the pressure a gas would exert if it occupied the container alone
- ☐ partial pressure (in general), $p_J = x_J p$

Real gases

1.3 Molecular interactions

- ☐ compression factor, $Z = V_m/V_m^\circ$
- ☐ virial equation of state, $pV_m = RT(1 + B/V_m + C/V_m^2 + \cdots)$
- ☐ virial coefficient, the coefficients B, C, ...
- ☐ Boyle temperature, the temperature at which $dZ/dp = 0$ as $p \to 0$
- ☐ vapour pressure, the pressure of a vapour in equilibrium with its condensed phase
- ☐ critical point, the point at which the volumes at each end of the horizontal part of the isotherm have merged to a single point
- ☐ critical temperature, the temperature at the critical point
- ☐ critical pressure, the pressure at the critical point
- ☐ critical molar volume, the molar volume at the critical point
- ☐ critical constants, p_c, V_c, and T_c
- ☐ supercritical fluid, a dense fluid phase above its critical temperature and pressure

1.4 The van der Waals equation

- ☐ van der Waals equation, $p = nRT/(V - nb) - a(n/V)^2$
- ☐ van der Waals coefficients, the parameters a and b
- ☐ van der Waals loops, unrealistic oscillations in the van der Waals isotherms
- ☐ Maxwell construction, the equal-area rule for the replacement of the van der Waals loops by straight lines
- ☐ critical compression factor, the value of Z at the critical point

1.5 The principle of corresponding states

- ☐ reduced variables, the actual variable divided by the corresponding critical constant
- ☐ principle of corresponding states: Real gases at the same reduced volume and reduced temperature exert the same reduced pressure

Further reading

Articles of general interest

J.M. Alvaiño, J. Veguillas, and S. Valasco, Equations of state, collisional energy transfer, and chemical equilibrium in gases. *J. Chem. Educ.* **66**, 157 (1989).

J.G. Eberhart, The many faces of van der Waals's equation of state. *J. Chem. Educ.* **66**, 906 (1989).

J.G. Eberhart, A least-squares technique for determining the van der Waals parameters from the critical constants. *J. Chem. Educ.* **69**, 220 (1992).

E.F. Meyer and T.P. Meyer, Supercritical fluid: liquid, gas, both, or neither? A different approach. *J. Chem. Educ.* **63**, 463 (1986).

I.M. Mills, The choice of names and symbols for quantities in chemistry. *J. Chem. Educ.* **66**, 887 (1989).

J.L. Pauley and E.H. Davis, *P–V–T* isotherms of real gases: Experimental versus calculated values. *J. Chem. Educ.* **63**, 466 (1986).

G. Rhodes, Does a one-molecule gas obey Boyle's law? *J. Chem. Educ.* **69**, 16 (1992).

M. Ross, Equations of state. In *Encyclopedia of applied physics* (ed. G.L. Trigg), **6**, 291. VCH, New York (1993).

J.B. Ott, J.R. Coates, and H.T. Hall, Comparison of equations of state. *J. Chem. Educ.* **48**, 515 (1971).

Texts and sources of data and information

J.H. Dymond and E.B. Smith, *The virial coefficients of pure gases and mixtures*. Oxford University Press (1980).

A.D. McNaught and A. Wilkinson, *Compendium of chemical terminology*. Blackwell Scientific, Oxford (1997).

J.O. Hirschfelder, C.F. Curtiss, and R.B. Bird, *The molecular theory of gases and liquids*. Wiley, New York (1954).

I.M. Mills (ed.), *Quantities, units, and symbols in physical chemistry*. Blackwell Scientific, Oxford (1993).

B.W. Rossiter and R.C. Baetzgold (ed.) *Physical methods of chemistry*, VI. Wiley-Interscience, New York (1992).

A.J. Walton, *Three phases of matter*. Oxford University Press (1983).

Exercises

Discussion questions

1.1 (a) Explain how the experiments of Boyle, Charles (and others), and Avogadro led to the formulation of the perfect gas equation of state.

1.1 (b) Explain the term 'partial pressure' and explain why Dalton's law is a limiting law.

1.2 (a) Explain how the compression factor varies with pressure and temperature and describe how it reveals information about intermolecular interactions in real gases.

1.2 (b) What is the significance of the critical constants?

1.3 (a) Describe the formulation of the van der Waals equation and suggest a rationale for one other equation of state in Table 1.6.

1.3 (b) Explain how the van der Waals equation accounts for critical behaviour.

Numerical exercises

1.4 (a) A sample of air occupies 1.0 L at 25°C and 1.00 atm. What pressure is needed to compress it to 100 cm^3 at this temperature?

1.4 (b) A sample of carbon dioxide gas occupies 2.0 dm^3 at 20°C and 104 kPa. What pressure is needed to compress it to 250 cm^3 at this temperature?

1.5 (a) (a) Could 131 g of xenon gas in a vessel of volume 1.0 L exert a pressure of 20 atm at 25°C if it behaved as a perfect gas? If not, what pressure would it exert? (b) What pressure would it exert if it behaved as a van der Waals gas?

1.5 (b) (a) Could 25 g of argon gas in a vessel of volume 1.5 L exert a pressure of 2.0 bar at 30°C if it behaved as a perfect gas? If not, what pressure would it exert? (b) What pressure would it exert if it behaved as a van der Waals gas?

1.6 (a) A perfect gas undergoes isothermal compression, which reduces its volume by 2.20 L. The final pressure and volume of the gas are 3.78×10^3 Torr and 4.65 L, respectively. Calculate the original pressure of the gas in (a) Torr, (b) atm.

1.6 (b) A perfect gas undergoes isothermal compression, which reduces its volume by 1.80 dm^3. The final pressure and volume of the gas are 1.48×10^3 Torr and 2.14 dm^3, respectively. Calculate the original pressure of the gas in (a) Torr, (b) bar.

1.7 (a) To what temperature must a 1.0 L sample of a perfect gas be cooled from 25°C in order to reduce its volume to 100 cm^3?

1.7 (b) To what temperature must a sample of a perfect gas of volume 500 mL be cooled from 35°C in order to reduce its volume to 150 cm^3?

1.8 (a) A car tyre (that is, an automobile tire) was inflated to a pressure of 24 lb in^{-2} (1.00 atm = 14.7 lb in^{-2}) on a winter's day when the temperature was −5°C. What pressure will be found, assuming no leaks have occurred and that the volume is constant, on a subsequent summer's day when the temperature is 35°C? What complications should be taken into account in practice?

1.8 (b) A sample of hydrogen gas was found to have a pressure of 125 kPa when the temperature was 23°C. What can its pressure be expected to be when the temperature is 11°C?

1.9 (a) A sample of 255 mg of neon occupies 3.00 L at 122 K. Use the perfect gas law to calculate the pressure of the gas.

1.9 (b) A homeowner uses 4.00×10^3 m^3 of natural gas in a year to heat a home. Assume that natural gas is all methane, CH_4, and that methane is a perfect gas for the conditions of this problem, which are 1.00 atm and 20°C. What is the mass of gas used?

1.10 (a) In an attempt to determine an accurate value of the gas constant, R, a student heated a 20.000 L container filled with 0.251 32 g of helium gas to 500°C and measured the pressure as 206.402 cm of water in a manometer at 25°C. Calculate the value of R from these data. (The density of water at 25°C is 0.997 07 g cm^{-3}.)

1.10 (b) The following data have been obtained for oxygen gas at 273.15 K. Calculate the best value of the gas constant R from them and the best value of the molar mass of O_2.

p/atm	0.750 000	0.500 000	0.250 000
V_m/(L mol^{-1})	29.9649	44.8090	89.6384
ρ/(g L^{-1})	1.071 44	0.714 110	0.356 975

1.11 (a) At 500°C and 699 Torr, the mass density of sulfur vapour is 3.71 g L^{-1}. What is the molecular formula of sulfur under these conditions?

1.11 (b) At 100°C and 120 Torr, the mass density of phosphorus vapour is 0.6388 kg m^{-3}. What is the molecular formula of phosphorus under these conditions?

1.12 (a) Calculate the mass of water vapour present in a room of volume 400 m^3 that contains air at 27°C on a day when the relative humidity is 60 per cent.

1.12 (b) Calculate the mass of water vapour present in a room of volume 250 m^3 that contains air at 23°C on a day when the relative humidity is 53 per cent.

1.13 (a) Given that the density of air at 740 Torr and 27°C is 1.146 g L^{-1}, calculate the mole fraction and partial pressure of nitrogen and oxygen assuming that (a) air consists only of these two gases, (b) air also contains 1.0 mole per cent Ar.

1.13 (b) A gas mixture consists of 320 mg of methane, 175 mg of argon, and 225 mg of neon. The partial pressure of neon at 300 K is 66.5 Torr. Calculate (a) the volume and (b) the total pressure of the mixture.

1.14 (a) The density of a gaseous compound was found to be 1.23 g L^{-1} at 330 K and 150 Torr. What is the molar mass of the compound?

1.14 (b) In an experiment to measure the molar mass of a gas, 250 cm^3 of the gas was confined in a glass vessel. The pressure was 152 Torr at 298 K and, after correcting for buoyancy effects, the mass of the gas was 33.5 mg. What is the molar mass of the gas?

1.15 (a) The densities of air at −85°C, 0°C, and 100°C are 1.877 g L^{-1}, 1.294 g L^{-1}, and 0.946 g L^{-1}, respectively. From these data, and assuming that air obeys Charles's law, determine a value for the absolute zero of temperature in degrees Celsius.

1.15 (b) A certain sample of a gas has a volume of 20.00 L at 0°C and 1.000 atm. A plot of the experimental data of its volume against the Celsius temperature, θ, at constant p, gives a straight line of slope 0.0741 L (°C)$^{-1}$. From these data alone (without making use of the perfect gas law), determine the absolute zero of temperature in degrees Celsius.

1.16 (a) Calculate the pressure exerted by 1.0 mol C_2H_6 behaving as (a) a perfect gas, (b) a van der Waals gas when it is confined under the following conditions: (i) at 273.15 K in 22.414 L, (ii) at 1000 K in 100 cm^3. Use the data in Table 1.5.

1.16 (b) Calculate the pressure exerted by 1.0 mol H_2S behaving as (a) a perfect gas, (b) a van der Waals gas when it is confined under the following conditions: (i) at 273.15 K in 22.414 L, (ii) at 500 K in 150 cm^3. Use the data in Table 1.5.

1.17 (a) Estimate the critical constants of a gas with van der Waals parameters $a = 0.751$ atm L^2 mol^{-2} and $b = 0.0226$ L mol^{-1}.

1.17 (b) Estimate the critical constants of a gas with van der Waals parameters $a = 1.32$ atm L^2 mol^{-2} and $b = 0.0436$ L mol^{-1}.

1.18 (a) A gas at 250 K and 15 atm has a molar volume 12 per cent smaller than that calculated from the perfect gas law. Calculate (a) the compression factor under these conditions and (b) the molar volume of the gas. Which are dominating in the sample, the attractive or the repulsive forces?

1.18 (b) A gas at 350 K and 12 atm has a molar volume 12 per cent larger than that calculated from the perfect gas law. Calculate (a) the compression factor under these conditions and (b) the molar volume of the gas. Which are dominating in the sample, the attractive or the repulsive forces?

1.19 (a) In an industrial process, nitrogen is heated to 500 K at a constant volume of 1.000 m^3. The gas enters the container at 300 K and 100 atm. The mass of the gas is 92.4 kg. Use the van der Waals equation to determine the approximate pressure of the gas at its working temperature of 500 K. For nitrogen, $a = 1.352$ L^2 atm mol^{-2}, $b = 0.0387$ L mol^{-1}.

1.19 (b) Cylinders of compressed gas are typically filled to a pressure of 200 bar. For oxygen, what would be the molar volume at this pressure and 25°C based on (a) the perfect gas equation, (b) the van der Waals equation. For oxygen, $a = 1.364$ L^2 atm mol^{-2}, $b = 3.19 \times 10^{-2}$ L mol^{-1}.

1.20 (a) The density of water vapour at 327.6 atm and 776.4 K is 133.2 g dm^{-3}. (a) Determine the molar volume, V_m, of water and the compression factor, Z, from these data. (b) Calculate Z from the van der Waals equation with $a = 5.464$ L^2 atm mol^{-2} and $b = 0.0305$ L mol^{-1}.

1.20 (b) The density of water vapour at 1.00 bar and 383 K is 950.99 kg m^{-3}. (a) Determine the molar volume, V_m, of water and the compression factor, Z, from these data. (b) Calculate Z from the van der Waals equation with $a = 5.464$ L^2 atm mol^{-2} and $b = 0.0305$ L mol^{-1}.

1.21 (a) Suppose that 10.0 mol C_2H_6(g) is confined to 4.860 L at 27°C. Predict the pressure exerted by the ethane from (a) the perfect gas and (b) the van der Waals equations of state. Calculate the compression factor based on these calculations. For ethane, $a = 5.507$ L^2 atm mol^{-2}, $b = 0.0651$ L mol^{-1}.

1.21 (b) At 300 K and 20 atm, the compression factor of a gas is 0.86. Calculate (a) the volume occupied by 8.2 mmol of the gas under these conditions and (b) an approximate value of the second virial coefficient B at 300 K.

1.22 (a) A vessel of volume 22.4 L contains 2.0 mol H_2 and 1.0 mol N_2 at 273.15 K. Calculate (a) the mole fractions of each component, (b) their partial pressures, and (c) their total pressure.

1.22 (b) A vessel of volume 22.4 L contains 1.5 mol H_2 and 2.5 mol N_2 at 273.15 K. Calculate (a) the mole fractions of each component, (b) their partial pressures, and (c) their total pressure.

1.23 (a) The critical constants of methane are $p_c = 45.6$ atm, $V_c = 98.7$ cm^3 mol^{-1}, and $T_c = 190.6$ K. Calculate the van der Waals parameters of the gas and estimate the radius of the molecules.

1.23 (b) The critical constants of ethane are $p_c = 48.20$ atm, $V_c = 148$ cm^3 mol^{-1}, and $T_c = 305.4$ K. Calculate the van der Waals parameters of the gas and estimate the radius of the molecules.

1.24 (a) Use the van der Waals parameters for chlorine to calculate approximate values of (a) the Boyle temperature of chlorine and (b) the radius of a Cl_2 molecule regarded as a sphere.

1.24 (b) Use the van der Waals parameters for hydrogen sulfide to calculate approximate values of (a) the Boyle temperature of the gas and (b) the radius of a H_2S molecule regarded as a sphere.

1.25 (a) Suggest the pressure and temperature at which 1.0 mol of (a) NH_3, (b) Xe, (c) He will be in states that correspond to 1.0 mol H_2 at 1.0 atm and 25°C.

1.25 (b) Suggest the pressure and temperature at which 1.0 mol of (a) H_2S, (b) CO_2, (c) Ar will be in states that correspond to 1.0 mol N_2 at 1.0 atm and 25°C.

1.26 (a) A certain gas obeys the van der Waals equation with $a = 0.50$ m^6 Pa mol^{-2}. Its volume is found to be 5.00×10^{-4} m^3 mol^{-1} at 273 K and 3.0 MPa. From this information calculate the van der Waals constant b. What is the compression factor for this gas at the prevailing temperature and pressure?

1.26 (b) A certain gas obeys the van der Waals equation with $a = 0.76$ m^6 Pa mol^{-2}. Its volume is found to be 4.00×10^{-4} m^3 mol^{-1} at 288 K and 4.0 MPa. From this information calculate the van der Waals constant b. What is the compression factor for this gas at the prevailing temperature and pressure?

Problems†

Numerical problems

1.1 A diving bell has an air space of 3.0 m^3 when on the deck of a boat. What is the volume of the air space when the bell has been lowered to a depth of 50 m? Take the mean density of sea water to be 1.025 g cm^{-3} and assume that the temperature is the same as on the surface.

1.2 What pressure difference must be generated across the length of a 15 cm vertical drinking straw in order to drink a water-like liquid of density 1.0 g cm^{-3}?

1.3 Recent communications with the inhabitants of Neptune have revealed that they have a Celsius-type temperature scale, but based on the melting point (0°N) and boiling point (100°N) of their most common substance, hydrogen. Further communications have revealed that the Neptunians know about perfect gas behaviour and they find that, in the limit of zero pressure, the value of pV is 28 L atm at 0°N and 40 L atm at 100°N. What is the value of the absolute zero of temperature on their temperature scale?

1.4 A meteorological balloon had a radius of 1.0 m when released at sea level at 20°C and expanded to a radius of 3.0 m when it had risen to its maximum altitude where the temperature was −20°C. What is the pressure inside the balloon at that altitude?

1.5 Deduce the relation between the pressure and mass density, ρ, of a perfect gas of molar mass M. Confirm graphically, using the following data on dimethyl ether at 25°C, that perfect behaviour is reached at low pressures and find the molar mass of the gas.

p/Torr	91.74	188.98	277.3	452.8	639.3	760.0
ρ/(g L^{-1})	0.225	0.456	0.664	1.062	1.468	1.734

1.6 Charles's law is sometimes expressed in the form $V = V_0(1 + \alpha\theta)$, where θ is the Celsius temperature, α is a constant, and V_0 is the volume of the sample at 0°C. The following values for α have been reported for nitrogen at 0°C:

p/Torr	749.7	599.6	333.1	98.6
$10^3\alpha$/(°C)$^{-1}$	3.6717	3.6697	3.6665	3.6643

For these data calculate the best value for the absolute zero of temperature on the Celsius scale.

1.7 The molar mass of a newly synthesized fluorocarbon was measured in a gas microbalance. This device consists of a glass bulb forming one end of a beam, the whole surrounded by a closed container. The beam is pivoted, and the balance point is attained by raising the pressure of gas in the container, so increasing the buoyancy of the enclosed bulb. In one experiment, the balance point was reached when the fluorocarbon pressure was 327.10 Torr; for the same setting of the pivot, a balance was reached when CHF_3 (M = 70.014 g mol^{-1}) was introduced at 423.22 Torr. A repeat of the experiment with a different setting of the pivot required a pressure of 293.22 Torr of the fluorocarbon and 427.22 Torr of the CHF_3. What is the molar mass of the fluorocarbon? Suggest a molecular formula.

1.8 A constant-volume perfect gas thermometer indicates a pressure of 50.2 Torr at the triple point temperature of water (273.16 K). (a) What change of pressure indicates a change of 1.00 K at this temperature? (b) What pressure indicates a temperature of 100.00°C? (c) What change of pressure indicates a change of 1.00 K at the latter temperature?

1.9 A vessel of volume 22.4 L contains 2.0 mol H_2 and 1.0 mol N_2 at 273.15 K initially. All the H_2 reacted with sufficient N_2 to form NH_3. Calculate the partial pressures and the total pressure of the final mixture.

1.10 Calculate the molar volume of chlorine gas at 350 K and 2.30 atm using (a) the perfect gas law and (b) the van der Waals equation. Use the answer to (a) to calculate a first approximation to the correction term for attraction and then use successive approximations to obtain a numerical answer for part (b).

1.11 At 273 K measurements on argon gave $B = -21.7$ cm^3 mol^{-1} and $C = 1200$ cm^6 mol^{-2}, where B and C are the second and third virial coefficients in the expansion of Z in powers of $1/V_m$. Assuming that the perfect gas law holds sufficiently well for the estimation of the second and third terms of the expansion, calculate the compression factor of argon at 100 atm and 273 K. From your result, estimate the molar volume of argon under these conditions.

1.12 Calculate the volume occupied by 1.00 mol N_2 using the van der Waals equation in the form of a virial expansion at (a) its critical temperature, (b) its Boyle temperature, and (c) its inversion temperature. Assume that the pressure is 10 atm throughout. At what temperature is the gas most perfect? Use the following data: $T_c = 126.3$ K, $a = 1.352$ L^2 atm mol^{-2}, $b = 0.0387$ L mol^{-1}.

1.13‡ The second virial coefficient of methane can be approximated by the empirical equation $B'(T) = a + be^{-c/T^2}$, where $a = -0.1993$ bar^{-1}, $b = 0.2002$ bar^{-1}, and $c = 1131$ K^2 with 300 K < T < 600 K. (a) What is the value of the Boyle temperature for methane? (b) Compare the molar volume of methane at 50 bar as predicted by both the virial equation and the perfect gas equation at (i) 298 K, (ii) 373 K.

1.14 The mass density of water vapour at 327.6 atm and 776.4 K is 1.332×10^2 g L^{-1}. Given that for water $T_c = 647.4$ K, $p_c = 218.3$ atm, $a = 5.464$ L^2 atm mol^{-2}, $b = 0.0305$ L mol^{-1}, and $M = 18.02$ g mol^{-1}, calculate (a) the molar volume. Then calculate the compression factor (b) from the data, (c) from the virial expansion of the van der Waals equation.

1.15 The critical volume and critical pressure of a certain gas are 160 cm^3 mol^{-1} and 40 atm, respectively. Estimate the critical temperature by assuming that the gas obeys the Berthelot equation of state. Estimate the radii of the gas molecules on the assumption that they are spheres.

1.16 Estimate the coefficients a and b in the Dieterici equation of state from the critical constants of xenon. Calculate the pressure exerted by 1.0 mol Xe when it is confined to 1.0 L at 25°C.

Theoretical problems

1.17 Show that the van der Waals equation leads to values of $Z < 1$ and $Z > 1$, and identify the conditions for which these values are obtained.

† Problems denoted with the symbol ‡ were supplied by Charles Trapp, Carmen Giunta, and Marshall Cady.

1.18 Express the van der Waals equation of state as a virial expansion in powers of $1/V_m$ and obtain expressions for B and C in terms of the parameters a and b. The expansion you will need is $(1-x)^{-1} = 1 + x + x^2 + \cdots$. Measurements on argon gave $B = -21.7\ \text{cm}^3\ \text{mol}^{-1}$ and $C = 1200\ \text{cm}^6\ \text{mol}^{-2}$ for the virial coefficients at 273 K. What are the values of a and b in the corresponding van der Waals equation of state?

1.19‡ Derive the relation between the critical constants and the Dieterici equation parameters. Show that $Z_c = 2e^{-2}$ and derive the reduced form of the Dieterici equation of state. Compare the van der Waals and Dieterici predictions of the critical compression factor. Which is closer to typical experimental values?

1.20 A scientist proposed the following equation of state:

$$p = \frac{RT}{V_m} - \frac{B}{V_m^2} + \frac{C}{V_m^3}$$

Show that the equation leads to critical behaviour. Find the critical constants of the gas in terms of B and C and an expression for the critical compression factor.

1.21 Equations 1.21 and 1.22 are expansions in p and $1/V_m$, respectively. Find the relation between B, C and B', C''.

1.22 The second virial coefficient B' can be obtained from measurements of the density ρ of a gas at a series of pressures. Show that the graph of p/ρ against p should be a straight line with slope proportional to B'. Use the data on dimethyl ether in Problem 1.5 to find the values of B' and B at 25°C.

1.23 The equation of state of a certain gas is given by $p = RT/V_m + (a + bT)/V_m^2$, where a and b are constants. Find $(\partial V/\partial T)_p$.

1.24 The following equations of state are occasionally used for approximate calculations on gases: (gas A) $pV_m = RT(1 + b/V_m)$, (gas B) $p(V_m - b) = RT$. Assuming that there were gases that actually obeyed these equations of state, would it be possible to liquefy either gas A or B? Would they have a critical temperature? Explain your answer.

1.25 Derive an expression for the compression factor of a gas that obeys the equation of state $p(V - nb) = nRT$, where b and R are constants. If the pressure and temperature are such that $V_m = 10b$, what is the numerical value of the compression factor?

1.26‡ Amedeo Avogadro (*Journal de Physique* (1811)) noted that two volumes of hydrogen combine with one volume of oxygen to form two volumes of water vapour. He also gave the density of water vapour relative to that of air as 0.625 and hydrogen as 0.0732 (in units in which air had unit density). Use this information and Avogadro's principle to compute the molar masses of water vapour and oxygen relative to that of hydrogen.

1.27‡ The discovery of the element argon by Lord Rayleigh and Sir William Ramsay had its origins in Rayleigh's measurements of the density of nitrogen with an eye toward accurate determination of its molar mass. Rayleigh prepared some samples of nitrogen by chemical reaction of nitrogen-containing compounds; under his standard conditions, a glass globe filled with this 'chemical nitrogen' had a mass of 2.2990 g. He prepared other samples by removing oxygen, carbon dioxide, and water vapour from atmospheric air; under the same conditions, this 'atmospheric nitrogen' had a mass of 2.3102 g (Lord Rayleigh, *Royal Institution Proceedings* **14**, 524 (1895)). With the hindsight of knowing accurate values for the molar masses of nitrogen and argon, compute the mole fraction of argon in the latter sample on the assumption that the former was pure nitrogen and the latter a mixture of nitrogen and argon.

1.28‡ A substance as elementary and well-known as argon still receives research attention. Stewart and Jacobsen have published a review of thermodynamic properties of argon (R.B. Stewart and R.T. Jacobsen, *J. Phys. Chem. Ref. Data* **18**, 639 (1989)) that included the following 300 K isotherm.

p/MPa	0.4000	0.5000	0.6000	0.8000	1.000
V_m/(L mol^{-1})	6.2208	4.9736	4.1423	3.1031	2.4795
p/MPa	1.500	2.000	2.500	3.000	4.000
V_m/(L mol^{-1})	1.6483	1.2328	0.983 57	0.817 46	0.609 98

(a) Compute the second virial coefficient, B, at this temperature. (b) Use nonlinear curve-fitting software to compute the third virial coefficient, C, at this temperature.

1.29‡ Show that the compression factor, Z, of a van der Waals gas can be expressed as $Z = V_r'/(V_r' - \frac{1}{8}) - 27/64T_rV_r'$, where $V_r' = p_cV_m/RT_c$ is the 'pseudoreduced volume', or alternatively as the solution of the cubic equation $Z^3 - \{(p_r/8T_r) + 1\}Z^2 + (27p_r/64T_r^2)Z - 27p_r^2/512T_r^3 = 0$. Solve this equation for Z for nitrogen, methane, propane, and ethene at $T_r = 1.2$ and $p_r = 3.0$ and compare to the value given in Fig. 1.18.

Applications: to atmospheric science

1.30 Balloons are still used to deploy sensors that monitor meteorological phenomena and the chemistry of the atmosphere. It is possible to investigate some of the technicalities of ballooning by using the perfect gas law. Suppose your balloon has a radius of 3.0 m and that it is spherical. (a) What amount of H_2 (in moles) is needed to inflate it to 1.0 atm in an ambient temperature of 25°C at sea level? (b) What mass can the balloon lift at sea level, where the density of air is 1.22 kg m^{-3}? (c) What would be the payload if He were used instead of H_2?

1.31‡ The preceding problem is most readily solved (see the *Solutions manual*) with the use of the Archimedes principle, which states that the lifting force is equal to the difference between the weight of the displaced air and the weight of the balloon. Prove the Archimedes principle for the atmosphere from the barometric formula. *Hint.* Assume a simple shape for the balloon, perhaps a right circular cylinder of cross-sectional area A and height h.

1.32‡ Chlorofluorocarbons such as CCl_3F and CCl_2F_2 have been linked to ozone depletion in Antarctica. As of 1994, these gases were found in quantities of 261 and 509 parts per trillion (10^{12}) by volume (World Resources Institute, *World Resources* 1996–97). Compute the molar concentration of these gases under conditions typical of (a) the mid-latitude troposphere (10°C and 1.0 atm) and (b) the Antarctic stratosphere (200 K and 0.050 atm).

1.33‡ The composition of the atmosphere is approximately 80 per cent nitrogen and 20 per cent oxygen by mass. At what height above the surface of the earth would the atmosphere become 90 per cent nitrogen and 10 per cent oxygen by mass? Assume that the temperature of the atmosphere is constant at 25°C. What is the pressure of the atmosphere at that height?

2 The First Law: the concepts

This chapter introduces some of the basic concepts of thermodynamics. It concentrates on the conservation of energy—the experimental observation that energy can be neither created nor destroyed—and shows how the principle of the conservation of energy can be used to assess the energy changes that accompany physical and chemical processes. Much of this chapter examines the means by which a system can exchange energy with its surroundings in terms of the work it may do or the heat that it may produce. The target concept of the chapter is enthalpy, which is a very useful book-keeping property for keeping track of the heat output (or requirements) of physical processes and chemical reactions at constant pressure.

The release of energy can be used to provide heat when a fuel burns in a furnace, to produce mechanical work when a fuel burns in an engine, and to generate electrical work when a chemical reaction pumps electrons through a circuit. In chemistry, we encounter reactions that can be harnessed to provide heat and work, reactions that liberate energy, which is squandered (often to the detriment of the environment), but that give products we require, and reactions that constitute the processes of life. **Thermodynamics**, the study of the transformations of energy, enables us to discuss all these matters quantitatively and to make useful predictions.

The basic concepts

For the purposes of physical chemistry, the universe is divided into two parts, the system and its surroundings. The **system** is the part of the world in which we have a special interest. It may be a reaction vessel, an engine, an electrochemical cell, a biological cell, and so on. The **surroundings** comprise the region outside the system and are where we make our measurements. The type of system depends on the characteristics of the boundary that divides it from the surroundings (Fig. 2.1). If matter can be transferred through the boundary between the system and its surroundings the system is classified as **open**. If matter cannot pass through the boundary the system is classified as **closed**. Both open and closed systems can exchange energy with their surroundings. For example, a closed system can expand and thereby raise a weight in the surroundings; it may also transfer energy to them if they are at a lower temperature. An **isolated system** is a closed system that has neither mechanical nor thermal contact with its surroundings.

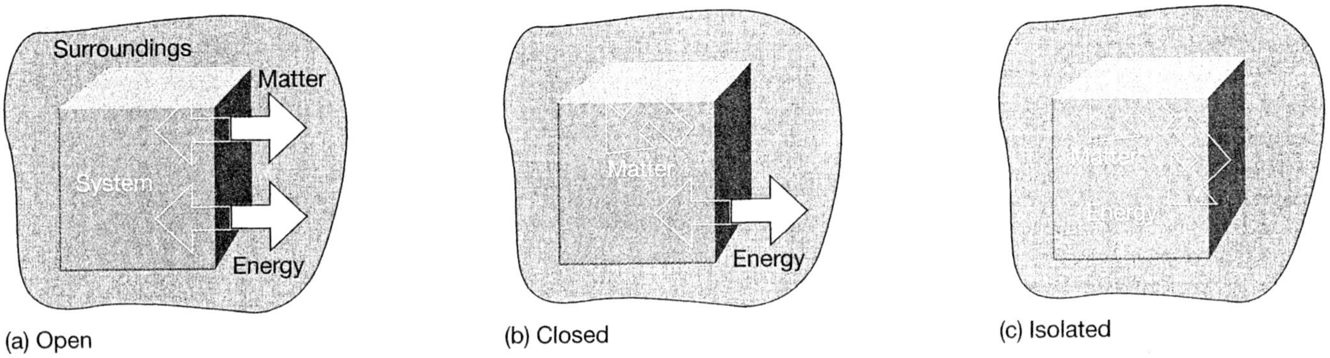

2.1 (a) An open system can exchange matter and energy with its surroundings. (b) A closed system can exchange energy with its surroundings, but it cannot exchange matter. (c) An isolated system can exchange neither energy nor matter with its surroundings.

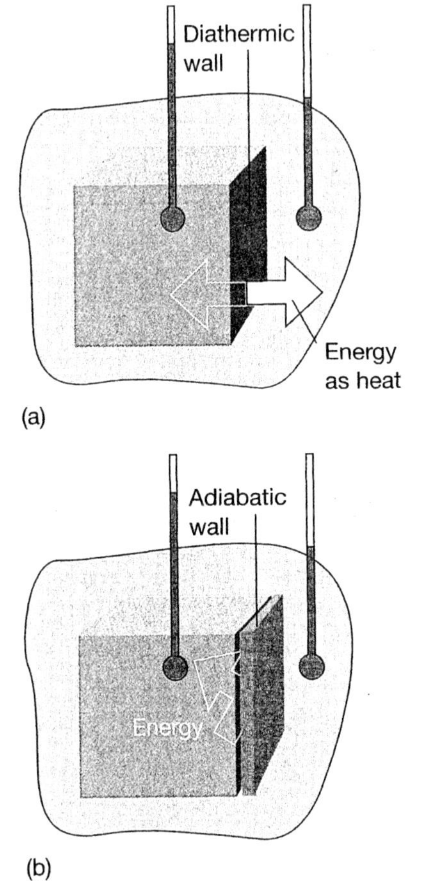

2.2 (a) A diathermic system is one that allows energy to escape as heat through its boundary if there is a difference in temperature between the system and its surroundings. (b) An adiabatic system is one that does not permit the passage of energy as heat through its boundary even if there is a temperature difference between the system and its surroundings.

2.1 Work, heat, and energy

The fundamental physical property in thermodynamics is work: **work** is done when an object is moved against an opposing force. Doing work is equivalent to raising a weight somewhere in the surroundings. An example of doing work is the expansion of a gas that pushes out a piston and raises a weight. A chemical reaction that drives an electric current through a resistance also does work, because the same current could be driven through a motor and used to raise a weight.

The **energy** of a system is its capacity to do work. When work is done on an otherwise isolated system (for instance, by compressing a gas or winding a spring), the capacity of the system to do work is increased, so the energy of the system is increased. When the system does work (when the piston moves out or the spring unwinds), the energy of the system is reduced and it can do less work than before.

Experiments have shown that the energy of a system may be changed by means other than work itself. When the energy of a system changes as a result of a temperature difference between the system and its surroundings we say that energy has been transferred as **heat**. When a heater is immersed in a beaker of water (the system), the capacity of the system to do work increases because hot water can be used to do more work than cold water. Not all boundaries permit the transfer of energy even though there is a temperature difference between the system and its surroundings. As we have seen (Section 1.1c), a boundary that does permit energy transfer as heat (such as steel and glass) is called **diathermic** and a boundary that does not permit energy transfer as heat is called **adiabatic** (Fig. 2.2).

An **exothermic process** is a process that releases energy as heat. All combustion reactions are exothermic. An **endothermic process** is a process that absorbs energy as heat. An example of an endothermic process is the vaporization of water. An endothermic process in a diathermic container results in energy flowing into the system as heat. An exothermic process in a similar diathermic container results in a release of energy as heat into the surroundings. When an endothermic process takes place in an adiabatic container, it results in a lowering of temperature of the system; an exothermic process results in a rise of temperature. These features are summarized in Fig. 2.3.

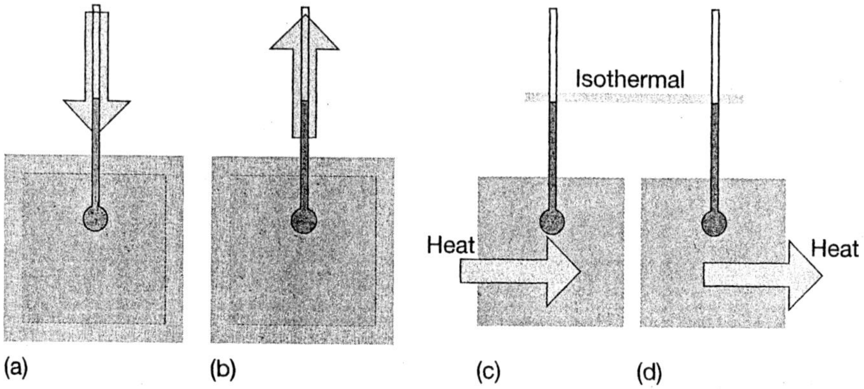

2.3 (a) When an endothermic process occurs in an adiabatic system, the temperature falls; (b) if the process is exothermic, then the temperature rises. (c) When an endothermic process occurs in a diathermic container, energy enters as heat from the surroundings, and the system remains at the same temperature. (d) If the process is exothermic, then energy leaves as heat, and the process is isothermal.

Molecular interpretation 2.1 Heat, work, and energy

In molecular terms, heat is the transfer of energy that makes use of *disorderly molecular motion*. The disorderly motion of molecules is called **thermal motion**. The thermal motion of the molecules in the hot surroundings stimulates the molecules in the cooler system to move more vigorously and, as a result, the energy of the system is increased. When a system heats its surroundings, molecules of the system stimulate the thermal motion of the molecules in the surroundings (Fig. 2.4).

In contrast, *work is the transfer of energy that makes use of organized motion* (Fig. 2.5). When a weight is raised or lowered, its atoms move in an organized way. The atoms

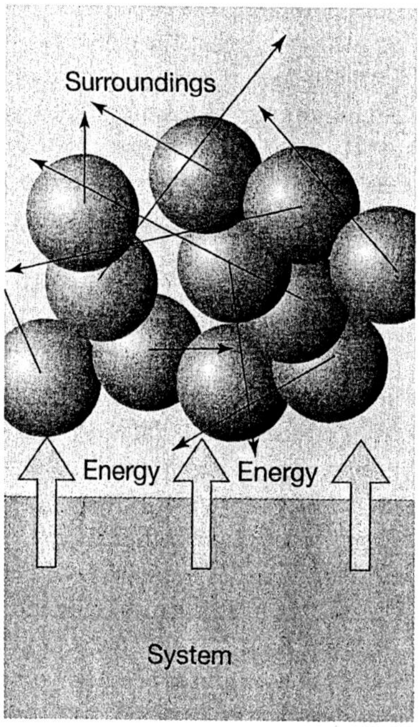

2.4 When energy is transferred to the surroundings as heat, the transfer stimulates disordered motion of the atoms in the surroundings. Transfer of energy from the surroundings to the system makes use of disordered motion (thermal motion) in the surroundings.

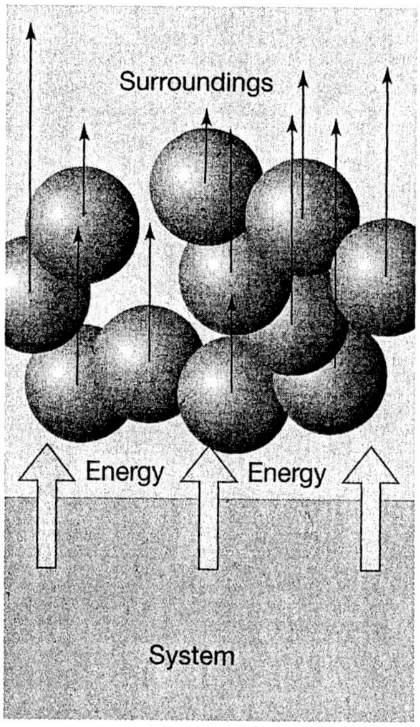

2.5 When a system does work, it stimulates orderly motion in the surroundings. For instance, the atoms shown here may be part of a weight that is being raised. The ordered motion of the atoms in a falling weight does work on the system.

in a spring move in an orderly way when it is wound; the electrons in an electric current move in an orderly direction when it flows. When a system does work it causes atoms or electrons in its surroundings to move in an organized way. Likewise, when work is done on a system, molecules in the surroundings are used to transfer energy to it in an organized way, as the atoms in a weight are lowered or a current of electrons is passed.

The distinction between work and heat is made in the surroundings. The fact that a falling weight may stimulate thermal motion in the system is irrelevant to the distinction between heat and work: work is identified as energy transfer making use of the organized motion of atoms in the surroundings, and heat is identified as energy transfer making use of thermal motion in the surroundings. In the compression of a gas, for instance, work is done as the particles of the compressing weight descend in an orderly way, but the effect of the incoming piston is to accelerate the gas molecules to higher average speeds. Because collisions between molecules quickly randomize their directions, the orderly motion of the atoms of the weight is in effect stimulating thermal motion in the gas. We observe the falling weight, the orderly descent of its atoms, and report that work is being done even though it is stimulating thermal motion.

The great revolution in physics that occurred in the opening years of the twentieth century was the realization that energy is *quantized*; that is, a particle can possess only certain energies, called its 'energy levels'. Broadly speaking, the heavier the particle and the greater its spatial freedom, the closer together are its energy levels. The number of particles, N_i, in a sample of N particles, that will be found in a state with an energy E_i when it is part of a system in thermal equilibrium at a temperature T is expressed by the *Boltzmann distribution*:

$$N_i = \frac{Ne^{-E_i/kT}}{q} \qquad q = \sum_i e^{-E_i/kT}$$

where k is *Boltzmann's constant*, a fundamental constant with the value 1.381×10^{-23} J K^{-1}. Boltzmann's constant is related to the gas constant by $R = N_A k$, where N_A is Avogadro's constant. One reason why the gas constant appears in so many expressions, including those apparently wholly unrelated to gases, is that it is really the fundamental constant k appearing in disguise. Broadly speaking, the Boltzmann distribution implies that the population of states decays exponentially with increasing energy, so most particles will be found in states of low energy and only a few will be found in states of high energy. Care must be taken with the exact interpretation though, because more than one state may correspond to the same energy: that is, an energy level may be degenerate and consist of several states. The function q is called the *partition function*. It plays a central role in the calculation of thermodynamic properties (Chapter 19).

2.2 The First Law

In thermodynamics, the total energy of a system is called its **internal energy**, U. The internal energy is the total kinetic and potential energy of the molecules in the system.[1] We denote by ΔU the change in internal energy when a system changes from an initial state i with internal energy U_i to a final state f of internal energy U_f:

$$\Delta U = U_f - U_i \qquad [2.1]$$

1. The internal energy does not include the kinetic energy arising from the motion of the system as a whole, such as its kinetic energy as it accompanies the Earth on its orbit round the Sun.

The internal energy is a **state function** in the sense that its value depends only on the current state of the system and is independent of how that state has been prepared. In other words, it is a function of the properties that determine the current state of the system. Changing any one of the state variables, such as the pressure, results in a change in internal energy. The internal energy is an extensive property.

Internal energy, heat, and work are all measured in the same units, the joule (J). The joule, which is named after the nineteenth-century scientist J.P. Joule, is defined as

$$1\ \text{J} = 1\ \text{kg m}^2\ \text{s}^{-2}$$

A joule is quite a small unit of energy: for instance, each beat of the human heart consumes about 1 J. Changes in molar internal energy, ΔU_m, are typically expressed in kilojoules per mole (kJ mol^{-1}). Certain other energy units are also used, but are more common in fields other than thermodynamics. Thus, 1 electronvolt (1 eV) is defined as the kinetic energy acquired when an electron is accelerated from rest through a potential difference of 1 V; the relation between electronvolts and joules is 1 eV ≈ 0.16 aJ (where 1 aJ = 10^{-18} J). Many processes in chemistry have an energy of several electronvolts. Thus, the energy to remove an electron from a sodium atom is close to 5 eV. Calories (cal) and kilocalories (kcal) are still encountered:

$$1\ \text{cal} = 4.184\ \text{J exactly}$$

An energy of 1 cal is enough to raise the temperature of 1 g of water by 1°C.

Molecular interpretation 2.2 The internal energy of a gas

The key to relating the internal energy to the properties of individual molecules is the *equipartition theorem* of classical mechanics. There are two parts to this theorem, one qualitative, the other quantitative. First, we need to know that a 'quadratic contribution' to the energy means a contribution that can be expressed as the square of a variable, such as the position or the velocity. For example, the kinetic energy of an atom of mass m as it moves through space is

$$E_K = \tfrac{1}{2}mv_x^2 + \tfrac{1}{2}mv_y^2 + \tfrac{1}{2}mv_z^2$$

and there are three quadratic contributions to its energy. The equipartition theorem then states that, for a collection of particles at thermal equilibrium at a temperature T, *the average value of each quadratic contribution to the energy is the same and equal to* $\tfrac{1}{2}kT$, where k is Boltzmann's constant (see *Molecular interpretation 2.1*). The equipartition theorem is a conclusion from classical mechanics, and is applicable only when the effects of quantization can be ignored. In practice, it can be used for molecular translation and rotation but not vibration. At 25°C, $\tfrac{1}{2}kT = 2$ zJ (where 1 zJ = 10^{-21} J), or about 13 meV.

According to the equipartition theorem, the average energy of each term in the expression above is $\tfrac{1}{2}kT$. Therefore, the mean energy of the atoms is $\tfrac{3}{2}kT$ and the total energy of the gas (there being no potential energy contribution) is $\tfrac{3}{2}NkT$, or $\tfrac{3}{2}nRT$. We can therefore write

$$U_m = U_m(0) + \tfrac{3}{2}RT$$

where $U_m(0)$ is the molar internal energy at $T = 0$, when all translational motion has ceased and the sole contribution to the internal energy arises from the internal structure of the atoms. This equation shows that the internal energy of a perfect gas increases linearly with temperature. At 25°C, $\tfrac{3}{2}RT$ = 3.7 kJ mol^{-1}, so translational motion contributes about 4 kJ mol^{-1} to the molar internal energy of a monatomic gas (the remaining contribution arises from the internal structure of the atoms).

Because the potential energy of interactions between the atoms (or molecules) of a perfect gas is zero, the internal energy is independent of how close they are together. Therefore, the internal energy is independent of the volume occupied by a perfect gas. A formal statement of this property is

$$\left(\frac{\partial U}{\partial V}\right)_T = 0$$

The internal energy of interacting molecules in condensed phases also has a contribution from the potential energy of their interaction. However, no simple expressions can be written down in general. Nevertheless, the crucial molecular point is that, as the temperature of a system is raised, the internal energy increases as the various modes of motion become more highly excited.

Self-test 2.1 Estimate the contribution of the rotational and translational motion of a nonlinear molecule to its molar internal energy. (see *Further information 2*) [$U_m = U_m(0) + 3RT$; $3RT = 7.4$ kJ mol^{-1}]

(a) The conservation of energy

It has been found experimentally that the internal energy of a system may be changed either by doing work on the system or by heating it. Whereas we may know how the energy transfer has occurred (because we can see if a weight has been raised or lowered in the surroundings, indicating transfer of energy by doing work, or if ice has melted in the surroundings, indicating transfer of energy as heat), the system is blind to the mode employed. *Heat and work are equivalent ways of changing a system's internal energy.* A system is like a bank: it accepts deposits in either currency, but stores its reserves as internal energy. It is also found experimentally that, if a system is isolated from its surroundings, then no change in internal energy takes place. This summary of observations is now known as the **First Law of thermodynamics** and expressed as follows:

The internal energy of an isolated system is constant.

We cannot use a system to do work, leave it isolated for a month, and then come back expecting to find it restored to its original state and ready to do the same work again. The evidence for this property is that no 'perpetual motion machine' (a machine that does work without consuming fuel or some other source of energy) has ever been built.

These remarks may be summarized as follows. If we write w for the work done on a system, q for the energy transferred as heat to a system, and ΔU for the resulting change in internal energy, then it follows that

$$\Delta U = q + w \tag{2.2}$$

Equation 2 is the mathematical statement of the First Law, for it summarizes the equivalence of heat and work and the fact that the internal energy is constant in an isolated system (for which $q = 0$ and $w = 0$). The equation states that the change in internal energy of a closed system is equal to the energy that passes through its boundary as heat or work. It employs the 'acquisitive convention', in which $w > 0$ or $q > 0$ if energy is transferred to the system as work or heat and $w < 0$ or $q < 0$ if energy is lost from the system as work or heat. In other words, we view the flow of energy as work or heat from the system's perspective.

Illustration 2.1

If an electric motor produced 15 kJ of energy each second as mechanical work and lost 2 kJ as heat to the surroundings, then the change in the internal energy of the motor each second is

$$\Delta U = -2\ \text{kJ} - 15\ \text{kJ} = -17\ \text{kJ}$$

Suppose that, when a spring was wound, 100 J of work was done on it but 15 J escaped to the surroundings as heat. The change in internal energy of the spring is

$$\Delta U = +100\ \text{kJ} - 15\ \text{kJ} = +85\ \text{kJ}$$

(b) The formal statement of the First Law

The expression of the First Law that we have given is adequate for most purposes in thermodynamics. However, there are several unsatisfactory features about it, such as how we define and measure 'heat'. This section gives a more sophisticated version of the law, and shows how eqn 2 can be put on a firmer foundation. As the remainder of the text does not depend on this material, it is possible to omit it and go immediately to the following section ('Work and heat').

We begin by pretending that we do not know what we mean by 'energy'. We pretend that we know only what is meant by work, because we can observe a weight being raised or lowered in the surroundings. We also know how to measure work by noting the height through which the weight is raised. Throughout this section, work will be the fundamental, measurable quantity, and we define energy, heat, and the First Law in terms of work alone. We shall employ terms that have been established by the Zeroth Law of thermodynamics (Section 1.1), namely, state and temperature and the concepts of adiabatic and diathermic walls.

In an adiabatic system of a given composition, it is known experimentally that the same increase in temperature is brought about by the same quantity of any kind of work we do on the system. Thus, if 1 kJ of mechanical work is done on the system (by stirring it with rotating paddles, for instance), or 1 kJ of electrical work is done (by passing an electric current through a heater), and so on, then the same rise in temperature is produced. The following statement of the First Law of thermodynamics is a summary of a large number of observations of this kind:

The work needed to change an adiabatic system from one specified state to another specified state is the same however the work is done.

This form of the law looks completely different from the form we gave before, but we shall now see how it implies eqn 2.

Suppose we do work w_{ad} on an adiabatic system to change it from an initial state i to a final state f. The work may be of any kind (mechanical or electrical) and may take the system through different intermediate states (different temperatures and pressures, for instance). We might (in ignorance of the First Law) think that we need to label w_{ad} with the path and to write w_{ad}(mechanical) or w_{ad}(electrical). However, the First Law tells us that w_{ad} is the same for all paths and depends only on the initial and final states. This conclusion is analogous to climbing a mountain: the height we must climb between any two points is independent of the path we take (Fig. 2.6). In mountain climbing we can attach a number, the altitude, A, to each point on the mountain and express the height, h, of the climb as a difference in altitudes:

$$h = A_f - A_i = \Delta A$$

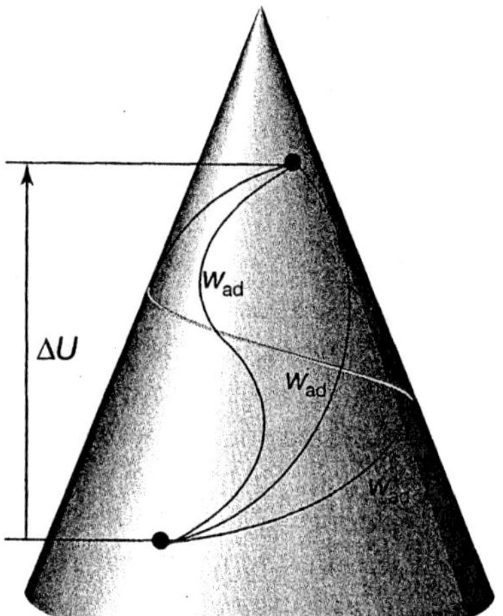

2.6 It is found that the same quantity of work must be done on an adiabatic system to achieve the same change of state even though different means of achieving that work may be used. This path independence implies the existence of a state function, the internal energy. The change in internal energy is like the change in altitude when climbing a mountain: its value is independent of path.

That is, in mountain climbing, the observation that h is independent of the path implies the existence of the state function A. The First Law has exactly the same implication. The fact that w_{ad} is independent of the path implies that to each state of the system we can attach a value of a quantity—we call it the 'internal energy', U—and express the work as a difference in internal energies:

$$\Delta w_{ad} = U_f - U_i = \Delta U \qquad [2.3]$$

This equation also shows that we can measure the change in the internal energy of a system by measuring the work needed to bring about the change in an adiabatic system.

(c) The mechanical definition of heat

Suppose we strip away the thermal insulation around the system and make it diathermic. The system is now in thermal contact with its surroundings as we drive it from the same initial state to the same final state. The change in internal energy is the same as before, because U is a state function, but we might find that the work we must do is not the same as before. Thus, whereas we might have needed to do 42 kJ of work when the system was in an adiabatic container, to achieve the same change of state we might now have to do 50 kJ of work. The difference between the work done in the two cases is *defined* as the heat absorbed by the system in the process:

$$q = w_{ad} - w \qquad [2.4]$$

In the present case, we would conclude that $q = 42\ \text{kJ} - 50\ \text{kJ} = -8\ \text{kJ}$, and report that 8 kJ of energy had left the system as heat. We see that we now have a purely mechanical definition of heat in terms of work. We know how to measure work in terms of the height through which a weight falls, so we now also have a method for measuring heat in terms of work.

Finally, we can express eqn 4 in a more familiar way. Because we already know that ΔU is (by definition) equal to w_{ad}, the expression for the energy transferred to the system as heat is $q = \Delta U - w$. However, this expression is equivalent to eqn 2, the mathematical form of the First Law that we saw earlier.

Work and heat

The way can now be opened to powerful methods of calculation by switching attention to infinitesimal changes of state (such as infinitesimal change in temperature) and infinitesimal changes in the internal energy dU. Then, if the work done on a system is dw and the energy supplied to it as heat is dq, in place of eqn 2 we have

$$dU = dq + dw \qquad (2.5)$$

To use eqn 5 we must be able to relate dq and dw to events taking place in the surroundings.

2.3 Expansion work

We begin by discussing **expansion work**, the work arising from a change in volume. This type of work includes the work done by a gas as it expands and drives back the atmosphere. Many chemical reactions result in the generation or consumption of

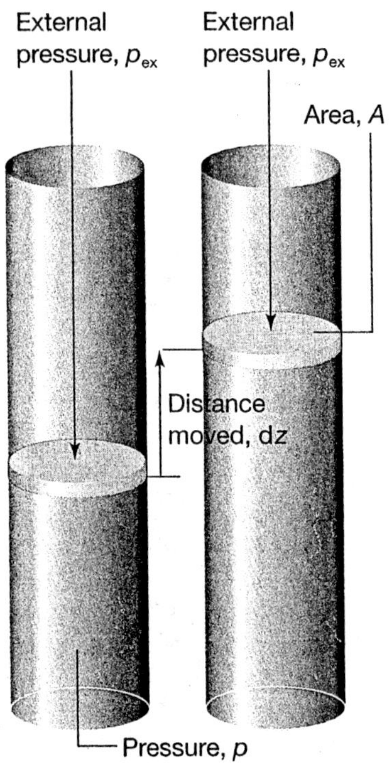

2.7 When a piston of area A moves out through a distance dz, it sweeps out a volume $\mathrm{d}V = A\,\mathrm{d}z$. The external pressure p_{ex} is equivalent to a weight pressing on the piston, and the force opposing expansion is $F = p_{ex}A$.

gases (for instance, the thermal decomposition of calcium carbonate or the combustion of octane), and the thermodynamic characteristics of a reaction depend on the work it can do. The term 'expansion work' also includes work associated with negative changes of volume, that is, compression.

(a) The general expression for work

The calculation of expansion work starts from the definition used in physics, which states that the work required to move an object a distance dz against an opposing force of magnitude F is

$$\mathrm{d}w = -F\,\mathrm{d}z \qquad [2.6]$$

The negative sign tells us that, when the system moves an object against an opposing force, the internal energy of the system doing the work will decrease. Now consider the arrangement shown in Fig. 2.7, in which one wall of a system is a massless, frictionless, rigid, perfectly fitting piston of area A. If the external pressure is p_{ex}, the magnitude of the force acting on the outer face of the piston is $F = p_{ex}A$. When the system expands through a distance dz against an external pressure p_{ex}, it follows that the work done is $\mathrm{d}w = -p_{ex}A\,\mathrm{d}z$. But $A\,\mathrm{d}z$ is the change in volume, dV, in the course of the expansion. Therefore, the work done when the system expands by dV against a pressure p_{ex} is

$$\mathrm{d}w = -p_{ex}\,\mathrm{d}V \qquad (2.7)$$

To obtain the total work done when the volume changes from V_i to V_f we integrate this expression between the initial and final volumes:

$$w = -\int_{V_i}^{V_f} p_{ex}\,\mathrm{d}V \qquad (2.8)$$

The force acting on the piston, $p_{ex}A$, is equivalent to a weight that is raised as the system expands.

If the system is compressed instead, then the same weight is lowered in the surroundings and eqn 8 can still be used, but now $V_f < V_i$. It is important to note that it is still the external pressure that determines the magnitude of the work. This somewhat perplexing conclusion seems to be inconsistent with the fact that the gas *inside* the container is opposing the compression. However, when a gas is compressed, the ability of the *surroundings* to do work is diminished by an amount determined by the weight that is lowered, and it is this energy that is transferred into the system.

Other types of work (for example, electrical work), which we shall call either **non-expansion work** or **additional work**, have analogous expressions, with each one the product of an intensive factor (the pressure, for instance) and an extensive factor (the change in volume). Some are collected in Table 2.1. For the present we continue with the work associated with changing the volume, the expansion work, and see what we can extract from eqn 7.

(b) Free expansion

By **free expansion** we mean expansion against zero opposing force. It occurs when $p_{ex} = 0$. According to eqn 7, $\mathrm{d}w = 0$ for each stage of the expansion. Hence, overall:

$$w = 0 \qquad (2.9)$$

That is, no work is done when a system expands freely. Expansion of this kind occurs when a system expands into a vacuum.

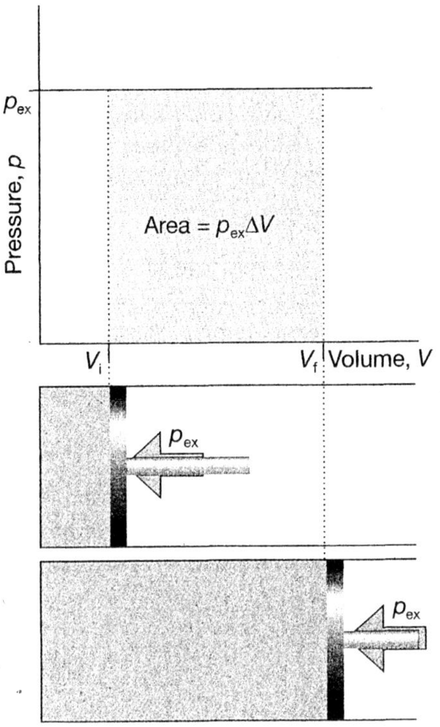

2.8 The work done by a gas when it expands against a constant external pressure, p_{ex}, is equal to the shaded area in this example of an indicator diagram.

The value of the integral $\int_a^b f(x)\,dx$ is equal to the area under the graph of $f(x)$ between $x = a$ and $x = b$.

Table 2.1 Varieties of work*

Type of work	dw	Comments	Units†
Expansion	$-p_{ex}\,dV$	p_{ex} is the external pressure dV is the change in volume	Pa m^3
Surface expansion	$\gamma\,d\sigma$	γ is the surface tension $d\sigma$ is the change in area	$N\,m^{-1}$ m^2
Extension	$f\,dl$	f is the tension dl is the change of length	N m
Electrical	$\phi\,dq$	ϕ is the electric potential dq is the change in charge	V C

* In general, the work done on a system can be expressed in the form $dw = -F\,dz$, where F is a 'generalized force' and dz is a 'generalized displacement'.
† For work in joules (J). Note that 1 N m = 1 J and 1 V C = 1 J.

(c) Expansion against constant pressure

Now suppose that the external pressure is constant throughout the expansion. For example, the piston may be pressed on by the atmosphere, which exerts the same pressure throughout the expansion. A chemical example of this condition is the expansion of a gas formed in a chemical reaction. We can evaluate eqn 8 by taking the constant p_{ex} outside the integral:

$$w = -p_{ex}\int_{V_i}^{V_f} dV = -p_{ex}(V_f - V_i)$$

Therefore, if we write the change in volume as $\Delta V = V_f - V_i$,

$$w = -p_{ex}\Delta V \qquad (2.10)$$

This result is illustrated graphically in Fig. 2.8, which makes use of the fact that an integral can be interpreted as an area. The magnitude of w, denoted $|w|$, is equal to the area beneath the horizontal line at $p = p_{ex}$ lying between the initial and final volumes. A p,V-graph used to compute expansion work is called an **indicator diagram**; James Watt first used one to indicate aspects of the operation of his steam engine.

(d) Reversible expansion

A **reversible change** in thermodynamics is a change that can be reversed by an infinitesimal modification of a variable. The key word 'infinitesimal' sharpens the everyday meaning of the word 'reversible' as something that can change direction. We say that a system is in **equilibrium** with its surroundings if an infinitesimal change in the conditions in opposite directions results in opposite changes in its state. One example of reversibility that we have encountered already is the thermal equilibrium of two systems with the same temperature. The transfer of energy as heat between the two is reversible because, if the temperature of either system is lowered infinitesimally, then energy flows into the system with the lower temperature. If the temperature of either system at thermal equilibrium is raised infinitesimally, then energy flows out of the hotter system.

Suppose a gas is confined by a piston and that the external pressure, p_{ex}, is set equal to the pressure, p, of the confined gas. Such a system is in mechanical equilibrium with its surroundings (as illustrated in Section 1.1) because an infinitesimal change in the external pressure in either direction causes changes in volume in opposite directions. If the external pressure is reduced infinitesimally, then the gas expands slightly. If the external pressure is increased infinitesimally, then the gas contracts slightly. In either case the change is reversible in the thermodynamic sense. If, on the other hand, the external pressure differs measurably from the internal pressure, then changing p_{ex} infinitesimally will not decrease it below the pressure of the gas and so will not change the direction of the process. Such a system is not in mechanical equilibrium with its surroundings and the expansion is thermodynamically irreversible.

To achieve reversible expansion we set p_{ex} equal to p at each stage of the expansion. In practice, this equalization could be achieved by gradually removing weights from the piston so that the downward force due to the weights always matches the changing upward force due to the pressure of the gas. When we set $p_{ex} = p$, eqn 7 becomes

$$dw = -p_{ex}\,dV = -p\,dV \qquad (2.11)_{rev}$$

(Equations valid only for reversible processes are labelled with a subscript 'rev'.) Although the pressure inside the system appears in this expression for the work, it does so only because p_{ex} has been set equal to p to ensure reversibility. The total work of reversible expansion is therefore

$$w = -\int_{V_i}^{V_f} p\,dV \qquad (2.12)_{rev}$$

We can evaluate the integral once we know how the pressure of the confined gas depends on its volume. Equation 12 is the link with the material covered in Chapter 1 for, if we know the equation of state of the gas, then we can express p in terms of V and evaluate the integral.

(e) Isothermal reversible expansion

Consider the isothermal, reversible expansion of a perfect gas. The expansion is made isothermal by keeping the system in thermal contact with its surroundings (which may be a constant-temperature bath). Because the equation of state is $pV = nRT$, we know that at each stage $p = nRT/V$, with V the volume at that stage of the expansion. The temperature T is constant in an isothermal expansion, so (together with n and R) it may be taken outside the integral. It follows that the work of reversible isothermal expansion of a perfect gas from V_i to V_f at a temperature T is

$$w = -nRT\int_{V_i}^{V_f} \frac{dV}{V} = -nRT \ln \frac{V_f}{V_i} \qquad (2.13)^{\circ}_{rev}$$

When the final volume is greater than the initial volume, as in an expansion, the logarithm in eqn 13 is positive and hence $w < 0$. In this case, the system has done work on the surroundings and the internal energy of the system has decreased as a result of the work it has done.[2] The equations also show that more work is done for a given change of volume when the temperature is increased. The greater pressure of the confined gas then needs a higher opposing pressure to ensure reversibility.

We can express the result of the calculation as an indicator diagram, for the magnitude of the work done is equal to the area under the isotherm $p = nRT/V$ (Fig. 2.9). Superimposed on the diagram is the rectangular area obtained for irreversible expansion against constant external pressure fixed at the same final value as that reached in the reversible expansion. More work is obtained when the expansion is reversible (the area is greater) because matching the external pressure to the inter-

2. We shall see later that there is a compensating influx of energy as heat, so overall the internal energy is constant for the isothermal expansion of a perfect gas.

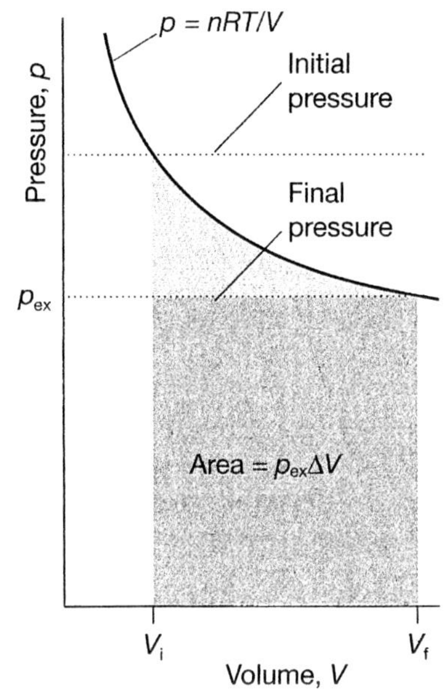

2.9 The work done by a perfect gas when it expands reversibly and isothermally is equal to the area under the isotherm $p = nRT/V$. The work done during the irreversible expansion against the same final pressure is equal to the rectangular area shown slightly darker. Note that the reversible work is greater than the irreversible work.

nal pressure at each stage of the process ensures that none of the system's pushing power is wasted. We cannot obtain more work than for the reversible process because increasing the external pressure even infinitesimally at any stage results in compression. We may infer from this discussion that, because some pushing power is wasted when $p > p_{ex}$, the maximum work available from a system operating between specified initial and final states and passing along a specified path is obtained when the change takes place reversibly.

We have introduced the connection between reversibility and maximum work for the special case of a perfect gas undergoing expansion. Later (in Section 4.5c) we shall see that it applies to all substances and to all kinds of work.

Example 2.1 Calculating the work of gas production

Calculate the work done when 50 g of iron reacts with hydrochloric acid to produce hydrogen gas in (a) a closed vessel of fixed volume, (b) an open beaker at 25°C.

Method We need to judge the magnitude of the volume change and then to decide how the process occurs. If there is no change in volume, there is no expansion work however the process takes place. If the system expands against a constant external pressure, the work can be calculated from eqn 10. A general feature of processes in which a condensed phase changes into a gas is that the volume of the former may usually be neglected relative to that of the gas it forms.

Answer In (a) the volume cannot change, so no expansion work is done and $w = 0$. In (b) the gas drives back the atmosphere and therefore $w = -p_{ex}\Delta V$. We can neglect the initial volume because the final volume (after the production of gas) is so much larger and $\Delta V = V_f - V_i \approx V_f = nRT/p_{ex}$, where n is the amount of H_2 produced. Therefore,

$$w = -p_{ex}\Delta V \approx -p_{ex} \times \frac{nRT}{p_{ex}} = -nRT$$

Because the reaction is $Fe(s) + 2\,HCl(aq) \rightarrow FeCl_2(aq) + H_2(g)$, we know that 1 mol H_2 is generated when 1 mol Fe is consumed, and n can be taken as the amount of Fe atoms that react. Because the molar mass of Fe is 55.85 g mol^{-1}, it follows that

$$w \approx -\frac{50\ \text{g}}{55.85\ \text{g mol}^{-1}} \times (8.314\,51\ \text{J K}^{-1}\ \text{mol}^{-1}) \times (298.15\ \text{K})$$
$$\approx -2.2\ \text{kJ}$$

The system (the reaction mixture) does 2.2 kJ of work driving back the atmosphere.

Comment Note that (for this perfect gas system) the external pressure does not affect the final result: the lower the pressure, the larger the volume occupied by the gas, so the effects cancel.

Self-test 2.2 Calculate the expansion work done when 50 g of water is electrolysed under constant pressure at 25°C. [−10 kJ]

2.4 Heat transactions

In general, the change in internal energy of a system is

$$dU = dq + dw_{exp} + dw_e \tag{2.14}$$

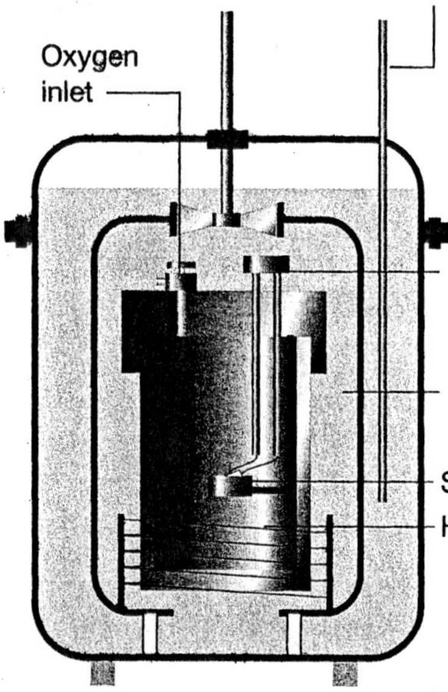

2.10 A constant-volume bomb calorimeter. The 'bomb' is the central vessel, which is strong enough to withstand high pressures. The calorimeter (for which the heat capacity must be known) is the entire assembly shown here. To ensure adiabaticity, the calorimeter is immersed in a water bath with a temperature continuously readjusted to that of the calorimeter at each stage of the combustion.

where dw_e is work in addition (e for 'extra') to the expansion work, dw_{exp}. For instance, dw_e might be the electrical work of driving a current through a circuit. A system kept at constant volume can do no expansion work, so $dw_{exp} = 0$. If the system is also incapable of doing any other kind of work (if it is not, for instance, an electrochemical cell connected to an electric motor), then $dw_e = 0$ too. Under these circumstances:

$$dU = dq \quad \text{(at constant volume, no additional work)} \tag{2.15}$$

We express this relation by writing $dU = dq_V$, where the subscript implies a change at constant volume. For a measurable change,

$$\Delta U = q_V \tag{2.16}$$

It follows that, by measuring the energy supplied to a constant-volume system as heat ($q > 0$) or obtained from it as heat ($q < 0$) when it undergoes a change of state, we are in fact measuring the change in its internal energy.

(a) Calorimetry

Calorimetry is the study of heat transfer during physical and chemical processes. A **calorimeter** is a device for measuring the heat transferred. The most common device for measuring ΔU is an **adiabatic bomb calorimeter** (Fig. 2.10). The process we wish to study—which may be a chemical reaction—is initiated inside a constant-volume container, the 'bomb'. The bomb is immersed in a stirred water bath, and the whole device is the calorimeter. The calorimeter is also immersed in an outer water bath. The water in the calorimeter and that of the outer bath are both monitored and adjusted to the same temperature. This arrangement ensures that there is no net loss of heat from the calorimeter to the surroundings (the bath) and hence that the calorimeter is adiabatic.

The change in temperature, ΔT, of the calorimeter is proportional to the heat that the reaction releases or absorbs. Therefore, by measuring ΔT we can determine q_V and hence find ΔU. The conversion of ΔT to q_V is best achieved by calibrating the calorimeter using a process of known energy output and determining the **calorimeter constant**, the constant C in the relation

$$q = C\Delta T \tag{2.17}$$

The calorimeter constant may be measured electrically by passing a current, I, from a source of known potential, $\mathcal{V}$, through a heater for a known period of time, t:

$$q = I\mathcal{V}t \tag{2.18}$$

Alternatively, C may be determined by burning a known mass of substance (benzoic acid is often used) that has a known heat output. With C known, it is simple to interpret an observed temperature rise as a release of heat.

Illustration 2.2

If we pass a current of 10.0 A from a 12 V supply for 300 s, then from eqn 18 the energy supplied as heat is

$$q = (10.0\ \text{A}) \times (12\ \text{V}) \times (300\ \text{s}) = 3.6 \times 10^4\ \text{A V s} = 36\ \text{kJ}$$

because 1 A V s = 1 J. If the observed rise in temperature is 5.5 K, then the calorimeter constant is $C = (36\ \text{kJ})/(5.5\ \text{K}) = 6.5\ \text{kJ K}^{-1}$.

(b) Heat capacity

The internal energy of a substance increases when its temperature is raised. The increase depends on the conditions under which the heating takes place and for the present we shall suppose that the sample is confined to a constant volume. For example, the sample may be a gas in a container of fixed volume. If the internal energy is plotted against temperature, then a curve like that in Fig. 2.11 may be obtained. The slope of the tangent to the curve at any temperature is called the **heat capacity** of the system at that temperature. The **heat capacity at constant volume** is denoted C_V and is defined formally as[3]

$$C_V = \left(\frac{\partial U}{\partial T}\right)_V \qquad [2.19]$$

Partial derivatives are reviewed in *Further information 1*.

The notation is that of a 'partial derivative'. A partial derivative is a slope calculated with all variables except one held constant. In this case, the internal energy varies with the temperature and the volume of the sample, but we are interested only in its variation with the temperature, the volume being held constant (Fig. 2.12).

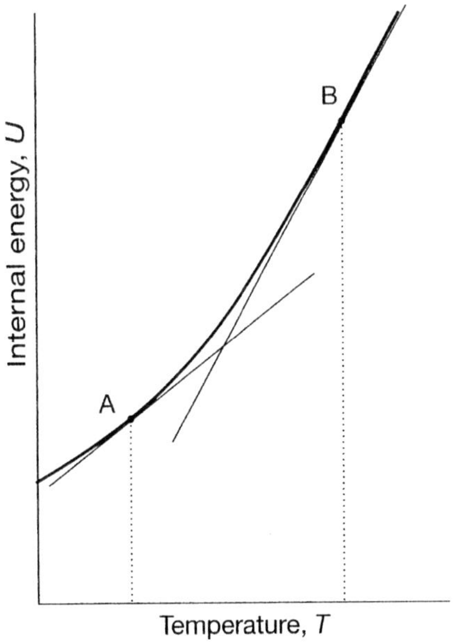

2.11 The internal energy of a system increases as the temperature is raised; this graph shows its variation as the system is heated at constant volume. The slope of the tangent to the curve at any temperature is the heat capacity at constant volume at that temperature. Note that, for the system illustrated, the heat capacity is greater at B than at A.

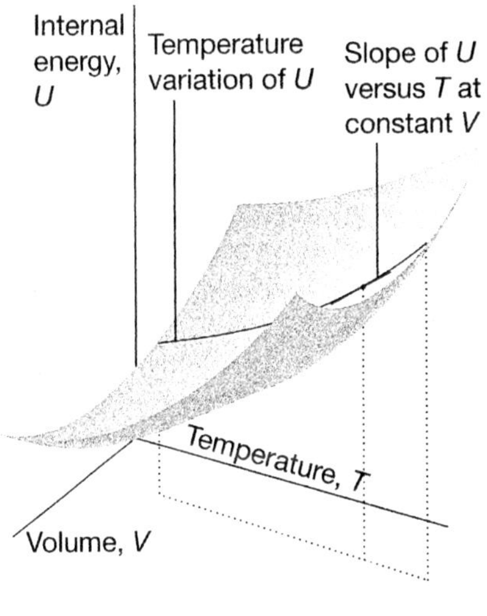

2.12 The internal energy of a system varies with volume and temperature, perhaps as shown here by the surface. The variation of the internal energy with temperature at one particular constant volume is illustrated by the curve drawn parallel to T. The slope of this curve at any point is the partial derivative $(\partial U/\partial T)_V$.

3. If the system can change its composition, it is necessary to distinguish between equilibrium and fixed-composition values of C_V. All applications in this chapter refer to a single substance, so this complication can be ignored.

Illustration 2.3

The heat capacity of a monatomic perfect gas can be calculated by inserting the expression for the internal energy derived in *Molecular interpretation 2.2*. There we saw that $U_m = U_m(0) + \frac{3}{2}RT$, so from eqn 19

$$C_{V,m} = \frac{\partial}{\partial T}(U_m(0) + \tfrac{3}{2}RT) = \tfrac{3}{2}R \qquad (2.20)$$

The numerical value is 12.47 J K^{-1} mol^{-1}.

Heat capacities are extensive properties: 100 g of water, for instance, has 100 times the heat capacity of 1 g of water (and therefore requires 100 times the heat to bring about the same rise in temperature). The **molar heat capacity at constant volume,** $C_{V,m}$, is the heat capacity per mole of material, and is an intensive property (all molar quantities are intensive). Typical values of $C_{V,m}$ for polyatomic gases are close to 25 J K^{-1} mol^{-1}. For certain applications it is useful to know the **specific heat capacity** (more informally, the 'specific heat') of a substance, which is the heat capacity of the sample divided by the mass, usually in grams. The specific heat capacity of water at room temperature is close to 4 J K^{-1} g^{-1}. In general, heat capacities depend on the temperature and decrease at low temperatures. However, over small ranges of temperature at and above room temperature, the variation is quite small and for approximate calculations heat capacities can be treated as almost independent of temperature.

The heat capacity can be used to relate a change in internal energy to a change in temperature of a constant-volume system. It follows from eqn 19 that

$$dU = C_V\,dT \qquad \text{(at constant volume)} \qquad (2.21a)$$

That is, an infinitesimal change in temperature brings about an infinitesimal change in internal energy, and the constant of proportionality is the heat capacity at constant volume. If the heat capacity is independent of temperature over the range of temperatures of interest, a measurable change of temperature, ΔT, brings about a measurable increase in internal energy, ΔU, where

$$\Delta U = C_V\Delta T \qquad \text{(at constant volume)} \qquad (2.21b)$$

Because a change in internal energy can be identified with the heat supplied at constant volume (eqn 16), the last equation can be written

$$q_V = C_V\Delta T \qquad (2.22)$$

This relation provides a simple way of measuring the heat capacity of a sample: a measured quantity of heat is supplied to the sample (electrically, for example), and the resulting increase in temperature is monitored. The ratio of the heat supplied to the temperature rise it causes is the heat capacity of the sample.

A large heat capacity implies that, for a given quantity of heat, there will be only a small increase in temperature (the sample has a large capacity for heat). An infinite heat capacity implies that there will be no increase in temperature however much heat is supplied. At a phase transition, such as at the boiling point of water, the temperature of a substance does not rise as heat is supplied: the energy is used to drive the endothermic phase transition, in this case to vaporize the water, rather than to increase its temperature. Therefore, at the temperature of a phase transition, the heat capacity of a sample is infinite. The properties of heat capacities close to phase transitions are treated more fully in Section 6.7.

2.5 Enthalpy

The change in internal energy is not equal to the heat supplied when the system is free to change its volume. Under these circumstances some of the energy supplied as heat to the system is returned to the surroundings as expansion work (Fig. 2.13), so $\mathrm{d}U$ is less than $\mathrm{d}q$. However, we shall now show that in this case the heat supplied at constant pressure is equal to the change in another thermodynamic property of the system, the enthalpy.

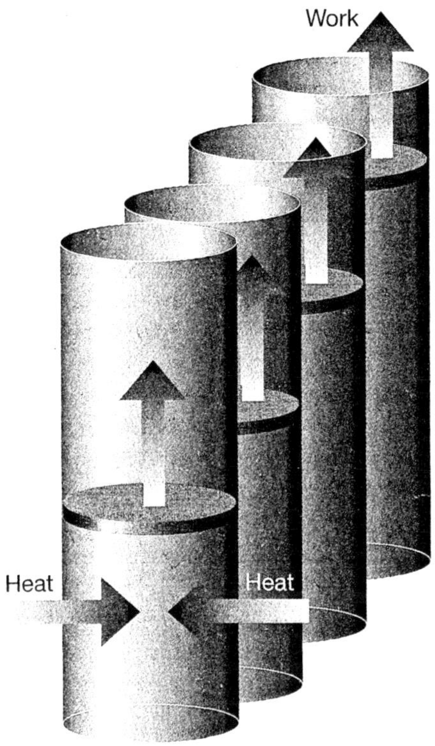

2.13 When a system is subjected to constant pressure and is free to change its volume, some of the energy supplied as heat may escape back into the surroundings as work. In such a case, the change in internal energy is smaller than the energy supplied as heat.

(a) The definition of enthalpy

The **enthalpy**, H, is defined as

$$H = U + pV \qquad [2.23]$$

where p is the pressure of the system and V is its volume. Because U, p, and V are all state functions, the enthalpy is a state function too. As is true of any state function, the change in enthalpy, ΔH, between any pair of initial and final states is independent of the path between them.

Although the definition of enthalpy may appear arbitrary, it has important implications for thermochemistry. For instance, we show in the following *Justification* that eqn 23 implies that *the change in enthalpy is equal to the heat supplied at constant pressure* (so long as the system does no additional work):

$$\mathrm{d}H = \mathrm{d}q \quad \text{(at constant pressure, no additional work)} \qquad (2.24a)$$

For a measurable change,

$$\Delta H = q_p \qquad (2.24b)$$

Justification 2.1

For a general infinitesimal change in the state of the system, U changes to $U + \mathrm{d}U$, p changes to $p + \mathrm{d}p$, and V changes to $V + \mathrm{d}V$, so from the definition in eqn 23, H changes from $U + pV$ to

$$H + \mathrm{d}H = (U + \mathrm{d}U) + (p + \mathrm{d}p)(V + \mathrm{d}V) = U + \mathrm{d}U + pV + p\,\mathrm{d}V + V\,\mathrm{d}p + \mathrm{d}p\,\mathrm{d}V$$

The last term is the product of two infinitesimally small quantities and can therefore be neglected. As a result, after recognizing $U + pV = H$ on the right, we find that H changes to

$$H + \mathrm{d}H = H + \mathrm{d}U + p\,\mathrm{d}V + V\,\mathrm{d}p$$

and hence that

$$\mathrm{d}H = \mathrm{d}U + p\,\mathrm{d}V + V\,\mathrm{d}p$$

This conclusion is the basis of the general 'product rule' for the differentiation of the product of two functions f and g: $\mathrm{d}(fg) = f\,\mathrm{d}g + g\,\mathrm{d}f$.

If we now substitute $\mathrm{d}U = \mathrm{d}q + \mathrm{d}w$ into this expression, we get

$$\mathrm{d}H = \mathrm{d}q + \mathrm{d}w + p\,\mathrm{d}V + V\,\mathrm{d}p$$

If the system is in mechanical equilibrium with its surroundings at a pressure p and does only expansion work, we can write $\mathrm{d}w = -p\,\mathrm{d}V$ and obtain

$$\mathrm{d}H = \mathrm{d}q + V\,\mathrm{d}p$$

Now we impose the condition that the heating occurs at constant pressure by writing $\mathrm{d}p = 0$. Then

$$\mathrm{d}H = \mathrm{d}q \quad \text{(at constant pressure, no additional work)}$$

as in eqn 24a.

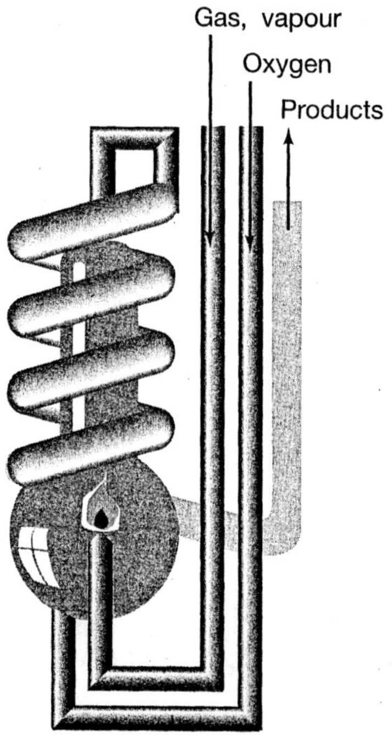

2.14 An adiabatic flame calorimeter consists of this component immersed in a stirred water bath. Combustion occurs as a known amount of reactant is passed through to fuel the flame, and the rise of temperature is monitored.

The result expressed in eqn 24 states that, when a system is subjected to a constant pressure, and only expansion work can occur, the change in enthalpy is equal to the energy supplied as heat. For example, if we supply 36 kJ of energy through an electric heater immersed in an open beaker of water, then the enthalpy of the water increases by 36 kJ and we write $\Delta H = +36$ kJ.

(b) The measurement of an enthalpy change

An enthalpy change can be measured calorimetrically by monitoring the temperature change that accompanies a physical or chemical change occurring at constant pressure. A calorimeter for studying processes at constant pressure is called an **isobaric calorimeter**. A simple example is a thermally insulated vessel open to the atmosphere: the heat released in the reaction is monitored by measuring the change in temperature of the contents. For a combustion reaction an **adiabatic flame calorimeter** may be used to measure ΔT when a given amount of substance burns in a supply of oxygen (Fig. 2.14). Another route to ΔH is to measure the internal energy change using a bomb calorimeter, and then to convert ΔU to ΔH. Because solids and liquids have small molar volumes, for them pV_m is so small that the molar enthalpy and molar internal energy are almost identical ($H_m = U_m + pV_m \approx U_m$). Consequently, if a process involves only solids or liquids, the values of ΔH and ΔU are almost identical. Physically, such processes are accompanied by a very small change in volume, the system does negligible work on the surroundings when the process occurs, so the energy supplied as heat stays entirely within the system. The most sophisticated way, however, is to use a **differential scanning calorimeter** (DSC), as explained in Box 2.1.

Box 2.1 Differential scanning calorimetry

A *differential scanning calorimeter* (DSC) measures the heat transferred to or from a sample at constant pressure during a physical or chemical change. The term 'differential' refers to the fact that the behaviour of the sample is compared to that of a reference material that does not undergo a physical or chemical change during the analysis. The term 'scanning' refers to the fact that the temperatures of the sample and reference material are increased, or scanned, systematically during the analysis.

A DSC consists of two small compartments that are heated electrically at a constant rate. The temperature, T, at time t during a linear scan is

$$T = T_0 + \alpha t$$

where T_0 is the initial temperature and α is the temperature scan rate (in kelvins per second, K s^{-1}). A computer controls the electrical power output in order to maintain the same temperature in the sample and reference compartments throughout the analysis (see the first illustration).

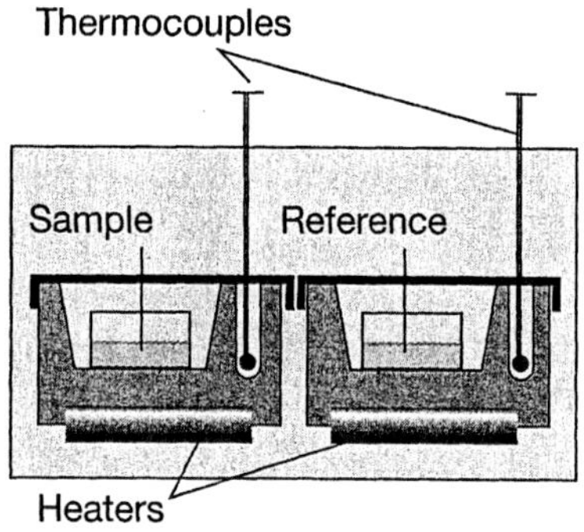

A differential scanning calorimeter. The sample and a reference material are heated in separate but identical metal heat sinks. The output is the difference in power needed to maintain the heat sinks at equal temperatures as the temperature rises.

The temperature of the sample changes significantly relative to that of the reference material if a chemical or physical process involving heat transfer occurs in the sample during the scan. To maintain the same temperature in both compartments, excess heat is transferred to or from the sample during the process. For example, an endothermic process lowers the temperature of the sample relative to that of the reference and, as a result, the sample must be heated more strongly than the reference in order to maintain equal temperatures. It is customary to interpret the excess heat transferred to the sample, $q_{p,ex}$, in terms of an apparent change in the heat capacity at constant pressure of the sample, C_p, during the temperature scan. We shall see that this interpretation leads to a simple calculation of the enthalpy change associated with the process that occurs in the sample.

If no physical or chemical change occurs in the sample at temperature T, we use eqn 29 to write the heat transferred to the sample as $q_p = C_p\Delta T$, where $\Delta T = T - T_0$ and we have assumed that C_p is independent of temperature. The chemical or physical process requires the transfer of $q_p + q_{p,ex}$ to attain the same change in temperature of the sample. If we write the heat capacity of the sample as $C_p + C_{p,ex}$, it follows that

$$q_p + q_{p,ex} = (C_p + C_{p,ex})\Delta T$$

It follows that

$$C_{p,ex} = \frac{q_{p,ex}}{\Delta T} = \frac{q_{p,ex}}{\alpha t} = \frac{P_{ex}}{\alpha}$$

where $P_{ex} = q_{p,ex}/t$ is the excess electrical power necessary to equalize the temperature of the sample and reference compartments.

A DSC trace, also called a *thermogram*, consists of a plot of P_{ex} or $C_{p,ex}$ versus T (see the second illustration). Broad peaks in the thermogram indicate processes requiring heat transfer. From eqn 27, the enthalpy change associated with the process is

$$\Delta H = \int_{T_1}^{T_2} C_{p,ex}\, dT$$

where T_1 and T_2 are, respectively, the temperatures at which the process begins and ends. This relation shows that the enthalpy change is then the area under the curve of $C_{p,ex}$ versus T. With a DSC, enthalpy changes may be determined in samples of masses as low as 0.5 mg, which is a significant advantage over bomb or flame calorimeters, which require several grams of material.

Differential scanning calorimetry is used in the chemical industry to characterize polymers and in the biochemistry laboratory to assess the stability of proteins, nucleic acids, and membranes. Large molecules, such as synthetic or biological polymers, attain complex three-dimensional structures due to intra- and inter-molecular interactions, such as hydrogen bonding and hydrophobic interactions (Chapter 22). Disruption of these interactions is an endothermic process

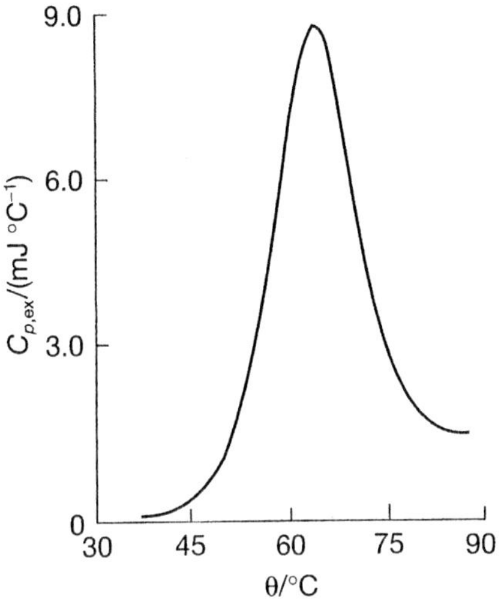

A thermogram for the protein ubiquitin at pH = 2.45. The protein retains its native structure up to about 45°C and then undergoes an endothermic conformational change. (Adapted from B. Chowdhry and S. LeHarne, *J. Chem. Educ.* **74**, 236 (1997).)

that can be studied with a DSC. For example, the thermogram shown in the illustration indicates that the protein ubiquitin retains its native structure up to about 45°C. At higher temperatures, the protein undergoes an endothermic conformational change that results in the loss of its three-dimensional structure. The same principles also apply to the study of structural integrity and stability of synthetic polymers, such as plastics.

Discussion questions

1. In many experimental thermograms, such as that shown in the illustration, the baseline below T_1 is at a different level from that above T_2. Explain this observation.

2. You have at your disposal a sample of pure polymer P and a sample of P that has just been synthesized in a large chemical reactor and that may contain impurities. Describe how you would use differential scanning calorimetry to determine the mole percentage composition of P in the allegedly impure sample.

Further reading

B. Chowdhry and S. LeHarne, Simulation and analysis of differential scanning calorimetry output: Protein unfolding studies. *J. Chem. Educ.* **74**, 236 (1997).

D.A. Skoog, F.J. Holler, and T.A. Nieman, *Principles of instrumental analysis*. Harcourt, Philadelphia (1998).

Example 2.2 Relating ΔH and ΔU

The internal energy change when 1.0 mol $CaCO_3$ in the form of calcite converts to aragonite is +0.21 kJ. Calculate the difference between the enthalpy change and the change in internal energy when the pressure is 1.0 bar given that the densities of the solids are 2.71 g cm^{-3} and 2.93 g cm^{-3}, respectively.

Method The starting point for the calculation is the relation between the enthalpy of a substance and its internal energy (eqn 23). The difference between the two quantities can be expressed in terms of the pressure and the difference of their molar volumes, and the latter can be calculated from their molar masses, M, and their mass densities, ρ, by using $\rho = M/V_m$.

Answer The change in enthalpy when the transition occurs is

$$\begin{aligned}\Delta H &= H(\text{aragonite}) - H(\text{calcite})\\ &= \{U(\text{a}) + pV(\text{a})\} - \{U(\text{c}) + pV(\text{c})\}\\ &= \Delta U + p\{V(\text{a}) - V(\text{c})\} = \Delta U + p\Delta V\end{aligned}$$

The volume of 1.0 mol $CaCO_3$ (100 g) as aragonite is 34 cm^3, and that of 1.0 mol $CaCO_3$ as calcite is 37 cm^3. Therefore,

$$p\Delta V = (1.0 \times 10^5 \text{ Pa}) \times (34 - 37) \times 10^{-6} \text{ m}^3 = -0.3 \text{ J}$$

(because 1 Pa m^3 = 1 J). Hence,

$$\Delta H - \Delta U = -0.3 \text{ J}$$

which is only 0.1 per cent of the value of ΔU.

Comment It is usually justifiable to ignore the difference between the enthalpy and internal energy of condensed phases, except at very high pressures, when pV is no longer negligible.

Self-test 2.3 Calculate the difference between ΔH and ΔU when 1.0 mol of grey tin (density 5.75 g cm^{-3}) changes to white tin (density 7.31 g cm^{-3}) at 10.0 bar. At 298 K, ΔH = +2.1 kJ. [$\Delta H - \Delta U = -4.4$ J]

The enthalpy of a perfect gas is related to its internal energy by using $pV = nRT$ in the definition of H:

$$H = U + pV = U + nRT \qquad (2.25)^{\circ}$$

This relation implies that the change of enthalpy in a reaction that produces or consumes gas is

$$\Delta H = \Delta U + \Delta n_g RT \qquad (2.26)^{\circ}$$

where Δn_g is the change in the amount of gas molecules in the reaction.

Illustration 2.4

In the reaction $2\,H_2(g) + O_2(g) \rightarrow 2\,H_2O(l)$, 3 mol of gas-phase molecules is replaced by 2 mol of liquid-phase molecules so $\Delta n_g = -3$ mol. Therefore, at 298 K the enthalpy and internal energy changes taking place in the system are related by

$$\Delta H - \Delta U = (-3 \text{ mol}) \times RT \approx -7.4 \text{ kJ}$$

Note that the difference is expressed in kilojoules, not joules as in Example 2.2. The enthalpy change is smaller than the change in internal energy because, although heat escapes from the system when the reaction occurs, the system contracts when the liquid is formed, so energy is restored to it from the surroundings.

Example 2.3 Calculating a change in enthalpy

Water is heated to boiling under a pressure of 1.0 atm. When an electric current of 0.50 A from a 12 V supply is passed for 300 s through a resistance in thermal contact with boiling water, it is found that 0.798 g of water is vaporized. Calculate the molar internal energy and enthalpy changes at the boiling point (373.15 K).

Method Because the vaporization occurs at constant pressure, the enthalpy change is equal to the heat supplied by the heater. Therefore, the strategy is to calculate the heat supplied (from $q = I\mathcal{V}t$), express that as an enthalpy change, and then convert the result to a molar enthalpy change by division by the amount of H_2O molecules vaporized. To convert from enthalpy change to internal energy change, we assume that the vapour is a perfect gas and use eqn 26.

Answer The enthalpy change is

$$\Delta H = q_p = (0.50\ \text{A}) \times (12\ \text{V}) \times (300\ \text{s}) = +(0.50 \times 12 \times 300)\ \text{J}$$

Here we have used 1 A V s = 1 J (see inside front cover). Because 0.798 g of water is (0.798 g)/(18.02 g mol^{-1}) = (0.798/18.02) mol H_2O, the enthalpy of vaporization per mole of H_2O is

$$\Delta H_{\text{m}} = +\frac{0.50 \times 12 \times 300\ \text{J}}{(0.798/18.02)\ \text{mol}} = +41\ \text{kJ mol}^{-1}$$

In the process $H_2O(l) \rightarrow H_2O(g)$ the change in the amount of gas molecules is $\Delta n_g = +1$ mol, so

$$\Delta U_{\text{m}} = \Delta H_{\text{m}} - RT = +38\ \text{kJ mol}^{-1}$$

Comment The plus sign is added to positive quantities to emphasize that they represent an increase in internal energy or enthalpy. Notice that the internal energy change is smaller than the enthalpy change because energy has been used to drive back the surrounding atmosphere to make room for the vapour.

Self-test 2.4 The molar enthalpy of vaporization of benzene at its boiling point (353.25 K) is 30.8 kJ mol^{-1}. What is the molar internal energy change? For how long would the same 12 V source need to supply a 0.50 A current in order to vaporize a 10 g sample? [+27.9 kJ mol^{-1}, 660 s]

Enthalpy and internal energy changes may also be measured by noncalorimetric methods (see Chapters 9 and 10).

(c) The variation of enthalpy with temperature

The enthalpy of a substance increases as its temperature is raised. The relation between the increase in enthalpy and the increase in temperature depends on the

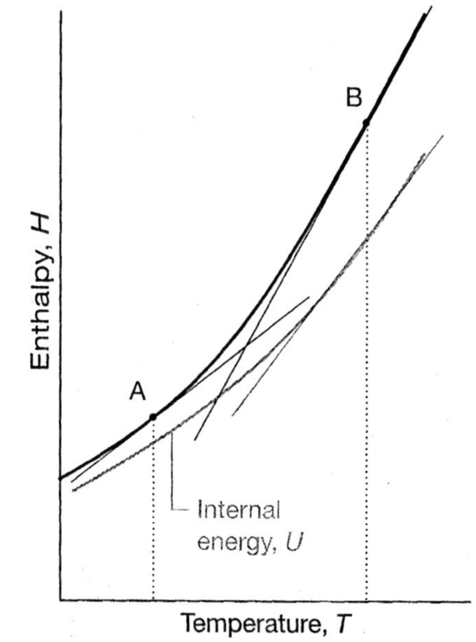

2.15 The slope of the tangent to a curve of the enthalpy of a system subjected to a constant pressure plotted against temperature is the constant-pressure heat capacity. The slope may change with temperature, in which case the heat capacity varies with temperature. Thus, the heat capacities at A and B are different. For gases, at a given temperature the slope of enthalpy versus temperature is steeper than that of internal energy versus temperature, and $C_{p,m}$ is larger than $C_{V,m}$.

conditions (for example, constant pressure or constant volume). The most important condition is constant pressure, and the slope of the tangent to a plot of enthalpy against temperature at constant pressure is called the **heat capacity at constant pressure**, C_p, at a given temperature (Fig. 2.15). More formally:

$$C_p = \left(\frac{\partial H}{\partial T}\right)_p \quad [2.27]$$

The heat capacity at constant pressure is the analogue of the heat capacity at constant volume, and is an extensive property.[4] The **molar heat capacity at constant pressure**, $C_{p,m}$, is the heat capacity per mole of material; it is an intensive property.

The heat capacity at constant pressure is used to relate the change in enthalpy to a change in temperature. For infinitesimal changes of temperature,

$$dH = C_p\, dT \quad \text{(at constant pressure)} \quad (2.28a)$$

If the heat capacity is constant over the range of temperatures of interest, then for a measurable increase in temperature

$$\Delta H = C_p \Delta T \quad \text{(at constant pressure)} \quad (2.28b)$$

Because an increase in enthalpy can be equated with the heat supplied at constant pressure, the practical form of the latter equation is

$$q_p = C_p \Delta T \quad (2.29)$$

This expression shows us how to measure the heat capacity of a sample: a measured quantity of heat is supplied under conditions of constant pressure (as in a sample exposed to the atmosphere and free to expand), and the temperature rise is monitored.

The variation of heat capacity with temperature can sometimes be ignored if the temperature range is small; this approximation is highly accurate for a monatomic perfect gas (one of the noble gases). However, when it is necessary to take the variation into account, a convenient approximate empirical expression is

$$C_{p,m} = a + bT + \frac{c}{T^2} \quad (2.30)$$

The empirical parameters a, b, and c are independent of temperature (Table 2.2).

Table 2.2* Temperature variation of molar heat capacities, $C_{p,m}/(\mathrm{J\,K^{-1}\,mol^{-1}}) = a + bT + c/T^2$

	a	$b/(10^{-3}\,\mathrm{K^{-1}})$	$c/(10^5\,\mathrm{K^2})$
C(s, graphite)	16.86	4.77	−8.54
CO_2(g)	44.22	8.79	−8.62
H_2O(l)	75.29	0	0
N_2(g)	28.58	3.77	−0.50

* More values are given in the *Data section* at the end.

4. As in the case of C_V, if the system can change its composition it is necessary to distinguish between equilibrium and fixed-composition values. All applications in this chapter refer to pure substances, so this complication can be ignored.

Example 2.4 Evaluating an increase in enthalpy with temperature

What is the change in molar enthalpy of N_2 when it is heated from 25°C to 100°C? Use the heat capacity information in Table 2.2.

Method The heat capacity of N_2 changes with temperature, so we cannot use eqn 28b (which assumes that the heat capacity of the substance is constant). Therefore, we must use eqn 28a, substitute eqn 30 for the temperature dependence of the heat capacity, and integrate the resulting expression from 25°C to 100°C.

Answer For convenience, we denote the two temperatures T_1 (298 K) and T_2 (373 K). The integrals we require are

$$\int_{H(T_1)}^{H(T_2)} \mathrm{d}H = \int_{T_1}^{T_2} \left(a + bT + \frac{c}{T^2}\right)\mathrm{d}T$$

Notice how the limits of integration correspond on each side of the equation: the integration over H on the left ranges from $H(T_1)$, the value of H at T_1, up to $H(T_2)$, the value of H at T_2, while on the right the integration over the temperature ranges from T_1 to T_2. Now we use the integrals

Integrals commonly encountered in physical chemistry are listed inside the front cover.

$$\int \mathrm{d}x = x + \text{constant} \qquad \int x\,\mathrm{d}x = \tfrac{1}{2}x^2 + \text{constant} \qquad \int \frac{\mathrm{d}x}{x^2} = -\frac{1}{x} + \text{constant}$$

to obtain

$$H(T_2) - H(T_1) = a(T_2 - T_1) + \tfrac{1}{2}b(T_2^2 - T_1^2) - c\left(\frac{1}{T_2} - \frac{1}{T_1}\right)$$

Substitution of the numerical data results in

$$H(373\ \mathrm{K}) = H(298\ \mathrm{K}) + 2.20\ \mathrm{kJ\ mol^{-1}}$$

If we had assumed a constant heat capacity of 29.14 J K^{-1} mol^{-1} (the value given by eqn 30 at 25°C), we would have found that the two enthalpies differed by 2.19 kJ mol^{-1}.

Self-test 2.5 At very low temperatures the heat capacity of a solid is proportional to T^3, and we can write $C_p = aT^3$. What is the change in enthalpy of such a substance when it is heated from 0 to a temperature T (with T close to 0)? [$\Delta H = \frac{1}{4}aT^4$]

(d) The relation between heat capacities

Most systems expand when heated at constant pressure. Such systems do work on the surroundings and therefore some of the energy supplied to them as heat escapes back to the surroundings. As a result, the temperature of the system rises less than when the heating occurs at constant volume. A smaller increase in temperature implies a larger heat capacity, so we conclude that in most cases the heat capacity at constant pressure of a system is larger than its heat capacity at constant volume.

There is a simple relation between the two heat capacities of a perfect gas:

$$C_p - C_V = nR \qquad (2.31)^\circ$$

(This relation is derived in Section 3.5.) It follows that the molar heat capacity of a perfect gas is about 8 J K^{-1} mol^{-1} larger at constant pressure than at constant

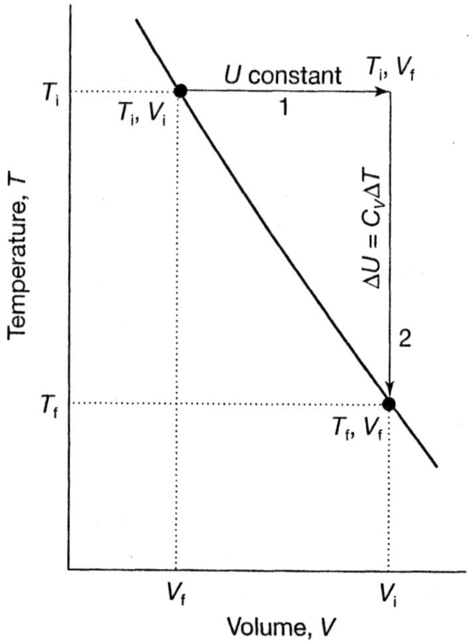

2.16 To achieve a change of state from one temperature and volume to another temperature and volume, we may consider the overall change as composed of two steps. In the first step, the system expands at constant temperature; there is no change in internal energy if the system consists of a perfect gas. In the second step, the temperature of the system is reduced at constant volume. The overall change in internal energy is the sum of the changes for the two steps.

volume. Because the heat capacity at constant volume of a monatomic gas is about 12 J K^{-1} mol^{-1}, the difference is highly significant and must be taken into account.

2.6 Adiabatic changes

We are now equipped to deal with the changes that occur when a perfect gas expands adiabatically. A decrease in temperature should be expected: because work is done, the internal energy falls, and therefore the temperature of the working gas also falls. In molecular terms, the kinetic energy of the molecules falls as work is done, so their average speed decreases, and hence the temperature falls.

(a) The work of adiabatic change

The change in internal energy of a perfect gas when the temperature is changed from T_i to T_f and the volume is changed from V_i to V_f can be expressed as the sum of two steps (Fig. 2.16). In the first step, only the volume changes and the temperature is held constant at its initial value. However, because the internal energy of a perfect gas is independent of the volume the molecules occupy, the overall change in internal energy arises solely from the second step, the change in temperature at constant volume. Provided the heat capacity is independent of temperature, this change is

$$\Delta U = C_V(T_f - T_i) = C_V \Delta T \qquad (2.32)$$

Because the expansion is adiabatic, we know that $q = 0$; because $\Delta U = q + w$, it then follows that $\Delta U = w_{ad}$. The subscript 'ad' denotes an adiabatic process. Therefore, by equating the two values we have obtained for ΔU, we obtain

$$w_{ad} = C_V \Delta T \qquad (2.33)^\circ$$

That is, the work done during an adiabatic expansion of a perfect gas is proportional to the temperature difference between the initial and final states. That is exactly what we expect on molecular grounds, because the mean kinetic energy is proportional to T, so a change in internal energy arising from temperature alone is also expected to be proportional to ΔT.

To use eqn 33, we need to relate the change in temperature to the change in volume (which we know). The most important type of adiabatic expansion (and the only kind that we need for later) is reversible adiabatic expansion, in which the external pressure is matched to the internal pressure throughout. For adiabatic, reversible expansion the initial and final temperatures are related by

$$V_f T_f^c = V_i T_i^c \qquad c = C_{V,m}/R \qquad (2.34)^\circ_{rev}$$

This expression is derived in the following *Justification* and the temperature dependence it implies is shown in Fig. 2.17. All we need do at this stage to find the work done by the system is to solve this relation for T_f,

$$T_f = T_i \left(\frac{V_i}{V_f}\right)^{1/c} \qquad (2.35)^\circ_{rev}$$

calculate ΔT, and substitute into eqn 33.

Justification 2.2

Consider a stage in the expansion when the pressure inside and out is p (the expansion is reversible, so the internal and external pressures are equal at all stages). The work done when the gas expands by dV is

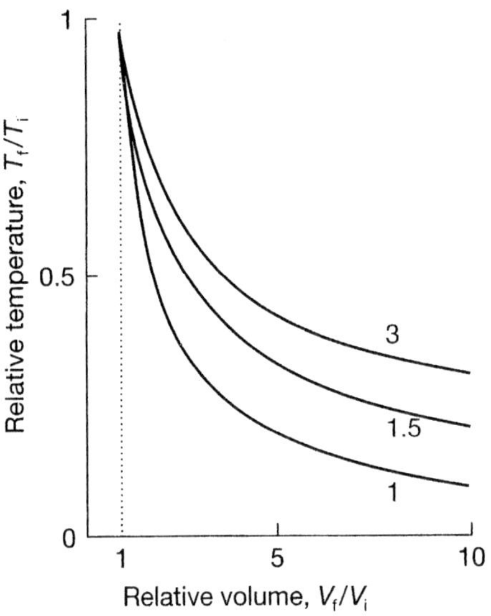

2.17 The variation of temperature as a perfect gas is expanded reversibly and adiabatically. The curves are labelled with different values of $c = C_{V,m}/R$. Note that the temperature falls most steeply for gases with low molar heat capacity.

For the final stages of this derivation we use the relations $a \ln x = \ln x^a$ and $-\ln(x/y) = \ln(y/x)$.

$$dw = -p\, dV$$

However, for a perfect gas,

$$dU = C_V\, dT$$

(by the same argument as in the text for a macroscopic change). Therefore, because for an adiabatic change ($dq = 0$)

$$dU = dw + dq = dw$$

we can equate these two expressions for dU and write

$$C_V\, dT = -p\, dV$$

We are dealing with a perfect gas, so we can replace p by nRT/V and obtain

$$\frac{C_V\, dT}{T} = -\frac{nR\, dV}{V}$$

To integrate this expression we note that T is equal to T_i when V is equal to V_i, and is equal to T_f when V is equal to V_f at the end of the expansion. Therefore,

$$C_V \int_{T_i}^{T_f} \frac{dT}{T} = -nR \int_{V_i}^{V_f} \frac{dV}{V}$$

(We are taking C_V to be independent of temperature.) Then, because $\int dx/x = \ln x +$ constant, we obtain

$$C_V \ln \frac{T_f}{T_i} = -nR \ln \frac{V_f}{V_i}$$

With $c = C_V/nR$ we obtain

$$\ln\left(\frac{T_f}{T_i}\right)^c = \ln\left(\frac{V_i}{V_f}\right)$$

which implies eqn 34.

Illustration 2.5

Consider the adiabatic, reversible expansion of 0.020 mol Ar, initially at 25°C, from 0.50 L to 1.00 L. The molar heat capacity of argon at constant volume is 12.48 J K^{-1} mol^{-1}, so $c = 1.501$. Therefore, from eqn 35,

$$T_f = (298\text{ K}) \times \left(\frac{0.50\text{ L}}{1.00\text{ L}}\right)^{1/1.501} = 188\text{ K}$$

It follows that $\Delta T = -110$ K, and therefore, from eqn 33, that

$$w_{ad} = C_V \Delta T = (0.020\text{ mol}) \times (12.48\text{ J K}^{-1}\text{ mol}^{-1}) \times (-110\text{ K}) = -27\text{ J}$$

Note that temperature change is independent of the mass of gas but the work is not.

Self-test 2.6 Calculate the final temperature, the work done, and the change of internal energy when ammonia is used in a reversible adiabatic expansion from 0.50 L to 2.00 L, the other initial conditions being the same.

[195 K, −56 J, −56 J]

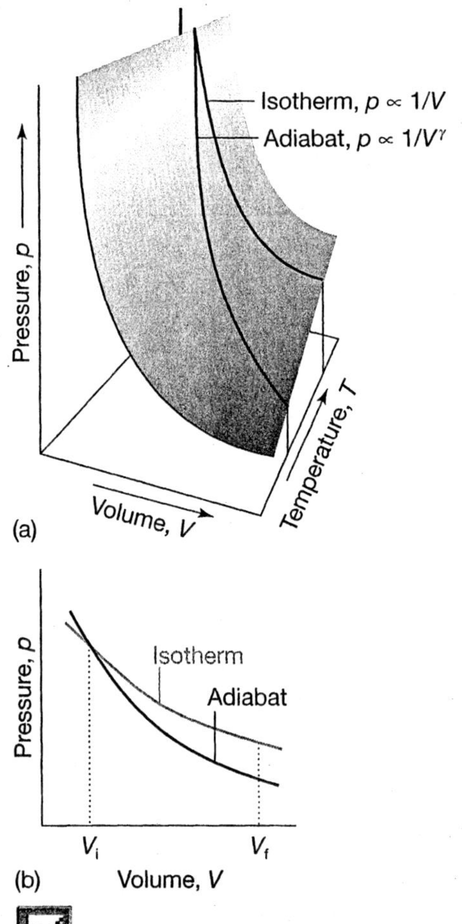

2.18 An adiabat depicts the variation of pressure with volume when a gas expands adiabatically. (a) An adiabat for a perfect gas undergoing reversible expansion. (b) Note that the pressure declines more steeply for an adiabat than it does for an isotherm because the temperature decreases in the former.

(b) Heat capacity ratio and adiabats

Now we consider the change in pressure that results from an adiabatic, reversible expansion of a perfect gas. We show in the *Justification* below that

$$pV^{\gamma} = \text{constant} \qquad (2.36)^{\circ}_{\text{rev}}$$

where the **heat capacity ratio**, γ, of a substance is defined as

$$\gamma = \frac{C_{p,\text{m}}}{C_{V,\text{m}}} \qquad [2.37]$$

Because the heat capacity at constant pressure is greater than the heat capacity at constant volume, $\gamma > 1$. For a perfect gas, it follows from eqn 31 that

$$\gamma = \frac{C_{V,\text{m}} + R}{C_{V,\text{m}}} = 1 + \frac{R}{C_{V,\text{m}}} \qquad (2.38)^{\circ}$$

For a monatomic perfect gas, $C_{V,\text{m}} = \frac{3}{2}R$ (see *Illustration* 2.3), so $\gamma = \frac{5}{3}$. For a gas of nonlinear polyatomic molecules (which can rotate as well as translate), $C_{V,\text{m}} = 3R$, so $\gamma = \frac{4}{3}$.

Justification 2.3

The initial and final states of a perfect gas satisfy the perfect gas law regardless of how the change of state takes place, so we can use $pV = nRT$ to write

$$\frac{p_i V_i}{p_f V_f} = \frac{T_i}{T_f}$$

However, we know from (eqn 35) that for a reversible adiabatic change the temperature changes so as to satisfy

$$\frac{T_i}{T_f} = \left(\frac{V_f}{V_i}\right)^{1/c} = \left(\frac{V_f}{V_i}\right)^{\gamma-1}$$

where we use the definition of the heat capacity ratio (eqn 37). Then we combine the two expressions, to obtain

$$p_i V_i^{\gamma} = p_f V_f^{\gamma}$$

It follows that $pV^{\gamma} = $ constant, as used in the text.

The curves of pressure versus volume for adiabatic change are known as **adiabats**, and one for a reversible path is illustrated in Fig. 2.18. Because $\gamma > 1$, an adiabat falls more steeply ($p \propto 1/V^{\gamma}$) than the corresponding isotherm ($p \propto 1/V$). The physical reason for the difference is that, in an isothermal expansion, energy flows into the system as heat and maintains the temperature; as a result, the pressure does not fall as much as in an adiabatic expansion.

Illustration 2.6

When a sample of argon (for which $\gamma = \frac{5}{3}$) at 100 kPa expands reversibly and adiabatically to twice its initial volume the final pressure will be

$$p_f = p_i\left(\frac{V_i}{V_f}\right)^{\gamma} = (100\text{ kPa}) \times (\tfrac{1}{2})^{5/3} = 31.5\text{ kPa}$$

For an isothermal doubling of volume, the final pressure would be 50 kPa.

Thermochemistry

The study of the heat produced or required by chemical reactions is called **thermochemistry**. Thermochemistry is a branch of thermodynamics because a reaction vessel and its contents form a system, and chemical reactions result in the exchange of energy between the system and the surroundings. Thus we can use calorimetry to measure the heat produced or absorbed by a reaction, and can identify q with a change in internal energy (if the reaction occurs at constant volume) or a change in enthalpy (if the reaction occurs at constant pressure). Conversely, if we know the ΔU or ΔH for a reaction, we can predict the heat the reaction can produce.

We have already remarked that a process that releases heat is classified as exothermic and one that absorbs heat is classified as endothermic. Because the release of heat signifies a decrease in the enthalpy of a system (at constant pressure), we can now see that an exothermic process at constant pressure is one for which $\Delta H < 0$. Conversely, because the absorption of heat results in an increase in enthalpy, an endothermic process at constant pressure has $\Delta H > 0$.

2.7 Standard enthalpy changes

Changes in enthalpy are normally reported for processes taking place under a set of standard conditions. In most of our discussions we shall consider the **standard enthalpy change**, $\Delta H^{\ominus}$, the change in enthalpy for a process in which the initial and final substances are in their standard states:

The standard state of a substance at a specified temperature is its pure form at 1 bar.[5]

For example, the standard state of liquid ethanol at 298 K is pure liquid ethanol at 298 K and 1 bar; the standard state of solid iron at 500 K is pure iron at 500 K and 1 bar. The standard enthalpy change for a reaction or a physical process is the difference between the products in their standard states and the reactants in their standard states, all at the same specified temperature.

As an example of a standard enthalpy change, the *standard enthalpy of vaporization*, $\Delta_{vap}H^{\ominus}$, is the enthalpy change per mole when a pure liquid at 1 bar vaporizes to a gas at 1 bar, as in

$$H_2O(l) \longrightarrow H_2O(g) \qquad \Delta_{vap}H^{\ominus}(373\text{ K}) = +40.66\text{ kJ mol}^{-1}$$

As implied by the examples, standard enthalpies may be reported for any temperature. However, the conventional temperature for reporting thermodynamic data is 298.15 K (corresponding to 25.00°C). Unless otherwise mentioned, all thermodynamic data in this text will refer to this conventional temperature.

(a) Enthalpies of physical change

The standard enthalpy change that accompanies a change of physical state is called the **standard enthalpy of transition** and is denoted $\Delta_{trs}H^{\ominus}$ (Table 2.3). The **standard enthalpy of vaporization**, $\Delta_{vap}H^{\ominus}$, is one example. Another is the **standard enthalpy of fusion**, $\Delta_{fus}H^{\ominus}$, the standard enthalpy change accompanying the conversion of a solid to a liquid, as in

$$H_2O(s) \longrightarrow H_2O(l) \qquad \Delta_{fus}H^{\ominus}(273\text{ K}) = +6.01\text{ kJ mol}^{-1}$$

5. The definition of standard state is more sophisticated for a real gas (Section 5.5) and for solutions (Section 10.2).

Table 2.3* Standard enthalpies of fusion and vaporization at the transition temperature, $\Delta_{trs}H^{\ominus}/(kJ\ mol^{-1})$

	T_f/K	Fusion	T_b/K	Vaporization
Ar	83.81	1.188	87.29	6.506
C_6H_6	278.61	10.59	353.2	30.8
H_2O	273.15	6.008	373.15	40.656 44.016 at 298 K
He	3.5	0.021	4.22	0.084

* More values are given in the *Data section*.

As in this case, it is sometimes convenient to know the standard enthalpy change at the transition temperature as well as at the conventional temperature.

Because enthalpy is a state function, a change in enthalpy is independent of the path between the two states. This feature is of great importance in thermochemistry, for it implies that the same value of $\Delta H^{\ominus}$ will be obtained however the change is brought about (so long as the initial and final states are the same). For example, we can picture the conversion of a solid to a vapour either as occurring by sublimation (the direct conversion from solid to vapour),

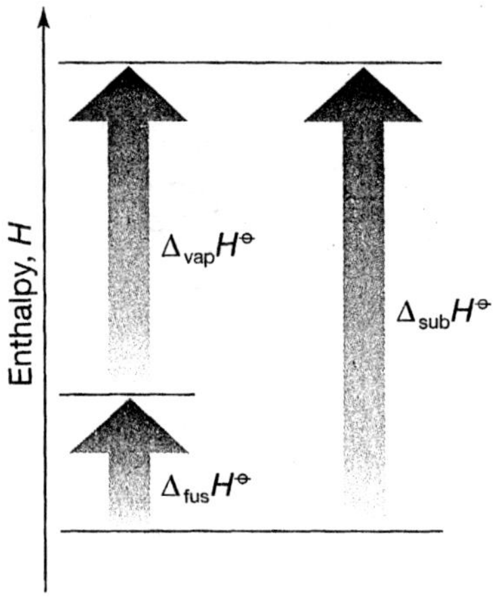

1

$$H_2O(s) \longrightarrow H_2O(g) \qquad \Delta_{sub}H^{\ominus}$$

or as occurring in two steps, first fusion (melting) and then vaporization of the resulting liquid:

$$\begin{array}{lll} & H_2O(s) \longrightarrow H_2O(l) & \Delta_{fus}H^{\ominus} \\ & H_2O(l) \longrightarrow H_2O(g) & \Delta_{vap}H^{\ominus} \\ \text{Overall:} & H_2O(s) \longrightarrow H_2O(g) & \Delta_{fus}H^{\ominus} + \Delta_{vap}H^{\ominus} \end{array}$$

Because the overall result of the indirect path is the same as that of the direct path, the overall enthalpy change is the same in each case (**1**), and we can conclude that (for processes occurring at the same temperature)

$$\Delta_{sub}H^{\ominus} = \Delta_{fus}H^{\ominus} + \Delta_{vap}H^{\ominus} \tag{2.39}$$

An immediate conclusion is that, because all enthalpies of fusion are positive, the enthalpy of sublimation of a substance is greater than its enthalpy of vaporization (at a given temperature).

Another consequence of H being a state function is that the standard enthalpy changes of a forward process and its reverse must differ only in sign (**2**):

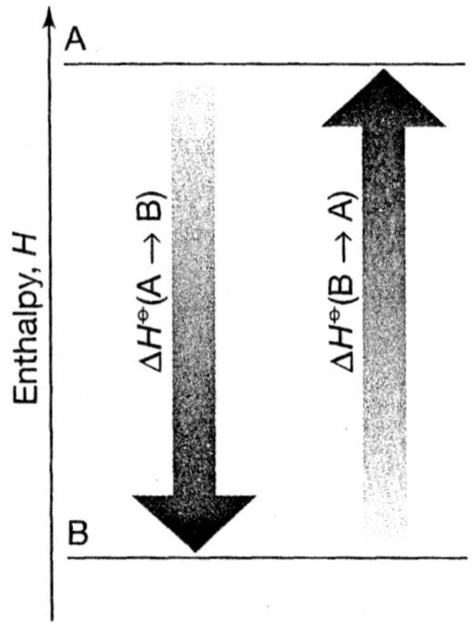

2

$$\Delta H^{\ominus}(A \rightarrow B) = -\Delta H^{\ominus}(B \rightarrow A) \tag{2.40}$$

For instance, because the enthalpy of vaporization of water is +44 kJ mol^{-1} at 298 K, its enthalpy of condensation at that temperature is −44 kJ mol^{-1}.

The different types of enthalpies encountered in thermochemistry are summarized in Table 2.4. We shall meet them again in various locations throughout the text.

(b) Enthalpies of chemical change

Now we consider enthalpy changes that accompany chemical reactions. There are two ways of reporting the change in enthalpy that accompanies a chemical reaction.

Table 2.4 Enthalpies of transition

Transition	Process	Symbol*
Transition	Phase $\alpha \longrightarrow$ phase β	$\Delta_{trs}H$
Fusion	$s \longrightarrow l$	$\Delta_{fus}H$
Vaporization	$l \longrightarrow g$	$\Delta_{vap}H$
Sublimation	$s \longrightarrow g$	$\Delta_{sub}H$
Mixing of fluids	Pure $\longrightarrow$ mixture	$\Delta_{mix}H$
Solution	Solute $\longrightarrow$ solution	$\Delta_{sol}H$
Hydration	$X^{\pm}(g) \longrightarrow X^{\pm}(aq)$	$\Delta_{hyd}H$
Atomization	Species(s, l, g) $\longrightarrow$ atoms(g)	$\Delta_{at}H$
Ionization	$X(g) \longrightarrow X^{+}(g) + e^{-}(g)$	$\Delta_{ion}H$
Electron gain	$X(g) + e^{-}(g) \longrightarrow X^{-}(g)$	$\Delta_{eg}H$
Reaction	Reactants $\longrightarrow$ products	$\Delta_{r}H$
Combustion	Compound(s, l, g) + $O_2(g) \longrightarrow CO_2(g), H_2O(l, g)$	$\Delta_{c}H$
Formation	Elements $\longrightarrow$ compound	$\Delta_{f}H$
Activation	Reactants $\longrightarrow$ activated complex	$\Delta^{\ddagger}H$

* IUPAC recommendations. In common usage, the transition subscript is often attached to ΔH, as in ΔH_{trs}.

One is to write the **thermochemical equation**, a combination of a chemical equation and the corresponding change in standard enthalpy:

$$CH_4(g) + 2\,O_2(g) \longrightarrow CO_2(g) + 2\,H_2O(l) \qquad \Delta H^{\ominus} = -890\text{ kJ}$$

$\Delta H^{\ominus}$ is the change in enthalpy when reactants in their standard states change to products in their standard states:

Pure, unmixed reactants in their standard states

$\longrightarrow$ pure, separated products in their standard states

Except in the case of ionic reactions in solution, the enthalpy changes accompanying mixing and separation are insignificant in comparison with the contribution from the reaction itself. For the combustion of methane, the standard value refers to the reaction in which 1 mol CH_4 in the form of pure methane gas at 1 bar reacts completely with 2 mol O_2 in the form of pure oxygen gas at 1 bar to produce 1 mol CO_2 as pure carbon dioxide at 1 bar and 2 mol H_2O as pure liquid water at 1 bar; the numerical value is for the reaction at 298 K.

Alternatively, we write the chemical equation and then report the **standard reaction enthalpy**, $\Delta_r H^{\ominus}$. Thus, for the combustion of methane, we write

$$CH_4(g) + 2\,O_2(g) \longrightarrow CO_2(g) + 2\,H_2O(l) \qquad \Delta_r H^{\ominus} = -890\text{ kJ mol}^{-1}$$

For the reaction

$$2\,A + B \longrightarrow 3\,C + D$$

the standard reaction enthalpy is

$$\Delta_r H^{\ominus} = \{3H_m^{\ominus}(C) + H_m^{\ominus}(D)\} - \{2H_m^{\ominus}(A) + H_m^{\ominus}(B)\}$$

where $H_m^{\ominus}(J)$ is the standard molar enthalpy of species J at the temperature of interest. Note how the 'per mole' of $\Delta_r H^{\ominus}$ comes directly from the fact that molar enthalpies appear in this expression. We interpret the 'per mole' by noting the stoichiometric coefficients in the chemical equation. In this case 'per mole' in $\Delta_r H^\ominus$ means 'per 2 mole A', 'per mole B', 'per 3 mole C', or 'per mole D'. In general,

$$\Delta_r H^\ominus = \sum_{\text{Products}} \nu H_m^\ominus - \sum_{\text{Reactants}} \nu H_m^\ominus \tag{2.41}$$

where in each case the molar enthalpies of the species are multiplied by their stoichiometric coefficients, ν.[6]

Some standard reaction enthalpies have special names and a particular significance. For instance, the **standard enthalpy of combustion**, $\Delta_c H^\ominus$, is the standard reaction enthalpy for the complete oxidation of an organic compound to CO_2 gas and liquid H_2O if the compound contains C, H, and O, and to N_2 gas if N is also present. An example is the combustion of glucose:

$$C_6H_{12}O_6(s) + 6\,O_2(g) \longrightarrow 6\,CO_2(g) + 6\,H_2O(l) \qquad \Delta_c H^\ominus = -2808 \text{ kJ mol}^{-1}$$

The value quoted shows that 2808 kJ of heat is released when 1 mol $C_6H_{12}O_6$ burns under standard conditions (at 298 K). Some further values are listed in Table 2.5. We consider the thermochemistry of fuels and foods in Box 2.2.

(c) Hess's law

Standard enthalpies of individual reactions can be combined to obtain the enthalpy of another reaction. This application of the First Law is called **Hess's law**:

The standard enthalpy of an overall reaction is the sum of the standard enthalpies of the individual reactions into which a reaction may be divided.

The individual steps need not be realizable in practice: they may be hypothetical reactions, the only requirement being that their chemical equations should balance. The thermodynamic basis of the law is the path-independence of the value of $\Delta_r H^\ominus$ and the implication that we may take the specified reactants, pass through any (possibly hypothetical) set of reactions to the specified products, and overall obtain the same change of enthalpy. The importance of Hess's law is that information about a reaction of interest, which may be difficult to determine directly, can be assembled from information on other reactions.

Table 2.5* Standard enthalpies of formation and combustion of organic compounds at 298 K

	$\Delta_f H^\ominus/(\text{kJ mol}^{-1})$	$\Delta_c H^\ominus/(\text{kJ mol}^{-1})$
Benzene, $C_6H_6(l)$	+49.0	−3268
Ethane, $C_2H_6(g)$	−84.7	−1560
Glucose, $C_6H_{12}O_6(s)$	−1274	−2808
Methane, $CH_4(g)$	−74.8	−890
Methanol, $CH_3OH(l)$	−238.7	−726

* More values are given in the *Data section*.

6. In this and similar expressions, all stoichiometric coefficients are positive. For a more sophisticated way of writing eqn 41, see Section 9.2.

Box 2.2 Food and energy reserves

The thermochemical properties of fuels and foods are commonly discussed in terms of their *specific enthalpy*, the magnitude of the enthalpy of combustion per gram of material or the *enthalpy density*, the magnitude of the enthalpy of combustion per litre of material. Thus, if the standard enthalpy of combustion is $\Delta_c H^{\ominus}$ and the molar mass of the compound is M, then the specific enthalpy is $|\Delta_c H^{\ominus}|/M$. The table lists the specific enthalpies of several fuels.

A typical 18–20 year old man requires a daily input of about 12 MJ; a woman of the same age needs about 9 MJ. If the entire consumption were in the form of glucose (which has a specific enthalpy of 16 kJ g^{-1}), that would require the consumption of 750 g of glucose for a man and 560 g for a woman. In fact, digestible carbohydrates have a slightly higher specific enthalpy (17 kJ g^{-1}) than glucose itself, so a carbohydrate diet is slightly less daunting than a pure glucose diet, as well as being more appropriate in the form of fibre, the indigestible cellulose that helps move digestion products through the intestine.

The specific enthalpy of fats, which are long-chain esters like tristearin (beef fat), is much greater than that of carbohydrates, at around 38 kJ g^{-1}, slightly less than the value for the hydrocarbon oils used as fuel (48 kJ g^{-1}). Fats are commonly used as an energy store, to be used only when the more readily accessible carbohydrates have fallen into short supply. In Arctic species, the stored fat also acts as a layer of insulation; in desert species (such as the camel), the fat is also a source of water, one of its oxidation products.

Proteins are also used as a source of energy, but their components, the amino acids, are often too valuable to squander in this way, and are used to construct other proteins instead. When proteins are oxidized (to urea, $CO(NH_2)_2$), the equivalent enthalpy density is comparable to that of carbohydrates.

The heat released by the oxidation of foods needs to be discarded in order to maintain body temperature within its typical range of 35.6–37.8°C. A variety of mechanisms contribute to this aspect of homeostasis. The general uniformity of temperature throughout the body is maintained largely by the flow of blood. When heat needs to be dissipated rapidly, warm blood is allowed to flow through the capillaries of the skin, so producing flushing. Radiation is one means of discarding heat; another is evaporation and the energy demands of the enthalpy of vaporization of water. Evaporation removes about 2.4 kJ per gram of water perspired. When vigorous exercise promotes sweating (through the influence of heat selectors in the hypothalamus), 1–2 L of perspired water can be produced per hour, corresponding to a heat loss of 2.4–5.0 MJ h^{-1}.

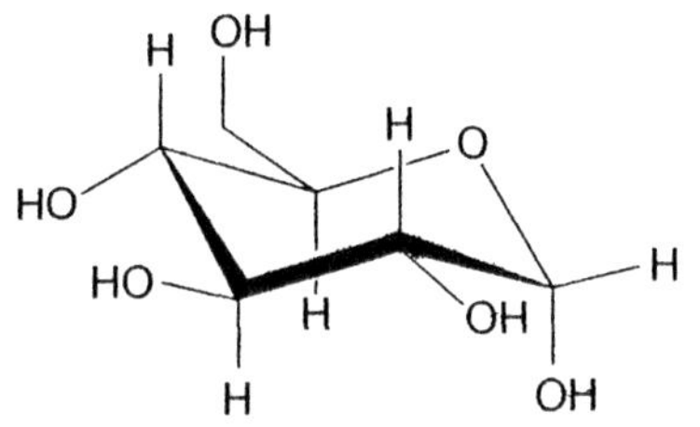

α-D-glucose

Discussion questions

1. An average human produces about 10 MJ of heat each day through metabolic activity. If a human body were an isolated system of mass 65 kg with the heat capacity of water, what temperature rise would the body experience? Human bodies are actually open systems, and the main mechanism of heat loss is through the evaporation of water. What mass of water should be evaporated each day to maintain constant temperature?

2. The mean bond enthalpies of the C—C, C—H, C=O, and O—H bonds are 348, 412, 743, and 463 kJ mol^{-1}, respectively. The combustion of a fuel such as octane is exothermic because relatively weak bonds break to form relatively strong bonds. Use this information to justify why glucose (see the illustration) has a lower specific enthalpy than the lipid decanoic acid even though these compounds have similar molar masses.

Further reading

V. Smil, *Energies: An illustrated guide to the biosphere and civilization.* MIT Press, Cambridge, Mass. (1999).

O.R. Fennema (ed.), *Food chemistry.* Marcel Dekker, New York (1996).

Thermochemical properties of some fuels

Fuel	Combustion equation	$\Delta_c H^{\ominus}$/(kJ mol^{-1})	Specific enthalpy/(kJ g^{-1})	Enthalpy density/(kJ L^{-1})
Hydrogen	$H_2(g) + \frac{1}{2}O_2(g) \longrightarrow H_2O(l)$	−286	142	13
Methane	$CH_4(g) + 2\,O_2(g) \longrightarrow CO_2(g) + 2\,H_2O(l)$	−890	55	40
Octane	$C_8H_{18}(l) + \frac{25}{2}O_2(g) \longrightarrow 8\,CO_2(g) + 9\,H_2O(l)$	−5471	48	3.8×10^4
Methanol	$CH_3OH(l) + \frac{3}{2}O_2(g) \longrightarrow CO_2(g) + 2\,H_2O(l)$	−726	23	1.8×10^4

Example 2.5 Using Hess's law

The standard reaction enthalpy for the hydrogenation of propene,

$$CH_2{=}CHCH_3(g) + H_2(g) \longrightarrow CH_3CH_2CH_3(g)$$

is −124 kJ mol−1. The standard reaction enthalpy for the combustion of propane,

$$CH_3CH_2CH_3(g) + 5\,O_2(g) \longrightarrow 3\,CO_2(g) + 4\,H_2O(l)$$

is −2220 kJ mol−1. Calculate the standard enthalpy of combustion of propene.

Method Add or subtract the reactions given, together with any others needed, so as to reproduce the reaction required. Then add or subtract the reaction enthalpies in the same way. Additional data are in Table 2.5.

Answer The combustion reaction we require is

$$C_3H_6(g) + \tfrac{9}{2}O_2(g) \longrightarrow 3\,CO_2(g) + 3\,H_2O(l)$$

This reaction can be recreated from the following sum:

	$\Delta_r H^{\ominus}/(\mathrm{kJ\ mol^{-1}})$
$C_3H_6(g) + H_2(g) \longrightarrow C_3H_8(g)$	−124
$C_3H_8(g) + 5\,O_2(g) \longrightarrow 3\,CO_2(g) + 4\,H_2O(l)$	−2220
$H_2O(l) \longrightarrow H_2(g) + \tfrac{1}{2}O_2(g)$	+286
$C_3H_6(g) + \tfrac{9}{2}O_2(g) \longrightarrow 3\,CO_2(g) + 3\,H_2O(l)$	−2058

Comment The skill to develop is the ability to assemble a given thermochemical equation from others.

Self-test 2.7 Calculate the enthalpy of hydrogenation of benzene from its enthalpy of combustion and the enthalpy of combustion of cyclohexane.

[−205 kJ mol−1]

2.8 Standard enthalpies of formation

The **standard enthalpy of formation**, $\Delta_f H^{\ominus}$, of a substance is the standard reaction enthalpy for the formation of the compound from its elements in their reference states. The **reference state** of an element is its most stable state at the specified temperature and 1 bar. For example, at 298 K the reference state of nitrogen is a gas of N_2 molecules, that of mercury is liquid mercury, that of carbon is graphite, and that of tin is the white (metallic) form. There is one exception to this general prescription of reference states: the reference state of phosphorus is taken to be white phosphorus despite this allotrope not being the most stable form but simply the more reproducible form of the element. Standard enthalpies of formation are expressed as enthalpies per mole of the compound. The standard enthalpy of formation of liquid benzene at 298 K, for example, refers to the reaction

$$6\,C(s,\text{graphite}) + 3\,H_2(g) \longrightarrow C_6H_6(l)$$

and is +49.0 kJ mol−1. The standard enthalpies of formation of elements in their reference states are zero at all temperatures because they are the enthalpies of such 'null' reactions as $N_2(g) \rightarrow N_2(g)$. Some enthalpies of formation are listed in Tables 2.5 and 2.6.

Table 2.6* Standard enthalpies of formation of inorganic compounds, $\Delta_f H^{\ominus}/(\mathrm{kJ\ mol^{-1}})$, at 298 K

$H_2O(l)$	−285.83	$H_2O_2(l)$	−187.78
$NH_3(g)$	−46.11	$N_2H_4(l)$	+50.63
$NO_2(g)$	+33.18	$N_2O_4(g)$	+9.16
NaCl(s)	−411.15	KCl(s)	−436.75

* More values are given in the *Data section*.

The web site contains links to online databases of thermochemical data.

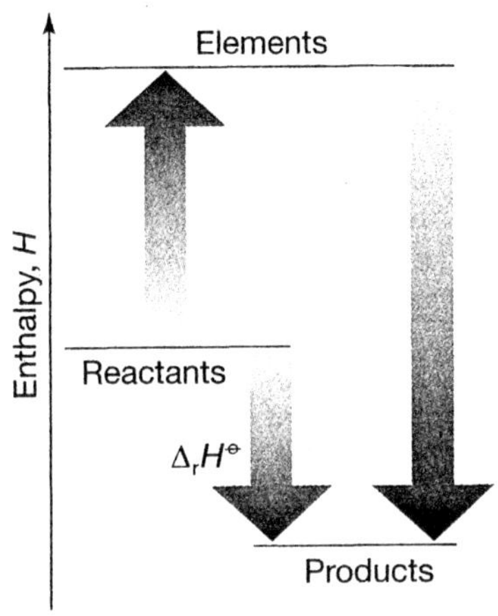

3

(a) The reaction enthalpy in terms of enthalpies of formation

Conceptually, we can regard a reaction as proceeding by decomposing the reactants into their elements and then forming those elements into the products (3). The value of $\Delta_r H^\ominus$ for the overall reaction is the sum of these 'unforming' and forming enthalpies. Because 'unforming' is the reverse of forming, the enthalpy of an unforming step is the negative of the enthalpy of formation. Hence, in the enthalpies of formation of substances, we have enough information to calculate the enthalpy of any reaction by using

$$\Delta_r H^\ominus = \sum_{\text{Products}} \nu \Delta_f H^\ominus - \sum_{\text{Reactants}} \nu \Delta_f H^\ominus \tag{2.42}$$

where in each case the enthalpies of formation of the species that occur are multiplied by their stoichiometric coefficients.

Illustration 2.7

The standard reaction enthalpy of $2\,HN_3(l) + 2\,NO(g) \rightarrow H_2O_2(l) + 4\,N_2(g)$ is calculated as follows:

$$\begin{aligned}\Delta_r H^\ominus &= \{\Delta_f H^\ominus(H_2O_2, l) + 4\Delta_f H^\ominus(N_2, g)\} - \{2\Delta_f H^\ominus(HN_3, l) + 2\Delta_f H^\ominus(NO, g)\}\\ &= \{-187.78 + 4(0)\}\ \text{kJ mol}^{-1} - \{2(264.0) + 2(90.25)\}\ \text{kJ mol}^{-1}\\ &= -896.3\ \text{kJ mol}^{-1}\end{aligned}$$

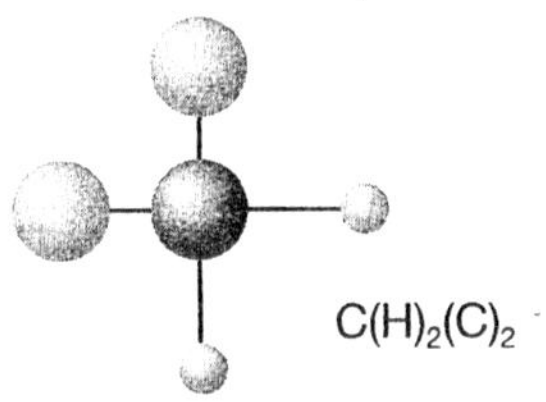

4

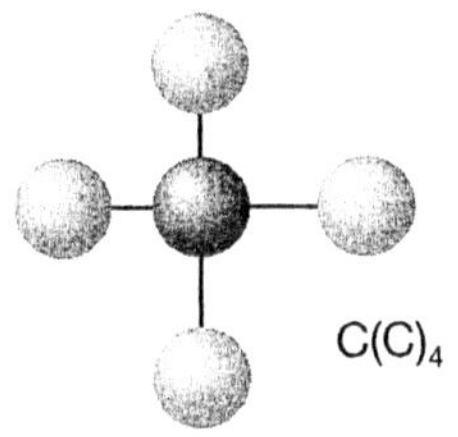

5

(b) Group contributions

We have seen how to construct standard reaction enthalpies by combining standard enthalpies of formation. The question that now arises is whether we can construct standard enthalpies of formation from a knowledge of the chemical constitution of the species. The short answer is that there is no thermodynamically exact way of expressing enthalpies of formation in terms of contributions from individual atoms and bonds. In the past, approximate procedures based on **mean bond enthalpies**, ΔH(A—B), the average enthalpy change associated with the breaking of a specific A—B bond,

$$\text{A—B(g)} \longrightarrow \text{A(g)} + \text{B(g)} \qquad \Delta H(\text{A—B})$$

have been used. However, this procedure is notoriously unreliable, in part because the ΔH(A—B) are average values for a series of related compounds. Nor does the approach distinguish between geometrical isomers, where the same atoms and bonds may be present but experimentally the enthalpies of formation might be significantly different.

Somewhat more reliable is an approach in which a molecule is regarded as being built up of **thermochemical groups**, an atom or physical group of atoms bound to at least two other atoms: two examples are shown as (4) and (5). The enthalpy of formation of the compound is then expressed (approximately, at least) as the sum of the contributions associated with all the thermochemical groups into which the molecule can be divided. A list of the appropriate values is given in Table 2.7 and the method is illustrated in the following example. Table 2.7 also contains information on heat capacity, which is approximately additive in a similar sense.

Table 2.7* Benson thermochemical groups

Group	$\Delta_f H^\ominus$/(kJ mol^{-1})	$C^\ominus_{p,m}$/(J K^{-1} mol^{-1})
$C(H)_3(C)$	−42.17	25.9
$C(H)_2(C)_2$	−20.7	22.8
$C(H)(C)_3$	−6.91	18.7
$C(C)_4$	+8.16	18.2

* More values are given in the *Data section*.

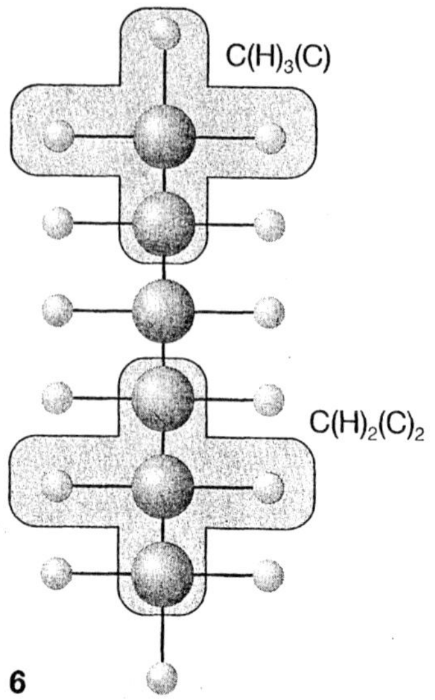

6

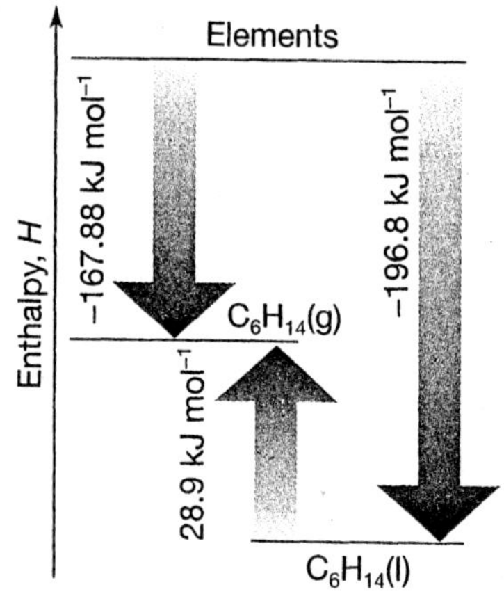

7

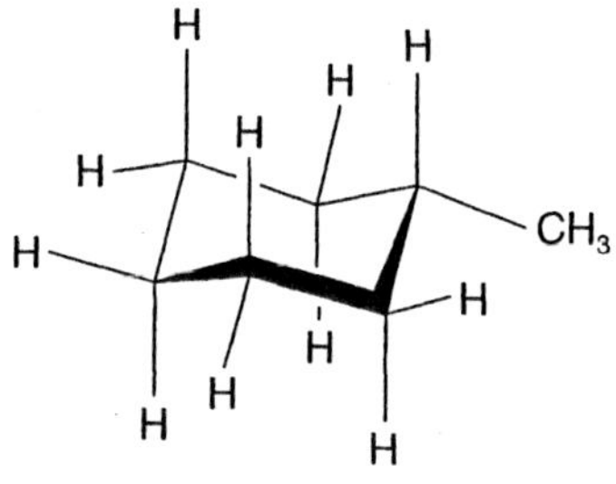

8

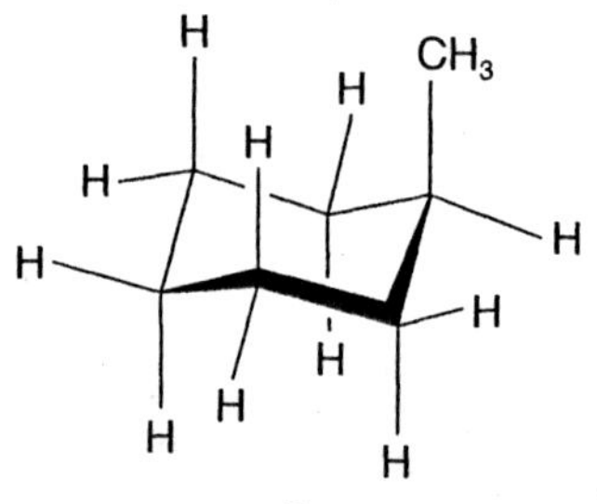

9

Example 2.6 **Using the thermochemical group approach**

Estimate the standard enthalpy of formation at 298 K of hexane in (a) the gas phase, (b) the liquid phase.

Method First, identify the thermochemical groups present in the molecule. Then add together the appropriate values in Table 2.7. To obtain the enthalpy of formation of the liquid, include the enthalpy of condensation of the vapour; that is, the negative of the enthalpy of vaporization of the liquid (Table 2.3).

Answer The decomposition of the molecule into groups is depicted in (**6**). There are two $C(H)_3(C)$ groups and four $C(H)_2(C)_2$ groups; therefore

$$\Delta_f H^{\ominus}(C_6H_{14}, g) = 2(-42.17 \text{ kJ mol}^{-1}) + 4(-20.7 \text{ kJ mol}^{-1}) = -167.1 \text{ kJ mol}^{-1}$$

The enthalpy of vaporization of hexane is 33.1 kJ mol^{-1}; therefore (**7**),

$$\Delta_f H^{\ominus}(C_6H_{14}, l) = (-167.1 \text{ kJ mol}^{-1}) - (33.1 \text{ kJ mol}^{-1}) = -200.2 \text{ kJ mol}^{-1}$$

The experimental value is −198.7 kJ mol^{-1}.

Self-test 2.8 Estimate the standard enthalpy of formation of gaseous 2,2-dimethylpropane. [−162.6 kJ mol^{-1}]

(c) Enthalpies of formation and molecular modelling

It is difficult to estimate standard enthalpies of formation of conformational isomers. For example, it has been observed experimentally that the equatorial (**8**) and axial (**9**) conformers of methylcyclohexane have different standard enthalpies of formation even though they consist of the same thermochemical groups. This difference arises from the steric repulsions in the axial conformer, which raise its energy relative to that of the equatorial conformer.

Although it is possible to modify the thermochemical group approach to take steric interactions into account, computer-aided molecular modelling is becoming the technique of choice for estimating standard enthalpies of formation of molecules with complex three-dimensional structures and the 'group approach' is regarded as primitive and old-fashioned. Commercial software packages use the principles developed in Chapter 14 to calculate the standard enthalpy of formation of a conformer drawn on the computer screen. The difference between calculated standard enthalpies of formation of two conformers is then an estimate of the conformational energy difference. In the case of methylcyclohexane, the calculated conformational energy difference ranges from 5.9 to 7.9 kJ mol^{-1}, with the equatorial conformer having the lower standard enthalpy of formation. These estimates compare favourably with the experimental value of 7.5 kJ mol^{-1}. However, good agreement between calculated and experimental values is relatively rare. Computational methods almost always predict correctly which conformer is more stable but do not always predict the correct magnitude of the conformational energy difference.

2.9 The temperature dependence of reaction enthalpies

The standard enthalpies of many important reactions have been measured at different temperatures, and for serious work these accurate data must be used. However, in the absence of this information, standard reaction enthalpies at different

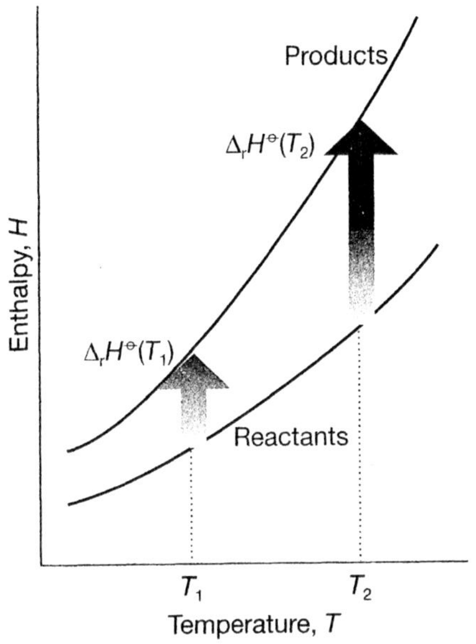

2.19 An illustration of the content of Kirchhoff's law. When the temperature is increased, the enthalpy of the products and the reactants both increase, but may do so to different extents. In each case, the change in enthalpy depends on the heat capacities of the substances. The change in reaction enthalpy reflects the difference in the changes of the enthalpies.

temperatures may be estimated from heat capacities and the reaction enthalpy at some other temperature (Fig. 2.19).

It follows from eqn 27 that, when a substance is heated from T_1 to T_2, its enthalpy changes from $H(T_1)$ to

$$H(T_2) = H(T_1) + \int_{T_1}^{T_2} C_p \, dT \tag{2.43}$$

(We have assumed that no phase transition takes place in the temperature range of interest.) Because this equation applies to each substance in the reaction, the standard reaction enthalpy changes from $\Delta_r H^\ominus(T_1)$ to

$$\Delta_r H^\ominus(T_2) = \Delta_r H^\ominus(T_1) + \int_{T_1}^{T_2} \Delta_r C_p^\ominus \, dT \tag{2.44}$$

where $\Delta_r C_p^\ominus$ is the difference of the molar heat capacities of products and reactants under standard conditions weighted by the stoichiometric numbers that appear in the chemical equation:

$$\Delta_r C_p^\ominus = \sum_{\text{Products}} \nu C_{p,\text{m}}^\ominus - \sum_{\text{Reactants}} \nu C_{p,\text{m}}^\ominus \qquad [2.45]$$

Equation 44 is known as **Kirchhoff's law**. It is normally a good approximation to assume that $\Delta_r C_p$ is independent of the temperature, at least over reasonably limited ranges, as illustrated in the following example. Although the individual heat capacities may vary, their difference varies less significantly. In some cases the temperature dependence of heat capacities is taken into account by using eqn 30.

Example 2.7 Using Kirchhoff's law

The standard enthalpy of formation of gaseous H_2O at 298 K is −241.82 kJ mol^{-1}. Estimate its value at 100°C given the following values of the molar heat capacities at constant pressure: H_2O(g): 33.58 J K^{-1} mol^{-1}; H_2(g): 28.84 J K^{-1} mol^{-1}; O_2(g): 29.37 J K^{-1} mol^{-1}. Assume that the heat capacities are independent of temperature.

Method When $\Delta C_p^\ominus$ is independent of temperature in the range T_1 to T_2, the integral in eqn 44 evaluates to $(T_2 - T_1)\Delta_r C_p^\ominus$. Therefore,

$$\Delta_r H^\ominus(T_2) = \Delta_r H^\ominus(T_1) + (T_2 - T_1)\Delta_r C_p^\ominus$$

To proceed, write the chemical equation, identify the stoichiometric coefficients, and calculate $\Delta_r C_p^\ominus$ from the data.

Answer The reaction is $H_2(g) + \frac{1}{2} O_2(g) \rightarrow H_2O(g)$, so

$$\Delta_r C_p^\ominus = C_{p,\text{m}}^\ominus(H_2O, g) - \{C_{p,\text{m}}^\ominus(H_2, g) + \tfrac{1}{2} C_{p,\text{m}}^\ominus(O_2, g)\}$$
$$= -9.94 \text{ J K}^{-1} \text{ mol}^{-1}$$

It then follows that

$$\Delta_f H^\ominus(373 \text{ K}) = -241.82 \text{ kJ mol}^{-1} + (75 \text{ K}) \times (-9.94 \text{ J K}^{-1} \text{ mol}^{-1})$$
$$= -242.6 \text{ kJ mol}^{-1}$$

Self-test 2.9 Estimate the standard enthalpy of formation of cyclohexene at 400 K from the data in Table 2.5. [−163 kJ mol^{-1}]

Checklist of key ideas

☐ thermodynamics, the study of the transformations of energy

The basic concepts

☐ system, the part of the world in which we have a special interest

☐ surroundings, the region outside the system where we make our measurements

☐ open system, a system with a boundary through which matter can be transferred

☐ closed system, a system with a boundary through which matter cannot be transferred

☐ isolated system, a system with a boundary through which neither matter nor energy can be transferred

2.1 Work, heat, and energy

☐ work, the transfer of energy by motion against an opposing force

☐ energy, the capacity to do work

☐ heat, the transfer of energy as a result of a temperature difference

☐ diathermic boundary, a boundary that permits energy transfer as heat

☐ adiabatic boundary, a boundary that does not permit energy transfer as heat

☐ exothermic process, a process that releases energy as heat

☐ endothermic process, a process that absorbs energy as heat

☐ the molecular interpretation of heat and work, as random and ordered motion, respectively

☐ thermal motion, the random motion of molecules

2.2 The First Law

☐ internal energy, the total energy of a system

☐ state function, a property that depends only on the current state of the system and is independent of how that state has been prepared

☐ First Law of thermodynamics: The internal energy of an isolated system is constant, $\Delta U = q + w$

☐ alternative statement of First Law: The work needed to change an adiabatic system from one specified state to another specified state is the same however the work is done

☐ mechanical definition of heat, $q = w_{ad} - w$

Work and heat

2.3 Expansion work

☐ work, motion against an opposing force, $dw = -F\,dz$

☐ expansion work, work of expansion (or compression), $dw = -p_{ex}\,dV$

☐ non-expansion work or additional work, work other than expansion work

☐ free expansion, expansion into a vacuum, $w = 0$

☐ expansion against constant pressure, $w = -p_{ex}\Delta V$

☐ indicator diagram, a p,V-graph used to compute expansion work

☐ reversible change, a change that can be reversed by an infinitesimal modification of a variable

☐ equilibrium, a condition in which an infinitesimal change in a variable in opposite directions results in opposite changes in its state

☐ work of isothermal reversible expansion, $w = -nRT\ln(V_f/V_i)$

☐ maximum work, the work achieved in a reversible change

2.4 Heat transactions

☐ internal energy change and heat transfer at constant volume, $\Delta U = q_V$

☐ calorimetry, the study of heat transfer during physical and chemical processes

☐ calorimeter, a device for measuring the heat transferred

☐ adiabatic bomb calorimeter, a calorimeter in which there is no net loss of heat from the calorimeter to the surroundings

☐ calorimeter constant, the constant C in the relation $q = C\Delta T$

☐ heat capacity at constant volume, $C_V = (\partial U/\partial T)_V$

☐ molar heat capacity, the heat capacity divided by the amount of substance

☐ specific heat capacity, the heat capacity divided by the mass

☐ internal energy change and temperature rise, $dU = C_V\,dT$

2.5 Enthalpy

☐ enthalpy, $H = U + pV$

☐ enthalpy change and heat transfer at constant pressure, $\Delta H = q_p$

☐ relation between ΔU and ΔH, $\Delta H = \Delta U + \Delta n_g RT$

☐ heat capacity at constant pressure, $C_p = (\partial H/\partial T)_p$

☐ isobaric calorimeter, a calorimeter for studying processes at constant pressure

☐ adiabatic flame calorimeter, an isobaric calorimeter used to study the combustion of volatile compounds

☐ differential scanning calorimeter (DSC), an isobaric calorimeter in which the behaviour of a sample material is compared to the behaviour of a reference material

☐ relation between heat capacities of a perfect gas, $C_p - C_V = nR$

2.6 Adiabatic changes

☐ work of adiabatic change, $w_{ad} = C_V\Delta T$

☐ temperature change accompanying adiabatic change of a perfect gas, $T_f = T_i(V_i/V_f)^{1/c}$, $c = C_{V,m}/R$

☐ heat capacity ratio, $\gamma = C_{p,m}/C_{V,m}$

☐ pressure–volume relation for adiabatic change, $pV^{\gamma} = \text{constant}$

☐ adiabat, curves of pressure versus volume for adiabatic, reversible changes

Thermochemistry

☐ thermochemistry, the study of the heat produced or required by chemical reactions

2.7 Standard enthalpy changes

☐ standard enthalpy change, the change in enthalpy for a process in which the initial and final substances are in their standard states

☐ standard state, the pure substance at 1 bar

☐ standard enthalpy of transition ($\Delta_{trs}H^{\ominus}$), the standard enthalpy change that accompanies a change of physical state

☐ standard enthalpy of vaporization, the standard enthalpy change accompanying the conversion of a solid or liquid to a vapour

- □ standard enthalpy of fusion, the standard enthalpy change accompanying the conversion of a solid to a liquid
- □ addition of enthalpy changes, as in $\Delta_{sub}H^{\ominus} = \Delta_{fus}H^{\ominus} + \Delta_{vap}H^{\ominus}$
- □ enthalpy change for reverse process, $\Delta H^{\ominus}(A \rightarrow B) = -\Delta H^{\ominus}(B \rightarrow A)$
- □ thermochemical equation, a combination of a chemical equation and the corresponding change in standard enthalpy
- □ standard reaction enthalpy, $\Delta_r H^{\ominus} = \sum_{Products} \nu H_m^{\ominus} - \sum_{Reactants} \nu H_m^{\ominus}$
- □ standard enthalpy of combustion, the standard reaction enthalpy for the complete oxidation of an organic compound to CO_2 gas and liquid H_2O if the compound contains C, H, and O, and to N_2 gas if N is also present
- □ Hess's law: The standard enthalpy of an overall reaction is the sum of the standard enthalpies of the individual reactions into which a reaction may be divided

2.8 Standard enthalpies of formation

- □ standard enthalpy of formation ($\Delta_f H^{\ominus}$), the standard reaction enthalpy for the formation of the compound from its elements in their reference states
- □ reference state, the most stable state of an element at the specified temperature and 1 bar
- □ combining enthalpies of formation, $\Delta_r H^{\ominus} = \sum_{Products} \nu\Delta_f H^{\ominus} - \sum_{Reactants} \nu\Delta_f H^{\ominus}$
- □ mean bond enthalpy, the average enthalpy change associated with the breaking of a specific A—B bond
- □ thermochemical group, an atom or physical group of atoms bound to at least two other atoms used in thermochemical calculations

2.9 The temperature dependence of reaction enthalpies

- □ Kirchhoff's law for the temperature dependence of reaction enthalpies, $\Delta_r H^{\ominus}(T_2) = \Delta_r H^{\ominus}(T_1) + \int_{T_1}^{T_2} \Delta_r C_p^{\ominus}\, dT$

Further reading

Articles of general interest

E.R. Boyko and J.F. Belliveau, Simplification of some thermochemical equations. *J. Chem. Educ.* **67**, 743 (1990).

F. Freeman, Z.M. Tsegai, M.L. Kasner, and W.J. Hehre, A comparison of the ab initio calculated and experimental conformational energies of alkylcyclohexanes. *J. Chem. Ed.* **77**, 661 (2000).

E.A. Gislason and N.C. Craig, General definitions of work and heat in thermodynamic processes. *J. Chem. Educ.* **64**, 670 (1987).

J. Güèmez, S. Velasco, and M.A. Matías, Thermal coefficients and heat capacities in systems with chemical reaction: the Le Chatelier-Braun principle. *J. Chem. Educ.* **72**, 199 (1995).

I.H. Segel and L.D. Segel, Energetics of biological processes. In *Encyclopedia of applied physics* (ed. G.L. Trigg), **6**, 207. VCH, New York (1993).

T. Solomon, Standard enthalpies of formation of ions in solution. *J. Chem. Educ.* **68**, 41 (1991).

R.S. Treptow, Bond energies and enthalpies: an often neglected difference. *J. Chem. Educ.* **72**, 497 (1995).

R.S. Treptow, How thermodynamic data and equilibrium constants changed when the standard-state pressure became 1 bar. *J. Chem. Educ.* **76**, 212 (1999).

Texts and sources of data and information

S.W. Benson, *Thermochemical kinetics*. Wiley, New York (1976).

M.W. Chase, Jr. (ed.), *NIST-JANAF thermochemical tables*. Published as *J. Phys. Chem. Reference Data, Monograph no. 9*. American Institute of Physics, New York (1998).

M. Zabransky, V. Ruzicka Jr., V. Majer, and E.S. Domalski, *Heat capacity of liquids*. Published as *J. Phys. Chem. Reference Data, Monograph no. 6*. American Institute of Physics, New York (1996).

J.D. Cox and G. Pilcher, *Thermochemistry of organic and organometallic compounds*. Academic Press, New York (1970).

J.D. Cox, D.D. Wagman, and V.A. Medvedev, *CODATA key values for thermodynamics*. Hemisphere Publishing Corp., New York (1989).

W.E. Dasent, *Inorganic energetics*. Cambridge University Press (1982).

J.A. Dean (ed.), *Handbook of organic chemistry*. McGraw-Hill, New York (1987).

J. Emsley, *The elements*. Oxford University Press (1998).

E. Fermi, *Thermodynamics*. Dover, New York (1956).

E.F.G. Harrington, Thermodynamic data and their uses. *Chemical thermodynamics: Specialist Periodical Reports*. **1**, 31 (1973).

A.M. James and M.P. Lord (ed.), *Macmillan's chemical and physical data*. Macmillan, London (1992).

D.A. Johnson, *Some thermodynamic aspects of inorganic chemistry*. Cambridge University Press (1982).

I.M. Klotz and R.M. Rosenberg, *Chemical thermodynamics: basic theory and methods*. Wiley-Interscience, New York (2000).

G.N. Lewis and M. Randall, *Thermodynamics*. Revised by K.S. Pitzer and L. Brewer. McGraw-Hill, New York (1961).

M.L. McGlashan, *Chemical thermodynamics*. Academic Press, London (1979).

J.B. Pedley, R.D. Naylor, and S.P. Kirby, *Thermochemical data of organic compounds*. Chapman and Hall, London (1986).

B.W. Rossiter and R.C. Baetzgold (ed.) *Physical methods of chemistry*, **VI**. Wiley-Interscience (1992).

V. Smil, *Energies*. MIT Press, Cambridge (1999).

D.R. Stull and G.C. Sinke, *Thermodynamic properties of the elements*. American Chemical Society, Washington (1956).

D.R. Stull, E.F. Westrum, and G.C. Sinke, *The chemical thermodynamics of organic compounds*. Wiley-Interscience, New York (1969).

H.C. Van Ness, *Understanding thermodynamics*. Dover, New York (1983).

D.B. Wagman, W.H. Evans, V.B. Parker, R.H. Schumm, I. Halow, S.M. Bailey, K.L. Churney, and R.L. Nuttall, *The NBS tables of chemical thermodynamic properties*. Published as *J. Phys. Chem. Reference Data* **11**, Supplement 2 (1982).

D.R. Lide (ed.), *Handbook of chemistry and physics*, Vol. 81. CRC Press, Boca Raton (2000).

Exercises

Discussion questions

2.1 (a) Provide mechanical definitions of work and heat.

2.1 (b) Provide molecular interpretations for work and heat.

2.2 (a) Distinguish between expansion work against constant pressure and work of reversible expansion.

2.2 (b) Consider the reversible expansion of a perfect gas. Provide a physical interpretation for the fact that pV^{γ} = constant for an adiabatic change, whereas pV = constant for an isothermal change.

2.3 (a) Explain the difference between the change in internal energy and the change in enthalpy of a chemical or physical process.

2.3 (b) Describe at least two methods by which standard reaction enthalpies may be predicted. Discuss the advantages and disadvantages of each method.

Numerical exercises

Assume all gases are perfect unless stated otherwise. Note that 1 atm = 1.013 25 bar exactly. Unless otherwise stated, thermochemical data are for 298.15 K.

2.4 (a) Calculate the work done to raise a mass of 1.0 kg through 10 m on the surface of (a) the Earth ($g = 9.81$ m s^{-2}) and (b) the moon ($g = 1.60$ m s^{-2}).

2.4 (b) Calculate the work done to raise a mass of 5.0 kg through 100 m on the surface of (a) the Earth ($g = 9.81$ m s^{-2}) and (b) Mars ($g = 3.73$ m s^{-2}).

2.5 (a) Calculate the work needed for a 65 kg person to climb through 4.0 m on the surface of the Earth.

2.5 (b) Calculate the work needed for a bird of mass 120 g to fly to a height of 50 m from the surface of the Earth.

2.6 (a) A chemical reaction takes place in a container of cross-sectional area 100 cm^2. As a result of the reaction, a piston is pushed out through 10 cm against an external pressure of 1.0 atm. Calculate the work done by the system.

2.6 (b) A chemical reaction takes place in a container of cross-sectional area 50.0 cm^2. As a result of the reaction, a piston is pushed out through 15 cm against an external pressure of 121 kPa. Calculate the work done by the system.

2.7 (a) A sample consisting of 1.00 mol Ar is expanded isothermally at 0°C from 22.4 L to 44.8 L (a) reversibly, (b) against a constant external pressure equal to the final pressure of the gas, and (c) freely (against zero external pressure). For the three processes calculate q, w, ΔU, and ΔH.

2.7 (b) A sample consisting of 2.00 mol He is expanded isothermally at 22°C from 22.8 L to 31.7 L (a) reversibly, (b) against a constant external pressure equal to the final pressure of the gas, and (c) freely (against zero external pressure). For the three processes calculate q, w, ΔU, and ΔH.

2.8 (a) A sample consisting of 1.00 mol of monatomic perfect gas, for which $C_{V,\mathrm{m}} = \frac{3}{2}R$, initially at $p_1 = 1.00$ atm and $T_1 = 300$ K, is heated reversibly to 400 K at constant volume. Calculate the final pressure, ΔU, q, and w.

2.8 (b) A sample consisting of 2.00 mol of perfect gas molecules, for which $C_{V,\mathrm{m}} = \frac{5}{2}R$, initially at $p_1 = 111$ kPa and $T_1 = 277$ K, is heated reversibly to 356 K at constant volume. Calculate the final pressure, ΔU, q, and w.

2.9 (a) A sample of 4.50 g of methane occupies 12.7 L at 310 K. (a) Calculate the work done when the gas expands isothermally against a constant external pressure of 200 Torr until its volume has increased by 3.3 L. (b) Calculate the work that would be done if the same expansion occurred reversibly.

2.9 (b) A sample of argon of mass 6.56 g occupies 18.5 L at 305 K. (a) Calculate the work done when the gas expands isothermally against a constant external pressure of 7.7 kPa until its volume has increased by 2.5 L. (b) Calculate the work that would be done if the same expansion occurred reversibly.

2.10 (a) In the isothermal reversible compression of 52.0 mmol of a perfect gas at 260 K, the volume of the gas is reduced to one-third its initial value. Calculate w for this process.

2.10 (b) In the isothermal reversible compression of 1.77 mmol of a perfect gas at 273 K, the volume of the gas is reduced to 0.224 of its initial value. Calculate w for this process.

2.11 (a) A sample of 1.00 mol H_2O(g) is condensed isothermally and reversibly to liquid water at 100°C. The standard enthalpy of vaporization of water at 100°C is 40.656 kJ mol^{-1}. Find w, q, ΔU, and ΔH for this process.

2.11 (b) A sample of 2.00 mol CH_3OH(g) is condensed isothermally and reversibly to liquid at 64°C. The standard enthalpy of vaporization of methanol at 64°C is 35.3 kJ mol^{-1}. Find w, q, ΔU, and ΔH for this process.

2.12 (a) A strip of magnesium of mass 15 g is dropped into a beaker of dilute hydrochloric acid. Calculate the work done by the system as a result of the reaction. The atmospheric pressure is 1.0 atm and the temperature 25°C.

2.12 (b) A piece of zinc of mass 5.0 g is dropped into a beaker of dilute hydrochloric acid. Calculate the work done by the system as a result of the reaction. The atmospheric pressure is 1.1 atm and the temperature 23°C.

2.13 (a) Calculate the heat required to melt 750 kg of sodium metal at 371 K. The enthalpy of fusion of sodium is 2.601 kJ mol^{-1}.

2.13 (b) Calculate the heat required to melt 500 kg of potassium metal at 336 K. The enthalpy of fusion of potassium is 2.35 kJ mol^{-1}.

2.14 (a) A sample of a perfect gas was found to vary with temperature according to the expression $C_p/(\mathrm{J\,K^{-1}}) = 20.17 + 0.3665(T/\mathrm{K})$. Calculate q, w, ΔU, and ΔH when the temperature is raised from 25°C to 200°C (a) at constant pressure, (b) at constant volume.

2.14 (b) The constant-pressure heat capacity of a sample of a perfect gas was found to vary with temperature according to the expression $C_p/(\mathrm{J\,K^{-1}}) = 20.17 + 0.4001(T/\mathrm{K})$. Calculate q, w, ΔU, and ΔH when the temperature is raised from 0°C to 100°C (a) at constant pressure, (b) at constant volume.

2.15 (a) Calculate the final temperature of a sample of argon of mass 12.0 g that is expanded reversibly and adiabatically from 1.0 L at 273.15 K to 3.0 L.

2.15 (b) Calculate the final temperature of a sample of carbon dioxide of mass 16.0 g that is expanded reversibly and adiabatically from 500 mL at 298.15 K to 2.00 L.

2.16 (a) A sample of carbon dioxide of mass 2.45 g at 27.0°C is allowed to expand reversibly and adiabatically from 500 mL to 3.00 L. What is the work done by the gas?

2.16 (b) A sample of nitrogen of mass 3.12 g at 23.0°C is allowed to expand reversibly and adiabatically from 400 mL to 2.00 L. What is the work done by the gas?

2.17 (a) Calculate the final pressure of a sample of carbon dioxide that expands reversibly and adiabatically from 57.4 kPa and 1.0 L to a final volume of 2.0 L. Take $\gamma = 1.4$.

2.17 (b) Calculate the final pressure of a sample of water vapour that expands reversibly and adiabatically from 87.3 Torr and 500 mL to a final volume of 3.0 L. Take $\gamma = 1.3$.

2.18 (a) Calculate the final pressure of a sample of carbon dioxide of mass 2.4 g that expands reversibly and adiabatically from an initial temperature of 278 K and volume 1.0 L to a final volume of 2.0 L. Take $\gamma = 1.4$.

2.18 (b) Calculate the final pressure of a sample of water vapour of mass 1.4 g that expands reversibly and adiabatically from an initial temperature of 300 K and volume 1.0 L to a final volume of 3.0 L. Take $\gamma = 1.3$.

2.19 (a) Calculate the standard enthalpy of formation of butane at 25°C from its standard enthalpy of combustion.

2.19 (b) Calculate the standard enthalpy of formation of hexane at 25°C from its standard enthalpy of combustion.

2.20 (a) When 229 J of energy is supplied as heat to 3.0 mol Ar(g), the temperature of the sample increases by 2.55 K. Calculate the molar heat capacities at constant volume and constant pressure of the gas.

2.20 (b) When 178 J of energy is supplied as heat to 1.9 mol of gas molecules, the temperature of the sample increases by 1.78 K. Calculate the molar heat capacities at constant volume and constant pressure of the gas.

2.21 (a) A sample of a liquid of mass 25 g is cooled from 290 K to 275 K at constant pressure by the extraction of 1.2 kJ of energy as heat. Calculate q and ΔH and estimate the heat capacity of the sample.

2.21 (b) A sample of a liquid of mass 30.5 g is cooled from 288 K to 275 K at constant pressure by the extraction of 2.3 kJ of energy as heat. Calculate q and ΔH and estimate the heat capacity of the sample.

2.22 (a) When 3.0 mol O_2 is heated at a constant pressure of 3.25 atm, its temperature increases from 260 K to 285 K. Given that the molar heat capacity of O_2 at constant pressure is 29.4 J K^{-1} mol^{-1}, calculate q, ΔH, and ΔU.

2.22 (b) When 2.0 mol CO_2 is heated at a constant pressure of 1.25 atm, its temperature increases from 250 K to 277 K. Given that the molar heat capacity of CO_2 at constant pressure is 37.11 J K^{-1} mol^{-1}, calculate q, ΔH, and ΔU.

2.23 (a) A sample of 4.0 mol O_2 is originally confined in 20 L at 270 K and then undergoes adiabatic expansion against a constant pressure of 600 Torr until the volume has increased by a factor of 3.0. Calculate q, w, ΔT, ΔU, and ΔH. (The final pressure of the gas is not necessarily 600 Torr.)

2.23 (b) A sample of 5.0 mol CO_2 is originally confined in 15 L at 280 K and then undergoes adiabatic expansion against a constant pressure of 78.5 kPa until the volume has increased by a factor of 4.0. Calculate q, w, ΔT, ΔU, and ΔH. (The final pressure of the gas is not necessarily 78.5 kPa.)

2.24 (a) A sample consisting of 3.0 mol of perfect gas molecules at 200 K and 2.00 atm is compressed reversibly and adiabatically until the temperature reaches 250 K. Given that its molar constant-volume heat capacity is 27.5 J K^{-1} mol^{-1}, calculate q, w, ΔU, ΔH, and the final pressure and volume.

2.24 (b) A sample consisting of 2.5 mol of gas molecules at 220 K and 200 kPa is compressed reversibly and adiabatically until the temperature reaches 255 K. Given that its molar constant-volume heat capacity is 27.6 J K^{-1} mol^{-1}, calculate q, w, ΔU, ΔH, and the final pressure and volume.

2.25 (a) A sample consisting of 1.0 mol of perfect gas molecules with $C_V = 20.8$ J K^{-1} is initially at 3.25 atm and 310 K. It undergoes reversible adiabatic expansion until its pressure reaches 2.50 atm. Calculate the final volume and temperature and the work done.

2.25 (b) A sample consisting of 1.5 mol of perfect gas molecules with $C_{p,m} = 20.8$ J K^{-1} mol^{-1} is initially at 230 kPa and 315 K. It undergoes reversible adiabatic expansion until its pressure reaches 170 kPa. Calculate the final volume and temperature and the work done.

2.26 (a) Estimate the changes in volume that occur when a sample of mercury of volume 1.0 cm^3 is heated through 5.0 K at room temperature.

2.26 (b) Estimate the changes in volume that occur when a sample of iron of volume 5.0 cm^3 is heated through 10.0 K at room temperature.

2.27 (a) Consider a system consisting of 2.0 mol CO_2 (assumed to be a perfect gas) at 25°C confined to a cylinder of cross-section 10 cm^2 at 10 atm. The gas is allowed to expand adiabatically and irreversibly against a constant pressure of 1.0 atm. Calculate w, q, ΔU, ΔH, and ΔT when the piston has moved 20 cm.

2.27 (b) Consider a system consisting of 3.0 mol O_2 (assumed to be a perfect gas) at 25°C confined to a cylinder of cross-section 22 cm^2 at 820 kPa. The gas is allowed to expand adiabatically and irreversibly against a constant pressure of 110 kPa. Calculate w, q, ΔU, ΔH, and ΔT when the piston has moved 15 cm.

2.28 (a) A sample consisting of 65.0 g of xenon is confined in a container at 2.00 atm and 298 K and then allowed to expand adiabatically (a) reversibly to 1.00 atm, (b) against a constant pressure of 1.00 atm. Calculate the final temperature in each case.

2.28 (b) A sample consisting of 15.0 g of nitrogen is confined in a container at 220 kPa and 200 K and then allowed to expand adiabatically (a) reversibly to 110 kPa, (b) against a constant pressure of 110 kPa. Calculate the final temperature in each case.

2.29 (a) A certain liquid has $\Delta_{vap}H^{\ominus} = 26.0$ kJ mol^{-1}. Calculate q, w, ΔH, and ΔU when 0.50 mol is vaporized at 250 K and 750 Torr.

2.29 (b) A certain liquid has $\Delta_{vap}H^{\ominus} = 32.0$ kJ mol^{-1}. Calculate q, w, ΔH, and ΔU when 0.75 mol is vaporized at 260 K and 765 Torr.

2.30 (a) The standard enthalpy of formation of ethylbenzene is −12.5 kJ mol^{-1}. Calculate its standard enthalpy of combustion.

2.30 (b) The standard enthalpy of formation of phenol is −165.0 kJ mol^{-1}. Calculate its standard enthalpy of combustion.

2.31 (a) Calculate the standard enthalpy of hydrogenation of 1-hexene to hexane given that the standard enthalpy of combustion of 1-hexene is -4003 kJ mol^{-1}.

2.31 (b) Calculate the standard enthalpy of hydrogenation of 1-butene to butane given that the standard enthalpy of combustion of 1-butene is -2717 kJ mol^{-1}.

2.32 (a) The standard enthalpy of combustion of cyclopropane is -2091 kJ mol^{-1} at 25°C. From this information and enthalpy of formation data for $CO_2(g)$ and $H_2O(g)$, calculate the enthalpy of formation of cyclopropane. The enthalpy of formation of propene is +20.42 kJ mol^{-1}. Calculate the enthalpy of isomerization of cyclopropane to propene.

2.32 (b) From the following data, determine $\Delta_f H^\ominus$ for diborane, $B_2H_6(g)$, at 298 K:

(1) $B_2H_6(g) + 3\,O_2(g) \longrightarrow B_2O_3(s) + 3\,H_2O(g)$ $\quad \Delta_r H^\ominus = -1941$ kJ mol^{-1}

(2) $2\,B(s) + \frac{3}{2}O_2(g) \longrightarrow B_2O_3(s)$ $\quad \Delta_r H^\ominus = -2368$ kJ mol^{-1}

(3) $H_2(g) + \frac{1}{2}O_2(g) \longrightarrow H_2O(g)$ $\quad \Delta_r H^\ominus = -241.8$ kJ mol^{-1}

2.33 (a) Calculate the standard internal energy of formation of liquid methyl acetate from its standard enthalpy of formation, which is -442 kJ mol^{-1}.

2.33 (b) Calculate the standard internal energy of formation of urea from its standard enthalpy of formation, which is -333.51 kJ mol^{-1}.

2.34 (a) The temperature of a bomb calorimeter rose by 1.617 K when a current of 3.20 A was passed for 27.0 s from a 12.0 V source. Calculate the calorimeter constant.

2.34 (b) The temperature of a bomb calorimeter rose by 1.712 K when a current of 2.86 A was passed for 22.5 s from a 12.0 V source. Calculate the calorimeter constant.

2.35 (a) When 120 mg of naphthalene, $C_{10}H_8(s)$, was burned in a bomb calorimeter the temperature rose by 3.05 K. Calculate the calorimeter constant. By how much will the temperature rise when 10 mg of phenol, $C_6H_5OH(s)$, is burned in the calorimeter under the same conditions?

2.35 (b) When 2.25 mg of anthracene, $C_{14}H_{10}(s)$, was burned in a bomb calorimeter the temperature rose by 1.35 K. Calculate the calorimeter constant. By how much will the temperature rise when 135 mg of phenol, $C_6H_5OH(s)$, is burned in the calorimeter under the same conditions?

2.36 (a) Calculate the standard enthalpy of solution of AgCl(s) in water from the enthalpies of formation of the solid and the aqueous ions.

2.36 (b) Calculate the standard enthalpy of solution of AgBr(s) in water from the enthalpies of formation of the solid and the aqueous ions.

2.37 (a) The standard enthalpy of decomposition of the yellow complex H_3NSO_2 into NH_3 and SO_2 is +40 kJ mol^{-1}. Calculate the standard enthalpy of formation of H_3NSO_2.

2.37 (b) Given that the standard enthalpy of combustion of graphite is -393.51 kJ mol^{-1} and that of diamond is -395.41 kJ mol^{-1}, calculate the enthalpy of the graphite-to-diamond transition.

2.38 (a) The standard enthalpy of combustion of propane gas is -2220 kJ mol^{-1} and the standard enthalpy of vaporization of propane liquid is +15 kJ mol^{-1}. Calculate (a) the standard enthalpy and (b) the standard internal energy of combustion of liquid propane.

2.38 (b) The standard enthalpy of combustion of butane gas is -2878 kJ mol^{-1} and the standard enthalpy of vaporization of butane liquid is +21.0 kJ mol^{-1}. Calculate (a) the standard enthalpy and (b) the standard internal energy of combustion of liquid butane.

2.39 (a) Use standard enthalpies of formation to calculate the standard enthalpies of the following reactions:

(a) $2\,NO_2(g) \longrightarrow N_2O_4(g)$

(b) $NH_3(g) + HCl(g) \longrightarrow NH_4Cl(s)$

2.39 (b) Use standard enthalpies of formation to calculate the standard enthalpies of the following reactions:

(a) Cyclopropane(g) $\longrightarrow$ propene(g)

(b) $HCl(aq) + NaOH(aq) \longrightarrow NaCl(aq) + H_2O(l)$

2.40 (a) Given the reactions (1) and (2) below, determine (a) $\Delta_r H^\ominus$ and $\Delta_r U^\ominus$ for reaction (3), (b) $\Delta_f H^\ominus$ for both HCl(g) and $H_2O(g)$ all at 298 K. Assume all gases are perfect.

(1) $H_2(g) + Cl_2(g) \longrightarrow 2\,HCl(g)$ $\quad \Delta_r H^\ominus = -184.62$ kJ mol^{-1}

(2) $2\,H_2(g) + O_2(g) \longrightarrow 2\,H_2O(g)$ $\quad \Delta_r H^\ominus = -483.64$ kJ mol^{-1}

(3) $4\,HCl(g) + O_2(g) \longrightarrow Cl_2(g) + 2\,H_2O(g)$

2.40 (b) Given the reactions (1) and (2) below, determine (a) $\Delta_r H^\ominus$ and $\Delta_r U^\ominus$ for reaction (3), (b) $\Delta_f H^\ominus$ for both HI(g) and $H_2O(g)$ all at 298 K. Assume all gases are perfect.

(1) $H_2(g) + I_2(s) \longrightarrow 2\,HI(g)$ $\quad \Delta_r H^\ominus = +52.96$ kJ mol^{-1}

(2) $2\,H_2(g) + O_2(g) \longrightarrow 2\,H_2O(g)$ $\quad \Delta_r H^\ominus = -483.64$ kJ mol^{-1}

(3) $4\,HI(g) + O_2(g) \longrightarrow 2\,I_2(s) + 2\,H_2O(g)$

2.41 (a) For the reaction $C_2H_5OH(l) + 3\,O_2(g) \rightarrow 2\,CO_2(g) + 3\,H_2O(g)$, $\Delta_r U^\ominus = -1373$ kJ mol^{-1} at 298 K. Calculate $\Delta_r H^\ominus$.

2.41 (b) For the reaction $2\,C_6H_5COOH(s) + 13\,O_2(g) \rightarrow 12\,CO_2(g) + 6\,H_2O(g)$, $\Delta_r U^\ominus = -772.7$ kJ mol^{-1} at 298 K. Calculate $\Delta_r H^\ominus$.

2.42 (a) Calculate the standard enthalpies of formation of (a) $KClO_3(s)$ from the enthalpy of formation of KCl, (b) $NaHCO_3(s)$ from the enthalpies of formation of CO_2 and NaOH together with the following information:

$2\,KClO_3(s) \longrightarrow 2\,KCl(s) + 3\,O_2(g)$ $\quad \Delta_r H^\ominus = -89.4$ kJ mol^{-1}

$NaOH(s) + CO_2(g) \longrightarrow NaCO_3(s)$ $\quad \Delta_r H^\ominus = -127.5$ kJ mol^{-1}

2.42 (b) Calculate the standard enthalpy of formation of NOCl(g) from the enthalpy of formation of NO given in Table 2.5, together with the following information:

$2\,NOCl(g) \longrightarrow 2\,NO(g) + Cl_2(g)$ $\quad \Delta_r H^\ominus = +75.5$ kJ mol^{-1}

2.43 (a) Use the information in Table 2.5 to predict the standard reaction enthalpy of $2\,NO_2(g) \rightarrow N_2O_4(g)$ at 100°C from its value at 25°C.

2.43 (b) Use the information in Table 2.5 to predict the standard reaction enthalpy of $2\,H_2(g) + O_2(g) \rightarrow 2\,H_2O(l)$ at 100°C from its value at 25°C.

2.44 (a) From the data in Table 2.5, calculate $\Delta_r H^\ominus$ and $\Delta_r U^\ominus$ at (a) 298 K, (b) 378 K for the reaction C(graphite) + $H_2O(g) \rightarrow CO(g) + H_2(g)$. Assume all heat capacities to be constant over the temperature range of interest.

2.44 (b) Calculate $\Delta_r H^\ominus$ and $\Delta_r U^\ominus$ at 298 K and $\Delta_r H^\ominus$ at 348 K for the hydrogenation of ethyne (acetylene) to ethene (ethylene) from the enthalpy of combustion and heat capacity data in Tables 2.5 and 2.6. Assume the heat capacities to be constant over the temperature range involved.

2.45 (a) Set up a thermodynamic cycle for determining the enthalpy of hydration of Mg^{2+} ions using the following data: enthalpy of sublimation of Mg(s), +167.2 kJ mol^{-1}; first and second ionization enthalpies of Mg(g), 7.646 eV and 15.035 eV; dissociation enthalpy of Cl_2(g), +241.6 kJ mol^{-1}; electron-gain enthalpy of Cl(g), −3.78 eV; enthalpy of solution of $MgCl_2$(s), −150.5 kJ mol^{-1}; enthalpy of hydration of Cl^-(g), −383.7 kJ mol^{-1}.

2.45 (b) Set up a thermodynamic cycle for determining the enthalpy of hydration of Ca^{2+} ions using the following data: enthalpy of sublimation of Ca(s), +178.2 kJ mol^{-1}; first and second ionization enthalpies of Ca(g), 589.7 kJ mol^{-1} and 1145 kJ mol^{-1}; enthalpy of vaporization of bromine, +30.91 kJ mol^{-1}; dissociation enthalpy of Br_2(g), +192.9 kJ mol^{-1}; electron-gain enthalpy of Br(g), −331.0 kJ mol^{-1}; enthalpy of solution of $CaBr_2$(s), −103.1 kJ mol^{-1}; enthalpy of hydration of Br^-(g), −337 kJ mol^{-1}.

2.46 (a) Use the thermochemical groups in Table 2.7 to estimate the standard enthalpy of formation of (a) cyclohexane, (b) 1,4-dimethylhexane.

2.46 (b) Use the thermochemical groups in Table 2.7 to estimate the standard enthalpy of formation of (a) 2,2,4-trimethylpentane, (b) 2,2-dimethylpropane.

Problems†

Assume all gases are perfect unless stated otherwise. Note that 1 atm = 1.013 25 bar. Unless otherwise stated, thermochemical data are for 298.15 K.

Numerical problems

2.1 Calculate the heat needed to raise the temperature of the air in a house from 20°C to 25°C. Assume that the house contains 600 m^3 of air, which should be taken to be a perfect diatomic gas. The density of air is 1.21 kg m^{-3} at 20°C. Calculate ΔU and ΔH for the heating of the air.

2.2 Consider a perfect gas contained in a cylinder and separated by a frictionless adiabatic piston into two sections, A and B; section B is in contact with a water bath that maintains it at constant temperature. Initially, $T_A = T_B = 300$ K, $V_A = V_B = 2.00$ L, and $n_A = n_B = 2.00$ mol. Heat is supplied to Section A and the piston moves to the right reversibly until the final volume of Section B is 1.00 L. Calculate (a) the work done by the gas in Section A, (b) ΔU for the gas in Section B, (c) q for the gas in B, (d) ΔU for the gas in A, and (e) q for the gas in A. Assume $C_{V,m} = 20.0$ J K^{-1} mol^{-1}.

2.3 A sample consisting of 1 mol of a monatomic perfect gas (for which $C_{V,m} = \frac{3}{2}R$) is taken through the cycle shown in Fig. 2.20. (a) Determine the temperature at 1, 2, and 3. (b) Calculate q, w, ΔU, and ΔH for each step and for the overall cycle. If a numerical answer cannot be obtained from the information given, then write in +, −, 0, or ? as appropriate.

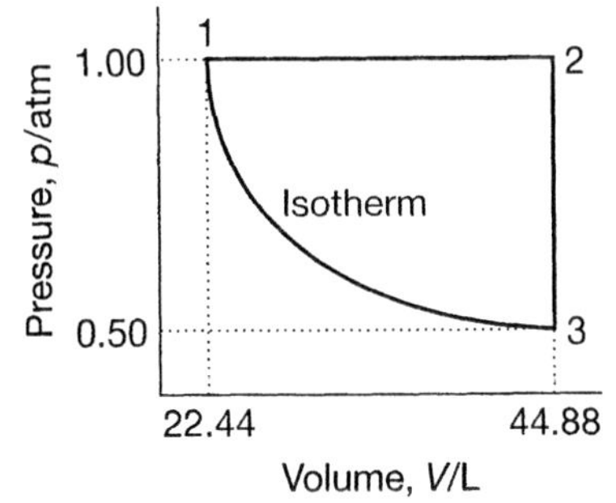

Fig. 2.20

2.4 A 5.0 g block of solid carbon dioxide is allowed to evaporate in a vessel of volume 100 cm^3 maintained at 25°C. Calculate the work done when the system expands (a) isothermally against a pressure of 1.0 atm, and (b) isothermally and reversibly to the same volume as in (a).

2.5 A sample consisting of 1.0 mol $CaCO_3$(s) was heated to 800°C, when it decomposed. The heating was carried out in a container fitted with a piston that was initially resting on the solid. Calculate the work done during complete decomposition at 1.0 atm. What work would be done if instead of having a piston the container was open to the atmosphere?

2.6 A sample consisting of 2.0 mol CO_2 occupies a fixed volume of 15.0 L at 300 K. When it is supplied with 2.35 kJ of energy as heat its temperature increases to 341 K. Assume that CO_2 is described by the van der Waals equation of state, and calculate w, ΔU, and ΔH.

2.7 A sample of 70 mmol Kr(g) expands reversibly and isothermally at 373 K from 5.25 cm^3 to 6.29 cm^3, and the internal energy of the sample is known to increase by 83.5 J. Use the virial equation of state up to the second coefficient $B = -28.7$ cm^3 mol^{-1} to calculate w, q, and ΔH for this change of state.

2.8 A piston exerting a pressure of 1.00 atm rests on the surface of water at 25°C. The pressure is reduced infinitesimally, and as a result 10.0 g of water evaporated and absorbed 22.2 kJ of heat. Calculate w, q, ΔU, and ΔH.

2.9 A new fluorocarbon of molar mass 102 g mol^{-1} was placed in an electrically heated vessel. When the pressure was 650 Torr, the liquid boiled at 78°C. After the boiling point had been reached, it was found that a current of 0.232 A from a 12.0 V supply passed for 650 s vaporized 1.871 g of the sample. Calculate the (molar) enthalpy and internal energy of vaporization.

2.10 A sample of 1.00 mol perfect gas molecules with $C_{p,m} = \frac{7}{2}R$ is put through the following cycle: (a) constant-volume heating to twice its initial temperature, (b) reversible, adiabatic expansion back to its initial temperature, (c) reversible isothermal compression back to 1.00 atm. Calculate q, w, ΔU, and ΔH for each step and overall.

2.11 An object is cooled by the evaporation of liquid methane at its normal boiling point (112 K). What volume of gas at 1.00 atm

† Problems denoted with the symbol ‡ were supplied by Charles Trapp, Carmen Giunta, and Marshall Cady.

pressure must be formed from the liquid in order to remove 32.5 kJ of energy as heat from the object?

2.12 The molar heat capacity of ethane is represented in the temperature range 298 K to 400 K by the empirical expression $C_{p,m}/(\text{J K}^{-1}\text{ mol}^{-1}) = 14.73 + 0.1272(T/\text{K})$. The corresponding expressions for C(s) and H_2(g) are given in Table 2.2. Calculate the standard enthalpy of formation of ethane at 350 K from its value at 298 K.

2.13 A sample of the sugar D-ribose ($C_5H_{10}O_5$) of mass 0.727 g was placed in a calorimeter and then ignited in the presence of excess oxygen. The temperature rose by 0.910 K. In a separate experiment in the same calorimeter, the combustion of 0.825 g of benzoic acid, for which the internal energy of combustion is −3251 kJ mol⁻¹, gave a temperature rise of 1.940 K. Calculate the internal energy of combustion of D-ribose and its enthalpy of formation.

2.14 The standard enthalpy of formation of the metallocene bis(benzene)chromium was measured in a calorimeter. It was found for the reaction $Cr(C_6H_6)_2(s) \rightarrow Cr(s) + 2\,C_6H_6(g)$ that $\Delta_r U^{\ominus}(583\text{ K}) = +8.0\text{ kJ mol}^{-1}$. Find the corresponding reaction enthalpy and estimate the standard enthalpy of formation of the compound at 583 K. The constant-pressure molar heat capacity of benzene is 140 J K^{-1} mol^{-1} in its liquid range and 28 J K^{-1} mol^{-1} as a gas.

2.15‡ When 1.3584 g of sodium acetate trihydrate was mixed into 100.0 mL of 0.2000 M HCl(aq) at 25°C in a solution calorimeter, its temperature fell by 0.397°C. This cooling occurred on account of the reaction:

$$H_3O^+(aq) + NaCH_3CO_2\cdot 3H_2O(s) \longrightarrow Na^+(aq) + CH_3COOH(aq) + 4\,H_2O(l)$$

The heat capacity of the calorimeter is 91.0 J K^{-1} and the heat capacity density of the acid solution is 4.144 J K^{-1} mL^{-1}. Determine the standard enthalpy of formation of the aqueous sodium cation. The standard enthalpy of formation of sodium acetate trihydrate is −1604 kJ mol^{-1}.

2.16‡ Since their discovery in 1985, fullerenes have received the attention of many chemical researchers. Kolesov *et al.* have recently reported the standard enthalpy of combustion and of formation of crystalline C_{60} based on calorimetric measurements (V.P. Kolesov, S.M. Pimenova, V.K. Pavlovich, N.B. Tamm, and A.A. Kurskaya, *J. Chem. Thermodynamics* **28**, 1121 (1996)). In one of their runs, they found the standard specific internal energy of combustion to be −36.0334 kJ g^{-1} at 298.15 K Compute $\Delta_c H^{\ominus}$ and $\Delta_f H^{\ominus}$ of C_{60}.

2.17‡ A thermodynamic study of $DyCl_3$ (E.H.P. Cordfunke, A.S. Booji, and M. Yu. Furkaliouk, *J. Chem. Thermodynamics* **28**, 1387 (1996)) determined its standard enthalpy of formation from the following information

(1) $DyCl_3(s) \longrightarrow DyCl_3(aq,\text{ in }4.0\text{ M HCl(aq)})$ $\quad \Delta_r H^{\ominus} = -180.06\text{ kJ mol}^{-1}$

(2) $Dy(s) + 3\,HCl(aq, 4.0\text{ M}) \longrightarrow DyCl_3(aq,\text{ in }4.0\text{ M HCl(aq)}) + \frac{3}{2}H_2(g)$ $\quad \Delta_r H^{\ominus} = -699.43\text{ kJ mol}^{-1}$

(3) $\frac{1}{2}H_2(g) + \frac{1}{2}Cl_2(g) \longrightarrow HCl(aq, 4.0\text{ M})$ $\quad \Delta_r H^{\ominus} = -158.31\text{ kJ mol}^{-1}$

Determine $\Delta_f H^{\ominus}(DyCl_3, s)$ from these data.

2.18‡ Silylene (SiH_2) is a key intermediate in the thermal decomposition of silicon hydrides such as silane (SiH_4) and disilane (Si_2H_6). Moffat *et al.* (H.K. Moffat, K.F. Jensen, and R.W. Carr, *J. Phys. Chem.* **95**, 145 (1991)) report $\Delta_f H^{\ominus}(SiH_2) = +274\text{ kJ mol}^{-1}$. If $\Delta_f H^{\ominus}(SiH_4) = +34.3\text{ kJ mol}^{-1}$ and $\Delta_f H^{\ominus}(Si_2H_6) = +80.3\text{ kJ mol}^{-1}$ (*CRC Handbook* (2000)), compute the standard enthalpies of the following reactions:

(a) $SiH_4(g) \longrightarrow SiH_2(g) + H_2(g)$

(b) $Si_2H_6(g) \longrightarrow SiH_2(g) + SiH_4(g)$

2.19‡ Silanone (SiH_2O) and silanol (SiH_3OH) are species believed to be important in the oxidation of silane (SiH_4). These species are much more elusive than their carbon counterparts. C.L. Darling and H.B. Schlegel (*J. Phys. Chem.* **97**, 8207 (1993)) report the following values (converted from calories) from a computational study: $\Delta_f H^{\ominus}(SiH_2O) = -98.3\text{ kJ mol}^{-1}$ and $\Delta_f H^{\ominus}(SiH_3OH) = -282\text{ kJ mol}^{-1}$. Compute the standard enthalpies of the following reactions:

(a) $SiH_4(g) + \frac{1}{2}O_2(g) \longrightarrow SiH_3OH(g)$

(b) $SiH_4(g) + O_2(g) \longrightarrow SiH_2O(g) + H_2O(l)$

(c) $SiH_3OH(g) \longrightarrow SiH_2O(g) + H_2(g)$

Note that $\Delta_f H^{\ominus}(SiH_4, g) = +34.3\text{ kJ mol}^{-1}$ (*CRC Handbook* (2000)).

2.20‡ *Polytropic processes* are defined as those that satisfy the condition $pV^n = C$, where C is a constant. In one experiment, 1.00 mol of 'air molecules' is compressed from 1.00 bar to 10.0 bar at 25°C by two different combinations of reversible polytropic processes: (1) heating at constant volume to the final pressure, followed by cooling at constant pressure, (2) adiabatic compression to the final volume, followed by cooling at constant volume. (a) Sketch these processes on a pV diagram and identify the value of n for each step in each process. (b) Calculate q, w, ΔU, and ΔH for each step in the processes and for the overall process. Note that the overall process can be accomplished in one isothermal reversible step. Assume air is a perfect diatomic gas with $C_{p,m} = \frac{7}{2}R$.

2.21‡ For reversible polytropic processes described by the general relation $pV^n = C$, derive the following expressions for work and heat

$$w = \frac{RT_1}{n-1}\left\{\left(\frac{p_2}{p_1}\right)^{(n-1)/n} - 1\right\}$$

$$q = \frac{(n-\gamma)RT_1}{(n-1)(\gamma-1)}\left\{\left(\frac{p_2}{p_1}\right)^{(n-1)/n} - 1\right\}$$

Show that these expressions reduce to already familiar expressions for $n = 0, 1, \gamma$, and ∞.

2.22‡ Ammonia is compressed in a piston-cylinder apparatus from an initial state of 30°C and 500 kPa to a final pressure of 1400 kPa. The following data were obtained during the process:

p/kPa	500	653	802	945	1100	1248	1400
V/L	1.25	1.08	0.96	0.84	0.72	0.60	0.50

(a) Is this a polytropic process? (See Problem 2.20 for the definition of a polytropic process.) If so, what is n? (b) Calculate the work done on the ammonia. (c) What is the final temperature?

Theoretical problems

2.23 In a machine of a particular design, the opposing force acting on a mass m varies as $F\sin(\pi x/a)$. Calculate the work needed to move the mass (a) from $x = 0$ to $x = a$ and (b) from $x = 0$ to $x = 2a$.

2.24 Calculate the work done during the isothermal reversible expansion of a gas that satisfies the virial equation of state, eqn 1.22. Evaluate (a) the work for 1.0 mol Ar at 273 K (for data, see Table 1.3) and (b) the same amount of a perfect gas. Let the expansion be from 500 cm³ to 1000 cm³ in each case.

2.25 With reference to Fig. 2.21, and assuming perfect gas behaviour, calculate: (a) the amount of gas molecules (in moles) in this system and its volume in states B and C, (b) the work done on the gas along the paths ACB and ADB, (c) The work done on the gas

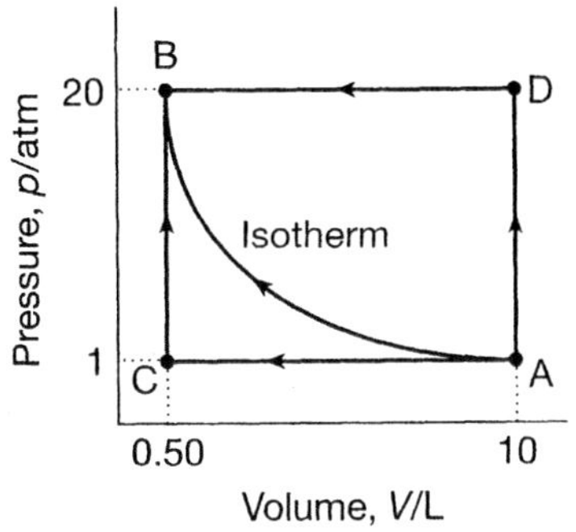

Fig. 2.21

along the isotherm AB, (d) q and ΔU for each of the three paths. Take $C_{V,m} = \frac{3}{2}R$ and $T = 313$ K.

2.26 Calculate the work done during the isothermal reversible expansion of a van der Waals gas. Account physically for the way in which the coefficients a and b appear in the final expression. Plot on the same graph the indicator diagrams for the isothermal reversible expansion of (a) a perfect gas, (b) a van der Waals gas in which $a = 0$ and $b = 5.11 \times 10^{-2}$ L mol^{-1}, and (c) $a = 4.2$ L^2 atm mol^{-2} and $b = 0$. (Use other values too if you have a computer available.) The values selected exaggerate the imperfections but give rise to significant effects on the indicator diagrams. Take $V_i = 1.0$ L, $n = 1.0$ mol, and $T = 298$ K.

2.27 When a system is taken from state A to state B along the path ACB in Fig. 2.22, 80 J of heat flows into the system and the system does 30 J of work. (a) How much heat flows into the system along path ADB if the work done is 10 J? (b) When the system is returned from state B to A along the curved path, the work done on the system is 20 J. Does the system absorb or liberate heat, and how much? (c) If $U_D - U_A = +40$ J, find the heat absorbed in the processes AD and DB.

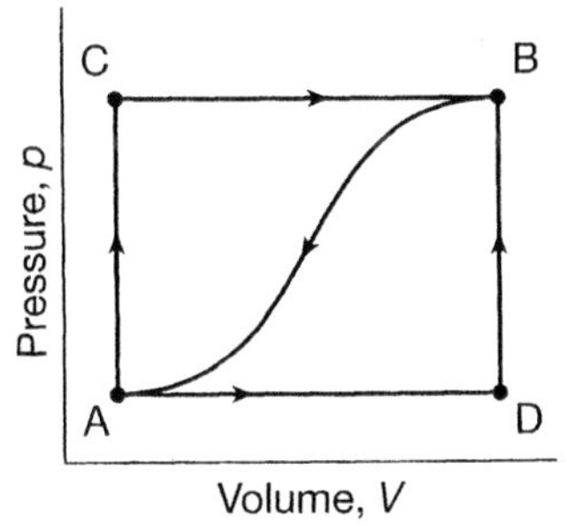

Fig. 2.22

2.28 Show that the value of ΔH for the adiabatic expansion of a perfect gas may be calculated by integration of $dH = V\,dp$, and evaluate the integral for reversible adiabatic expansion.

2.29 Express the work of isothermal reversible expansion of a van der Waals gas in reduced variables and find a definition of reduced work that makes the overall expression independent of the identity of the gas. Calculate the work of isothermal reversible expansion along the critical isotherm from V_c to xV_c.

Applications: to biology, geology, and chemical engineering

2.30 Glucose and fructose are simple sugars with the molecular formula $C_6H_{12}O_6$. Sucrose, or table sugar, is a complex sugar with molecular formula $C_{12}H_{22}O_{11}$ that consists of a glucose unit covalently bound to a fructose unit (a water molecule is released as a result of the reaction between glucose and fructose to form sucrose). (a) Calculate the energy released as heat when a typical table sugar cube of mass 1.5 g is burned in air. (b) To what height could you climb on the energy a table sugar cube provides assuming 25 per cent of the energy is available for work? (c) The mass of a typical glucose tablet is 2.5 g. Calculate the energy released as heat when a glucose tablet is burned in air. (d) To what height could you climb on the energy a cube provides assuming that 25 per cent of the energy is available for work?

2.31 In biological cells that have a plentiful supply of O_2, glucose is oxidized completely to CO_2 and H_2O by a process called *aerobic oxidation* (see Box 9.1). Muscle cells may be deprived of O_2 during vigorous exercise and, in that case, one molecule of glucose is converted to two molecules of lactic acid ($CH_3CH(OH)COOH$) by a process called *anaerobic glycolysis* (see Box 9.1). (a) When 0.3212 g of glucose was burned in a bomb calorimeter of calorimeter constant 641 J K^{-1} the temperature rose by 7.793 K. Calculate (i) the standard molar enthalpy of combustion, (ii) the standard internal energy of combustion, and (iii) the standard enthalpy of formation of glucose. (b) What is the biological advantage (in kilojoules per mole of energy released as heat) of complete aerobic oxidation compared with anaerobic glycolysis to lactic acid?

2.32 Geological conditions are sometimes so extreme that quantities neglected in normal laboratory experiments take on an overriding importance. For example, consider the formation of diamond under geophysically typical conditions. The density of graphite is 2.27 g cm^{-3} and that of diamond is 3.52 g cm^{-3} at a certain temperature and 500 kbar. By how much does $\Delta_{trs}U$ differ from $\Delta_{trs}H$ for the graphite-to-diamond transition?

2.33 Several characteristics of a hydrocarbon must be considered when it is being considered as a fuel. Among them are the specific enthalpy, the heat evolved divided by the mass of the sample, for the advantage of a high molar enthalpy of combustion may be eliminated if a large mass of fuel is to be transported. Use the data in Table 2.5 to calculate the following information on butane, pentane, and octane: (a) the heat output per mole, (b) the heat output per gram.

2.34‡ From the enthalpy of combustion data in Table 2.5 for the alkanes methane through octane, test the extent to which the relation $\Delta_c H^\ominus = k\{(M/(\text{g mol}^{-1})\}^n$ holds and find the numerical values for k and n. Predict $\Delta_c H^\ominus$ for decane and compare to the known value.

2.35 It is possible to investigate the thermochemical properties of hydrocarbons with molecular modelling methods. (a) Use electronic structure software to predict $\Delta_c H^\ominus$ values for the alkanes methane through pentane. To calculate $\Delta_c H^\ominus$ values, estimate the standard enthalpy of formation of $C_nH_{2(n+1)}$(g) by performing semi-empirical calculations (for example, AM1 or PM3 methods) and use experimental standard enthalpy of formation values for CO_2(g) and H_2O(l). (b) Compare your estimated values with the experimental values of $\Delta_c H^\ominus$ (Table 2.5) and comment on the reliability of the molecular modelling method. (c) Test the extent to which the relation $\Delta_c H^\ominus = k\{(M/(\text{g mol}^{-1})\}^n$ holds and find the numerical values for k and n.

2.36‡ Alkyl radicals are important intermediates in the combustion and atmospheric chemistry of hydrocarbons. Seakins *et al.* (P.W. Seakins, M.J. Pilling, J.T. Niiranen, D. Gutman, and L.N. Krasnoperov, *J. Phys. Chem.* **96**, 9847 (1992)) report $\Delta_f H^{\ominus}$ for a variety of alkyl radicals in the gas phase—information that is applicable to studies of pyrolysis and oxidation reactions of hydrocarbons. This information can be combined with thermodynamic data on alkenes to determine the reaction enthalpy for possible fragmentation of a large alkyl radical into smaller radicals and alkenes. Use the following set of data to compute the standard reaction enthalpies for three possible fates of the *tert*-butyl radical, namely, (a) *tert*-$C_4H_9 \rightarrow$ *sec*-C_4H_9, (b) *tert*-$C_4H_9 \rightarrow C_3H_6 + CH_3$, (c) *tert*-$C_4H_9 \rightarrow C_2H_4 + C_2H_5$.

Species:	C_2H_5	*sec*-C_4H_9	*tert*-C_4H_9
$\Delta_f H^{\ominus}/(\text{kJ mol}^{-1})$	+121.0	+67.5	+51.3

2.37‡ N. Cohen (*J. Phys. Chem.* **96**, 9052 (1992)) has reported group additivity tables for the thermochemistry of alkyl radicals in the gas phase. He suggests computing enthalpies of formation based on the following bond dissociation energies ($\Delta U^{\ominus}$) for C—H bonds: primary (—(H)C(H)—H): 420.5 kJ mol^{-1}; secondary (—(C)C(H)—H): 410.5 kJ mol^{-1}; tertiary (—(C)C(C)—H): 398.3 kJ mol^{-1}. Estimate $\Delta_f H^{\ominus}$ of (a) C_2H_5, (b) *sec*-C_4H_9, and (c) *tert*-C_4H_9. ($\Delta_f H^{\ominus}$(2-methylpropane, g) = −134.2 kJ mol^{-1}.)

3 The First Law: the machinery

In this chapter we begin to unfold some of the power of thermodynamics by showing how to establish relations between different properties of a system. The procedure we use is based on the experimental fact that the internal energy and the enthalpy are state functions. We shall see that one very useful aspect of thermodynamics is that a property can be measured indirectly by measuring others and then combining their values. The relations we derive also enable us to discuss the liquefaction of gases and to establish the relation between the heat capacities of a substance at constant pressure and constant volume.

We saw in Section 2.2 that a 'state function' is a property that is independent of how a sample is prepared. In general, such properties are functions of variables that define the current state of the system, such as pressure and temperature. The internal energy and enthalpy are examples of state functions, for they depend on the current state of the system and are independent of its previous history. Properties that relate to the preparation of the state are called **path functions**. Examples of path functions are the work that is done in preparing a state and the energy transferred as heat. We do not speak of a system in a particular state as possessing work or heat. In each case, the energy transferred as work or heat relates to the path being taken, not the current state itself.

State functions and exact differentials

We can use the mathematical properties of state functions to draw far-reaching conclusions about the relations between physical properties and establish connections that may be completely unexpected. The practical importance of these results is that we can combine measurements of different properties to obtain the value of a property we require.

3.1 State and path functions

Consider a system undergoing the changes depicted in Fig. 3.1. The initial state of the system is i and in this state the internal energy is U_i. Work is done by the system as it expands adiabatically to a state f. In this state the system has an internal energy U_f and the work done on the system as it changes along Path 1 from i to f is w. Notice our use of language: U is a property of the state; w is a property of the path. Now

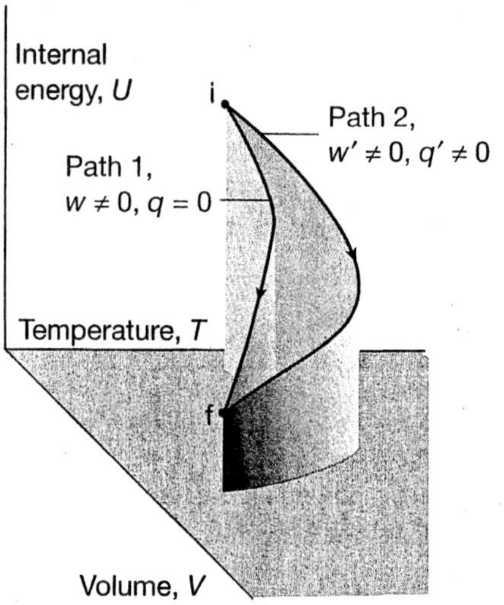

3.1 As the volume and temperature of a system are changed, the internal energy changes. An adiabatic and a nonadiabatic path are shown as Path 1 and Path 2, respectively: they correspond to different values of q and w but to the same value of ΔU.

consider another process, Path 2, in which the initial and final states are the same as those in Path 1 but in which the expansion is not adiabatic. The internal energy of both the initial and the final states are the same as before (because U is a state function). However, in the second path an energy q' enters the system as heat and the work w' is not the same as w. The work and the heat are path functions. In terms of the mountaineering analogy in Section 2.2, the change in altitude (a state function) is independent of the path, but the distance travelled (a path function) does depend on the path taken between the fixed endpoints.

3.2 Exact and inexact differentials

If a system is taken along a path (for example, by heating it), U changes from U_i to U_f, and the overall change is the sum (integral) of all the infinitesimal changes along the path:

$$\Delta U = \int_i^f \mathrm{d}U \tag{3.1}$$

The value of ΔU depends on the initial and final states of the system but is independent of the path between them. This path independence of the integral is expressed by saying that $\mathrm{d}U$ is an 'exact differential'. In general, an **exact differential** is an infinitesimal quantity that, when integrated, gives a result that is independent of the path between the initial and final states.

When a system is heated, the total energy transferred as heat is the sum of all individual contributions at each point of the path:

$$q = \int_{i,\text{path}}^f \mathrm{d}q \tag{3.2}$$

Notice the difference between this equation and eqn 1. First, we do not write Δq, because q is not a state function and the energy supplied as heat cannot be expressed as $q_f - q_i$. Secondly, we must specify the path of integration because q depends on the path selected (for example, an adiabatic path has $q = 0$, whereas a nonadiabatic path between the same two states would have $q \neq 0$). This path dependence is expressed by saying that $\mathrm{d}q$ is an 'inexact differential'. In general, an **inexact differential** is an infinitesimal quantity that, when integrated, gives a result that depends on the path between the initial and final states. Often $\mathrm{d}q$ is written $\text{đ}q$ to emphasize that it is inexact.

The work done on a system to change it from one state to another depends on the path taken between the two specified states; for example, in general the work is different if the change takes place adiabatically and nonadiabatically. It follows that $\mathrm{d}w$ is an inexact differential. It is often written $\text{đ}w$.

Example 3.1 Calculating work, heat, and internal energy

Consider a perfect gas inside a cylinder fitted with a piston. Let the initial state be T, V_i and the final state be T, V_f. The change of state can be brought about in many ways, of which the two simplest are the following: Path 1, in which there is free expansion against zero external pressure; Path 2, in which there is reversible, isothermal expansion. Calculate w, q, and ΔU for each process.

Method To find a starting point for a calculation in thermodynamics, it is often a good idea to go back to first principles, and to look for a way of expressing the quantity we are asked to calculate in terms of other quantities that are easier to calculate. Because the internal energy of a perfect gas arises only from the kinetic energy of its molecules, it is independent of the volume those molecules occupy; therefore, for any isothermal change, $\Delta U = 0$. We also know that in general $\Delta U = q + w$. The question depends on being able to combine the two expressions. We derived a number of expressions for the work done in a variety of processes in Chapter 2, and here we need to select the appropriate ones.

Answer Because $\Delta U = 0$ for both paths and $\Delta U = q + w$, in each case $q = -w$. The work of free expansion is zero (Section 2.3b); so in Path 1, $w = 0$ and $q = 0$. For Path 2, the work is given by eqn 2.13, so $w = -nRT \ln(V_f/V_i)$ and consequently $q = nRT \ln(V_f/V_i)$.

Self-test 3.1 Calculate the values of q, w, and ΔU for an irreversible isothermal expansion of a perfect gas against a constant nonzero external pressure. $[q = p_{ex}\Delta V,\ w = -p_{ex}\Delta V,\ \Delta U = 0]$

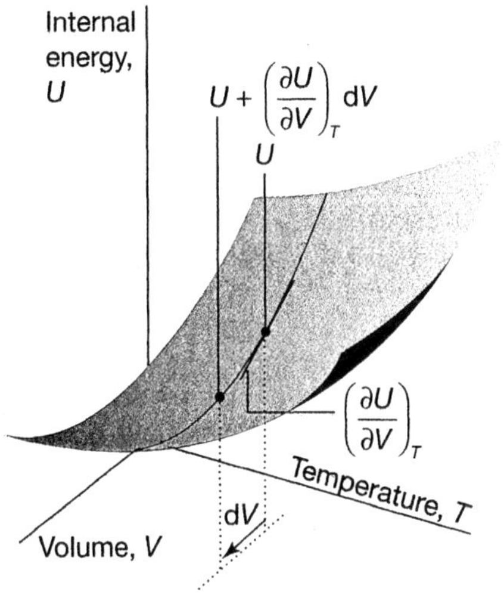

3.2 The partial derivative $(\partial U/\partial V)_T$ is the slope of U with respect to V with the temperature T held constant.

Partial derivatives are reviewed in *Further information 1*.

Thermodynamic consequences

The fact that a function is a state function has profound—and very useful—consequences. This chapter illustrates how to use the properties of state functions to establish unexpected relations between the physical properties of a system.

3.3 Changes in internal energy

We begin to unfold the consequences of dU being an exact differential by noting that, for a closed system of constant composition (the only type of system considered in this chapter), U is a function of volume and temperature.[1]

(a) General considerations

When V changes to $V + dV$ at constant temperature, U changes to

$$U' = U + \left(\frac{\partial U}{\partial V}\right)_T dV$$

The coefficient $(\partial U/\partial V)_T$, the slope of a plot of U against V at constant temperature, is the partial derivative of U with respect to V (Fig. 3.2). If, instead, T changes to $T + dT$ at constant volume (Fig. 3.3), the internal energy changes to

$$U' = U + \left(\frac{\partial U}{\partial T}\right)_V dT$$

1. U could be regarded as a function of V, T, and p but, because there is an equation of state, it is possible to express p in terms of V and T, so p is not an independent variable. We could choose p, T or p, V as independent variables, but V, T fit our purpose.

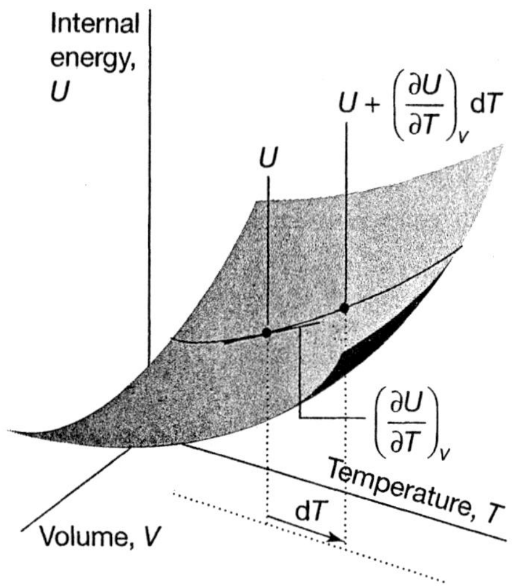

3.3 The partial derivative $(\partial U/\partial T)_V$ is the slope of U with respect to T with the volume V held constant.

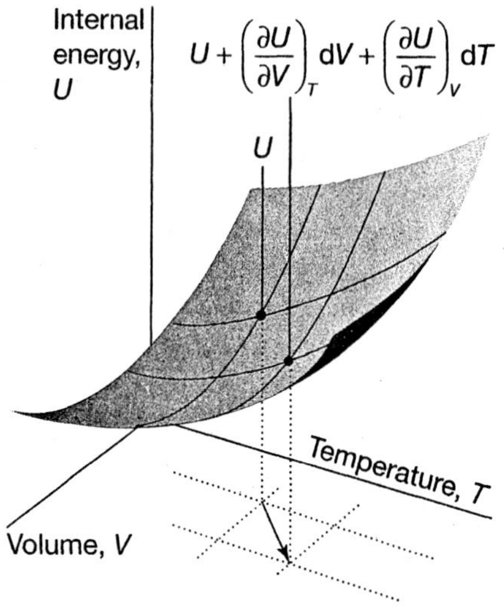

3.4 An overall change in U, which is denoted $\mathrm{d}U$, arises when both V and T are allowed to change. If second-order infinitesimals are ignored, the overall change is the sum of changes for each variable separately.

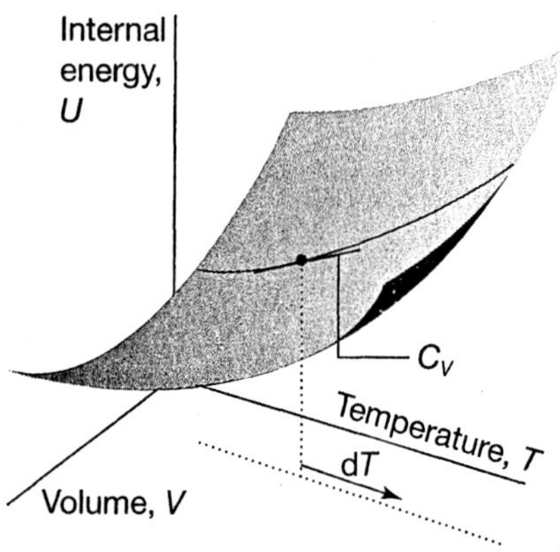

3.5 The constant-volume heat capacity, C_V, is the slope of U with respect to T with the volume V held constant.

Now suppose that V and T both change infinitesimally (Fig. 3.4). The new internal energy, neglecting second-order infinitesimals (those proportional to $\mathrm{d}V\,\mathrm{d}T$), is

$$U' = U + \left(\frac{\partial U}{\partial V}\right)_T \mathrm{d}V + \left(\frac{\partial U}{\partial T}\right)_V \mathrm{d}T$$

As a result of the infinitesimal changes in conditions, the internal energy U' differs from U by the infinitesimal amount $\mathrm{d}U$. Therefore, from the last equation we obtain the very important result that

$$\mathrm{d}U = \left(\frac{\partial U}{\partial V}\right)_T \mathrm{d}V + \left(\frac{\partial U}{\partial T}\right)_V \mathrm{d}T \qquad (3.3)$$

The interpretation of this equation is that, in a closed system of constant composition, any infinitesimal change in the internal energy is proportional to the infinitesimal changes of volume and temperature, the coefficients of proportionality being the partial derivatives.

A partial derivative is the slope of a graph of the property of interest against one of the variables on which it depends (recall Fig. 2.12), all the other variables being held constant. In many cases the slopes have a straightforward physical interpretation, and thermodynamics gets shapeless and difficult only when that meaning is not kept in sight. In the present case, we have already met $(\partial U/\partial T)_V$ in eqn 2.19, where we saw that it is the constant-volume heat capacity, C_V (Fig. 3.5). Therefore, we can write

$$\mathrm{d}U = \left(\frac{\partial U}{\partial V}\right)_T \mathrm{d}V + C_V\,\mathrm{d}T \qquad (3.4)$$

The other coefficient, $(\partial U/\partial V)_T$, plays a major role in thermodynamics because it is a measure of the variation of the internal energy of a substance as its volume is changed at constant temperature (Fig. 3.6). We shall denote it π_T (because it has the same dimensions as pressure), and call it the **internal pressure**:

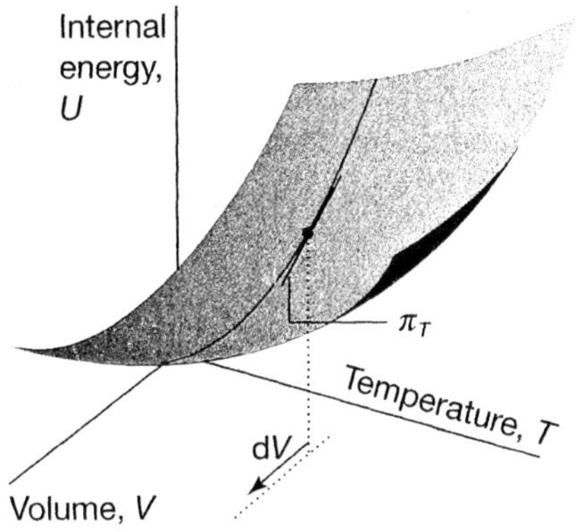

3.6 The internal pressure, π_T, is the slope of U with respect to V with the temperature T held constant.

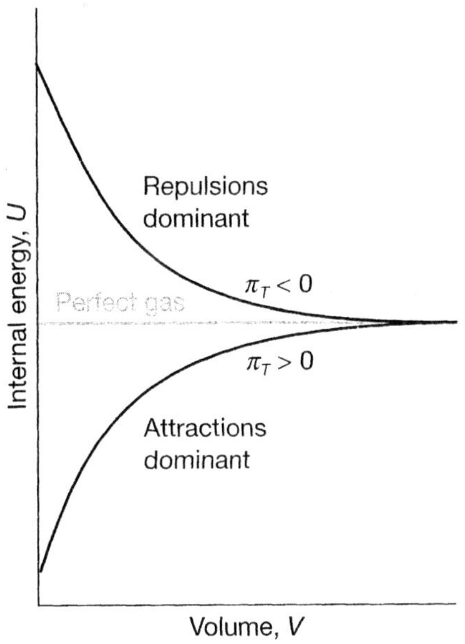

3.7 For a perfect gas, the internal energy is independent of the volume (at constant temperature). If attractions are dominant in a real gas, the internal energy increases with volume because the molecules become further apart on average. If repulsions are dominant, the internal energy decreases as the gas expands.

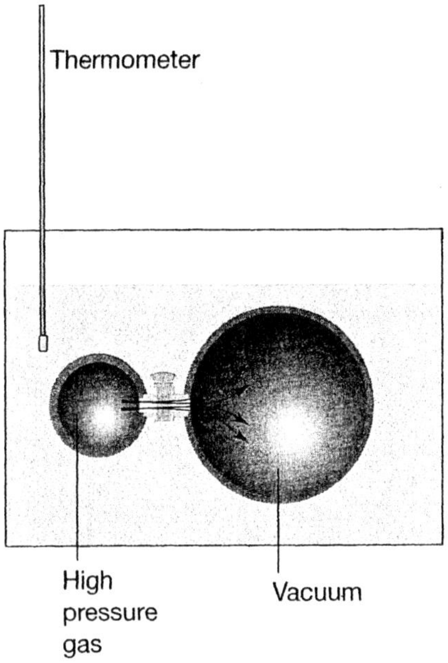

3.8 A schematic diagram of the apparatus used by Joule in an attempt to measure the change in internal energy when a gas expands isothermally. The heat absorbed by the gas is proportional to the change in temperature of the bath.

$$\pi_T = \left(\frac{\partial U}{\partial V}\right)_T \qquad [3.5]$$

We shall see that the internal pressure is a measure of the strength of the cohesive forces in the sample. Then

$$dU = \pi_T\, dV + C_V\, dT \qquad (3.6)$$

If the internal energy increases ($dU > 0$) as the volume of the sample expands isothermally ($dV > 0$), which is the case when there are attractive forces between the particles, a graph of internal energy against volume slopes upwards and $\pi_T > 0$ (Fig. 3.7). When there are no interactions between the molecules, the internal energy is independent of their separation and hence independent of the volume of the sample. We can therefore write

$$\text{For a perfect gas} \quad \left(\frac{\partial U}{\partial V}\right)_T = 0 \qquad (3.7)°$$

or, more succinctly, $\pi_T = 0$. The statement $\pi_T = 0$ (that is, the internal energy is independent of the volume occupied by the sample) can be taken to be the definition of a perfect gas, for later we shall see that it implies the equation of state $pV = nRT$.

(b) The Joule experiment

James Joule thought that he could measure π_T by observing the change in temperature of a gas when it is allowed to expand into a vacuum. He used two metal vessels immersed in a water bath (Fig. 3.8). One was filled with air at about 22 atm and the other was evacuated. He then tried to measure the change in temperature of the water of the bath when a stopcock was opened and the air expanded into a vacuum. He observed no change in temperature.

The thermodynamic implications of the experiment are as follows. No work was done in the expansion into a vacuum, so $w = 0$. No heat entered or left the system (the gas) because the temperature of the bath did not change, so $q = 0$. Consequently, within the accuracy of the experiment, $\Delta U = 0$. It follows that U does not change much when a gas expands isothermally and therefore that $\pi_T = 0$.

Joule's experiment was crude. In particular, the heat capacity of the apparatus was so large that the temperature change that gases do in fact cause was too small to measure. His experiment was on a par with Boyle's: he extracted an essential limiting property of a gas, a property of a perfect gas, without detecting the small deviations characteristic of real gases.

Illustration 3.1

For ammonia, $\pi_{T,\text{m}} = 840$ Pa at 300 K and 1.0 bar where $\pi_{T,\text{m}} = (\partial U_\text{m}/\partial V)_T$, and $C_{V,\text{m}} = 27.32\ \text{J K}^{-1}\ \text{mol}^{-1}$. The change in molar internal energy of ammonia when it is heated through 2.0 K and compressed through 100 cm^3 is approximately

$$\Delta U_\text{m} \approx (840\ \text{J m}^{-3}\ \text{mol}^{-1}) \times (-100 \times 10^{-6}\ \text{m}^3) + (27.32\ \text{J K}^{-1}\ \text{mol}^{-1}) \times (2.0\ \text{K})$$
$$\approx -0.084\ \text{J mol}^{-1} + 55\ \text{J mol}^{-1} = 55\ \text{J mol}^{-1}$$

Note that the contribution of the heating term greatly dominates that of the compression term for this gas.

(c) Changes in internal energy at constant pressure

Partial derivatives have many useful properties and some that we shall draw on frequently are reviewed in *Further information 1*. Skilful use of them can often turn some unfamiliar quantity into a quantity that can be recognized, interpreted, or measured.

As an example, suppose we want to find out how the internal energy varies with temperature when the pressure of the system is kept constant. If we divide both sides of eqn 6 by dT and impose the condition of constant pressure on the resulting differentials, so dU/dT on the left becomes $(\partial U/\partial T)_p$, we obtain

$$\left(\frac{\partial U}{\partial T}\right)_p = \pi_T \left(\frac{\partial V}{\partial T}\right)_p + C_V \qquad (3.8)$$

It is usually sensible in thermodynamics to inspect the output of a manipulation like this to see if it contains any recognizable physical quantity. The differential coefficient on the right in this expression is the slope of the plot of volume against temperature (at constant pressure). This property is normally tabulated as the **expansion coefficient**, α, of a substance,[2] which is defined as

$$\alpha = \frac{1}{V}\left(\frac{\partial V}{\partial T}\right)_p \qquad [3.9]$$

A large value of α means that the volume of the sample responds strongly to changes in temperature. Table 3.1 lists some experimental values.

Table 3.1* Expansion coefficients (α) and isothermal compressibilities (κ_T)

Substance	$\alpha/(10^{-4}\ \text{K}^{-1})$	$\kappa_T/(10^{-6}\ \text{atm}^{-1})$
Benzene	12.4	92.1
Diamond	0.030	0.187
Lead	0.861	2.21
Water	2.1	49.6

* More values are given in the *Data section* at the end of this volume.

2. As for heat capacities, the expansion coefficient of a mixture depends on whether or not the composition is allowed to change. Throughout this chapter, we deal only with pure substances, so this complication can be disregarded.

Example 3.2 **Using the expansion coefficient of a gas**

Derive an expression for the expansion coefficient for a perfect gas.

Method The expansion coefficient is defined in eqn 9. To use this expression, substitute the expression for V in terms of T obtained from the equation of state for the gas. As implied by eqn 9, the pressure, p, is treated as a constant.

Answer Because $pV = nRT$, we can write

$$\alpha = \frac{1}{V}\left(\frac{\partial(nRT/p)}{\partial T}\right)_p = \frac{nR}{pV} = \frac{1}{T}$$

The higher the temperature, the less responsive is its volume to a change in temperature.

Self-test 3.2 Evaluate α for a gas for which the equation of state is $p = nRT/(V - nb)$. $[\alpha = (1 - b/V_m)/T]$

When we introduce the definition of α into the equation for $(\partial U/\partial T)_p$, we obtain

$$\left(\frac{\partial U}{\partial T}\right)_p = \alpha\pi_T V + C_V \qquad (3.10)$$

This equation is entirely general (provided the system is closed and its composition is constant). It expresses the dependence of the internal energy on the temperature at constant pressure in terms of C_V, which can be measured in one experiment, in terms of α, which can be measured in another, and in terms of the quantity π_T. For a perfect gas, $\pi_T = 0$, so then

$$\left(\frac{\partial U}{\partial T}\right)_p = C_V \qquad (3.11)^\circ$$

That is, the constant-volume heat capacity of a perfect gas is equal to the slope of a plot of internal energy against temperature at constant pressure as well as (by definition) to the slope at constant volume.

3.4 The temperature dependence of the enthalpy

We can carry out a similar set of operations on the enthalpy, $H = U + pV$. The quantities U, p, and V are all state functions; therefore H is also a state function and dH is an exact differential.

(a) Changes in the enthalpy at constant volume

The variation of enthalpy with temperature at constant pressure is simply the constant-pressure heat capacity, C_p. The simplicity of this relation strongly suggests that H will prove to be a useful thermodynamic function when the pressure is under our control. We saw a sign of that in the relation $\Delta H = q_p$ (eqn 2.24). We shall therefore regard H as a function of p and T, and adapt the argument in Section 3.3 to find an expression for the variation of H with temperature at constant volume. This relation will prove useful for relating the heat capacities at constant pressure and volume and for a discussion of the liquefaction of gases.

By the same argument as for U (but with p in place of V) we find that for a closed system of constant composition,

$$dH = \left(\frac{\partial H}{\partial p}\right)_T dp + \left(\frac{\partial H}{\partial T}\right)_p dT \tag{3.12}$$

We recognize the second coefficient as the definition of the constant-pressure heat capacity, C_p, so

$$dH = \left(\frac{\partial H}{\partial p}\right)_T dp + C_p\, dT \tag{3.13}$$

The manipulation of this expression is slightly more involved than before, but we show in the following *Justification* that it implies that

$$\left(\frac{\partial H}{\partial T}\right)_V = \left(1 - \frac{\alpha\mu}{\kappa_T}\right)C_p \tag{3.14}$$

where the **isothermal compressibility**, κ_T (kappa), is defined as

$$\kappa_T = -\frac{1}{V}\left(\frac{\partial V}{\partial p}\right)_T \qquad [3.15]$$

and the **Joule–Thomson coefficient**, μ (mu), is defined as

$$\mu = \left(\frac{\partial T}{\partial p}\right)_H \qquad [3.16]$$

Equation 14 applies to any substance. Because all the quantities that appear in it can be measured in suitable experiments, we now know how H varies with T when the volume of the sample is held constant.

Justification 3.1

First, divide eqn 13 through by dT and impose constant volume:

$$\left(\frac{\partial H}{\partial T}\right)_V = \left(\frac{\partial H}{\partial p}\right)_T \left(\frac{\partial p}{\partial T}\right)_V + C_p$$

The third differential coefficient looks like something we ought to recognize, and is perhaps related to $(\partial V/\partial T)_p$, the expansion coefficient. It follows from Euler's chain relation that

$$\left(\frac{\partial p}{\partial T}\right)_V \left(\frac{\partial T}{\partial V}\right)_p \left(\frac{\partial V}{\partial p}\right)_T = -1$$

and therefore that

$$\left(\frac{\partial p}{\partial T}\right)_V = -\frac{1}{(\partial T/\partial V)_p(\partial V/\partial p)_T}$$

Unfortunately, $(\partial T/\partial V)_p$ occurs instead of $(\partial V/\partial T)_p$. However, the reciprocal identity allows us to invert partial derivatives and to write

$$\left(\frac{\partial p}{\partial T}\right)_V = -\frac{(\partial V/\partial T)_p}{(\partial V/\partial p)_T} = \frac{\alpha}{\kappa_T}$$

The Euler chain relation states that, for a differentiable function $z = z(x, y)$,

$$\left(\frac{\partial y}{\partial x}\right)_z \left(\frac{\partial x}{\partial z}\right)_y \left(\frac{\partial z}{\partial y}\right)_x = -1$$

The reciprocal identity states that

$$\left(\frac{\partial y}{\partial x}\right)_z = \frac{1}{(\partial x/\partial y)_z}$$

Next, we must change $(\partial H/\partial p)_T$ into something recognizable. The chain relation lets us write this partial derivative as

$$\left(\frac{\partial H}{\partial p}\right)_T = -\frac{1}{(\partial p/\partial T)_H(\partial T/\partial H)_p}$$

Both derivatives may be brought up into the numerator:

$$\left(\frac{\partial H}{\partial p}\right)_T = -\left(\frac{\partial T}{\partial p}\right)_H\left(\frac{\partial H}{\partial T}\right)_p$$

and we can recognize both the constant-pressure heat capacity, C_p, and the Joule–Thomson coefficient μ as defined in eqn 16. Therefore,

$$\left(\frac{\partial H}{\partial p}\right)_T = -\mu C_p$$

When we use this expression and the expression for $(\partial p/\partial T)_V$ in the first equation in this *Justification*, we obtain eqn 14.

(b) The isothermal compressibility

The negative sign in the definition of κ_T, eqn 15, ensures that the compressibility is a positive quantity, because an increase of pressure, implying a positive dp, brings about a reduction of volume, a negative dV. The isothermal compressibility is obtained from the slope of the plot of volume against pressure at constant temperature (that is, it is proportional to the slope of an isotherm). Table 3.1 lists some values of κ_T. Its value for a perfect gas is obtained by substitution of the perfect gas equation of state into eqn 15, which gives

$$\kappa_T = -\frac{1}{V}\left(\frac{\partial(nRT/p)}{\partial p}\right)_T = -\frac{nRT}{V}\left(-\frac{1}{p^2}\right) = \frac{1}{p} \qquad (3.17)^\circ$$

This expression shows that, the higher the pressure of the gas, the lower its compressibility.

Example 3.3 Using the isothermal compressibility

The isothermal compressibility of water at 20°C and 1 atm is 4.96×10^{-5} atm^{-1}. What change of volume occurs when a sample of volume 50 cm^3 is subjected to an additional 1000 atm at constant temperature?

Method We know from the definition of κ_T that, for an infinitesimal change of pressure at constant temperature, the volume changes by

$$\mathrm{d}V = \left(\frac{\partial V}{\partial p}\right)_T \mathrm{d}p = -\kappa_T V\,\mathrm{d}p$$

For a finite change in pressure, we need to integrate both sides. When confronted by an integration, it is often a useful first approximation (for substances other than gases) to suppose that the integrand is constant over the range of integration.

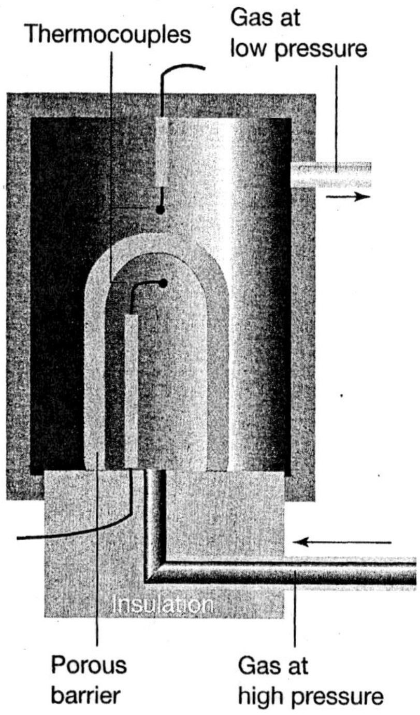

3.9 The apparatus used for measuring the Joule–Thomson effect. The gas expands through the porous barrier, which acts as a throttle, and the whole apparatus is thermally insulated. As explained in the text, this arrangement corresponds to an isenthalpic expansion (expansion at constant enthalpy). Whether the expansion results in a heating or a cooling of the gas depends on the conditions.

Answer The integrals we need to evaluate are

$$\int_{V_i}^{V_f} dV = -\int_{p_i}^{p_f} \kappa_T V\,dp$$

The integral on the left is ΔV. If we suppose that κ_T and V are approximately constant over the range of pressures of interest, we can write

$$\Delta V = -\kappa_T V \int_{p_i}^{p_f} dp = -\kappa_T V\,\Delta p$$

Substitution of the data into the last expression then gives

$$\Delta V = -(4.96\times10^{-5}\ \text{atm}^{-1})\times(50\ \text{cm}^3)\times(1000\ \text{atm}) = -2.5\ \text{cm}^3$$

Comment Because the compression results in a decrease in volume of only 5 per cent, the assumption of constant V and κ_T is probably acceptable as a first approximation. Note that very high pressures are needed to bring about significant changes of volume.

Self-test 3.3 A sample of copper of volume 50 cm^3 is subjected to a additional pressure of 100 atm and a temperature increase of 5.0 K. Estimate the total change in volume. [8.8 mm^3]

(c) The Joule–Thomson effect

The analysis of the Joule–Thomson coefficient is central to the technological problems associated with the liquefaction of gases. We need to be able to interpret it physically and to measure it.

As shown in the *Justification* below, the cunning required to impose the constraint of constant enthalpy, so that the process is **isenthalpic**, was supplied by Joule and William Thomson (later Lord Kelvin). They let a gas expand through a porous barrier from one constant pressure to another, and monitored the difference of temperature that arose from the expansion (Fig. 3.9). The whole apparatus was insulated so that the process was adiabatic. They observed a lower temperature on the low pressure side, the difference in temperature being proportional to the pressure difference they maintained. This cooling by adiabatic expansion is now called the **Joule–Thomson effect**.

Justification 3.2

Here we show that the experimental arrangement results in expansion at constant enthalpy. Because all changes to the gas occur adiabatically,

$$q = 0, \qquad \text{which implies} \quad \Delta U = w$$

Consider the work done as the gas passes through the barrier. We focus on the passage of a fixed amount of gas from the high pressure side, where the pressure is p_i, the temperature T_i, and the gas occupies a volume V_i (Fig. 3.10). The gas emerges on the low pressure side, where the same amount of gas has a pressure p_f, a temperature T_f, and occupies a volume V_f. The gas on the left is compressed isothermally by the upstream gas acting as a piston. The relevant pressure is p_i and the volume changes from V_i to 0; therefore, the work done on the gas is

$$w_1 = -p_i(0 - V_i) = p_iV_i$$

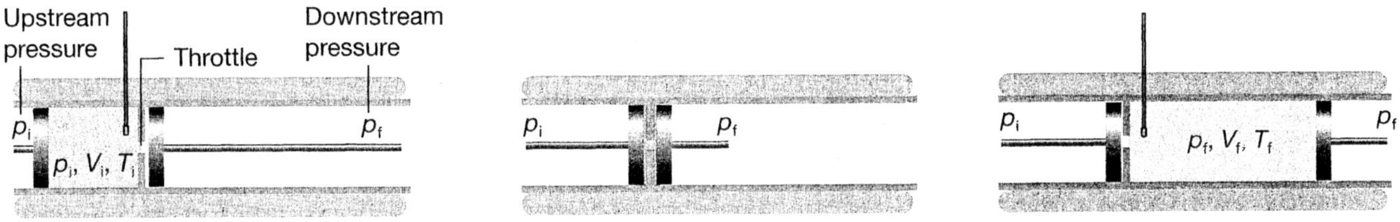

3.10 The thermodynamic basis of the Joule–Thomson expansion. The pistons represent the upstream and downstream gases, which maintain constant pressures either side of the throttle. The transition from the left-hand diagram to the right-hand diagram, which represents the passage of a given amount of gas through the throttle, occurs without change of enthalpy.

The gas expands isothermally on the right of the barrier (but possibly at a different constant temperature) against the pressure p_f provided by the downstream gas acting as a piston to be driven out. The volume changes from 0 to V_f, so the work done on the gas in this stage is

$$w_2 = -p_f(V_f - 0) = -p_fV_f$$

The total work done on the gas is the sum of these two quantities, or

$$w = w_1 + w_2 = p_iV_i - p_fV_f$$

It follows that the change of internal energy of the gas as it moves adiabatically from one side of the barrier to the other is

$$U_f - U_i = w = p_iV_i - p_fV_f$$

Reorganization of this expression gives

$$U_f + p_fV_f = U_i + p_iV_i, \qquad \text{or} \quad H_f = H_i$$

Therefore, the expansion occurs without change of enthalpy.

The property measured in the experiment is the ratio of the temperature change to the change of pressure, $\Delta T/\Delta p$. Adding the constraint of constant enthalpy and taking the limit of small Δp implies that the thermodynamic quantity measured is $(\partial T/\partial p)_H$, which is the Joule–Thomson coefficient, μ. In other words, the physical interpretation of μ is that it is the ratio of the change in temperature to the change in pressure when a gas expands under adiabatic conditions.

The modern method of measuring μ is indirect, and involves measuring the **isothermal Joule–Thomson coefficient**, the quantity

$$\mu_T = \left(\frac{\partial H}{\partial p}\right)_T \qquad [3.18]$$

which is the slope of enthalpy against pressure at constant temperature (Fig. 3.11). As shown by the last equation in *Justification* 3.1, the two coefficients are related by:

$$\mu_T = -C_p\mu \qquad (3.19)$$

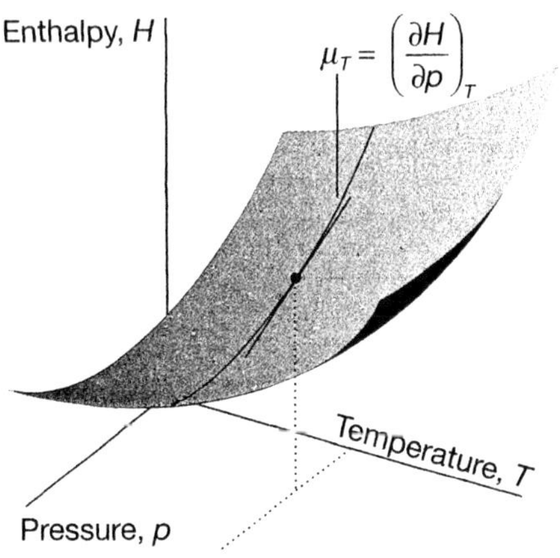

3.11 The isothermal Joule–Thomson coefficient is the slope of the enthalpy with respect to changing pressure, the temperature being held constant.

To measure μ_T, the gas is pumped continuously at a steady pressure through a heat exchanger (which brings it to the required temperature), and then through a porous plug inside a thermally insulated container. The steep pressure drop is measured, and the cooling effect is exactly offset by an electric heater placed immediately after

Table 3.2* Inversion temperatures (T_I), normal freezing (T_f) and boiling (T_b) points, and Joule–Thomson coefficients (μ) at 1 atm and 298 K

	T_I/K	T_f/K	T_b/K	$\mu/(K\ atm^{-1})$
Ar	723	83.8	87.3	
CO_2	1500		194.7	1.11
He	40		4.2	−0.060
N_2	621	63.3	77.4	0.25

* More values are given in the *Data section*.

the plug (Fig. 3.12). The energy provided by the heater is monitored. Because the heat can be identified with the value of ΔH for the gas (because $\Delta H = q_p$), and the pressure change Δp is known, we can find μ_T from the limiting value of $\Delta H/\Delta p$ as $\Delta p \rightarrow 0$, and then convert it to μ. Table 3.2 lists some values obtained in this way.

Real gases have nonzero Joule–Thomson coefficients. Depending on the identity of the gas, the pressure, the relative magnitudes of the attractive and repulsive intermolecular forces, and the temperature, the sign of the coefficient may be either positive or negative (Fig. 3.13). A positive sign implies that dT is negative when dp is negative, in which case the gas cools on expansion. Gases that show a heating effect ($\mu < 0$) at one temperature show a cooling effect ($\mu > 0$) when the temperature is below their upper **inversion temperature**, T_I (Table 3.2, Fig. 3.14). As indicated in Fig. 3.14, a gas typically has two inversion temperatures, one at high temperature and the other at low.

The 'Linde refrigerator' makes use of Joule–Thompson expansion to liquefy gases (Fig. 3.15). The gas at high pressure is allowed to expand through a throttle; it cools and is circulated past the incoming gas. That gas is cooled, and its subsequent expansion cools it still further. There comes a stage when the circulating gas becomes so cold that it condenses to a liquid.

For a perfect gas, $\mu = 0$; hence, the temperature of a perfect gas is unchanged by Joule–Thomson expansion.[3] This characteristic points clearly to the involvement of intermolecular forces in determining the size of the effect. However, the Joule–

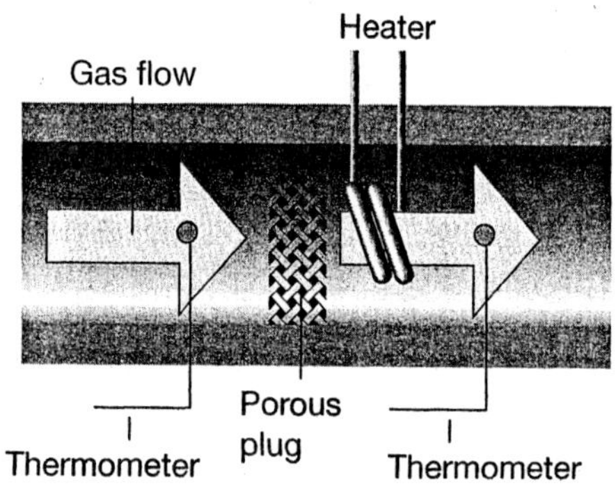

3.12 A schematic diagram of the apparatus used for measuring the isothermal Joule–Thomson coefficient. The electrical heating required to offset the cooling arising from expansion is interpreted as ΔH and used to calculate $(\partial H/\partial p)_T$ which is then converted to μ as explained in the text.

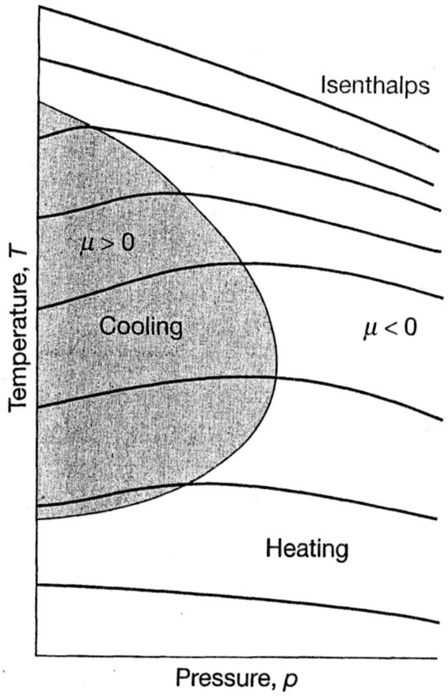

3.13 The sign of the Joule–Thomson coefficient, μ, depends on the conditions. Inside the boundary, the shaded area, it is positive and outside it is negative. The temperature corresponding to the boundary at a given pressure is the 'inversion temperature' of the gas at that pressure. For a given pressure, the temperature must be below a certain value if cooling is required, but if it becomes too low, the boundary is crossed again and heating occurs. Reduction of pressure under adiabatic conditions moves the system along one of the isenthalps, or curves of constant enthalpy. The inversion temperature curve runs through the points of the isenthalps where their slope changes from negative to positive.

3. Simple adiabatic expansion does cool a perfect gas, because the gas does work; recall Section 2.6.

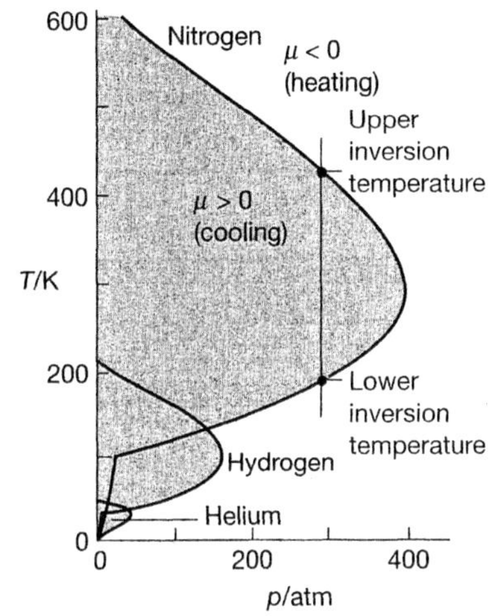

3.14 The inversion temperatures for three real gases, nitrogen, hydrogen, and helium.

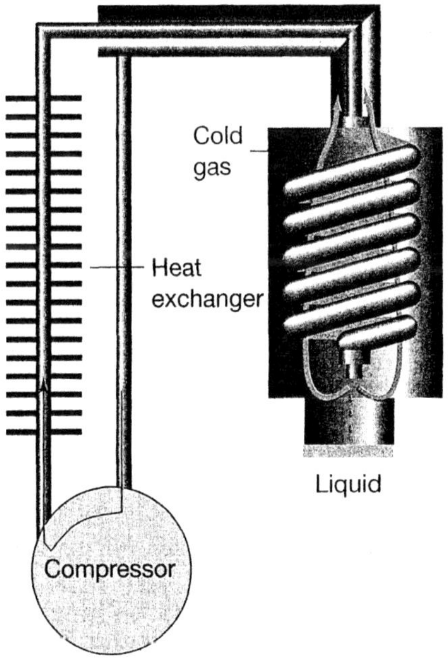

3.15 The principle of the Linde refrigerator is shown in this diagram. The gas is recirculated, and so long as it is beneath its inversion temperature it cools on expansion through the throttle. The cooled gas cools the high-pressure gas, which cools still further as it expands. Eventually liquefied gas drips from the throttle.

Thomson coefficient of a real gas does not necessarily approach zero as the pressure is reduced even though the equation of state of the gas approaches that of a perfect gas. The coefficient behaves like the properties discussed in Section 1.3b in the sense that it depends on derivatives and not on p, V, and T themselves.

3.5 The relation between C_V and C_p

The constant-pressure heat capacity C_p differs from the constant-volume heat capacity C_V by the work needed to change the volume of the system. This work arises in two ways. One is the work of driving back the atmosphere; the other is the work of stretching the bonds in the material, including any weak intermolecular interactions. In the case of a perfect gas, the second makes no contribution. We shall now derive a general relation between the two heat capacities, and show that it reduces to the perfect gas result in the absence of intermolecular forces.

First, we carry through the calculation for a perfect gas. In this special case, we can use eqn 11 to express both heat capacities in terms of derivatives at constant pressure:

$$C_p - C_V = \left(\frac{\partial H}{\partial T}\right)_p - \left(\frac{\partial U}{\partial T}\right)_p \tag{3.20}$$

Then we introduce

$$H = U + pV = U + nRT$$

into the first term, which results in

$$C_p - C_V = \left(\frac{\partial U}{\partial T}\right)_p + nR - \left(\frac{\partial U}{\partial T}\right)_p = nR$$

This is the formal derivation of eqn 2.31.

In the *Justification* below we show that the general relation between the two heat capacities for any pure substance is

$$C_p - C_V = \frac{\alpha^2 TV}{\kappa_T} \tag{3.21}$$

This formula is a thermodynamic expression, which means that it applies to any substance (that is, that it is 'universally true'). It reduces to eqn 2.31 for a perfect gas when we set $\alpha = 1/T$ and $\kappa_T = 1/p$.

Justification 3.3

A useful rule when doing a problem in thermodynamics is to go back to first principles. In the present problem we do this twice, first by expressing C_p and C_V in terms of their definitions and then by inserting the definition $H = U + pV$:

$$C_p - C_V = \left(\frac{\partial H}{\partial T}\right)_p - \left(\frac{\partial U}{\partial T}\right)_V = \left(\frac{\partial U}{\partial T}\right)_p + \left(\frac{\partial (pV)}{\partial T}\right)_p - \left(\frac{\partial U}{\partial T}\right)_V$$

We have already calculated the difference of the first and third terms on the right, and eqn 9 lets us write this difference as $\alpha\pi_T V$. The factor αV gives the change in volume when the temperature is raised, and $\pi_T = (\partial U/\partial V)_T$ converts this change in volume into a change in internal energy. We can simplify the remaining term by noting that, because p is constant,

$$\left(\frac{\partial (pV)}{\partial T}\right)_p = p\left(\frac{\partial V}{\partial T}\right)_p = \alpha pV$$

The middle term of this expression identifies it as the contribution to the work of pushing back the atmosphere: $(\partial V/\partial T)_p$ is the change of volume caused by a change of temperature, and multiplication by p converts this expansion into work.

Collecting the two contributions gives

$$C_p - C_V = \alpha(p + \pi_T)V$$

As just remarked, the first term on the right (αpV) is a measure of the work needed to push back the atmosphere; the second term on the right, $\alpha\pi_T V$, is the work required to separate the molecules composing the system.

At this point we can go further by using the result we prove in *Justification 5.1* that

$$\pi_T = T\left(\frac{\partial p}{\partial T}\right)_V - p$$

When this expression is inserted in the last equation we obtain

$$C_p - C_V = \alpha TV\left(\frac{\partial p}{\partial T}\right)_V$$

The coefficient that appears here was encountered in *Justification 3.1*, where we saw that it is equal to α/κ_T. Therefore, this expression turns into eqn 21.

Because thermal expansivities α of liquids and solids are small, it is tempting to deduce from eqn 21 that for them $C_p \approx C_V$. But this is not always so, because the compressibility κ_T might also be small, so α^2/κ_T might be large. That is, although only a little work need be done to push back the atmosphere, a great deal of work may have to be done to pull atoms apart from one another as the solid expands. As an illustration, for water at 25°C, eqn 21 gives $C_{p,\mathrm{m}} = 75.3\ \mathrm{J\ K^{-1}\ mol^{-1}}$ compared with $C_{V,\mathrm{m}} = 74.8\ \mathrm{J\ K^{-1}\ mol^{-1}}$. In some cases, the two heat capacities differ by as much as 30 per cent.

Checklist of key ideas

☐ path function, a property that depends on the path between two states

State functions and exact differentials

3.1 State and path functions

3.2 Exact and inexact differentials

☐ exact differential, an infinitesimal quantity that, when integrated, gives a result that is independent of the path between the initial and final states

☐ inexact differential, an infinitesimal quantity that, when integrated, gives a result that depends on the path between the initial and final states

Thermodynamic consequences

3.3 Changes in internal energy

☐ general change in internal energy for a closed system, $dU = (\partial U/\partial V)_T\,dV + (\partial U/\partial T)_V\,dT$

☐ internal pressure, $\pi_T = (\partial U/\partial V)_T$

☐ internal energy and isothermal expansion of perfect gas, $(\partial U/\partial V)_T = 0$

☐ Joule experiment, the failure to observe a change in temperature when a gas expands freely into a vacuum

☐ expansion coefficient, $\alpha = (1/V)(\partial V/\partial T)_p$

☐ temperature dependence of internal energy at constant pressure, $(\partial U/\partial T)_p = \alpha\pi_T V + C_V$

3.4 The temperature dependence of the enthalpy

☐ isothermal compressibility, $\kappa_T = -(1/V)(\partial V/\partial p)_T$

☐ Joule-Thomson coefficient, $\mu = (\partial T/\partial p)_H$

☐ isenthalpic process, a process at constant enthalpy

☐ Joule-Thomson effect, the cooling of a gas by isenthalpic adiabatic expansion

☐ isothermal Joule-Thomson coefficient, $\mu_T = (\partial H/\partial p)_T = -C_p\mu$

☐ inversion temperature, the temperature at which the Joule-Thomson coefficient changes sign

3.5 The relation between C_V and C_p

☐ relation between C_V and C_p, in general $C_p - C_V = \alpha^2 TV/\kappa_T$

Further reading

Articles of general interest

R.A. Alberty, Legendre transforms in chemical thermodynamics. *Chem. Rev.* **94**, 1457 (1994).

E.W. Anacker, S.E. Anacker, and W.J. Swartz, Some comments on partial derivatives in thermodynamics. *J. Chem. Educ.* **64**, 674 (1987).

S.M. Blinder, Mathematical methods in elementary thermodynamics. *J. Chem. Educ.* **43**, 85 (1966).

G.A. Estèvez, K. Yang, and B.B. Dasgupta, Thermodynamic partial derivatives and experimentally measurable quantities. *J. Chem. Educ.* **66**, 890 (1989).

J. Wisniak, The Joule–Thomson coefficient for pure gases and their mixtures. *Chem. Educ.* **4**, 51 (1999).

Texts and sources of data and information

J.R. Barrante, *Applied mathematics for physical chemistry*. Prentice Hall, Upper Saddle River (1998).

I.M. Klotz and R.M. Rosenberg, *Chemical thermodynamics: basic theory and methods*. Wiley-Interscience, New York (2000).

M.L. McGlashan, *Chemical thermodynamics*. Academic Press, London (1979).

R.G. Mortimer, *Mathematics for physical chemistry*. Academic Press, San Diego (1999).

E. Steiner, *The chemistry maths book*. Oxford University Press (1996).

Exercises

Discussion questions

3.1 (a) Explain the significance of a physical observable being a state function.

3.1 (b) Compile a list of as many state functions as you can identify.

3.2 (a) Explain why the internal pressure of a perfect gas is zero.

3.2 (b) Suggest how the internal energy of a van der Waals gas should vary with volume at constant temperture.

3.3 (a) Explain the significance of the Joule experiment. What would Joule observe in a more sensitive apparatus?

3.3 (b) Explain the origin and implications of the Joule–Thomson effect.

Numerical exercises

Assume that all gases are perfect and that all data refer to 298.15 K unless stated otherwise.

3.4 (a) Show that the following functions have exact differentials: (a) $x^2y + 3y^2$, (b) $x \cos xy$.

3.4 (b) Show that the following functions have exact differentials: (a) x^3y^2, (b) $t(t + e^s) + s$.

3.5 (a) Let $z = ax^2y^3$. Find dz.

3.5 (b) Let $z = x/(1 + y)^2$. Find dz.

3.6 (a) (a) What is the total differential of $z = x^2 + 2y^2 - 2xy + 2x - 4y - 8$? (b) Show that $\partial^2 z/\partial y\partial x = \partial^2 z/\partial x\partial y$ for this function.

3.6 (b) (a) What is the total differential of $z = x^3 - 2xy^2 + 15$? (b) Show that $\partial^2 z/\partial y\partial x = \partial^2 z/\partial x\partial y$ for this function.

3.7 (a) Let $z = xy - y + \ln x + 2$. Find dz and show that it is exact.

3.7 (b) Let $z = x^2y + xy^2$. Find dz and show that it is exact.

3.8 (a) Express $(\partial C_V/\partial V)_T$ as a second-derivative of U and find its relation to $(\partial U/\partial V)_T$. From this relation show that $(\partial C_V/\partial V)_T = 0$ for a perfect gas.

3.8 (b) Express $(\partial C_p/\partial p)_T$ as a second-derivative of H and find its relation to $(\partial H/\partial p)_T$. From this relation show that $(\partial C_p/\partial p)_T = 0$ for a perfect gas.

3.9 (a) By direct differentiation of $H = U + pV$, obtain a relation between $(\partial H/\partial U)_p$ and $(\partial U/\partial V)_p$.

3.9 (b) Confirm that $(\partial H/\partial U)_p = 1 + p(\partial V/\partial U)_p$ by expressing $(\partial H/\partial U)_p$ as the ratio of two derivatives with respect to volume and then using the definition of enthalpy.

3.10 (a) Write an expression for dV given that V is a function of p and T. Deduce an expression for d ln V in terms of the expansion coefficient and the isothermal compressibility.

3.10 (b) Write an expression for dp given that p is a function of V and T. Deduce an expression for d ln p in terms of the expansion coefficient and the isothermal compressibility.

3.11 (a) The internal energy of a perfect monatomic gas relative to its value at $T = 0$ is $\frac{3}{2}nRT$. Calculate $(\partial U/\partial V)_T$ and $(\partial H/\partial V)_T$ for the gas.

3.11 (b) The internal energy of a perfect monatomic gas relative to its value at $T = 0$ is $\frac{3}{2}nRT$. Calculate $(\partial H/\partial p)_T$ and $(\partial U/\partial p)_T$ for the gas.

3.12 (a) The coefficient of thermal expansion, α, is defined in eqn 8 and the isothermal compressibility, κ_T, is defined in eqn 14. Starting from the expression for the total differential dV in terms of T and p, show that $(\partial p/\partial T)_V = \alpha/\kappa_T$.

3.12 (b) Evaluate α and κ_T for a perfect gas.

3.13 (a) When a certain freon used in refrigeration was expanded adiabatically from an initial pressure of 32 atm and 0°C to a final pressure of 1.00 atm, the temperature fell by 22 K. Calculate the Joule–Thomson coefficient, μ, at 0°C, assuming it remains constant over this temperature range.

3.13 (b) A vapour at 22 atm and 5°C was allowed to expand adiabatically to a final pressure of 1.00 atm; the temperature fell by 10 K. Calculate the Joule–Thomson coefficient, μ, at 5°C, assuming it remains constant over this temperature range.

3.14 (a) For a van der Waals gas, $\pi_T = a/V_m^2$. Calculate ΔU_m for the isothermal reversible expansion of nitrogen gas from an initial volume of 1.00 L to 24.8 L at 298 K. What are the values of q and w?

3.14 (b) For a van der Waals gas, $\pi_T = a/V_m^2$. Calculate ΔU_m for the isothermal reversible expansion of argon from an initial volume of 1.00 L to 22.1 L at 298 K. What are the values of q and w?

3.15 (a) The volume of a certain liquid varies with temperature as

$$V = V'\{0.75 + 3.9 \times 10^{-4}(T/\mathrm{K}) + 1.48 \times 10^{-6}(T/\mathrm{K})^2\}$$

where V' is its volume at 300 K. Calculate its expansion coefficient, α, at 320 K.

3.15 (b) The volume of a certain liquid varies with temperature as

$$V = V'\{0.77 + 3.7 \times 10^{-4}(T/\mathrm{K}) + 1.52 \times 10^{-6}(T/\mathrm{K})^2\}$$

where V' is its volume at 298 K. Calculate its expansion coefficient, α, at 310 K.

3.16 (a) The isothermal compressibility of copper at 293 K is 7.35×10^{-7} atm^{-1}. Calculate the pressure that must be applied in order to increase its density by 0.08 per cent.

3.16 (b) The isothermal compressibility of lead at 293 K is 2.21×10^{-6} atm^{-1}. Calculate the pressure that must be applied in order to increase its density by 0.08 per cent.

3.17 (a) Given that $\mu = 0.25$ K atm^{-1} for nitrogen, calculate the value of its isothermal Joule–Thomson coefficient. Calculate the energy that must be supplied as heat to maintain constant temperature when 15.0 mol N_2 flows through a throttle in an isothermal Joule–Thomson experiment and the pressure drop is 75 atm.

3.17 (b) Given that $\mu = 1.11$ K atm^{-1} for carbon dioxide, calculate the value of its isothermal Joule–Thomson coefficient. Calculate the energy that must be supplied as heat to maintain constant temperature when 12.0 mol CO_2 flows through a throttle in an isothermal Joule–Thomson experiment and the pressure drop is 55 atm.

3.18 (a) To design a particular kind of refrigerator we need to know the temperature reduction brought about by adiabatic expansion of the refrigerant gas. For one type of freon, $\mu = 1.2$ K atm^{-1}. What pressure difference is needed to produce a temperature drop of 5.0 K?

3.18 (b) For another type of freon (see previous exercise), $\mu = 13.3$ mK kPa^{-1}. What pressure difference is needed to produce a temperature drop of 4.5 K?

Problems†

Assume all gases are perfect unless stated otherwise. Unless otherwise stated, thermochemical data are for 298.15 K.

Numerical problems

3.1 The isothermal compressibility of lead is 2.21×10^{-6} atm^{-1}. Express this value in Pa^{-1}. A cube of lead of side 10 cm at 25°C was to be inserted in the keel of an underwater exploration TV camera, and its designers needed to know the stresses in the equipment. Calculate the change of volume of the cube at a depth of 1.000 km (disregarding the effects of temperature). Take the mean density of sea water as 1.03 g cm^{-3}. Given that the expansion coefficient of lead is 8.61×10^{-5} K^{-1} and that the temperature where the camera operates is −5°C, calculate the volume of the block taking the temperature into account too.

3.2 Calculate the change in (a) the molar internal energy and (b) the molar enthalpy of water when its temperature is raised by 10 K. Account for the difference between the two quantities.

3.3 The constant-volume heat capacity of a gas can be measured by observing the decrease in temperature when it expands adiabatically and reversibly. If the decrease in pressure is also measured, we can use it to infer the value of γ (the ratio of heat capacities, C_p/C_V) and hence, by combining the two values, deduce the constant-pressure heat capacity. A fluorocarbon gas was allowed to expand reversibly and adiabatically to twice its volume; as a result, the temperature fell from 298.15 K to 248.44 K and its pressure fell from 1522.2 Torr to 613.85 Torr. Evaluate C_p.

3.4 A sample consisting of 1.00 mol of a van der Waals gas is compressed from 20.0 L to 10.0 L at 300 K. In the process, 20.2 kJ of work is done on the gas. Given that $\mu = \{(2a/RT) - b\}/C_{p,m}$, with $C_{p,m} = 38.4$ J K^{-1} mol^{-1}, $a = 3.60$ L^2 atm mol^{-2}, and $b = 0.44$ L mol^{-1}, calculate ΔH for the process.

3.5 Estimate γ (the ratio of heat capacities) for xenon at 100°C and 1.00 atm on the assumption that it is a van der Waals gas.

3.6‡ A cylindrical container of fixed total volume is divided into three sections, S_1, S_2, and S_3. The sections S_1 and S_2 are separated by an adiabatic piston whereas S_2 and S_3 are separated by a diathermic (heat-conducting) piston. The pistons can slide along the walls of the cylinder without friction. Each section of the cylinder contains 1.00 mol of a perfect diatomic gas. Initially, the gas pressure in all three sections is 1.00 bar and the temperature is 298 K. The gas in S_1 is heated slowly until the temperature of the gas in S_3 reaches 348 K. Find the final temperature, pressure, and volume, as well as the change in internal energy for each section. Determine the total energy supplied to the gas in S_1.

3.7‡ Solve Problem 3.6 for the case where both pistons are (a) adiabatic, (b) diathermic.

3.8‡ A gas obeying the equation of state $p(V - nb) = nRT$ is subjected to a Joule–Thomson expansion. Will the temperature increase, decrease, or remain the same?

† Problems denoted with the symbol ‡ were supplied by Charles Trapp and Carmen Giunta.

Theoretical problems

3.9 Determine whether or not $dz = xy\,dx + xy\,dy$ is exact by integrating it around the closed curve formed by the paths $y = x$ and $y = x^2$ between the points (0, 0) and (1, 1).

3.10 Decide whether $dq = (RT/p)\,dp - R\,dT$ is exact. Then determine whether multiplication of dq by $1/T$ is exact. Comment on the significance of your result.

3.11 Obtain the total differential of the function $w = xy + yz + xz$. Demonstrate that dw is exact by integration between the points (0, 0, 0) and (1, 1, 1) along the two different paths (1) $z = y = x$ and (2) $z = y = x^2$.

3.12 Derive the relation $C_V = -(\partial U/\partial V)_T(\partial V/\partial T)_U$ from the expression for the total differential of $U(T, V)$.

3.13 Starting from the expression for the total differential of $H(T, p)$, express $(\partial H/\partial p)_T$ in terms of C_p and the Joule–Thomson coefficient, μ.

3.14 Starting from the expression $C_p - C_V = T(\partial p/\partial T)_V(\partial V/\partial T)_p$, use the appropriate relations between partial derivatives to show that

$$C_p - C_V = \frac{T(\partial V/\partial T)_p^2}{(\partial V/\partial T)_T}$$

Evaluate $C_p - C_V$ for a perfect gas.

3.15 From an analysis of Joule's free expansion experiment, demonstrate that it is possible to calculate the change in internal energy of a perfect gas for any process by knowing only C_V and ΔT.

3.16 By the consideration of a suitable cycle involving a perfect gas, demonstrate that dq is an inexact differential and, therefore, that q is not a state function.

3.17 Use the fact that $(\partial U/\partial V)_T = a/V_m^2$ for a van der Waals gas to show that $\mu C_{p,m} \approx (2a/RT) - b$ by using the definition of μ and appropriate relations between partial derivatives. (*Hint.* Use the approximation $pV_m \approx RT$ when it is justifiable to do so.)

3.18 Obtain the expression for the total differential dp for a van der Waals gas in terms of dT and dV. Also obtain $(\partial V/\partial T)_p$. Demonstrate that dp is not an exact differential by integrating it from (T_1, V_1) to (T_2, V_2) along the two different paths, namely (1) $(T_1, V_1) \rightarrow (T_2, V_1) \rightarrow (T_2, V_2)$ and (2) $(T_1, V_1) \rightarrow (T_1, V_2) \rightarrow (T_2, V_2)$.

3.19 Take nitrogen to be a van der Waals gas with a = 1.352 L^2 atm mol^{-2} and b = 0.0387 L mol^{-1}, and calculate ΔH_m when the pressure on the gas is decreased from 500 atm to 1.00 atm at 300 K. For a van der Waals gas, $\mu = \{(2a/RT) - b\}/C_{p,m}$. Assume $C_{p,m} = \frac{7}{2}R$.

3.20 The pressure of a given amount of a van der Waals gas depends on T and V. Find an expression for dp in terms of dT and dV.

3.21 Rearrange the van der Waals equation of state to give an expression for T as a function of p and V (with n constant). Calculate $(\partial T/\partial p)_V$ and confirm that $(\partial T/\partial p)_V = 1/(\partial p/\partial T)_V$. Go on to confirm Euler's chain relation.

3.22 Calculate the isothermal compressibility and the expansion coefficient of a van der Waals gas. Show, using Euler's chain relation, that $\kappa_T R = \alpha(V_m - b)$.

3.23 Given that $\mu C_p = T(\partial V/\partial T)_p - V$, derive an expression for μ in terms of the van der Waals parameters a and b, and express it in terms of reduced variables. Evaluate μ at 25°C and 1.0 atm, when the molar volume of the gas is 24.6 L mol^{-1}. Use the expression obtained to derive a formula for the inversion temperature of a van der Waals gas in terms of reduced variables, and evaluate it for the xenon sample.

3.24 The thermodynamic equation of state $(\partial U/\partial V)_T = T(\partial p/\partial T)_V - p$ was quoted in the chapter. Derive its partner

$$\left(\frac{\partial H}{\partial p}\right)_T = -T\left(\frac{\partial V}{\partial T}\right)_p + V$$

from it and the general relations between partial differentials.

3.25 Show that for a van der Waals gas,

$$C_{p,m} - C_{V,m} = \lambda R \qquad \frac{1}{\lambda} = 1 - \frac{(3V_r - 1)^2}{4V_r^3 T_r}$$

and evaluate the difference for xenon at 25°C and 10.0 atm.

3.26 The speed of sound, c_s, in a gas of molar mass M is related to the ratio of heat capacities γ by $c_s = (\gamma RT/M)^{1/2}$. Show that $c_s = (\gamma p/\rho)^{1/2}$, where ρ is the mass density of the gas. Calculate the speed of sound in argon at 25°C.

3.27‡ A gas obeys the equation of state $V_m = RT/p + aT^2$ and its constant-pressure heat capacity is given by $C_{p,m} = A + BT + Cp$, where a, A, B, and C are constants independent of T and p. Obtain expressions for (a) the Joule–Thomson coefficient and (b) its constant-volume heat capacity.

Applications: to environmental science

3.28‡ In 1995, the Intergovernmental Panel on Climate Change considered a global average temperature rise of 1.0–3.5°C likely by the year 2100, with 2.0°C its best estimate (IPCC Second Assessment Synthesis of Scientific-Technical Information Relevant to Interpreting Article 2 of the UN Framework Convention on Climate Change (1995)). Predict the average rise in sea level due to thermal expansion of sea water based on temperature rises of 1.0°C, 2.0°C, and 3.5°C given that the volume of the Earth's oceans is 1.37×10^9 km^3 and their surface area is 361×10^6 km^2, and state the approximations that go into the estimates.

3.29‡ Concerns over the harmful effects of chlorofluorocarbons on stratospheric ozone have motivated a search for new refrigerants. One such alternative is 2,2-dichloro-1,1,1-trifluoroethane (refrigerant 123). B.A. Younglove and M. McLinden published a compendium of thermophysical properties of this substance (*J. Phys. Chem. Ref. Data* **23**, 7 (1994)), from which properties such as the Joule–Thomson coefficient μ can be computed. (a) Compute μ at 1.00 bar and 50°C given that $(\partial H/\partial p)_T = -3.29 \times 10^3$ J MPa^{-1} mol^{-1} and $C_{p,m}$ = 110.0 J K^{-1} mol^{-1}. (b) Compute the temperature change that would accompany adiabatic expansion of 2.0 mol of this refrigerant from 1.5 bar to 0.5 bar at 50°C.

3.30‡ Another alternative refrigerant (see preceding problem) is 1,1,1,2-tetrafluoroethane (refrigerant HFC-134a). R. Tillner-Roth and H.D. Baehr published a compendium of thermophysical properties of this substance (*J. Phys. Chem. Ref. Data* **23**, 657 (1994)), from which properties such as the Joule–Thomson coefficient μ can be computed. (a) Compute μ at 0.100 MPa and 300 K from the following data (all referring to 300 K):

p/MPa	0.080	0.100	0.12
specific enthalpy/(kJ kg^{-1})	426.48	426.12	425.76

(The specific constant-pressure heat capacity is 0.7649 kJ K^{-1} kg^{-1}.) (b) Compute μ at 1.00 MPa and 350 K from the following data (all referring to 350 K):

p/MPa	0.80	1.00	1.2
specific enthalpy/(kJ kg^{-1})	461.93	459.12	456.15

(The specific constant-pressure heat capacity is 1.0392 kJ K^{-1} kg^{-1}.)

4 The Second Law: the concepts

The purpose of this chapter is to explain the origin of the spontaneity of physical and chemical change. We examine two simple processes and show how to define, measure, and use a property, the entropy, to discuss spontaneous changes quantitatively. The chapter also introduces a major subsidiary thermodynamic property, the Gibbs energy. The introduction of the Gibbs energy enables the spontaneity of a process to be expressed solely in terms of the properties of a system (instead of having to consider entropy changes in the system and its surroundings). The Gibbs energy also enables us to predict the maximum non-expansion work that a process can do.

Some things happen naturally; some things don't. A gas expands to fill the available volume, a hot body cools to the temperature of its surroundings, and a chemical reaction runs in one direction rather than another. Some aspect of the world determines the **spontaneous** direction of change, the direction of change that does not require work to be done to bring the change about. We can confine a gas to a smaller volume, we can cool an object with a refrigerator, and we can force some reactions to go in reverse (as in the electrolysis of water). However, none of these processes happens spontaneously; each one must be brought about by doing work.

The recognition of two classes of process, spontaneous and non-spontaneous, is summarized by the **Second Law of thermodynamics**. This law may be expressed in a variety of equivalent ways. One statement was formulated by Kelvin:

No process is possible in which the sole result is the absorption of heat from a reservoir and its complete conversion into work.

For example, it has proved impossible to construct an engine like that shown in Fig. 4.1, in which heat is drawn from a hot reservoir and completely converted into work. All real heat engines have both a hot source and a cold sink; some heat is always discarded into the cold sink and not converted into work. The Kelvin statement is a generalization of another everyday observation, that a ball at rest on a surface has never been observed to leap spontaneously upwards. An upward leap of the ball would be equivalent to the conversion of heat from the surface into work.

The direction of spontaneous change

What determines the direction of spontaneous change? It is not the total energy of the isolated system. The First Law of thermodynamics states that energy is conserved in any process, and we cannot disregard that law now and say that

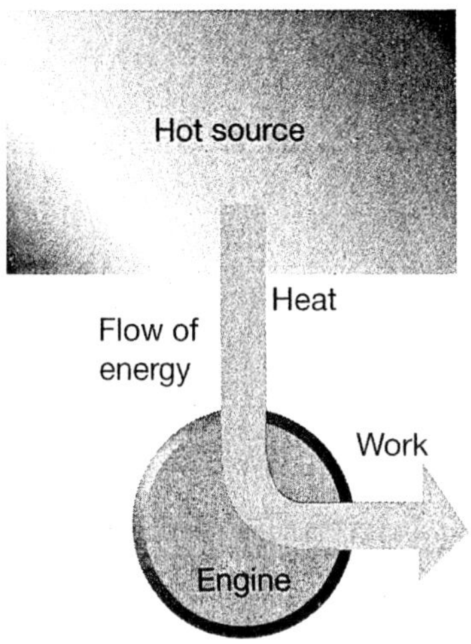

4.1 The Kelvin statement of the Second Law denies the possibility of the process illustrated here, in which heat is changed completely into work, there being no other change. The process is not in conflict with the First Law because energy is conserved.

everything tends towards a state of lower energy: the total energy of an isolated system is constant.

Is it perhaps the energy of the *system* that tends towards a minimum? Two arguments show that this cannot be so. First, a perfect gas expands spontaneously into a vacuum, yet its internal energy remains constant as it does so. Secondly, if the energy of a system does happen to decrease during a spontaneous change, the energy of its surroundings must increase by the same amount (by the First Law). The increase in energy of the surroundings is just as spontaneous a process as the decrease in energy of the system.

When a change occurs, the total energy of an isolated system remains constant but it is parcelled out in different ways. Can it be, therefore, that the direction of change is related to the *distribution* of energy? We shall see that this idea is the key, and that spontaneous changes are always accompanied by a dispersal of energy into a more disordered form.

4.1 The dispersal of energy

We can begin to understand the role of the distribution of energy by thinking about a ball (the system) bouncing on a floor (the surroundings). The ball does not rise as high after each bounce because there are inelastic losses in the materials of the ball and floor and the kinetic energy of the ball's overall motion is converted into the energy of thermal motion. The direction of spontaneous change is towards a state in which the ball is at rest with all its energy degraded into the thermal motion of the atoms of the virtually infinite floor (Fig. 4.2).

A ball resting on a warm floor has never been observed to start bouncing. For bouncing to begin, something rather special would need to happen. In the first place, some of the thermal motion of the atoms in the floor would have to accumulate in a single, small object, the ball. This accumulation requires a spontaneous localization of energy from the myriad of vibrations of the atoms of the floor into the much smaller number of atoms that constitute the ball (Fig. 4.3). Furthermore, whereas the thermal motion is disorderly, for the ball to move upwards its atoms must all move in the same direction. The localization of random motion as orderly motion is so unlikely that we can dismiss it as virtually impossible.[1]

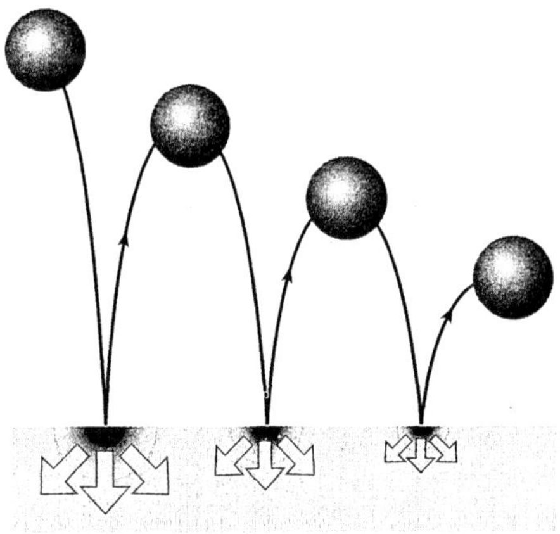

4.2 The direction of spontaneous change for a ball bouncing on a floor. On each bounce some of its energy is degraded into the thermal motion of the atoms of the floor, and that energy disperses. The reverse has never been observed to take place on a macroscopic scale.

We appear to have found the signpost of spontaneous change: *we look for the direction of change that leads to more disorderly dispersal of the total energy of the isolated system.* This principle accounts for the direction of change of the bouncing ball, because its energy is dissipated as thermal motion of the atoms of the floor. The reverse process is not spontaneous because it is highly improbable that the disorderly distribution of energy will become organized into localized, uniform motion. A gas does not contract spontaneously, because to do so the disordered motion of its molecules would have to take them all into the same region of the container. The opposite change, spontaneous expansion, is a natural consequence of increasing disorder. An object does not spontaneously become warmer than its surroundings because it is highly improbable that the jostling of randomly vibrating atoms in the surroundings will lead to the accumulation of excess thermal motion in the object. The opposite change, the spreading of the object's energy into the surroundings as thermal motion, is natural.

It may seem very puzzling that collapse into disorder can account for the formation of such ordered substances as crystals or proteins. Nevertheless, in due course we shall see that organized structures can emerge as energy and matter disperse. We shall see, in fact, that collapse into disorder accounts for change in all its forms.

1. Orderly motion, but on a much smaller scale, is observed as *Brownian motion*.

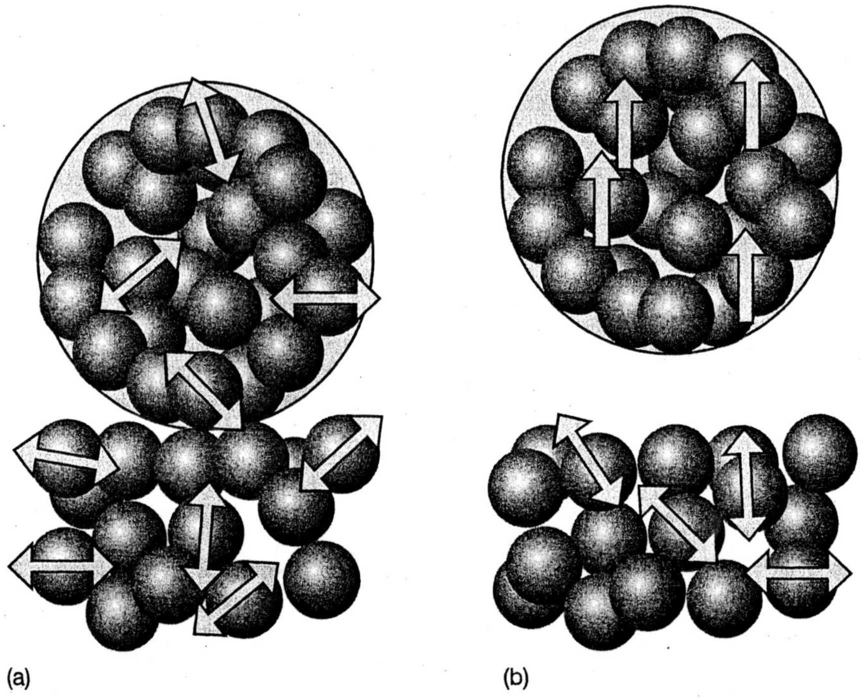

4.3 The molecular interpretation of the irreversibility expressed by the Second Law. (a) A ball resting on a warm surface; the atoms are undergoing thermal motion (chaotic vibration, in this instance), as indicated by the arrows. (b) For the ball to fly upwards, some of the random vibrational motion would have to change into coordinated, directed motion. Such a conversion is highly improbable.

4.2 Entropy

The First Law of thermodynamics led to the introduction of the internal energy, U. The internal energy is a state function that lets us assess whether a change is permissible: only those changes may occur for which the internal energy of an isolated system remains constant. The law that is used to identify the signpost of spontaneous change, the Second Law of thermodynamics, may also be expressed in terms of another state function, the **entropy**, S. We shall see that the entropy (which we shall define shortly, but is a measure of the molecular disorder of a system) lets us assess whether one state is accessible from another by a spontaneous change. The First Law uses the internal energy to identify *permissible* changes; the Second Law uses the entropy to identify the *spontaneous changes* among those permissible changes.

The Second Law of thermodynamics can be expressed in terms of the entropy:

The entropy of an isolated system increases in the course of a spontaneous change:

$$\Delta S_{tot} > 0 \tag{4.1}$$

where S_{tot} is the total entropy of the system and its surroundings. Thermodynamically irreversible processes (like cooling to the temperature of the surroundings and the free expansion of gases) are spontaneous processes, and hence must be accompanied by an increase in total entropy.

(a) The thermodynamic definition of entropy

The thermodynamic definition of entropy concentrates on the change in entropy, dS, that occurs as a result of a physical or chemical change (in general, as a result of a 'process'). The definition is motivated by the idea that a change in the extent to which energy is dispersed in a disorderly manner depends on how much energy is transferred as heat. As we have remarked, heat stimulates disorderly motion in the surroundings. Work, which stimulates uniform motion of atoms in the surroundings, does not change the degree of disorder and so does not change the entropy.

The thermodynamic definition of entropy is based on the expression

$$dS = \frac{dq_{rev}}{T} \qquad [4.2]$$

For a measurable change between two states i and f this expression integrates to

$$\Delta S = \int_i^f \frac{dq_{rev}}{T} \qquad (4.3)$$

That is, to calculate the difference in entropy between any two states of a system, we find a reversible path between them, and integrate the heat supplied at each stage of the path divided by the temperature at which the heat is supplied.

Molecular interpretation 4.1

The molecules in a system at high temperature are highly disorganized, either in terms of their locations or in terms of the occupation of their available translational, rotational, and vibrational energy states. A small additional transfer of energy will result in a relatively small additional disorder, much as sneezing in a busy street may be barely noticed. In contrast, the molecules in a system at low temperature have access to far fewer energy states (at $T = 0$, only the lowest state is accessible), and the same quantity of heat will have a pronounced effect on the degree of disorder, much as sneezing in a quiet library can be very disruptive. Hence, the change in entropy when a given quantity of heat is transferred will be greater when it is transferred to a cold body than when it is transferred to a hot body. This argument suggests that the change in entropy should be inversely proportional to the temperature at which the transfer takes place, as in eqn 2.

According to eqn 2, when the heat transferred is expressed in joules and the temperature is in kelvins, the units of entropy are joules per kelvin ($J K^{-1}$). Molar entropy, the entropy divided by the amount of substance, is expressed in joules per kelvin per mole ($J K^{-1} mol^{-1}$).[2]

Example 4.1 **Calculating the entropy change for the isothermal expansion of a perfect gas**

Calculate the entropy change of a sample of perfect gas when it expands isothermally from a volume V_i to a volume V_f.

Method The definition of entropy instructs us to find the heat absorbed for a reversible path between the stated initial and final states regardless of the actual manner in which the process takes place. A simplification is that the expansion is isothermal, so the temperature is a constant and may be taken outside the integral in eqn 3. The heat absorbed during a reversible isothermal expansion of a perfect gas can be calculated from $\Delta U = q + w$ and $\Delta U = 0$, which implies that $q = -w$ in general and therefore that $q_{rev} = -w_{rev}$ for a reversible change. The work of reversible isothermal expansion was calculated in Section 2.3 (eqn 2.13).

2. These are the same units as those of the gas constant, R, and molar heat capacities.

Answer Because the temperature is constant, eqn 3 becomes

$$\Delta S = \frac{1}{T}\int_i^f dq_{rev} = \frac{q_{rev}}{T}$$

From eqn 2.13, we know that

$$q_{rev} = -w_{rev} = nRT \ln \frac{V_f}{V_i}$$

Therefore, it follows that

$$\Delta S = nR \ln \frac{V_f}{V_i}$$

Comment As an illustration of this formula, when the volume of 1.00 mol of any perfect gas is doubled at any constant temperature

$$\Delta S = (1.00 \text{ mol}) \times (8.3145 \text{ J K}^{-1} \text{ mol}^{-1}) \times \ln 2 = +5.76 \text{ J K}^{-1}$$

Self-test 4.1 Calculate the change in entropy when the pressure of a perfect gas is changed isothermally from p_i to p_f. [$\Delta S = nR \ln(p_i/p_f)$]

We can use the definition in eqn 2 to formulate an expression for the change in entropy of the surroundings, ΔS_{sur}. Consider an infinitesimal transfer of heat dq_{sur} to the surroundings. The surroundings consist of a reservoir of constant volume, so the heat supplied to them can be identified with the change in their internal energy, dU_{sur}.[3] The internal energy is a state function, and dU_{sur} is an exact differential. As we have seen, these properties imply that dU_{sur} is independent of how the change is brought about and in particular is independent of whether the process is reversible or irreversible. The same remarks therefore apply to dq_{sur}, to which dU_{sur} is equal. Therefore, we can adapt the definition of entropy change in eqn 2 to write

$$dS_{sur} = \frac{dq_{sur,rev}}{T_{sur}} = \frac{dq_{sur}}{T_{sur}} \tag{4.4}$$

Furthermore, because the temperature of the surroundings is constant whatever the change, for a measurable change

$$\Delta S_{sur} = \frac{q_{sur}}{T_{sur}} \tag{4.5}$$

That is, regardless of how the change is brought about in the system, reversibly or irreversibly, we can calculate the change of entropy of the surroundings by dividing the heat transferred by the temperature at which the transfer takes place.

Equation 5 makes it very simple to calculate the changes in entropy of the surroundings that accompany any process. For instance, for any adiabatic change, $q_{sur} = 0$, so

$$\Delta S_{sur} = 0 \tag{4.6}$$

This expression is true however the change takes place, reversibly or irreversibly, provided no local hot spots are formed in the surroundings. That is, it is true so long as the surroundings remain in internal equilibrium. If hot spots do form, then the localized energy may subsequently disperse spontaneously and hence generate more entropy.

3. Alternatively, the surroundings can be regarded as being at constant pressure, in which case we could equate dq_{sur} to dH_{sur}.

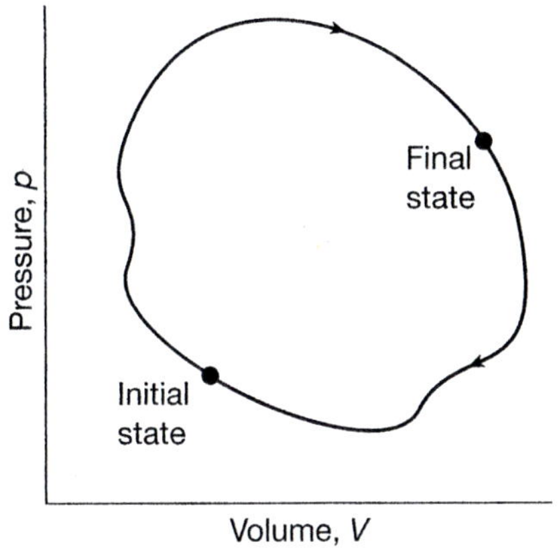

4.4 In a thermodynamic cycle, the overall change in a state function (from the initial state to the final state and then back to the initial state again) is zero.

Illustration 4.1

To calculate the entropy change in the surroundings when 1.00 mol $H_2O(l)$ is formed from its elements under standard conditions at 298 K, we use $\Delta H^{\ominus} = -286$ kJ from Table 2.6. The heat released is supplied to the surroundings, now regarded as being at constant pressure, so $q_{sur} = +286$ kJ. Therefore,

$$\Delta S_{sur} = \frac{2.86 \times 10^5\ \text{J}}{298\ \text{K}} = +960\ \text{J K}^{-1}$$

This strongly exothermic reaction results in an increase in the entropy of the surroundings as heat is released into them.

Self-test 4.2 Calculate the entropy change in the surroundings when 1.00 mol $N_2O_4(g)$ is formed from 2.00 mol $NO_2(g)$ under standard conditions at 298 K. [$+192\ \text{J K}^{-1}$]

(b) The entropy as a state function

Entropy is a state function. To prove this assertion, we need to show that the integral of dS is independent of path. To do so, it is sufficient to prove that the integral of eqn 2 around an arbitrary cycle is zero, for that guarantees that the entropy is the same at the initial and final states of the system regardless of the path taken between them (Fig. 4.4). That is, we need to show that

$$\oint \frac{dq_{rev}}{T} = 0 \tag{4.7}$$

where the symbol $\oint$ denotes integration around a closed path.

To prove eqn 7, we first consider the special cycle shown in Fig. 4.5. A **Carnot cycle**, which is named after the French engineer Sadi Carnot, consists of four reversible stages:

1 Reversible isothermal expansion from A to B at T_h; the entropy change is q_h/T_h, where q_h is the heat supplied to the system from the hot source.

2 Reversible adiabatic expansion from B to C. No heat leaves the system, so the change in entropy is zero. In the course of this expansion, the temperature falls from T_h to T_c, the temperature of the cold sink.

3 Reversible isothermal compression from C to D at T_c. Heat is released to the cold sink; the change in entropy of the system is q_c/T_c; in this expression q_c is negative.

4 Reversible adiabatic compression from D to A. No heat enters the system, so the change in entropy is zero. The temperature rises from T_c to T_h.

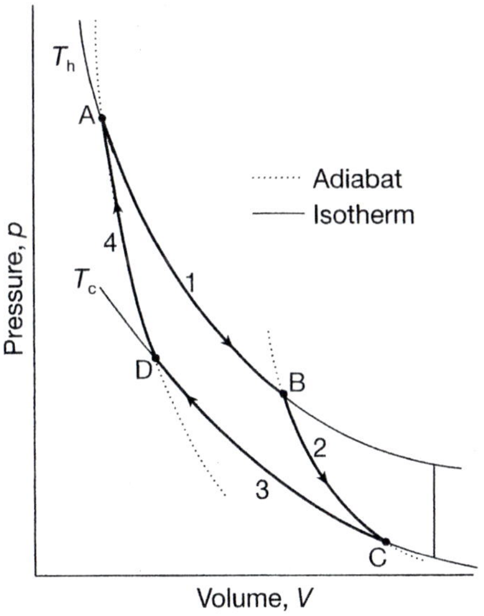

4.5 The basic structure of a Carnot cycle. In step 1, there is isothermal reversible expansion at the temperature T_h. Step 2 is a reversible adiabatic expansion in which the temperature falls from T_h to T_c. In Step 3 there is an isothermal reversible compression at T_c, and that isothermal step is followed by an adiabatic reversible compression, which restores the system to its initial state.

The total change in entropy around the cycle is

$$\oint dS = \frac{q_h}{T_h} + \frac{q_c}{T_c}$$

However, we show in the *Justification* below that

$$\frac{q_h}{q_c} = -\frac{T_h}{T_c} \tag{4.8}$$

Substitution of this relation into the preceding equation gives zero on the right, which is what we wanted to prove.

Justification 4.1

This *Justification* is based on the fact that the two temperatures in eqn 4.8 lie on the same adiabat in Fig. 4.5. As explained in Example 4.1, for a perfect gas:

$$q_h = nRT_h \ln \frac{V_B}{V_A} \qquad q_c = nRT_c \ln \frac{V_D}{V_C}$$

From the relations between temperature and volume for reversible adiabatic processes (eqn 2.34):

$$V_A T_h^c = V_D T_c^c \qquad V_C T_c^c = V_B T_h^c$$

Multiplication of the first of these expressions by the second gives

$$V_A V_C T_h^c T_c^c = V_D V_B T_h^c T_c^c$$

which simplifies to

$$\frac{V_A}{V_B} = \frac{V_D}{V_C}$$

Consequently,

$$q_c = nRT_c \ln \frac{V_D}{V_C} = nRT_c \ln \frac{V_A}{V_B} = -nRT_c \ln \frac{V_B}{V_A}$$

and therefore

$$\frac{q_h}{q_c} = \frac{nRT_h \ln(V_B/V_A)}{-nRT_c \ln(V_B/V_A)} = -\frac{T_h}{T_c}$$

as in eqn 8.

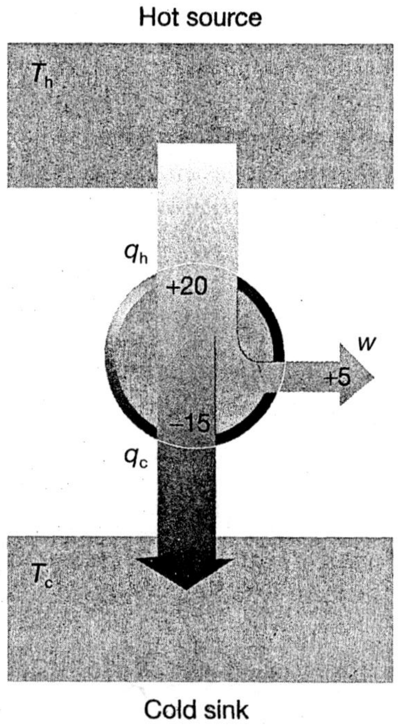

4.6 Suppose an energy q_h (e.g. 20 kJ) is supplied to the engine and q_c is lost from the engine (e.g. $q_c = -15$ kJ) and discarded into the cold reservoir. The work done by the engine is equal to $q_h + q_c$ (e.g. 20 kJ + (−15 kJ) = 5 kJ). The efficiency is the work done divided by the heat supplied from the hot source.

Now we need to show that the same conclusion applies to any material, not just a perfect gas. To do so, consider the **efficiency**, ε (epsilon), of a heat engine:

$$\varepsilon = \frac{\text{work performed}}{\text{heat absorbed}} = \frac{|w|}{q_h} \qquad [4.9]$$

The definition implies that, the greater the work output for a given supply of heat from the hot reservoir, the greater is the efficiency of the engine. We can express the definition in terms of the heat transactions alone, because (as shown in Fig. 4.6) the work supplied by the engine is the difference between the heat supplied by the hot reservoir and returned to the cold reservoir:

$$\varepsilon = \frac{q_h + q_c}{q_h} = 1 + \frac{q_c}{q_h} \qquad (4.10)$$

(Remember that $q_c < 0$.) It then follows that for a Carnot engine

$$\varepsilon_{rev} = 1 - \frac{T_c}{T_h} \qquad (4.11)$$

Now, the Second Law of thermodynamics implies that *all reversible engines have the same efficiency regardless of their construction.* To see the truth of this statement, suppose two reversible engines are coupled together and run between the same two reservoirs (Fig. 4.7). The working substances and details of construction of the two engines are entirely arbitrary. Initially, suppose that engine A is more efficient than engine B, and that we choose a setting of the controls that causes engine B to acquire

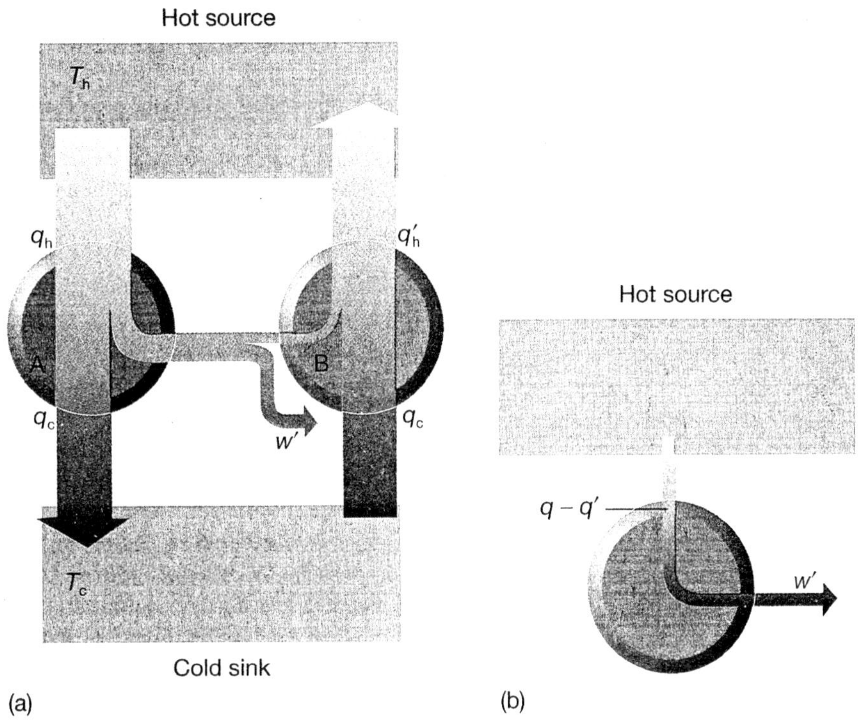

4.7 (a) The demonstration of the equivalence of the efficiencies of all reversible engines working between the same thermal reservoirs is based on the flow of energy represented in this diagram. (b) The net effect of the processes is the conversion of heat into work without there being a need for a cold sink: this is contrary to the Kelvin statement of the Second Law.

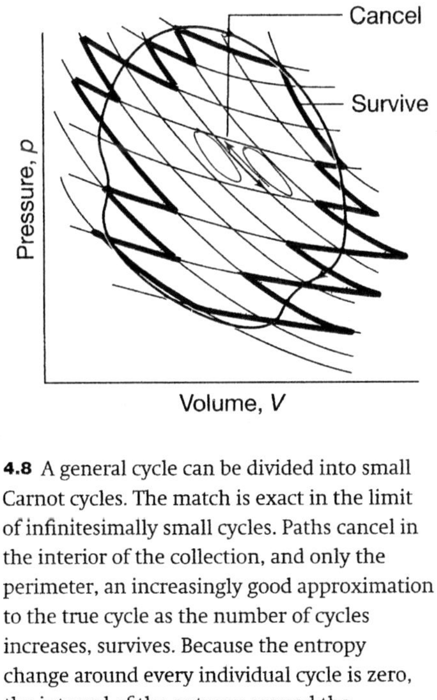

4.8 A general cycle can be divided into small Carnot cycles. The match is exact in the limit of infinitesimally small cycles. Paths cancel in the interior of the collection, and only the perimeter, an increasingly good approximation to the true cycle as the number of cycles increases, survives. Because the entropy change around every individual cycle is zero, the integral of the entropy around the perimeter is zero too.

the heat q_c from the cold reservoir and to release a certain quantity of heat into the hot reservoir. However, because engine A is more efficient than engine B, not all the work that A produces is needed for this process, and the difference can be used to do work. The net result is that the cold reservoir is unchanged, work has been produced, and the hot reservoir has lost a certain amount of energy. This outcome is contrary to the Kelvin statement of the Second Law, because some heat has been converted directly into work. In molecular terms, the disordered thermal motion of the hot reservoir has been converted into ordered motion characteristic of work. Because the conclusion is contrary to experience, the initial assumption that engines A and B can have different efficiencies must be false. It follows that the relation between the heat transfers and the temperatures (eqn 11) must also be independent of the working material, and therefore that eqn 8 is always true for any substance involved in a Carnot cycle.

To complete the argument, we note that any reversible cycle can be approximated as a collection of Carnot cycles and the cyclic integral around an arbitrary path is the sum of the integrals around each of the Carnot cycles (Fig. 4.8). This approximation becomes exact as the individual cycles are allowed to become infinitesimal. The entropy change around each individual cycle is zero (as demonstrated above), so the sum of entropy changes for all the cycles is zero. However, in the sum, the entropy change along any individual path is cancelled by the entropy change along the path it shares with the neighbouring cycle. Therefore, all the entropy changes cancel except for those along the perimeter of the overall cycle. That is,

$$\sum_{\text{all}} \frac{q_{\text{rev}}}{T} = \sum_{\text{perimeter}} \frac{q_{\text{rev}}}{T} = 0$$

In the limit of infinitesimal cycles, the non-cancelling edges of the Carnot cycles match the overall cycle exactly, and the sum becomes an integral. Equation 7 then follows immediately. This result implies that dS is an exact differential and therefore that S is a state function.

(c) The thermodynamic temperature

Suppose we have an engine that is working reversibly between a hot source at a temperature T_h and a cold sink at a temperature T; then we know from eqn 11 that

$$T = (1 - \varepsilon)T_h \qquad (4.12)$$

This expression enabled Kelvin to define the **thermodynamic temperature scale** in terms of the efficiency of a heat engine. The zero of the scale occurs for a Carnot efficiency of 1. The size of the unit is entirely arbitrary, but on the Kelvin scale is defined by setting the temperature of the triple point of water as 273.16 K exactly. Then, if the heat engine has a hot source at the triple point of water, the temperature of the cold sink (the object we want to measure) is found by measuring the efficiency of the engine. This result is independent of the working substance (Box 4.1).

An additional point is that, as we saw in Section 2.2c, heat transferred can, in principle, be measured mechanically (in terms of the location of a weight). Therefore, it is possible, in principle at least, to use the distance moved by a weight to measure temperature. Kelvin's definition of the thermodynamic temperature scale puts the measurement of temperature on to a purely mechanical basis.

(d) The Clausius inequality

So far, we have verified that the entropy as defined in eqn 2 is a state function. We now need to verify that the entropy is a signpost of spontaneous change in the sense that $dS_{tot} \geq 0$ for any spontaneous change (with the equality applying for a reversible process).

Consider a system in thermal and mechanical contact with its surroundings at the same temperature, T. The system and the surroundings are not necessarily in mechanical equilibrium; for instance, a gas might have a greater pressure than its surroundings. Any change of state is accompanied by a change in entropy of the system, dS, and of the surroundings, dS_{sur}. Because the process might be irreversible, the total entropy will increase when a process occurs in the system, so we can write

$$dS + dS_{sur} \geq 0, \qquad \text{or} \qquad dS \geq -dS_{sur}$$

Box 4.1 The thermodynamics of refrigeration

The discussion of the text is the basis of the thermodynamic assessment of the power needed to cool objects in refrigerators. First, we consider the work required to cool an object, and refer to the illustration.

When an energy $|q_c|$ is removed from a cool source at a temperature T_c and then deposited in a warmer sink at a temperature T_h, as in a typical refrigerator, the change in entropy is

$$\Delta S = -\frac{|q_c|}{T_c} + \frac{|q_c|}{T_h} < 0$$

The process is not spontaneous because not enough entropy is generated in the warm sink to overcome the entropy loss from the hot source. To generate more entropy, energy must be added to the stream that enters the warm sink. Our task is to find the minimum energy that needs to be supplied. The outcome is expressed as the *coefficient of performance*, *c*:

$$c = \frac{\text{heat transferred}}{\text{work done}} = \frac{|q_c|}{|w|}$$

The less the work that is required to achieve a given transfer, the greater the coefficient of performance and the more efficient the refrigerator.

Because $|q_c|$ is removed from the cold source, and the work $|w|$ is added to the energy stream, the heat deposited in the hot sink is $|q_h| = |q_c| + |w|$. Therefore,

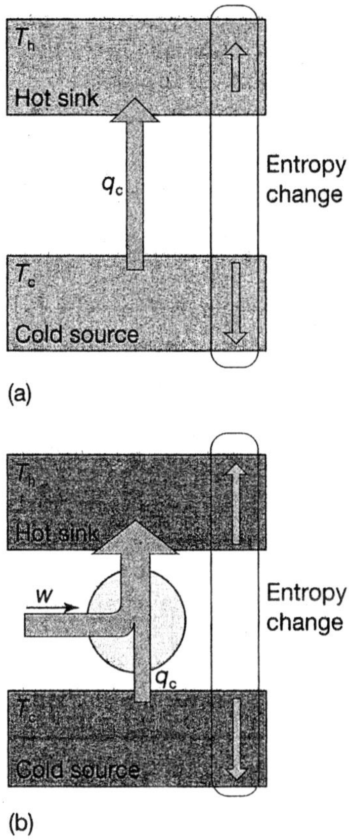

(a) The flow of energy as heat from a cold source to a hot sink is not spontaneous. As shown here, the entropy increase of the hot sink is smaller than the entropy increase of the cold source, so there is a net decrease in entropy. (b) The process becomes feasible if work is provided to add to the energy stream. Then the increase in entropy of the hot sink can be made to cancel the entropy decrease of the hot source.

$$\frac{1}{c} = \frac{|q_h| - |q_c|}{|q_c|} = \frac{|q_h|}{|q_c|} - 1$$

We can now use the result established in the text to express this result in terms of the temperatures alone, which is possible if the transfer is performed reversibly. This substitution leads to

$$c = \frac{T_c}{T_h - T_c}$$

for the thermodynamically optimum coefficient of performance. For a refrigerator withdrawing heat from ice-cold water ($T_c = 273$ K) in a typical environment ($T_h = 293$ K), $c = 14$, so, to remove 10 kJ (enough to freeze 30 g of water), requires an input of at least 0.71 kJ of work. Practical refrigerators, of course, have a lower coefficient of performance.

The work to *maintain* a low temperature is also relevant to the design of refrigerators. No thermal insulation is perfect, so there is always a flow of heat into the sample at a rate proportional to the temperature difference. If the rate at which heat leaks in is written $A(T_h - T_c)$, where A is a constant that depends on the size of the sample and the details of the insulation, then the minimum power, P, required to maintain the original temperature difference by pumping out that heat is

$$P = \frac{1}{c} \times A(T_h - T_c) = A \times \frac{(T_h - T_c)^2}{T_c}$$

We see that the power increases as the square of the temperature difference we are trying to maintain. For this reason, air-conditioners are much more expensive to run on hot days than on mild days.

Discussion questions

1. The calculation of the work that has to be done in order to lower the temperature of an object is a bit involved because the coefficient of performance changes with the temperature of the object. (a) Find an expression for the work needed to cool an object from T_i to T_f when the refrigerator is in a room at a temperature T_h. *Hint.* Write $dw = dq/c(T)$, relate dq to dT through the heat capacity C_p, and integrate the resulting expression. Assume that the heat capacity is independent of temperature in the range of interest. (b) Use the result in part (a) to calculate the work needed to freeze 250 g of water put into a refrigerator at 293 K. How long will it take when the refrigerator operates at 100 W?

2. The same expressions that apply to the treatment of refrigerators also describe the behaviour of heat pumps, where warmth is obtained from the back of a refrigerator while its front is being used to cool the outside world. Heat pumps are popular home heating devices because they are very efficient, as we can see by performing the following calculation. Compare the heat delivered to a room at 295 K by each of two methods: (a) direct conversion of 1.00 kJ of electrical energy in an electrical heater, and (b) use of 1.00 kJ of electrical energy to run a reversible heat pump with the outside at 260 K. Discuss the origin of the difference in the heat delivered to the interior of the house by the two methods.

Further reading

R.J. Dossat, *Principles of refrigeration*. Prentice-Hall, New Jersey (1997).

F.J. Hale, Heat engines and refrigerators. In *Encyclopedia of applied physics* (ed. G.L. Trigg), **7**, 303. VCH, New York (1993).

Because eqn 4 implies that $dS_{sur} = -dq/T$, where dq is the heat supplied to the system during the process (that is, $dq_{sur} = -dq$, because the heat that enters the system comes from the surroundings), it follows that for any change

$$dS \geq \frac{dq}{T} \tag{4.13}$$

This expression is the **Clausius inequality**.

Suppose the system is isolated from its surroundings. Then $dq = 0$, and the Clausius inequality implies that

$$dS \geq 0 \tag{4.14}$$

This is exactly the characteristic we need for the entropy to be the signpost of spontaneous change, for it tells us that in an *isolated* system the entropy cannot decrease when a spontaneous change takes place.

We can illustrate the content of the Clausius inequality in two simple cases. First, suppose a system undergoes an irreversible adiabatic change. Then $dq = 0$ and, by eqn 13, $dS \geq 0$. That is, for this type of spontaneous change, the entropy of the system has increased. Because no heat flows into the surroundings, their entropy remains constant and $dS_{sur} = 0$. Therefore, the total entropy of the system and its surroundings obeys $dS_{tot} \geq 0$.

Now consider irreversible isothermal expansion of a perfect gas. As we saw in Example 4.1, for such a change $dq = -dw$ (because $dU = 0$). If the gas expands freely into a vacuum, it does no work and $dw = 0$, which implies that $dq = 0$ too. Therefore, according to the Clausius inequality, $dS \geq 0$. Next, consider the surroundings. No heat is transferred into the surroundings, so $dS_{sur} = 0$. Therefore, in this case too $dS_{tot} \geq 0$.

Another type of irreversible process is spontaneous cooling. Consider a transfer of energy as heat from one system—the hot source—at a temperature T_h to another system—the cold sink—at a temperature T_c (Fig. 4.9). When $|dq|$ leaves the hot source, the entropy changes by $-|dq|/T_h$ (a decrease). When $|dq|$ enters the cold sink the entropy changes by $+|dq|/T_c$ (an increase). The overall change in entropy is therefore

$$dS = \frac{|dq|}{T_c} - \frac{|dq|}{T_h} = |dq|\left(\frac{1}{T_c} - \frac{1}{T_h}\right) \tag{4.15}$$

which is positive (because $T_h \geq T_c$). Hence, cooling (the transfer of heat from hot to cold) is spontaneous, as we know from experience. When the temperatures of the two systems are equal, $dS_{tot} = 0$: the two systems are then at thermal equilibrium.

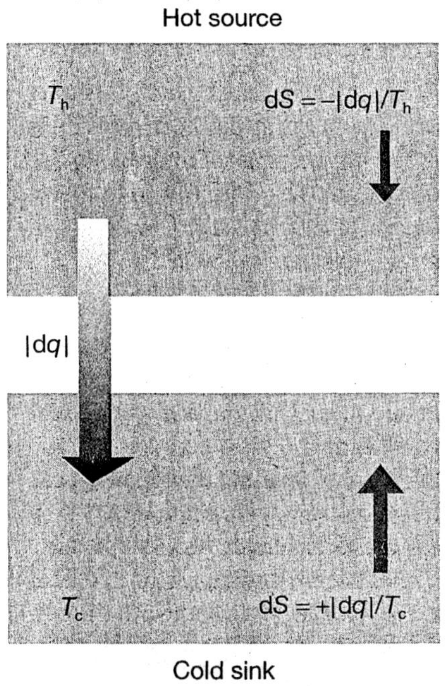

4.9 When energy leaves a hot reservoir as heat, the entropy of the reservoir decreases. When the same quantity of energy enters a cooler reservoir, the entropy increases by a larger amount. Hence, overall there is an increase in entropy and the process is spontaneous. Relative changes in entropy are indicated by the sizes of the arrows.

4.3 Entropy changes accompanying specific processes

We now see how to calculate the entropy changes that accompany a variety of basic processes.

(a) The entropy of phase transition at the transition temperature

Because a change in the degree of molecular order occurs when a substance freezes or boils, we should expect the transition to be accompanied by a change in entropy. For example, when a substance vaporizes, a compact condensed phase changes into a widely dispersed gas, and we can expect the entropy of the substance to increase

Table 4.1* Standard entropies (and temperatures) of phase transitions, $\Delta_{trs}S^{\ominus}/(\mathrm{J\,K^{-1}\,mol^{-1}})$

	Fusion (at T_f)	Vaporization (at T_b)
Argon, Ar	14.17 (at 83.8 K)	74.53 (at 87.3 K)
Benzene, C_6H_6	38.00 (at 279 K)	87.19 (at 353 K)
Water, H_2O	22.00 (at 273.15 K)	109.0 (at 373.15 K)
Helium, He	4.8 (at 1.8 K and 30 bar)	19.9 (at 4.22 K)

* More values are given in the *Data section*.

considerably. The entropy of a solid substance increases when it melts to a liquid. The entropy also increases when the liquid phase turns into a gas.

Consider a system and its surroundings at the **normal transition temperature**, T_{trs}, the temperature at which two phases are in equilibrium at 1 atm. This temperature is 0°C (273 K) for ice in equilibrium with liquid water at 1 atm, and 100°C (373 K) for liquid water in equilibrium with its vapour at 1 atm. At the transition temperature, any transfer of heat between the system and its surroundings is reversible because the two phases in the system are in equilibrium. Because at constant pressure $q = \Delta_{trs}H$, the change in molar entropy of the system is[4]

$$\Delta_{trs}S = \frac{\Delta_{trs}H}{T_{trs}} \tag{4.16}$$

If the phase transition is exothermic ($\Delta_{trs}H < 0$, as in freezing or condensing), then the entropy change is negative. This decrease in entropy is consistent with the system becoming more ordered when a solid forms from a liquid. If the transition is endothermic ($\Delta_{trs}H > 0$, as in melting), then the entropy change is positive, which is consistent with the system becoming more disordered. Melting and vaporizing are endothermic processes, so both are accompanied by an increase in the system's entropy. This increase is consistent with liquids being more disordered than solids, and gases more disordered than liquids. Table 4.1 lists some experimental entropies of transition.

Table 4.2 lists in more detail the standard entropies of vaporization of several liquids at their boiling points. An interesting feature of the data is that a wide range of liquids give approximately the same standard entropy of vaporization (about 85 J K^{-1} mol^{-1}): this empirical observation is called **Trouton's rule**.

Molecular interpretation 4.2

The explanation of Trouton's rule is that a comparable amount of disorder is generated when any liquid evaporates and becomes a gas. Hence, all liquids can be expected to have similar standard entropies of vaporization.

Liquids that show significant deviations from Trouton's rule do so on account of the molecules in the liquid being arranged in a partially orderly manner. In such

4. Recall from Section 2.7 that $\Delta_{trs}H$ is an enthalpy change per mole of substance; so $\Delta_{trs}S$ is also a molar quantity.

Table 4.2* The standard entropies of vaporization of liquids

	$\Delta_{vap}H^{\ominus}/(kJ\ mol^{-1})$	$\theta_b/°C$	$\Delta_{vap}S^{\ominus}/(J\ K^{-1}\ mol^{-1})$
Benzene	+30.8	80.1	+87.2
Carbon tetrachloride	+30.00	76.7	+85.8
Cyclohexane	+30.1	80.7	+85.1
Hydrogen sulfide	+18.7	−60.4	+87.9
Methane	+8.18	−161.5	+73.2
Water	+40.7	100.0	+109.1

* More values are given in the *Data section*.

cases, a greater change of disorder occurs when the liquid evaporates than if the molecules were highly disordered in the liquid. An example is water, where the large entropy of vaporization reflects the presence of structure arising from hydrogen-bonding in the liquid. Hydrogen bonds tend to organize the molecules in the liquid so that they are less random than, for example, the molecules in liquid hydrogen sulfide (which is not hydrogen bonded).

Methane has an unusually low entropy of vaporization. A part of the reason is that the entropy of the gas itself is slightly low (186 J K^{-1} mol^{-1} at 298 K); the entropy of N_2 under the same conditions is 192 J K^{-1} mol^{-1}. As we shall see in Chapter 19, light molecules are difficult to excite into rotation; as a result, only a few rotational states are accessible at room temperature, and the disorder associated with the population of rotational states is low.

Example 4.2 Using Trouton's rule

Predict the standard molar enthalpy of vaporization of bromine given that it boils at 59.2°C.

Method We need to judge whether there is the likelihood of anomalous structural organization in the liquid phase or some anomaly in the gas phase. If there is not, it is permissible to use Trouton's rule in the form

$$\Delta_{vap}H^{\ominus} = T_b \times (85\ J\ K^{-1}\ mol^{-1})$$

Answer There is no hydrogen bonding in liquid bromine and Br_2 is a heavy molecule that is unlikely to display unusual behaviour in the gas phase, so it is probably safe to use Trouton's rule. Substitution of the data then gives

$$\Delta_{vap}H^{\ominus} = (332.4\ K) \times (85\ J\ K^{-1}\ mol^{-1}) = 2.8 \times 10^4\ J\ mol^{-1} = 28\ kJ\ mol^{-1}$$

The experimental value is 29.45 kJ mol^{-1}.

Self-test 4.3 Predict the enthalpy of vaporization of ethane from its boiling point, −88.6°C. [16 kJ mol^{-1}]

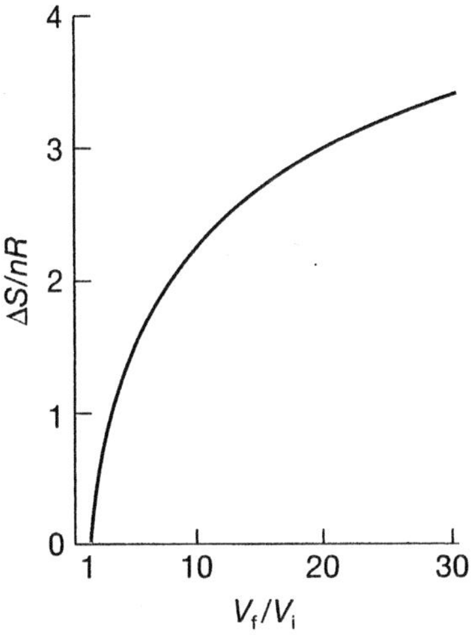

4.10 The logarithmic increase in entropy of a perfect gas as it expands isothermally.

(b) The expansion of a perfect gas

We established in Example 4.1 that the change in entropy of a perfect gas that expands isothermally from V_i to V_f is

$$\Delta S = nR\ln\frac{V_f}{V_i} \tag{4.17°}$$

Because S is a state function, the value of ΔS is independent of the path between the initial and final states, so this expression applies whether the change of state occurs reversibly or irreversibly. The logarithmic dependence of entropy on volume is illustrated in Fig. 4.10.

The *total* change in entropy, however, does depend on how the expansion takes place. If the change is reversible, the entropy change in the surroundings must be such as to give $\Delta S_{tot} = 0$. Therefore, in this case, ΔS_{sur} must be the negative of the expression in eqn 17. If the isothermal expansion occurs freely ($w = 0$) and irreversibly, then $q = 0$ (because $\Delta U = 0$). Consequently, $\Delta S_{sur} = 0$, and the total entropy change is given by eqn 17.

(c) The variation of entropy with temperature

We can use eqn 3 to calculate the entropy of a system at a temperature T_f from a knowledge of its entropy at a temperature T_i and the heat supplied to change its temperature from one value to the other:

$$S(T_f) = S(T_i) + \int_i^f \frac{dq_{rev}}{T} \tag{4.18}$$

We shall be particularly interested in the entropy change when the system is subjected to constant pressure (such as from the atmosphere) during the heating. Then, from the definition of constant-pressure heat capacity (eqn 2.27),

$$dq_{rev} = C_p\,dT$$

so long as the system is doing no non-expansion work. Consequently, at constant pressure:

$$S(T_f) = S(T_i) + \int_i^f \frac{C_p\,dT}{T} \tag{4.19}$$

The same expression applies at constant volume, but with C_p replaced by C_V. When C_p is independent of temperature in the temperature range of interest, we obtain

$$S(T_f) = S(T_i) + C_p\int_i^f \frac{dT}{T} = S(T_i) + C_p\ln\frac{T_f}{T_i} \tag{4.20}$$

with a similar expression for heating at constant volume. The logarithmic dependence of entropy on temperature is illustrated in Fig. 4.11.

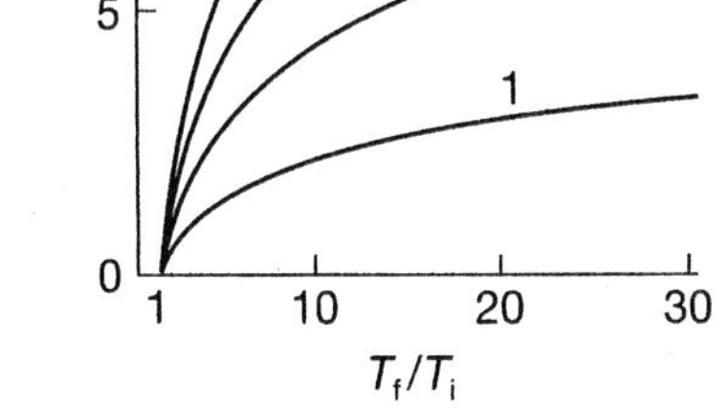

4.11 The logarithmic increase in entropy of a substance as it is heated at constant volume. Different curves correspond to different values of the constant-volume heat capacity (which is assumed constant over the temperature range) expressed as $C_{V,m}/R$.

Example 4.3 Calculating the entropy change

Calculate the entropy change when argon at 25°C and 1.00 atm in a container of volume 500 cm^3 is allowed to expand to 1000 cm^3 and is simultaneously heated to 100°C.

Method Because S is a state function, we are free to choose the most convenient path from the initial state. One such path is reversible isothermal expansion to the final volume, followed by reversible heating at constant volume to the final temperature. The entropy change in the first step is given by eqn 17 and that of the second step, provided C_V is independent of temperature, by eqn 20 (with C_V in place of C_p). In each case we need to know n, the amount of gas, and can calculate it from the perfect gas equation and the data for the initial state. The heat capacity at constant volume can be obtained from the value of $C_{p,\mathrm{m}}$ in Table 2.6 and the relation $C_{p,\mathrm{m}} - C_{V,\mathrm{m}} = R$.

Answer The amount of Ar present (from $n = pV/RT$) is 0.0204 mol. The entropy change in the first step (expansion from 500 cm^3 to 1000 cm^3 at 298 K) is

$$\Delta S = nR \ln 2.00 = +0.118\ \mathrm{J\,K^{-1}}$$

The entropy change in the second step, from 298 K to 373 K at constant volume, is

$$\Delta S = (0.0204\ \mathrm{mol}) \times (12.48\ \mathrm{J\,K^{-1}\,mol^{-1}}) \times \ln\left(\frac{373\ \mathrm{K}}{298\ \mathrm{K}}\right)$$

$$= +0.057\ \mathrm{J\,K^{-1}}$$

The overall entropy change, the sum of these two changes, is $\Delta S = +0.175\ \mathrm{J\,K^{-1}}$.

Self-test 4.4 Calculate the entropy change when the same initial sample is compressed to 50.0 cm^3 and cooled to −25°C. [−0.44 J K^{-1}]

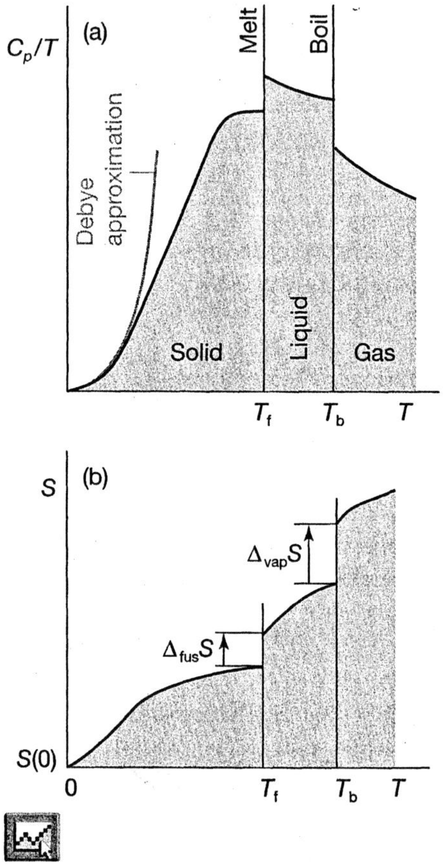

4.12 The calculation of an entropy from heat capacity data. (a) The variation of C_p/T with the temperature for a sample. (b) The entropy, which is equal to the area beneath the upper curve up to the corresponding temperature, plus the entropy of each phase transition passed.

(d) The measurement of entropy

The entropy of a system at a temperature T is related to its entropy at $T = 0$ by measuring its heat capacity C_p at different temperatures and evaluating the integral in eqn 20. The entropy of transition ($\Delta_{\mathrm{trs}}H/T_{\mathrm{trs}}$) must be added for each phase transition between $T = 0$ and the temperature of interest. For example, if a substance melts at T_f and boils at T_b, then its entropy above its boiling temperature is given by

$$S(T) = S(0) + \int_0^{T_f} \frac{C_p(\mathrm{s})\mathrm{d}T}{T} + \frac{\Delta_{\mathrm{fus}}H}{T_f} + \int_{T_f}^{T_b} \frac{C_p(\mathrm{l})\mathrm{d}T}{T} + \frac{\Delta_{\mathrm{vap}}H}{T_b} + \int_{T_b}^{T} \frac{C_p(\mathrm{g})\mathrm{d}T}{T} \qquad (4.21)$$

All the properties required, except $S(0)$, can be measured calorimetrically, and the integrals can be evaluated either graphically or, as is now more usual, by fitting a polynomial to the data and integrating the polynomial analytically. The procedure is illustrated in Fig. 4.12: the area under the curve of C_p/T against T is the integral required. Because $\mathrm{d}T/T = \mathrm{d}\ln T$, an alternative procedure is to evaluate the area under a plot of C_p against $\ln T$.

One problem with the measurement of entropy is the difficulty of measuring heat capacities near $T = 0$. There are good theoretical grounds for assuming that the heat capacity is proportional to T^3 when T is low (see Section 11.1c), and this dependence is the basis of the **Debye extrapolation**. In this method, C_p is measured down to as low a temperature as possible, and a curve of the form aT^3 is fitted to the data. That fit determines the value of a, and the expression $C_p = aT^3$ is assumed valid down to $T = 0$.

Illustration 4.2

The standard molar entropy of nitrogen gas at 25°C has been calculated from the following data:

	$S_m^{\ominus}/(\mathrm{J\,K^{-1}\,mol^{-1}})$
Debye extrapolation	1.92
Integration, from 10 K to 35.61 K	25.25
Phase transition at 35.61 K	6.43
Integration, from 35.61 K to 63.14 K	23.38
Fusion at 63.14 K	11.42
Integration, from 63.14 K to 77.32 K	11.41
Vaporization at 77.32 K	72.13
Integration, from 77.32 K to 298.15 K	39.20
Correction for gas imperfection	0.92
Total	192.06

Therefore,

$$S_m(298.15\,\mathrm{K}) = S_m(0) + 192.1\,\mathrm{J\,K^{-1}\,mol^{-1}}$$

Example 4.4 Calculating the entropy at low temperatures

The molar constant-pressure heat capacity of a certain solid at 10 K is $0.43\,\mathrm{J\,K^{-1}\,mol^{-1}}$. What is its molar entropy at that temperature?

Method Because the temperature is so low, we can assume that the heat capacity varies with temperature as aT^3, in which case we can use eqn 19 to calculate the entropy at a temperature T in terms of the entropy at $T = 0$ and the constant a. When the integration is carried out, it turns out that the result can be expressed in terms of the heat capacity at the temperature T, so the data can be used directly to calculate the entropy.

Answer The integration required is

$$S(T) = S(0) + \int_0^T \frac{aT^3\,dT}{T}$$

$$= S(0) + a\int_0^T T^2\,dT = S(0) + \tfrac{1}{3}aT^3$$

However, because aT^3 is the heat capacity at the temperature T,

$$S(T) = S(0) + \tfrac{1}{3}C_p(T)$$

from which it follows that

$$S_m(10\,\mathrm{K}) = S_m(0) + 0.14\,\mathrm{J\,K^{-1}\,mol^{-1}}$$

Self-test 4.5 For metals, there is also a contribution to the heat capacity from the electrons that is linearly proportional to T when the temperature is low. Find its contribution to the entropy at low temperatures.

[$S(T) = S(0) + C_p(T)$]

4.4 The Third Law of thermodynamics

At $T = 0$, all energy of thermal motion has been quenched, and in a perfect crystal all the atoms or ions are in a regular, uniform array. The absence of both spatial disorder and thermal motion suggests that such materials also have zero entropy. This conclusion is consistent with the molecular interpretation of entropy, because $S = 0$ if there is only one way of arranging the molecules.

(a) The Nernst heat theorem

The experimental observation that turns out to be consistent with the view that the entropy of a regular array of molecules is zero at $T = 0$ is summarized by the **Nernst heat theorem**:

The entropy change accompanying any physical or chemical transformation approaches zero as the temperature approaches zero: $S \to 0$ as $T \to 0$ provided all the substances involved are perfectly ordered.

As an example of the experimental evidence for this theorem, consider the entropy of the transition between orthorhombic sulfur, S(α), and monoclinic sulfur, S(β), which can be calculated from the transition enthalpy ($-402\ \text{J mol}^{-1}$) at the transition temperature (369 K):

$$\Delta_{\text{trs}}S = S_{\text{m}}(\alpha) - S_{\text{m}}(\beta) = \frac{(-402\ \text{J mol}^{-1})}{369\ \text{K}} = -1.09\ \text{J K}^{-1}\ \text{mol}^{-1}$$

The two individual entropies can also be determined by measuring the heat capacities from $T = 0$ up to $T = 369$ K. It is found that

$$S_{\text{m}}(\alpha) = S_{\text{m}}(\alpha, 0) + 37\ \text{J K}^{-1}\ \text{mol}^{-1} \qquad S_{\text{m}}(\beta) = S_{\text{m}}(\beta, 0) + 38\ \text{J K}^{-1}\ \text{mol}^{-1}$$

These two values imply that at the transition temperature

$$\Delta_{\text{trs}}S = S_{\text{m}}(\alpha, 0) - S_{\text{m}}(\beta, 0) - 1\ \text{J K}^{-1}\ \text{mol}^{-1}$$

On comparing this value with the one above, we conclude that

$$S_{\text{m}}(\alpha, 0) - S_{\text{m}}(\beta, 0) \approx 0$$

in accord with the theorem.

It follows from the Nernst theorem that, if we arbitrarily ascribe the value zero to the entropies of elements in their perfect crystalline form at $T = 0$, then all perfect crystalline compounds also have zero entropy at $T = 0$ (because the change in entropy that accompanies the formation of the compounds, like the entropy of all transformations at that temperature, is zero). This conclusion is summarized by the **Third Law of thermodynamics**:

The entropy of all perfect crystalline substances is zero at $T = 0$.

As far as thermodynamics is concerned, choosing this common value as zero is then a matter of convenience. The molecular interpretation of entropy, however, justifies the value $S = 0$ at $T = 0$. The Third Law has important implications for the attainment of very low temperatures (Box 4.2).

(b) Third-Law entropies

Entropies reported on the basis that $S(0) = 0$ are called **Third-Law entropies** (and often just 'entropies'). When the substance is in its standard state at the temperature T, the **standard (Third-Law) entropy** is denoted $S^{\ominus}(T)$. A list of values at 298 K is given in Table 4.3.

Table 4.3* Standard Third-Law entropies at 298 K

	$S_{\text{m}}^{\ominus}/(\text{J K}^{-1}\ \text{mol}^{-1})$
Solids:	
Graphite, C(s)	5.7
Diamond, C(s)	2.4
Sucrose, $C_{12}H_{22}O_{11}$(s)	360.2
Iodine, I_2(s)	116.1
Liquids:	
Benzene, C_6H_6(l)	173.3
Water, H_2O(l)	69.9
Mercury, Hg(l)	76.0
Gases:	
Methane, CH_4(g)	186.3
Carbon dioxide, CO_2(g)	213.7
Hydrogen, H_2(g)	130.7
Helium, He(g)	126.2
Ammonia, NH_3(g)	192.4

* More values are given in the *Data section*.

Box 4.2 Reaching very low temperatures

The world record low temperature stands at about 280 pK. Gases may be cooled by Joule–Thomson expansion below their inversion temperature, and temperatures lower than 4 K (the boiling point of helium) can be reached by the evaporation of liquid helium by pumping rapidly through large diameter pipes. Temperatures as low as about 1 K can be reached in this way, but at lower temperatures helium is insufficiently volatile for this procedure to be effective; moreover, the superfluid phase begins to interfere with the cooling process by creeping round the apparatus.

Common methods used to reach very low temperatures include laser cooling (see *Discussion question 2*) and *adiabatic demagnetization*. In the absence of a magnetic field, the unpaired electrons of a paramagnetic material are orientated at random, but in the presence of a magnetic field there are more β spins ($m_s = -\frac{1}{2}$; these terms are explained in Section 12.8) than α spins ($m_s = +\frac{1}{2}$). In thermodynamic terms, the application of a magnetic field lowers the entropy of a sample (see the illustration), and at a given temperature, the entropy of a sample is lower when the field is on than when it is off.

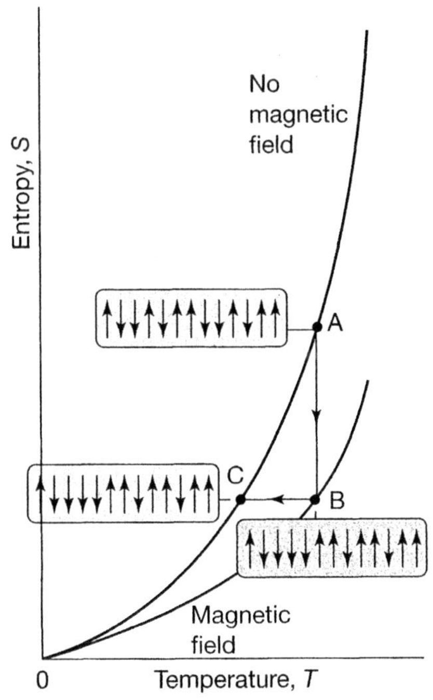

The technique of adiabatic demagnetization is used to attain very low temperatures. The upper curve shows the variation of the entropy of a paramagnetic system in the absence of an applied field. The lower curve shows the variation in entropy when a field is applied and has made the electron magnets more orderly. The isothermal magnetization step is from A to B; the adiabatic demagnetization step (at constant entropy) is from B to C.

A sample of paramagnetic material, such as a *d*- or *f*-metal complex, is cooled to about 1 K by using helium. Gadolinium(III) sulfate octahydrate, $Gd_2(SO_4)_3 \cdot 8H_2O$, has been used because each gadolinium ion carries several unpaired electrons but is separated from its neighbours by a coordination sphere of hydrating H_2O molecules. The sample is then exposed to a strong magnetic field while it is surrounded by helium, which provides thermal contact with the cold reservoir. This magnetization step is isothermal, and heat leaves the sample as the electron spins adopt the lower energy state (AB in the illustration). Thermal contact between the sample and the surroundings is now broken by pumping away the helium and the magnetic field is reduced to zero. This step is adiabatic and effectively reversible, so the state of the sample changes from B to C. At the end of this step the sample is the same as it was at A except that it now has a lower entropy. That lower entropy in the absence of a magnetic field corresponds to a lower temperature. That is, adiabatic demagnetization has cooled the sample.

Even lower temperatures can be reached if nuclear spins (which also behave like small magnets) are used instead of electron spins in the technique of *adiabatic nuclear demagnetization*. This technique was used to reach the current world record (in silver) of 280 pK.

Discussion questions

1. Suppose the two curves in the illustration could be expressed as aT^3 (field off) and bT^3 (field on). How many magnetization–demagnetization cycles would be needed to reduce the temperature of a sample of mass 0.10 g from (a) 1 K to 1 mK, (b) 1 K to 1 nK? Take $a = 6.2$ kJ K^{-4} g^{-1} and $b = 5.1$ kJ K^{-4} g^{-1}.

2. A group of 2000 rubidium atoms has been cooled to 20 nK. In the optical trapping technique, atoms in the gas phase are cooled by inelastic collisions with photons from intense laser beams, which act as walls of a very small container. (a) Assuming perfect gas behaviour, calculate the root mean square speed of and the pressure exerted by 1000 rubidium atoms trapped by laser beams in a volume of 1.0×10^{-15} m^3 at 0.10 mK (see Section 24.1). (b) What is the de Broglie wavelength of a rubidium atom travelling at the root mean square speed calculated in part (a) (see Section 11.2)?

Further reading

A.T. Serlemitsos, B.A. Warner, and S. Castles, Adiabatic demagnetization refrigerator for space use. *Advances in Cryogenics* **35B**, 1431 (1990).

C.N. Cohen-Tannoudji and W.D. Phillips, New mechanisms for laser cooling. *Physics Today*, 33 (October 1990).

The **standard reaction entropy**, $\Delta_r S^\ominus$, is defined, like the standard reaction enthalpy, as the difference between the molar entropies of the pure, separated products and the pure, separated reactants, all substances being in their standard states at the specified temperature:

$$\Delta_r S^\ominus = \sum_{\text{Products}} \nu S_m^\ominus - \sum_{\text{Reactants}} \nu S_m^\ominus \tag{4.22}$$

In this expression, each term is weighted by the appropriate stoichiometric coefficient. Standard reaction entropies are likely to be positive if there is a net formation of gas in a reaction, and are likely to be negative if there is a net consumption of gas.

Illustration 4.3

To calculate the standard reaction entropy of $H_2(g) + \frac{1}{2}O_2(g) \rightarrow H_2O(l)$ at 25°C, we use the data in Table 2.6 of the *Data section* to write

$$\begin{aligned}\Delta_r S^\ominus &= S_m^\ominus(H_2O, l) - \{S_m^\ominus(H_2, g) + \tfrac{1}{2}S_m^\ominus(O_2, g)\}\\ &= 69.9\ \text{J K}^{-1}\ \text{mol}^{-1} - \{130.7 + \tfrac{1}{2}(205.1)\}\ \text{J K}^{-1}\ \text{mol}^{-1}\\ &= -163.4\ \text{J K}^{-1}\ \text{mol}^{-1}\end{aligned}$$

The negative value is consistent with the conversion of two gases to a compact liquid.

Self-test 4.6 Calculate the standard reaction entropy for the combustion of methane to carbon dioxide and liquid water at 25°C. [$-243\ \text{J K}^{-1}\ \text{mol}^{-1}$]

Concentrating on the system

Entropy is the basic concept for discussing the direction of natural change, but to use it we have to analyse changes in both the system and its surroundings. We have seen that it is always very simple to calculate the entropy change in the surroundings, and we shall now see that it is possible to devise a simple method for taking that contribution into account automatically. This approach focuses our attention on the system and simplifies discussions. Moreover, it is the foundation of all the applications of chemical thermodynamics that follow.

4.5 The Helmholtz and Gibbs energies

Consider a system in thermal equilibrium with its surroundings at a temperature T. When a change in the system occurs and there is a transfer of energy as heat between the system and the surroundings, the Clausius inequality, eqn 13, reads

$$dS - \frac{dq}{T} \geq 0 \tag{4.23}$$

We can develop this inequality in two ways according to the conditions (of constant volume or constant pressure) under which the process occurs.

(a) Criteria for spontaneity

First, consider heat transfer at constant volume. Then, in the absence of non-expansion work, we can write $dq_V = dU$; consequently

$$dS - \frac{dU}{T} \geq 0 \qquad (4.24)$$

The importance of the inequality in this form is that it expresses the criterion for spontaneous change solely in terms of the state functions of the system. The inequality is easily rearranged to

$$T\,dS \geq dU \quad \text{(constant } V\text{, no additional work)}^{5} \qquad (4.25)$$

At either constant internal energy ($dU = 0$) or constant entropy ($dS = 0$), this expression becomes, respectively,

$$dS_{U,V} \geq 0 \qquad dU_{S,V} \leq 0 \qquad (4.26)$$

where the subscripts indicate the constant conditions.

Equation 26 expresses the criteria for spontaneous change in terms of properties relating to the system. The first inequality states that, in a system at constant volume and constant internal energy (such as an isolated system), the entropy increases in a spontaneous change. That statement is essentially the content of the Second Law. The second inequality is less obvious, for it says that, if the entropy and volume of the system are constant, then the internal energy must decrease in a spontaneous change. Do not interpret this criterion as a tendency of the system to sink to lower energy. It is a disguised statement about entropy, and should be interpreted as implying that, if the entropy of the system is unchanged, then there must be an increase in entropy of the surroundings, which can be achieved only if the energy of the system decreases as energy flows out as heat.

When heat is transferred at constant pressure, and there is no work other than expansion work, we can write $dq_p = dH$ and obtain

$$T\,dS \geq dH \quad \text{(constant } p\text{, no additional work)} \qquad (4.27)$$

At either constant enthalpy or constant entropy this inequality becomes, respectively,

$$dS_{H,p} \geq 0 \qquad dH_{S,p} \leq 0 \qquad (4.28)$$

The interpretations of these inequalities are similar to those of eqn 26. The entropy of the system at constant pressure must increase if its enthalpy remains constant (for there can then be no change in entropy of the surroundings). Alternatively, the enthalpy must decrease if the entropy of the system is constant, for then it is essential to have an increase in entropy of the surroundings.

Because eqns 25 and 27 have the forms $dU - T\,dS \leq 0$ and $dH - T\,dS \leq 0$, respectively, they can be expressed more simply by introducing two more thermodynamic quantities. One is the **Helmholtz energy**, A, which is defined as

$$A = U - TS \qquad [4.29]$$

The other is the **Gibbs energy**, G:

$$G = H - TS \qquad [4.30]$$

All the symbols in these two definitions refer to the system.

5. Recall that 'additional work' is work other than expansion work.

When the state of the system changes at constant temperature, the two properties change as follows:

$$\text{(a)}\ dA = dU - T\,dS \qquad \text{(b)}\ dG = dH - T\,dS \tag{4.31}$$

When we introduce eqns 25 and 27, respectively, we obtain the criteria of spontaneous change as

$$\text{(a)}\ dA_{T,V} \le 0 \qquad \text{(b)}\ dG_{T,p} \le 0 \tag{4.32}$$

These inequalities are the most important conclusions from thermodynamics for chemistry. They are developed in subsequent sections and chapters.

(b) Some remarks on the Helmholtz energy

A change in a system at constant temperature and volume is spontaneous if $dA_{T,V} \le 0$. That is, a change under these conditions is spontaneous if it corresponds to a decrease in the Helmholtz energy. Such systems move spontaneously towards states of lower A if a path is available. The criterion of equilibrium, when neither the forward nor reverse process has a tendency to occur, is

$$dA_{T,V} = 0 \tag{4.33}$$

The expressions $dA = dU - T\,dS$ and $dA < 0$ are sometimes interpreted as follows. A negative value of dA is favoured by a negative value of dU and a positive value of $T\,dS$. This observation suggests that the tendency of a system to move to lower A is due to its tendency to move towards states of lower internal energy and higher entropy. However, this interpretation is false (even though it is a good rule of thumb for remembering the expression for dA) because the tendency to lower A is solely a tendency towards states of greater overall entropy. Systems change spontaneously if in doing so the total entropy of the system and its surroundings increases, not because they tend to lower internal energy. The form of dA may give the impression that systems favour lower energy, but that is misleading: dS is the entropy change of the system, $-dU/T$ is the entropy change of the surroundings (when the volume of the system is constant), and their total tends to a maximum.

(c) Maximum work

It turns out that A carries a greater significance than being simply a signpost of spontaneous change: *the change in the Helmholtz function is equal to the maximum work accompanying a process*:

$$dw_{max} = dA \tag{4.34}$$

As a result, A is sometimes called the 'maximum work function', or the 'work function'.[6]

Justification 4.2

First we prove that a system does maximum work when it is working reversibly. (This conclusion was demonstrated in Section 2.3e for the expansion of a perfect gas; now we prove its universal validity.) We combine the Clausius inequality $dS \ge dq/T$ in the form $T\,dS \ge dq$ with the First Law, $dU = dq + dw$, and obtain

$$dU \le T\,dS + dw$$

6. *Arbeit* is the German word for work; hence the symbol A.

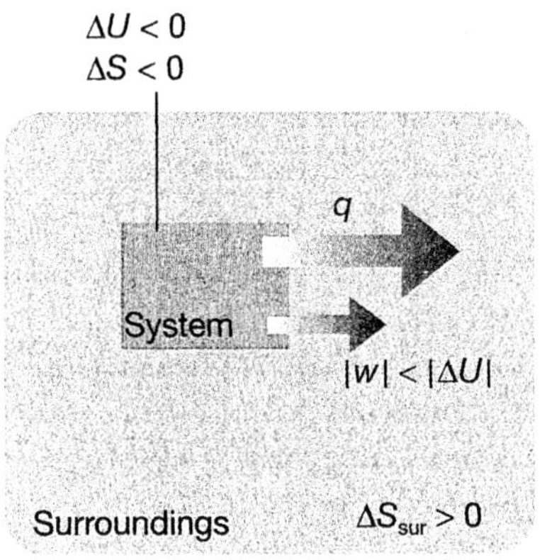

4.13 In a system not isolated from its surroundings, the work done may be different from the change in internal energy. Moreover, the process is spontaneous if overall the entropy of the global, isolated system (the system plus its surroundings) increases. In the process depicted here, the entropy of the system decreases, so that of the surroundings must increase in order for the process to be spontaneous, which means that energy must pass from the system to the surroundings as heat. Therefore, less work than ΔU can be obtained.

(dU is smaller than the term of the right because we are replacing dq by $T\,dS$, which in general is larger.) This expression rearranges to

$$dw \geq dU - T\,dS$$

It follows that the most negative value of dw, and therefore the maximum energy that can be obtained from the system as work, is given by

$$dw_{max} = dU - T\,dS$$

and that this work is done only when the path is traversed reversibly (because then the equality applies). Because at constant temperature $dA = dU - T\,dS$, we conclude that $dw_{max} = dA$.

When a macroscopic isothermal change takes place in the system, eqn 34 becomes

$$w_{max} = \Delta A \tag{4.35}$$

with

$$\Delta A = \Delta U - T\Delta S \tag{4.36}$$

This expression shows that in some cases, depending on the sign of $T\Delta S$, not all the change in internal energy may be available for doing work. If the change occurs with a decrease in entropy (of the system), in which case $T\Delta S < 0$, then the right-hand side of this equation is not as negative as ΔU itself, and consequently the maximum work is less than ΔU. For the change to be spontaneous, some of the energy must escape as heat in order to generate enough entropy in the surroundings to overcome the reduction in entropy in the system (Fig. 4.13). In this case, Nature is demanding a tax on the internal energy as it is converted into work. This is the origin of the alternative name 'Helmholtz free energy' for A, because ΔA is that part of the change in internal energy that we are free to use to do work.

Molecular interpretation 4.3

Further insight into the relation between the work that a system can do and the Helmholtz energy is to recall that work is energy transferred to the surroundings as the uniform motion of atoms. We can interpret the expression $A = U - TS$ as showing that A is the total internal energy of the system, U, less a contribution that is stored in a disordered manner (the quantity TS). Because energy stored in a disordered manner cannot be used to achieve uniform motion in the surroundings, only the part of U that is not stored in a disordered way, the quantity $U - TS$, is available for conversion into work.

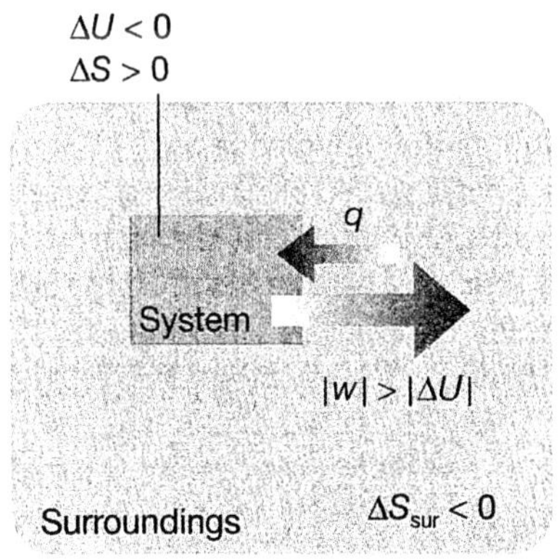

4.14 In this process, the entropy of the system increases; hence we can afford to lose some entropy of the surroundings. That is, some of their energy may be lost as heat to the system. This energy can be returned to them as work. Hence the work done can exceed ΔU.

If the change occurs with an increase of entropy of the system (in which case $T\Delta S > 0$), the right-hand side of the equation is more negative than ΔU. In this case, the maximum work that can be obtained from the system is greater than ΔU. The explanation of this apparent paradox is that the system is not isolated and energy may flow in as heat as work is done. Because the entropy of the system increases, we can afford a reduction of the entropy of the surroundings yet still have, overall, a spontaneous process. Therefore, some heat (no more than the value of $T\Delta S$) may leave the surroundings and contribute to the work the change is generating (Fig. 4.14). Nature is now providing a tax refund.

Example 4.5 **Calculating the maximum available work**

When 1.000 mol $C_6H_{12}O_6$ (glucose) is oxidized to carbon dioxide and water at 25°C according to the equation $C_6H_{12}O_6(s) + 6\,O_2(g) \rightarrow 6\,CO_2(g) + 6\,H_2O(l)$, calorimetric measurements give $\Delta_r U^\ominus = -2808\ \text{kJ mol}^{-1}$ and $\Delta_r S = +182.4\ \text{J K}^{-1}\ \text{mol}^{-1}$ at 25°C. How much of this energy change can be extracted as (a) heat at constant pressure, (b) work?

Method We know that the heat released at constant pressure is equal to the value of ΔH, so we need to relate $\Delta_r H^\ominus$ to $\Delta_r U^\ominus$, which is given. To do so, we suppose that all the gases involved are perfect, and use eqn 2.26 in the form $\Delta_r H = \Delta_r U + \Delta \nu_g RT$. For the maximum work available from the process we use eqn 35.

Answer (a) Because $\Delta \nu_g = 0$, we know that $\Delta_r H^\ominus = \Delta_r U^\ominus = -2808\ \text{kJ mol}^{-1}$. Therefore, at constant pressure, the energy available as heat is 2808 kJ mol^{-1}. (b) Because $T = 298$ K, the value of $\Delta_r A^\ominus$ is

$$\Delta_r A^\ominus = \Delta_r U^\ominus - T\Delta_r S^\ominus = -2862\ \text{kJ mol}^{-1}$$

Therefore, the combustion of 1.000 mol $C_6H_{12}O_6$ can be used to produce up to 2862 kJ of work.

Comment The maximum work available is greater than the change in internal energy on account of the positive entropy of reaction (which is partly due to the generation of a large number of small molecules from one big one). The system can therefore draw in energy from the surroundings (so reducing their entropy) and make it available for doing work.

Self-test 4.7 Repeat the calculation for the combustion of 1.000 mol $CH_4(g)$ under the same conditions, using data from Table 2.5 and Self-test 4.6.

[$|q_p| = 890$ kJ, $|w_{max}| = 813$ kJ]

ΔH negative
= exothermic
ΔS positive
= more disorder
= " spont.
ΔG negative
spontaneous

(d) Some remarks on the Gibbs energy

The Gibbs energy (the 'free energy') is more common in chemistry than the Helmholtz energy because, at least in laboratory chemistry, we are usually more interested in changes occurring at constant pressure than at constant volume. The criterion $dG_{T,p} \le 0$ carries over into chemistry as the observation that, *at constant temperature and pressure, chemical reactions are spontaneous in the direction of decreasing Gibbs energy*. Therefore, if we want to know whether a reaction is spontaneous, the pressure and temperature being constant, we assess the change in the Gibbs energy. If G decreases as the reaction proceeds, then the reaction has a spontaneous tendency to convert the reactants into products. If G increases, then the reverse reaction is spontaneous.

The existence of spontaneous endothermic reactions provides an illustration of the role of G. In such reactions, H increases, the system rises spontaneously to states of higher enthalpy, and $dH > 0$. Because the reaction is spontaneous we know that $dG < 0$ despite $dH > 0$; it follows that the entropy of the system increases so much that $T\,dS$ is strongly positive and outweighs dH in $dG = dH - T\,dS$. Endothermic reactions are therefore driven by the increase of entropy of the system, and this entropy change overcomes the reduction of entropy brought about in the surroundings by the inflow of heat into the system ($dS_{sur} = -dH/T$ at constant pressure).

(e) Maximum non-expansion work

The analogue of the maximum work interpretation of ΔA, and the origin of the name free energy, can be found for ΔG. In the *Justification* below, we show that, at constant temperature and pressure, the maximum additional (non-expansion) work, $w_{add,max}$, is given by the change in Gibbs energy:

$$dw_{add,max} = dG \tag{4.37}$$

The corresponding expression for a measurable change is

$$w_{add,max} = \Delta G \tag{4.38}$$

This expression is particularly useful for assessing the electrical work that may be produced by fuel cells and electrochemical cells, and we shall see many applications of it.

Justification 4.3

Because $H = U + pV$, for a general change in conditions, the change in enthalpy is

$$dH = dq + dw + d(pV)$$

The corresponding change in Gibbs energy ($G = H - TS$) is

$$dG = dH - T\,dS - S\,dT = dq + dw + d(pV) - T\,dS - S\,dT$$

When the change is isothermal,

$$dG = dq + dw + d(pV) - T\,dS$$

When the change is reversible, $dw = dw_{rev}$ and $dq = dq_{rev} = T\,dS$, so for a reversible, isothermal process

$$dG = T\,dS + dw_{rev} + d(pV) - T\,dS = dw_{rev} + d(pV)$$

The work consists of expansion work, which for a reversible change is given by $-p\,dV$, and possibly some other kind of work (for instance, the electrical work of pushing electrons through a circuit or of raising a column of liquid); this additional work we denote dw_{add}. Therefore, with $d(pV) = p\,dV + V\,dp$,

$$dG = (-p\,dV + dw_{add,rev}) + p\,dV + V\,dp = dw_{add,rev} + V\,dp$$

If the change occurs at constant pressure (as well as constant temperature), the last term disappears and $dG = dw_{add,rev}$. Therefore, at constant temperature and pressure, $dw_{add,rev} = dG$. However, because the process is reversible, the work done must now have its maximum value, so eqn 37 follows.

Example 4.6 **Calculating the maximum non-expansion work of a reaction**

How much energy is available for sustaining muscular and nervous activity from the combustion of 1.00 mol of glucose molecules under standard conditions at 37°C (blood temperature)? The standard entropy of reaction is $+182.4\ \mathrm{J\,K^{-1}\,mol^{-1}}$.

Method The non-expansion work available from the reaction is equal to the change in standard Gibbs energy for the reaction ($\Delta_r G^\ominus$, a quantity defined more fully below). To calculate this quantity, it is legitimate to ignore the temperature dependence of the reaction enthalpy, to obtain $\Delta_r H^\ominus$ from Table 2.5, and to substitute the data into $\Delta_r G^\ominus = \Delta_r H^\ominus - T\Delta_r S^\ominus$.

Answer Because the standard reaction enthalpy is −2808 kJ mol^{-1}, it follows that the standard reaction Gibbs energy is

$$\Delta_r G^\ominus = -2808\ \text{kJ mol}^{-1} - (310\ \text{K}) \times (182.4\ \text{J K}^{-1}\ \text{mol}^{-1}) = -2865\ \text{kJ mol}^{-1}$$

Therefore, $w_{\text{add,max}} = -2865$ kJ for the combustion of 1 mol glucose molecules, and the reaction can be used to do up to 2865 kJ of non-expansion work.

Comment A person of mass 70 kg would need to do 2.1 kJ of work to climb vertically through 3.0 m; therefore, at least 0.13 g of glucose is needed to complete the task (and in practice significantly more).

Self-test 4.8 How much non-expansion work can be obtained from the combustion of 1.00 mol CH_4(g) under standard conditions at 298 K? Use $\Delta_r S^\ominus = -243\ \text{J K}^{-1}\ \text{mol}^{-1}$. [818 kJ]

4.6 Standard molar Gibbs energies

Standard entropies and enthalpies of reaction can be combined to obtain the **standard Gibbs energy of reaction, $\Delta_r G^\ominus$**:

$$\Delta_r G^\ominus = \Delta_r H^\ominus - T\Delta_r S^\ominus \qquad [4.39]$$

The standard Gibbs energy of reaction is the difference in standard molar Gibbs energies of the products and reactants in their standard states at the temperature specified for the reaction as written. As in the case of standard reaction enthalpies, it is convenient to define the **standard Gibbs energy of formation**, $\Delta_f G^\ominus$ (or 'standard reaction Gibbs energy'), the standard reaction Gibbs energy for the formation of a compound from its elements in their reference states.[7] Standard free energies of formation of the elements in their reference states are zero, because their formation is a 'null' reaction. A selection of values for compounds is given in Table 4.4. From the values there, it is a simple matter to obtain the standard Gibbs energy of reaction by taking the appropriate combination:

$$\Delta_r G^\ominus = \sum_{\text{Products}} \nu\Delta_f G^\ominus_m - \sum_{\text{Reactants}} \nu\Delta_f G^\ominus_m \qquad (4.40)$$

with each term weighted by the appropriate stoichiometric coefficient.

Table 4.4* Standard Gibbs energies of formation at 298 K

	$\Delta_f G^\ominus/(\text{kJ mol}^{-1})$
Diamond, C(s)	+2.9
Benzene, C_6H_6(l)	+124.3
Methane, CH_4(g)	−50.7
Carbon dioxide, CO_2(g)	−394.4
Water, H_2O(l)	−237.1
Ammonia, NH_3(g)	−16.5
Sodium chloride, NaCl(s)	−384.1

* More values are given in the *Data section*.

Illustration 4.4

To calculate the standard Gibbs energy of the reaction $CO(g) + \frac{1}{2}O_2(g) \rightarrow CO_2(g)$ at 25°C, we write

$$\begin{aligned}\Delta_r G^\ominus &= \Delta_f G^\ominus(CO_2, g) - \{\Delta_f G^\ominus(CO, g) + \tfrac{1}{2}\Delta_f G^\ominus(O_2, g)\}\\ &= -394.4\ \text{kJ mol}^{-1} - \{(-137.2) + \tfrac{1}{2}(0)\}\ \text{kJ mol}^{-1} = -257.2\ \text{kJ mol}^{-1}\end{aligned}$$

Self-test 4.9 Calculate the standard reaction Gibbs energy for the combustion of CH_4(g) at 298 K. [−818 kJ mol^{-1}]

7. The reference state of an element was defined in Section 2.8.

The web site is a good source of links to online databases of thermochemical data.

Calorimetry (for ΔH directly, and for S via heat capacities) is only one of the ways of determining the values of Gibbs energies. They may also be obtained from equilibrium constants (Chapter 9) and electrochemical measurements (Chapter 10), and they may be calculated using data from spectroscopic observations (Chapter 20). The information in Tables 2.5 and 2.6 of the *Data section*, however, together with the machinery we shall now construct, is all we need in order to draw far-reaching conclusions about reactions and other processes of interest in chemistry.

Checklist of key ideas

- ☐ spontaneous direction of change, the direction of change that does not require work to be done to bring the change about
- ☐ Kelvin statement of the Second Law of thermodynamics: No process is possible in which the sole result is the absorption of heat from a reservoir and its complete conversion into work

The direction of spontaneous change

4.1 The dispersal of energy

- ☐ Spontaneity and the rise of disorder
- ☐ Collapse into disorder as the driving force of change

4.2 Entropy

- ☐ Second Law in terms of entropy: The entropy of an isolated system increases in the course of a spontaneous change: $\Delta S_{tot} > 0$
- ☐ thermodynamic definition of entropy, $dS = dq_{rev}/T$
- ☐ entropy change in the surroundings, $\Delta S_{sur} = q_{sur}/T_{sur}$
- ☐ entropy change for an adiabatic process, $\Delta S_{sur} = 0$
- ☐ Carnot cycle, a cycle composed of alternating isothermal and adiabatic reversible expansions and compressions
- ☐ efficiency of a heat engine, $\varepsilon = |w|/q_h$
- ☐ Carnot efficiency, $\varepsilon_{rev} = 1 - T_c/T_h$
- ☐ proof that entropy is a state function by showing that $\oint (1/T)dq_{rev} = 0$
- ☐ thermodynamic temperature scale, $T = (1 - \varepsilon)T_h$
- ☐ Kelvin scale, a thermodynamic temperature scale in which the triple point of water is set at 273.16 K
- ☐ Clausius inequality, $dS \geq dq/T$

4.3 Entropy changes accompanying specific processes

- ☐ normal transition temperature, T_{trs}, the temperature at which two phases are in equilibrium at 1 atm
- ☐ Entropy of transition at the transition temperature, $\Delta_{trs}S = \Delta_{trs}H/T_{trs}$
- ☐ Trouton's rule: many normal liquids have approximately the same standard entropy of vaporization (about 85 J K^{-1} mol^{-1})
- ☐ Entropy of isothermal expansion of a perfect gas, $\Delta S = nR \ln (V_f/V_i)$
- ☐ Variation of entropy with temperature,

$$S(T_f) = S(T_i) + \int_{T_i}^{T_f} (C_p/T)dT$$

- ☐ Measurement of entropy, from area under graph of C_p/T against T
- ☐ Debye extrapolation, $C_p = aT^3$ as $T \to 0$

4.4 The Third Law of thermodynamics

- ☐ Nernst heat theorem: The entropy change accompanying any physical or chemical transformation approaches zero as the temperature approaches zero: $\Delta S \to 0$ as $T \to 0$ provided all the substances involved are perfectly ordered
- ☐ Third Law of thermodynamics: The entropy of all perfect crystalline substances is zero at $T = 0$
- ☐ Third-Law entropy, an entropy reported on the basis that $S(0) = 0$
- ☐ standard reaction entropy, $\Delta_r S^\ominus = \sum_{Products} \nu S_m^\ominus - \sum_{Reactants} \nu S_m^\ominus$

Concentrating on the system

4.5 The Helmholtz and Gibbs energies

- ☐ criteria of spontaneity, $dS_{U,V} \geq 0$ and $dU_{S,V} \leq 0$
- ☐ Helmholtz energy, $A = U - TS$
- ☐ Gibbs energy, $G = H - TS$
- ☐ criteria of spontaneity, $dA_{T,V} \leq 0$ and $dG_{T,p} \leq 0$
- ☐ criterion of equilibrium at constant temperature and volume, $dA_{T,V} = 0$
- ☐ maximum work and the Helmholtz energy, $w_{max} = \Delta A$
- ☐ the criterion of equilibrium at constant temperature and pressure, $dG_{T,p} = 0$
- ☐ maximum additional (non-expansion) work and the Gibbs energy, $w_{add,max} = \Delta G$

4.6 Standard molar Gibbs energies

- ☐ standard Gibbs energy of reaction, $\Delta_r G^\ominus = \Delta_r H^\ominus - T\Delta_r S^\ominus = \sum_{Products} \nu G_m^\ominus - \sum_{Reactants} \nu G_m^\ominus$
- ☐ standard Gibbs energy of formation ($\Delta_f G^\ominus$), the standard reaction Gibbs energy for the formation of a compound from its elements in their reference states
- ☐ expressing $\Delta_r G^\ominus$ in terms of $\Delta_f G^\ominus$, $\Delta_r G^\ominus = \sum_{Products} \nu \Delta_f G^\ominus - \sum_{Reactants} \nu \Delta_f G^\ominus$

Further reading

Articles of general interest

M. Barón, With Clausius from energy to entropy. *J. Chem. Educ.* **66**, 1001 (1989).

N.C. Craig, Entropy analyses of four familiar processes. *J. Chem. Educ.* **65**, 760 (1988).

N.C. Craig, Entropy diagrams. *J. Chem. Educ.* **73**, 710 (1996).

W.H. Cropper, Walther Nernst and the last law. *J. Chem. Educ.* **64**, 3 (1987).

P. Djurdjevic and I. Gutman, A simple method for showing that entropy is a function of state. *J. Chem. Educ.* **65**, 399 (1988).

H. Frohlich, On the entropy change of surroundings of finite and infinite size. *J. Chem. Educ.* **73**, 716 (1996).

D.F.R. Gilson, Order end disorder and entropies of fusion. *J. Chem. Educ.* **69**, 23 (1992).

L. Glasser, Order, chaos, and all that! *J. Chem. Educ.* **66**, 997 (1989).

H.B. Hollinger and M.J. Zenzen, Thermodynamic irreversibility: 1. What is it? *J. Chem. Educ.* **68**, 31 (1991).

F.L. Lambert, Shuffled cards, messy desks, and disorderly dorm rooms—examples of entropy increase? Nonsense! *J. Chem. Educ.* **76**, 1385 (1999).

E.F. Meyer, The Carnot cycle revisited. *J. Chem. Educ.* **65**, 873 (1988).

P.G. Nelson, Derivation of the Second Law of thermodynamics from Boltzmann's distribution law. *J. Chem. Educ.* **65**, 390 (1988).

R.S. Ochs, Thermodynamics and spontaneity. *J. Chem. Educ.* **73**, 952 (1996).

K. Seidman and T.R. Michalik, The efficiency of reversible heat engines: The possible misinterpretation of a corollary to Carnot's theorem. *J. Chem. Educ.* **68**, 208 (1991).

J. Waser and V. Schomaker, A note on thermodynamic inequalities. *J. Chem. Educ.* **65**, 393 (1988).

Texts and sources of data and information

P.W. Atkins, *The second law*. Scientific American Books, New York (1984).

K.E. Bett, J.S. Rowlinson, and G. Saville, *Thermodynamics for chemical engineers*. MIT Press, Cambridge (1975).

J.B. Fenn, *Engines, energy, and entropy*. W.H. Freeman and Co., New York (1982).

J. Harte, *Consider a spherical cow: a course in environmental problem solving*. University Science Books, Sausalito (1988).

D.R. Gaskell, *Introduction to the thermodynamics of materials*. Taylor and Francis, London (1995).

D. Kondepudi and I. Prigogine, *Modern thermodynamics: from heat engines to dissipative structures*. Wiley, New York (1998).

H.-G. Lee, *Chemical thermodynamics for metals and materials*. Imperial College Press, London (1999).

J.M. Smith and H.C. Van Ness, *Introduction to chemical engineering thermodynamics*. McGraw–Hill, New York (1987).

Exercises

Discussion questions

4.1 (a) The evolution of life requires the organization of a very large number of molecules into biological cells. Does the formation of living organisms violate the Second Law of thermodynamics? State your conclusion clearly and present detailed arguments to support it.

4.1 (b) Justify Trouton's rule.

4.2 (a) Distinguish between a heat engine, a refrigerator, and a heat pump.

4.2 (b) You received an unsolicited proposal from a self-declared inventor who is seeking investors for the development of his latest idea: a device that uses heat extracted from the ground by a heat pump to boil water into steam that is used to heat a home and to power a steam engine that drives the heat pump. This procedure is potentially very lucrative because, after an initial extraction of energy from the ground, no fossil fuels would be required to keep the device running indefinitely. Would you invest in this idea? State your conclusion clearly and present detailed arguments to support it.

4.3 (a) The following expressions have been used to establish criteria for spontaneous change: $\Delta S_{tot} > 0$, $dS_{U,V} > 0$ and $dU_{S,V} < 0$, $dA_{T,V} < 0$, and $dG_{T,p} < 0$. Discuss the origin, significance, and applicability of each criterion.

4.3 (b) The following expressions have been used to establish criteria for equilibrium: $dA_{T,V} = 0$ and $dG_{T,p} = 0$. Discuss the origin, significance, and applicability of each criterion.

Numerical exercises

Assume that all gases are perfect and that data refer to 298.15 K unless otherwise stated.

4.4 (a) Calculate the change in entropy when 25 kJ of energy is transferred reversibly and isothermally as heat to a large block of iron at (a) 0°C, (b) 100°C.

4.4 (b) Calculate the change in entropy when 50 kJ of energy is transferred reversibly and isothermally as heat to a large block of copper at (a) 0°C, (b) 70°C.

4.5 (a) Calculate the molar entropy of a constant-volume sample of neon at 500 K given that it is 146.22 J K^{-1} mol^{-1} at 298 K.

4.5 (b) Calculate the molar entropy of a constant-volume sample of argon at 250 K given that it is 154.84 J K^{-1} mol^{-1} at 298 K.

4.6 (a) A sample consisting of 1.00 mol of a monatomic perfect gas with $C_{V,m} = \frac{3}{2}R$ is heated from 100°C to 300°C at constant pressure. Calculate ΔS (for the system).

4.6 (b) A sample consisting of 1.00 mol of a diatomic perfect gas with $C_{V,m} = \frac{5}{2}R$ is heated from 0°C to 100°C at constant pressure. Calculate ΔS (for the system).

4.7 (a) Calculate ΔS (for the system) when the state of 3.00 mol of a monatomic perfect gas, for which $C_{p,m} = \frac{5}{2}R$, is changed from 25°C and 1.00 atm to 125°C and 5.00 atm. How do you rationalize the sign of ΔS?

4.7 (b) Calculate ΔS (for the system) when the state of 2.00 mol of a diatomic perfect gas, for which $C_{p,m} = \frac{7}{2}R$, is changed from 25°C and 1.50 atm to 135°C and 7.00 atm. How do you rationalize the sign of ΔS?

4.8 (a) A sample consisting of 3.00 mol of a diatomic perfect gas at 200 K is compressed reversibly and adiabatically until its temperature reaches 250 K. Given that $C_{V,m} = 27.5$ J K^{-1} mol^{-1}, calculate q, w, ΔU, ΔH, and ΔS.

4.8 (b) A sample consisting of 2.00 mol of a diatomic perfect gas at 250 K is compressed reversibly and adiabatically until its temperature reaches 300 K. Given that $C_{V,m} = 27.5$ J K^{-1} mol^{-1}, calculate q, w, ΔU, ΔH, and ΔS.

4.9 (a) Calculate the increase in entropy when 1.00 mol of a monatomic perfect gas with $C_{p,m} = \frac{5}{2}R$ is heated from 300 K to 600 K and simultaneously expanded from 30.0 L to 50.0 L.

4.9 (b) Calculate the increase in entropy when 3.50 mol of a monatomic perfect gas with $C_{p,m} = \frac{5}{2}R$ is heated from 250 K to 700 K and simultaneously expanded from 20.0 L to 60.0 L.

4.10 (a) A system undergoes a process in which the entropy change is +2.41 J K^{-1}. During the process, 1.00 kJ of heat is added to the system at 500 K. Is the process thermodynamically reversible? Explain your reasoning.

4.10 (b) A system undergoes a process in which the entropy change is +5.51 J K^{-1}. During the process, 1.50 kJ of heat is added to the system at 350 K. Is the process thermodynamically reversible? Explain your reasoning.

4.11 (a) A sample of aluminium of mass 1.75 kg is cooled at constant pressure from 300 K to 265 K. Calculate (a) the energy that must be removed as heat and (b) the change in entropy of the sample.

4.11 (b) A sample of copper of mass 2.75 kg is cooled at constant pressure from 330 K to 275 K. Calculate (a) the energy that must be removed as heat and (b) the change in entropy of the sample.

4.12 (a) A sample of methane gas of mass 25 g at 250 K and 18.5 atm expands isothermally until its pressure is 2.5 atm. Calculate the change in entropy of the gas.

4.12 (b) A sample of nitrogen gas of mass 35 g at 230 K and 21.1 atm expands isothermally until its pressure is 4.3 atm. Calculate the change in entropy of the gas.

4.13 (a) A sample of perfect gas that initially occupies 15.0 L at 250 K and 1.00 atm is compressed isothermally. To what volume must the gas be compressed to reduce its entropy by 5.0 J K^{-1}?

4.13 (b) A sample of perfect gas that initially occupies 11.0 L at 270 K and 1.20 atm is compressed isothermally. To what volume must the gas be compressed to reduce its entropy by 3.0 J K^{-1}?

4.14 (a) Calculate the change in entropy when 50 g of water at 80°C is poured into 100 g of water at 10°C in an insulated vessel given that $C_{p,m} = 75.5$ J K^{-1} mol^{-1}.

4.14 (b) Calculate the change in entropy when 25 g of ethanol at 50°C is poured into 70 g of ethanol at 10°C in an insulated vessel, given that $C_{p,m} = 111.5$ J K^{-1} mol^{-1}.

4.15 (a) Calculate ΔH and ΔS_{tot} when two copper blocks, each of mass 10.0 kg, one at 100°C and the other at 0°C, are placed in contact in an isolated container. The specific heat capacity of copper is 0.385 J K^{-1} g^{-1} and may be assumed constant over the temperature range involved.

4.15 (b) Calculate ΔH and ΔS_{tot} when two iron blocks, each of mass 1.00 kg , one at 200°C and the other at 25°C, are placed in contact in an isolated container. The specific heat capacity of iron is 0.449 J K^{-1} g^{-1} and may be assumed constant over the temperature range involved.

4.16 (a) Consider a system consisting of 2.0 mol CO_2(g), initially at 25°C and 10 atm and confined to a cylinder of cross-section 10.0 cm^2. It is allowed to expand adiabatically against an external pressure of 1.0 atm until the piston has moved outwards through 20 cm. Assume that carbon dioxide may be considered a perfect gas with $C_{V,m} = 28.8$ J K^{-1} mol^{-1} and calculate (a) q, (b) w, (c) ΔU, (d) ΔT, (e) ΔS.

4.16 (b) Consider a system consisting of 1.5 mol CO_2(g), initially at 15°C and 9.0 atm and confined to a cylinder of cross-section 100.0 cm^2. The sample is allowed to expand adiabatically against an external pressure of 1.5 atm until the piston has moved outwards through 15 cm. Assume that carbon dioxide may be considered a perfect gas with $C_{V,m} = 28.8$ J K^{-1} mol^{-1}, and calculate (a) q, (b) w, (c) ΔU, (d) ΔT, (e) ΔS.

4.17 (a) The enthalpy of vaporization of chloroform ($CHCl_3$) is 29.4 kJ mol^{-1} at its normal boiling point of 334.88 K. Calculate (a) the entropy of vaporization of chloroform at this temperature and (b) the entropy change of the surroundings.

4.17 (b) The enthalpy of vaporization of methanol is 35.27 kJ mol^{-1} at its normal boiling point of 64.1°C. Calculate (a) the entropy of vaporization of methanol at this temperature and (b) the entropy change of the surroundings.

4.18 (a) Calculate the standard reaction entropy at 298 K of

(a) $2\,CH_3CHO(g) + O_2(g) \longrightarrow 2\,CH_3COOH(l)$

(b) $2\,AgCl(s) + Br_2(l) \longrightarrow 2\,AgBr(s) + Cl_2(g)$

(c) $Hg(l) + Cl_2(g) \longrightarrow HgCl_2(s)$

4.18 (b) Calculate the standard reaction entropy at 298 K of

(a) $Zn(s) + Cu^{2+}(aq) \longrightarrow Zn^{2+}(aq) + Cu(s)$

(b) $C_{12}H_{22}O_{11}(s) + 12\,O_2(g) \longrightarrow 12\,CO_2(g) + 11\,H_2O(l)$

4.19 (a) Combine the reaction entropies calculated in Exercise 4.18a with the reaction enthalpies, and calculate the standard reaction Gibbs energies at 298 K.

4.19 (b) Combine the reaction entropies calculated in Exercise 4.18b with the reaction enthalpies, and calculate the standard reaction Gibbs energies at 298 K.

4.20 (a) Use standard Gibbs energies of formation to calculate the standard reaction Gibbs energies at 298 K of the reactions in Exercise 4.18a.

4.20 (b) Use standard Gibbs energies of formation to calculate the standard reaction Gibbs energies at 298 K of the reactions in Exercise 4.18b.

4.21 (a) Calculate the standard Gibbs energy of the reaction $4\,HCl(g) + O_2(g) \rightarrow 2\,Cl_2(g) + 2\,H_2O(l)$ at 298 K, from the standard entropies and enthalpies of formation given in Table 2.6.

4.21 (b) Calculate the standard Gibbs energy of the reaction $CO(g) + CH_3OH(l) \rightarrow CH_3COOH(l)$ at 298 K, from the standard entropies and enthalpies of formation given in Tables 2.5 and 2.6.

4.22 (a) The standard enthalpy of combustion of solid phenol (C_6H_5OH) is -3054 kJ mol^{-1} at 298 K and its standard molar entropy is 144.0 J K^{-1} mol^{-1}. Calculate the standard Gibbs energy of formation of phenol at 298 K.

4.22 (b) The standard enthalpy of combustion of solid urea ($CO(NH_2)_2$) is -632 kJ mol^{-1} at 298 K and its standard molar entropy is 104.60 J K^{-1} mol^{-1}. Calculate the standard Gibbs energy of formation of urea at 298 K.

4.23 (a) Calculate the change in the entropies of the system and the surroundings, and the total change in entropy, when a sample of nitrogen gas of mass 14 g at 298 K and 1.00 bar doubles its volume in (a) an isothermal reversible expansion, (b) an isothermal irreversible expansion against $p_{ex} = 0$, and (c) an adiabatic reversible expansion.

4.23 (b) Calculate the change in the entropies of the system and the surroundings, and the total change in entropy, when the volume of a sample of argon gas of mass 21 g at 298 K and 1.50 bar increases from 1.20 L to 4.60 L in (a) an isothermal reversible expansion, (b) an isothermal irreversible expansion against $p_{ex} = 0$, and (c) an adiabatic reversible expansion.

4.24 (a) Calculate the change in entropy when a monatomic perfect gas is compressed to half its volume and simultaneously heated to twice its initial temperature.

4.24 (b) Calculate the change in entropy when a diatomic perfect gas is compressed to one-third its volume and simultaneously heated to three times its initial temperature.

4.25 (a) Calculate the maximum non-expansion work per mole that may be obtained from a fuel cell in which the chemical reaction is the combustion of methane at 298 K.

4.25 (b) Calculate the maximum non-expansion work per mole that may be obtained from a fuel cell in which the chemical reaction is the combustion of propane at 298 K.

4.26 (a) Calculate the Carnot efficiency of a primitive steam engine operating on steam at 100°C and discharging at 60°C. (b) Repeat the calculation for a modern steam turbine that operates with steam at 300°C and discharges at 80°C.

4.26 (b) A certain heat engine operates between 1000 K and 500 K. (a) What is the maximum efficiency of the engine? (b) Calculate the maximum work that can be done for each 1.0 kJ of heat supplied by the hot source. (c) How much heat is discharged into the cold sink in a reversible process for each 1.0 kJ supplied by the hot source?

Problems†

Assume that all gases are perfect and that data refer to 298 K unless otherwise stated.

Numerical problems

4.1 Calculate the difference in molar entropy (a) between liquid water and ice at −5°C, (b) between liquid water and its vapour at 95°C and 1.00 atm. The differences in heat capacities on melting and on vaporization are 37.3 J K^{-1} mol^{-1} and −41.9 J K^{-1} mol^{-1}, respectively. Distinguish between the entropy changes of the sample, the surroundings, and the total system, and discuss the spontaneity of the transitions at the two temperatures.

4.2 The heat capacity of chloroform (trichloromethane, $CHCl_3$) in the range 240 K to 330 K is given by $C_{p,m}/(\text{J K}^{-1}\,\text{mol}^{-1}) = 91.47 + 7.5 \times 10^{-2}\,(T/\text{K})$. In a particular experiment, 1.00 mol $CHCl_3$ is heated from 273 K to 300 K. Calculate the change in molar entropy of the sample.

4.3 A block of copper of mass 2.00 kg ($C_{p,m} = 0.385$ J K^{-1} mol^{-1}) and temperature 0°C is introduced into an insulated container in which there is 1.00 mol $H_2O(g)$ at 100°C and 1.00 atm. (a) Assuming all the steam is condensed to water, what will be the final temperature of the system, the heat transferred from water to copper, and the entropy change of the water, copper, and the total system? (b) In fact, some water vapour is present at equilibrium. From the vapour pressure of water at the temperature calculated in (a), and assuming that the heat capacities of both gaseous and liquid water are constant and given by their values at that temperature, obtain an improved value of the final temperature, the heat transferred, and the various entropies. (*Hint.* You will need to make plausible approximations.)

4.4 Consider a perfect gas contained in a cylinder and separated by a frictionless adiabatic piston into two sections A and B. All changes in B are isothermal; that is, a thermostat surrounds B to keep its temperature constant. There is 2.00 mol of the gas in each section. Initially, $T_A = T_B = 300$ K, $V_A = V_B = 2.00$ L. Heat is added to Section A and the piston moves to the right reversibly until the final volume of Section B is 1.00 L. Calculate (a) ΔS_A and ΔS_B, (b) ΔA_A and ΔA_B, (c) ΔG_A and ΔG_B, (d) ΔS of the total system and its surroundings. If numerical values cannot be obtained, indicate whether the values should be positive, negative, or zero or are indeterminate from the information given. (Assume $C_{V,m} = 20$ J K^{-1} mol^{-1}.)

4.5 A Carnot cycle uses 1.00 mol of a monatomic perfect gas as the working substance from an initial state of 10.0 atm and 600 K. It expands isothermally to a pressure of 1.00 atm (step 1), and then adiabatically to a temperature of 300 K (step 2). This expansion is followed by an isothermal compression (step 3), and then an adiabatic compression (step 4) back to the initial state. Determine

† Problems denoted with the symbol ‡ were supplied by Charles Trapp, Carmen Giunta, and Marshall Cady.

the values of q, w, ΔU, ΔH, ΔS, ΔS_{tot}, and ΔG for each stage of the cycle and for the cycle as a whole. Express your answer as a table of values.

4.6 1.00 mol of a perfect gas at 27°C is expanded isothermally from an initial pressure of 3.00 atm to a final pressure of 1.00 atm in two ways: (a) reversibly, and (b) against a constant external pressure of 1.00 atm. Determine the values of q, w, ΔU, ΔH, ΔS, ΔS_{sur} and ΔS_{tot} for each path.

4.7 A sample of 1.00 mol of a perfect gas at 27°C and 1.00 atm is expanded adiabatically in two ways: (a) reversibly to 0.50 atm, and (b) against a constant external pressure of 0.50 atm. Determine the values of q, w, ΔU, ΔH, ΔS, ΔS_{sur} and ΔS_{tot} for each path where the data permit. Take $C_{V,m} = \frac{3}{2}R$.

4.8 A sample of 1.00 mol of a monatomic perfect gas with $C_{V,m} = \frac{3}{2}R$, initially at 298 K and 10 L, is expanded, with the surroundings maintained at 298 K, to a final volume of 20 L, in three ways: (a) isothermally and reversibly, (b) isothermally against a constant external pressure of 0.50 atm, (c) adiabatically against a constant external pressure of 0.50 atm. Calculate ΔS, ΔS_{sur}, ΔH, ΔT, ΔA, and ΔG for each path where the data permit.

4.9 The standard molar entropy of NH_3(g) is 192.45 J K^{-1} mol^{-1} at 298 K, and its heat capacity is given by eqn 2.30 with the coefficients given in Table 2.2. Calculate the standard molar entropy at (a) 100°C and (b) 500°C.

4.10 A block of copper of mass 500 g and initially at 293 K is in thermal contact with an electric heater of resistance 1.00 kΩ and negligible mass. A current of 1.00 A is passed for 15.0 s. Calculate the change in entropy of the copper, taking $C_{p,m} = 24.4$ J K^{-1} mol^{-1}. The experiment is then repeated with the copper immersed in a stream of water that maintains its temperature at 293 K. Calculate the change in entropy of the copper and the water in this case.

4.11 Calculate the standard Helmholtz energy of formation, $\Delta_f A$, of CH_3OH(l) at 298 K from the standard Gibbs energy of formation and the assumption that H_2 and O_2 are perfect gases.

4.12 Calculate the change in entropy when 200 g of (a) water at 0°C, (b) ice at 0°C is added to 200 g of water at 90°C in an insulated container.

4.13 Calculate (a) the maximum work and (b) the maximum non-expansion work that can be obtained from the freezing of supercooled water at −5°C and 1.0 atm. The densities of water and ice are 0.999 g cm^{-3} and 0.917 g cm^{-3}, respectively, at −5°C.

4.14 The molar heat capacity of lead varies with temperature as follows:

T/K	10	15	20	25	30	50
$C_{p,m}$/(J K^{-1} mol^{-1})	2.8	7.0	10.8	14.1	16.5	21.4
T/K	70	100	150	200	250	298
$C_{p,m}$/(J K^{-1} mol^{-1})	23.3	24.5	25.3	25.8	26.2	26.6

Calculate the standard Third-Law entropy of lead at (a) 0°C and (b) 25°C.

4.15 From standard enthalpies of formation, standard entropies, and standard heat capacities available from tables in the *Data section*, calculate: (a) the standard enthalpies and entropies at 298 K and 398 K for the reaction $CO_2(g) + H_2(g) \rightarrow CO(g) + H_2O(g)$. Assume that the heat capacities are constant over the temperature range involved.

4.16 The standard reaction Gibbs energy of

$$K_4[Fe(CN)_6]\cdot 3H_2O(s) \longrightarrow 4\,K^+(aq) + [Fe(CN)_6]^{4-}(aq) + 3\,H_2O(l)$$

is +26.120 kJ mol^{-1} (I.R. Malcolm, L.A.K. Staveley, and R.D. Worswick, *J. Chem. Soc. Faraday Trans. I*, 1532 (1973)). The enthalpy of solution of the trihydrate is +55.000 kJ mol^{-1}. Calculate (a) the standard molar entropy of the hexacyanoferrate(II) ion in water and (b) the standard reaction entropy given that the standard molar entropy of the solid trihydrate is 599.7 J K^{-1} mol^{-1} and that of the K^+ ion in water is 102.5 J K^{-1} mol^{-1}.

4.17 The heat capacity of anhydrous potassium hexacyanoferrate(II) varies with temperature as follows:

T/K	$C_{p,m}$/(J K^{-1} mol^{-1})	T/K	$C_{p,m}$/(J K^{-1} mol^{-1})
10	2.09	100	179.6
20	14.43	110	192.8
30	36.44	150	237.6
40	62.55	160	247.3
50	87.03	170	256.5
60	111.0	180	265.1
70	131.4	190	273.0
80	149.4	200	280.3
90	165.3		

Calculate the molar enthalpy relative to its value at $T = 0$ and the Third-Law entropy at each of these temperatures.

4.18 The compound 1,3,5-trichloro-2,4,6-trifluorobenzene is an intermediate in the conversion of hexachlorobenzene to hexafluorobenzene, and its thermodynamic properties have been examined by measuring its heat capacity over a wide temperature range (R.L. Andon and J.F. Martin, *J. Chem. Soc. Faraday Trans. I.* 871 (1973)). Some of the data are as follows:

T/K	14.14	16.33	20.03	31.15	44.08	64.81
$C_{p,m}$/(J K^{-1} mol^{-1})	9.492	12.70	18.18	32.54	46.86	66.36
T/K	100.90	140.86	183.59	225.10	262.99	298.06
$C_{p,m}$/(J K^{-1} mol^{-1})	95.05	121.3	144.4	163.7	180.2	196.4

Calculate the molar enthalpy relative to its value at $T = 0$ and the Third-Law entropy of the compound at these temperatures.

4.19‡ Given that $S_m^{\ominus} = 29.79$ J K^{-1} mol^{-1} for bismuth at 100 K and the following tabulated heat capacities data (D.G. Archer, *J. Chem. Eng. Data* **40**, 1015 (1995)), compute the standard molar entropy of bismuth at 200 K.

T/K	100	120	140	150	160	180	200
$C_{p,m}$/(J K^{-1} mol^{-1})	23.00	23.74	24.25	24.44	24.61	24.89	25.11

Compare the value to the value that would be obtained by taking the heat capacity to be constant at 24.44 J K^{-1} mol^{-1} over this range.

Theoretical problems

4.20 Show that the integral of dq_{rev}/T round a Carnot cycle is zero. Then show that the integral is negative if the isothermal reversible expansion stage is replaced by an isothermal irreversible expansion.

4.21 Prove that two reversible adiabatic paths can never cross. Assume that the energy of the system under consideration is a function of temperature only. (*Hint.* Suppose that two such paths can intersect, and complete a cycle with the two paths plus one isothermal path. Consider the changes accompanying each stage of the cycle and show that they conflict with the Kelvin statement of the Second Law.)

4.22 Represent the Carnot cycle on a temperature–entropy diagram and show that the area enclosed by the cycle is equal to the work done.

4.23 Find an expression for the change in entropy when two blocks of the same substance and of equal mass, one at the temperature T_h and the other at T_c, are brought into thermal contact and allowed to reach equilibrium. Evaluate the change for two blocks of copper, each of mass 500 g, with $C_{p,m} = 24.4\ \mathrm{J\ K^{-1}\ mol^{-1}}$, taking $T_h = 500$ K and $T_c = 250$ K.

4.24 A gaseous sample consisting of 1.00 mol molecules is described by the equation of state $pV_m = RT(1 + Bp)$. Initially at 373 K, it undergoes Joule–Thomson expansion from 100 atm to 1.00 atm. Given that $C_{p,m} = \frac{5}{2}R$, $\mu = 0.21\ \mathrm{K\ atm^{-1}}$, $B = -0.525(\mathrm{K}/T)\ \mathrm{atm^{-1}}$, and that these are constant over the temperature range involved, calculate ΔT and ΔS for the gas.

4.25 Prove that the perfect gas temperature scale and the thermodynamic temperature scale based on the Second Law of thermodynamics differ from each other by at most a constant numerical factor.

4.26 The definitions of the enthalpy, Gibbs energy, and Helmholtz energy have all been of the form $g = f + yz$. Show that the addition of the product yz is a general way of converting a function of x and y to a function of x and z in the sense that, if $\mathrm{d}f = a\,\mathrm{d}x - z\,\mathrm{d}y$, then $\mathrm{d}g = a\,\mathrm{d}x + y\,\mathrm{d}z$.

4.27‡ It is found that the molar heat capacity of β-$Ce_2Si_2O_7$ at 1 bar in the range of temperatures 0–30 K is described by

$$C_{p,m}(T) = a(T/\mathrm{K})^3(1 - e^{-b/T^2})$$

where $a = 5.597 \times 10^{-4}\ \mathrm{J\ mol^{-1}\ K^{-4}}$ and $b = 1701.8\ \mathrm{K^2}$ are empirical parameters. In the high-temperature range, 500–900 K, the molar heat capacity is described by

$$C_{p,m}(T)/(\mathrm{J\ mol^{-1}\ K^{-1}}) = 222.3716 + 0.076\,790\,42(T/\mathrm{K}) - 3\,631\,886(T/\mathrm{K})^{-2}$$

No phase transition has been detected below 900 K. (a) Draw graphs of $S_m(T)$, $H_m(T) - H_m(0)$, $G_m(T) - G_m(0)$ in the range $0 \le T \le 30$ K. (b) According to the law of Dulong and Petit, the high-temperature heat capacity per mole of atoms of a crystalline material is $3R$. How does the classical prediction compare with the experimental? (*Hint.* Use mathematical software or a spreadsheet to draw the graphs.)

4.28‡ Polytropic processes are those that satisfy the relation $pV^n = C$. Make schematic plots of polytropic processes on a pV and TS diagram for $n = 0, \pm1, \gamma$ (the heat capacity ratio), and $\pm\infty$. (*Hint.* Use mathematical software or a spreadsheet to draw the graphs.)

Applications: to chemical engineering and atmospheric science

4.29‡ Alkyl radicals are important intermediates in the combustion and atmospheric chemistry of hydrocarbons. N. Cohen (*J. Phys. Chem.* **96**, 9052 (1992)) reports group additivity tables for the thermochemistry of alkyl free radicals. A portion of the table follows. Use the table to estimate the standard molar entropies of C_2H_5, *sec*-C_4H_9, and *tert*-C_4H_9. (For the structures of these species, see the problems in Chapter 3.) Note that $S_m^{\ominus} = S_{int}^{\ominus} - R \ln s$, where $S_{int}^{\ominus}$ is the so-called intrinsic molar entropy, computed by group additivity, and s is a symmetry number: $s = 6$ for C_2H_5, 3^2 for *sec*-C_4H_9, and 3^4 for *tert*-C_4H_9.

Group	$S_{int}^{\ominus}/(\mathrm{J\ K^{-1}\ mol^{-1}})$
C—(C)(H)$_3$	126.8
·C—(C)(H)$_2$	135.9
·C—(C)$_2$(H)	59.3
·C—(C)$_3$	−29.2
C—(·C)(H)$_3$	126.8
C—(·C)(C)(H)$_2$	42.0

4.30‡ Use the following enthalpies of formation reported by Seakins *et al.* (P.W. Seakins, M.J. Pilling, J.T. Niiranen, D. Gutman, and L.N. Krasnoperov, *J. Phys. Chem.* **96**, 9847 (1992)) and entropies based on group additivity tables of Cohen (N. Cohen, *J. Phys. Chem.* **96**, 9052 (1992)) to compute $\Delta_rG^{\ominus}$ for three possible fates of the *tert*-butyl radical at 700 K, namely (a) *tert*-$C_4H_9 \rightarrow$ *sec*-C_4H_9, (b) *tert*-$C_4H_9 \rightarrow C_3H_6 + CH_3$, (c) *tert*-$C_4H_9 \rightarrow C_2H_4 + C_2H_5$.

Species	$\Delta_fH^{\ominus}/(\mathrm{kJ\ mol^{-1}})$	$S_m^{\ominus}/(\mathrm{J\ K^{-1}\ mol^{-1}})$
C_2H_5	+121.0	247.8
sec-C_4H_9	+67.5	336.6
tert-C_4H_9	+51.3	314.6

4.31 Suppose that an internal combustion engine runs on octane, for which the enthalpy of combustion is $-5512\ \mathrm{kJ\ mol^{-1}}$ and take the mass of 1 gallon of fuel as 3 kg. What is the maximum height, neglecting all forms of friction, to which a car of mass 1000 kg can be driven on 1.00 gallon of fuel given that the engine cylinder temperature is 2000°C and the exit temperature is 800°C?

4.32 The cycle involved in the operation of an internal combustion engine is called the *Otto cycle*. Air can be considered to be the working substance and can be assumed to be a perfect gas. The cycle consists of the following steps: (1) reversible adiabatic compression from A to B, (2) reversible constant-volume pressure increase from B to C due to the combustion of a small amount of fuel, (3) reversible adiabatic expansion from C to D, and (4) reversible and constant-volume pressure decrease back to state A. Determine the change in entropy (of the system and of the surroundings) for each step of the cycle and determine an expression for the efficiency of the cycle, assuming that the heat is supplied in Step 2. Evaluate the efficiency for a compression ratio of 10:1. Assume that in state A, $V = 4.00$ L, $p = 1.00$ atm, and $T = 300$ K, that $V_A = 10V_B$, $p_C/p_B = 5$, and that $C_{p,m} = \frac{7}{2}R$.

4.33‡ Consider a Carnot engine operating in outer space between the temperatures T_h and T_c. The only way that the engine can discard heat at T_c is by radiation. The power radiated by the engine at T_c follows the Stefan–Boltzmann law (see Section 11.1), which for our purposes here is written $\mathrm{d}q_c/\mathrm{d}t = kAT_c^4$, where k is a constant related to the Stefan–Boltzmann constant. Find the ratio T_c/T_h that corresponds to a minimum area A of the radiator for a fixed power output and constant T_h.

5 The Second Law: the machinery

One of the principal applications of thermodynamics is to find relations between properties that might not be thought to be related. Several relations of this kind can be established by making use of the fact that the Gibbs energy is a state function. We also see how to derive expressions for the variation of the Gibbs energy with temperature and pressure and how to formulate expressions that are valid for real gases. These expressions will prove useful later when we need to discuss the effect of temperature and pressure on equilibrium constants.

The Gibbs energy, G, is of central importance to chemistry. In this chapter it begins to move to the centre of the stage. We have seen that, at constant temperature and pressure, a system tends to adjust so as to minimize the value of G. Therefore, the equilibrium state of a system corresponds to the state of minimum Gibbs energy. It follows that to know how the equilibrium state responds to changes in the conditions, such as how a melting temperature responds to changes in pressure or the equilibrium composition of a reaction mixture responds to changes in temperature, we need to know how G varies. To do so, we first consider how our new information about entropy can be combined with our knowledge about the internal energy that came from the First Law.

Combining the First and Second Laws

We have treated the First and Second Laws of thermodynamics as distinct statements. Both are relevant to the behaviour of matter, so we can bring the whole force of thermodynamics to bear on a problem by setting up a formulation that combines concepts from the two laws.

5.1 The fundamental equation

We have seen that the First Law of thermodynamics may be written

$$dU = dq + dw \tag{5.1}$$

For a reversible change in a closed system of constant composition, and in the absence of any additional (non-expansion) work,

$$dw_{rev} = -p\,dV \qquad dq_{rev} = T\,dS$$

The second of these two relations, of course, comes from the definition of entropy introduced in the discussion of the Second Law. Therefore, for a reversible change in a closed system,

$$dU = T\,dS - p\,dV \tag{5.2}$$

However, because dU is an exact differential, its value is independent of path. Therefore, the same value of dU is obtained whether the change is brought about irreversibly or reversibly. Consequently, *eqn 2 applies to any change—reversible or irreversible—of a closed system that does no additional (non-expansion) work*. We shall call this combination of the First and Second Laws the **fundamental equation**.

The fact that the fundamental equation applies to both reversible and irreversible changes may be puzzling at first sight. The reason is that only in the case of a reversible change may $T\,dS$ be identified with dq and $-p\,dV$ with dw. When the change is irreversible, $T\,dS > dq$ (the Clausius inequality) and $-p\,dV < dw$. The sum of dw and dq remains equal to the sum of $T\,dS$ and $-p\,dV$, provided the composition is constant.

5.2 Properties of the internal energy

Equation 2 shows that the internal energy of a closed system changes in a simple way when either S or V is changed ($dU \propto dS$ and $dU \propto dV$). These simple proportionalities suggest that U should be regarded as a function of S and V. We could regard U as a function of other variables, such as S and p or T and V, because they are all interrelated; but the simplicity of the fundamental equation suggests that $U(S, V)$ is the best choice.

The *mathematical* consequence of U being a function of S and V is that we can express an infinitesimal change dU in terms of changes dS and dV by

Partial derivatives are reviewed in *Further information 1*.

$$dU = \left(\frac{\partial U}{\partial S}\right)_V dS + \left(\frac{\partial U}{\partial V}\right)_S dV \tag{5.3}$$

The two coefficients are the slopes of the plots of U against S and V, respectively. When this expression is compared to the *thermodynamic* relation, eqn 2, we see that, for systems of constant composition,

$$\left(\frac{\partial U}{\partial S}\right)_V = T \qquad \left(\frac{\partial U}{\partial V}\right)_S = -p \tag{5.4}$$

The first of these two equations is a purely thermodynamic definition of temperature as the ratio of the changes in the internal energy and entropy of a constant-volume, closed, constant-composition system. We are beginning to generate relations between the properties of a system and to discover the power of thermodynamics for establishing unexpected relations.

(a) The Maxwell relations

The properties of differentials are reviewed in *Further information 1*.

Suppose we can write an infinitesimal change in a function f as $df = g\,dx + h\,dy$ where g and h are functions of x and y. The mathematical criterion for df being an exact differential (in the sense that its integral is independent of path) is that

$$\left(\frac{\partial g}{\partial y}\right)_x = \left(\frac{\partial h}{\partial x}\right)_y \tag{5.5}$$

Because the fundamental equation, eqn 2, is an expression for an exact differential, the functions multiplying dS and dV (namely T and $-p$) must pass this test. Therefore, it must be the case that

Table 5.1 The Maxwell relations

$\left(\frac{\partial T}{\partial V}\right)_S = -\left(\frac{\partial p}{\partial S}\right)_V$
$\left(\frac{\partial T}{\partial p}\right)_S = \left(\frac{\partial V}{\partial S}\right)_p$
$\left(\frac{\partial p}{\partial T}\right)_V = \left(\frac{\partial S}{\partial V}\right)_T$
$\left(\frac{\partial V}{\partial T}\right)_p = -\left(\frac{\partial S}{\partial p}\right)_T$

$$\left(\frac{\partial T}{\partial V}\right)_S = -\left(\frac{\partial p}{\partial S}\right)_V \qquad (5.6)$$

We have generated a relation between quantities that, at first sight, would not seem to be related.

Equation 6 is an example of a **Maxwell relation**. However, apart from being unexpected, it does not look particularly interesting. Nevertheless, it does suggest that there may be other similar relations that are more useful. Indeed, we can use the fact that H, G, and A are all state functions to derive three more Maxwell relations. The argument to obtain them runs in the same way in each case: because H, G, and A are state functions, the expressions for dH, dG, and dA satisfy relations like eqn 6. All four relations are listed in Table 5.1 and we put them to work later in the chapter.

(b) The variation of internal energy with volume

The coefficient

$$\pi_T = \left(\frac{\partial U}{\partial V}\right)_T \qquad [5.7]$$

which represents how the internal energy changes as the volume of a system is changed isothermally played a central role in the manipulation of the First Law, and in *Justification 3.3* we used the relation

$$\pi_T = T\left(\frac{\partial p}{\partial T}\right)_V - p \qquad (5.8)$$

This relation is called a **thermodynamic equation of state** because it is an expression for pressure in terms of a variety of thermodynamic properties of the system. We are now ready to derive it by using a Maxwell relation.

Justification 5.1

We obtain an expression for the coefficient π_T by dividing both sides of eqn 3 by dV, imposing the constraint of constant temperature, which gives

$$\left(\frac{\partial U}{\partial V}\right)_T = \left(\frac{\partial U}{\partial S}\right)_V\left(\frac{\partial S}{\partial V}\right)_T + \left(\frac{\partial U}{\partial V}\right)_S$$

Next, we introduce the two relations in eqn 4, to obtain

$$\left(\frac{\partial U}{\partial V}\right)_T = T\left(\frac{\partial S}{\partial V}\right)_T - p$$

This equation is already beginning to look like the expression we want. The third Maxwell relation in Table 5.1 turns $(\partial S/\partial V)_T$ into $(\partial p/\partial T)_V$, which completes the proof of eqn 8

Example 5.1 Deriving a thermodynamic relation

Show thermodynamically that $\pi_T = 0$ for a perfect gas, and compute its value for a van der Waals gas.

Method Proving a result 'thermodynamically' means basing it entirely on general thermodynamic relations and equations of state, without drawing on molecular arguments (such as the existence of intermolecular forces). We know that for a perfect gas, $p = nRT/V$, so this relation should be used in eqn 8. Similarly, the van der Waals equation is given in Table 1.6, and for the second part of the question it should be used in eqn 8.

Answer Because $(\partial p/\partial T)_V = nR/V$ for a perfect gas (by differentiation of the equation of state), eqn 8 becomes

$$\pi_T = \frac{nRT}{V} - p = 0$$

The equation of state of a van der Waals gas is

$$p = \frac{nRT}{V - nb} - a\frac{n^2}{V^2}$$

Therefore, because a and b are independent of temperature, we can write

$$\left(\frac{\partial p}{\partial T}\right)_V = \frac{nR}{V - nb}$$

That is,

$$\pi_T = \frac{nRT}{V - nb} - \frac{nRT}{V - nb} + a\frac{n^2}{V^2} = a\frac{n^2}{V^2}$$

Comment This result for π_T implies that the internal energy of a van der Waals gas increases when it expands isothermally (that is, $(\partial U/\partial V)_T > 0$), and that the increase is related to the parameter a, which models the attractive interactions between the particles. A larger molar volume, corresponding to a greater average separation between molecules, implies weaker mean intermolecular attractions, so the total energy is greater.

Self-test 5.1 Calculate π_T for a gas that obeys the virial equation of state.
[$\pi_T = RT^2(\partial B/\partial T)_V/V_m^2 + \cdots$]

Properties of the Gibbs energy

The same arguments that we have used for U can be used for the Gibbs energy $G = H - TS$ and we can derive important expressions showing how G varies with pressure and temperature.

5.3 General considerations

When the system undergoes a change of state, G may change because H, T, and S change. For infinitesimal changes in each property,

$$dG = dH - T\,dS - S\,dT$$

Because $H = U + pV$, we know that

$$dH = dU + p\,dV + V\,dp$$

For a closed system doing no additional work, we can replace dU by the fundamental equation $dU = T\,dS - p\,dV$. The result of these steps is

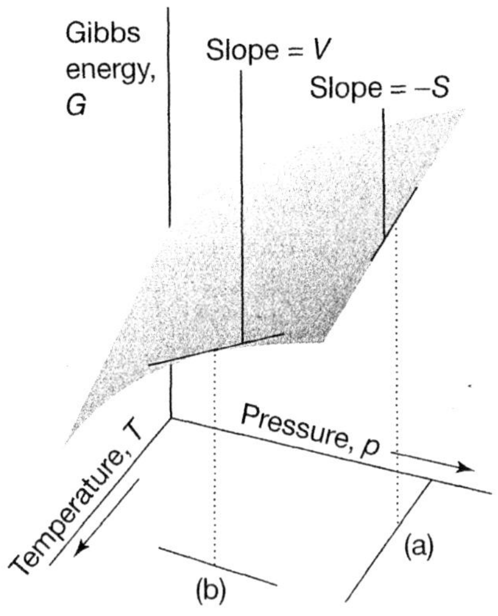

5.1 The variation of the Gibbs energy of a system with (a) temperature at constant pressure and (b) pressure at constant temperature. The slope of the former is equal to the negative of the entropy of the system and that of the latter is equal to the volume.

$$dG = (T\,dS - p\,dV) + p\,dV + V\,dp - T\,dS - S\,dT$$

That is, for a closed system in the absence of non-expansion work and at constant composition

$$dG = V\,dp - S\,dT \qquad (5.9)$$

This expression, which shows that a change in G is proportional to a change in p or T, suggests that G may be best regarded as a function of p and T. It confirms that G is an important quantity in chemistry because the pressure and temperature are usually the variables under our control. In other words, G carries around the combined consequences of the First and Second Laws in a way that makes it particularly suitable for chemical applications.

The same argument that led to eqn 4, when applied to the exact differential dG, now gives

$$\left(\frac{\partial G}{\partial T}\right)_p = -S \qquad \left(\frac{\partial G}{\partial p}\right)_T = V \qquad (5.10)$$

These relations show how the Gibbs energy varies with temperature and pressure (Fig. 5.1). Because S is positive, it follows that G decreases when the temperature is raised at constant pressure and composition. Moreover, the relation shows that G decreases most sharply when the entropy of the system is large. Therefore, the Gibbs energy of the gaseous phase of a substance, which has a high molar entropy, is more sensitive to temperature than its liquid and solid phases (Fig. 5.2). Because V is positive, G always increases when the pressure of the system is increased at constant temperature and composition. Because the molar volumes of gases are large, G is more sensitive to pressure for the gaseous phase of a substance than for its liquid and solid phases (Fig. 5.3).

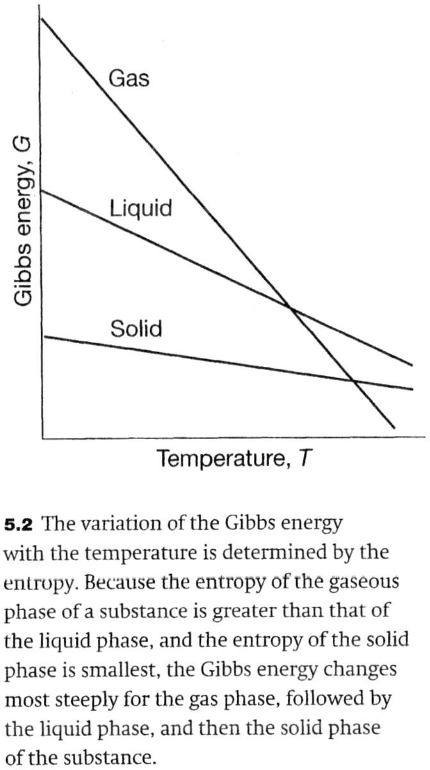

5.2 The variation of the Gibbs energy with the temperature is determined by the entropy. Because the entropy of the gaseous phase of a substance is greater than that of the liquid phase, and the entropy of the solid phase is smallest, the Gibbs energy changes most steeply for the gas phase, followed by the liquid phase, and then the solid phase of the substance.

Example 5.2 Calculating the effect of pressure on the Gibbs energy

Calculate the change in the molar Gibbs energy of (a) liquid water treated as an incompressible fluid and (b) water vapour treated as a perfect gas, when the pressure is increased isothermally from 1.0 bar to 2.0 bar at 298 K.

Method In each case, we obtain the change in molar Gibbs energy by integration of eqn 9 with the temperature held constant (that is, setting dT = 0):

$$G_m(p_f) - G_m(p_i) = \int_{p_i}^{p_f} V_m\,dp$$

For an incompressible fluid, the molar volume is independent of the pressure, so V_m is a constant. For a perfect gas, the molar volume varies with pressure as $V_m = RT/p$, so this expression must be used in the integrand and the integration performed with RT a constant.

Answer For the incompressible liquid, V_m is constant at 18.0 cm^3 mol^{-1}, so

$$G_m(p_f) - G_m(p_i) = V_m \int_{p_i}^{p_f} dp = V_m \times (p_f - p_i)$$

$$= (18.0 \times 10^{-6}\ \text{m}^3\ \text{mol}^{-1}) \times (1.0 \times 10^5\ \text{Pa})$$

$$= +1.8\ \text{J mol}^{-1}$$

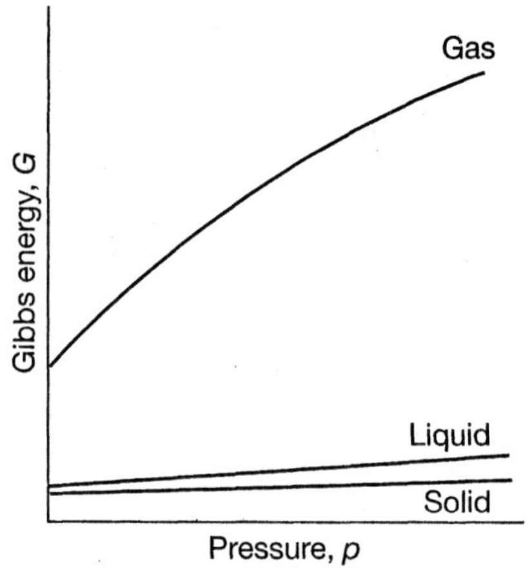

5.3 The variation of the Gibbs energy with the pressure is determined by the volume of the sample. Because the volume of the gaseous phase of a substance is greater than that of the same amount of liquid phase, and the entropy of the solid phase is smallest (for most substances), the Gibbs energy changes most steeply for the gas phase, followed by the liquid phase, and then the solid phase of the substance. Because the volumes of the solid and liquid phases of a substance are similar, they vary by similar amounts as the pressure is changed.

(because 1 Pa m^3 = 1 N m = 1 J). For a perfect gas:

$$G_m(p_f) - G_m(p_i) = \int_{p_i}^{p_f} \frac{RT}{p}\,dp = RT \ln \frac{p_f}{p_i}$$

$$= (2.48 \text{ kJ mol}^{-1}) \times \ln 2.0 = +1.7 \text{ kJ mol}^{-1}$$

Comment Note that G increases in both cases, and that the increase for a gas is 1000 times greater than for the liquid.

Self-test 5.2 Calculate the change in G_m for ice at −10°C, with density 0.917 g cm^{-3}, when the pressure is increased from 1.0 bar to 2.0 bar.
[+2.0 J mol^{-1}]

5.4 The variation of the Gibbs energy with temperature

As we remarked in the introduction, because the equilibrium composition of a system depends on the Gibbs energy, to discuss the response of the composition to temperature we need to know how G varies with temperature.

The first expression in eqn 10 is our starting point for this discussion. Although it expresses the variation of G in terms of the entropy, we can express it in terms of the enthalpy by using the definition of G to write $S = (H - G)/T$. Then

$$\left(\frac{\partial G}{\partial T}\right)_p = \frac{G - H}{T} \tag{5.11}$$

We shall see later that the equilibrium constant of a reaction is related to G/T rather than to G itself,[1] and it turns out that the variation of this quantity with temperature is simpler than the temperature variation of G alone. In fact, it is easy to deduce from the last equation (see the *Justification* below) that

$$\left(\frac{\partial}{\partial T}\frac{G}{T}\right)_p = -\frac{H}{T^2} \tag{5.12a}$$

This expression is called the **Gibbs–Helmholtz equation**. It shows that, if we know the enthalpy of the system, then we know how G/T varies with temperature.

Justification 5.2

First, we write eqn 11 as

$$\left(\frac{\partial G}{\partial T}\right)_p - \frac{G}{T} = -\frac{H}{T}$$

Then we simplify the expression on the left by noting that

1. In Section 9.2b we derive the result that the equilibrium constant for a reaction is related to its standard reaction free energy by $\Delta_r G^{\ominus}/T = -R \ln K$.

$$\left(\frac{\partial}{\partial T}\frac{G}{T}\right)_p = \frac{1}{T}\left(\frac{\partial G}{\partial T}\right)_p + G\frac{\mathrm{d}}{\mathrm{d}T}\frac{1}{T} = \frac{1}{T}\left(\frac{\partial G}{\partial T}\right)_p - \frac{G}{T^2} = \frac{1}{T}\left\{\left(\frac{\partial G}{\partial T}\right)_p - \frac{G}{T}\right\}$$

When we substitute eqn 11 into this expression, we obtain eqn 12a.

Now note that we can use the chain rule of differentiation to write the left-hand side of eqn 12a as

The *chain rule* states that for a function $f = f(g)$, where $g = g(t)$,

$$\frac{\mathrm{d}f}{\mathrm{d}t} = \frac{\mathrm{d}f}{\mathrm{d}g}\frac{\mathrm{d}g}{\mathrm{d}t}$$

$$\left(\frac{\partial(G/T)}{\partial T}\right)_p = \left(\frac{\partial(G/T)}{\partial(1/T)}\right)_p \frac{\mathrm{d}(1/T)}{\mathrm{d}T} = \left(\frac{\partial(G/T)}{\partial(1/T)}\right)_p \times \left(-\frac{1}{T^2}\right)$$

Substitution of this result into eqn 12a and multiplication of both sides by $-T^2$ gives an alternative form of the Gibbs–Helmholtz equation:

$$\left(\frac{\partial(G/T)}{\partial(1/T)}\right)_p = H \qquad (5.12\text{b})$$

The Gibbs–Helmholtz equation is most useful when it is applied to changes, including changes of physical state and chemical reactions at constant pressure. Then, because $\Delta G = G_f - G_i$ for the change of Gibbs energy between the final and initial states and because the equation applies to both G_f and G_i, we can write

$$\left(\frac{\partial}{\partial T}\frac{\Delta G}{T}\right)_p = -\frac{\Delta H}{T^2} \qquad (5.13)$$

This equation shows that, if we know the change in enthalpy of a system undergoing a change, then we know how the change in Gibbs energy varies with temperature. As we shall see, this is a crucial piece of information in chemistry.

5.5 The variation of the Gibbs energy with pressure

To find the Gibbs energy at one pressure in terms of its value at another pressure, the temperature being constant, we set $\mathrm{d}T = 0$ in eqn 9 and integrate the remaining expression:

$$G(p_f) = G(p_i) + \int_{p_i}^{p_f} V\,\mathrm{d}p \qquad (5.14)$$

This expression is applicable to any phase of matter.

(a) Liquids and solids

For a liquid or solid, the volume changes only slightly as the pressure changes (Fig. 5.4), so we can treat V as a constant and take it outside the integral. Then, for molar quantities,

$$G_m(p_f) = G_m(p_i) + V_m(p_f - p_i) = G_m(p_i) + V_m\Delta p \qquad (5.15)$$

where $\Delta p = p_f - p_i$. Under normal laboratory conditions $V_m\Delta p$ is very small, as we saw in Example 5.2, and may be neglected. Hence, we may usually suppose that the Gibbs energies of solids and liquids are independent of pressure. However, if we are interested in geophysical problems, then, because pressures in the Earth's interior are huge, their effect on the Gibbs energy cannot be ignored. If the pressures are so great that there are substantial volume changes, we must use the complete expression, eqn 14.

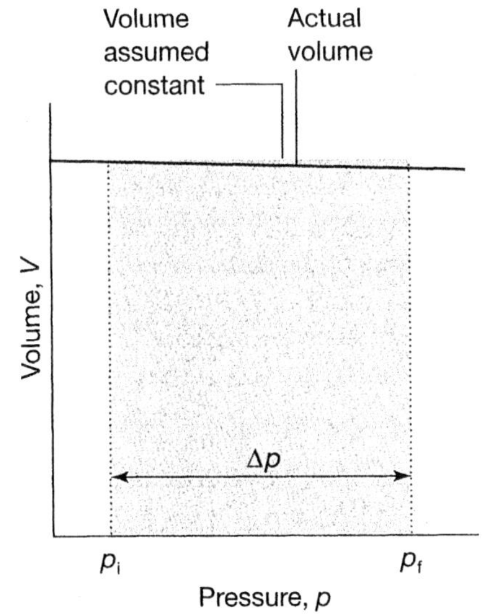

5.4 The difference in Gibbs energy of a solid or liquid at two pressures is equal to the rectangular area shown. We have assumed that the variation of volume with pressure is negligible.

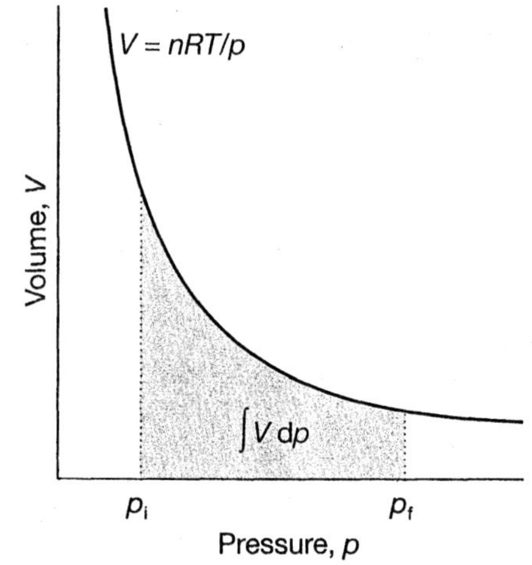

5.5 The difference in Gibbs energy for a perfect gas at two pressures is equal to the area shown below the perfect-gas isotherm.

Illustration 5.1

Suppose that for a certain phase transition of a solid $\Delta_{trs}V = +1.0\ cm^3\ mol^{-1}$. Then, for an increase in pressure to 3.0 Mbar, the Gibbs energy of the transition changes from $\Delta_{trs}G(1\ bar)$ to

$$\Delta_{trs}G(3\ Mbar) = \Delta_{trs}G(1\ bar) + (1.0\times10^{-6}\ m^3\ mol^{-1})\times(3.0\times10^{11}\ Pa - 1.0\times10^5\ Pa)$$
$$= \Delta_{trs}G(1\ bar) + 3.0\times10^2\ kJ\ mol^{-1}$$

where we have used $1\ Pa\ m^3 = 1\ J$.

Self-test 5.3 Calculate the difference in molar Gibbs energy between the top and bottom of a column of mercury in a barometer of height 760 mm. The mass density of mercury is $13.6\ g\ cm^{-3}$. [$+1.49\ J\ mol^{-1}$]

(b) Gases

The molar volumes of gases are large, so the Gibbs energy of a gas depends strongly on the pressure. Furthermore, because the volume also varies markedly with the pressure, we cannot treat it as a constant in the integral in eqn 14 (Fig. 5.5). For a perfect gas we substitute $V = nRT/p$ into the integral, and find

$$G(p_f) = G(p_i) + nRT\int_{p_i}^{p_f}\frac{dp}{p} = G(p_i) + nRT\ln\frac{p_f}{p_i} \qquad (5.16)^\circ$$

This expression shows that, when the pressure is increased tenfold at room temperature, the molar Gibbs energy increases by about $6\ kJ\ mol^{-1}$. It also follows from this equation that, if we set $p_i = p^\ominus$ (the standard pressure of 1 bar), then the Gibbs energy of a perfect gas at a pressure p is related to its standard value by

$$G(p) = G^\ominus + nRT\ln\frac{p}{p^\ominus} \qquad (5.17)^\circ$$

The logarithmic dependence of the molar Gibbs energy on the pressure predicted by this expression is illustrated in Fig. 5.6.

At various stages in the development of physical chemistry it is necessary to switch from a consideration of idealized systems to real systems. In many cases it is desirable to preserve the form of the expressions that have been derived for an idealized system. Then deviations from the idealized behaviour can be expressed most simply. For instance, the pressure dependence of the molar Gibbs energy of a real gas might resemble that shown in Fig. 5.7. To adapt eqn 17 to this case, we replace the true pressure, p, by an effective pressure, called the **fugacity**,[2] f, and write

$$G_m = G_m^\ominus + RT\ln\frac{f}{p^\ominus} \qquad [5.18]$$

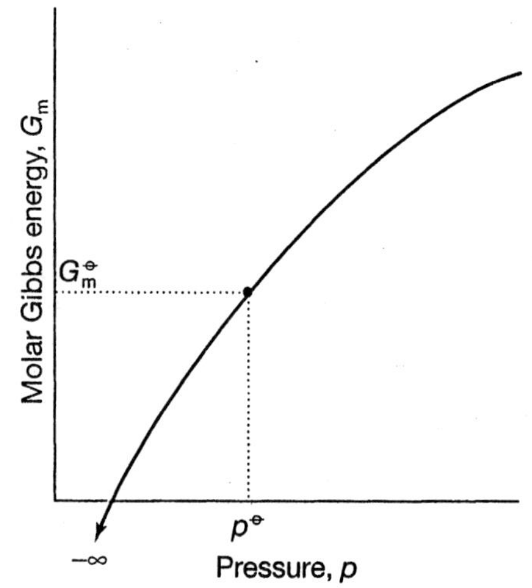

5.6 The molar Gibbs energy of a perfect gas is proportional to ln p, and the standard state is reached at $p^\ominus$. Note that as $p\to 0$, the molar Gibbs energy becomes negatively infinite.

The fugacity, a function of the pressure and temperature, is defined so that this relation is exactly true. In later chapters we derive thermodynamically exact expressions in terms of the Gibbs energy, and therefore in terms of fugacities. For example, from elementary chemistry we know that the equilibrium constant for a reaction such as $H_2(g) + Br_2(g) \rightleftharpoons 2\,HBr(g)$ should be written

$$K = \frac{p_{HBr}^2}{p_{H_2}p_{Br_2}}$$

2. The name 'fugacity' comes from the Latin for 'fleetness' in the sense of 'escaping tendency'; fugacity has the same dimensions as pressure.

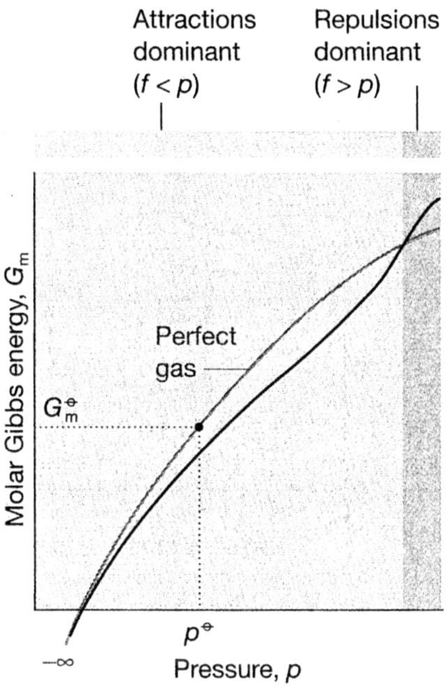

5.7 The molar Gibbs energy of a real gas. As $p \to 0$, the molar Gibbs energy coincides with the value for a perfect gas (shown by the pale line). When attractive forces are dominant (at intermediate pressures), the molar Gibbs energy is less than that of a perfect gas and the molecules have a lower 'escaping tendency'. At high pressures, when repulsive forces are dominant, the molar Gibbs energy of a real gas is greater than that of a perfect gas. Then the 'escaping tendency' is increased.

where p_J is the partial pressure of substance J; however, this expression is only an approximation. The thermodynamically exact expression is

$$K = \frac{f_{HBr}^2}{f_{H_2} f_{Br_2}}$$

where f_J is the fugacity of J.

Although expressions in terms of fugacities are exact, they are useful only if we know how to interpret fugacities in terms of actual pressures. We show in the *Justification* below that, if we write the fugacity as

$$f = \phi p \qquad [5.19]$$

where ϕ is the dimensionless **fugacity coefficient**, then ϕ is related to the compression factor, Z (Section 1.3a), of the gas between $p = 0$ and the pressure of interest by

$$\ln \phi = \int_0^p \frac{Z-1}{p} \mathrm{d}p \qquad (5.20)$$

Justification 5.3

Equation 14 is true for all gases whether real or perfect. Expressing it in terms of molar quantities and then using eqn 18 gives

$$\int_{p'}^{p} V_m \,\mathrm{d}p = G_m - G'_m = RT \ln \frac{f}{f'}$$

In this expression, f is the fugacity when the pressure is p and f' is the fugacity when the pressure is p'. If the gas were perfect, we would write

$$\int_{p'}^{p} V_{perfect,m} \,\mathrm{d}p = RT \ln \frac{p}{p'}$$

The difference between the two equations is

$$\int_{p'}^{p} (V_m - V_{perfect,m})\mathrm{d}p = RT\left(\ln \frac{f}{f'} - \ln \frac{p}{p'}\right)$$

which can be rearranged into

$$\ln\left(\frac{f}{p} \times \frac{p'}{f'}\right) = \frac{1}{RT}\int_{p'}^{p} (V_m - V_{perfect,m})\mathrm{d}p$$

When $p' \to 0$, the gas behaves perfectly and f' becomes equal to the pressure, p'. Therefore, $f'/p' \to 1$ as $p' \to 0$. If we take the limit, which means setting $f'/p' = 1$ on the left and $p' = 0$ on the right, the last equation becomes

$$\ln \frac{f}{p} = \frac{1}{RT}\int_0^p (V_m - V_{perfect,m})\mathrm{d}p$$

Then, with $\phi = f/p$,

$$\ln \phi = \frac{1}{RT}\int_0^p (V_m - V_{perfect,m})\mathrm{d}p$$

For a perfect gas, $V_{perfect,m} = RT/p$. For a real gas, $V_m = RTZ/p$. With these two substitutions, we obtain eqn 20.

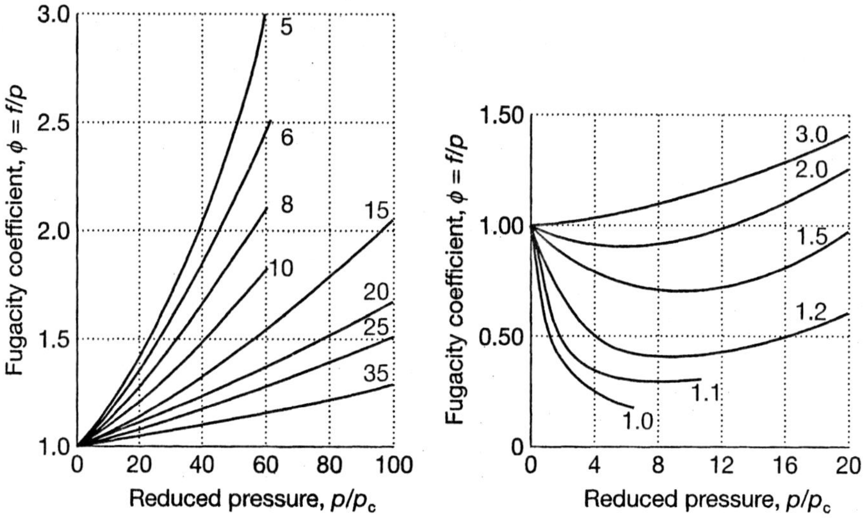

5.8 The fugacity coefficient of a van der Waals gas plotted using the reduced variables of the gas. The curves are labelled with the reduced temperature $T_r = T/T_c$.

Table 5.2* The fugacity of nitrogen at 273 K

p/atm	f/atm
1	0.999 55
10	9.9560
100	97.03
1000	1839

* More values are given in the *Data section*.

We see from Fig. 1.13 that for most gases $Z < 1$ up to moderate pressures, but that $Z > 1$ at higher pressures. If $Z < 1$ throughout the range of integration, then the integrand in eqn 20 is negative and $\phi < 1$. This implies that $f < p$ (the molecules tend to stick together) and that the molar Gibbs energy of the gas is less than that of a perfect gas. At higher pressures, the range over which $Z > 1$ may dominate the range over which $Z < 1$. The integral is then positive, $\phi > 1$, and $f > p$ (the repulsive interactions are dominant and tend to drive the particles apart). Now the molar Gibbs energy of the gas is greater than that of the perfect gas at the same pressure. Figure 5.8, which has been calculated using the full van der Waals equation of state, shows how the fugacity coefficient depends on the pressure in terms of the reduced variables (Section 1.5). Because critical constants are available in Table 1.4, the graphs can be used for quick estimates of the fugacities of a wide range of gases. Table 5.2 gives some explicit values for nitrogen.

Checklist of key ideas

Combining the First and Second Laws

5.1 The fundamental equation

- ☐ fundamental equation, $dU = T\,dS - p\,dV$

5.2 Properties of the internal energy

- ☐ Maxwell relations (Table 5.1)
- ☐ thermodynamic equation of state, an expression for pressure in terms of thermodynamic quantities, $\pi_T = T(\partial p/\partial T)_V - p$

Properties of the Gibbs energy

5.3 General considerations

- ☐ general change in Gibbs energy, $dG = V\,dp - S\,dT$
- ☐ the variation of Gibbs energy with pressure, $(\partial G/\partial p)_T = V$
- ☐ the variation of Gibbs energy with temperature, $(\partial G/\partial T)_p = -S$

5.4 The variation of the Gibbs energy with temperature

- ☐ Gibbs–Helmholtz equation, $(\partial(G/T)/\partial T)_p = -H/T^2$

5.5 The variation of the Gibbs energy with pressure

- ☐ for a condensed phase, $G_m(p_f) = G_m(p_i) + V_m\Delta p$
- ☐ for a perfect gas, $G(p_f) = G(p_i) + nRT\ln(p_f/p_i)$
- ☐ fugacity, the effective pressure of a gas, $G_m = G_m^{\ominus} + RT\ln(f/p^{\ominus})$
- ☐ fugacity coefficient, $\phi = f/p$, with $\ln\phi = \int_0^p \{(Z-1)/p\}dp$

Further reading

Articles of general interest

L.L. Combs, An alternative view of fugacity. *J. Chem. Educ.* **69**, 218 (1992).

L. D'Alessio, On the fugacity of a van der Waals gas: An approximate expression that separates attractive and repulsive forces. *J. Chem. Educ.* **70**, 96 (1993).

R.M. Noyes, Thermodynamics of a process in a rigid container. *J. Chem. Educ.* **69**, 470 (1992).

R.M. Noyes, Application of the Gibbs function to chemical systems and subsystems. *J. Chem. Educ.* **73**, 404 (1996).

R.J. Tykodi, Spontaneity, accessibility, irreversibility, 'useful work': the availability function, the Helmholtz function, and the Gibbs function. *J. Chem. Educ.* **72**, 103 (1995).

R.J. Tykodi, The Gibbs function, spontaneity, and walls. *J. Chem. Educ.* **73**, 398 (1996).

J.S. Winn, The fugacity of van der Waals gas. *J. Chem. Educ.* **65**, 772 (1988).

S.E. Wood and R. Battino, The Gibbs function controversy. *J. Chem. Educ.* **73**, 408 (1996).

Texts and sources of data and information

I.M. Klotz and R.M. Rosenberg, *Chemical thermodynamics: basic theory and methods*. Wiley-Interscience, New York (2000).

G.N. Lewis and M. Randall, *Thermodynamics* (revised by K.S. Pitzer and L. Brewer). McGraw-Hill, New York (1961).

R.G. Mortimer, *Mathematics for physical chemistry*. Academic Press, San Diego (1999).

J.M. Smith and H.C. Van Ness, *Introduction to chemical engineering thermodynamics*. McGraw-Hill, New York (1987).

M.A. White, *Properties of materials*. Oxford University Press, Oxford (1999).

B.D. Wood, *Applications of thermodynamics*. Addison-Wesley, New York (1982).

Exercises

Discussion questions

5.1 (a) Discuss the physical interpretation of any one Maxwell relation.

5.1 (b) Account for the dependence of π_T of a van der Waals gas in terms of the significance of the parameters a and b.

5.2 (a) Suggest a physical interpretation of the dependence of the Gibbs energy on the pressure.

5.2 (b) Suggest a physical interpretation of the dependence of the Gibbs energy on the temperature.

5.3 (a) Explain the significance of the fugacity of a gas.

5.3 (b) Suggest a physical interpretation of the variation of the fugacity coefficient of a van der Waals gas as depicted in Fig. 5.8.

Numerical exercises

Assume all gases are perfect and that the temperature is 298.15 K unless stated otherwise.

5.4 (a) Express $(\partial S/\partial V)_T = (\partial p/\partial T)_V$ in terms of α and κ_T (see eqns 3.8 and 3.14 for definitions).

5.4 (b) Express $(\partial S/\partial p)_T = -(\partial V/\partial T)_p$ in terms of α (see eqn 3.8 for definition).

5.5 (a) Suppose that 3.0 mmol N_2(g) occupies 36 cm^3 at 300 K and expands to 60 cm^3. Calculate ΔG for the process.

5.5 (b) Suppose that 2.5 mmol Ar(g) occupies 72 dm^3 at 298 K and expands to 100 dm^3. Calculate ΔG for the process.

5.6 (a) The change in the Gibbs energy of a certain constant-pressure process was found to fit the expression $\Delta G/\mathrm{J} = -85.40 + 36.5(T/\mathrm{K})$. Calculate the value of ΔS for the process.

5.6 (b) The change in the Gibbs energy of a certain constant-pressure process was found to fit the expression $\Delta G/\mathrm{J} = -73.1 + 42.8(T/\mathrm{K})$. Calculate the value of ΔS for the process.

5.7 (a) Calculate the change in Gibbs energy of 35 g of ethanol (mass density 0.789 g cm^{-3}) when the pressure is increased isothermally from 1 atm to 3000 atm.

5.7 (b) Calculate the change in Gibbs energy of 25 g of methanol (mass density 0.791 g cm^{-3}) when the pressure is increased isothermally from 100 kPa to 100 MPa.

5.8 (a) When 2.00 mol of a gas at 330 K and 3.50 atm is subjected to isothermal compression, its entropy decreases by 25.0 J K^{-1}. Calculate (a) the final pressure of the gas and (b) ΔG for the compression.

5.8 (b) When 3.00 mol of a gas at 230 K and 150 kPa is subjected to isothermal compression, its entropy decreases by 15.0 J K^{-1}. Calculate (a) the final pressure of the gas and (b) ΔG for the compression.

5.9 (a) Calculate the change in molar Gibbs energy of a perfect gas when its pressure is increased isothermally from 1.8 atm to 29.5 atm at 40°C.

5.9 (b) Calculate the change in molar Gibbs energy of a perfect gas when its pressure is increased isothermally from 92.0 kPa to 252.0 kPa at 50°C.

5.10 (a) The fugacity coefficient of a certain gas at 200 K and 50 bar is 0.72. Calculate the difference of its chemical potential from that of a perfect gas in the same state.

5.10 (b) The fugacity coefficient of a certain gas at 290 K and 2.1 MPa is 0.68. Calculate the difference of its chemical potential from that of a perfect gas in the same state.

5.11 (a) At 373 K, the second virial coefficient B of xenon is $-81.7\ \mathrm{cm^3\ mol^{-1}}$. Calculate the value of B' and hence estimate the fugacity coefficient of xenon at 50 atm and 373 K.

5.11 (b) At 100 K, the second virial coefficient B of nitrogen is $-160.0\ \mathrm{cm^3\ mol^{-1}}$. Calculate the value of B' and hence estimate the fugacity coefficient of nitrogen at 62 MPa and 100 K.

5.12 (a) Estimate the change in the Gibbs energy of 1.0 L of benzene when the pressure acting on it is increased from 1.0 atm to 100 atm.

5.12 (b) Estimate the change in the Gibbs energy of 1.0 L of water when the pressure acting on it is increased from 100 kPa to 300 kPa.

5.13 (a) Calculate the change in the molar Gibbs energy of hydrogen gas when its pressure is increased isothermally from 1.0 atm to 100.0 atm at 298 K.

5.13 (b) Calculate the change in the molar Gibbs energy of oxygen when its pressure is increased isothermally from 50.0 kPa to 100.0 kPa at 500 K.

5.14 (a) The molar Helmholtz energy of a certain gas is given by:

$$A_m = -\frac{a}{V_m} - RT\ln(V_m - b) + f(T)$$

where a and b are constants and $f(T)$ is a function of temperature only. Obtain the equation of state of the gas.

5.14 (b) The molar Gibbs energy of a certain gas is given by:

$$G_m = RT\ln p + A + Bp + \tfrac{1}{2}Cp^2 + \tfrac{1}{3}Dp^3$$

where A, B, C, and D are constants. Obtain the equation of state of the gas.

5.15 (a) Evaluate $(\partial S/\partial V)_T$ for a van der Waals gas. For an isothermal expansion, will ΔS be greater for a perfect gas or a van der Waals gas? Explain your conclusion.

5.15 (b) Evaluate $(\partial S/\partial V)_T$ for a Dieterici gas (Table 1.6). For an isothermal expansion, will ΔS be greater for a perfect gas or a Dieterici gas? Explain your conclusion.

Problems†

Numerical problems

5.1 Calculate $\Delta_r G^\ominus(375\ \mathrm{K})$ for the reaction $2\,CO(g) + O_2(g) \rightarrow 2\,CO_2(g)$ from the value of $\Delta_r G^\ominus$ (298 K), $\Delta_r H^\ominus$ (298 K), and the Gibbs–Helmholtz equation.

5.2 Estimate the standard reaction Gibbs energy of $N_2(g) + 3\,H_2(g) \rightarrow 2\,NH_3(g)$ at (a) 500 K, (b) 1000 K from their values at 298 K.

5.3 At 200 K, the compression factor of oxygen varies with pressure as shown below. Evaluate the fugacity of oxygen at this temperature and 100 atm.

p/atm	1.0000	4.000 00	7.000 00	10.0000	40.00	70.00	100.0
Z	0.9971	0.987 96	0.978 80	0.969 56	0.8734	0.7764	0.6871

5.4‡ In an investigation of thermophysical properties of toluene (*J. Phys. Chem. Ref. Data* **18**, 1565 (1989)), R.D. Goodwin tabulated (among other quantities) the compression factor, Z, at several temperatures and pressures. From the following information, compute the fugacity coefficient of toluene at 600 K and (a) 30.0 bar and (b) 1000 bar.

p/bar	0.500	1.013	2.00	3.00	5.00	
Z	0.994 12	0.988 96	0.979 42	0.969 95	0.951 33	
p/bar	10.00	20.0	30.0	42.4	50.0	
Z	0.905 69	0.812 27	0.701 77	0.471 98	0.223 76	
p/bar	70.0	100.0	200	300	500	1000
Z	0.265 20	0.349 20	0.623 62	0.882 88	1.371 09	2.488 36

Theoretical problems

5.5 Show that, for a perfect gas, $(\partial U/\partial S)_V = T$ and $(\partial U/\partial V)_S = -p$.

5.6 Two of the four Maxwell relations were derived in the text, but two were not. Complete their derivation by showing that $(\partial S/\partial V)_T = (\partial p/\partial T)_V$ and $(\partial T/\partial p)_S = (\partial V/\partial S)_p$.

5.7 Use the Maxwell relations to express the derivatives $(\partial S/\partial V)_T$ and $(\partial V/\partial S)_p$ in terms of the expansion coefficient α and the isothermal compressibility κ_T.

5.8 Use the Maxwell relations and Euler's chain relation to express $(\partial p/\partial S)_V$ and $(\partial V/\partial S)_p$ in terms of the heat capacities, the expansion coefficient, and the isothermal compressibility.

5.9 Use the Maxwell relations to show that the entropy of a perfect gas depends on the volume as $S \propto R\ln V$.

5.10 Derive the thermodynamic equation of state

$$\left(\frac{\partial H}{\partial p}\right)_T = V - T\left(\frac{\partial V}{\partial T}\right)_p$$

Derive an expression for $(\partial H/\partial p)_T$ for (a) a perfect gas and (b) a van der Waals gas. In the latter case, estimate its value for 1.0 mol Ar(g) at 298 K and 10 atm. By how much does the enthalpy of the argon change when the pressure is increased isothermally to 11 atm?

5.11 Prove the following relation:

$$\left(\frac{\partial H}{\partial V}\right)_T = -V^2\left(\frac{\partial p}{\partial T}\right)_V\left(\frac{\partial (T/V)}{\partial V}\right)_p$$

† Problems denoted with the symbol ‡ were supplied by Charles Trapp, Carmen Giunta, and Marshall Cady.

5.12 Show that, if $B(T)$ is the second virial coefficient of a gas, and $\Delta B = B(B'') - B(T')$, $\Delta T = T'' - T'$, and T is the mean of T'' and T', then

$$\pi_T \approx \frac{RT^2\Delta B}{V_m^2 \Delta T}$$

Estimate π_T for argon given that $B(250\text{ K}) = -28.0\text{ cm}^3\text{ mol}^{-1}$ and $B(300\text{ K}) = -15.6\text{ cm}^3\text{ mol}^{-1}$ at 275 K at (a) 1.0 atm, (b) 10.0 atm.

5.13 (a) Prove that the heat capacities C_V and C_p of a perfect gas are independent of both volume and pressure. May they depend on the temperature? (b) Deduce an expression for the dependence of C_V on volume of a gas that is described by the equation of state $pV_m/RT = 1 + B/V_m$.

5.14 The Joule coefficient, μ_J, is defined as $\mu_J = (\partial T/\partial V)_U$. Show that $\mu_J C_V = p - (\alpha T/\kappa_T)$.

5.15 Evaluate π_T for a Dieterici gas (Table 1.6). Justify physically the form of the expression obtained.

5.16‡ The second virial coefficient of methane can be approximated by the empirical equation $B'(T) = a + be^{-c/T^2}$, where $a = -0.1993\text{ bar}^{-1}$, $b = 0.2002\text{ bar}^{-1}$, and $c = 1131\text{ K}^2$ with $300\text{ K} < T < 600\text{ K}$. Derive an equation for the computation of each of the following properties for methane: (a) π_T at 25°C and 100 bar, (b) ΔS_m, ΔH_m, and ΔG_m, when the gas is compressed isothermally from 1 bar to 100 bar at 25°C. Assume that $B(T) = RTB'(T)$.

5.17 Instead of assuming that the volume of a condensed phase is constant when pressure is applied, assume only that the compressibility is constant. Show that when the pressure is changed isothermally by Δp, G changes to

$$G' = G + V_m\Delta p(1 - \tfrac{1}{2}\kappa_T\Delta p)$$

Assess the error in assuming that a solid is incompressible by applying this expression to the compression of copper when $\Delta p = 500$ atm. (For copper at 25°C, $\kappa_T = 0.8 \times 10^{-6}\text{ atm}^{-1}$ and $\rho = 8.93\text{ g cm}^{-3}$.)

5.18 Derive an expression for the standard reaction Gibbs energy $\Delta_r G^\ominus$ (the analogue of the standard reaction enthalpy) at a temperature T' in terms of its value $\Delta_r G^\ominus$ at T by using the Gibbs–Helmholtz equation and (a) assuming that $\Delta_r H^\ominus$ does not vary with temperature, (b) assuming instead that $\Delta_r C_p$ does not vary with temperature and using Kirchhoff's law.

5.19 The adiabatic compressibility, κ_S, is defined like κ_T (eqn 3.14) but at constant entropy. Show that for a perfect gas $p\gamma\kappa_S = 1$ (where γ is the ratio of heat capacities).

5.20 Show that, if S is regarded as a function of T and p, then

$$T\,dS = C_V\,dT + T\left(\frac{\partial p}{\partial T}\right)_V dV$$

Calculate the energy that must be transferred as heat to a van der Waals gas that expands reversibly and isothermally from V_i to V_f.

5.21 Suppose that S is regarded as a function of p and T. Show that

$$T\,dS = C_p\,dT - \alpha TV\,dp$$

Hence, show that the energy transferred as heat when the pressure on an incompressible liquid or solid is increased by Δp is equal to $-\alpha TV\Delta p$. Evaluate q when the pressure acting on 100 cm³ of mercury at 0°C is increased by 1.0 kbar. ($\alpha = 1.82 \times 10^{-4}\text{ K}^{-1}$.)

5.22‡ At 1 atm, liquid water has its maximum density at 4°C. What can be concluded about the variation of the entropy of liquid water with pressure at constant temperature at 3°C, 4°C, and 5°C?

5.23‡ You are told that the differential of pressure consistent with an equation of state is given by *one* of the following two expressions. Determine the equation of state.

$$dp = \frac{2nRT\,dV}{(V - nb)^2} + \frac{R(V - nb)\,dT}{nb^2}$$

$$dp = -\frac{nRT\,dV}{(V - nb)^2} + \frac{nR\,dT}{V - nb}$$

5.24 Suppose that the attractive interactions between gas particles can be neglected, and find an expression for the fugacity of a van der Waals gas in terms of the pressure. Estimate its value for ammonia at 10.00 atm and 298.15 K.

5.25 Find an expression for the fugacity coefficient when the attractive interaction is dominant in a van der Waals gas, and the pressure is low enough to make the approximation $4ap/(RT)^2 \ll 1$. Evaluate the fugacity for ammonia at 10.00 atm and 298.15 K.

5.26 Find an expression for the fugacity coefficient of a gas that obeys the equation of state

$$\frac{pV_m}{RT} = 1 + \frac{B}{V_m} + \frac{C}{V_m^2}$$

Use the resulting expression to estimate the fugacity of argon at 1.00 atm and 100 K using $B = -21.13\text{ cm}^3\text{ mol}^{-1}$ and $C = 1054\text{ cm}^6\text{ mol}^{-2}$.

5.27 Derive an expression for the fugacity coefficient of a gas that obeys the equation of state

$$\frac{pV_m}{RT} = 1 + \frac{qT}{V_m}$$

where q is a constant, and plot ϕ against $4pq/R$.

Applications: to biology, environmental science, and polymer science

5.28 At 298 K the standard enthalpy of combustion of sucrose is -5797 kJ mol^{-1} and the standard Gibbs energy of the reaction is -6333 kJ mol^{-1}. Estimate the additional (non-expansion) work that may be obtained by raising the temperature to blood temperature, 37°C.

5.29‡ In 1995, the Intergovernmental Panel on Climate Change considered a global average temperature rise of 1.0–3.5°C likely by the year 2100, with 2.0°C its best estimate (IPCC Second Assessment Synthesis of Scientific–Technical Information Relevant to Interpreting Article 2 of the UN Framework Convention on Climate Change (1995)). Because water vapour is itself a greenhouse gas, the increase in water vapour content of the atmosphere is of some concern to climate change experts. Predict the relative increase in water vapour in the atmosphere based on a temperature rise of 2.0 K, assuming that the relative humidity remains constant. (The present global mean temperature is 290 K, and the equilibrium vapour pressure of water at that temperature is 0.0189 bar.)

5.30‡ Protocols have outlawed the use of certain CFC refrigerants. It has been suggested that sodium oxalate, $Na_2(CO_2)_2$, at elevated temperatures can be used to destroy existing CFC stockpiles. The mineralization of freon-12, for example, is

$$CF_2Cl_2(g) + 2\,Na_2(CO_2)_2(s) \longrightarrow 2\,NaF(s) + 2\,NaCl(s) + C(s) + 4\,CO_2(g)$$

(a) Without performing a calculation, what is your expectation concerning the spontaneity of the reaction? Why is the high temperature required? Why should reaction mixtures not be heated to temperatures much higher than 270°C? (b) Use library, internet, or database resources to find thermodynamic data that will make possible the calculation of $\Delta_r G^{\ominus}(T)$. Estimate the value of any thermodynamic property that cannot be found. Consult the web site of this text for a list of internet resources.

5.31‡ Nitric acid hydrates have received much attention as possible catalysts for heterogeneous reactions which bring about the Antarctic ozone hole. Worsnop *et al.* investigated the thermodynamic stability of these hydrates under conditions typical of the polar winter stratosphere (D.R. Worsnop, L.E. Fox, M.S. Zahniser, and S.C. Wofsy, *Science* **259**, 71 (1993)). They report thermodynamic data for the sublimation of mono-, di-, and trihydrates to nitric acid and water vapours, $HNO_3 \cdot nH_2O(s) \rightarrow HNO_3(g) + n\,H_2O(g)$, for $n = 1, 2$, and 3. Given $\Delta_r G^{\ominus}$ and $\Delta_r H^{\ominus}$ for these reactions at 220 K, use the Gibbs–Helmholtz equation to compute $\Delta_r G^{\ominus}$ at 190 K.

n	1	2	3
$\Delta_r G^{\ominus}/(\text{kJ mol}^{-1})$	46.2	69.4	93.2
$\Delta_r H^{\ominus}/(\text{kJ mol}^{-1})$	127	188	237

5.32‡ J. Gao and J.H. Weiner in their study of the origin of stress on the atomic level in dense polymer systems (*Science* **266**, 748 (1994)) observe that the tensile force required to maintain the length, l, of a long linear chain of N freely jointed links each of length, a, can be interpreted as arising from an entropic spring. For such a chain, $S(l) = -3kl^2/2Na^2 + C$, where k is the Boltzmann constant and C is a constant. Using thermodynamic relations of this and previous chapters, show that the tensile force obeys Hooke's law, $f = -k_f l$, if we assume that the energy U is independent of l.

5.33 The volume of a newly synthesized polymer was found to depend exponentially on the pressure as $V = V_0 e^{-p/p^*}$, where p is the excess pressure and p^* is a constant. Deduce an expression for the Gibbs energy of the polymer as a function of excess pressure. What is the natural direction of change of the compressed material when the pressure is relaxed?

6

Physical transformations of pure substances

The simplest applications of thermodynamics to chemically significant systems are to the discussion of the phase transitions of pure substances. We shall see that a phase diagram is a map of the pressures and temperatures at which each phase of a substance is the most stable. First, we describe the interpretation of empirically determined phase diagrams for a selection of materials. Then we turn to a consideration of the factors that determine the positions and shapes of the boundaries between the regions on a phase diagram. The practical importance of the expressions we derive is that they show how the vapour pressure of a substance varies with temperature and how the melting point varies with pressure. We shall see that the transitions between phases can be classified by noting how various thermodynamic functions change when the transition occurs. This chapter also introduces the chemical potential, a property that is at the centre of discussions of phase transitions and chemical reactions. Finally, we consider the thermodynamic discussion of liquid surfaces, such as surface tension and capillary action.

Vaporization, melting, and the conversion of graphite to diamond are all examples of changes of phase without change of chemical composition. In this chapter we describe such processes thermodynamically, using as the guiding principle the tendency of systems at constant temperature and pressure to minimize their Gibbs energy.

Phase diagrams

One of the most succinct ways of presenting the physical changes of state that a substance can undergo is in terms of its phase diagram. We present the concept in this section.

6.1 The stabilities of phases

A **phase** of a substance is a form of matter that is uniform throughout in chemical composition and physical state. Thus, we speak of solid, liquid, and gas phases of a substance, and of its various solid phases, such as the white and black allotropes of

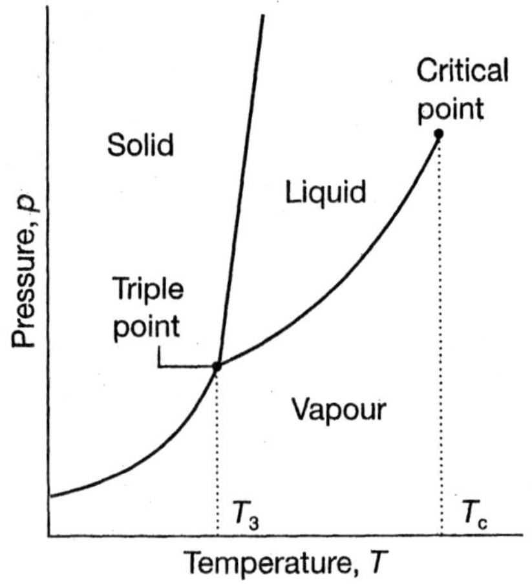

6.1 The general regions of pressure and temperature where solid, liquid, or gas is stable (that is, has minimum molar Gibbs energy) are shown on this phase diagram. For example, the solid phase is the most stable phase at low temperatures and high pressures. In the following paragraphs we locate the precise boundaries between the regions.

phosphorus. A **phase transition**, the spontaneous conversion of one phase into another phase, occurs at a characteristic temperature for a given pressure. Thus, at 1 atm, ice is the stable phase of water below 0°C, but above 0°C liquid water is more stable. This difference indicates that below 0°C the Gibbs energy decreases as liquid water changes into ice and that above 0°C the Gibbs energy decreases as ice changes into liquid water. The **transition temperature**, T_{trs}, is the temperature at which the two phases are in equilibrium and the Gibbs energy is minimized at the prevailing pressure.

We must distinguish between the thermodynamic description of a phase transition and the rate at which the transition occurs. A transition that is predicted from thermodynamics to be spontaneous may occur too slowly to be significant in practice. For instance, at normal temperatures and pressures the molar Gibbs energy of graphite is lower than that of diamond, so there is a thermodynamic tendency for diamond to change into graphite. However, for this transition to take place, the C atoms must change their locations, which is an immeasurably slow process in a solid except at high temperatures. The rate of attainment of equilibrium is a kinetic problem and is outside the range of thermodynamics. In gases and liquids the mobilities of the molecules allow phase transitions to occur rapidly, but in solids thermodynamic instability may be frozen in. Thermodynamically unstable phases that persist because the transition is kinetically hindered are called **metastable phases**. Diamond is a metastable phase of carbon under normal conditions.

6.2 Phase boundaries

The **phase diagram** of a substance shows the regions of pressure and temperature at which its various phases are thermodynamically stable (Fig. 6.1). The lines separating the regions, which are called **phase boundaries**, show the values of p and T at which two phases coexist in equilibrium.

Consider a liquid sample of a pure substance in a closed vessel. The pressure of a vapour in equilibrium with the liquid is called the **vapour pressure** of the substance (Fig. 6.2). Therefore, the liquid–vapour phase boundary in a phase diagram shows how the vapour pressure of the liquid varies with temperature. Similarly, the solid–vapour phase boundary shows the temperature variation of the **sublimation vapour pressure**, the vapour pressure of the solid phase. The vapour pressure of a substance increases with temperature because at higher temperatures the Boltzmann distribution populates more heavily the states of higher energy corresponding to molecules that have escaped from their neighbours.

The web site is a good source of links to online databases of data on phase transitions.

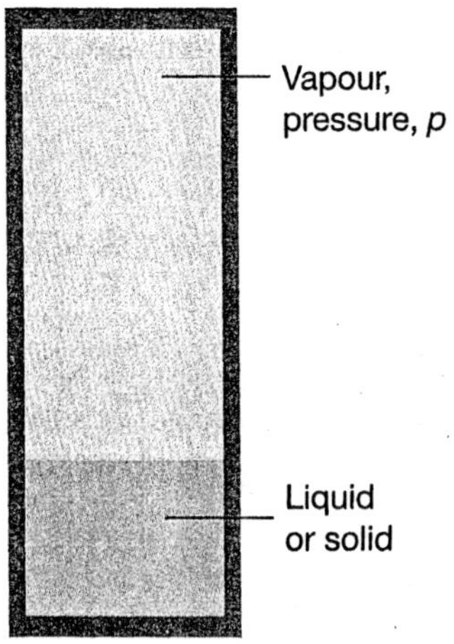

6.2 The vapour pressure of a liquid or solid is the pressure exerted by the vapour in equilibrium with the condensed phase.

(a) Critical points and boiling points

When a liquid is heated in an open vessel, the liquid vaporizes from its surface. At the temperature at which its vapour pressure would be equal to the external pressure, vaporization can occur throughout the bulk of the liquid and the vapour can expand freely into the surroundings. The condition of free vaporization throughout the liquid is called **boiling**. The temperature at which the vapour pressure of a liquid is equal to the external pressure is called the **boiling temperature** at that pressure. For the special case of an external pressure of 1 atm, the boiling temperature is called the **normal boiling point**, T_b. With the replacement of 1 atm by 1 bar as standard pressure, there is some advantage in using the **standard boiling point** instead: this is the temperature at which the vapour pressure reaches 1 bar. Because 1 bar is slightly less than 1 atm (1.00 bar = 0.987 atm), the standard boiling point of a liquid is slightly lower than its normal boiling point. The normal boiling point of water is 100.0°C; its standard boiling point is 99.6°C.

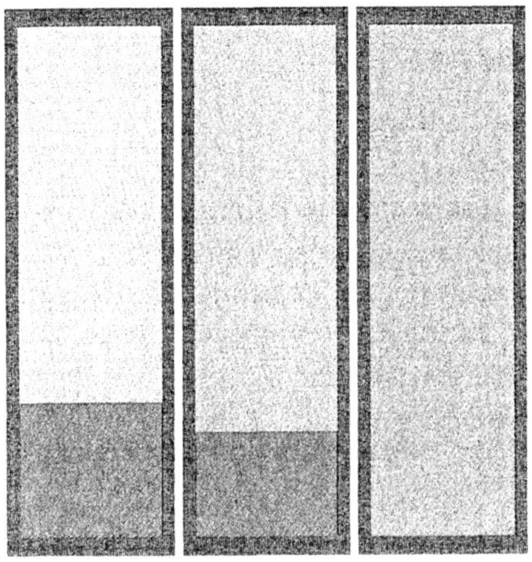

6.3 (a) A liquid in equilibrium with its vapour. (b) When a liquid is heated in a sealed container, the density of the vapour phase increases and that of the liquid decreases slightly. There comes a stage, (c), at which the two densities are equal and the interface between the fluids disappears. This disappearance occurs at the critical temperature. The container needs to be strong: the critical temperature of water is 374°C and the vapour pressure is then 218 atm.

Boiling does not occur when a liquid is heated in a closed vessel. Instead, the vapour pressure, and hence the density of the vapour, rise continuously as the temperature is raised (Fig. 6.3). At the same time, the density of the liquid decreases slightly as a result of its expansion. There comes a stage when the density of the vapour is equal to that of the remaining liquid and the surface between the two phases disappears. The temperature at which the surface disappears is the **critical temperature**, T_c, of the substance. We first encountered this property in Section 1.3d. The vapour pressure at the critical temperature is called the **critical pressure**, p_c. At and above the critical temperature, a single uniform phase called a **supercritical fluid** fills the container and an interface no longer exists. That is, above the critical temperature, the liquid phase of the substance does not exist (Box 6.1).

(b) Melting points and triple points

The temperature at which, under a specified pressure, the liquid and solid phases of a substance coexist in equilibrium is called the **melting temperature**. Because a substance melts at exactly the same temperature as that at which it freezes, the melting temperature of a substance is the same as its **freezing temperature**. The freezing temperature when the pressure is 1 atm is called the **normal freezing point**, T_f, and its freezing point when the pressure is 1 bar is called the **standard freezing point**. The normal and standard freezing points are negligibly different for most purposes. The normal freezing point is also called the **normal melting point**.

There is a set of conditions under which three different phases of a substance (typically solid, liquid, and vapour) all simultaneously coexist in equilibrium. These conditions are represented by the **triple point**, a point at which the three phase boundaries meet. The temperature at the triple point is denoted T_3. The triple point of a pure substance is outside our control: it occurs at a single definite pressure and

Box 6.1 Supercritical fluids

Supercritical carbon dioxide, $scCO_2$, is the centre of attention for an increasing number of solvent-based processes. The critical temperature 304.2 K (31.0°C) and pressure (72.9 atm) are readily accessible and carbon dioxide is cheap. The density of $scCO_2$ at its critical point is 0.45 g cm^{-3}. However, the transport properties of any supercritical fluid depend strongly on its density, which in turn is sensitive to the pressure and temperature. For instance, densities may be adjusted from a gas-like 0.1 g cm^{-3} to a liquid-like 1.2 g cm^{-3}. A useful rule of thumb is that the solubility of a solute is an exponential function of the density of the supercritical fluid, so small increases in pressure, particularly close to the critical point, can have very large effects on solubility.

A great advantage of $scCO_2$ is that there are no noxious residues once the solvent has been allowed to evaporate, so, coupled with its low critical temperature, $scCO_2$ is ideally suited to food processing and the production of pharmaceuticals. It is used, for instance, to remove caffeine from coffee. The supercritical fluid is also increasingly being used for dry cleaning, which avoids the use of carcinogenic and environmentally deleterious chlorinated hydrocarbons.

Supercritical CO_2 has been used since the 1960s as a mobile phase in *supercritical fluid chromatography* (SFC), but it fell out of favour when the more convenient technique of high-performance liquid chromatography (HPLC) was introduced. However, interest in SFC has returned, and there are separations possible in SFC that cannot easily be achieved by HPLC, such as the separation of lipids and of phospholipids. Samples as small as 1 pg can be analysed. The essential advantage of SFC is that diffusion coefficients in supercritical fluids are an order of magnitude greater than in liquids, so there is less resistance to the transfer of solutes through the column, with the result that separations may be effected rapidly or with high resolution.

The principal problem with $scCO_2$, though, is that the fluid is not a very good solvent and surfactants are needed to induce many potentially interesting solutes to dissolve. Indeed, $scCO_2$-based dry cleaning depends on the availability of cheap surfactants; so too does the use of $scCO_2$ as a solvent for homogeneous catalysts, such as metal complexes. There appear to be two principal approaches to solving the solubilizing problem. One solution is to use fluorinated and siloxane-based polymeric stabilizers, which allow polymerization reactions to proceed in $scCO_2$. The disadvantage of these stabilizers for commercial use is their great expense. An alternative and much cheaper approach is poly(ether-carbonate) copolymers. The copolymers can be made more soluble in $scCO_2$ by adjusting the ratio of ether and carbonate groups.

The critical temperature of water is 374°C and its pressure is 218 atm. The conditions for using scH_2O are therefore much more demanding than for $scCO_2$ and the properties of the fluid are highly sensitive to pressure. Thus, as the density of scH_2O decreases, the characteristics of a solution change from those of an aqueous solution through those of a non-aqueous solution and eventually to those of a gaseous solution. One consequence is that reaction mechanisms may change from ionic to radical.

Discussion questions

1.[1] The use of supercritical fluids as mobile phases in SFC depends on their properties as nonpolar solvents. The solubility parameter, δ, is defined as $(\Delta U_{\text{cohesive}}/V_{\text{m}})^{1/2}$, where $\Delta U_{\text{cohesive}}$ is the cohesive energy of the solvent, the energy per mole needed to increase the volume isothermally to an infinite value. Diethyl ether, carbon tetrachloride, and dioxane have solubility parameter ranges of 7–8, 8–9, and 10–11, respectively. (a) Derive a practical equation for the computation of the isotherms for the reduced internal energy change, $\Delta U_{\text{r}}(T_{\text{r}}, V_{\text{r}})$ defined as

$$\Delta U_{\text{r}}(T_{\text{r}}, V_{\text{r}}) = \frac{U(T_{\text{r}}, V_{\text{r}}) - U(T_{\text{r}}, \infty)}{p_{\text{c}}V_{\text{c}}}$$

(b) Draw a graph of ΔU_{r} against p_{r} for the isotherms $T_{\text{r}} = 1, 1.2$, and 1.5 in the reduced pressure range for which $0.7 \le V_{\text{r}} \le 2$. (c) Draw a graph of δ against p_{r} for the carbon dioxide isotherms $T_{\text{r}} = 1$ and 1.5 in the reduced pressure range for which $1 \le V_{\text{r}} \le 3$. In what pressure range at $T_{\text{f}} = 1$ will carbon dioxide have solvent properties similar to those of liquid carbon tetrachloride? (*Hint.* Use mathematical software or a spreadsheet.)

2. The use of supercritical fluids for the extraction of a component from a complicated mixture is not confined to the decaffeination of coffee. Consult library and internet resources and prepare a discussion of the principles, advantages, disadvantages, and current uses of supercritical fluid extraction technology. (*Hint.* see *Further reading* below for ideas.)

1. Provided by Marshal Cady and Charles Trapp.

Further reading

C.L. Phelps, N.G. Smart, and C.M. Wai, Past, present, and future applications of supercritical fluid extraction technology. *J. Chem. Educ.* **73**, 1163 (1996).

T.L. Chester, J.D. Pinkston, and D.E. Raynie, Supercritical fluid chromatography and extraction. *Analytical Chemistry* **70**, 301 (1998).

temperature characteristic of the substance. The triple point of water lies at 273.16 K and 611 Pa (6.11 mbar, 4.58 Torr), and the three phases of water (ice, liquid water, and water vapour) coexist in equilibrium at no other combination of pressure and temperature. This invariance of the triple point is the basis of its use in the definition of the thermodynamic temperature scale (Section 4.2c).

As we can see from Fig. 6.1, the triple point marks the lowest pressure at which a liquid phase of a substance can exist. If (as is common) the slope of the solid–liquid phase boundary is as shown in the diagram, then the triple point also marks the lowest temperature at which the liquid can exist; the critical temperature is the upper limit.

6.3 Three typical phase diagrams

We shall now see how these general features appear in the phase diagrams of pure substances.

(a) Water

Figure 6.4 is the phase diagram for water. The liquid–vapour boundary in the phase diagram summarizes how the vapour pressure of liquid water varies with temperature. It also summarizes how the boiling temperature varies with pressure: we simply read off the temperature at which the vapour pressure is equal to the prevailing atmospheric pressure. The solid–liquid boundary shows how the melting temperature

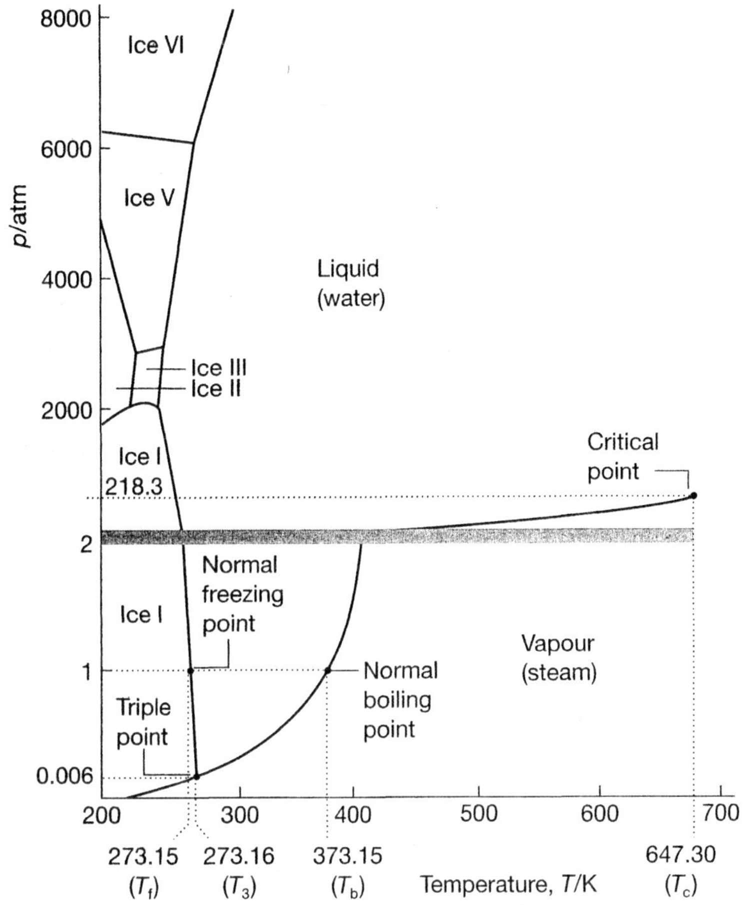

6.4 The experimental phase diagram for water showing the different solid phases. Note the change of vertical scale at 2 atm.

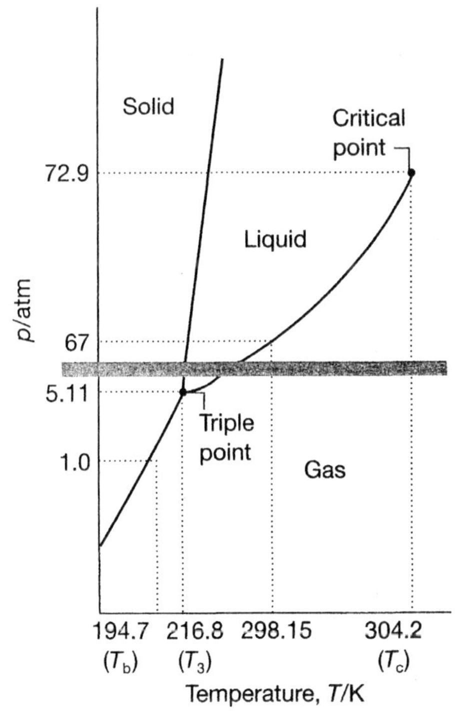

6.5 The experimental phase diagram for carbon dioxide. Note that, as the triple point lies at pressures well above atmospheric, liquid carbon dioxide does not exist under normal conditions (a pressure of at least 5.1 atm must be applied).

varies with the pressure. Its very steep slope indicates that enormous pressures are needed to bring about significant changes. Notice that the line has a negative slope up to 2 kbar, which means that the melting temperature falls as the pressure is raised. The reason for this almost unique behaviour can be traced to the decrease in volume that occurs on melting, and hence it being more favourable for the solid to transform into the liquid as the pressure is raised. The decrease in volume is a result of the very open molecular structure of ice: the H_2O molecules are held apart (as well as together) by the hydrogen bonds between them, but the structure partially collapses on melting, and the liquid is denser than the solid.

At high pressures, different structural forms of ice come into stability as the hydrogen bonds between molecules are modified by the stress. Some of these phases (which are called ice II, III, V, VI, and VII) melt at high temperatures.[2] Ice VII, for instance, melts at 100°C but exists only above 25 kbar. Note that five more triple points occur in the diagram other than the one where vapour, liquid, and ice I coexist. Each one occurs at a definite pressure and temperature that cannot be changed.

(b) Carbon dioxide

The phase diagram for carbon dioxide is shown in Fig. 6.5. The features to notice include the positive slope of the solid–liquid boundary (the direction of this line is

2. Ice IV was an illusion, or possibly a transient, not thermodynamically stable, phase.

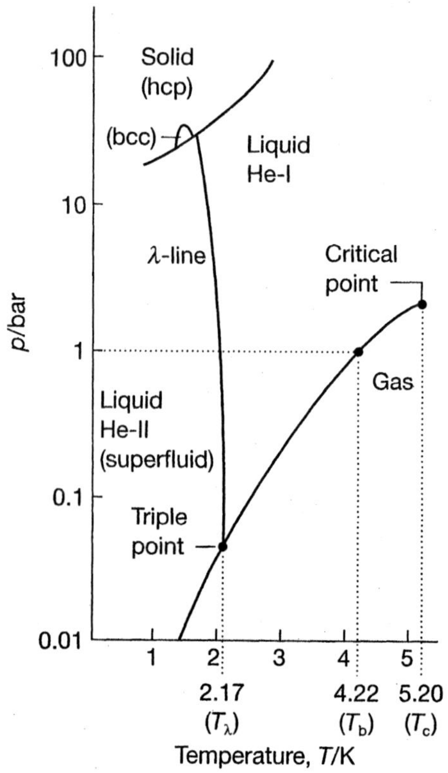

6.6 The phase diagram for helium (^{4}He). The λ-line marks the conditions under which the two liquid phases are in equilibrium. Helium-II is the superfluid phase. Note that a pressure of over 20 bar must be exerted before solid helium can be obtained. The labels hcp and bcc denote different solid phases in which the atoms pack together differently: hcp denotes hexagonal closed packing and bcc denotes body-centred cubic (see Section 21.1 for a description of these structures).

characteristic of most substances), which indicates that the melting temperature of solid carbon dioxide rises as the pressure is increased. Notice also that, as the triple point lies above 1 atm, the liquid cannot exist at normal atmospheric pressures whatever the temperature, and the solid sublimes when left in the open (hence the name 'dry ice'). To obtain the liquid, it is necessary to exert a pressure of at least 5.11 atm. Cylinders of carbon dioxide generally contain the liquid or compressed gas; at 25°C that implies a vapour pressure of 67 atm if both gas and liquid are present in equilibrium. When the gas squirts through the throttle it cools by the Joule–Thomson effect so, when it emerges into a region where the pressure is only 1 atm, it condenses into a finely divided snow-like solid.

(c) Helium

The phase diagram of helium is shown in Fig. 6.6. Helium behaves unusually at low temperatures. For instance, the solid and gas phases are never in equilibrium however low the temperature: the He atoms are so light that they vibrate with a large-amplitude motion even at very low temperatures, and the solid simply shakes itself apart. Solid helium can be obtained, but only by holding the atoms together by applying pressure.

When considering helium at low temperatures it is necessary to distinguish between the isotopes ^{3}He and ^{4}He, because quantum mechanical effects become important and these two isotopes differ not only in mass but also in the spin of their nuclei.[3] At this stage nuclear spin can be visualized as a spinning motion of the nucleus. A ^{4}He nucleus has zero spin; a ^{3}He nucleus has nonzero spin. As a result of quantum mechanical effects, pure helium-4 has a liquid–liquid phase transition at its **λ-line** (lambda line); the reason for this name is explained in Section 6.7. The liquid phase marked He-I behaves like a normal liquid. The other phase, He-II, is a **superfluid**. It is so called because it flows without viscosity. The phase diagram of helium-3 differs from the phase diagram of helium-4, but it also possesses a superfluid phase. Helium-3 is unusual in that the entropy of the liquid is lower than that of the solid, and melting is exothermic.

Phase stability and phase transitions

We shall now see how thermodynamic considerations can account for the features of the phase diagrams we have just described. All our considerations will be based on the Gibbs energy of a substance, and in particular on its molar Gibbs energy, G_m. In fact, this quantity will play such an important role in this chapter and the rest of the text, that we give it a special name and symbol, the **chemical potential**, μ (mu). For a one-component system, 'molar Gibbs energy' and 'chemical potential' are synonyms, so $\mu = G_m$, but in Chapter 7 we see that chemical potential has a broader significance and a more general definition. The name 'chemical potential' is also instructive: as we develop the concept, we shall see that μ is a measure of the potential that a substance has for bringing about change in a system. In this chapter, it reflects the potential of a substance to bring about physical change. In Chapter 9 we see that μ is the potential of a substance to bring about chemical change.

3. Nuclear spin is a quantum mechanical property that will be encountered in more detail in Part 2.

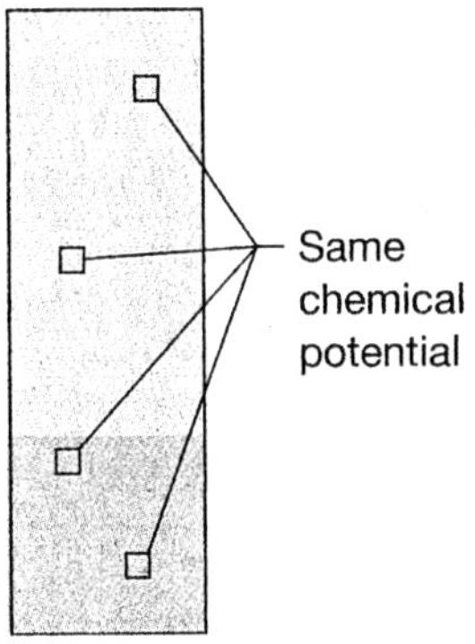

6.7 When two or more phases are in equilibrium, the molar Gibbs energy of a substance (and, in a mixture, a component) is the same in each phase and is the same at all points in each phase.

6.4 The thermodynamic criterion of equilibrium

We base our discussion on the following consequence of the Second Law: *at equilibrium, the chemical potential of a substance is the same throughout a sample, regardless of how many phases are present.* When the liquid and solid phases of a substance are in equilibrium, the chemical potential of the substance is the same throughout the system (Fig. 6.7).

To see the validity of this remark, consider a system in which the chemical potential of a substance is μ_1 at one location and μ_2 at another location. The locations may be in the same or in different phases. When an amount dn of the substance is transferred from one location to the other, the Gibbs energy of the system changes by $-\mu_1 dn$ when material is removed from location 1, and it changes by $+\mu_2 dn$ when that material is added to location 2. The overall change is therefore $dG = (\mu_2 - \mu_1)\, dn$. If the chemical potential at location 1 is higher than that at location 2, the transfer is accompanied by a decrease in G, and so has a spontaneous tendency to occur. Only if $\mu_1 = \mu_2$ is there no change in G, and only then is the system at equilibrium. We conclude that the transition temperature, T_{trs}, is the temperature at which the chemical potentials of two phases are equal.

6.5 The dependence of stability on the conditions

At low temperatures, the solid phase of a substance has the lowest chemical potential and, provided the pressure is not too low, is usually the most stable at low temperatures. However, the chemical potentials of phases change with temperature in different ways, and as the temperature is raised the chemical potential of another phase (perhaps another solid phase, a liquid, or a gas) will fall below that of the solid. When that happens, a phase transition occurs if it is kinetically feasible to do so.

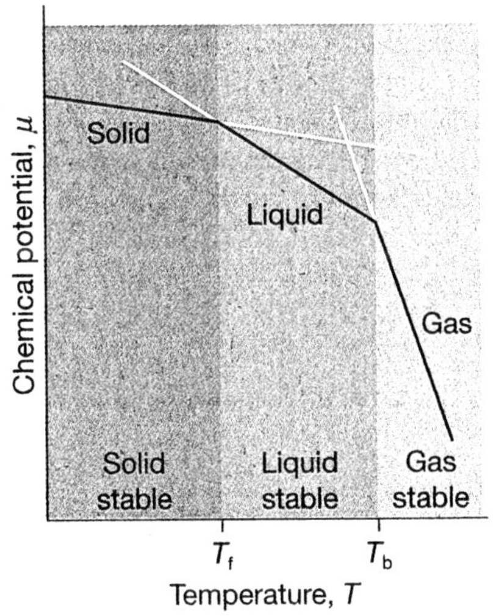

6.8 The schematic temperature dependence of the chemical potential of the solid, liquid, and gas phases of a substance (in practice, the lines are curved). The phase with the lowest molar Gibbs energy at a specified temperature is the most stable one at that temperature. The transition temperatures, the melting and boiling temperatures, are the temperatures at which the chemical potentials of the two phases are equal.

(a) The temperature dependence of phase stability

The temperature dependence of the Gibbs energy is expressed in terms of the entropy of the system by eqn 5.10 ($(\partial G/\partial T)_p = -S$). Because the chemical potential of a pure substance is equal to the molar Gibbs energy of that substance, it follows that

$$\left(\frac{\partial \mu}{\partial T}\right)_p = -S_m \tag{6.1}$$

This relation shows that, as the temperature is raised, the chemical potential of a pure substance decreases: $S_m > 0$ for all substances, so the slope of a plot of μ against T is negative.

Equation 1 implies that the slope of a plot of μ against temperature is steeper for gases than for liquids, because $S_m(g) > S_m(l)$. The slope is also steeper for a liquid than the corresponding solid, because $S_m(l) > S_m(s)$ almost always, on account of the greater disorder of a liquid. These features are illustrated in Fig. 6.8. The steep negative slope of $\mu(l)$ results in its falling below $\mu(s)$ when the temperature is high enough, and then the liquid becomes the stable phase: the solid melts. The chemical potential of the gas phase plunges steeply downwards as the temperature is raised (because the molar entropy of the vapour is so high), and there comes a temperature at which it lies lowest. Then the gas is the stable phase and the liquid vaporizes. When we bring about a phase transition, what we are actually doing is modifying the relative values of the chemical potentials of the phases. The easiest way of doing that is by changing the temperature of the sample.

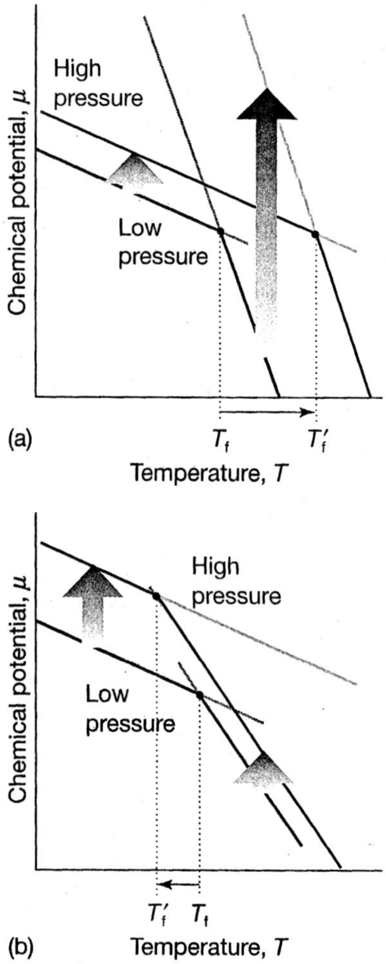

6.9 The pressure dependence of the chemical potential of a substance depends on the molar volume of the phase. The lines show schematically the effect of increasing pressure on the chemical potential of the solid and liquid phases (in practice, the lines are curved), and the corresponding effects on the freezing temperatures. (a) In this case the molar volume of the solid is less than that of the liquid and $\mu(s)$ increases less than $\mu(l)$. As a result, the freezing temperature rises. (b) Here the molar volume is greater for the solid than the liquid (as for water), $\mu(s)$ increases more strongly than $\mu(l)$, and the freezing temperature is lowered.

(b) The response of melting to applied pressure

Most substances melt at a higher temperature when subjected to pressure. It is as though the pressure is preventing the formation of the less dense liquid phase. Exceptions to this behaviour include water, for which the liquid is denser than the solid. Application of pressure to water encourages the formation of the liquid phase. That is, water freezes at a lower temperature when it is under pressure.

We can rationalize the response of melting temperatures to pressure as follows. The variation of the chemical potential with pressure is expressed (from eqn. 5.10) by

$$\left(\frac{\partial \mu}{\partial p}\right)_T = V_m \qquad (6.2)$$

This equation shows that the slope of a plot of chemical potential against pressure is equal to the molar volume of the substance. An increase in pressure raises the chemical potential of any pure substance (because $V_m > 0$). In most cases, $V_m(l) > V_m(s)$ and the equation predicts that an increase in pressure increases the chemical potential of the liquid more than that of the solid. As shown in Fig. 6.9a, the effect of pressure in such a case is to raise the melting temperature slightly. For water, however, $V_m(l) < V_m(s)$, and an increase in pressure increases the chemical potential of the solid more than that of the liquid. In this case, the melting temperature is lowered slightly (Fig. 6.9b).

Example 6.1 Assessing the effect of pressure on the chemical potential

Calculate the effect on the chemical potentials of ice and water of increasing the pressure from 1.00 bar to 2.00 bar at 0°C. The density of ice is 0.917 g cm^{-3} and that of liquid water is 0.999 g cm^{-3} under these conditions.

Method From eqn 2, we know that the change in chemical potential of an incompressible substance when the pressure is changed by Δp is $\Delta\mu = V_m\Delta p$. Therefore, to answer the question, we need to know the molar volumes of the two phases of water. These values are obtained from the mass density, ρ, and the molar mass, M, by using $V_m = M/\rho$. We therefore use the expression

$$\Delta\mu = \frac{M\Delta p}{\rho}$$

Convert all data to SI units and use 1 Pa m^3 = 1 J.

Answer The molar mass of water is 18.02 g mol^{-1} (1.802×10^{-2} kg mol^{-1}); therefore,

$$\Delta\mu(\text{ice}) = \frac{(1.802\times10^{-2}\ \text{kg mol}^{-1})\times(1.00\times10^{5}\ \text{Pa})}{917\ \text{kg m}^{-3}} = +1.97\ \text{J mol}^{-1}$$

$$\Delta\mu(\text{water}) = \frac{(1.802\times10^{-2}\ \text{kg mol}^{-1})\times(1.00\times10^{5}\ \text{Pa})}{999\ \text{kg m}^{-3}} = +1.80\ \text{J mol}^{-1}$$

Comment The chemical potential of ice rises more sharply than that of water; so, if they are initially in equilibrium at 1 bar, there will be a tendency for the ice to melt at 2 bar.

Self-test 6.1 Calculate the effect of an increase in pressure of 1.00 bar on the liquid and solid phases of carbon dioxide (of molar mass 44.0 g mol^{-1}) in equilibrium with densities 2.35 g cm^{-3} and 2.50 g cm^{-3}, respectively.

[$\Delta\mu(l) = +1.87$ J mol^{-1}, $\Delta\mu(s) = +1.76$ J mol^{-1}; solid forms]

(c) The effect of applied pressure on vapour pressure

When pressure is applied to a condensed phase, its vapour pressure rises: in effect, molecules are squeezed out of the phase and escape as a gas. Pressure can be exerted on the condensed phases mechanically or by subjecting it to the applied pressure of an inert gas (Fig. 6.10); in the latter case, the vapour pressure is the partial pressure of the vapour in equilibrium with the condensed phase, and we speak of the **partial vapour pressure** of the substance. One complication (which we ignore here) is that if the condensed phase is a liquid, then the pressurizing gas might dissolve and change the properties of the liquid. Another complication is that the gas-phase molecules might attract molecules out of the liquid by the process of **gas solvation,** the attachment of molecules to gas-phase species.

The quantitative relation between the vapour pressure, p, when a pressure ΔP is applied and the vapour pressure, p^*, of the liquid in the absence of an additional pressure is

$$p = p^* e^{V_m \Delta P/RT} \qquad (6.3a)$$

This equation shows how the vapour pressure increases when the pressure acting on the condensed phase is increased. If $V_m\Delta P/RT \ll 1$, where V_m is the molar volume of the liquid, the exponential function on the right of eqn 3 may be approximated by $1 + V_m\Delta P/RT$:

Calculations involving exponentials often become simpler if we note that $e^x = 1 + x + \frac{1}{2}x^2 + \cdots$. If $x \ll 1$, a good approximation is $e^x \approx 1 + x$.

$$p \approx p^*\left(1 + \frac{V_m \Delta P}{RT}\right)$$

which rearranges to

$$\frac{p - p^*}{p^*} \approx \frac{V_m \Delta P}{RT} \qquad (6.3b)$$

Justification 6.1

We calculate the vapour pressure of a pressurized liquid by using the fact that at equilibrium the chemical potentials of the liquid and its vapour are equal: $\mu(\mathrm{l}) = \mu(\mathrm{g})$. It follows that, for any change that preserves equilibrium, the resulting change in $\mu(\mathrm{l})$ must be equal to the change in $\mu(\mathrm{g})$; therefore, we can write $d\mu(\mathrm{g}) = d\mu(\mathrm{l})$. When the pressure P on the liquid is increased by dP, the chemical potential of the liquid changes by $d\mu(\mathrm{l}) = V_m(\mathrm{l})\,dP$. The chemical potential of the vapour changes by $d\mu(\mathrm{g}) = V_m(\mathrm{g})\,dp$ where dp is the change in the vapour pressure we are trying to find. If we treat the vapour as a perfect gas, the molar volume can be replaced by $V_m(\mathrm{g}) = RT/p$, and we obtain

$$d\mu(\mathrm{g}) = \frac{RT\,dp}{p}$$

Next, we equate the changes in chemical potentials of the vapour and the liquid:

$$\frac{RT\,dp}{p} = V_m(\mathrm{l})\,dP$$

We can integrate this expression once we know the limits of integration.

When there is no additional pressure acting on the liquid, P (the pressure experienced by the liquid) is equal to the normal vapour pressure p^*; so, when $P = p^*$, $p = p^*$ too. When there is an additional pressure ΔP on the liquid, with the result that $P = p + \Delta P$, the vapour pressure is p (the value we want to find). The effect of pressure on the vapour pressure is so small that it is a good approximation to replace the p in

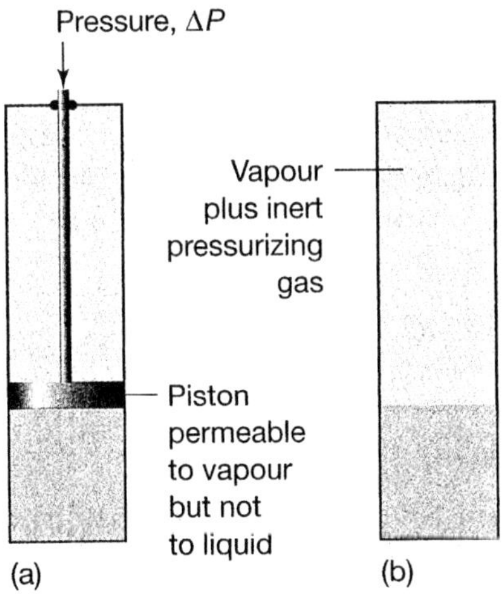

6.10 Pressure may be applied to a condensed phase either (a) by compressing the condensed phase or (b) by subjecting it to an inert pressurizing gas. When pressure is applied, the vapour pressure of the condensed phase increases.

$p + \Delta P$ by p^* itself, and to set the upper limit of the integral to $p^* + \Delta P$. The integrations required are therefore as follows:

$$RT\int_{p^*}^{p}\frac{dp}{p} = \int_{p^*}^{p^*+\Delta p} V_m(l)\,dP$$

We now assume that the molar volume of the liquid is the same throughout the small range of pressures involved. Then both integrations are straightforward, and lead to

$$RT\ln\frac{p}{p^*} = V_m(l)\,\Delta P$$

which rearranges to eqn 3.

Illustration 6.1

For water, which has density 0.997 g cm^{-3} at 25°C and therefore molar volume 18.1 cm^3 mol^{-1}, when subjected to an increase in pressure of 10 bar,

$$\frac{V_m\Delta P}{RT} = \frac{(1.81\times10^{-5}\ \text{m}^3\,\text{mol}^{-1})\times(1.0\times10^{6}\ \text{Pa})}{(8.3145\ \text{J K}^{-1}\,\text{mol}^{-1})\times(298\ \text{K})} = 7.3\times10^{-3}$$

where we have used 1 J = 1 Pa m^3. Because $V_m\Delta P/RT \ll 1$, eqn 3b can be used, and we obtain $(p - p^*)/p^* = 7.3\times10^{-3}$, an increase of 0.73 per cent.

Self-test 6.2 Calculate the effect of an increase in pressure of 100 bar on the vapour pressure of benzene at 25°C, which has density 0.879 g cm^{-3}.

[43 per cent; because the change is so large, use eqn 3a]

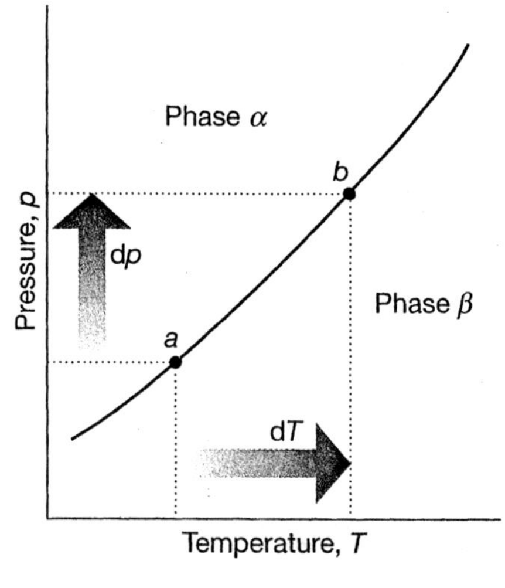

6.11 When pressure is applied to a system in which two phases are in equilibrium (at *a*), the equilibrium is disturbed. It can be restored by changing the temperature, so moving the state of the system to *b*. It follows that there is a relation between d*p* and d*T* that ensures that the system remains in equilibrium as either variable is changed.

6.6 The location of phase boundaries

We can find the precise locations of the phase boundaries—the pressures and temperatures at which two phases can coexist—by making use of the fact that, when two phases are in equilibrium, their chemical potentials must be equal. Therefore, where the phases α and β are in equilibrium,

$$\mu_\alpha(p,T) = \mu_\beta(p,T) \qquad (6.4)$$

By solving this equation for p in terms of T, we get an equation for the phase boundary.

(a) The slopes of the phase boundaries

It turns out to be simplest to discuss the phase boundaries in terms of their slopes, dp/dT.

Let p and T be changed infinitesimally, but in such a way that the two phases α and β remain in equilibrium. The chemical potentials of the phases are initially equal (the two phases are in equilibrium). They remain equal when the conditions are changed to another point on the phase boundary, where the two phases continue to be in equilibrium (Fig. 6.11). Therefore, the changes in the chemical potentials of the two phases must be equal and we can write $d\mu_\alpha = d\mu_\beta$. Because, from eqn 5.9, we know that

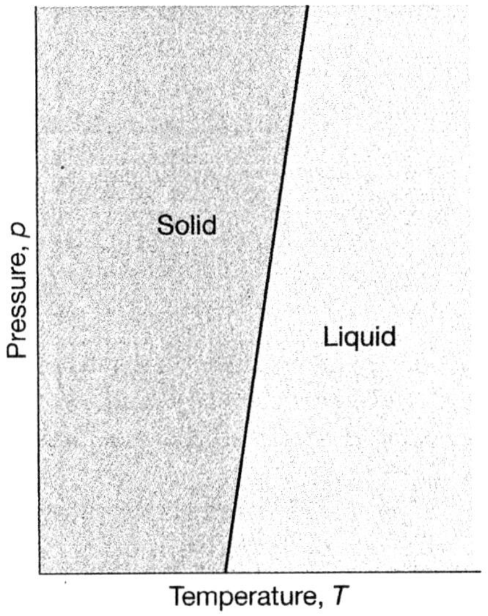

6.12 A typical solid–liquid phase boundary slopes steeply upwards. This slope implies that, as the pressure is raised, the melting temperature rises. Most substances behave in this way.

$$d\mu = -S_m\, dT + V_m\, dp$$

for each phase, it follows that

$$-S_{\alpha,m}\, dT + V_{\alpha,m}\, dp = -S_{\beta,m}\, dT + V_{\beta,m}\, dp$$

where $S_{\alpha,m}$ and $S_{\beta,m}$ are the molar entropies of the phases and $V_{\alpha,m}$ and $V_{\beta,m}$ are their molar volumes. Hence

$$(V_{\beta,m} - V_{\alpha,m})dp = (S_{\beta,m} - S_{\alpha,m})dT \tag{6.5}$$

which rearranges into the **Clapeyron equation**:

$$\frac{dp}{dT} = \frac{\Delta_{trs}S}{\Delta_{trs}V} \tag{6.6}$$

In this expression $\Delta_{trs}S = S_{\beta,m} - S_{\alpha,m}$ and $\Delta_{trs}V = V_{\beta,m} - V_{\alpha,m}$ are the entropy and volume of transition. The Clapeyron equation is an exact expression for the slope of the phase boundary and applies to any phase equilibrium of any pure substance. Its importance is that we can use thermodynamic data to predict the appearance of phase diagrams and to understand their form. A more practical application is to the prediction of the response of freezing and boiling points to the application of pressure.

(b) The solid–liquid boundary

Melting (fusion) is accompanied by a molar enthalpy change $\Delta_{fus}H$ and occurs at a temperature T. The molar entropy of melting at T is therefore $\Delta_{fus}H/T$ (Section 4.3a), and the Clapeyron equation becomes

$$\frac{dp}{dT} = \frac{\Delta_{fus}H}{T\Delta_{fus}V} \tag{6.7}$$

where $\Delta_{fus}V$ is the change in molar volume that occurs on melting. The enthalpy of melting is positive (the only exception is helium-3) and the volume change is usually positive and always small. Consequently, the slope dp/dT is steep and usually positive (Fig. 6.12).

We can obtain the formula for the phase boundary by integrating dp/dT, assuming that $\Delta_{fus}H$ and $\Delta_{fus}V$ change so little with temperature and pressure that they can be treated as constant. If the melting temperature is T^* when the pressure is p^*, and T when the pressure is p, the integration required is

$$\int_{p^*}^{p} dp = \frac{\Delta_{fus}H}{\Delta_{fus}V}\int_{T^*}^{T} \frac{dT}{T}$$

Therefore, the approximate equation of the solid–liquid boundary is

$$p \approx p^* + \frac{\Delta_{fus}H}{\Delta_{fus}V} \ln \frac{T}{T^*} \tag{6.8}$$

This equation was originally obtained by yet another Thomson—James, the brother of William, Lord Kelvin. When T is close to T^*, the logarithm can be approximated by using

$$\ln \frac{T}{T^*} = \ln\left(1 + \frac{T - T^*}{T^*}\right) \approx \frac{T - T^*}{T^*}$$

Therefore,

$$p \approx p^* + \frac{\Delta_{fus}H}{T^*\Delta_{fus}V}(T - T^*) \tag{6.9}$$

Calculations involving natural logarithms often become simpler if we note that $\ln(1 + x) = x - \frac{1}{2}x^2 + \frac{1}{3}x^3 \cdots$. If $x \ll 1$, a good approximation is $\ln(1 + x) \approx x$.

This expression is the equation of a steep straight line when p is plotted against T (as in Fig. 6.12).

(c) The liquid–vapour boundary

The entropy of vaporization at a temperature T is equal to $\Delta_{vap}H/T$; the Clapeyron equation for the liquid–vapour boundary is therefore

$$\frac{dp}{dT} = \frac{\Delta_{vap}H}{T\Delta_{vap}V} \tag{6.10}$$

The enthalpy of vaporization is positive; $\Delta_{vap}V$ is large and positive. Therefore, dp/dT is positive, but it is much smaller than for the solid–liquid boundary. It follows that dT/dp is large, and hence that the boiling temperature is more responsive to pressure than the freezing temperature.

As an illustration of eqn 10, consider the cooking of food in water in a covered pot fitted with a valve that opens when the vapour pressure reaches a higher than ambient value. Upon heating, the vapour pressure of the water increases and boiling takes place only when the temperature reaches a value that opens the valve. This temperature is higher than for a liquid in a vessel at the ambient pressure. As a result, the food can be cooked at temperatures that are significantly higher than 100°C, and the process takes less time.[4]

Example 6.2 Estimating the effect of pressure on the boiling temperature

Estimate the typical size of the effect of increasing pressure on the boiling point of a liquid.

Method To use eqn 10 we need to estimate the right-hand side. At the boiling point, the term $\Delta_{vap}H/T$ is Trouton's constant (Section 4.3a). Because the molar volume of a gas is so much greater than the molar volume of a liquid, we can write

$$\Delta_{vap}V = V_m(g) - V_m(l) \approx V_m(g)$$

and take for $V_m(g)$ the molar volume of a perfect gas (at low pressures, at least).

Answer Trouton's constant has the value 85 J K^{-1} mol^{-1}. The molar volume of a perfect gas is about 25 L mol^{-1} at 1 atm and near but above room temperature. Therefore,

$$\frac{dp}{dT} \approx \frac{85\ \text{J K}^{-1}\ \text{mol}^{-1}}{2.5 \times 10^{-2}\ \text{m}^3\ \text{mol}^{-1}} = 3.4 \times 10^3\ \text{Pa K}^{-1}$$

We have used 1 J = 1 Pa m^3. This value corresponds to 0.034 atm K^{-1}, and hence to dT/dp = 29 K atm^{-1}. Therefore, a change of pressure of +0.1 atm can be expected to change a boiling temperature by about +3 K.

Self-test 6.3 Estimate dT/dp for water at its normal boiling point using the information in Table 4.2 and $V_m(g) = RT/p$. [28 K atm^{-1}]

4. Like most chemical reactions, those associated with cooking of food occur at higher rates as the temperature increases. We discuss the temperature dependence of rates of chemical reactions in Chapter 25.

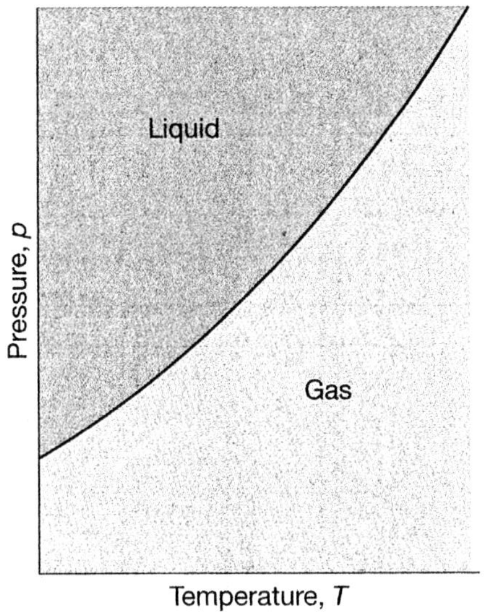

6.13 A typical liquid–vapour phase boundary. The boundary can be regarded as a plot of the vapour pressure against the temperature. Note that, in some depictions of phase diagrams in which a logarithmic pressure scale is used, the phase boundary has the opposite curvature (see Fig. 6.6). This phase boundary terminates at the critical point (not shown).

Because the molar volume of a gas is so much greater than the molar volume of a liquid, we can write $\Delta_{vap}V \approx V_m(g)$ (see the example above). Moreover, if the gas behaves perfectly, $V_m(g) = RT/p$. These two approximations turn the exact Clapeyron equation into

$$\frac{dp}{dT} = \frac{\Delta_{vap}H}{T(RT/p)}$$

which rearranges into the **Clausius–Clapeyron equation** for the variation of vapour pressure with temperature:

$$\frac{d\ln p}{dT} = \frac{\Delta_{vap}H}{RT^2} \qquad (6.11)°$$

(We have used $dx/x = d\ln x$.) Like the Clapeyron equation, the Clausius–Clapeyron equation is important for understanding the appearance of phase diagrams, particularly the location and shape of the liquid–vapour and solid–vapour phase boundaries. Its practical significance is that it lets us predict how the vapour pressure varies with temperature and how the boiling temperature varies with pressure. For instance, if we also assume that the enthalpy of vaporization is independent of temperature, this equation integrates to

$$p = p^* e^{-\chi} \qquad \chi = \frac{\Delta_{vap}H}{R}\left(\frac{1}{T} - \frac{1}{T^*}\right) \qquad (6.12)°$$

where p^* is the vapour pressure when the temperature is T^* and p the vapour pressure when the temperature is T. Equation 12 is the curve plotted as the liquid–vapour boundary in Fig. 6.13. The line does not extend beyond the critical temperature T_c, because above this temperature the liquid does not exist.

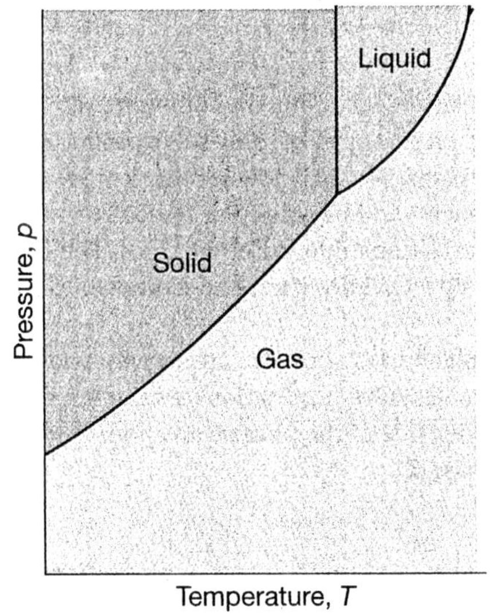

6.14 Near the point where they coincide (at the triple point), the solid–gas boundary has a steeper slope than the liquid–gas boundary because the enthalpy of sublimation is greater than the enthalpy of vaporization and the temperatures that occur in the Clausius-Clapeyron equation for the slope have similar values.

Illustration 6.2

Equation 12 can be used to estimate the vapour pressure of a liquid at any temperature from its normal boiling point, the temperature at which the vapour pressure is 1 atm (760 Torr). Thus, because the normal boiling point of benzene is 80°C (353 K) and (from Table 2.3), $\Delta_{vap}H^{\ominus} = 30.8\ \mathrm{kJ\ mol^{-1}}$, to calculate the vapour pressure at 20°C (293 K), we write

$$\chi = \frac{3.08 \times 10^4\ \mathrm{J\,mol^{-1}}}{8.3145\ \mathrm{J\,K^{-1}\,mol^{-1}}}\left(\frac{1}{293\ \mathrm{K}} - \frac{1}{353\ \mathrm{K}}\right)$$

$$= \frac{3.08 \times 10^4}{8.3145}\left(\frac{1}{293} - \frac{1}{353}\right)$$

and substitute this value into eqn 12 with $p^* = 760$ Torr. The result is 89 Torr. The experimental value is 75 Torr.

(d) The solid–vapour boundary

The only difference between this case and the last is the replacement of the enthalpy of vaporization by the enthalpy of sublimation, $\Delta_{sub}H$. Because the enthalpy of sublimation is greater than the enthalpy of vaporization, the equation predicts a steeper slope for the sublimation curve than for the vaporization curve at similar temperatures, which is near where they meet (Fig. 6.14).

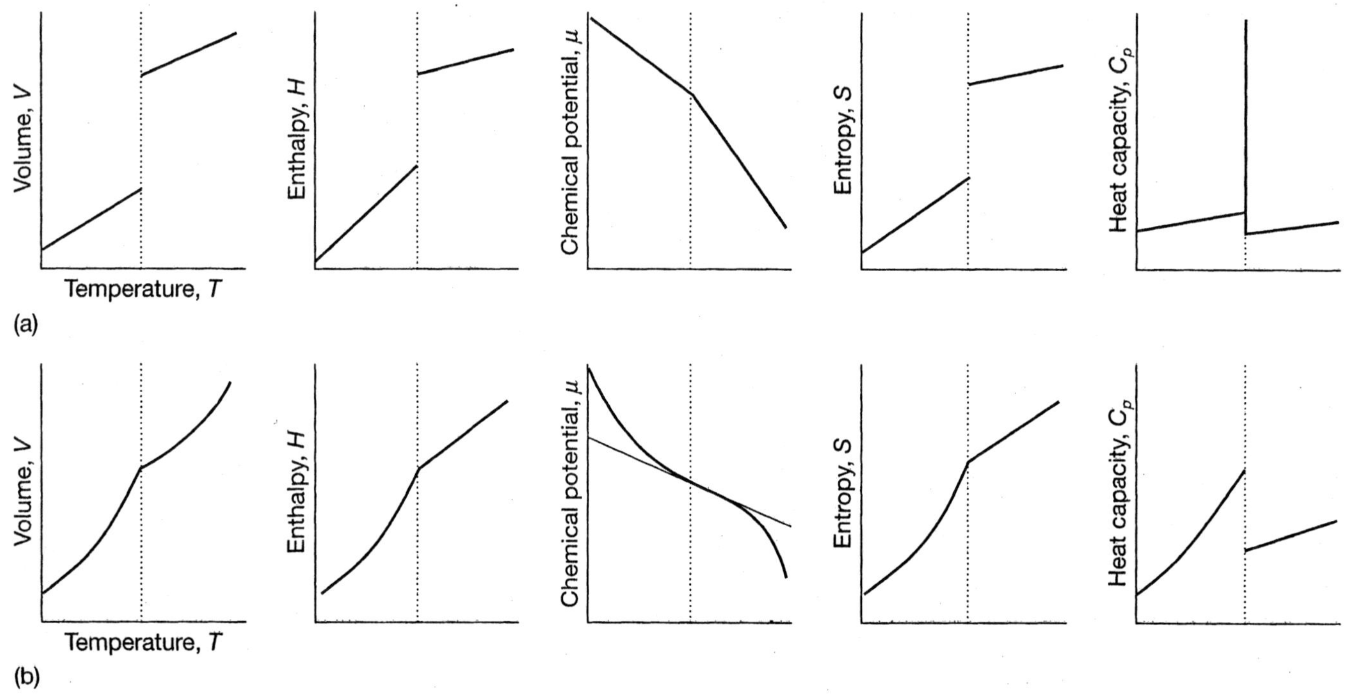

6.15 The changes in thermodynamic properties accompanying (a) first-order and (b) second-order phase transitions.

6.7 The Ehrenfest classification of phase transitions

There are many different types of phase transition, including the common examples of fusion and vaporization and the less common examples of solid–solid, conducting–superconducting, and fluid–superfluid transitions. We shall now see that it is possible to use thermodynamic properties of substances, and in particular the behaviour of the chemical potential, to classify phase transitions into different types. The classification scheme was originally proposed by Paul Ehrenfest, and is known as the **Ehrenfest classification.**

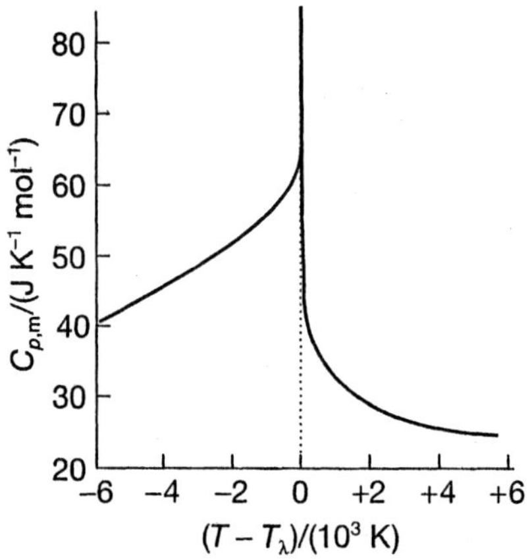

6.16 The λ-curve for helium, where the heat capacity rises to infinity. The shape of this curve is the origin of the name λ-transition.

Many familiar phase transitions, like fusion and vaporization, are accompanied by changes of enthalpy and volume. These changes have implications for the slopes of the chemical potentials of the phases at either side of the phase transition. Thus, at the transition from a phase α to another phase β,

$$\left(\frac{\partial \mu_\beta}{\partial p}\right)_T - \left(\frac{\partial \mu_\alpha}{\partial p}\right)_T = V_{\beta,\mathrm{m}} - V_{\alpha,\mathrm{m}} = \Delta_{\mathrm{trs}}V$$

$$\left(\frac{\partial \mu_\beta}{\partial T}\right)_p - \left(\frac{\partial \mu_\alpha}{\partial T}\right)_p = -S_{\beta,\mathrm{m}} + S_{\alpha,\mathrm{m}} = -\Delta_{\mathrm{trs}}S = -\frac{\Delta_{\mathrm{trs}}H}{T_{\mathrm{trs}}} \tag{6.13}$$

Because $\Delta_{\mathrm{trs}}V$ and $\Delta_{\mathrm{trs}}H$ are nonzero for melting and vaporization, it follows that for such transitions the slopes of the chemical potential plotted against either pressure or temperature are different on either side of the transition (Fig. 6.15a). In other words, the first derivatives of the chemical potentials with respect to pressure and temperature are discontinuous at the transition.

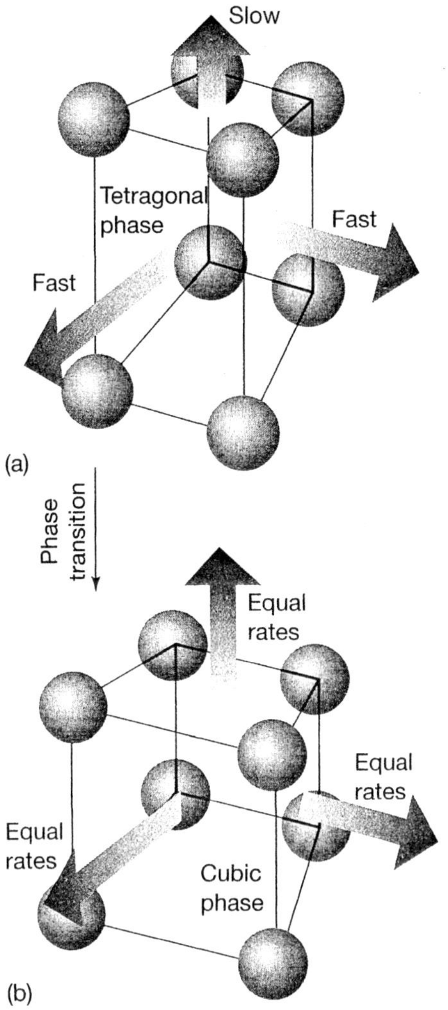

6.17 One version of a second-order phase transition in which (a) a tetragonal phase expands more rapidly in two directions than a third, and hence becomes a cubic phase, which (b) expands uniformly in three directions as the temperature is raised. There is no rearrangement of atoms at the transition temperature, and hence no enthalpy of transition.

A transition for which the first derivative of the chemical potential with respect to temperature is discontinuous is classified as a **first-order phase transition**. The constant-pressure heat capacity, C_p, of a substance is the slope of a plot of the enthalpy with respect to temperature. At a first-order phase transition, H changes by a finite amount for an infinitesimal change of temperature. Therefore, at the transition temperature the heat capacity is infinite. The physical reason is that heating drives the transition rather than raising the temperature. For example, boiling water stays at the same temperature even though heat is being supplied.

A **second-order phase transition** in the Ehrenfest sense is one in which the first derivative of μ with respect to temperature is continuous but its second derivative is discontinuous. A continuous slope of μ (a graph with the same slope on either side of the transition) implies that the volume and entropy (and hence the enthalpy) do not change at the transition (Fig. 6.15b). The heat capacity is discontinuous at the transition but does not become infinite there. An example of a second-order transition is the conducting–superconducting transition in metals at low temperatures.

The term **λ-transition** is applied to a phase transition that is not first-order yet the heat capacity becomes infinite at the transition temperature. Typically, the heat capacity of a system that shows such a transition begins to increase well before the transition (Fig. 6.16), and the shape of the heat capacity curve resembles the Greek letter lambda. This type of transition includes order–disorder transitions in alloys, the onset of ferromagnetism, and the fluid–superfluid transition of liquid helium.

Molecular interpretation 6.1

One type of second-order transition is associated with a change in symmetry of the crystal structure of a solid. Thus, suppose the arrangement of atoms in a solid is like that represented in Fig. 6.17a, with one dimension (technically, of the unit cell) longer than the other two, which are equal. Such a crystal structure is classified as tetragonal (see Section 23.1). Moreover, suppose the two shorter dimensions increase more than the long dimension when the temperature is raised. There may come a stage when the three dimensions become equal. At that point the crystal has cubic symmetry (Fig. 6.17b), and at higher temperatures it will expand equally in all three directions (because there is no longer any distinction between them). The tetragonal → cubic phase transition has occurred but, as it has not involved a discontinuity in the interaction energy between the atoms or the volume they occupy, the transition is not first-order.

The order–disorder transition in β-brass (CuZn) is an example of a λ-transition. The low-temperature phase is an orderly array of alternating Cu and Zn atoms. The high-temperature phase is a random array of the atoms (Fig. 6.18). At $T = 0$ the order is perfect, but islands of disorder appear as the temperature is raised. The islands form because the transition is cooperative in the sense that, once two atoms have exchanged locations, it is easier for their neighbours to exchange their locations. The islands grow in extent, and merge throughout the crystal at the transition temperature (742 K). The heat capacity increases as the transition temperature is approached because the cooperative nature of the transition means that it is increasingly easy for the heat supplied to drive the phase transition rather than to be stored as thermal motion.

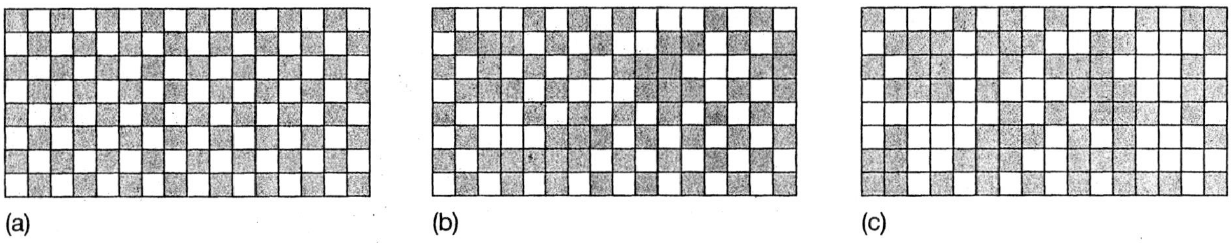

6.18 An order–disorder transition. (a) At $T = 0$, there is perfect order, with different kinds of atoms occupying alternate sites. (b) As the temperature is increased, atoms exchange locations and islands of each kind of atom form in regions of the solid. Some of the original order survives. (c) At and above the transition temperature, the islands occur at random throughout the sample.

The physical liquid surface

So far, we have concentrated on the properties of the boundaries in the phase diagram of a substance. However, the physical boundary between phases, such as the surface where solid is in contact with liquid or liquid is in contact with vapour, has interesting properties. In this section we concentrate on the liquid–vapour interface, which is interesting because it is so mobile. Chapter 28 deals with solid surfaces and their important role in catalysis.

6.8 Surface tension

Liquids tend to adopt shapes that minimize their surface area, for then the maximum number of molecules are in the bulk and hence surrounded by and interacting with neighbours. Droplets of liquids therefore tend to be spherical, because a sphere is the shape with the smallest surface-to-volume ratio. However, there may be other forces present that compete against the tendency to form this ideal shape and, in particular, gravity may flatten spheres into puddles or oceans.

Surface effects may be expressed in the language of Helmholtz and Gibbs energies. The link between these quantities and the surface area is the work needed to change the area by a given amount, and the fact that dA and dG are equal (under different conditions) to the work done in changing the energy of a system. The work needed to change the surface area, σ, of a sample by an infinitesimal amount $d\sigma$ is proportional to $d\sigma$, and we write

$$dw = \gamma d\sigma \qquad [6.14]$$

The constant of proportionality, γ, is called the **surface tension**; its dimensions are energy/area and its units are typically joules per metre squared (J m^{-2}). However, as in Table 6.1, values of γ are usually reported in newtons per metre (N m^{-1}, because 1 J = 1 N m). The work of surface formation at constant volume and temperature can be identified with the change in the Helmholtz energy, and we can write

$$dA = \gamma d\sigma \qquad (6.15)$$

Because the Helmholtz energy decreases ($dA < 0$) if the surface area decreases ($d\sigma < 0$), surfaces have a natural tendency to contract. This is a more formal way of expressing what we have already described.

Table 6.1* Surface tensions of liquids at 293 K

	γ/(mN m^{-1})†
Benzene	28.88
Mercury	472
Methanol	22.6
Water	72.75

* More values are given in the *Data section*.
† Note that 1 N m^{-1} = 1 J m^{-2}.

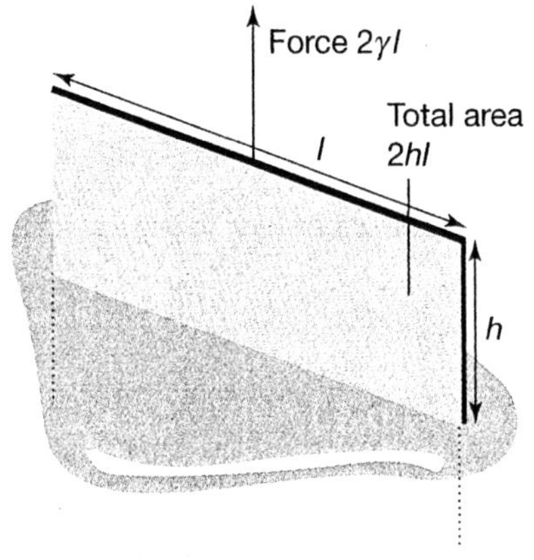

6.19 The model used for calculating the work of forming a liquid film when a wire of length l is raised and pulls the surface with it through a height h.

Example 6.3 Using the surface tension

Calculate the work needed to raise a wire of length l and to stretch the surface of a liquid through a height h in the arrangement shown in Fig. 6.19. Disregard gravitational potential energy.

Method According to eqn 14, the work required to create a surface area given that the surface tension does not vary as the surface is formed is $w = \gamma\sigma$. Therefore, all we need do is to calculate the surface area of the two-sided rectangle formed as the frame is withdrawn from the liquid.

Answer When the wire of length l is raised through a height h it increases the area of the liquid by twice the area of the rectangle (because there is a surface on each side). The total increase is therefore $2lh$ and the work done is $2\gamma lh$.

Comment The work can be expressed as a force × distance by writing it as $2\gamma l \times h$, and identifying γl as the opposing force on the wire of length l. This is why γ is called a tension and why its units are often chosen to be newtons per metre (N m^{-1}, so γl is a force in newtons).

Self-test 6.4 Calculate the work of creating a spherical cavity of radius r in a liquid of surface tension γ. [$4\pi r^2\gamma$]

6.9 Curved surfaces

The minimization of the surface area of a liquid may result in the formation of a curved surface, as in a bubble. We shall now see that there are two consequences of curvature, and hence of the surface tension, that are relevant to the properties of liquids. One is that the vapour pressure of a liquid depends on the curvature of its surface. The other is the capillary rise (or fall) of liquids in narrow tubes.

(a) Bubbles, cavities, and droplets

A **bubble** is a region in which vapour (and possibly air too) is trapped by a thin film; a **cavity** is a vapour-filled hole in a liquid. What are widely called 'bubbles' in liquids are therefore strictly cavities. True bubbles have two surfaces (one on each side of the film); cavities have only one. The treatments of both are similar, but a factor of 2 is required for bubbles to take into account the doubled surface area. A **droplet** is a small volume of liquid at equilibrium surrounded by its vapour (and possibly also air).

The pressure on the concave side of an interface, p_{in}, is always greater than the pressure on the convex side, p_{out}. This relation is expressed by the **Laplace equation**, which is derived in the following *Justification*:

$$p_{\text{in}} = p_{\text{out}} + \frac{2\gamma}{r} \qquad (6.16)$$

Justification 6.2

The cavities in a liquid are at equilibrium when the tendency for their surface area to decrease is balanced by the rise of internal pressure which would then result. When the pressure inside a cavity is p_{in} and its radius is r, the outward force is

$$\text{pressure} \times \text{area} = 4\pi r^2 p_{\text{in}}$$

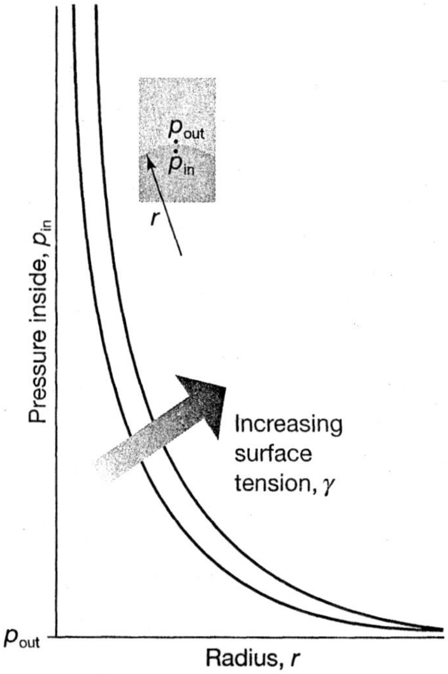

6.20 The dependence of the pressure inside a curved surface on the radius of the surface, for two different values of the surface tension.

The force inwards arises from the external pressure and the surface tension. The former has magnitude $4\pi r^2 p_{out}$. The latter is calculated as follows. The change in surface area when the radius of a sphere changes from r to $r + dr$ is

$$d\sigma = 4\pi(r + dr)^2 - 4\pi r^2 = 8\pi r\, dr$$

(The second-order infinitesimal, $(dr)^2$, is ignored.) The work done when the surface is stretched by this amount is therefore

$$dw = 8\pi\gamma r\, dr$$

As force × distance is work, the force opposing stretching through a distance dr when the radius is r is

$$F = 8\pi\gamma r$$

The total inward force is therefore $4\pi r^2 p_{out} + 8\pi\gamma r$. At equilibrium, the outward and inward forces are balanced, so we can write

$$4\pi r^2 p_{in} = 4\pi r^2 p_{out} + 8\pi\gamma r$$

which rearranges into eqn 16.

The Laplace equation shows that the difference in pressure decreases to zero as the radius of curvature becomes infinite (when the surface is flat, Fig. 6.20). Small cavities have small radii of curvature, so the pressure difference across their surface is quite large. For instance, a 'bubble' (actually, a cavity) of radius 0.10 mm in champagne implies a pressure difference of 1.5 kPa, which is enough to sustain a column of water of height 15 cm.

We saw in Section 6.5 that the vapour pressure of a liquid depends on the pressure applied to the liquid. Because curving a surface gives rise to a pressure differential of $2\gamma/r$, we can expect the vapour pressure above a curved surface to be different from that above a flat surface. By substituting this value of the pressure difference into eqn 3a we obtain the **Kelvin equation** for the vapour pressure of a liquid when it is dispersed as droplets of radius r:

$$p = p^* e^{2\gamma V_m/rRT} \qquad (6.17)$$

The analogous expression for the vapour pressure inside a cavity can be written at once. The pressure of the liquid outside the cavity is less than the pressure inside, so the only change is in the sign of the exponent in the last expression.

(b) Nucleation

For droplets of water of radius 1 μm and 1 nm the ratios p/p^* at 25°C are about 1.001 and 3, respectively. The second figure, although quite large, is unreliable because at that radius the droplet is less than about 10 molecules in diameter and the basis of the calculation is suspect. The first figure shows that the effect is usually small; nevertheless it may have important consequences.

Consider, for example, the formation of a cloud. Warm, moist air rises into the cooler regions higher in the atmosphere. At some altitude the temperature is so low that the vapour becomes thermodynamically unstable with respect to the liquid and we expect it to condense into a cloud of liquid droplets. The initial step can be imagined as a swarm of water molecules congregating into a microscopic droplet. Because the initial droplet is so small it has an enhanced vapour pressure. Therefore, instead of growing it evaporates. This effect effectively stabilizes the vapour because an initial tendency to condense is overcome by a heightened tendency to evaporate.

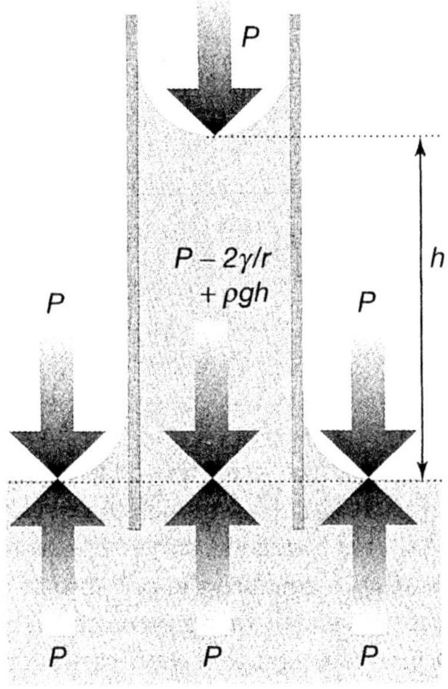

6.21 When a capillary tube is first stood in a liquid, the latter climbs up the walls, so curving the surface. The pressure just under the meniscus is less than that arising from the atmosphere by $2\gamma/r$. The pressure is equal at equal heights throughout the liquid provided the hydrostatic pressure (which is equal to ρgh) cancels the pressure difference arising from the curvature.

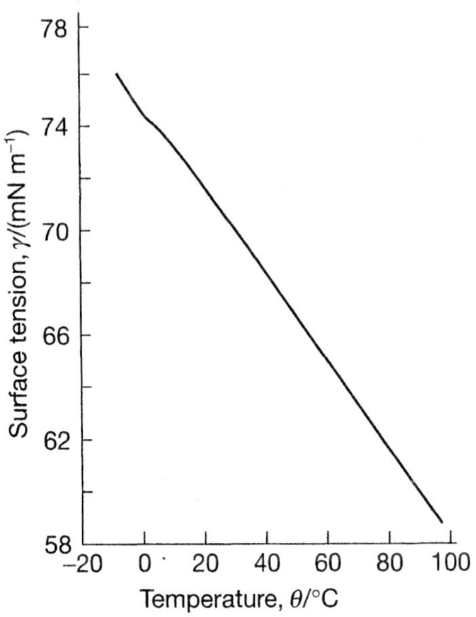

6.22 The variation of the surface tension of water with temperature.

The vapour phase is then said to be **supersaturated**. It is thermodynamically unstable with respect to the liquid but not unstable with respect to the small droplets that need to form before the bulk liquid phase can appear, so the formation of the latter by a simple, direct mechanism is hindered.

Clouds do form, so there must be a mechanism. Two processes are responsible. The first is that a sufficiently large number of molecules might congregate into a droplet so big that the enhanced evaporative effect is unimportant. The chance of one of these **spontaneous nucleation centres** forming is low, and in rain formation it is not a dominant mechanism. The more important process depends on the presence of minute dust particles or other kinds of foreign matter. These **nucleate** the condensation (that is, provide centres at which it can occur) by providing surfaces to which the water molecules can attach.

Liquids may be **superheated** above their boiling temperatures and **supercooled** below their freezing temperatures. In each case the thermodynamically stable phase is not achieved on account of the kinetic stabilization that occurs in the absence of nucleation centres. For example, superheating occurs because the vapour pressure inside a cavity is artificially low, so any cavity that does form tends to collapse. This instability is encountered when an unstirred beaker of water is heated, for its temperature may be raised above its boiling point. Violent bumping often ensues as spontaneous nucleation leads to bubbles big enough to survive. To ensure smooth boiling at the true boiling temperature, nucleation centres, such as small pieces of sharp-edged glass or bubbles (cavities) of air, should be introduced. A bubble chamber, a device for tracking elementary particles, depends on the nucleation of the evaporation of superheated liquid hydrogen by ionizing radiation.

6.10 Capillary action

The tendency of liquids to rise up capillary tubes (tubes of narrow bore), which is called **capillary action**, is a consequence of surface tension. Consider what happens when a glass capillary tube is first immersed in water or any liquid that has a tendency to adhere to the walls. The energy is lowest when a thin film covers as much of the glass as possible. As this film creeps up the inside wall it has the effect of curving the surface of the liquid inside the tube. This curvature implies that the pressure just beneath the curving meniscus is less than the atmospheric pressure by approximately $2\gamma/r$, where r is the radius of the tube and we assume a hemispherical surface. The pressure immediately under the flat surface outside the tube is p, the atmospheric pressure; but inside the tube under the curved surface it is only $p - 2\gamma/r$. The excess external pressure presses the liquid up the tube until hydrostatic equilibrium (equal pressures at equal depths) has been reached (Fig. 6.21).

(a) Capillary rise

The pressure exerted by a column of liquid of mass density ρ and height h is

$$p = \rho gh \qquad (6.18)$$

This hydrostatic pressure matches the pressure difference $2\gamma/r$ at equilibrium. Therefore, the height of the column at equilibrium is obtained by equating $2\gamma/r$ and ρgh, which gives

$$h = \frac{2\gamma}{\rho gr} \qquad (6.19)$$

This simple expression provides a reasonably accurate way of measuring the surface tension of liquids. Surface tension decreases with increasing temperature (Fig. 6.22).

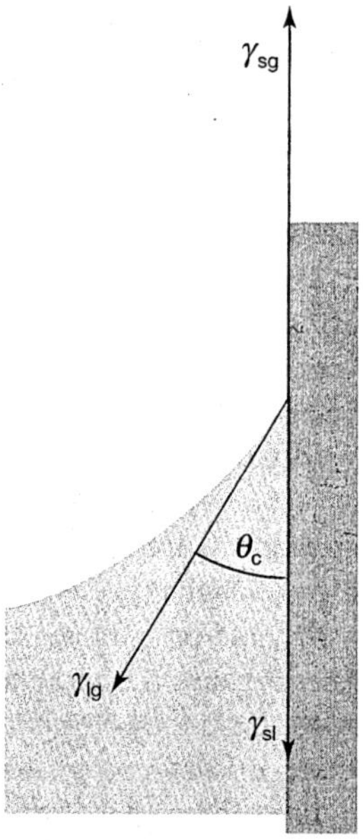

6.23 The balance of forces that results in a contact angle, θ_c.

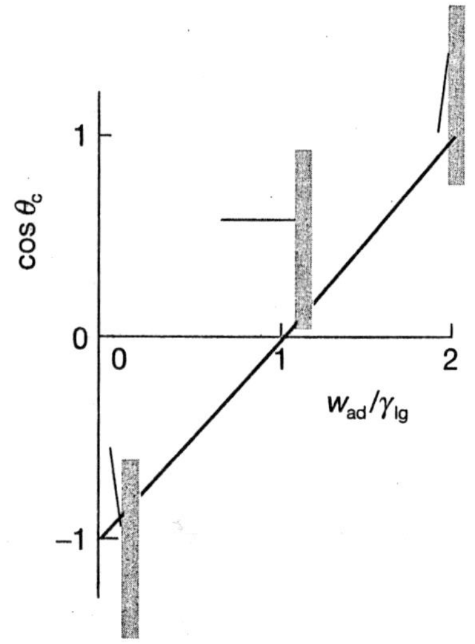

6.24 The variation of contact angle (shown by the semaphore-like object) as the ratio w_{ad}/γ_{lg} changes.

Illustration 6.3

If water at 25°C rises through 7.36 cm in a capillary of radius 0.20 mm, its surface tension at that temperature is

$$\gamma = \tfrac{1}{2}\rho g h r$$
$$= \tfrac{1}{2} \times (997.1\ \text{kg m}^{-3}) \times (9.81\ \text{m s}^{-2}) \times (7.36 \times 10^{-2}\ \text{m}) \times (2.0 \times 10^{-4}\ \text{m})$$
$$= 72\ \text{mN m}^{-1}$$

where we have used $1\ \text{kg m s}^{-2} = 1\ \text{N}$.

When the adhesive forces between the liquid and the material of the capillary wall are weaker than the cohesive forces within the liquid (as for mercury in glass), the liquid in the tube retracts from the walls. This retraction curves the surface with the concave, high-pressure side downwards. To equalize the pressure at the same depth throughout the liquid the surface must fall to compensate for the heightened pressure arising from its curvature. This compensation results in a capillary depression.

(b) The contact angle

In many cases there is a nonzero angle between the edge of the meniscus and the wall. If this contact angle is θ_c, then eqn 19 should be modified by multiplying the right hand side by $\cos\theta_c$.

The origin of the contact angle can be traced to the balance of forces at the line of contact between the liquid and the solid (Fig. 6.23). If the solid–gas, solid–liquid, and liquid–gas surface tensions (essentially the energy needed to create unit area of each of the interfaces) are denoted γ_{sg}, γ_{sl}, and γ_{lg}, respectively, then the vertical forces are in balance if

$$\gamma_{sg} = \gamma_{sl} + \gamma_{lg}\cos\theta_c \qquad (6.20)$$

This expression solves to

$$\cos\theta_c = \frac{\gamma_{sg} - \gamma_{sl}}{\gamma_{lg}} \qquad (6.21)$$

If we note that the work of adhesion of the liquid to the solid (per unit area of contact) is

$$w_{ad} = \gamma_{sg} + \gamma_{lg} - \gamma_{sl} \qquad (6.22)$$

eqn 21 can be written

$$\cos\theta_c = \frac{w_{ad}}{\gamma_{lg}} - 1 \qquad (6.23)$$

We now see that the liquid 'wets' (spreads over) the surface, corresponding to $0 < \theta_c < 90°$, when $1 < w_{ad}/\gamma_{lg} < 2$ (Fig. 6.24). The liquid does not wet the surface, corresponding to $90° < \theta_c < 180°$, when $0 < w_{ad}/\gamma_{lg} < 1$. For mercury in contact with glass, $\theta_c = 140°$, which corresponds to $w_{ad}/\gamma_{lg} = 0.23$, indicating a relatively low work of adhesion of the mercury to glass on account of the strong cohesive forces within mercury.

Checklist of key ideas

Phase diagrams

6.1 The stabilities of phases

- ☐ phase, a form of matter that is uniform throughout in chemical composition and physical state
- ☐ phase transition, the spontaneous conversion of one phase into another phase
- ☐ transition temperature, the temperature at which the two phases are in equilibrium
- ☐ metastable phase, a thermodynamically unstable phase that persists because the transition is kinetically hindered

6.2 Phase boundaries

- ☐ phase diagram, a diagram showing the regions of pressure and temperature at which its various phases are thermodynamically stable
- ☐ phase boundary, lines separating the regions in a phase diagram showing the values of p and T at which two phases coexist in equilibrium
- ☐ vapour pressure, the pressure of a vapour in equilibrium with the condensed (normally liquid) phase
- ☐ sublimation vapour pressure, the vapour pressure of a solid
- ☐ boiling, the condition of free vaporization throughout the liquid
- ☐ boiling temperature, the temperature at which the vapour pressure of a liquid is equal to the external pressure
- ☐ normal boiling point, the boiling point when the external pressure is 1 atm
- ☐ standard boiling point, the boiling point when the external pressure is 1 bar
- ☐ critical temperature, the temperature at which a liquid surface disappears and above which a liquid does not exist whatever the pressure
- ☐ critical pressure, the vapour pressure at the critical temperature
- ☐ supercritical fluid, a dense fluid phase above the critical temperature
- ☐ melting temperature or freezing temperature, the temperature at which, under a specified pressure, the liquid and solid phases of a substance coexist in equilibrium
- ☐ normal melting point or normal freezing point, the freezing temperature when the pressure is 1 atm
- ☐ standard freezing point, the freezing temperature when the pressure is 1 bar
- ☐ triple point, a point on a phase diagram at which the three phase boundaries meet and all three phases are in mutual equilibrium

6.3 Three typical phase diagrams

- ☐ the interpretation of the phase diagram for water
- ☐ the interpretation of the phase diagram for carbon dioxide
- ☐ the interpretation of the phase diagram for helium
- ☐ λ-line, the phase boundary between normal and superfluid liquid helium
- ☐ superfluid helium, a liquid phase that flows without viscosity

Phase stability and phase transitions

- ☐ chemical potential (of a pure substance), the molar Gibbs energy, $\mu = G_m$

6.4 The thermodynamic criterion of equilibrium

- ☐ the uniformity of chemical potential throughout a system at equilibrium

6.5 The dependence of stability on the conditions

- ☐ the variation of chemical potential with temperature, $(\partial\mu/\partial T)_p = -S_m$
- ☐ the variation of chemical potential with pressure, $(\partial\mu/\partial p)_T = V_m$
- ☐ partial vapour pressure, the partial pressure of the vapour of a component of a mixture in equilibrium with the condensed phase
- ☐ gas solvation, the attachment of molecules to gas phase species
- ☐ vapour pressure in the presence of applied pressure, $p = p^* e^{V_m \Delta P/RT}$

6.6 The location of phase boundaries

- ☐ Clapeyron equation, $dp/dT = \Delta_{trs}S/\Delta_{trs}V$
- ☐ the slope of the solid–liquid boundary, $dp/dT = \Delta_{fus}H/T\Delta_{fus}V$
- ☐ the slope of the liquid–vapour boundary, $dp/dT = \Delta_{vap}H/T\Delta_{vap}V$
- ☐ Clausius–Clapeyron equation, $d \ln p/dT = \Delta_{vap}H/RT^2$

6.7 The Ehrenfest classification of phase transitions

- ☐ Ehrenfest classification, a classification of phase transitions based on the behaviour of the chemical potential
- ☐ first-order phase transition, a transition for which the first derivative of the chemical potential with respect to temperature is discontinuous; at such a transition the heat capacity is infinite
- ☐ second-order phase transition, a transition for which the first derivative of the chemical potential with respect to temperature is continuous but its second derivative is discontinuous
- ☐ λ-transition, a phase transition that is not first-order but for which the heat capacity becomes infinite at the transition temperature

The physical liquid surface

6.8 Surface tension

- ☐ work of forming a surface, $dw = \gamma d\sigma$
- ☐ surface tension γ, the constant of proportionality between the work required to form a surface and the area of the surface

6.9 Curved surfaces

- ☐ bubble, a region in which vapour (and possibly air too) is trapped by a thin film
- ☐ cavity, a vapour-filled hole in a liquid
- ☐ droplet, a small volume of liquid at equilibrium surrounded by its vapour (and possibly also air)
- ☐ Laplace equation for the vapour pressure at a curved surface, $p_{in} = p_{out} + 2\gamma/r$
- ☐ Kelvin equation for the vapour pressure of droplets, $p = p^* e^{2\gamma V_m/rRT}$
- ☐ supersaturated phase, a phase that is thermodynamically unstable with respect to the liquid
- ☐ spontaneous nucleation centre, a location at which a sufficiently large number of molecules congregate into a droplet
- ☐ nucleate, provide surfaces to which molecules can attach and thereby induce condensation

☐ superheated, a liquid that has not boiled but is above its boiling temperature

☐ supercooled, a liquid that has not frozen but is below its freezing temperature

6.10 Capillary action

☐ capillary action, the tendency of liquids to rise up capillary tubes

☐ capillary rise and surface tension, $h = 2\gamma/\rho g r$

☐ contact angle and interfacial tension, $\cos\theta_c = (\gamma_{sg} - \gamma_{sl})/\gamma_{lg}$

☐ the criteria for surface wetting, $1 < w_{ad}/\gamma_{lg} < 2$, with $w_{ad} = \gamma_{sg} + \gamma_{lg} - \gamma_{sl}$

Further reading

Articles of general interest

W.B. Daniels, High-pressure techniques. In *Encyclopedia of applied physics* (ed. G.L. Trigg), **7**, 495. VCH, New York (1995).

B.B. Earl, The direct relation between altitude and boiling point. *J. Chem. Educ.* **67**, 45 (1990).

J.M. Honig, Mean field theory of phase transitions. *J. Chem. Educ.* **76**, 848 (1999).

E.K.H. Salje, Phase transitions, structural. In *Encyclopedia of applied physics* (ed. G.L. Trigg), **13**, 373. VCH, New York (1995).

J.M. Sanchez, Order–disorder transitions. In *Encyclopedia of applied physics* (ed. G.L. Trigg), **13**, 1. VCH, New York (1995).

K.M. Scholsky, Supercritical phase transitions at very high pressure. *J. Chem. Educ.* **66**, 989 (1989).

Texts and sources of data and information

J.R. Anderson, Determination of boiling and condensation temperatures. In *Techniques of Chemistry* (A. Weissberger and B.W. Rossiter, ed.) **5**, 199. Wiley, New York (1971).

T. Boublik, V. Fried, and E. Hála, *The vapor pressures of pure substances*. Elsevier, Amsterdam (1984).

W.D. Callister, Jr., *Materials science and engineering, an introduction*. Wiley, New York (2000).

J.J. Jasper, The surface tension of pure liquid compounds. *J. Phys. Chem. Ref. Data* **1**, 841 (1972).

A.N. Nesmeyanov. *Vapour pressure of the elements*. Pergamon Press, New York (1963).

J.S. Rowlinson and B. Widom, *Molecular theory of capillarity*. Clarendon Press, Oxford (1982).

S.I. Sandler, *Chemical and engineering thermodynamics*. Wiley, New York (1998).

E.L. Skau and J.C. Arthur, Determination of melting and freezing temperatures. In *Techniques of Chemistry* (A. Weissberger and B.W. Rossiter, ed.) **5**, 105. Wiley, New York (1971).

H.E. Stanley, Introduction to phase transitions and critical phenomena. Clarendon Press, Oxford (1971).

M.A. White, *Properties of materials*. Oxford University Press, Oxford (1999).

Exercises

Discussion questions

6.1 (a) Discuss the implications for phase stability of the variation of chemical potential with temperature and pressure.

6.1 (b) Suggest a physical interpretation of the phenomena of superheating and supercooling.

6.2 (a) Refer to Fig. 6.4 and describe the changes that would be observed when water vapour at 1.0 atm and 400 K is cooled at constant pressure to 260 K. Suggest the appearance of the cooling curve, a plot of temperature against time.

6.2 (b) Use the phase diagram in Fig. 6.5 to state what would be observed when a sample of carbon dioxide, initially at 1.0 atm and 298 K, is subjected to the following cycle: (a) isobaric (constant-pressure) heating to 320 K, (b) isothermal compression to 100 atm, (c) isobaric cooling to 210 K, (d) isothermal decompression to 1.0 atm, (e) isobaric heating to 298 K.

6.3 (a) Explain the significance of the Clapeyron equation and of the Clausius–Clapeyron equation.

6.3 (b) Distinguish between a first-order phase transition, a second-order phase transition, and a λ-transition.

Numerical exercises

6.4 (a) The vapour pressure of dichloromethane at 24.1°C is 400 Torr and its enthalpy of vaporization is 28.7 kJ mol^{-1}. Estimate the temperature at which its vapour pressure is 500 Torr.

6.4 (b) The vapour pressure of a substance at 20.0°C is 58.0 kPa and its enthalpy of vaporization is 32.7 kJ mol^{-1}. Estimate the temperature at which its vapour pressure is 66.0 kPa.

6.5 (a) The molar volume of a certain solid is 161.0 cm^3 mol^{-1} at 1.00 atm and 350.75 K, its melting temperature. The molar volume of the liquid at this temperature and pressure is 163.3 cm^3 mol^{-1}. At 100 atm the melting temperature changes to 351.26 K. Calculate the enthalpy and entropy of fusion of the solid.

6.5 (b) The molar volume of a certain solid is 142.0 cm^3 mol^{-1} at 1.00 atm and 427.15 K, its melting temperature. The molar volume of the liquid at this temperature and pressure is 152.6 $cm^3\ mol^{-1}$. At 1.2 MPa the melting temperature changes to 429.26 K. Calculate the enthalpy and entropy of fusion of the solid.

6.6 (a) The vapour pressure of a liquid in the temperature range 200 K to 260 K was found to fit the expression $\ln(p/\text{Torr}) = 16.255 - 2501.8/(T/\text{K})$. Calculate the enthalpy of vaporization of the liquid.

6.6 (b) The vapour pressure of a liquid in the temperature range 200 K to 260 K was found to fit the expression $\ln(p/\text{Torr}) = 18.361 - 3036.8/(T/\text{K})$. Calculate the enthalpy of vaporization of the liquid.

6.7 (a) The vapour pressure of benzene between 10°C and 30°C fits the expression $\log(p/\text{Torr}) = 7.960 - 1780/(T/\text{K})$. Calculate (a) the enthalpy of vaporization and (b) the normal boiling point of benzene.

6.7 (b) The vapour pressure of a liquid between 15°C and 35°C fits the expression $\log(p/\text{Torr}) = 8.750 - 1625/(T/\text{K})$. Calculate (a) the enthalpy of vaporization and (b) the normal boiling point of the liquid.

6.8 (a) When benzene freezes at 5.5°C its density changes from 0.879 $g\ cm^{-3}$ to 0.891 $g\ cm^{-3}$. Its enthalpy of fusion is 10.59 $kJ\ mol^{-1}$. Estimate the freezing point of benzene at 1000 atm.

6.8 (b) When a certain liquid freezes at −3.65°C its density changes from 0.789 $g\ cm^{-3}$ to 0.801 $g\ cm^{-3}$. Its enthalpy of fusion is 8.68 $kJ\ mol^{-1}$. Estimate the freezing point of the liquid at 100 MPa.

6.9 (a) In July in Los Angeles, the incident sunlight at ground level has a power density of 1.2 $kW\ m^{-2}$ at noon. A swimming pool of area 50 m^2 is directly exposed to the sun. What is the maximum rate of loss of water assuming that all radiation is absorbed?

6.9 (b) Suppose the incident sunlight at ground level has a power density of 0.87 $kW\ m^{-2}$ at noon. What is the maximum rate of loss of water from a lake of area 1.0 ha? (1 ha = $10^4\ m^2$.)

6.10 (a) An open vessel containing (a) water, (b) benzene, (c) mercury stands in a laboratory measuring 5.0 m × 5.0 m × 3.0 m at 25°C. What mass of each substance will be found in the air if there is no ventilation? (The vapour pressures are (a) 24 Torr, (b) 98 Torr, (c) 1.7 mTorr.)

6.10 (b) On a cold, dry morning after a frost, the temperature was −5°C and the partial pressure of water in the atmosphere fell to 2.0 Torr. Will the frost sublime? What partial pressure of water would ensure that the frost remained?

6.11 (a) Naphthalene, $C_{10}H_8$, melts at 80.2°C. If the vapour pressure of the liquid is 10 Torr at 85.8°C and 40 Torr at 119.3°C, use the Clausius–Clapeyron equation to calculate (a) the enthalpy of vaporization, (b) the normal boiling point, and (c) the entropy of vaporization at the boiling point.

6.11 (b) The boiling point of hexane is 69.0°C. Estimate (a) its enthalpy of vaporization and (b) its vapour pressure at 25°C and 60°C.

6.12 (a) Calculate the melting point of ice under a pressure of 50 bar. Assume that the density of ice under these conditions is approximately 0.92 $g\ cm^{-3}$ and that of liquid water is 1.00 $g\ cm^{-3}$.

6.12 (b) Calculate the melting point of ice under a pressure of 10 MPa. Assume that the density of ice under these conditions is approximately 0.915 $g\ cm^{-3}$ and that of liquid water is 0.998 $g\ cm^{-3}$.

6.13 (a) What fraction of the enthalpy of vaporization of water is spent on expanding the water vapour?

6.13 (b) What fraction of the enthalpy of vaporization of ethanol is spent on expanding its vapour?

6.14 (a) Calculate the vapour pressure of a spherical droplet of water of radius 10 nm at 20°C. The vapour pressure of bulk water at that temperature is 2.3 kPa and its density is 0.9982 $g\ cm^{-3}$.

6.14 (b) Calculate the vapour pressure of a spherical droplet of water of radius 20.0 nm at 35.0°C. The vapour pressure of bulk water at that temperature is 5.623 kPa and its density is 994.0 $kg\ m^{-3}$.

6.15 (a) The contact angle for water on clean glass is close to zero. Calculate the surface tension of water at 20°C given that at that temperature water climbs to a height of 4.96 cm in a clean glass capillary tube of internal radius 0.300 mm. The density of water at 20°C is 998.2 $kg\ m^{-3}$.

6.15 (b) The contact angle for water on clean glass is close to zero. Calculate the surface tension of water at 30°C given that at that temperature water climbs to a height of 9.11 cm in a clean glass capillary tube of internal diameter 0.320 mm. The density of water at 30°C is 0.9956 $g\ cm^{-3}$.

6.16 (a) Calculate the pressure differential of water across the surface of a spherical droplet of radius 200 nm at 20°C.

6.16 (b) Calculate the pressure differential of ethanol across the surface of a spherical droplet of radius 220 nm at 20°C. The surface tension of ethanol at that temperature is 22.39 $mN\ m^{-1}$.

Problems†

Numerical problems

6.1 The temperature dependence of the vapour pressure of solid sulfur dioxide can be approximately represented by the relation $\log(p/\text{Torr}) = 10.5916 - 1871.2/(T/\text{K})$ and that of liquid sulfur dioxide by $\log(p/\text{Torr}) = 8.3186 - 1425.7/(T/\text{K})$. Estimate the temperature and pressure of the triple point of sulfur dioxide.

6.2 Prior to the discovery that freon-12 (CF_2Cl_2) was harmful to the Earth's ozone layer, it was frequently used as the dispersing agent in spray cans for hair spray, etc. Its enthalpy of vaporization at its normal boiling point of −29.2°C is 20.25 $kJ\ mol^{-1}$. Estimate the pressure that a can of hair spray using freon-12 had to withstand at 40°C, the temperature of a can that has been standing in sunlight.

† Problems denoted by the symbol ‡ were supplied by Charles Trapp, Carmen Giunta, and Marshall Cady.

Assume that $\Delta_{vap}H$ is a constant over the temperature range involved and equal to its value at −29.2°C.

6.3 The enthalpy of vaporization of a certain liquid is found to be 14.4 kJ mol^{-1} at 180 K, its normal boiling point. The molar volumes of the liquid and the vapour at the boiling point are 115 cm^3 mol^{-1} and 14.5 dm^3 mol^{-1}, respectively. (a) Estimate dp/dT from the Clapeyron equation and (b) the percentage error in its value if the Clausius–Clapeyron equation is used instead.

6.4 Calculate the difference in slope of the chemical potential against temperature on either side of (a) the normal freezing point of water and (b) the normal boiling point of water. (c) By how much does the chemical potential of water supercooled to −5.0°C exceed that of ice at that temperature?

6.5 Calculate the difference in slope of the chemical potential against pressure on either side of (a) the normal freezing point of water and (b) the normal boiling point of water. The densities of ice and water at 0°C are 0.917 g cm^{-3} and 1.000 g cm^{-3}, and those of water and water vapour at 100°C are 0.958 g cm^{-3} and 0.598 g L^{-1}, respectively. By how much does the chemical potential of water vapour exceed that of liquid water at 1.2 atm and 100°C?

6.6 The enthalpy of fusion of mercury is 2.292 kJ mol^{-1}, and its normal freezing point is 234.3 K with a change in molar volume of +0.517 cm^3 mol^{-1} on melting. At what temperature will the bottom of a column of mercury (density 13.6 g cm^{-3}) of height 10.0 m be expected to freeze?

6.7 50.0 L of dry air was slowly bubbled through a thermally insulated beaker containing 250 g of water initially at 25°C. Calculate the final temperature. (The vapour pressure of water is approximately constant at 23.8 Torr throughout, and its heat capacity is 75.5 J K^{-1} mol^{-1}. Assume that the air is not heated or cooled and that water vapour is a perfect gas.)

6.8 The vapour pressure, p, of nitric acid varies with temperature as follows:

θ/°C	0	20	40	50	70	80	90	100
p/Torr	14.4	47.9	133	208	467	670	937	1282

What are (a) the normal boiling point and (b) the enthalpy of vaporization of nitric acid?

6.9 The vapour pressure of the ketone carvone (M = 150.2 g mol^{-1}), a component of oil of spearmint, is as follows:

θ/°C	57.4	100.4	133.0	157.3	203.5	227.5
p/Torr	1.00	10.0	40.0	100	400	760

What are (a) the normal boiling point and (b) the enthalpy of vaporization of carvone?

6.10 Construct the phase diagram for benzene near its triple point at 36 Torr and 5.50°C using the following data: $\Delta_{fus}H$ = 10.6 kJ mol^{-1}, $\Delta_{vap}H$ = 30.8 kJ mol^{-1}, ρ(s) = 0.891 g cm^{-3}, ρ(l) = 0.879 g cm^{-3}.

6.11‡ In an investigation of the thermophysical properties of toluene (*J. Phys. Chem. Ref. Data* **18**, 1565 (1989)), R.D. Goodwin presented expressions for two coexistence curves. The solid–liquid coexistence curve is given by $p/\text{bar} = p_3/\text{bar} + 1000 \times (5.60 + 11.727x)x$, where $x = T/T_3 - 1$ and the triple point pressure and temperature are p_3 = 0.4362 μbar and T_3 = 178.15 K, respectively. The liquid–vapour curve is given by:

$$\ln(p/\text{bar}) = -10.418/y + 21.157 - 15.996y + 14.015y^2 - 5.0120y^3 + 4.7224(1-y)^{1.70}$$

where $y = T/T_c = T/593.95$ K. (a) Plot the solid–liquid and liquid–vapour phase boundaries. (b) Estimate the standard melting point of toluene. (c) Estimate the standard boiling point of toluene. (d) Compute the standard enthalpy of vaporization of toluene, given that the molar volumes of the liquid and vapour at the normal boiling point are 0.12 L mol^{-1} and 30.3 L mol^{-1}, respectively.

6.12‡ In a study of the vapour pressure of chloromethane (*J. Chem. Eng. Data* **40**, 869 (1995)), A. Bah and N. Dupont-Pavlovsky presented data for the vapour pressure over solid chloromethane at low temperatures. Some of that data is shown below:

T/K	145.94	147.96	149.93	151.94	153.97	154.94
p/Pa	13.07	18.49	25.99	36.76	50.86	59.56

Estimate the standard enthalpy of sublimation of chloromethane at 150 K. (Take the molar volume of the vapour to be that of a perfect gas, and that of the solid to be negligible.)

6.13‡ The standard enthalpy of combustion of 7,8-benzoquinoline (benz) is −6556 ± 4 kJ mol^{-1} for the pure solid at 25°C. The combustion reaction is:

$$C_{13}H_9N(s) + \tfrac{61}{4}O_2(g) \longrightarrow 13\,CO_2(g) + \tfrac{9}{2}H_2O(l) + \tfrac{1}{2}N_2(g)$$

A restricted molecular flow apparatus has been used to determine the very small vapour pressures, p, of the solid by measuring solid mass loss, Δm, in measured time periods Δt, at a fixed temperature. Theoretical analysis of the apparatus yields the relation

$$p/\text{Torr} = 17.4\left(\frac{f(\Delta m/\text{g})}{\alpha(A/\text{cm}^2)(\Delta t/\text{s})}\right)\left(\frac{T/\text{K}}{M/(\text{g mol}^{-1})}\right)^{1/2}$$

where A is the surface area of the sample (A = 3.64 cm^2). The apparatus constants are f = 8.99 and α = 1. The table below contains the experimental data. Use mathematical software or a spreadsheet to determine the enthalpy of sublimation and the entropy of sublimation for a vapour pressure of 1.00 Torr.

T/K	293.10	298.20	303.13	308.23	313.17	318.24	323.15
Δt/s	21 600	19 200	18 000	7140	4200	3600	1800
Δm/g	0.0481	0.0797	0.1275	0.0823	0.0801	0.1055	0.0877

Theoretical problems

6.14 Show that, for a transition between two incompressible solid phases, ΔG is independent of the pressure.

6.15 The change in enthalpy is given by $dH = C_p\,dT + V\,dp$. The Clapeyron equation relates dp and dT at equilibrium, and so in combination the two equations can be used to find how the enthalpy changes along a phase boundary as the temperature changes and the two phases remain in equilibrium. Show that $d(\Delta H/T) = \Delta C_p\,d\ln T$.

6.16 In the 'gas saturation method' for the measurement of vapour pressure, a volume V of gas (as measured at a temperature T and a pressure p) is bubbled slowly through the liquid, which is maintained at the temperature T, and a mass loss m is measured. Show that the vapour pressure, p, of the liquid is related to its molar mass, M, by $p = AmP/(1 + Am)$, where $A = RT/MPV$. The vapour pressure of geraniol (M = 154.2 g mol^{-1}), which is a component of oil of roses, was measured at 110°C. It was found that, when 5.00 L of nitrogen at 760 Torr was passed slowly through the heated liquid, the loss of mass was 0.32 g. Calculate the vapour pressure of geraniol.

6.17 Combine the barometric formula (stated in Box 1.1) for the dependence of the pressure on altitude with the Clausius–Clapeyron equation, and predict how the boiling temperature of a liquid depends on the altitude and the ambient temperature. Take the mean ambient temperature as 20°C and predict the boiling temperature of water at 3000 m.

6.18 Figure 6.8 gives a schematic representation of how the chemical potentials of the solid, liquid, and gaseous phases of a substance vary with temperature. All have a negative slope, but it is unlikely that they are truly straight lines as indicated in the illustration. Derive an expression for the curvatures (specifically, the second derivatives with respect to temperature) of these lines. Is there a restriction on the curvature of these lines? Which state of matter shows the greatest curvature?

6.19 The Clapeyron equation does not apply to second-order phase transitions, but there are two analogous equations, the *Ehrenfest equations*, that do. They are:

$$\frac{dp}{dT} = \frac{\alpha_2 - \alpha_1}{\kappa_{T,2} - \kappa_{T,1}}$$

$$\frac{dp}{dT} = \frac{C_{p,m2} - C_{p,m1}}{TV_m(\alpha_2 - \alpha_1)}$$

where α is the expansion coefficient, κ_T the isothermal compressibility, and the subscripts 1 and 2 refer to two different phases. Derive these two equations. Why does the Clapeyron equation not apply to second-order transitions?

6.20 For a first-order phase transition, to which the Clapeyron equation does apply, prove the relation

$$C_S = C_p - \frac{\alpha V \Delta_{trs}H}{\Delta_{trs}V}$$

where $C_S = (\partial q/\partial T)_S$ is the heat capacity along the coexistence curve of two phases.

Applications: to chemical engineering

6.21‡ A substance as well-known as methane still receives research attention because it is an important component of natural gas, a commonly used fossil fuel . Friend *et al.* have published a review of the thermophysical properties of methane (D.G. Friend, J.F. Ely, and H. Ingham, *J. Phys. Chem. Ref. Data* **18**, 583 (1989)), which included the following data describing the liquid–vapour phase boundary.

T/K	100	108	110	112	114	120	130	140
p/MPa	0.034	0.074	0.088	0.104	0.122	0.192	0.368	0.642

T/K	150	160	170	190
p/MPa	1.041	1.593	2.329	4.521

(a) Plot the liquid–vapour phase boundary. (b) Estimate the standard boiling point of methane. (c) Compute the standard enthalpy of vaporization of methane, given that the molar volumes of the liquid and vapour at the standard boiling point are 3.80×10^{-2} and 8.89 L mol^{-1}, respectively.

6.22‡ Diamond, an allotrope of carbon, is the hardest substance and the best conductor of heat yet characterized. For these reasons, diamond is used widely in industrial applications that require a strong abrasive. Unfortunately, it is difficult to synthesize diamond from the more readily available allotropes of carbon, such as graphite. To illustrate this point, calculated the pressure required to convert graphite into diamond at 25°C. The following data apply to 25°C and 100 kPa. Assume the specific volume, V_s, and κ_T are constant with respect to pressure changes.

	Graphite	Diamond
$\Delta_f G^{\ominus}$/(kJ mol^{-1})	0	+2.8678
V_s/(cm^3 g^{-1})	0.444	0.284
κ_T/kPa	3.04×10^{-8}	0.187×10^{-8}

7 Simple mixtures

This chapter begins by developing the concept of chemical potential to show that it is a particular case of a class of properties called partial molar quantities. Then it explores how to use the chemical potential of a substance to describe the physical properties of a mixture. The underlying principle to keep in mind is that at equilibrium the chemical potential of a species is the same in every phase. We see, by making use of the experimental observations known as Raoult's and Henry's laws, how to express the chemical potential of a substance in terms of its mole fraction in a mixture. With this result established, we can calculate the effect of a solute on certain thermodynamic properties of a solution. These properties include the lowering of vapour pressure of the solvent, the elevation of its boiling point, the depression of its freezing point, and the origin of osmotic pressure. Finally, we see how to express the chemical potential of a substance in a real mixture in terms of a property known as the activity. We see how the activity may be measured, and conclude with a brief discussion of how the standard states of solutes and solvents are defined.

Chemistry deals with mixtures, including mixtures of substances that can react together. Therefore, we need to generalize the concepts introduced so far to deal with substances that are mingled together. As a first step towards dealing with chemical reactions (which are treated in Chapter 9), here we consider mixtures of substances that do not react together. At this stage we deal mainly with binary mixtures (mixtures of two components, A and B). We shall therefore often be able to simplify equations by making use of the relation $x_A + x_B = 1$. Another restriction of this chapter is that we consider mainly non-electrolyte solutions, in which the solute is not present as ions.

The thermodynamic description of mixtures

We have already seen that the partial pressure, which is the contribution of one component to the total pressure, is used to discuss the properties of mixtures of gases. For a more general description of the thermodynamics of mixtures we need to introduce other analogous 'partial' properties.

7.1 Partial molar quantities

The easiest partial molar property to visualize is the 'partial molar volume', the contribution that a component of a mixture makes to the total volume of a sample.

(a) Partial molar volume

Imagine a huge volume of pure water at 25°C. When a further 1 mol H_2O is added, the volume increases by 18 cm^3 and we can report that 18 cm^3 mol^{-1} is the molar volume of pure water. However, when we add 1 mol H_2O to a huge volume of pure ethanol, the volume increases by only 14 cm^3. The reason for the different increase in volume is that the volume occupied by a given number of water molecules depends on the identity of the molecules that surround them. In the latter case there is so much ethanol present that each H_2O molecule is surrounded by ethanol molecules, and the packing of the molecules results in the H_2O molecules increasing the volume by only 14 cm^3. The quantity 14 cm^3 mol^{-1} is the partial molar volume of water in pure ethanol. In general, the **partial molar volume** of a substance A in a mixture is the change in volume per mole of A added to a large volume of the mixture.

The partial molar volumes of the components of a mixture vary with composition because the environment of each type of molecule changes as the composition changes from pure A to pure B. It is this changing molecular environment, and the consequential modification of the forces acting between molecules, that results in the variation of the thermodynamic properties of a mixture as its composition is changed. The partial molar volumes of water and ethanol across the full composition range at 25°C are shown in Fig. 7.1.

The partial molar volume, V_J, of a substance J at some general composition is defined formally as follows:

$$V_J = \left(\frac{\partial V}{\partial n_J}\right)_{p,T,n'} \qquad [7.1]$$

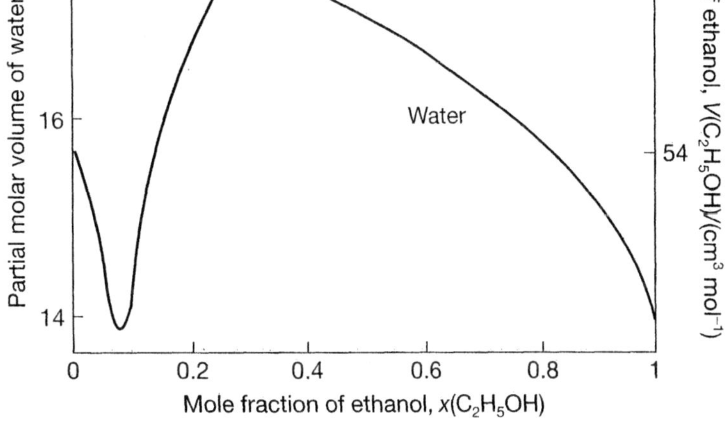

7.1 The partial molar volumes of water and ethanol at 25°C. Note the different scales (water on the left, ethanol on the right).

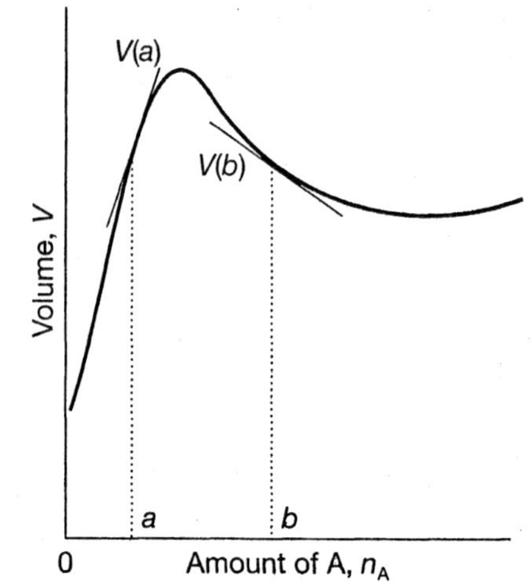

7.2 The partial molar volume of a substance is the slope of the variation of the total volume of the sample plotted against the composition. In general, partial molar quantities vary with the composition, as shown by the different slopes at the compositions *a* and *b*. Note that the partial molar volume at *b* is negative: the overall volume of the sample decreases as A is added.

where the subscript n' signifies that the amounts of all other substances present are constant.[1] The partial molar volume is the slope of the plot of the total volume as the amount of J is changed, the pressure, temperature, and amount of the other components being constant (Fig. 7.2). Its value depends on the composition, as we saw for water and ethanol. The definition in eqn 1 implies that, when the composition of the mixture is changed by the addition of dn_A of A and dn_B of B, then the total volume of the mixture changes by

$$dV = \left(\frac{\partial V}{\partial n_A}\right)_{p,T,n_B} dn_A + \left(\frac{\partial V}{\partial n_B}\right)_{p,T,n_A} dn_B = V_A\, dn_A + V_B\, dn_B \qquad (7.2)$$

Once the partial molar volumes of the two components of a mixture at the composition (and temperature) of interest are known, we can state the total volume, V, of the mixture by using

$$V = n_A V_A + n_B V_B \qquad (7.3)$$

Justification 7.1

Consider a very large sample of the mixture of the specified composition. Then, when an amount n_A of A is added to the mixture, the composition remains virtually unchanged, the partial molar volume V_A is constant, and the volume of the sample changes by $n_A V_A$. When n_B of B is added, the volume changes by $n_B V_B$, for the same reason. The total change of volume is therefore $n_A V_A + n_B V_B$. The sample now occupies a larger volume, but the proportions of the components are still the same. At this stage, scoop out of the enlarged volume a sample containing n_A of A and n_B of B. Its volume is $n_A V_A + n_B V_B$. Because V is a state function, the same sample could have been prepared simply by mixing the appropriate amounts of A and B. This justifies eqn 3.

Partial molar volumes can be measured in several ways. One method is to measure the dependence of the volume on the composition and to fit the observed volume to a function of the amount of the substance. Once the function has been found, its slope can be determined at any composition of interest by differentiation.

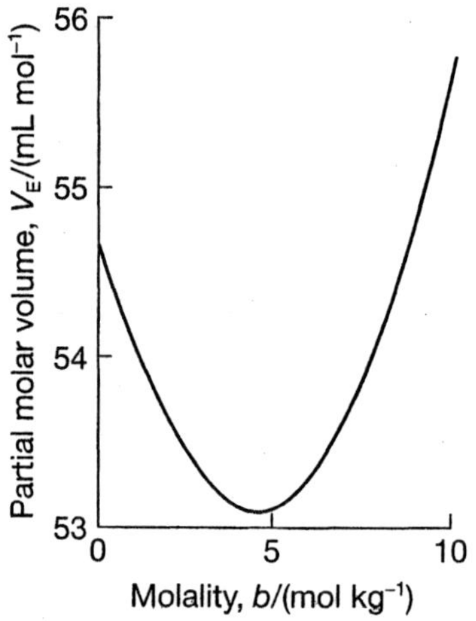

7.3 The partial molar volume of ethanol as expressed by the polynomial in *Illustration 7.1*.

Illustration 7.1

The total volume of an ethanol solution at 25°C containing 1.000 kg of water is found to be given by the expression

$$V/\text{mL} = 1002.93 + 54.6664b - 0.363\,94b^2 + 0.028\,256b^3$$

where b is the numerical value of the molality. Because the amount of CH_3CH_2OH in moles is equal to the numerical value of the molality in moles per kilogram, we can write the partial molar volume of ethanol, V_E, as

$$V_E/(\text{mL mol}^{-1}) = \left(\frac{\partial(V/\text{mL})}{\partial b}\right)_{p,T,n_W} = 54.6664 - 2(0.363\,94)b + 3(0.028\,256)b^2$$

Figure 7.3 is a graph of this function.

1. The IUPAC recommendation is to denote a partial molar quantity by $\bar{X}$, but only when there is the possibility of confusion with the quantity X. For instance, the partial molar volume of NaCl in water could be written $\bar{V}$(NaCl, aq) to distinguish it from the volume of the solution, V(NaCl, aq).

Self-test 7.1 At 25°C, the density of a 50 per cent by mass ethanol/water solution is 0.914 g cm^{-3}. Given that the partial molar volume of water in the solution is 17.4 cm^3 mol^{-1}, what is the partial molar volume of the ethanol?

[56.4 cm^3 mol^{-1}]

Molar volumes are always positive, but partial molar quantities need not be. For example, the limiting partial molar volume of $MgSO_4$ in water (its partial molar volume in the limit of zero concentration) is −1.4 cm^3 mol^{-1}, which means that the addition of 1 mol $MgSO_4$ to a large volume of water results in a decrease in volume of 1.4 cm^3. The contraction occurs because the salt breaks up the open structure of water as the ions become hydrated, and it collapses slightly.

(b) Partial molar Gibbs energies

The concept of a partial molar quantity can be extended to any extensive state function. For a pure substance, the chemical potential is just another name for the molar Gibbs energy. For a substance in a mixture, the chemical potential is *defined* as being the partial molar Gibbs energy:

$$\mu_J = \left(\frac{\partial G}{\partial n_J}\right)_{p,T,n'} \qquad [7.4]$$

That is, the chemical potential is the slope of a plot of Gibbs energy against the amount of the component J, with the pressure and temperature (and the amounts of the other substances) held constant (Fig. 7.4). By the same argument that led to eqn 3, it follows that the total Gibbs energy of a binary mixture is

$$G = n_A\mu_A + n_B\mu_B \qquad (7.5)$$

where μ_A and μ_B are the chemical potentials at the composition of the mixture. That is, the chemical potential of a substance in a mixture is the contribution of that substance to the total Gibbs energy of the mixture.

The Gibbs energy depends on the composition, pressure, and temperature of the system. Thus, G may change when p, T, and the composition change and, for a system of components A, B, etc., the equation $dG = V\,dp - S\,dT$ becomes

$$dG = V\,dp - S\,dT + \mu_A\,dn_A + \mu_B\,dn_B + \cdots \qquad (7.6)$$

This expression is the **fundamental equation of chemical thermodynamics**. Its implications and consequences are explored and developed in this and the next three chapters.

At constant pressure and temperature, eqn 6 simplifies to

$$dG = \mu_A\,dn_A + \mu_B\,dn_B + \cdots \qquad (7.7)$$

We saw in Section 4.5e that under the same conditions $dG = dw_{add,max}$. Therefore, at constant temperature and pressure,

$$dw_{add,max} = \mu_A\,dn_A + \mu_B\,dn_B + \cdots \qquad (7.8)$$

That is, additional (non-expansion) work can arise from the changing composition of a system. For instance, in an electrochemical cell, the chemical reaction is arranged to take place in two distinct sites (at the two electrodes). The electrical work the cell performs can be traced to its changing composition as products are formed from reactants.

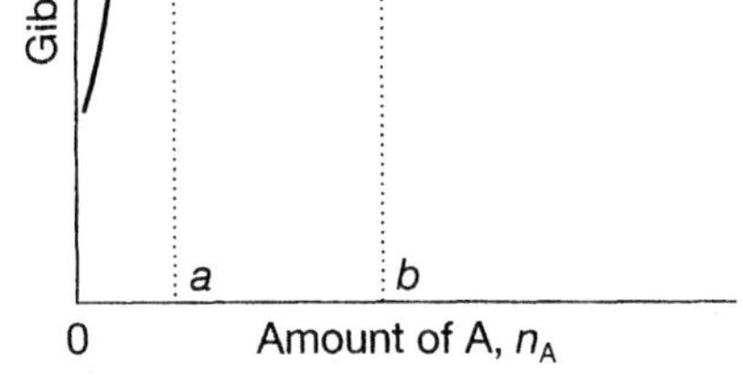

7.4 The chemical potential of a substance is the slope of the total Gibbs energy of a mixture with respect to the amount of substance of interest. In general, the chemical potential varies with composition, as shown for the two values at a and b. In this case, both chemical potentials are positive.

(c) The wider significance of the chemical potential

The chemical potential does more than show how G varies with composition. Because $G = U + pV - TS$, we can write a general infinitesimal change in U for a system of variable composition as

$$\begin{aligned} dU &= -p\,dV - V\,dp + S\,dT + T\,dS + dG \\ &= -p\,dV - V\,dp + S\,dT + T\,dS + (V\,dp - S\,dT + \mu_A\,dn_A + \mu_B\,dn_B + \cdots) \\ &= -p\,dV + T\,dS + \mu_A\,dn_A + \mu_B\,dn_B + \cdots \end{aligned}$$

This expression is the generalization of eqn 5.2 (that $dU = T\,dS - p\,dV$) to systems in which the composition may change. It follows that, at constant volume and entropy,

$$dU = \mu_A\,dn_A + \mu_B\,dn_B + \cdots$$

and hence that

$$\mu_J = \left(\frac{\partial U}{\partial n_J}\right)_{S,V,n'} \tag{7.9}$$

Therefore, not only does the chemical potential show how G changes when the composition changes, it also shows how the internal energy changes too (but under a different set of conditions). In the same way it is easy to deduce that

$$\text{(a)}\quad \mu_J = \left(\frac{\partial H}{\partial n_J}\right)_{S,p,n'} \qquad \text{(b)}\quad \mu_J = \left(\frac{\partial A}{\partial n_J}\right)_{V,T,n'} \tag{7.10}$$

Thus we see that the μ_J shows how all the extensive thermodynamic properties U, H, A, and G depend on the composition. This is why the chemical potential is so central to chemistry.

(d) The Gibbs–Duhem equation

Because the total Gibbs energy of a mixture is given by eqn 5 and the chemical potentials depend on the composition, when the compositions are changed infinitesimally we might expect G of a binary system to change by

$$dG = \mu_A\,dn_A + \mu_B\,dn_B + n_A\,d\mu_A + n_B\,d\mu_B$$

However, we have seen that at constant pressure and temperature a change in Gibbs energy is given by eqn 7. Because G is a state function, these two equations must be equal, which implies that at constant temperature and pressure

$$n_A\,d\mu_A + n_B\,d\mu_B = 0 \tag{7.11}$$

This equation is a special case of the **Gibbs–Duhem equation**:

$$\sum_J n_J\,d\mu_J = 0 \tag{7.12}$$

The significance of the Gibbs–Duhem equation is that the chemical potential of one component of a mixture cannot change independently of the chemical potentials of the other components. In a binary mixture, if one partial molar quantity increases, then the other must decrease:

$$d\mu_B = -\frac{n_A}{n_B}\,d\mu_A \tag{7.13}$$

The same line of reasoning applies to all partial molar quantities. We can see in Fig. 7.1, for example, that, where the partial molar volume of water increases, that of ethanol decreases. Moreover, as eqn 13 shows, and as we can see from Fig. 7.1, a small change in the partial molar volume of A corresponds to a large change in the partial molar volume of B if n_A/n_B is large, but the opposite is true when this ratio is small. In practice, the Gibbs–Duhem equation is used to determine the partial molar volume of one component of a binary mixture from measurements of the partial molar volume of the second component.

Example 7.1 Using the Gibbs-Duhem equation

The experimental values of the partial molar volume of K_2SO_4(aq) at 298 K are given by the expression

$$V_{K_2SO_4}/(\text{cm}^3\ \text{mol}^{-1}) = 32.280 + 18.216b^{1/2}$$

where b is the numerical value of the molality of K_2SO_4. Use the Gibbs-Duhem equation to derive an expression for the partial molar volume of water in the solution. The molar volume of pure water at 298 K is 18.079 cm^3 mol^{-1}.

Method Let A denote K_2SO_4 and B denote H_2O, the solvent. The Gibbs–Duhem equation for the partial molar volumes of two components is

$$n_A\, dV_A + n_B\, dV_B = 0$$

This relation implies that

$$dV_B = -\frac{n_A}{n_B}\, dV_A$$

Therefore, V_B can be found by integration:

$$V_B = V_B^* - \int \frac{n_A}{n_B}\, dV_A$$

where V_B^* is the molar volume of pure B.

The first step is to change the variable V_A to the molality b, and then to integrate the right-hand side between $b = 0$ (pure B) and the molality of interest.

Answer It follows from the information in the question that, with A = K_2SO_4,

$$\frac{dV_A}{db} = 9.108b^{-1/2}$$

Therefore, the integration required is

$$V_B = V_B^* - 9.108 \int_0^b \frac{n_A}{n_B}\, b^{-1/2}\, db$$

However, we need to note that the molality is related to the amounts of A and B by

$$b = \frac{n_A}{n_B M_B}$$

where M_B is the molar mass of water (in kilograms per mole). Therefore,

$$V_B = V_B^* - 9.108M_B \int_0^b b^{1/2}\, db$$

$$= V_B^* - \tfrac{2}{3}(9.108M_B b^{3/2})$$

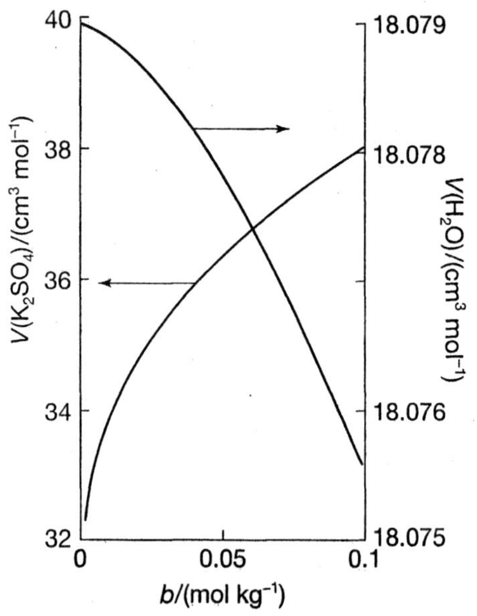

7.5 The partial molar volumes of the components of an aqueous solution of potassium sulfate.

It then follows, by substituting the data, that

$$V_B/(\text{cm}^3\ \text{mol}^{-1}) = 18.079 - 0.1094b^{3/2}$$

The partial molar volumes are plotted in Fig. 7.5.

Self-test 7.2 Repeat the calculation for a salt A for which $V_A/(\text{cm}^3\ \text{mol}^{-1}) = 6.218 + 5.146b - 7.147b^2$. [$V_B/(\text{cm}^3\ \text{mol}^{-1}) = 18.079 - 0.0464b^2 + 0.0859b^3$]

7.2 The thermodynamics of mixing

The dependence of the Gibbs energy of a mixture on its composition is given by eqn 5, and we know that at constant temperature and pressure systems tend towards lower Gibbs energy. This is the link we need in order to apply thermodynamics to the discussion of spontaneous changes of composition, as in the mixing of two substances. One simple example of a spontaneous mixing process is that of two gases introduced into the same container. The mixing is spontaneous, so it must correspond to a decrease in G. We shall now see how to express this idea quantitatively.

(a) The Gibbs energy of mixing

Let the amounts of two perfect gases in the two containers be n_A and n_B; both are at a temperature T and a pressure p. At this stage, the chemical potentials of the two gases have their 'pure' values, which are obtained by applying the definition $\mu = G_m$ to eqn 5.17:

$$\mu = \mu^{\ominus} + RT \ln \frac{p}{p^{\ominus}} \qquad (7.14)^{\circ}$$

where $\mu^{\ominus}$ is the **standard chemical potential**, the chemical potential of the pure gas at 1 bar.

The Gibbs energy of the total system is then given by eqn 5 as

$$G_i = n_A\mu_A + n_B\mu_B$$

$$= n_A\left(\mu_A^{\ominus} + RT \ln \frac{p}{p^{\ominus}}\right) + n_B\left(\mu_B^{\ominus} + RT \ln \frac{p}{p^{\ominus}}\right)$$

It will be much simpler notationally if we agree to let p denote the pressure relative to $p^{\ominus}$; that is, to replace $p/p^{\ominus}$ by p, for then we can write

$$G_i = n_A\{\mu_A^{\ominus} + RT \ln p\} + n_B\{\mu_B^{\ominus} + RT \ln p\} \qquad \{7.15\}^{\circ}$$

Equations for which this convention is used will be labelled {1}, {2}, ... ; to use the equations, we have to remember to replace p by $p/p^{\ominus}$ again. In practice, that simply means using the numerical value of p in bars. After mixing, the partial pressures of the gases are p_A and p_B, with $p_A + p_B = p$. The total Gibbs energy changes to

$$G_f = n_A\{\mu_A^{\ominus} + RT \ln p_A\} + n_B\{\mu_B^{\ominus} + RT \ln p_B\} \qquad \{7.16\}^{\circ}$$

The difference $G_f - G_i$, the Gibbs energy of mixing, $\Delta_{mix}G$, is therefore

$$\Delta_{mix}G = n_A RT \ln \frac{p_A}{p} + n_B RT \ln \frac{p_B}{p} \qquad (7.17)^{\circ}$$

At this point we may replace n_J by $x_J n$ and use Dalton's law (Section 1.2d) to write $p_J/p = x_J$ for each component, which gives

$$\Delta_{mix}G = nRT(x_A \ln x_A + x_B \ln x_B) \qquad (7.18)^{\circ}$$

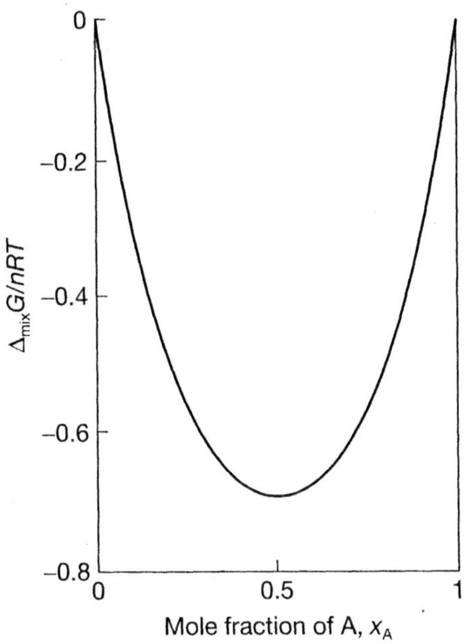

7.6 The Gibbs energy of mixing of two perfect gases and (as discussed later) of two liquids that form an ideal solution. The Gibbs energy of mixing is negative for all compositions and temperatures, so perfect gases mix spontaneously in all proportions.

Because mole fractions are never greater than 1, the logarithms in this equation are negative, and $\Delta_{mix}G < 0$ (Fig. 7.6). The conclusion that $\Delta_{mix}G$ is negative for all compositions confirms that perfect gases mix spontaneously in all proportions. However, the equation extends common sense by allowing us to discuss the process quantitatively. We see, for instance, that $\Delta_{mix}G$ is directly proportional to the temperature but is independent of the total pressure.

Example 7.2 Calculating a Gibbs energy of mixing

A container is divided into two equal compartments (Fig. 7.7). One contains 3.0 mol H_2 at 25°C; the other contains 1.0 mol N_2 at 25°C. Calculate the Gibbs energy of mixing when the partition is removed. Assume perfect behaviour.

Method We proceed by calculating the initial Gibbs energy from the chemical potentials. To do so, we need the pressure of each gas. Write the pressure of nitrogen as p; then the pressure of hydrogen as a multiple of p can be found from the gas laws. Next, calculate the Gibbs energy for the system when the partition is removed. The volume of each gas doubles, so its partial pressure falls by a factor of 2.

Answer Given that the pressure of nitrogen is p, the pressure of hydrogen is $3p$; therefore, the initial Gibbs energy is

$$G_i = (3.0\text{ mol})\{\mu^\ominus(H_2) + RT\ln 3p\} + (1.0\text{ mol})\{\mu^\ominus(N_2) + RT\ln p\}$$

The partial pressure of nitrogen falls to $\frac{1}{2}p$ and that of hydrogen falls to $\frac{3}{2}p$. Therefore, the Gibbs energy changes to

$$G_f = (3.0\text{ mol})\{\mu^\ominus(H_2) + RT\ln \tfrac{3}{2}p\} + (1.0\text{ mol})\{\mu^\ominus(N_2) + RT\ln \tfrac{1}{2}p\}$$

The Gibbs energy of mixing is the difference of these two quantities:

$$\begin{aligned}\Delta_{mix}G &= (3.0\text{ mol})RT\ln\frac{\frac{3}{2}p}{3p} + (1.0\text{ mol})RT\ln\frac{\frac{1}{2}p}{p}\\ &= -(3.0\text{ mol})RT\ln 2 - (1.0\text{ mol})RT\ln 2\\ &= -(4.0\text{ mol})RT\ln 2 = -6.9\text{ kJ}\end{aligned}$$

Comment In this example, the value of $\Delta_{mix}G$ is the sum of two contributions: the mixing itself, and the changes in pressure of the two gases to their final pressure, $2p$. When 3.0 mol H_2 mixes with 1.0 mol N_2 at the same pressure, the change of Gibbs energy is −5.6 kJ and is independent of the initial common pressure.

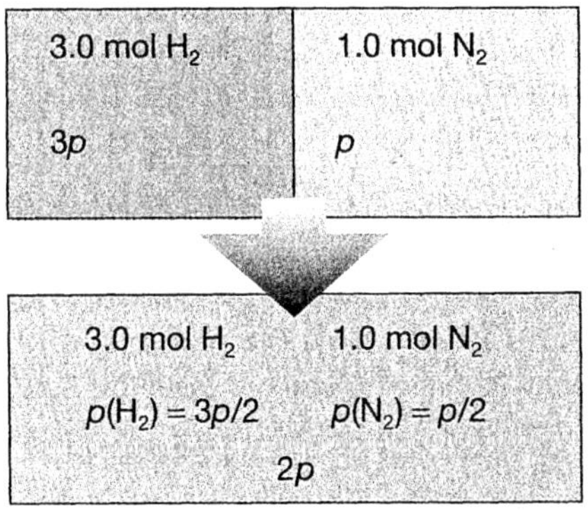

7.7 The initial and final states considered in the calculation of the Gibbs energy of mixing of gases at different initial pressures.

Self-test 7.3 Suppose that 2.0 mol H_2 at 2.0 atm and 25°C and 4.0 mol N_2 at 3.0 atm and 25°C are mixed at constant volume. Calculate $\Delta_{mix}G$. What would be the value of $\Delta_{mix}G$ had the pressures been identical initially?

[−9.7 kJ, −9.5 kJ]

(b) Other thermodynamic mixing functions

Because $(\partial G/\partial T)_{p,n} = -S$, it follows immediately from eqn 18 that, for a mixture of perfect gases, the **entropy of mixing**, $\Delta_{mix}S$, is

$$\Delta_{mix}S = -\left(\frac{\partial \Delta_{mix}G}{\partial T}\right)_{p,n_A,n_B} = -nR(x_A\ln x_A + x_B\ln x_B) \qquad (7.19)^\circ$$

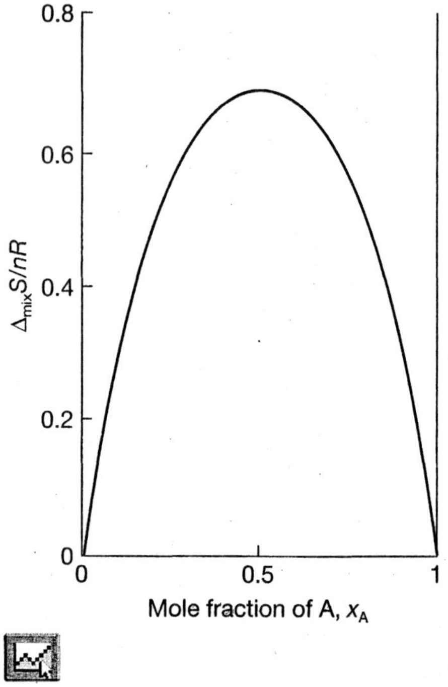

7.8 The entropy of mixing of two perfect gases and (as discussed later) of two liquids that form an ideal solution. The entropy increases for all compositions and temperatures, so perfect gases mix spontaneously in all proportions. Because there is no transfer of heat to the surroundings when perfect gases mix, the entropy of the surroundings is unchanged. Hence, the graph also shows the total entropy of the system plus the surroundings when perfect gases mix.

Because $\ln x < 0$, it follows that $\Delta_{mix}S > 0$ for all compositions (Fig. 7.8). This increase in entropy is what we expect when one gas disperses into the other and the system becomes more disordered. The Gibbs energy of mixing in Example 7.2 was calculated as $-(4.0\text{ mol})RT \ln 2$, so the corresponding entropy of mixing is $+(4.0\text{ mol})R \ln 2$, or $+23\text{ J K}^{-1}$.

We can calculate the isothermal, isobaric (constant pressure) **enthalpy of mixing**, $\Delta_{mix}H$, the enthalpy change accompanying mixing, of two perfect gases from $\Delta G = \Delta H - T\Delta S$. It follows from eqns 18 and 19 that

$$\Delta_{mix}H = 0 \qquad (7.20)^{\circ}$$

The enthalpy of mixing is zero, as we should expect for a system in which there are no interactions between the molecules forming the gaseous mixture. It follows that the whole of the driving force for mixing comes from the increase in entropy of the system, because the entropy of the surroundings is unchanged.

7.3 The chemical potentials of liquids

To discuss the equilibrium properties of liquid mixtures we need to know how the Gibbs energy of a liquid varies with composition. To calculate its value, we use the fact that, at equilibrium, the chemical potential of a substance present as a vapour must be equal to its chemical potential in the liquid.

(a) Ideal solutions

We shall denote quantities relating to pure substances by a superscript *, so the chemical potential of pure A is written μ_A^*, and as $\mu_A^*(\text{l})$ when we need to emphasize that A is a liquid. Because the vapour pressure of the pure liquid is p_A^*, it follows from eqn 14 that the chemical potential of A in the vapour is $\mu_A^{\ominus} + RT \ln p_A^*$ (with p_A to be interpreted as the relative pressure $p_A/p^{\ominus}$). These two chemical potentials are equal at equilibrium (Fig. 7.9), so we can write

$$\mu_A^* = \mu_A^{\ominus} + RT \ln p_A^* \qquad \{7.21\}$$

If another substance, a solute, is also present in the liquid, the chemical potential of A in the liquid is μ_A and its vapour pressure is p_A. In this case

$$\mu_A = \mu_A^{\ominus} + RT \ln p_A \qquad \{7.22\}$$

Next, we combine these two equations to eliminate the standard chemical potential of the gas, and obtain

$$\mu_A = \mu_A^* + RT \ln \frac{p_A}{p_A^*} \qquad (7.23)$$

In the final step we draw on additional experimental information about the relation between the ratio of vapour pressures and the composition of the liquid. In a series of experiments on mixtures of closely related liquids (such as benzene and methylbenzene), the French chemist François Raoult found that the ratio of the partial vapour pressure of each component to its vapour pressure as a pure liquid, p_A/p_A^*, is approximately equal to the mole fraction of A in the liquid mixture. That is, he established what we now call **Raoult's law**:

$$p_A = x_A p_A^* \qquad (7.24)^{\circ}$$

This law is illustrated in Fig. 7.10. Some mixtures obey Raoult's law very well, especially when the components are structurally similar (Fig. 7.11). Mixtures that obey the law throughout the composition range from pure A to pure B are called **ideal**

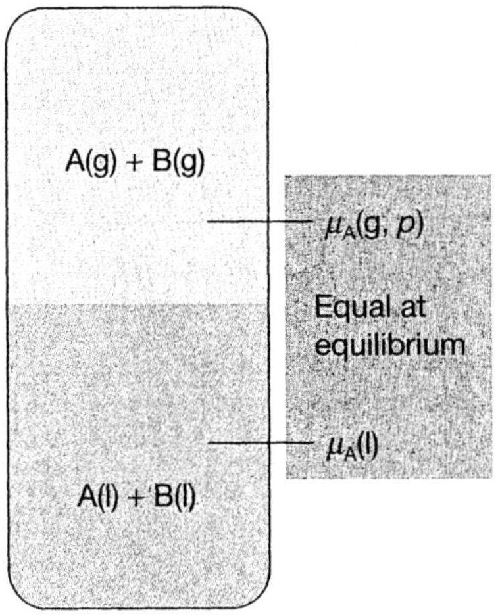

7.9 At equilibrium, the chemical potential of the gaseous form of a substance A is equal to the chemical potential of its condensed phase. The equality is preserved if a solute is also present. Because the chemical potential of A in the vapour depends on its partial vapour pressure, it follows that the chemical potential of liquid A can be related to its partial vapour pressure.

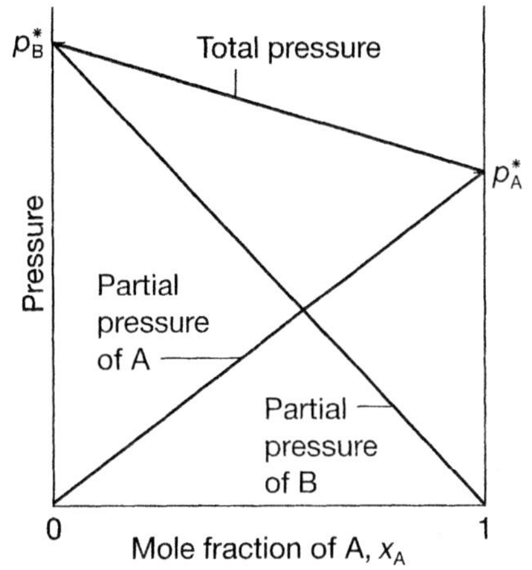

7.10 The total vapour pressure and the two partial vapour pressures of an ideal binary mixture are proportional to the mole fractions of the components.

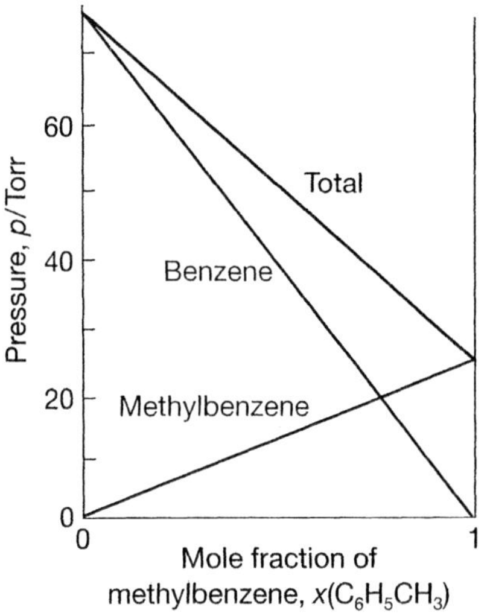

7.11 Two similar liquids, in this case benzene and methylbenzene (toluene), behave almost ideally, and the variation of their vapour pressures with composition resembles that for an ideal solution.

solutions. When we write equations that are valid only for ideal solutions, we shall label them with a superscript °, as in eqn 24.

For an ideal solution, it follows from eqns 23 and 24 that

$$\mu_A = \mu_A^* + RT \ln x_A \qquad (7.25)^\circ$$

This important equation can be used as the *definition* of an ideal solution (so that it implies Raoult's law rather than stemming from it). It is in fact a better definition than eqn 24 because it does not assume that the gas is perfect.

Molecular interpretation 7.1

The origin of Raoult's law can be understood in molecular terms by considering the rates at which molecules leave and return to the liquid. The law reflects the fact that the presence of a second component reduces the rate at which A molecules leave the surface of the liquid but does not inhibit the rate at which they return (Fig. 7.12).

The rate at which A molecules leave the surface is proportional to the number of them at the surface, which in turn is proportional to the mole fraction of A:

$$\text{rate of vaporization} = kx_A$$

where k is a constant of proportionality. The rate at which molecules condense is proportional to their concentration in the gas phase, which in turn is proportional to their partial pressure:

$$\text{rate of condensation} = k'p_A$$

At equilibrium, the rates of vaporization and condensation are equal, so

$$k'p_A = kx_A$$

It follows that

$$p_A = \frac{k}{k'} x_A$$

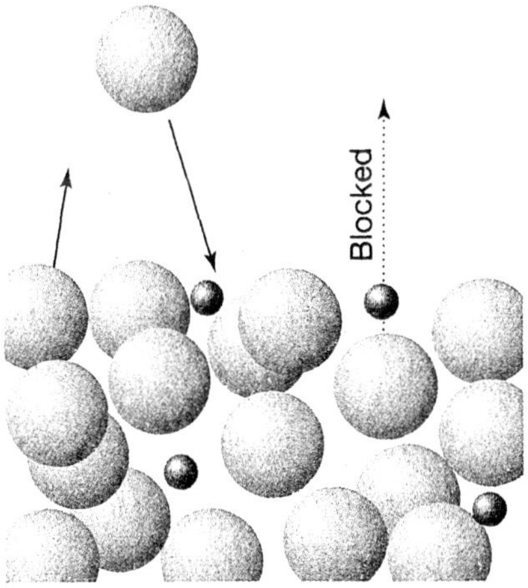

7.12 A pictorial representation of the molecular basis of Raoult's law. The large spheres represent solvent molecules at the surface of a solution (the uppermost line of spheres), and the small spheres are solute molecules. The latter hinder the escape of solvent molecules into the vapour, but do not hinder their return.

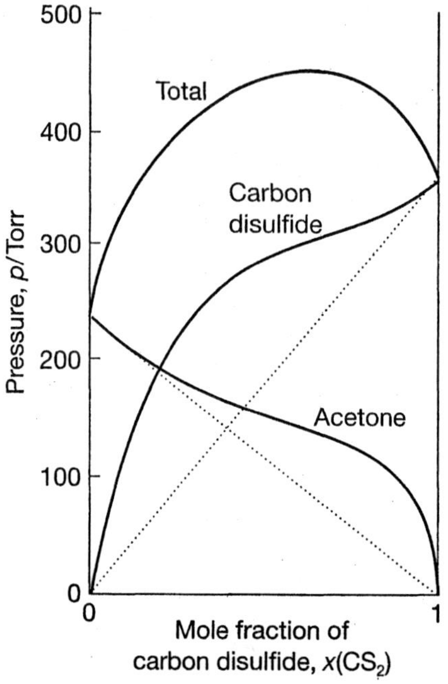

7.13 Strong deviations from ideality are shown by dissimilar liquids (in this case carbon disulfide and acetone (propanone)).

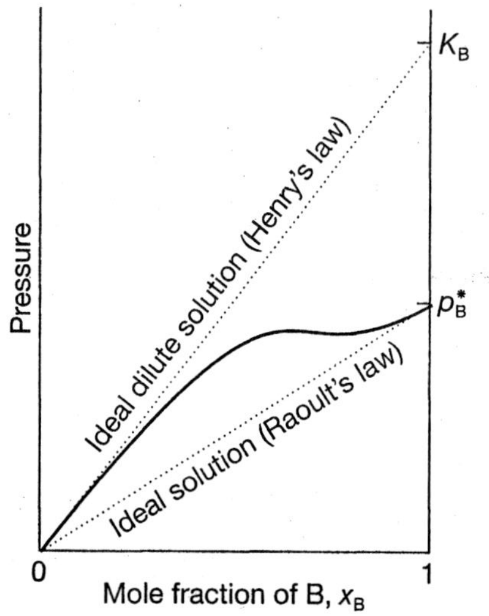

7.14 When a component (the solvent) is nearly pure, it has a vapour pressure that is proportional to the mole fraction with a slope p_B^* (Raoult's law). When it is the minor component (the solute) its vapour pressure is still proportional to the mole fraction, but the constant of proportionality is now K_B (Henry's law).

For the pure liquid, $x_A = 1$; so

$$p_A^* = \frac{k}{k'}$$

Equation 24 then follows by substitution of this relation into the previous one.

Some solutions depart significantly from Raoult's law (Fig. 7.13). Nevertheless, even in these cases the law is obeyed increasingly closely for the component in excess (the solvent) as it approaches purity. The law is therefore a good approximation for the properties of the solvent if the solution is dilute.

(b) Ideal-dilute solutions

In ideal solutions the solute, as well as the solvent, obeys Raoult's law. However, the English chemist William Henry found experimentally that, for real solutions at low concentrations, although the vapour pressure of the solute is proportional to its mole fraction, the constant of proportionality is not the vapour pressure of the pure substance (Fig. 7.14). **Henry's law** is:

$$p_B = x_B K_B \qquad (7.26)^\circ$$

In this expression x_B is the mole fraction of the solute and K_B is an empirical constant (with the dimensions of pressure) chosen so that the plot of the vapour pressure of B against its mole fraction is tangent to the experimental curve at $x_B = 0$.

7.15 In a dilute solution, the solvent molecules (the blue spheres) are in an environment that differs only slightly from that of the pure solvent. The solute particles, however, are in an environment totally unlike that of the pure solute.

Mixtures for which the solute obeys Henry's law and the solvent obeys Raoult's law are called **ideal-dilute solutions**. We shall also label equations with a superscript ° when they have been derived from Henry's law.

Molecular interpretation 7.2

The difference in behaviour of the solute and solvent at low concentrations (as expressed by Henry's and Raoult's laws, respectively) arises from the fact that in a dilute solution the solvent molecules are in an environment very much like the one they have in the pure liquid (Fig. 7.15). In contrast, the solute molecules are surrounded by solvent molecules, which is entirely different from their environment when pure. Thus, the solvent behaves like a slightly modified pure liquid, but the solute behaves entirely differently from its pure state unless the solvent and solute molecules happen to be very similar. In the latter case, the solute also obeys Raoult's law.

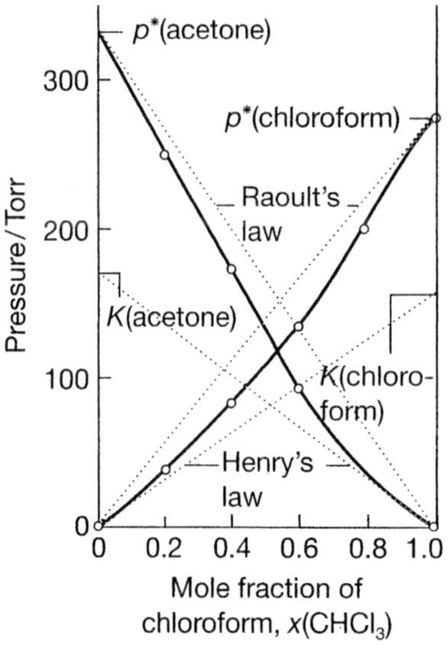

7.16 The experimental partial vapour pressures of a mixture of chloroform (trichloromethane) and acetone (propanone) based on the data in Example 7.3. The values of *K* are obtained by extrapolating the dilute solution vapour pressures as explained in the Example.

Example 7.3 Investigating the validity of Raoult's and Henry's laws

The vapour pressures of each component in a mixture of propanone (acetone, A) and trichloromethane (chloroform, C) were measured at 35°C with the following results:

x_C	0	0.20	0.40	0.60	0.80	1
p_C/Torr	0	35	82	142	200	273
p_A/Torr	347	250	175	92	37	0

Confirm that the mixture conforms to Raoult's law for the component in large excess and to Henry's law for the minor component. Find the Henry's law constants.

Method Both Raoult's and Henry's laws are statements about the form of the graph of partial vapour pressure against mole fraction. Therefore, plot the partial vapour pressures against mole fraction. Raoult's law is tested by comparing the data with the straight line $p_J = x_J p_J^*$ for each component in the region in which it is in excess (and acting as the solvent). Henry's law is tested by finding a straight line $p_J = x_J K_J$ that is tangent to each partial vapour pressure at low x, where the component can be treated as the solute.

Answer The data are plotted in Fig. 7.16 together with the Raoult's law lines. Henry's law requires $K = 175$ Torr for propanone and $K = 165$ Torr for trichloromethane.

Comment Notice how the system deviates from both Raoult's and Henry's laws even for quite small departures from $x = 1$ and $x = 0$, respectively. We deal with these deviations in Section 7.5.

Self-test 7.4 The vapour pressure of chloromethane at various mole fractions in a mixture at 25°C was found to be as follows:

x	0.005	0.009	0.019	0.024
p/Torr	205	363	756	946

Estimate Henry's law constant. [4×10^4 Torr]

The web site contains links to online databases of Henry's law constants.

Some Henry's law data are listed in Table 7.1. As well as providing a link between the mole fraction of solute and its partial pressure, the data in the table may also be used to calculate gas solubilities. The following example illustrates the procedure.

Table 7.1* Henry's law constants for gases in water at 298 K

	K/(10 MPa)	K/Torr
CO_2	0.167	1.25×10^6
H_2	7.12	5.34×10^7
N_2	8.68	6.51×10^7
O_2	4.40	3.30×10^7

* More values are given in the *Data section.*

Example 7.4 Using Henry's law

Estimate the molar solubility (the solubility in moles per litre) of oxygen in water at 25°C and a partial pressure of 160 Torr, its partial pressure in the atmosphere at sea level.

Method The mole fraction of solute is given by Henry's law as $x_J = p_J/K_J$, where p_J is the partial pressure of the gaseous solute J. All we need do is to calculate the mole fraction that corresponds to the stated partial pressure, and then interpret that mole fraction as a molar concentration. For the latter part of the calculation, we calculate the amount of O_2 dissolved in 1.00 kg of water (which corresponds to about 1.00 L water). The solution is dilute, so the expressions for the mole fraction can be simplified.

Answer Because the amount of O_2 dissolved is small, its mole fraction is

$$x_{O_2} = \frac{n_{O_2}}{n_{O_2} + n_{H_2O}} \approx \frac{n_{O_2}}{n_{H_2O}}$$

Hence,

$$n_{O_2} \approx x_{O_2} n_{H_2O} = \frac{p_{O_2} n_{H_2O}}{K} \approx \frac{(160\ \text{Torr}) \times (55.5\ \text{mol})}{3.30 \times 10^7\ \text{Torr}} = 2.69 \times 10^{-4}\ \text{mol}$$

The molality of the saturated solution is therefore 2.69×10^{-4} mol kg^{-1}, corresponding to a molar concentration of approximately 2.7×10^{-4} mol L^{-1}.

Self-test 7.5 Calculate the molar solubility of nitrogen in water exposed to air at 25°C; partial pressures were calculated in Example 1.5.

[0.51 mmol L^{-1}]

A knowledge of Henry's law constants for gases in blood and fats is important for the discussion of respiration, especially when the partial pressure of oxygen is abnormal, as in diving and mountaineering, and for the discussion of the action of gaseous anaesthetics. For example, the molar solubility of O_2 gas in blood decreases with decreasing atmospheric pressure and this impairs respiration at high altitudes. Conversely, the solubility of air in blood increases with increasing pressures. For this reason, a diver who has been underwater for a long period of time cannot be returned to the surface too rapidly: the sudden decrease in pressure lowers the solubility of gases and forms bubbles (emboli) in the bloodstream, with potentially fatal consequences.

The properties of solutions

In this section we consider the thermodynamics of mixing of liquids. First, we consider the simple case of mixtures of liquids that mix to form an ideal solution. In this way, we identify the thermodynamic consequences of molecules of one species mingling randomly with molecules of the second species. The calculation provides a background for discussing the deviations from ideal behaviour exhibited by real solutions.

7.4 Liquid mixtures

Thermodynamics can provide insight into the properties of liquid mixtures, and a few simple ideas can bring the whole field of study together.

(a) Ideal solutions

The Gibbs energy of mixing of two liquids to form an ideal solution is calculated in exactly the same way as for two gases (Section 7.2). The total Gibbs energy before liquids are mixed is

$$G_i = n_A\mu_A^* + n_B\mu_B^*$$

When they are mixed, the individual chemical potentials are given by eqn 25 and the total Gibbs energy is

$$G_f = n_A\{\mu_A^* + RT\ln x_A\} + n_B\{\mu_B^* + RT\ln x_B\}$$

Consequently, the Gibbs energy of mixing is

$$\Delta_{mix}G = nRT\{x_A\ln x_A + x_B\ln x_B\} \qquad (7.27)^\circ$$

where $n = n_A + n_B$. As for gases, it follows that the ideal entropy of mixing of two liquids is

$$\Delta_{mix}S = -nR\{x_A\ln x_A + x_B\ln x_B\} \qquad (7.28)^\circ$$

and the ideal enthalpy of mixing is zero.

Equation 27 is the same as that for two perfect gases and all the conclusions drawn there are valid here: the driving force for mixing is the increasing entropy of the system as the molecules mingle and the enthalpy of mixing is zero. It should be noted, however, that solution ideality means something different from gas perfection. In a perfect gas there are no interactions between molecules. In ideal solutions there are interactions, but the average A–B interactions in the mixture are the same as the average A–A and B–B interactions in the pure liquids. The variation of the Gibbs energy of mixing with composition is the same as that already depicted for gases in Fig. 7.6; the same is true of the entropy of mixing, Fig. 7.8.

Real solutions are composed of particles for which A–A, A–B, and B–B interactions are all different. Not only may there be an enthalpy change when liquids mix, but there may also be an additional contribution to the entropy arising from the way in which the molecules of one type might cluster together instead of mingling freely with the others. If the enthalpy change is large and positive or if the entropy change is adverse (because of a reorganization of the molecules that results in an orderly mixture), then the Gibbs energy might be positive for mixing. In that case, separation is spontaneous and the liquids may be immiscible. Alternatively, the liquids might be **partially miscible**, which means that they are miscible only over a certain range of compositions.

(b) Excess functions and regular solutions

The thermodynamic properties of real solutions are expressed in terms of the **excess functions**, X^E, the difference between the observed thermodynamic function of mixing and the function for an ideal solution. The **excess entropy**, S^E, for example, is defined as

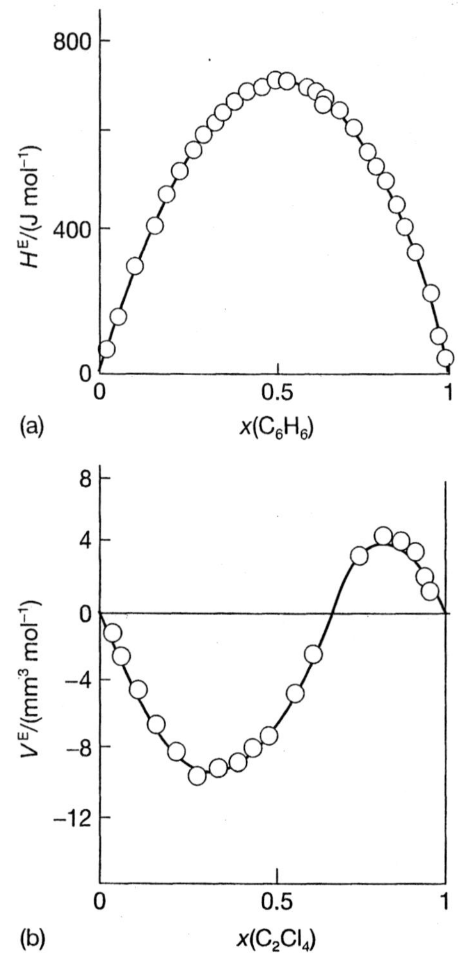

7.17 Experimental excess functions at 25°C. (a) H^E for benzene/cyclohexane; this graph shows that the mixing is endothermic (because $\Delta_{mix}H = 0$ for an ideal solution). (b) The excess volume, V^E, for tetrachloroethane/cyclopentane; this graph shows that there is a contraction at low tetrachloroethane mole fractions, but an expansion at high mole fractions (because $\Delta_{mix}V = 0$ for an ideal mixture).

$$S^E = \Delta_{mix}S - \Delta_{mix}S^{ideal} \qquad [7.29]$$

where $\Delta_{mix}S^{ideal}$ is given by eqn 28. The excess enthalpy and volume are both equal to the observed enthalpy and volume of mixing, because the ideal values are zero in each case.

Deviations of the excess energies from zero indicate the extent to which the solutions are nonideal. In this connection a useful model system is the **regular solution**, a solution for which $H^E \neq 0$ but $S^E = 0$. We can think of a regular solution as one in which the two kinds of molecules are distributed randomly (as in an ideal solution) but have different energies of interactions with each other. Figure 7.17 shows two examples of the composition dependence of excess functions.

We can make this discussion more quantitative by supposing that the excess enthalpy depends on composition as

$$H^E = n\beta RTx_Ax_B \qquad (7.30)$$

where β is a parameter. We let $\beta = w/RT$, where w is an indication of the energy of AB interactions relative to AA and BB interactions. The function given by eqn 30 is plotted in Fig. 7.18, and we see it resembles the experimental curve in Fig. 7.17. If $\beta < 0$, mixing is exothermic and the solute–solvent interactions are more favourable than the solvent–solvent and solute–solute interactions. If $\beta > 0$, then the mixing is endothermic. Because the entropy of mixing has its ideal value for a regular solution, the excess Gibbs energy is equal to the excess enthalpy, and the Gibbs energy of mixing is

$$\Delta_{mix}G = nRT\{x_A \ln x_A + x_B \ln x_B + \beta x_Ax_B\} \qquad (7.31)$$

Figure 7.19 shows how $\Delta_{mix}G$ varies with composition for different values of β. The important feature is that for $\beta > 2$ the graph shows two minima separated by a maximum. The implication of this observation is that, provided $\beta > 2$, then the system will separate spontaneously into two phases with compositions corresponding to the two minima, for that separation corresponds to a reduction in Gibbs energy. We take up this point in Sections 7.8 and 8.5.

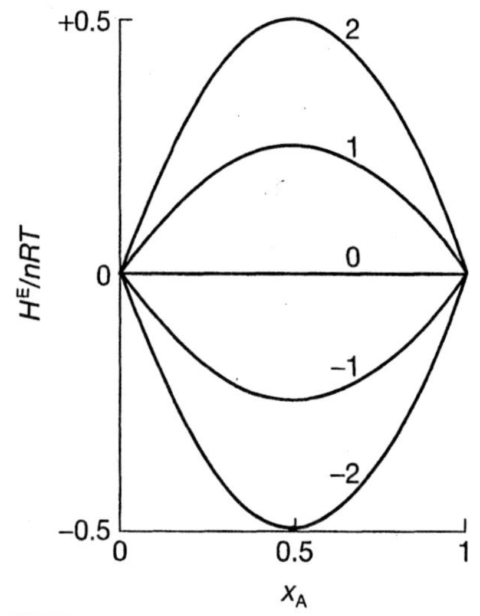

7.18 The excess enthalpy according to a model in which it is proportional to βx_Ax_B, for different values of the parameter β.

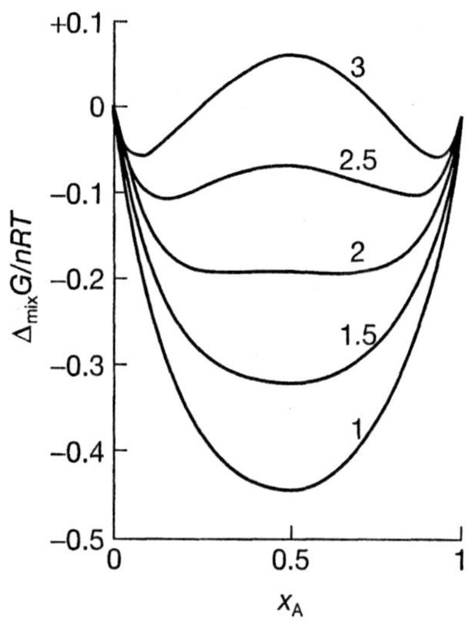

7.19 The Gibbs energy of mixing for different values of the parameter β.

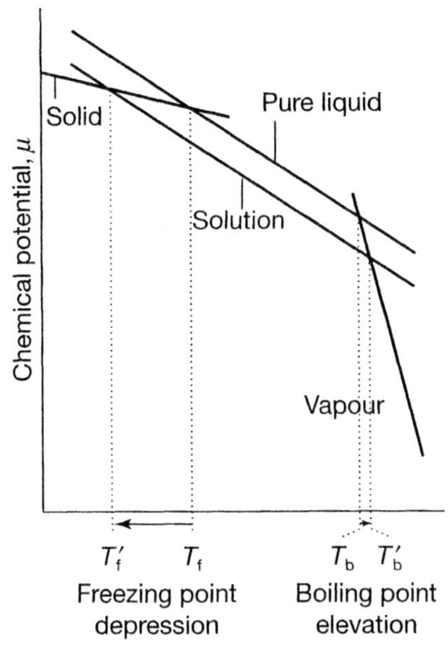

7.20 The chemical potential of a solvent in the presence of a solute. The lowering of the liquid's chemical potential has a greater effect on the freezing point than on the boiling point because of the angles at which the lines intersect (which are determined by entropies; recall Fig. 6.8).

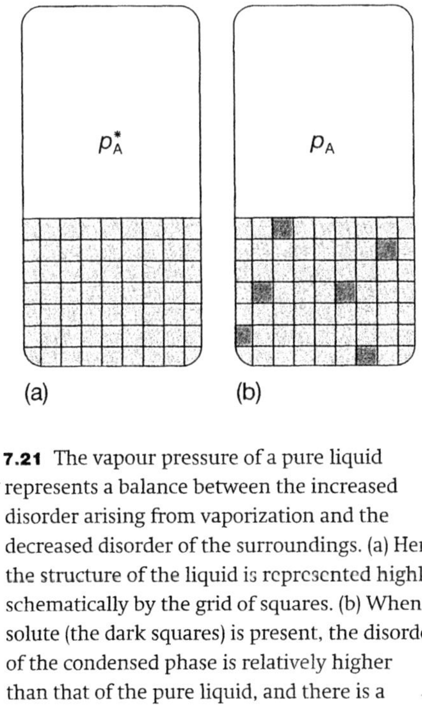

7.21 The vapour pressure of a pure liquid represents a balance between the increased disorder arising from vaporization and the decreased disorder of the surroundings. (a) Here the structure of the liquid is represented highly schematically by the grid of squares. (b) When solute (the dark squares) is present, the disorder of the condensed phase is relatively higher than that of the pure liquid, and there is a decreased tendency to acquire the disorder characteristic of the vapour.

7.5 Colligative properties

The properties we now consider are the elevation of boiling point, the depression of freezing point, and the osmotic pressure arising from the presence of a solute. In dilute solutions these properties depend only on the number of solute particles present, not their identity. For this reason, they are called **colligative properties** (denoting 'depending on the collection').

We assume throughout the following that the solute is not volatile, so it does not contribute to the vapour. We also assume that the solute does not dissolve in the solid solvent: that is, the pure solid solvent separates when the solution is frozen. The latter assumption is quite drastic, although it is true of many mixtures; it can be avoided at the expense of more algebra, but that introduces no new principles.

(a) The common features of colligative properties

All the colligative properties stem from the reduction of the chemical potential of the liquid solvent as a result of the presence of solute. The reduction is from μ_A^* for the pure solvent to $\mu_A^* + RT \ln x_A$ when a solute is present ($\ln x_A$ is negative because $x_A < 1$). There is no direct influence of the solute on the chemical potential of the solvent vapour and the solid solvent because the solute appears in neither the vapour nor the solid. As can be seen from Fig. 7.20, the reduction in chemical potential of the solvent implies that the liquid–vapour equilibrium occurs at a higher temperature (the boiling point is raised) and the solid–liquid equilibrium occurs at a lower temperature (the freezing point is lowered).

Molecular interpretation 7.3

The molecular origin of the lowering of the chemical potential is not the energy of interaction of the solute and solvent particles, because the lowering occurs even in an ideal solution (for which the enthalpy of mixing is zero). If it is not an enthalpy effect, it must be an entropy effect.

The pure liquid solvent has an entropy that reflects the disorder of its molecules. Its vapour pressure reflects the tendency of the solution towards greater entropy, which can be achieved if the liquid vaporizes to form a more disordered gas. When a solute is present, there is an additional contribution to the entropy of the liquid, even in an ideal solution. Because the entropy of the liquid is already higher than that of the pure liquid, there is a weaker tendency to form the gas (Fig. 7.21). The effect of the solute appears as a lowered vapour pressure, and hence a higher boiling point.

Similarly, the enhanced molecular randomness of the solution opposes the tendency to freeze. Consequently, a lower temperature must be reached before equilibrium between solid and solution is achieved. Hence, the freezing point is lowered.

The strategy for the quantitative discussion of the elevation of boiling point and the depression of freezing point is to look for the temperature at which, at 1 atm, one phase (the pure solvent vapour or the pure solid solvent) has the same chemical potential as the solvent in the solution. This is the new equilibrium temperature for the phase transition at 1 atm, and hence corresponds to the new boiling point or the new freezing point of the solvent.

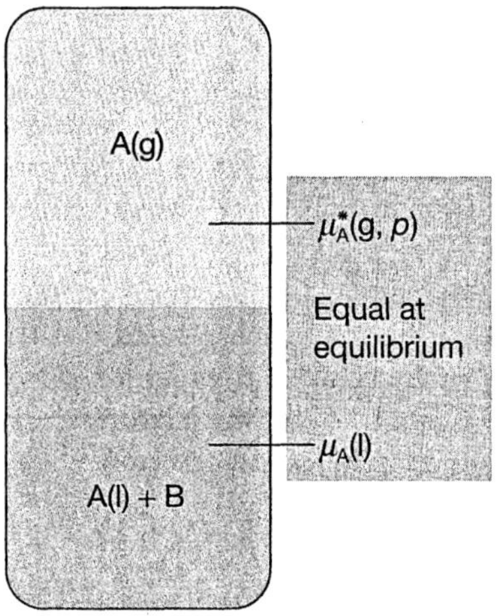

7.22 The heterogeneous equilibrium involved in the calculation of the elevation of boiling point is between A in the pure vapour and A in the mixture, A being the solvent and B an involatile solute.

(b) The elevation of boiling point

The heterogeneous equilibrium of interest when considering boiling is between the solvent vapour and the solvent in solution at 1 atm (Fig. 7.22). We denote the solvent by A and the solute by B. The equilibrium is established at a temperature for which

$$\mu_A^*(g) = \mu_A^*(l) + RT \ln x_A \qquad (7.32)$$

(The pressure of 1 atm is the same throughout, and will not be written explicitly.) We show in the *Justification* below that this equation implies that the presence of a solute at a mole fraction x_B causes an increase in normal boiling point from T^* to $T^* + \Delta T$, where

$$\Delta T = Kx_B \qquad K = \frac{RT^{*2}}{\Delta_{vap}H} \qquad (7.33)^\circ$$

Justification 7.2

Equation 32 rearranges into

$$\ln(1 - x_B) = \frac{\mu_A^*(g) - \mu_A^*(l)}{RT} = \frac{\Delta_{vap}G}{RT}$$

where $\Delta_{vap}G$ is the Gibbs energy of vaporization of the pure solvent (A) and x_B is the mole fraction of the solute; we have used $x_A + x_B = 1$. We now write

$$\Delta_{vap}G = \Delta_{vap}H - T\Delta_{vap}S$$

and ignore the small temperature dependence of $\Delta_{vap}H$ and $\Delta_{vap}S$. Then,

$$\ln(1 - x_B) = \frac{\Delta_{vap}H}{RT} - \frac{\Delta_{vap}S}{R}$$

When $x_B = 0$, the boiling point is that of pure liquid A, T^*, and

$$\ln 1 = \frac{\Delta_{vap}H}{RT^*} - \frac{\Delta_{vap}S}{R}$$

Because $\ln 1 = 0$, the difference of the two equations is

$$\ln(1 - x_B) = \frac{\Delta_{vap}H}{R}\left(\frac{1}{T} - \frac{1}{T^*}\right)$$

We now suppose that the amount of solute present is so small that $x_B \ll 1$. We can then write $\ln(1 - x_B) \approx -x_B$ and hence obtain

$$x_B = \frac{\Delta_{vap}H}{R}\left(\frac{1}{T^*} - \frac{1}{T}\right)$$

Finally, because $T \approx T^*$, it also follows that

$$\frac{1}{T^*} - \frac{1}{T} = \frac{T - T^*}{TT^*} \approx \frac{\Delta T}{T^{*2}}$$

with $\Delta T = T - T^*$. The previous equation then rearranges into eqn 33.

Because eqn 33 makes no reference to the identity of the solute, only to its mole fraction, we conclude that the elevation of boiling point is a colligative property. The

Table 7.2* Cryoscopic and ebullioscopic constants

	K_f/ (K/(mol kg^{-1}))	K_b/ (K/(mol kg^{-1}))
Benzene	5.12	2.53
Camphor	40	
Phenol	7.27	3.04
Water	1.86	0.51

* More values are given in the *Data section*.

value of ΔT does depend on the properties of the solvent, and the biggest changes occur for solvents with high boiling points.[2] For practical applications of eqn 33, we note that the mole fraction of B is proportional to its molality, b, in the solution, and write

$$\Delta T = K_b b \tag{7.34}$$

where K_b is the empirical **ebullioscopic constant** of the solvent (Table 7.2).

(c) The depression of freezing point

The heterogeneous equilibrium now of interest is between pure solid solvent A and the solution with solute present at a mole fraction x_B (Fig. 7.23). At the freezing point, the chemical potentials of A in the two phases are equal:

$$\mu_A^*(s) = \mu_A^*(l) + RT \ln x_A \tag{7.35}$$

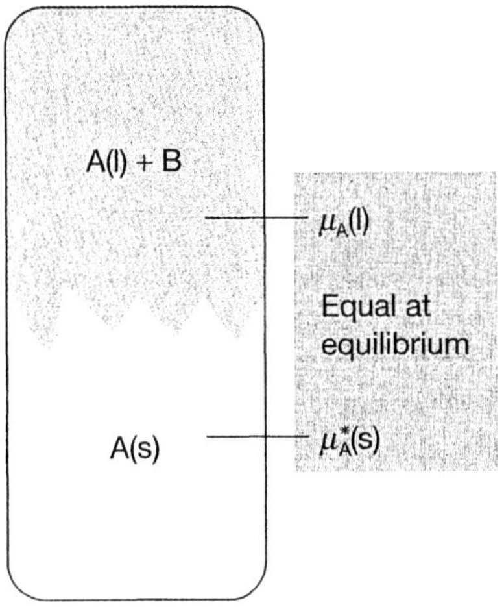

7.23 The heterogeneous equilibrium involved in the calculation of the lowering of freezing point is between A in the pure solid and A in the mixture, A being the solvent and B a solute that is insoluble in solid A.

The only difference between this calculation and the last is the appearance of the solid's chemical potential in place of the vapour's. Therefore we can write the result directly from eqn 33:

$$\Delta T = K'x_B \qquad K' = \frac{RT^{*2}}{\Delta_{fus}H} \tag{7.36}^\circ$$

where ΔT is the freezing point depression, $T^* - T$, and $\Delta_{fus}H$ is the enthalpy of fusion of the solvent. Larger depressions are observed in solvents with low enthalpies of fusion and high melting points. When the solution is dilute, the mole fraction is proportional to the molality of the solute, b, and it is common to write the last equation as

$$\Delta T = K_f b \tag{7.37}$$

where K_f is the empirical **cryoscopic constant** (Table 7.2). Once the cryoscopic constant of a solvent is known, the depression of freezing point may be used to measure the molar mass of a solute in the method known as cryoscopy; however, the technique is of little more than historical interest.

(d) Solubility

Although it is not strictly a colligative property, the solubility of a solute may be estimated by the same techniques as we have been using. When a solid solute is left in contact with a solvent, it dissolves until the solution is saturated. Saturation is a state of equilibrium, with the undissolved solute in equilibrium with the dissolved solute. Therefore, in a saturated solution the chemical potential of the pure solid solute, $\mu_B^*(s)$, and the chemical potential of B in solution, μ_B, are equal (Fig. 7.24). Because the latter is

$$\mu_B = \mu_B^*(l) + RT \ln x_B$$

we can write

$$\mu_B^*(s) = \mu_B^*(l) + RT \ln x_B \tag{7.38}$$

This expression is the same as the starting equation of the last section, except that the quantities refer to the solute B, not the solvent A.

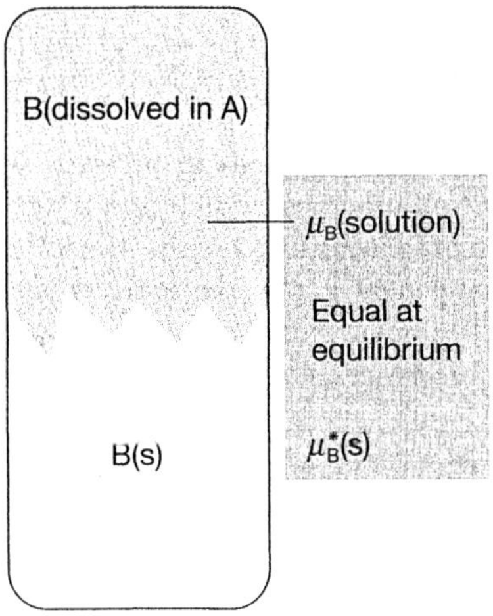

7.24 The heterogeneous equilibrium involved in the calculation of the solubility is between pure solid B and B in the mixture.

2. By Trouton's rule (Section 4.3a), $\Delta_{vap}H/T^*$ is a constant; therefore eqn 33 has the form $\Delta T \propto T^*$ and is independent of $\Delta_{vap}H$ itself.

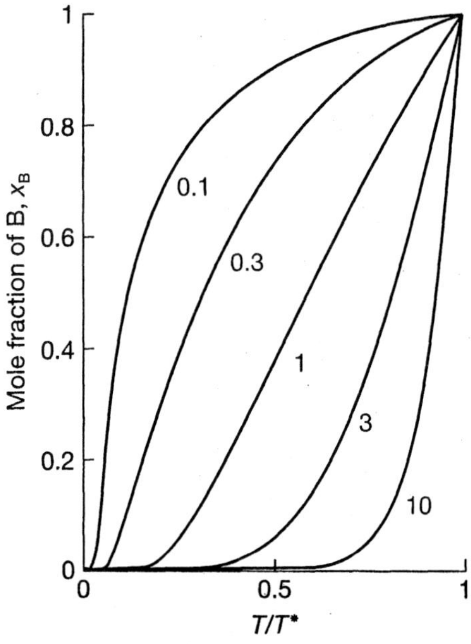

7.25 The variation of solubility (the mole fraction of solute in a saturated solution) with temperature (T^* is the freezing temperature of the solute). Individual curves are labelled with the value of $\Delta_{fus}H/RT^*$.

The starting point is the same but the aim is different. In the present case, we want to find the mole fraction of B in solution at equilibrium when the temperature is T. Therefore, we start by rearranging the last equation into

$$\ln x_B = \frac{\mu_B^*(s) - \mu_B^*(l)}{RT} = -\frac{\Delta_{fus}G}{RT}$$

$$= -\frac{\Delta_{fus}H}{RT} + \frac{\Delta_{fus}S}{R}$$

At the melting point of the solute, T^*, we know that $\Delta_{fus}G = 0$, so $\Delta_{fus}G/RT^* = 0$ too; consequently, this term may be added to the right-hand side to give

$$\ln x_B = -\frac{\Delta_{fus}H}{RT} + \frac{\Delta_{fus}S}{R} + \frac{\Delta_{fus}H}{RT^*} - \frac{\Delta_{fus}S}{R}$$

$$= -\frac{\Delta_{fus}H}{RT} + \frac{\Delta_{fus}H}{RT^*}$$

It then follows that

$$\ln x_B = \frac{\Delta_{fus}H}{R}\left(\frac{1}{T^*} - \frac{1}{T}\right) \qquad (7.39)^\circ$$

Equation 39 is plotted in Fig. 7.25. It shows that the solubility of B decreases exponentially as the temperature is lowered from its melting point. The illustration also shows that solutes with high melting points and large enthalpies of melting have low solubilities at normal temperatures. However, the detailed content of eqn 39 should not be treated too seriously because it is based on highly questionable approximations, such as the ideality of the solution. One aspect of its approximate character is that it fails to predict that solutes will have different solubilities in different solvents, for no solvent properties appear in the expression.

(e) Osmosis

The phenomenon of **osmosis** (from the Greek word for 'push') is the spontaneous passage of a pure solvent into a solution separated from it by a **semipermeable membrane**, a membrane permeable to the solvent but not to the solute (Fig. 7.26). The **osmotic pressure**, Π, is the pressure that must be applied to the solution to stop the influx of solvent. Important examples of osmosis include transport of fluids through cell membranes, dialysis (Box 7.1), and **osmometry**, the determination of molar mass by the measurement of osmotic pressure. Osmometry is widely used to determine the molar masses of macromolecules.

In the simple arrangement shown in Fig. 7.27, the opposing pressure arises from the head of solution that the osmosis itself produces. Equilibrium is reached when the hydrostatic pressure of the column of solution matches the osmotic pressure. The complicating feature of this arrangement is that the entry of solvent into the solution results in its dilution, and so it is more difficult to treat than the arrangement in Fig. 7.26, in which there is no flow and the concentrations remain unchanged.

The thermodynamic treatment of osmosis depends on noting that, at equilibrium, the chemical potential of the solvent must be the same on each side of the membrane. As shown in the *Justification* below, this equality implies that for dilute solutions the osmotic pressure is given by the **van 't Hoff equation**:

$$\Pi = [B]RT \qquad (7.40)^\circ$$

where $[B] = n_B/V$ is the molar concentration of the solute.

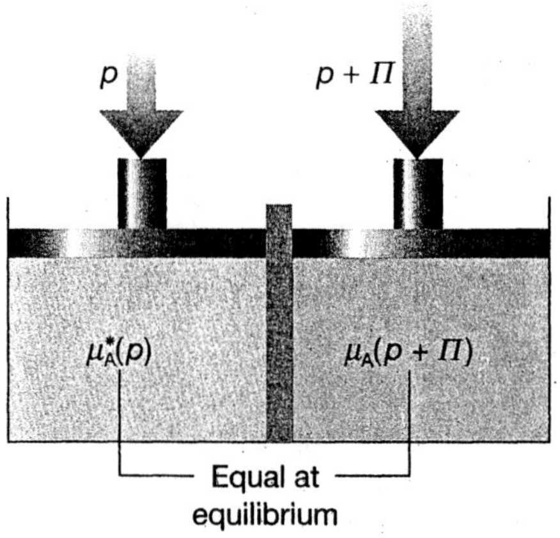

7.26 The equilibrium involved in the calculation of osmotic pressure, Π, is between pure solvent A at a pressure p on one side of the semipermeable membrane and A as a component of the mixture on the other side of the membrane, where the pressure is $p + \Pi$.

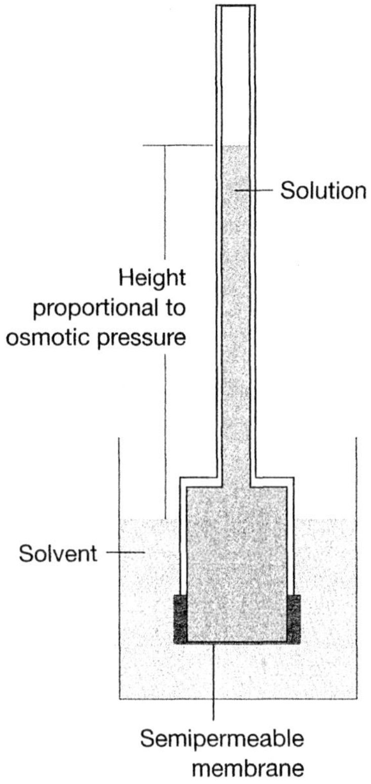

7.27 In a simple version of the osmotic pressure experiment, A is at equilibrium on each side of the membrane when enough has passed into the solution to cause a hydrostatic pressure difference.

Justification 7.3

On the pure solvent side the chemical potential of the solvent, which is at a pressure p, is $\mu_A^*(p)$. On the solution side, the chemical potential is lowered by the presence of the solute, which reduces the mole fraction of the solvent from 1 to x_A. However, the chemical potential of A is raised on account of the greater pressure, $p + \Pi$, that the solution experiences. At equilibrium the chemical potential of A is the same in both compartments, and we can write

$$\mu_A^*(p) = \mu_A(x_A, p + \Pi)$$

The presence of solute is taken into account in the normal way:

$$\mu_A(x_A, p + \Pi) = \mu_A^*(p + \Pi) + RT \ln x_A$$

We saw in Section 5.5 (eqn 5.14) how to take the effect of pressure into account:

$$\mu_A^*(p + \Pi) = \mu_A^*(p) + \int_p^{p+\Pi} V_m \, dp$$

where V_m is the molar volume of the pure solvent A. When these three equations are combined we get

$$-RT \ln x_A = \int_p^{p+\Pi} V_m \, dp$$

For dilute solutions, $\ln x_A$ may be replaced by $\ln(1 - x_B) \approx -x_B$. We may also assume that the pressure range in the integration is so small that the molar volume of the solvent is a constant. That being so, V_m may be taken outside the integral, giving

$$RTx_B = \Pi V_m$$

When the solution is dilute, $x_B \approx n_B/n_A$. Moreover, because $n_A V_m = V$, the total volume of the solvent, the equation simplifies to eqn 40.

Box 7.1 Dialysis and protein binding

In a dialysis experiment, a solution of macromolecules and smaller ions is placed in a bag made of a material that acts as a semipermeable membrane and the filled bag is immersed in a solvent. The membrane permits the passage of the small ions but not the macromolecules, so the former migrate through the membrane, leaving the macromolecules behind. Dialysis is used to study the binding of small molecules to macromolecules, such as an inhibitor to an enzyme, an antibiotic to DNA, and any other instance of cooperation or inhibition by small molecules attaching to large ones.

Suppose the molar concentration of the macromolecule M is [M] and the total concentration of small molecule A in the compartment containing the macromolecule is $[A]_{in}$. This total concentration is the sum of the concentrations of free A and bound A, which we write $[A]_{free}$ and $[A]_{bound}$, respectively. At equilibrium, the chemical potential of free A in the macromolecule solution is equal to the chemical potential of A in the solution on the other side of the membrane, where its concentration is $[A]_{out}$. The equality $\mu_{A,free} = \mu_{A,out}$ implies that $[A]_{free} = [A]_{out}$, provided the activity coefficient of A is the same in both solutions. Therefore, by measuring the concentration of A in the 'outside' solution, we can find the concentration of unbound A in the macromolecule solution and, from the difference $[A]_{in} - [A]_{free}$, which is equal to $[A]_{in} - [A]_{out}$, the concentration of bound A. The average number of A molecules bound to M molecules, ν, is then the ratio

$$\nu = \frac{[A]_{bound}}{[M]} = \frac{[A]_{in} - [A]_{out}}{[M]}$$

The bound and unbound A molecules are in equilibrium, $M + A \rightleftharpoons MA$, so their concentrations are related by an equilibrium constant K (see Chapter 9), where

$$K = \frac{[MA]}{[M]_{free}[A]_{free}} = \frac{[A]_{bound}}{([M] - [A]_{bound})[A]_{free}}$$

On division by [M], and replacement of $[A]_{free}$ by $[A]_{out}$, this expression becomes

$$K = \frac{\nu}{(1-\nu)[A]_{out}}$$

If there are N *identical* and *independent* binding sites on each macromolecule, each macromolecule behaves like N separate smaller macromolecules, with the same value of K for each site. The average number of A molecules per site is ν/N, so the last equation becomes

$$K = \frac{\nu/N}{\left(1-\frac{\nu}{N}\right)[A]_{out}}$$

It then follows that

$$\frac{\nu}{[A]_{out}} = KN - K\nu$$

This expression is the *Scatchard equation*. It implies that a plot of $\nu/[A]_{out}$ against ν should be a straight line of slope $-K$ and intercept KN at $\nu = 0$ (see the illustration). From these two quantities, we can find the equilibrium constant for binding, the enthalpy of binding (from the temperature dependence of K and the van 't Hoff equation, Chapter 9), and the number of binding sites on each macromolecule. If a straight line is not obtained we can conclude that the binding sites are not equivalent or independent.

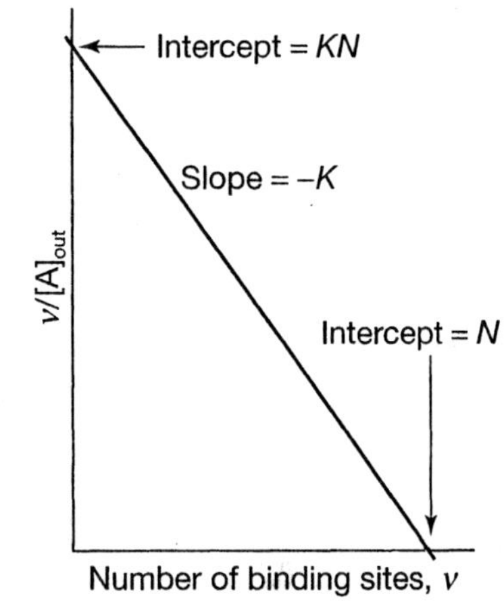

A Scatchard plot of $\nu/[A]_{out}$ against ν. The slope is $-K$ and the intercept at $\nu = 0$ is KN.

Discussion questions

1. The *Donnan equilibrium* determines the distribution of ions between two solutions in contact through a semipermeable membrane, in one of which there is a polyelectrolyte (a molecule bearing many charged groups, such as DNA). Consider the dialysis of a solution of a polyelectrolyte $Na_\nu P$, where the membrane is not permeable to the large anion $P^{\nu-}$. In such experiments, it is customary to add a high concentration of a salt such as NaCl to the solution inside and outside of the dialysis bag, so that the number of cations $Na_\nu P$ provides is insignificant in comparison with the number supplied by the additional salt. Show that at equilibrium

$$[Na^+]_{in} - [Na^+]_{out} = \frac{\nu[P^{\nu-}][Na^+]_{in}}{2[Cl^-] + \nu[P^{\nu-}]}$$

$$[Cl^-]_{in} - [Cl^-]_{out} = -\frac{\nu[P^{\nu-}][Cl^-]_{in}}{2[Cl^-]}$$

where $[Cl^-] = \frac{1}{2}([Cl^-]_{in} + [Cl^-]_{out})$. Note that Na^+ ions accumulate inside the dialysis bag, which contains $P^{\nu-}$, and from a measurement of the ion concentrations it is possible to determine the net charge of the anion. *Hint.* Electrical neutrality must be preserved in the bulk on both sides of the membrane, so that $[Na^+]_{in} = [Cl^-]_{in} + \nu[P^{\nu-}]$ and $[Na^+]_{out} = [Cl^-]_{out}$.

2. For the experiment described in the previous question, show that the osmotic pressure is

$$\Pi = RT[P^{\nu-}](1 + B[P^{\nu-}]) \qquad B = \frac{\nu^2[Cl^-]_{out}}{4[Cl^-] + 2\nu[Cl^-]_{in}[P^{\nu-}]}$$

Finally, show that, when $[Cl^-]_{in}$ and $[Cl^-]_{out}$ are large relative to $[P^{\nu-}]$, $\Pi = RT[P^{\nu-}]$. Note that the osmotic pressure is independent of the value of ν.

Further reading

K.E. van Holde, W.C. Johnson, and P.S. Ho, *Principles of physical biochemistry*. Prentice Hall, NJ (1998).

C.R. Cantor and P.R. Schimmel, *Biophysical chemistry, part III: The behavior of biological macromolecules*. W.H. Freeman, San Francisco (1980).

Because the effect of osmotic pressure is so readily measurable and large, one of the most common applications of osmometry is to the measurement of molar masses of macromolecules, such as proteins and synthetic polymers. As these huge molecules dissolve to produce solutions that are far from ideal, it is assumed that the van 't Hoff equation is only the first term of a virial-like expansion:

$$\Pi = [B]RT\{1 + B[B] + \cdots\} \qquad (7.41)$$

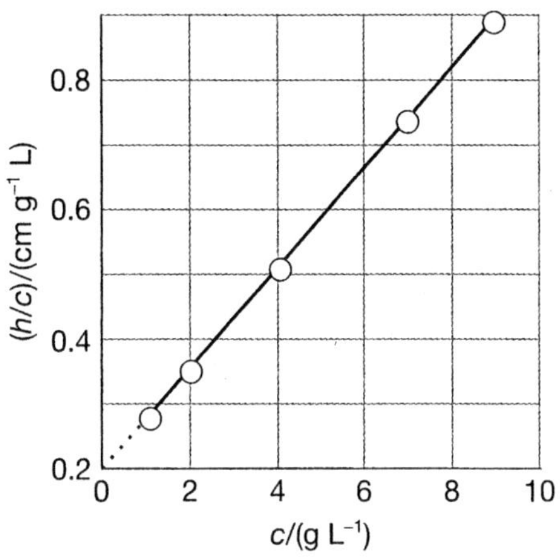

7.28 The plot involved in the determination of molar mass by osmometry. The molar mass is calculated from the intercept at $c = 0$; in Chapter 23 we shall see that additional information comes from the slope.

The additional terms take the nonideality into account; the empirical constant B is called the **osmotic virial coefficient**.

Example 7.5 **Using osmometry to determine the molar mass of a macromolecule**

The osmotic pressures of solutions of poly(vinyl chloride), PVC, in cyclohexanone at 298 K are given below. The pressures are expressed in terms of the heights of solution (of mass density $\rho = 0.980\ \mathrm{g\ cm^{-3}}$) in balance with the osmotic pressure. Determine the molar mass of the polymer.

$c/(\mathrm{g\ L^{-1}})$	1.00	2.00	4.00	7.00	9.00
h/cm	0.28	0.71	2.01	5.10	8.00

Method The osmotic pressure is measured at a series of mass concentrations, c, and a plot of Π/c against c is used to determine the molar mass of the polymer. We use eqn 41 with $[\mathrm{B}] = c/M$ where c is the mass concentration of the polymer and M is its molar mass. The osmotic pressure is related to the hydrostatic pressure by $\Pi = \rho g h$ (Example 1.2) with $g = 9.81\ \mathrm{m\ s^{-2}}$. With these substitutions, eqn 41 becomes

$$\frac{h}{c} = \frac{RT}{\rho g M}\left(1 + \frac{Bc}{M} + \cdots\right)$$

$$= \frac{RT}{\rho g M} + \left(\frac{RTB}{\rho g M^2}\right)c + \cdots$$

Therefore, to find M, plot h/c against c, and expect a straight line with intercept $RT/\rho g M$ at $c = 0$.

Answer The data give the following values for the quantities to plot:

$c/(\mathrm{g\ L^{-1}})$	1.00	2.00	4.00	7.00	9.00
$(h/c)/(\mathrm{cm\ g^{-1}\ L})$	0.28	0.36	0.503	0.729	0.889

The points are plotted in Fig. 7.28. The intercept is at 0.21. Therefore,

$$M = \frac{RT}{\rho g} \times \frac{1}{0.21\ \mathrm{cm\,g^{-1}\,L}}$$

$$= \frac{(8.3145\ \mathrm{J\,K^{-1}\,mol^{-1}}) \times (298\ \mathrm{K})}{(980\ \mathrm{kg\,m^{-3}}) \times (9.81\ \mathrm{m\,s^{-2}})} \times \frac{1}{2.1 \times 10^{-3}\ \mathrm{m^4\,kg^{-1}}}$$

$$= 1.2 \times 10^2\ \mathrm{kg\,mol^{-1}}$$

where we have used $1\ \mathrm{kg\ m^2\ s^{-2}} = 1\ \mathrm{J}$.

Comment Molar masses of macromolecules are often reported in daltons (Da), with $1\ \mathrm{Da} = 1\ \mathrm{g\ mol^{-1}}$. The macromolecule in this example has a molar mass of about 120 kDa. Modern osmometers give readings of osmotic pressure in pascals, so the analysis of the data is more straightforward and eqn 41 can be used directly.

Self-test 7.6 Estimate the depression of freezing point of the most concentrated of these solutions, taking K_f as about $10\ \mathrm{K/(mol\ kg^{-1})}$. [0.8 mK]

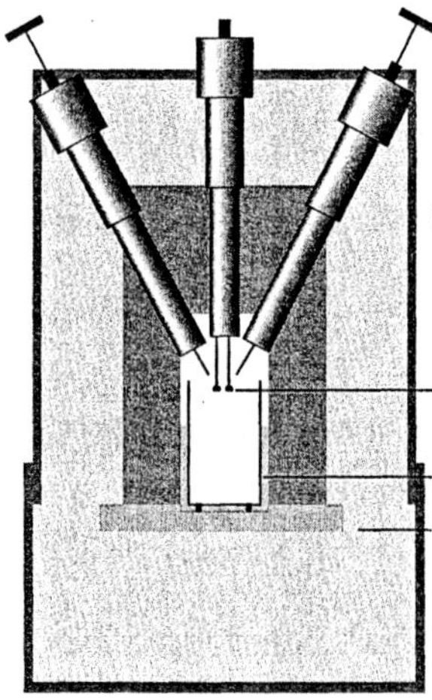

7.29 A vapour-phase osmometer. The syringes introduce droplets of solvent and solution, respectively, on to the thermistors, and the difference in temperature (which arises from the different rates of vaporization) is noted.

(f) Vapour-phase osmometry

In the technique of **vapour-phase osmometry**, a droplet of solution is placed on one thermistor (a semiconductor temperature probe) and a droplet of pure solvent is placed on another thermistor (Fig. 7.29). The two droplets are surrounded by an atmosphere of solvent vapour. Because the vapour pressure of a solvent in solution is lower than when it is pure, the net rate of condensation on to a droplet of solution is greater than the rate of condensation on to a droplet of pure solvent. It follows that more heat is liberated in the solution droplet, and the rise in temperature is greater there than in the droplet of pure solvent. The difference in temperature is measured for a series of concentrations and extrapolated to zero concentration. After calibration by using samples of known molar mass, the molar mass of the sample can be inferred from the temperature difference between the two thermistors. The method is not particularly sensitive, and its use is confined to samples of low molar mass.

Activities

Now we see how to adjust the expressions developed earlier in the chapter to take into account deviations from ideal behaviour. In Section 5.5b we saw how the fugacity was introduced to take into account the effects of gas imperfections in a manner that resulted in the least upset of the form of equations. Here we see how the expressions encountered in the treatment of ideal solutions can also be preserved almost intact by introducing the concept of activity. It is important to be aware of the different definitions of standard states and activities, and they are summarized in Table 7.3. We shall put them to work in the next few chapters, when we shall see that using them is much easier than defining them.

7.6 The solvent activity

The general form of the chemical potential of a real or ideal solvent is given by a straightforward modification of eqn 23:

$$\mu_A = \mu_A^* + RT\ln\frac{p_A}{p_A^*} \qquad (7.42)^\circ$$

Table 7.3 Standard states

Component	Basis	Standard state	Chemical potential	Limits
Solvent‡	Raoult	Pure solvent	$\mu = \mu^* + RT\ln a$ $a = p/p^*$ and $a = \gamma x$	$\gamma \to 1$ as $x \to 1$ (pure solvent)
Solute	Henry	(1) A hypothetical state of the pure solute	$\mu = \mu^\ominus + RT\ln a$ $a = p/K$ and $a = \gamma x$	$\gamma \to 1$ as $x \to 0$
		(2) A hypothetical state of the solute at molality $b^\ominus$	$\mu = \mu^\ominus + RT\ln a$ $a = \gamma b/b^\ominus$	$\gamma \to 1$ as $b \to 0$

‡ The implication of this entry is that the activity of a pure solid or liquid is 1. The pressure in each case is 1 bar.

where p_A^* is the vapour pressure of pure A and p_A is the vapour pressure of A when it is a component of a solution. For an ideal solution, the solvent obeys Raoult's law at all concentrations and we write

$$\mu_A = \mu_A^* + RT \ln x_A \qquad (7.43)°$$

The standard state of the solvent is the pure liquid (at 1 bar) and is obtained when $x_A = 1$. The form of the last equation can be preserved when the solution does not obey Raoult's law by writing

$$\mu_A = \mu_A^* + RT \ln a_A \qquad [7.44]$$

The quantity a_A is the **activity** of A, a kind of 'effective' mole fraction, just as the fugacity is an effective pressure.

Because eqn 42 is true for both real and ideal solutions (the only approximation being the use of pressures rather than fugacities), we can conclude by comparing it with eqn 44 that

$$a_A = \frac{p_A}{p_A^*} \qquad (7.45)$$

We see that there is nothing mysterious about the activity of a solvent: it can be determined experimentally simply by measuring the vapour pressure and then using eqn 45.

Illustration 7.2

The vapour pressure of 0.500 M KNO_3(aq) at 100°C is 749.7 Torr, so the activity of water in the solution at this temperature is

$$a_A = \frac{749.7 \text{ Torr}}{760.0 \text{ Torr}} = 0.9864$$

Because all solvents obey Raoult's law (that $p_A/p_A^* = x_A$) increasingly closely as the concentration of solute approaches zero, the activity of the solvent approaches the mole fraction as $x_A \to 1$:

$$a_A \to x_A \quad \text{as} \quad x_A \to 1 \qquad (7.46)$$

As in the case of real gases, a convenient way of expressing this convergence is to introduce the **activity coefficient**, γ, by the definition

$$a_A = \gamma_A x_A \qquad \gamma_A \to 1 \quad \text{as} \quad x_A \to 1 \qquad [7.47]$$

at all temperatures and pressures. The chemical potential of the solvent is then

$$\mu_A = \mu_A^* + RT \ln x_A + RT \ln \gamma_A \qquad (7.48)$$

The standard state of the solvent, the pure liquid solvent at 1 bar, is established when $x_A = 1$.

7.7 The solute activity

The problem with defining activity coefficients and standard states for solutes is that they approach ideal-dilute (Henry's law) behaviour as $x_B \to 0$, not as $x_B \to 1$ (corresponding to pure solute). We shall show how to set up the definitions for a solute that obeys Henry's law exactly, and then show how to allow for deviations.

(a) Ideal-dilute solutions

A solute B that satisfies Henry's law has a vapour pressure given by $p_B = K_B x_B$, where K_B is an empirical constant. In this case, the chemical potential of B is

$$\mu_B = \mu_B^* + RT \ln \frac{p_B}{p_B^*}$$
$$= \mu_B^* + RT \ln \frac{K_B}{p_B^*} + RT \ln x_B$$

Both K_B and p_B^* are characteristics of the solute, so the second term may be combined with the first to give a new standard chemical potential:

$$\mu_B^{\ominus} = \mu_B^* + RT \ln \frac{K_B}{p_B^*} \qquad [7.49]$$

It then follows that

$$\mu_B = \mu_B^{\ominus} + RT \ln x_B \qquad (7.50)^{\circ}$$

(b) Real solutes

We now permit deviations from ideal-dilute, Henry's law behaviour. For the solute, we introduce a_B in place of x_B in eqn 50, and obtain

$$\mu_B = \mu_B^{\ominus} + RT \ln a_B \qquad [7.51]$$

The standard state remains unchanged in this last stage, and all the deviations from ideality are captured in the activity a_B. The value of the activity at any concentration can be obtained in the same way as for the solvent, but in place of eqn 45 we use

$$a_B = \frac{p_B}{K_B} \qquad (7.52)$$

As for the solvent, it is sensible to introduce an activity coefficient through

$$a_B = \gamma_B x_B \qquad [7.53]$$

Now all the deviations from ideality are captured in the activity coefficient γ_B. Because the solute obeys Henry's law as its concentration goes to zero, it follows that

$$a_B \to x_B \quad \text{and} \quad \gamma_B \to 1 \quad \text{as} \quad x_B \to 0 \qquad (7.54)$$

at all temperatures and pressures. Deviations of the solute from ideality disappear as zero concentration is approached.

Example 7.6 Measuring activity

Use the information in Example 7.3 to calculate the activity and activity coefficient of chloroform in acetone at 25°C, treating it first as a solvent and then as a solute.

Method For the activity of chloroform as a solvent (the Raoult's law activity), form $a = p/p^*$ and $\gamma = a/x$. For its activity as a solute (the Henry's law activity), form $a = p/K$ and $\gamma = a/x$.

Answer Because $p^* = 273$ Torr and $K = 165$ Torr, we can construct the following table (with x the mole fraction of chloroform):

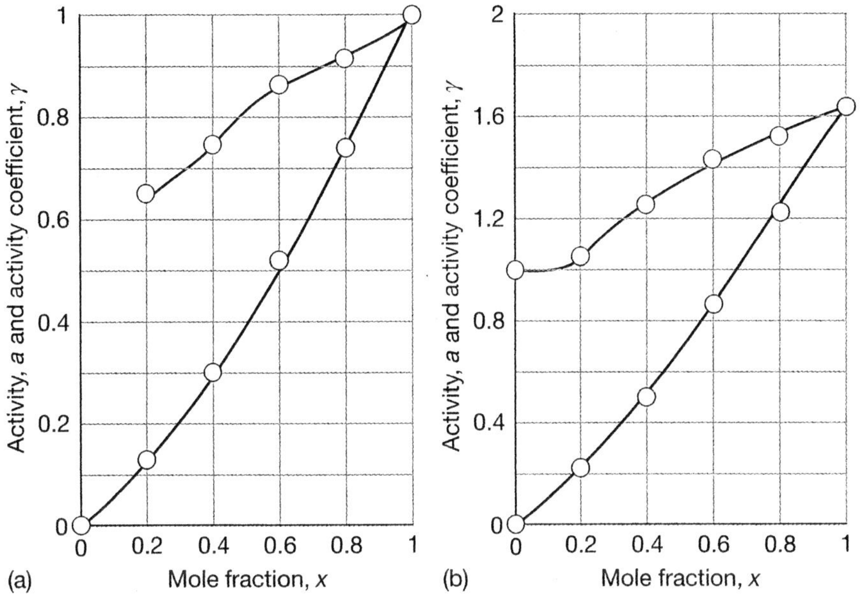

7.30 The variation of activity and activity coefficient of chloroform (trichloromethane) with composition according to (a) Raoult's law, (b) Henry's law.

x	0	0.20	0.40	0.60	0.80	1.00

From Raoult's law (chloroform regarded as the solvent):

a	0	0.13	0.30	0.52	0.73	1.00
γ		0.65	0.75	0.87	0.91	1.00

From Henry's law (chloroform regarded as the solute):

a	0	0.21	0.50	0.86	1.21	1.65
γ	1	1.05	1.25	1.43	1.51	1.65

These values are plotted in Fig. 7.30.

Comment Notice that $\gamma \to 1$ as $x \to 1$ in the Raoult's law case, but that $\gamma \to 1$ as $x \to 0$ in the Henry's law case.

Self-test 7.7 Calculate the activities and activity coefficients for acetone according to the two conventions.

[At $x_A = 0.60$, for instance, $a_R = 0.50$, $\gamma_R = 0.83$; $a_H = 1.00$, $\gamma_H = 1.67$]

(c) Activities in terms of molalities

The selection of a standard state is entirely arbitrary, so we are free to choose one that best suits our purpose and the description of the composition of the system. In chemistry, compositions are often expressed as molalities, b, in place of mole fractions. It therefore proves convenient to write

$$\mu_B = \mu_B^{\ominus} + RT \ln b_B \qquad \{7.55\}$$

where $\mu^{\ominus}$ has a different value from the standard values introduced earlier. According to this definition, the chemical potential of the solute has its standard value $\mu^{\ominus}$ when the molality of B is equal to $b^{\ominus}$ (that is, at 1 mol kg^{-1}). Note that as $b_B \to 0$, $\mu_B \to -\infty$; that is, as the solution becomes diluted, so the solute becomes increasingly stabilized. The practical consequence of this result is that it is very difficult to remove the last traces of a solute from a solution.

Now, as before, we incorporate deviations from ideality by introducing a dimensionless activity a_B, a dimensionless activity coefficient γ_B, and writing

$$a_B = \gamma_B \frac{b_B}{b^\ominus} \quad \text{where} \quad \gamma_B \to 1 \quad \text{as} \quad b_B \to 0 \qquad [7.56]$$

at all temperatures and pressures. The standard state remains unchanged in this last stage and, as before, all the deviations from ideality are captured in the activity coefficient γ_B. We then arrive at the following succinct expression for the chemical potential of a real solute at any molality:

$$\mu = \mu^\ominus + RT \ln a \qquad (7.57)$$

(d) The biological standard state

One important illustration of the ability to choose a standard state to suit the circumstances arises in biological applications. The conventional standard state of hydrogen ions (unit activity, corresponding to pH = 0)[3] is not appropriate to normal biological conditions. Therefore, in biochemistry it is common to adopt the **biological standard state**, in which pH = 7 (an activity of 10^{-7}, neutral solution) and to label the corresponding standard thermodynamic functions as $G^\oplus$, $H^\oplus$, $\mu^\oplus$, and $S^\oplus$ (some texts use $X^{\ominus\prime}$).

To find the relation between the thermodynamic and biological standard values of the chemical potential of hydrogen ions we need to note from eqn 57 that

We have used $\ln x = (\ln 10)\log x$.

$$\mu_{H^+} = \mu^\ominus_{H^+} + RT \ln a_{H^+} = \mu^\ominus_{H^+} - (RT \ln 10) \times \text{pH}$$

It follows that

$$\mu^\oplus_{H^+} = \mu^\ominus_{H^+} - 7RT \ln 10 \qquad (7.58)$$

At 298.15 K, $7RT \ln 10 = 39.96$ kJ mol^{-1}, so the two standard values differ by about 40 kJ mol^{-1}.

7.8 The activities of regular solutions

The material on regular solutions presented in Section 7.4 gives further insight into the origin of deviations from Raoult's law and its relation to activity coefficients. The starting point is the expression for the Gibbs energy of mixing for a regular solution (eqn 31). We show in the following *Justification* that eqn 31 implies that the activity coefficients are given by expressions of the form

$$\ln \gamma_A = \beta x_B^2 \qquad \ln \gamma_B = \beta x_A^2 \qquad (7.59)$$

These relations are called the **Margules equations.**

Justification 7.4

The Gibbs energy of mixing to form a nonideal solution is

$$\Delta_{mix}G = nRT\{x_A \ln a_A + x_B \ln a_B\}$$

This relation follows from the derivation of eqn 27 with activities in place of mole fractions. If each activity is replaced by γx, this expression becomes

$$\Delta_{mix}G = nRT\{x_A \ln x_A + x_B \ln x_B + x_A \ln \gamma_A + x_B \ln \gamma_B\}$$

3. pH = $-\log a_{H_3O^+}$ is introduced in Chapter 9; we presume that the concept is familiar from introductory chemistry courses.

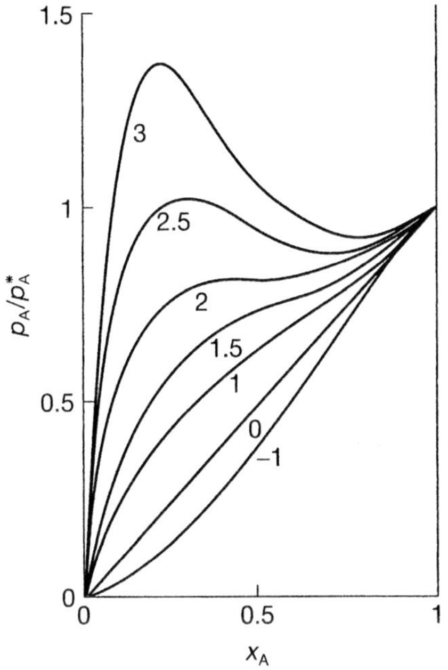

7.31 The vapour pressure of a mixture based on a model in which the excess enthalpy is proportional to $\beta x_A x_B$. An ideal solution corresponds to $\beta = 0$ and gives a straight line, in accord with Raoult's law. Positive values of β give vapour pressures higher than ideal. Negative values of β give a lower vapour pressure.

Now we introduce the two expressions in eqn 59, and use $x_A + x_B = 1$, which gives

$$\Delta_{mix}G = nRT\{x_A \ln x_A + x_B \ln x_B + \beta x_A x_B^2 + \beta x_B x_A^2\}$$
$$= nRT\{x_A \ln x_A + x_B \ln x_B + \beta x_A x_B(x_A + x_B)\}$$
$$= nRT\{x_A \ln x_A + x_B \ln x_B + \beta x_A x_B\}$$

as required by eqn 31. Note, moreover, that the activity coefficients behave correctly for dilute solutions: $\gamma_A \to 1$ as $x_B \to 0$ and $\gamma_B \to 1$ as $x_A \to 0$.

At this point we can use the Margules equations to write the activity of A as

$$a_A = \gamma_A x_A = x_A e^{\beta x_B^2} = x_A e^{\beta(1-x_A)^2} \quad (7.60)$$

with a similar expression for a_B. The activity of A, though, is just the ratio of the vapour pressure of A in the solution to the vapour pressure of pure A, so we can write

$$p_A = \{x_A e^{\beta(1-x_A)^2}\}p_A^* \quad (7.61)$$

This function is plotted in Fig. 7.31. We see that $\beta = 0$, corresponding to an ideal solution, gives a straight line, in accord with Raoult's law (indeed, when $\beta = 0$, eqn 61 becomes $p_A = x_A p_A^*$, which is Raoult's law). Positive values of β (endothermic mixing, unfavourable solute–solvent interactions) give vapour pressures higher than ideal. Negative values of β (exothermic mixing, favourable solute–solvent interactions) give a lower vapour pressure. All the curves approach linearity and coincide with the Raoult's law line as $x_A \to 1$ and the exponential function in eqn 61 approaches 1. When $x_A \ll 1$, eqn 61 approaches

$$p_A = x_A e^{\beta} p_A^* \quad (7.62)$$

This expression has the form of Henry's law once we identify K with $e^{\beta}p_A^*$, which is different for each solute–solvent system. It follows that β can be determined from the ratio K/p_A^*.

Checklist of key ideas

The thermodynamic description of mixtures

7.1 Partial molar quantities

- ☐ partial molar volume, the change in volume per mole of A added to a large volume of the mixture; $V_J = (\partial V/\partial n_J)_{p,T,n'}$
- ☐ total volume of a mixture, $V = n_A V_A + n_B V_B$
- ☐ partial molar Gibbs energy and the definition of the chemical potential, $\mu_J = (\partial G/\partial n_J)_{p,T,n'}$
- ☐ total Gibbs energy of a mixture, $G = n_A\mu_A + n_B\mu_B$
- ☐ fundamental equation of chemical thermodynamics $dG = V\,dp - S\,dT + \mu_A\,dn_A + \mu_B\,dn_B + \cdots$
- ☐ chemical potential in terms of U, $\mu_J = (\partial U/\partial n_J)_{S,V,n'}$
- ☐ chemical potential in terms of H, $\mu_J = (\partial H/\partial n_J)_{S,p,n'}$
- ☐ chemical potential in terms of A, $\mu_J = (\partial A/\partial n_J)_{T,V,n'}$
- ☐ Gibbs–Duhem equation, $\sum_J n_J\,d\mu_J = 0$

7.2 The thermodynamics of mixing

- ☐ standard chemical potential of a perfect gas, the chemical potential of the pure gas at 1 bar
- ☐ chemical potential of a perfect gas, $\mu = \mu^{\ominus} + RT\ln(p/p^{\ominus})$
- ☐ Gibbs energy of mixing of two perfect gases, $\Delta_{mix}G = nRT(x_A \ln x_A + x_B \ln x_B)$
- ☐ entropy of mixing of two perfect gases, $\Delta_{mix}S = -nR(x_A \ln x_A + x_B \ln x_B)$
- ☐ enthalpy of mixing, the enthalpy change accompanying mixing: $\Delta_{mix}H = 0$ for perfect gases

7.3 The chemical potential of liquids

- ☐ Raoult's law, $p_A = x_A p_A^*$
- ☐ ideal solution, a solution in which all components obeys Raoult's law throughout the composition range
- ☐ chemical potential of a component of an ideal solution, $\mu_A = \mu_A^* + RT\ln x_A$
- ☐ Henry's law, $p_B = x_B K_B$
- ☐ ideal-dilute solution, mixtures for which the solute obeys Henry's law and the solvent obeys Raoult's law

The properties of solutions

7.4 Liquid mixtures

- ☐ Gibbs energy of mixing of two liquids, $\Delta_{mix}G = nRT(x_A \ln x_A + x_B \ln x_B)$
- ☐ entropy of mixing, $\Delta_{mix}S = -nR(x_A \ln x_A + x_B \ln x_B)$
- ☐ partially miscible liquids, liquids that are miscible only over a range of compositions
- ☐ excess function (X^E), the difference between the observed thermodynamic function of mixing and the function for an ideal solution
- ☐ excess entropy, $S^E = \Delta_{mix}S - \Delta_{mix}S^{ideal}$
- ☐ regular solution, a solution for which $H^E \neq 0$ but $S^E = 0$

7.5 Colligative properties

- ☐ colligative property, a property that depends only on the number of solute particles present, not their identity
- ☐ the elevation of boiling point, $\Delta T = K_b b$
- ☐ ebullioscopic constant, the constant K_b in $\Delta T = K_b b$
- ☐ the depression of freezing point, $\Delta T = K_f b$
- ☐ cryoscopic constant, the constant K_f in $\Delta T = K_f b$
- ☐ ideal solubility, $\ln x_B = (\Delta_{fus}H/R)(1/T^* - 1/T)$
- ☐ osmosis, the spontaneous passage of a pure solvent into a solution separated from it by a semipermeable membrane
- ☐ semipermeable membrane, a membrane permeable to the solvent but not to the solute
- ☐ osmotic pressure, the pressure that must be applied to the solution to stop the influx of solvent
- ☐ osmometry, the determination of molar mass by the measurement of osmotic pressure
- ☐ van 't Hoff equation for the osmotic pressure, $\Pi = [B]RT$
- ☐ osmotic virial coefficient, B in $\Pi = [B]RT(1 + B[B] + \cdots)$
- ☐ vapour-phase osmometry, a technique for measuring osmotic pressure

Activities

7.6 The solvent activity

- ☐ activity, $a_A = p_A/p_A^*$
- ☐ solvent activity and chemical potential, $\mu_A = \mu_A^* + RT \ln a_A$
- ☐ Raoult's law basis of activity and activity coefficient, $\gamma = a/x$

7.7 The solute activity

- ☐ chemical potential of a solute in an ideal-dilute solution, $\mu_B = \mu_B^{\ominus} + RT \ln a_B$
- ☐ solute activity and the Henry's law basis of activity coefficient, $\gamma = a/x$
- ☐ chemical potential of a solute in terms of its molality, $\mu_B = \mu_B^{\ominus} + RT \ln b_B$
- ☐ biological standard state, $\mu_{H^+}^{\oplus} = \mu_{H^+}^{\ominus} - 7RT \ln 10$

7.8 The activities of regular solutions

- ☐ Margules equations, $\ln \gamma_J = \beta x_J^2$
- ☐ activity of regular solution, $a_A = x_A e^{\beta(1-x_A)^2}$

Further reading

Articles of general interest

S.F. Ackley, Sea ice. In *Encyclopedia of applied physics* (ed. G.L. Trigg), **17**, 81. VCH, New York (1996).

M.E. Cardinali and C. Giomini. Boiling temperature vs. composition: An almost-exact explicit equation for a binary mixture following Raoult's law. *J. Chem. Educ.* **66**, 549 (1989).

J.J. Carroll, Henry's law: A historical view. *J. Chem. Educ.* **70**, 91 (1993).

M.J. Clugston, A mathematical verification of the second law of thermodynamics from the entropy of mixing. *J. Chem. Educ.* **67**, 203 (1990).

H.F. Franzen, The freezing point depression law in physical chemistry. *J. Chem. Educ.* **65**, 1077 (1988).

B. Freeman, Osmosis. In *Encyclopedia of applied physics* (ed. G.L. Trigg), **13**, 59. VCH, New York (1995).

E.F. Meyer, Thermodynamics of 'mixing' of ideal gases: a persistent pitfall. *J. Chem. Educ.* **64**, 676 (1987).

S. Sattar, Thermodynamics of mixing real gases. *J. Chem. Educ.* **77**, 1361 (2000).

R.L. Scott, Models for phase equilibria in fluid mixtures. *Acc. Chem. Res.* **20**, 97 (1987).

A. Vincent, Osmotic pressure and the effect of gravity on solutions. *J. Chem. Educ.* **73**, 998 (1996).

Texts and sources of data and information

C.E. Carraher, Jr., *Seymour/Carraher's polymer chemistry*. Marcel Dekker, New York (2000).

M.R.J. Dack, Solutions and solubilities. In *Techniques of chemistry* (A. Weissberger and B.W. Rossiter, ed.), **8**. Wiley, New York (1975).

I.M. Klotz and R.M. Rosenberg, *Chemical thermodynamics: basic theory and methods*. Wiley–Interscience, New York (2000).

W.J. Mader and L.T. Brady, Determination of solubility. In *Techniques of chemistry* (A. Weissberger and B.W. Rossiter, ed.), **5**, 257. Wiley, New York (1971).

J.N. Murrell and A.D. Jenkins, *Properties of liquids and solutions*. Wiley–Interscience, New York (1994).

J.R. Overton, Determination of osmotic pressure. In *Techniques of chemistry* (A. Weissberger and B.W. Rossiter, ed.), **5**, 309. Wiley, New York (1971).

J.S. Rowlinson and F.L. Swinton, *Liquids and liquid mixtures*. Butterworths, London (1982).

S.I. Sandler, *Chemical and engineering thermodynamics*. Wiley, New York (1998).

Exercises

Discussion questions

7.1 (a) State and justify the thermodynamic criterion for solution–vapour equilibrium.

7.1 (b) How is Raoult's law modified so as to describe the vapour pressure of real solutions?

7.2 (a) Explain what is meant by a regular solution.

7.2 (b) Explain the origin of colligative properties.

7.3 (a) Explain how colligative properties are used to determine molar mass.

7.3 (b) What is meant by the activity of a solute?

Numerical exercises

7.4 (a) The partial molar volumes of acetone (propanone) and chloroform (trichloromethane) in a mixture in which the mole fraction of $CHCl_3$ is 0.4693 are 74.166 $cm^3 mol^{-1}$ and 80.235 $cm^3 mol^{-1}$, respectively. What is the volume of a solution of mass 1.000 kg?

7.4 (b) The partial molar volumes of two liquids A and B in a mixture in which the mole fraction of A is 0.3713 are 188.2 $cm^3 mol^{-1}$ and 176.14 $cm^3 mol^{-1}$, respectively. The molar masses of the A and B are 241.1 g mol^{-1} and 198.2 g mol^{-1}. What is the volume of a solution of mass 1.000 kg?

7.5 (a) At 25°C, the density of a 50 per cent by mass ethanol–water solution is 0.914 g cm^{-3}. Given that the partial molar volume of water in the solution is 17.4 $cm^3 mol^{-1}$, calculate the partial molar volume of the ethanol.

7.5 (b) At 20°C, the density of a 20 per cent by mass ethanol–water solution is 968.7 kg m^{-3}. Given that the partial molar volume of ethanol in the solution is 52.2 $cm^3 mol^{-1}$, calculate the partial molar volume of the water.

7.6 (a) At 300 K, the partial vapour pressures of HCl (that is, the partial pressure of the HCl vapour) in liquid $GeCl_4$ are as follows:

x_{HCl}	0.005	0.012	0.019
p_{HCl}/kPa	32.0	76.9	121.8

Show that the solution obeys Henry's law in this range of mole fractions, and calculate Henry's law constant at 300 K.

7.6 (b) At 310 K, the partial vapour pressures of a substance B dissolved in a liquid A are as follows:

x_B	0.010	0.015	0.020
p_B/kPa	82.0	122.0	166.1

Show that the solution obeys Henry's law in this range of mole fractions, and calculate Henry's law constant at 310 K.

7.7 (a) Predict the partial vapour pressure of HCl above its solution in liquid germanium tetrachloride of molality 0.10 mol kg^{-1}. For data, see Exercise 7.6a.

7.7 (b) Predict the partial vapour pressure of the component B above its solution in A in Exercise 7.6b when the molality of B is 0.25 mol kg^{-1}. The molar mass of A is 74.1 g mol^{-1}.

7.8 (a) Calculate the cryoscopic and ebullioscopic constants of tetrachloromethane.

7.8 (b) Calculate the cryoscopic and ebullioscopic constants of naphthalene.

7.9 (a) The vapour pressure of benzene is 400 Torr at 60.6°C, but it fell to 386 Torr when 19.0 g of an involatile organic compound was dissolved in 500 g of benzene. Calculate the molar mass of the compound.

7.9 (b) The vapour pressure of 2-propanol is 50.00 kPa at 338.8°C, but it fell to 49.62 kPa when 8.69 g of an involatile organic compound was dissolved in 250 g of 2-propanol. Calculate the molar mass of the compound.

7.10 (a) The addition of 100 g of a compound to 750 g of CCl_4 lowered the freezing point of the solvent by 10.5 K. Calculate the molar mass of the compound.

7.10 (b) The addition of 5.00 g of a compound to 250 g of naphthalene lowered the freezing point of the solvent by 0.780 K. Calculate the molar mass of the compound.

7.11 (a) The osmotic pressure of an aqueous solution at 300 K is 120 kPa. Calculate the freezing point of the solution.

7.11 (b) The osmotic pressure of an aqueous solution at 288 K is 99.0 kPa. Calculate the freezing point of the solution.

7.12 (a) Consider a container of volume 5.0 L that is divided into two compartments of equal size. In the left compartment there is nitrogen at 1.0 atm and 25°C; in the right compartment there is hydrogen at the same temperature and pressure. Calculate the entropy and Gibbs energy of mixing when the partition is removed. Assume that the gases are perfect.

7.12 (b) Consider a container of volume 250 mL that is divided into two compartments of equal size. In the left compartment there is argon at 100 kPa and 0°C; in the right compartment there is neon at the same temperature and pressure. Calculate the entropy and Gibbs energy of mixing when the partition is removed. Assume that the gases are perfect.

7.13 (a) Air is a mixture with a composition given in Self-test 1.5. Calculate the entropy of mixing when it is prepared from the pure (and perfect) gases.

7.13 (b) Calculate the Gibbs energy, entropy, and enthalpy of mixing when 1.00 mol C_6H_{14} (hexane) is mixed with 1.00 mol C_7H_{16} (heptane) at 298 K; treat the solution as ideal.

7.14 (a) What proportions of hexane and heptane should be mixed (a) by mole fraction, (b) by mass in order to achieve the greatest entropy of mixing?

7.14 (b) What proportions of benzene and ethylbenzene should be mixed (a) by mole fraction, (b) by mass in order to achieve the greatest entropy of mixing?

7.15 (a) Use Henry's law and the data in Table 7.1 to calculate the solubility (as a molality) of CO_2 in water at 25°C when its partial pressure is (a) 0.10 atm, (b) 1.00 atm.

7.15 (b) The mole fractions of N_2 and O_2 in air at sea level are approximately 0.78 and 0.21. Calculate the molalities of the solution formed in an open flask of water at 25°C.

7.16 (a) A water carbonating plant is available for use in the home and operates by providing carbon dioxide at 5.0 atm. Estimate the molar concentration of the soda water it produces.

7.16 (b) After some weeks of use, the pressure in the water carbonating plant mentioned in the previous exercise has fallen to 2.0 atm. Estimate the molar concentration of the soda water it produces at this stage.

7.17 (a) Calculate the freezing point of a glass of water of volume 250 cm^3 sweetened with 7.5 g of sucrose.

7.17 (b) Calculate the freezing point of a glass of water of volume 200 cm^3 in which 10 g of glucose has been dissolved.

7.18 (a) The enthalpy of fusion of anthracene is 28.8 kJ mol^{-1} and its melting point is 217°C. Calculate its ideal solubility in benzene at 25°C.

7.18 (b) Predict the ideal solubility of lead in bismuth at 280°C given that its melting point is 327°C and its enthalpy of fusion is 5.2 kJ mol^{-1}.

7.19 (a) The osmotic pressures of solutions of polystyrene in toluene were measured at 25°C and the pressure was expressed in terms of the height of the solvent of density 1.004 g cm^{-3}:

$c/(g\ L^{-1})$	2.042	6.613	9.521	12.602
h/cm	0.592	1.910	2.750	3.600

Calculate the molar mass of the polymer.

7.19 (b) The molar mass of an enzyme was determined by dissolving it in water, measuring the osmotic pressure at 20°C, and extrapolating the data to zero concentration. The following data were obtained:

$c/(mg\ cm^{-3})$	3.221	4.618	5.112	6.722
h/cm	5.746	8.238	9.119	11.990

Calculate the molar mass of the enzyme.

7.20 (a) Substances A and B are both volatile liquids with $p_A^* = 300$ Torr, $p_B^* = 250$ Torr, and $K_B = 200$ Torr (concentration expressed in mole fraction). When $x_A = 0.9$, $b_B = 2.22$ mol kg^{-1}, $p_A = 250$ Torr, and $p_B = 25$ Torr. Calculate the activities and activity coefficients of A and B. Use the mole fraction, Raoult's law basis system, for A and the Henry's law basis system (both mole fractions and molalities) for B.

7.20 (b) Given that $p^*(H_2O) = 0.023\,08$ atm and $p(H_2O) = 0.022\,39$ atm in a solution in which 0.122 kg of a non-volatile solute ($M = 241$ g mol^{-1}) is dissolved in 0.920 kg water at 293 K, calculate the activity and activity coefficient of water in the solution.

7.21 (a) A dilute solution of bromine in carbon tetrachloride behaves as an ideal-dilute solution. The vapour pressure of pure CCl_4 is 33.85 Torr at 298 K. The Henry's law constant when the concentration of Br_2 is expressed as a mole fraction is 122.36 Torr. Calculate the vapour pressure of each component, the total pressure, and the composition of the vapour phase when the mole fraction of Br_2 is 0.050, on the assumption that the conditions of the ideal-dilute solution are satisfied at this concentration.

7.21 (b) Benzene and toluene form nearly ideal solutions. The boiling point of pure benzene is 80.1°C. Calculate the chemical potential of benzene relative to that of pure benzene when $x_{benzene} = 0.30$ at its boiling point. If the activity coefficient of benzene in this solution were actually 0.93 rather than 1.00, what would be its vapour pressure?

7.22 (a) By measuring the equilibrium between liquid and vapour phases of an acetone (A)–ethanol (E) solution at 57.2°C at 1.00 atm, it was found that $x_A = 0.400$ when $y_A = 0.516$. Calculate the activities and activity coefficients of both components in this solution on the Raoult's law basis. The vapour pressures of the pure components at this temperature are: $p_A^* = 786$ Torr and $p_E^* = 551$ Torr. (x_A is the mole fraction in the liquid and y_A the mole fraction in the vapour.)

7.22 (b) By measuring the equilibrium between liquid and vapour phases of a solution at 30°C at 1.00 atm, it was found that $x_A = 0.220$ when $y_A = 0.314$. Calculate the activities and activity coefficients of both components in this solution on the Raoult's law basis. The vapour pressures of the pure components at this temperature are: $p_A^* = 73.0$ kPa and $p_B^* = 92.1$ kPa. (x_A is the mole fraction in the liquid and y_A the mole fraction in the vapour.)

Problems†

Numerical problems

7.1 The following table gives the mole fraction of methylbenzene (A) in liquid and gaseous mixtures with butanone at equilibrium at 303.15 K and the total pressure p. Take the vapour to be perfect and calculate the partial pressures of the two components. Plot them against their respective mole fractions in the liquid mixture and find the Henry's law constants for the two components.

x_A	0	0.0898	0.2476	0.3577	0.5194	0.6036
y_A	0	0.0410	0.1154	0.1762	0.2772	0.3393
p/kPa	36.066	34.121	30.900	28.626	25.239	23.402

x_A	0.7188	0.8019	0.9105	1
y_A	0.4450	0.5435	0.7284	1
p/kPa	20.6984	18.592	15.496	12.295

7.2 The volume of an aqueous solution of NaCl at 25°C was measured at a series of molalities b, and it was found that the volume fitted the expression

$$V/cm^3 = 1003 + 16.62b + 1.77b^{3/2} + 0.12b^2$$

where V is the volume of a solution formed from 1.000 kg of water and b is to be understood as $b/b^{\ominus}$. Calculate the partial molar volume of the components in a solution of molality 0.100 mol kg^{-1}.

† Problems denoted with the symbol ‡ were supplied by Charles Trapp, Carmen Giunta, and Marshall Cady.

7.3 At 18°C the total volume of a solution formed from $MgSO_4$ and 1.000 kg of water fits the expression

$$V/\text{cm}^3 = 1001.21 + 34.69(b - 0.070)^2$$

where b is to be understood as $b/b^{\ominus}$. Calculate the partial molar volumes of the salt and the solvent when in a solution of molality 0.050 mol kg^{-1}.

7.4 The densities of aqueous solutions of copper(II) sulfate at 20°C were measured as set out below. Determine and plot the partial molar volume of $CuSO_4$ in the range of the measurements.

$m(CuSO_4)$/g	5	10	15	20
ρ/(g cm^{-3})	1.051	1.107	1.167	1.230

where $m(CuSO_4)$ is the mass of $CuSO_4$ dissolved in 100 g of solution.

7.5 What proportions of ethanol and water should be mixed in order to produce 100 cm^3 of a mixture containing 50 per cent by mass of ethanol? What change in volume is brought about by adding 1.00 cm^3 of ethanol to the mixture? (Use data from Fig. 7.1.)

7.6 Potassium fluoride is very soluble in glacial acetic acid and the solutions have a number of unusual properties. In an attempt to understand them, freezing point depression data were obtained by taking a solution of known molality and then diluting it several times (J. Emsley, *J. Chem. Soc. A*, 2702 (1971)). The following data were obtained:

b/(mol kg^{-1})	0.015	0.037	0.077	0.295	0.602
ΔT/K	0.115	0.295	0.470	1.381	2.67

Calculate the apparent molar mass of the solute and suggest an interpretation. Use $\Delta_{\text{fus}}H = 11.4$ kJ mol^{-1} and $T_f^* = 290$ K.

7.7 In a study of the properties of an aqueous solution of $Th(NO_3)_4$ (by A. Apelblat, D. Azoulay, and A. Sahar, *J. Chem. Soc. Faraday Trans., I*, 1618, (1973)), a freezing point depression of 0.0703 K was observed for an aqueous solution of molality 9.6 mmol kg^{-1}. What is the apparent number of ions per formula unit?

7.8 The table below lists the vapour pressures of mixtures of iodoethane (I) and ethyl acetate (A) at 50°C. Find the activity coefficients of both components on (a) the Raoult's law basis, (b) the Henry's law basis with I as solute.

x_I	0	0.0579	0.1095	0.1918	0.2353	0.3718
p_I/Torr	0	20.0	52.7	87.7	105.4	155.4
p_A/Torr	280.4	266.1	252.3	231.4	220.8	187.9

x_I	0.5478	0.6349	0.8253	0.9093	1.0000
p_I/Torr	213.3	239.1	296.9	322.5	353.4
p_A/Torr	144.2	122.9	66.6	38.2	0

7.9 Plot the vapour pressure data for a mixture of benzene (B) and acetic acid (A) given below and plot the vapour pressure/composition curve for the mixture at 50°C. Then confirm that Raoult's and Henry's laws are obeyed in the appropriate regions. Deduce the activities and activity coefficients of the components on the Raoult's law basis and then, taking B as the solute, its activity and activity coefficient on a Henry's law basis. Finally, evaluate the excess Gibbs energy of the mixture over the composition range spanned by the data.

x_A	0.0160	0.0439	0.0835	0.1138	0.1714
p_A/Torr	3.63	7.25	11.51	14.2	18.4
p_B/Torr	262.9	257.2	249.6	244.8	231.8

x_A	0.2973	0.3696	0.5834	0.6604	0.8437	0.9931
p_A/Torr	24.8	28.7	36.3	40.2	50.7	54.7
p_B/Torr	211.2	195.6	153.2	135.1	75.3	3.5

7.10‡ Aminabhavi *et al.* examined mixtures of cyclohexane with various long-chain alkanes (T.M. Aminabhavi, V.B. Patil, M.I, Aralaguppi, J.D. Ortego, and K.C. Hansen, *J. Chem. Eng. Data* **41**, 526 (1996)). Among their data are the following measurements of the density of a mixture of cyclohexane and pentadecane as a function of the mole fraction of cyclohexane (x_c) at 298.15 K:

x_c	0.6965	0.7988	0.9004
ρ/(g cm^{-3})	0.7661	0.7674	0.7697

Compute the partial molar volume for each component in a mixture that has a mole fraction cyclohexane of 0.7988.

7.11‡ Comelli and Francesconi examined mixtures of propionic acid with various other organic liquids at 313.15 K (F. Comelli and R. Francesconi, *J. Chem. Eng. Data* **41**, 101 (1996)). They report the excess volume of mixing propionic acid with oxane as $V^E = x_1x_2\{a_0 + a_1(x_1 - x_2)\}$, where x_1 is the mole fraction of propionic acid, x_2 that of oxane, $a_0 = -2.4697$ cm^3 mol^{-1}, and $a_1 = 0.0608$ cm^3 mol^{-1}. The density of propionic acid at this temperature is 0.971 74 g cm^{-3}; that of oxane is 0.863 98 g cm^{-3}. (a) Derive an expression for the partial molar volume of each component at this temperature. (b) Compute the partial molar volume for each component in an equimolar mixture.

7.12‡ Francesconi, Lunelli, and Comelli studied the liquid–vapour equilibria of trichloromethane and 1,2-epoxybutane at several temperatures (R. Francesconi, B. Lunelli, and F. Comelli, *J. Chem. Eng. Data* **41**, 310 (1996)). Among their data are the following measurements of the mole fractions of trichloromethane in the liquid phase (x_T) and the vapour phase (y_T) at 298.15 K as a function of pressure.

p/kPa	23.40	21.75	20.25	18.75	18.15	20.25	22.50	26.30
x	0	0.129	0.228	0.353	0.511	0.700	0.810	1
y	0	0.065	0.145	0.285	0.535	0.805	0.915	1

Compute the activity coefficients of both components on the basis of Raoult's law.

7.13‡ Chen and Lee studied the liquid–vapour equilibria of cyclohexanol with several gases at elevated pressures (J.-T. Chen and M.-J. Lee, *J. Chem. Eng. Data* **41**, 339 (1996)). Among their data are the following measurements of the mole fractions of cyclohexanol in the vapour phase (y_{cyc}) and the liquid phase (x_{cyc}) at 393.15 K as a function of pressure.

p/bar	10.0	20.0	30.0	40.0	60.0	80.0
y_{cyc}	0.0267	0.0149	0.0112	0.009 47	0.008 35	0.009 21
x_{cyc}	0.9741	0.9464	0.9204	0.892	0.836	0.773

Determine the Henry's law constant of CO_2 in cyclohexanol, and compute the activity coefficient of CO_2.

7.14‡ Equation 39 indicates that solubility is an exponential function of temperature. The data in the table below give the solubility, S, of calcium acetate in water as a function of temperature.

θ/°C	0	20	40	60	80
S/(mol L^{-1})	36.4	34.9	33.7	32.7	31.7

Determine the extent to which the data fit the exponential $S = S_0e^{\tau/T}$ and obtain values for S_0 and τ. Express these constants in terms of properties of the solute.

7.15‡ The excess enthalpy of mixing data for 1-propanol and tripropylamine at 25°C is found to fit the expression

$$H^E/(\text{J mol}^{-1}) = x_1(1-x_1)\sum_{r=0}^{3} A_r(1-2x_1)^r$$

where $A_r/(\text{J mol}^{-1}) = -1131, 1355, -855, 1127$ for $r = 0, 1, 2, 3$.
(a) Draw a graph of H^E against x_1 and compare the graph shape to that expected for a solution in which $H^E = \beta RTx_1x_2$. What do the graphs imply about intermolecular interaction energies?
(b) At 25°C and infinite dilution the excess enthalpy of mixing 1-propanol in heptane is $H^E = +25.3$ kJ mol^{-1}, whereas for the mixing of tripropylamine in heptane $H^E = +121$ J mol^{-1}. Deduce the strength of the alcohol–amine hydrogen bond. Assume that intermolecular structural reorganization enthalpies cancel for the cycle. *Hint.* Use mathematical software.

7.16 The excess Gibbs energies of solutions of methylcyclohexane (MCH) and tetrahydrofuran (THF) at 303.15 K were found to fit the expression

$$G^E = RTx(1-x)\{0.4857 - 0.1077(2x-1) + 0.0191(2x-1)^2\}$$

where x is the mole fraction of the methylcyclohexane. Calculate the Gibbs energy of mixing when a mixture of 1.00 mol of MCH and 3.00 mol of THF is prepared.

Theoretical problems

7.17 The excess Gibbs energy of a certain binary mixture is equal to $gRTx(1-x)$ where g is a constant and x is the mole fraction of a solute A. Find an expression for the chemical potential of A in the mixture and sketch its dependence on the composition.

7.18 Use the Gibbs–Duhem equation to derive the Gibbs–Duhem–Margules equation

$$\left(\frac{\partial \ln f_A}{\partial \ln x_A}\right)_{p,T} = \left(\frac{\partial \ln f_B}{\partial \ln x_B}\right)_{p,T}$$

where f is the fugacity. Use the relation to show that, when the fugacities are replaced by pressures, if Raoult's law applies to one component in a mixture it must also apply to the other.

7.19 Use the Gibbs–Duhem equation to show that the partial molar volume (or any partial molar property) of a component B can be obtained if the partial molar volume (or other property) of A is known for all compositions up to the one of interest. Do this by proving that

$$V_B = V_B^* + \int_{V_A^*}^{V_A} \frac{x_A}{1-x_A}\,dV_A$$

Use the following data (which are for 298 K) to evaluate the integral graphically to find the partial molar volume of acetone at $x = 0.500$.

$x(\text{CHCl}_3)$	0	0.194	0.385	0.559	0.788	0.889	1.000
$V_m/(\text{cm}^3\,\text{mol}^{-1})$	73.99	75.29	76.50	77.55	79.08	79.82	80.67

7.20 Use the Gibbs–Helmholtz equation to find an expression for $d\ln x_A$ in terms of dT. Integrate $d\ln x_A$ from $x_A = 0$ to the value of interest, and integrate the right-hand side from the transition temperature for the pure liquid A to the value in the solution. Show that, if the enthalpy of transition is constant, then eqns 33 and 36 are obtained.

7.21 The 'osmotic coefficient', ϕ, is defined as $\phi = -(x_A/x_B)\ln a_A$. By writing $r = x_B/x_A$, and using the Gibbs–Duhem equation, show that we can calculate the activity of B from the activities of A over a composition range by using the formula

$$\ln\left(\frac{a_B}{r}\right) = \phi - \phi(0) + \int_0^r \left(\frac{\phi - 1}{r}\right)dr$$

7.22 Show that the osmotic pressure of a real solution is given by

$$\Pi V = -RT\ln a_A$$

Go on to show that, provided the concentration of the solution is low, this expression takes the form

$$\Pi V = \phi RT[\text{B}]$$

and hence that the osmotic coefficient, ϕ, (which is defined in Problem 7.21) may be determined from osmometry.

Applications: to biology and polymer science

7.23 Suppose that two equal volumes of 0.200 M NaCl(aq) are separated by a membrane and that an anionic protein P of mass 55 kg mol^{-1}, which cannot pass through the membrane, is added as its sodium salt Na_6P to a concentration of 50 g L^{-1} to the left-hand compartment of the apparatus. Calculate the molar concentrations of Na^+ and Cl^- in each compartment.

7.24‡ Polymer scientists often report their data in rather strange units. For example, in the determination of molar masses of polymers in solution by osmometry, osmotic pressures are often reported in grams per square centimetre (g cm^{-2}) and concentrations in grams per cubic centimetre (g cm^{-3}). (a) With these choices of units, what would be the units of R in the van 't Hoff equation? (b) The data in the table below on the concentration dependence of the osmotic pressure of polyisobutene in chlorobenzene at 25°C have been adapted from J. Leonard and H. Daoust (*J. Polymer Sci.* **57**, 53 (1962)). From these data, determine the molar mass of polyisobutene by plotting Π/c against c. (c) Theta solvents are solvents for which the second osmotic coefficient is zero; for 'poor' solvents the plot is linear and for good solvents the plot is nonlinear. From your plot, how would you classify chlorobenzene as a solvent for polyisobutene? Rationalize the result in terms of the molecular structure of the polymer and solvent. (d) Determine the second and third osmotic virial coefficients by fitting the curve to the virial form of the osmotic pressure equation. (e) Experimentally, it is often found that the virial expansion can be represented as

$$\Pi/c = RT/M(1 + B'c + gB'^2c'^2 + \cdots)$$

and, in good solvents, the parameter g is often about 0.25. With terms beyond the second power ignored, obtain an equation for $(\Pi/c)^{1/2}$ and plot this quantity against c. Determine the second and third virial coefficients from the plot and compare to the values from the first plot. Does this plot confirm the assumed value of g?

$10^{-2}(\Pi/c)/(\text{g cm}^{-2}/\text{g cm}^{-3})$	2.6	2.9	3.6	4.3	6.0	12.0
$c/(\text{g cm}^{-3})$	0.0050	0.010	0.020	0.033	0.057	0.10

$10^{-2}(\Pi/c)/(\text{g cm}^{-2}/\text{g cm}^{-3})$	19.0	31.0	38.0	52	63
$c/(\text{g cm}^{-3})$	0.145	0.195	0.245	0.27	0.29

7.25‡ K. Sato, F.R. Eirich, and J.E. Mark (*J. Polymer Sci., Polym. Phys.* **14**, 619 (1976)) have reported the data in the table below for the osmotic pressures of polychloroprene ($\rho = 1.25$ g cm^{-3}) in toluene ($\rho = 0.858$ g cm^{-3}) at 30°C. Determine the molar mass of polychloroprene and its second osmotic virial coefficient.

$c/(\text{mg cm}^{-3})$	1.33	2.10	4.52	7.18	9.87
$\Pi/(\text{N m}^{-2})$	30	51	132	246	390

8 Phase diagrams

Phase diagrams for pure substances were introduced in Chapter 6. Now we develop their use systematically and show how they are rich summaries of empirical information about a wide range of systems. To set the stage, we introduce the famous phase rule of Gibbs, which shows the extent to which various parameters can be varied yet the equilibrium between phases preserved. With the rule established, we see how it can be used to discuss the phase diagrams that we met in the two preceding chapters. The chapter then introduces systems of gradually increasing complexity. In each case we shall see how the phase diagram for the system summarizes empirical observations on the conditions under which the various phases of the system are stable.

In this chapter we describe a systematic way of discussing the physical changes mixtures undergo when they are heated or cooled and when their compositions are changed. In particular, we see how to use phase diagrams to judge whether two substances are mutually miscible, whether an equilibrium can exist over a range of conditions, or whether the system must be brought to a definite pressure, temperature, and composition before equilibrium is established. Phase diagrams are of considerable commercial and industrial significance, particularly for semiconductors, ceramics, steels, and alloys. They are also the basis of separation procedures in the petroleum industry and of the formulation of foods and cosmetic preparations.

Phases, components, and degrees of freedom

All phase diagrams can be discussed in terms of a relationship, the phase rule, derived by J.W. Gibbs. We shall derive this rule first, and then apply it to a wide variety of systems. The phase rule requires a careful use of terms, so we begin by presenting a number of definitions.

8.1 Definitions

The term **phase** was introduced at the start of Chapter 6, where we saw that it signifies a state of matter that is uniform throughout, not only in chemical composition but also in physical state.[1] Thus we speak of the solid, liquid, and gas

1. The words are Gibbs's.

(a)

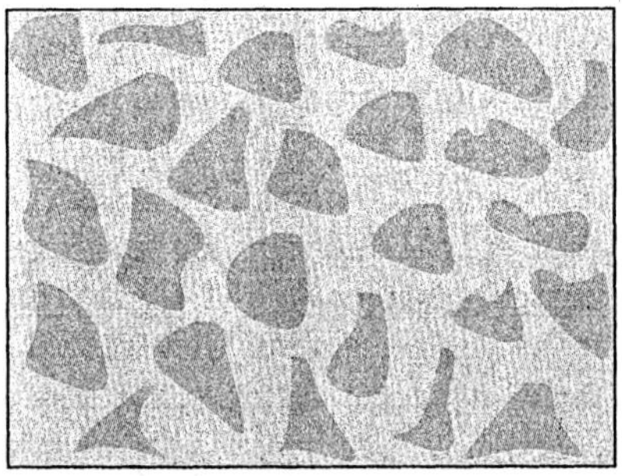
(b)

8.1 The difference between (a) a single-phase solution, in which the composition is uniform on a microscopic scale, and (b) a dispersion, in which regions of one component are embedded in a matrix of a second component.

phases of a substance, and of its various solid phases (as for black phosphorus and white phosphorus). The number of phases in a system is denoted P. A gas, or a gaseous mixture, is a single phase, a crystal is a single phase, and two totally miscible liquids form a single phase. A solution of sodium chloride in water is a single phase. Ice is a single phase ($P = 1$) even though it might be chipped into small fragments. A slurry of ice and water is a two-phase system ($P = 2$) even though it is difficult to map the boundaries between the phases. A system in which calcium carbonate undergoes thermal decomposition consists of two solid phases (one consisting of calcium carbonate and the other of calcium oxide) and one gaseous phase (consisting of carbon dioxide).

An alloy of two metals is a two-phase system ($P = 2$) if the metals are immiscible, but a single-phase system ($P = 1$) if they are miscible. This example shows that it is not always easy to decide whether a system consists of one phase or of two. A solution of solid A in solid B—a homogeneous mixture of the two substances—is uniform on a molecular scale. In a solution, atoms of A are surrounded by atoms of A and B, and any segment cut from the sample, however small, is representative of the composition of the whole.

A dispersion is uniform on a macroscopic scale but not on a microscopic scale, for it consists of grains or droplets of one substance in a matrix of the other. A small sample could come entirely from one of the minute grains of pure A and would not be representative of the whole (Fig. 8.1). Dispersions are important because, in many advanced materials (including steels), heat treatment cycles are used to achieve the precipitation of a fine dispersion of particles of one phase (such as a carbide phase) within a matrix formed by a saturated solid solution phase. The ability to control this microstructure resulting from phase equilibria makes it possible to tailor the mechanical properties of the materials to a particular application.

By a **constituent** of a system we mean a chemical species (an ion or a molecule) that is present. Thus, a mixture of ethanol and water has two constituents. A solution of sodium chloride has two constituents (sodium chloride and water) despite the fact that it contains Na^+ and Cl^- ions. The term constituent should be carefully distinguished from 'component', which has a more technical meaning. A **component** is a chemically independent constituent of a system. The number of components, C, in a system is the minimum number of independent species necessary to define the composition of all the phases present in the system.

When no reaction takes place, the number of components is equal to the number of constituents. Thus, pure water is a one-component system ($C = 1$), because we need only the species H_2O to specify its composition. Similarly, a mixture of ethanol and water is a two-component system ($C = 2$): we need the species H_2O and C_2H_5OH to specify its composition. When a reaction can occur under the conditions prevailing in the system, we need to decide the minimum number of species that, after allowing for reactions in which one species is synthesized from others, can be used to specify the composition of all the phases. Consider, for example, the equilibrium

$$\underset{\text{Phase 1}}{CaCO_3(s)} \rightleftharpoons \underset{\text{Phase 2}}{CaO(s)} + \underset{\text{Phase 3}}{CO_2(g)}$$

in which there are three phases. To specify the composition of the gas phase (Phase 3) we need the species CO_2, and to specify the composition of Phase 2 we need the species CaO. However, we do not need an additional species to specify the composition of Phase 1 because its identity ($CaCO_3$) can be expressed in terms of the other two constituents by making use of the stoichiometry of the reaction. Hence, the system has three constituents but only two components ($C = 2$).

Example 8.1 Counting components

How many components are present in a system in which ammonium chloride undergoes thermal decomposition?

Method Begin by writing down the chemical equation for the reaction and identifying the constituents of the system (all the species present) and the phases. Then decide whether, under the conditions prevailing in the system, any of the constituents can be prepared from any of the other constituents. The removal of these constituents leaves the number of independent constituents. Finally, identify the minimum number of these independent constituents that are needed to specify the composition of all the phases.

Answer The chemical reaction is

$$NH_4Cl(s) \rightleftharpoons NH_3(g) + HCl(g)$$

There are three constituents and two phases (one solid, one gas). However, NH_3 and HCl are formed in fixed stoichiometric proportions by the reaction. Therefore, the compositions of both phases can be expressed in terms of the single species NH_4Cl. It follows that there is only one component in the system ($C = 1$).

Comment If additional HCl (or NH_3) were supplied to the system, the decomposition of NH_4Cl would not give the correct composition of the gas phase and HCl (or NH_3) would have to be invoked as a second component. A system that consists of hydrogen, oxygen, and water at room temperature has three components, despite it being possible to form H_2O from H_2 and O_2: under the conditions prevailing in the system, hydrogen and oxygen do not react to form water, so they are independent constituents.

Self-test 8.1 Give the number of components in the following systems: (a) water, allowing for its autoprotolysis, (b) aqueous acetic acid, (c) magnesium carbonate in equilibrium with its decomposition products. [(a) 1, (b) 2, (c) 2]

The **variance**, F, of a system is the number of intensive variables that can be changed independently without disturbing the number of phases in equilibrium. In a single-component, single-phase system ($C = 1$, $P = 1$), the pressure and temperature may be changed independently without changing the number of phases, so $F = 2$. We say that such a system is **bivariant**, or that it has two **degrees of freedom**. On the other hand, if two phases are in equilibrium (a liquid and its vapour, for instance) in a single-component system ($C = 1$, $P = 2$), the temperature (or the pressure) can be changed at will, but the change in temperature (or pressure) demands an accompanying change in pressure (or temperature) to preserve the number of phases in equilibrium. That is, the variance of the system has fallen to 1.

8.2 The phase rule

In one of the most elegant calculations of the whole of chemical thermodynamics, J.W. Gibbs deduced the **phase rule**,[2] which is a general relation between the

2. Josiah Willard Gibbs spent most of his working life at Yale, and may justly be regarded as the originator of chemical thermodynamics. He reflected for years before publishing his conclusions, and then did so in precisely expressed papers in an obscure journal (*The Transactions of the Connecticut Academy of Arts and Sciences*). He needed interpreters before the power of his work was recognized and before it could be applied to industrial processes. He is regarded by many as the first great American theoretical scientist.

variance, F, the number of components, C, and the number of phases at equilibrium, P, for a system of any composition:

$$F = C - P + 2 \tag{8.1}$$

Justification 8.1

Consider first the special case of a one-component system. For two phases in equilibrium, we can write $\mu_J(\alpha) = \mu_J(\beta)$. Each chemical potential is a function of the pressure and temperature, so

$$\mu_J(\alpha, p, T) = \mu_J(\beta, p, T)$$

This is an equation relating p and T, so only one of these variables is independent (just as the equation $x + y = 2$ is a relation for y in terms of x: $y = 2 - x$). That conclusion is consistent with $F = 1$. For three phases in mutual equilibrium,

$$\mu_J(\alpha, p, T) = \mu_J(\beta, p, T) = \mu_J(\gamma, p, T)$$

This relation is actually two equations for two unknowns ($\mu_J(\alpha, p, T) = \mu_J(\beta, p, T)$ and $\mu_J(\beta, p, T) = \mu_J(\gamma, p, T)$), and therefore has a solution only for a single value of p and T (just as the pair of equations $x + y = 2$ and $3x - y = 4$ has the single solution $x = \frac{3}{2}$ and $y = \frac{1}{2}$). That conclusion is consistent with $F = 0$. Four phases cannot be in mutual equilibrium in a one-component system because the three equalities

$$\mu_J(\alpha, p, T) = \mu_J(\beta, p, T) \qquad \mu_J(\beta, p, T) = \mu_J(\gamma, p, T) \qquad \mu_J(\gamma, p, T) = \mu_J(\delta, p, T)$$

are three equations for two unknowns (p and T) and are not consistent (just as $x + y = 2$, $3x - y = 4$, and $x + 4y = 6$ has no solution).

Now consider the general case. We begin by counting the total number of intensive variables (properties that do not depend on the size of the system). The pressure, p, and temperature, T, count as 2. We can specify the composition of a phase by giving the mole fractions of $C - 1$ components. We need specify only $C - 1$ and not all C mole fractions because $x_1 + x_2 + \cdots + x_C = 1$, and all mole fractions are known if all except one are specified. Because there are P phases, the total number of composition variables is $P(C - 1)$. At this stage, the total number of intensive variables is $P(C - 1) + 2$.

At equilibrium, the chemical potential of a component J must be the same in every phase (Section 6.4):

$$\mu_J(\alpha) = \mu_J(\beta) = \cdots \quad \text{for } P \text{ phases}$$

That is, there are $P - 1$ equations of this kind to be satisfied for each component J. As there are C components, the total number of equations is $C(P - 1)$. Each equation reduces our freedom to vary one of the $P(C - 1) + 2$ intensive variables. It follows that the total number of degrees of freedom is

$$F = P(C - 1) + 2 - C(P - 1) = C - P + 2$$

which is eqn 1.

(a) One-component systems

For a one-component system, such as pure water, $F = 3 - P$. When only one phase is present, $F = 2$ and both p and T can be varied independently without changing the number of phases. In other words, a single phase is represented by an *area* on a phase

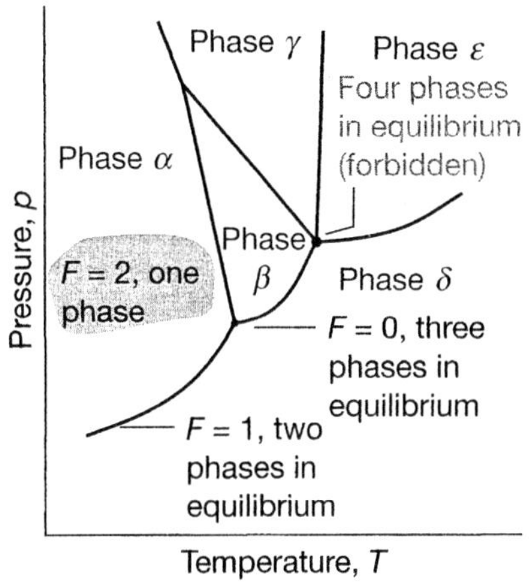

8.2 The typical regions of a one-component phase diagram. The lines represent conditions under which the two adjoining phases are in equilibrium. A point represents the unique set of conditions under which three phases coexist in equilibrium. Four phases cannot mutually coexist in equilibrium.

diagram. When two phases are in equilibrium $F = 1$, which implies that pressure is not freely variable if the temperature is set; indeed, at a given temperature, a liquid has a characteristic vapour pressure. It follows that the equilibrium of two phases is represented by a *line* in the phase diagram. Instead of selecting the temperature, we could select the pressure, but having done so the two phases would be in equilibrium at a single definite temperature. Therefore, freezing (or any other phase transition) occurs at a definite temperature at a given pressure.

When three phases are in equilibrium, $F = 0$ and the system is invariant. This special condition can be established only at a definite temperature and pressure that is characteristic of the substance and outside our control. The equilibrium of three phases is therefore represented by a *point*, the triple point, on a phase diagram. Four phases cannot be in equilibrium in a one-component system because F cannot be negative. These features are summarized in Fig. 8.2.

We can identify the features in Fig. 8.2 in the experimentally determined phase diagram for water (Fig. 8.3). This diagram summarizes the changes that take place as a sample, such as that at *a*, is cooled at constant pressure. The sample remains entirely gaseous until the temperature reaches *b*, when liquid appears. Two phases are now in equilibrium and $F = 1$. Because we have decided to specify the pressure, which uses up the single degree of freedom, the temperature at which this equilibrium occurs is not under our control. Lowering the temperature takes the system to *c* in the one-phase, liquid region. The temperature can now be varied around the point *c* at will, and only when ice appears at *d* does the variance become 1 again.

(b) Experimental procedures

Detecting a phase change is not always as simple as seeing water boil in a kettle, so special techniques have been developed. Two techniques are **thermal analysis**, which takes advantage of the effect of the enthalpy change during a first-order transition (Section 6.7), and differential scanning calorimetry (see Box 2.1). They are useful for solid–solid transitions, where simple visual inspection of the sample may be inadequate. In thermal analysis, a sample is allowed to cool and its temperature is monitored. At a first-order transition, heat is evolved and the cooling stops until the transition is complete. The cooling curve along the isobar *cde* in Fig. 8.3 therefore has the shape shown in Fig. 8.4. The transition temperature is obvious, and is used to mark point *d* on the phase diagram.

Modern work on phase transitions often deals with systems at very high pressures, and more sophisticated detection procedures must be adopted. Some of the highest pressures currently attainable are produced in a **diamond-anvil cell** like that illustrated in Fig. 8.5. The sample is placed in a minute cavity between two gemquality diamonds, and then pressure is exerted simply by turning the screw. The advance in design this represents is quite remarkable for, with a turn of the screw, pressures of up to about 1 Mbar can be reached which a few years ago could not be reached with equipment weighing tons.

The pressure is monitored spectroscopically by observing the shift of spectral lines in small pieces of ruby added to the sample, and the properties of the sample itself are observed optically through the diamond anvils. One application of the technique is to study the transition of covalent solids to metallic solids. Iodine, I_2, for instance, becomes metallic at around 200 kbar and makes a transition to a monatomic metallic solid at around 210 kbar. Studies such as these are relevant to the structure of material deep inside the Earth (at the centre of the Earth the pressure is around 5 Mbar) and in the interiors of the giant planets, where even hydrogen may be metallic.

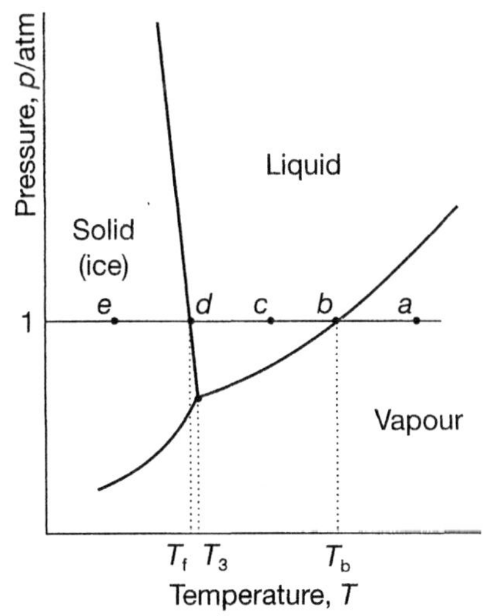

8.3 The phase diagram for water, a simplified version of Fig. 6.4. The label T_3 marks the temperature of the triple point, T_b the normal boiling point, and T_f the normal freezing point.

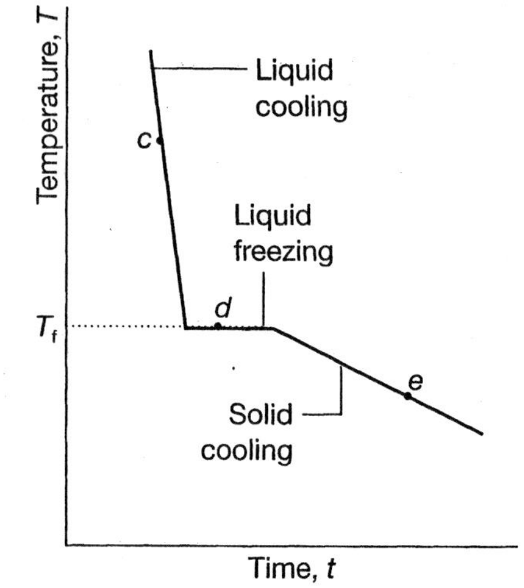

8.4 The cooling curve for the isobar *cde* in Fig. 8.3. The halt marked *d* corresponds to the pause in the fall of temperature while the first-order exothermic transition (freezing) occurs. This pause enables T_f to be located even if the transition cannot be observed visually.

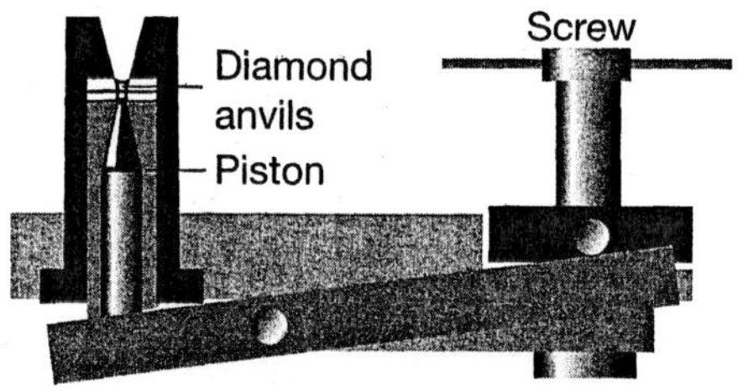

8.5 Ultrahigh pressures (up to about 2 Mbar) can be achieved using a diamond anvil. The sample, together with a ruby for pressure measurement and a drop of liquid for pressure transmission, are placed between two gem-quality diamonds. The principle of its action is like that of a nutcracker: the pressure is exerted by turning the screw by hand.

Two-component systems

When two components are present in a system, $C = 2$ and $F = 4 - P$. If the temperature is constant, the remaining variance is $F' = 3 - P$, which has a maximum value of 2. (The prime on F indicates that one of the degrees of freedom has been discarded; in this case the temperature.) One of these two remaining degrees of freedom is the pressure and the other is the composition (as expressed by the mole fraction of one component). Hence, one form of the phase diagram is a map of pressures and compositions at which each phase is stable. Alternatively, the pressure could be held constant and the phase diagram depicted in terms of temperature and composition.

8.3 Vapour pressure diagrams

The partial vapour pressures of the components of an ideal solution of two volatile liquids are related to the composition of the liquid mixture by Raoult's law (Section 7.3a)

$$p_A = x_A p_A^* \qquad p_B = x_B p_B^* \tag{8.2°}$$

where p_A^* is the vapour pressure of pure A and p_B^* that of pure B. The total vapour pressure p of the mixture is therefore

$$p = p_A + p_B = x_A p_A^* + x_B p_B^* = p_B^* + (p_A^* - p_B^*)x_A \tag{8.3°}$$

This expression shows that the total vapour pressure (at some fixed temperature) changes linearly with the composition from p_B^* to p_A^* as x_A changes from 0 to 1 (Fig. 8.6).

(a) The composition of the vapour

The compositions of the liquid and vapour that are in mutual equilibrium are not necessarily the same. Common sense suggests that the vapour should be richer in the more volatile component. This expectation can be confirmed as follows. The partial pressures of the components are given by eqn 2. It follows from Dalton's law that the mole fractions in the gas, y_A and y_B, are

$$y_A = \frac{p_A}{p} \qquad y_B = \frac{p_B}{p} \tag{8.4}$$

Provided the mixture is ideal, the partial pressures and the total pressure may be expressed in terms of the mole fractions in the liquid by using eqn 2 for p_J and eqn 3 for the total vapour pressure p, which gives

$$y_A = \frac{x_A p_A^*}{p_B^* + (p_A^* - p_B^*)x_A} \qquad y_B = 1 - y_A \tag{8.5°}$$

Figure 8.7 shows the composition of the vapour plotted against the composition of the liquid for various values of $p_A^*/p_B^* > 1$. We see that in all cases $y_A > x_A$; that is, the vapour is richer than the liquid in the more volatile component. Note that, if B is non-volatile, so that $p_B^* = 0$ at the temperature of interest, then it makes no contribution to the vapour ($y_B = 0$).

Equation 3 shows how the total vapour pressure of the mixture varies with the composition of the liquid. Because we can relate the composition of the liquid to the composition of the vapour through eqn 5, we can now also relate the total vapour pressure to the composition of the vapour:

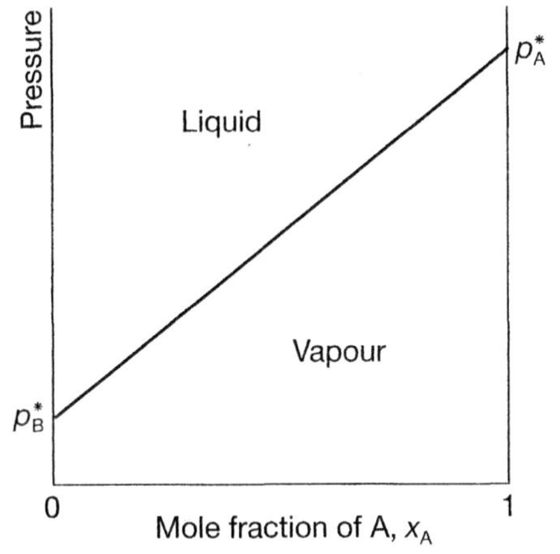

8.6 The variation of the total vapour pressure of a binary mixture with the mole fraction of A in the liquid when Raoult's law is obeyed.

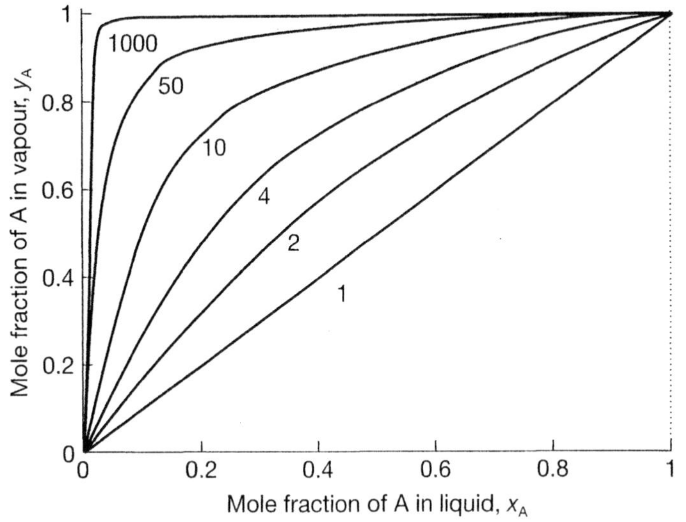

8.7 The mole fraction of A in the vapour of a binary ideal solution expressed in terms of its mole fraction in the liquid, calculated using eqn 5 for various values of p_A^*/p_B^* (the label on each curve) with A more volatile than B. In all cases the vapour is richer than the liquid in A.

$$p = \frac{p_A^* p_B^*}{p_A^* + (p_B^* - p_A^*)y_A} \qquad (8.6)°$$

This expression is plotted in Fig. 8.8.

(b) The interpretation of the diagrams

If we are interested in distillation, both the vapour and the liquid compositions are of equal interest. It is therefore sensible to combine the two preceding diagrams into one. The result is summarized in Fig. 8.9. The point *a* indicates the vapour pressure of a mixture of composition x_A, and the point *b* indicates the composition of the vapour that is in equilibrium with the liquid at that pressure. Note that when two phases are in equilibrium, $P = 2$ so $F' = 1$ (as usual, the prime indicating that one

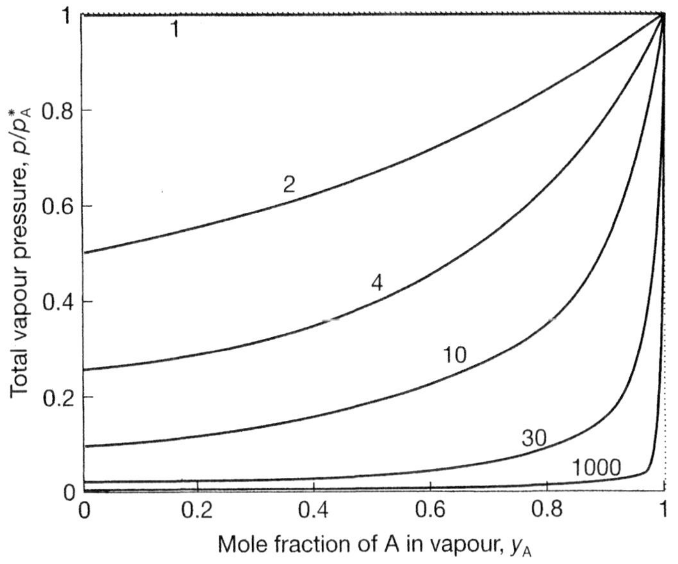

8.8 The dependence of the vapour pressure of the same system as in Fig. 8.7, but expressed in terms of the mole fraction of A in the vapour by using eqn 6. Individual curves are labelled with the value of p_A^*/p_B^*.

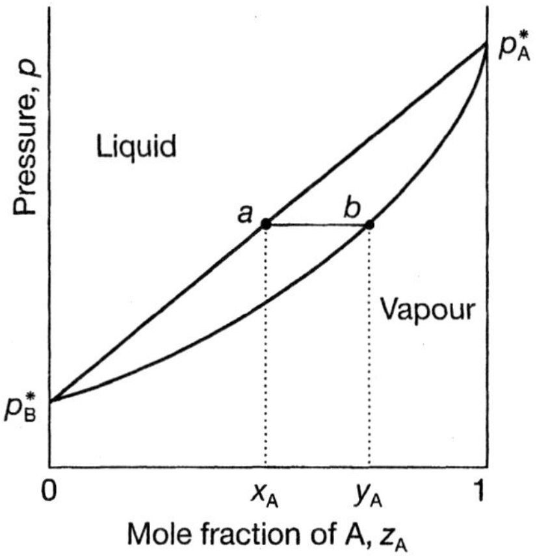

8.9 The dependence of the total vapour pressure of an ideal solution on the mole fraction of A in the entire system. A point between the two lines corresponds to both liquid and vapour being present; outside that region there is only one phase present. The mole fraction of A is denoted z_A, as explained below.

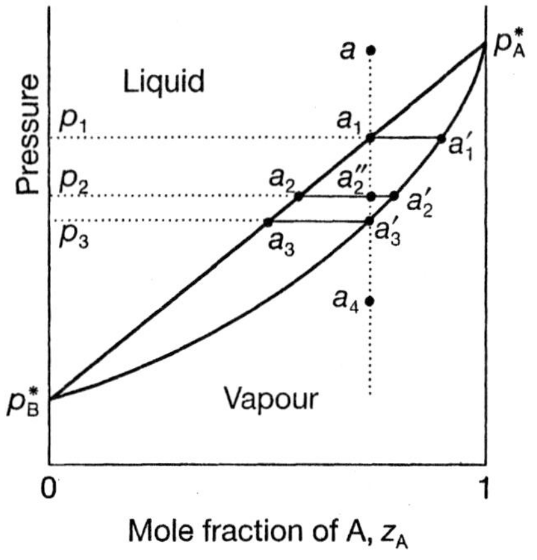

8.10 The points of the pressure–composition diagram discussed in the text. The vertical line through a is an *isopleth*, a line of constant composition of the entire system.

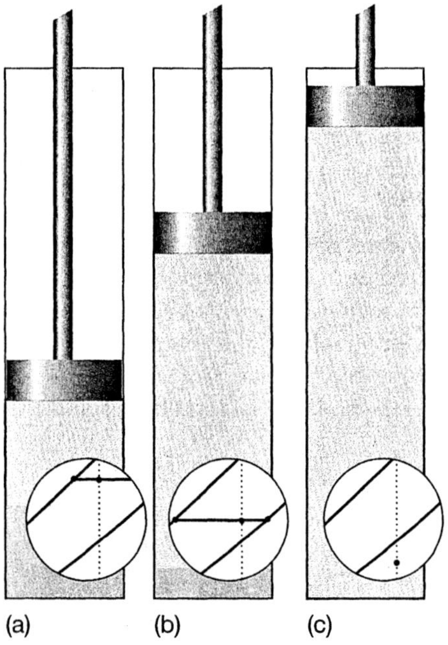

8.11 (a) A liquid in a container exists in equilibrium with its vapour. The superimposed fragment of the phase diagram shows the compositions of the two phases and their abundances (by the lever rule). (b) When the pressure is changed by drawing out a piston, the compositions of the phases adjust as shown by the tie line in the phase diagram. (c) When the piston is pulled so far out that all the liquid has vaporized and only the vapour is present, the pressure falls as the piston is withdrawn and the point on the phase diagram moves into the one-phase region.

degree of freedom, the temperature, has already been discarded). That is, if the composition is specified (so using up the only remaining degree of freedom), the pressure at which the two phases are in equilibrium is fixed.

A richer interpretation of the phase diagram is obtained if we interpret the horizontal axis as showing the *overall* composition, z_A, of the system. If the horizontal axis of the vapour pressure diagram is labelled with z_A, then all the points down to the solid diagonal line in the graph correspond to a system that is under such high pressure that it contains only a liquid phase (the applied pressure is higher than the vapour pressure), so $z_A = x_A$, the composition of the liquid. On the other hand, all points below the lower curve correspond to a system that is under such low pressure that it contains only a vapour phase (the applied pressure is lower than the vapour pressure), so $z_A = y_A$.

Points that lie between the two lines correspond to a system in which there are two phases present, one a liquid and the other a vapour. To see this interpretation, consider the effect of lowering the pressure on a liquid mixture of overall composition a in Fig. 8.10. The lowering of pressure can be achieved by drawing out a piston (Fig. 8.11). This degree of freedom is permitted by the phase rule because $F' = 2$ when $P = 1$, and even if the composition is selected one degree of freedom remains. The changes to the system do not affect the overall composition, so the state of the system moves down the vertical line that passes through a. This vertical line is called an **isopleth**, from the Greek words for 'equal abundance'. Until the point a_1 is reached (when the pressure has been reduced to p_1), the sample consists of a single liquid phase. At a_1 the liquid can exist in equilibrium with its vapour. As we have seen, the composition of the vapour phase is given by point a_1'. A line joining two points representing phases in equilibrium is called a **tie line**. The composition of the liquid is the same as initially (a_1 lies on the isopleth through a), so we have to conclude that at this pressure there is virtually no vapour present; however, the tiny amount of vapour that is present has the composition a_1'.

Now consider the effect of lowering the pressure to p_2, so taking the system to a pressure and overall composition represented by the point a_2''. This new pressure is below the vapour pressure of the original liquid, so it vaporizes until the vapour

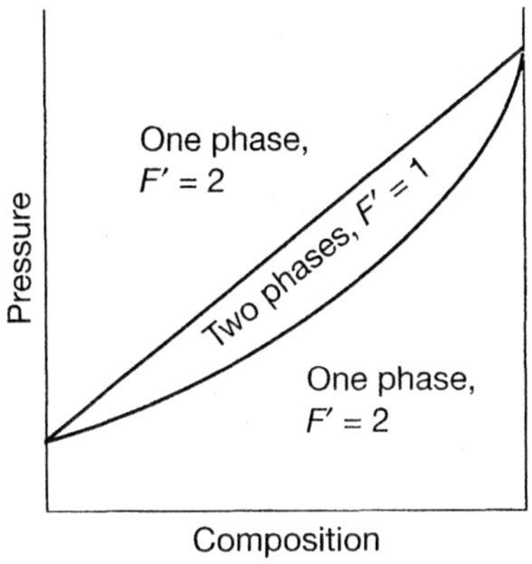

8.12 The general scheme of interpretation of a pressure–composition diagram (a vapour pressure diagram).

pressure of the remaining liquid falls to p_2. Now we know that the composition of such a liquid must be a_2. Moreover, the composition of the vapour in equilibrium with that liquid must be given by the point a_2' at the other end of the tie line. Note that two phases are now in equilibrium, so $F' = 1$ for all points between the two lines; hence, for a given pressure (such as at p_2) the variance is zero, and the vapour and liquid phases have fixed compositions (Fig. 8.12). If the pressure is reduced to p_3, a similar readjustment in composition takes place, and now the compositions of the liquid and vapour are represented by the points a_3 and a_3', respectively. The latter point corresponds to a system in which the composition of the vapour is the same as the overall composition, so we have to conclude that the amount of liquid present is now virtually zero, but the tiny amount of liquid present has the composition a_3. A further decrease in pressure takes the system to the point a_4; at this stage, only vapour is present and its composition is the same as the initial overall composition of the system (the composition of the original liquid).

(c) The lever rule

A point in the two-phase region of a phase diagram indicates not only qualitatively that both liquid and vapour are present, but represents quantitatively the relative amounts of each. To find the relative amounts of two phases α and β that are in equilibrium, we measure the distances l_α and l_β along the horizontal tie line, and then use the **lever rule** (Fig. 8.13):

$$n_\alpha l_\alpha = n_\beta l_\beta \qquad (8.7)$$

Here n_α is the amount of phase α and n_β the amount of phase β. In the case illustrated in Fig. 8.13, because $l_\beta \approx 2l_\alpha$, the amount of phase α is about twice the amount of phase β.

Justification 8.2

To prove the lever rule we write $n = n_\alpha + n_\beta$ and the overall amount of A as nz_A. The overall amount of A is also the sum of its amounts in the two phases:

$$nz_A = n_\alpha x_A + n_\beta y_A$$

Since also

$$nz_A = n_\alpha z_A + n_\beta z_A$$

by equating these two expressions it follows that

$$n_\alpha(x_A - z_A) = n_\beta(z_A - y_A)$$

which corresponds to eqn 7.

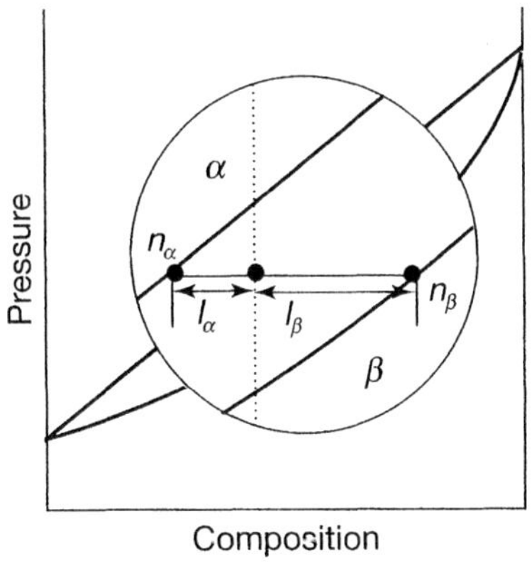

8.13 The lever rule. The distances l_α and l_β are used to find the proportions of the amounts of phases α (such as vapour) and β (for example, liquid) present at equilibrium. The lever rule is so called because a similar rule relates the masses at two ends of a lever to their distances from a pivot ($m_\alpha l_\alpha = m_\beta l_\beta$ for balance).

Illustration 8.1

At p_1 in Fig. 8.10, the ratio l_{vap}/l_{liq} is almost infinite for this tie line, so n_{liq}/n_{vap} is also almost infinite, and there is only a trace of vapour present. When the pressure is reduced to p_2, the value of l_{vap}/l_{liq} is about 0.3, so $n_{liq}/n_{vap} \approx 0.3$ and the amount of liquid is about 0.3 times the amount of vapour. When the pressure has been reduced to p_3, the sample is almost completely gaseous and because $l_{vap}/l_{liq} \approx 0$ we conclude that there is only a trace of liquid present.

The textbook's web site contains links to online databases of data on phase diagrams.

8.4 Temperature–composition diagrams

To discuss distillation we need a **temperature-composition diagram**, a phase diagram in which the boundaries show the composition of the phases that are in equilibrium at various temperatures (and a given pressure, typically 1 atm). An example is shown in Fig. 8.14. Note that the liquid phase now lies in the lower part of the diagram.

(a) The distillation of mixtures

The region between the lines in Fig. 8.14 is a two-phase region where $F' = 1$. As usual, the prime indicates that one degree of freedom has been discarded; in this case, the pressure is being kept fixed, and hence at a given temperature the compositions of the phases in equilibrium are fixed. The regions outside the phase lines correspond to a single phase, so $F' = 2$, and the temperature and composition are both independently variable.

Consider what happens when a liquid of composition a_1 is heated. It boils when the temperature reaches T_2. Then the liquid has composition a_2 (the same as a_1) and the vapour (which is present only as a trace) has composition a_2'. The vapour is richer in the more volatile component A (the component with the lower boiling point). From the location of a_2, we can state the vapour's composition at the boiling point, and from the location of the tie line joining a_2 and a_2' we can read off the boiling temperature (T_2) of the original liquid mixture.

In a **simple distillation**, the vapour is withdrawn and condensed. This technique is used to separate a volatile liquid from a non-volatile solute or solid. In **fractional distillation**, the boiling and condensation cycle is repeated successively. This technique is used to separate volatile liquids. We can follow the changes that occur by seeing what happens when the first condensate of composition a_3 is reheated. The phase diagram shows that this mixture boils at T_3 and yields a vapour of composition a_3', which is even richer in the more volatile component. That vapour is drawn off, and the first drop condenses to a liquid of composition a_4. The cycle can then be repeated until in due course almost pure A is obtained.

The efficiency of a fractionating column is expressed in terms of the number of **theoretical plates**, the number of effective vaporization and condensation steps that are required to achieve a condensate of given composition from a given distillate. Thus, to achieve the degree of separation shown in Fig. 8.15a, the fractionating

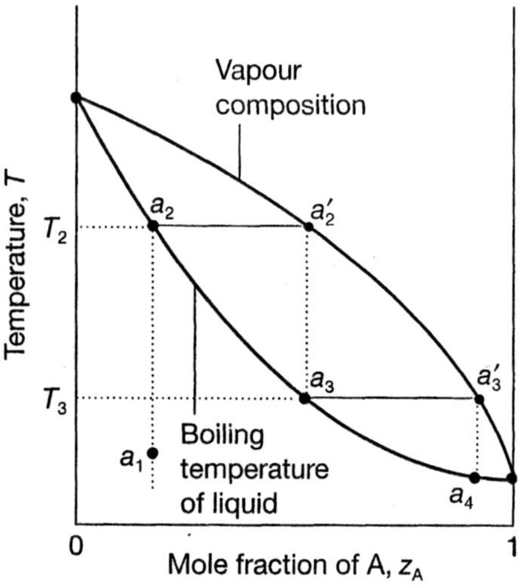

8.14 The temperature-composition diagram corresponding to an ideal mixture with the component A more volatile than component B. Successive boilings and condensations of a liquid originally of composition a_1 lead to a condensate that is pure A. The separation technique is called fractional distillation.

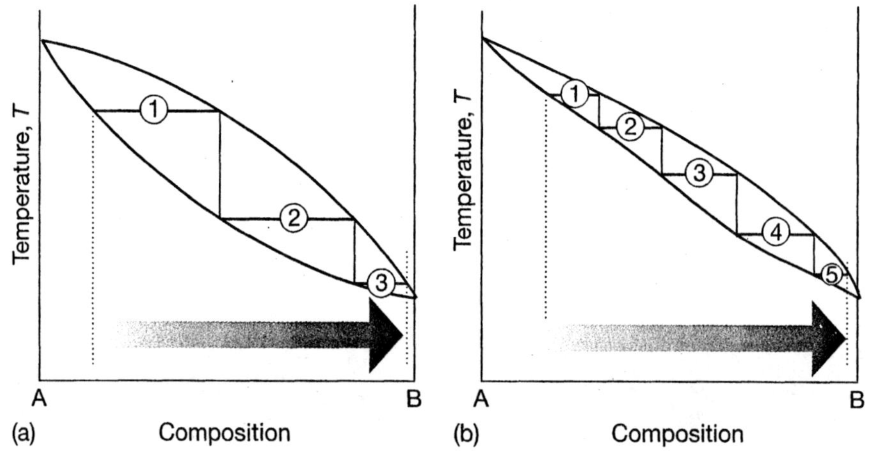

8.15 The number of theoretical plates is the number of steps needed to bring about a specified degree of separation of two components in a mixture. The two systems shown correspond to (a) 3, (b) 5 theoretical plates.

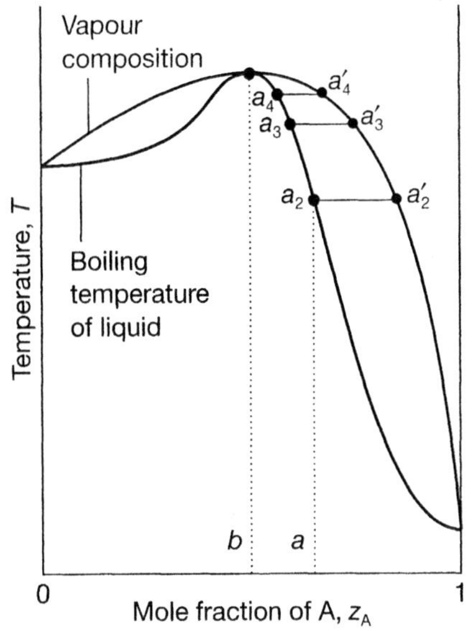

8.16 A high-boiling azeotrope. When the liquid of composition a is distilled, the composition of the remaining liquid changes towards b but no further.

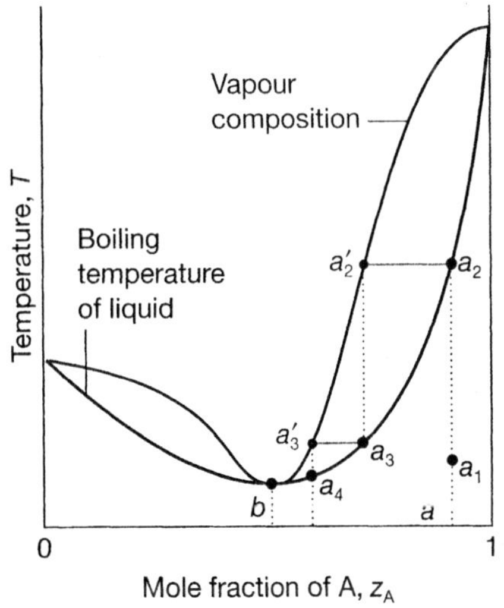

8.17 A low-boiling azeotrope. When the mixture at a is fractionally distilled, the vapour in equilibrium in the fractionating column moves towards b and then remains unchanged.

column must correspond to three theoretical plates. To achieve the same separation for the system shown in Fig. 8.15b, in which the components have more similar partial pressures, the fractionating column must be designed to correspond to five theoretical plates.

(b) Azeotropes

Although many liquids have temperature–composition phase diagrams resembling the ideal version in Fig. 8.14, in a number of important cases there are marked deviations. A maximum in the phase diagram (Fig. 8.16) may occur when the favourable interactions between A and B molecules reduce the vapour pressure of the mixture below the ideal value: in effect, the A–B interactions stabilize the liquid. In such cases the excess Gibbs energy, G^E (Section 7.4), is negative (more favourable to mixing than ideal). Examples of this behaviour include trichloromethane/propanone and nitric acid/water mixtures. Phase diagrams showing a minimum (Fig. 8.17) indicate that the mixture is destabilized relative to the ideal solution, the A–B interactions then being unfavourable. For such mixtures G^E is positive (less favourable to mixing than ideal), and there may be contributions from both enthalpy and entropy effects. Examples include dioxane/water and ethanol/water mixtures.

Deviations from ideality are not always so strong as to lead to a maximum or minimum in the phase diagram, but when they do there are important consequences for distillation. Consider a liquid of composition a on the right of the maximum in Fig. 8.16. The vapour (at a_2') of the boiling mixture (at a_2) is richer in A. If that vapour is removed (and condensed elsewhere), then the remaining liquid will move to a composition that is richer in B, such as that represented by a_3, and the vapour in equilibrium with this mixture will have composition a_3'. As that vapour is removed, the composition of the boiling liquid shifts to a point such as a_4, and the composition of the vapour shifts to a_4'. Hence, as evaporation proceeds, the composition of the remaining liquid shifts towards B as A is drawn off. The boiling point of the liquid rises, and the vapour becomes richer in B. When so much A has been evaporated that the liquid has reached the composition b, the vapour has the same composition as the liquid. Evaporation then occurs without change of composition. The mixture is said to form an **azeotrope**.[3] When the azeotropic composition has been reached, distillation cannot separate the two liquids because the condensate has the same composition as the azeotropic liquid. One example of azeotrope formation is hydrochloric acid/water, which is azeotropic at 80 per cent by mass of water and boils unchanged at 108.6°C.

The system shown in Fig. 8.17 is also azeotropic, but shows its azeotropy in a different way. Suppose we start with a mixture of composition a_1, and follow the changes in the composition of the vapour that rises through a fractionating column (essentially a vertical glass tube packed with glass rings to give a large surface area). The mixture boils at a_2 to give a vapour of composition a_2'. This vapour condenses in the column to a liquid of the same composition (now marked a_3). That liquid reaches equilibrium with its vapour at a_3', which condenses higher up the tube to give a liquid of the same composition, which we now call a_4. The fractionation therefore shifts the vapour towards the azeotropic composition at b, but not beyond, and the azeotropic vapour emerges from the top of the column. An example is ethanol/water, which boils unchanged when the water content is 4 per cent by mass and the temperature is 78°C.

3. The name comes from the Greek words for 'boiling without changing'.

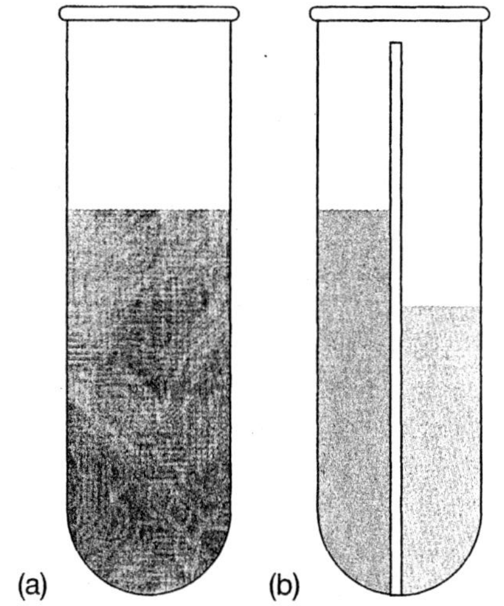

8.18 The distillation of (a) two immiscible liquids can be regarded as (b) the joint distillation of the separated components, and boiling occurs when the sum of the partial pressures equals the external pressure.

(c) Immiscible liquids

Finally, we consider the distillation of two immiscible liquids, such as octane and water. At equilibrium, there is a tiny amount of A dissolved in B, and similarly a tiny amount of B dissolved in A: both liquids are saturated with the other component (Fig. 8.18a). As a result, the total vapour pressure of the mixture is close to $p = p_A^* + p_B^*$. If the temperature is raised to the value at which this total vapour pressure is equal to the atmospheric pressure, boiling commences and the dissolved substances are purged from their solution. However, this boiling results in a vigorous agitation of the mixture, so each component is kept saturated in the other component, and the purging continues as the very dilute solutions are replenished. This intimate contact is essential: two immiscible liquids heated in a container like that shown in Fig. 8.18b would not boil at the same temperature. The presence of the saturated solutions means that the 'mixture' boils at a lower temperature than either component would alone because boiling begins when the total vapour pressure reaches 1 atm, not when either vapour pressure reaches 1 atm. This distinction is the basis of **steam distillation**, which enables some heat-sensitive, water-insoluble organic compounds to be distilled at a lower temperature than their normal boiling point. The only snag is that the composition of the condensate is in proportion to the vapour pressures of the components, so oils of low volatility distil in low abundance.

8.5 Liquid–liquid phase diagrams

Now we consider temperature–composition diagrams for systems that consist of pairs of **partially miscible** liquids, which are liquids that do not mix in all proportions at all temperatures. An example is hexane and nitrobenzene. The same principles of interpretation apply as to liquid–vapour diagrams. When $P = 2$, $F' = 1$ (the prime denoting the adoption of constant pressure), and the selection of a temperature implies that the compositions of the immiscible liquid phases are fixed. When $P = 1$ (corresponding to a system in which the two liquids are fully mixed), both the temperature and the composition may be adjusted.

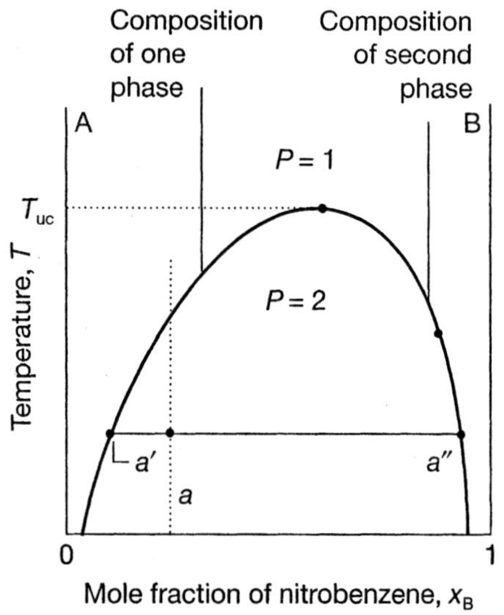

8.19 The temperature–composition diagram for hexane and nitrobenzene at 1 atm. The region below the curve corresponds to the compositions and temperatures at which the liquids are partially miscible. The upper critical temperature, T_{uc}, is the temperature above which the two liquids are miscible in all proportions.

(a) Phase separation

Suppose a small amount of a liquid B is added to a sample of another liquid A at a temperature T'. It dissolves completely, and the binary system remains a single phase. As more B is added, a stage comes at which no more dissolves. The sample now consists of two phases in equilibrium with each other ($P = 2$), the most abundant one consisting of A saturated with B, the minor one a trace of B saturated with A. In the temperature–composition diagram drawn in Fig. 8.19, the composition of the former is represented by the point a' and that of the latter by the point a''. The relative abundances of the two phases are given by the lever rule.

When more B is added, A dissolves in it slightly. The compositions of the two phases in equilibrium remain a' and a'' because $P = 2$ implies that $F' = 0$, and hence that the compositions of the phases are invariant at a fixed temperature and pressure. However, the amount of one phase increases at the expense of the other. A stage is reached when so much B is present that it can dissolve all the A, and the system reverts to a single phase. The addition of more B now simply dilutes the solution, and from then on it remains a single phase.

The composition of the two phases at equilibrium varies with the temperature. For hexane and nitrobenzene, raising the temperature increases their miscibility. The two-phase system therefore becomes less extensive, because each phase in equilibrium is richer in its minor component: the A-rich phase is richer in B and the B-rich phase is richer in A. We can construct the entire phase diagram by repeating

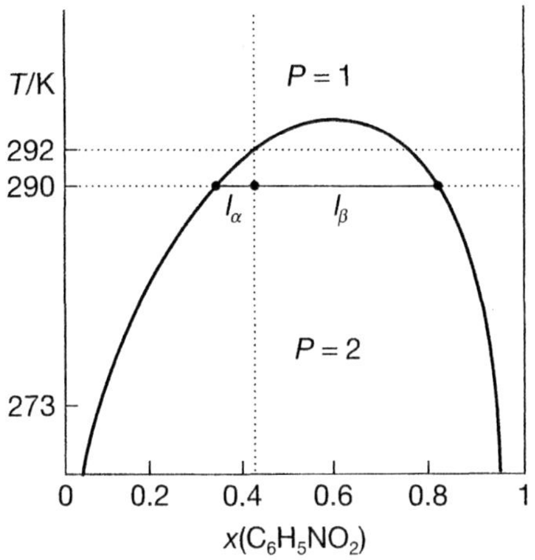

8.20 The temperature–composition diagram for hexane and nitrobenzene at 1 atm again, with the points and lengths discussed in the text.

the observations at different temperatures and drawing the envelope of the two-phase region.

Example 8.2 Interpreting a liquid–liquid phase diagram

A mixture of 50 g of hexane (0.59 mol C_6H_{14}) and 50 g nitrobenzene (0.41 mol $C_6H_5NO_2$) was prepared at 290 K. What are the compositions of the phases, and in what proportions do they occur? To what temperature must the sample be heated in order to obtain a single phase? Refer to Fig. 8.20.

Method The compositions of phases in equilibrium are given by the points where the tie line representing the temperature intersects the phase boundary. Their proportions are given by the lever rule (eqn 7). The temperature at which the components are completely miscible is found by following the isopleth upwards and noting the temperature at which it enters the one-phase region of the phase diagram.

Answer We denote hexane by H and nitrobenzene by N; refer to Fig. 8.20. The point $x_N = 0.41$, $T = 290$ K occurs in the two-phase region of the phase diagram. The horizontal tie line cuts the phase boundary at $x_N = 0.35$ and $x_N = 0.83$, so those are the compositions of the two phases. The ratio of amounts of each phase is equal to the ratio of the distances l_α and l_β:

$$\frac{n_\alpha}{n_\beta} = \frac{l_\beta}{l_\alpha} = \frac{0.83 - 0.41}{0.41 - 0.35} = \frac{0.42}{0.06} = 7$$

That is, there is about 7 times more hexane-rich phase than nitrobenzene-rich phase. Heating the sample to 292 K takes it into the single-phase region.

Comment Because the phase diagram has been constructed experimentally, these conclusions are not based on any assumptions about ideality. They would be modified if the system were subjected to a different pressure.

Self-test 8.2 Repeat the problem for 50 g of hexane and 100 g of nitrobenzene at 273 K. [$x_N = 0.09$ and 0.95 in ratio 1:1.3; 294 K]

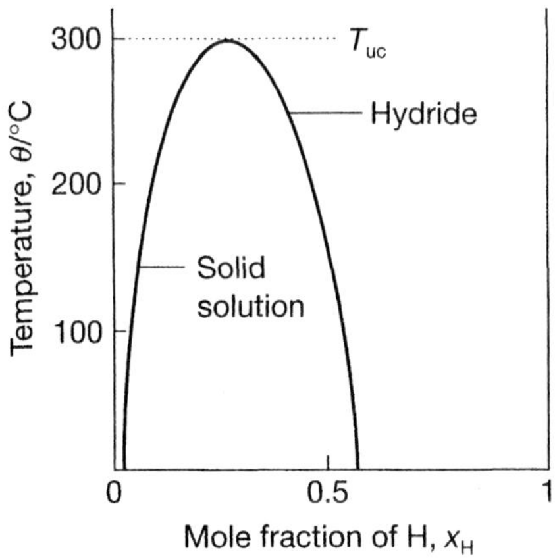

8.21 The phase diagram for palladium and palladium hydride, which has an upper critical temperature at 300°C.

(b) Critical solution temperatures

The **upper critical solution temperature**, T_{uc}, is the highest temperature at which phase separation occurs.[4] Above the upper critical temperature the two components are fully miscible. This temperature exists because the greater thermal motion overcomes any potential energy advantage in molecules of one type being close together. One example is the nitrobenzene/hexane system shown in Fig. 8.20; an example of a solid solution is the palladium/hydrogen system, which shows two phases, one a solid solution of hydrogen in palladium and the other a palladium hydride, up to 300°C but forms a single phase at higher temperatures (Fig. 8.21).

The thermodynamic interpretation of the upper critical solution temperature focuses on the Gibbs energy of mixing and its variation with temperature. We saw in Section 7.4 that a simple model of a real solution results in a Gibbs energy of

4. The upper critical solution temperature is also called the 'upper consolute temperature'.

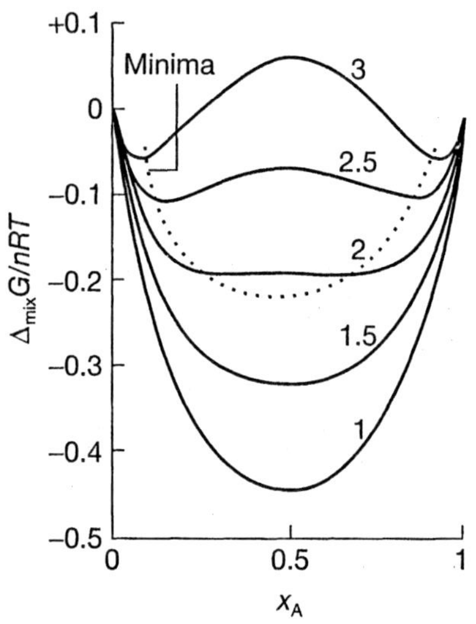

8.22 The temperature variation of the Gibbs energy of mixing of a system that is partially miscible at low temperatures. A system of composition in the region $P = 2$ forms two phases with compositions corresponding to the two local minima of the curve. This illustration is a duplicate of Fig. 7.19.

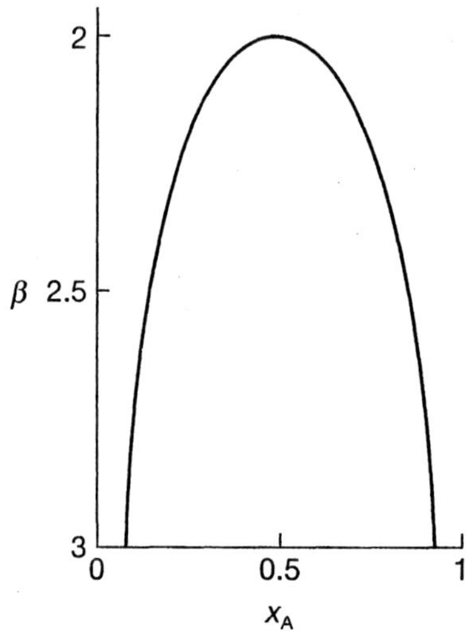

8.23 The location of the phase boundary as computed on the basis of the β-parameter model introduced in Section 7.4.

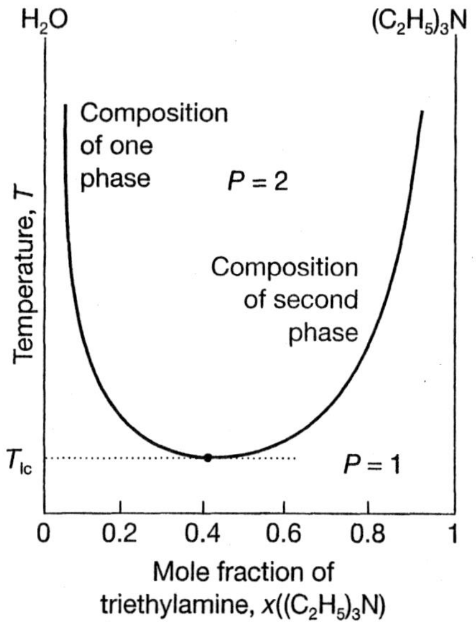

8.24 The temperature-composition diagram for water and triethylamine. This system shows a lower critical temperature at 292 K. The labels indicate the interpretation of the boundaries.

mixing that behaves as shown in Fig. 7.19. Provided the parameter $\beta = w/RT$ is greater than 2, the Gibbs energy of mixing has a double minimum (Fig. 8.22). As a result, for $\beta > 2$ we can expect phase separation to occur. The same model shows that the compositions corresponding to the minima are obtained by looking for the conditions at which $\partial\Delta_{mix}G/\partial x = 0$, and a simple manipulation of eqn 7.31 shows that we have to solve

$$\ln\frac{x}{1-x} + \beta(1-2x) = 0$$

This expression is an example of a *transcendental equation*, an equation that does not have a solution that can be expressed in a closed form. The solutions can be found numerically by using mathematical software or by plotting the first term against the second and identifying the points of intersection as β is changed.

The solutions are plotted in Fig. 8.23. We see that, as β decreases, which can be interpreted as an increase in temperature provided the intermolecular forces and w (the energy AB interactions relative to AA and BB interactions) remain constant, then the two minima move together and merge when $\beta = 2$. The upper critical solution temperature therefore corresponds to $w/2R$.

Some systems show a **lower critical solution temperature**, T_{lc}, below which they mix in all proportions and above which they form two phases.[5] An example is water and triethylamine (Fig. 8.24). In this case, at low temperatures the two components are more miscible because they form a weak complex; at higher temperatures the complexes break up and the two components are less miscible.

Some systems have both upper and lower critical solution temperatures. They occur because, after the weak complexes have been disrupted, leading to partial miscibility, the thermal motion at higher temperatures homogenizes the mixture again, just as in the case of ordinary partially miscible liquids. The most famous

5. The lower critical solution temperature is also called the 'lower consolute temperature'.

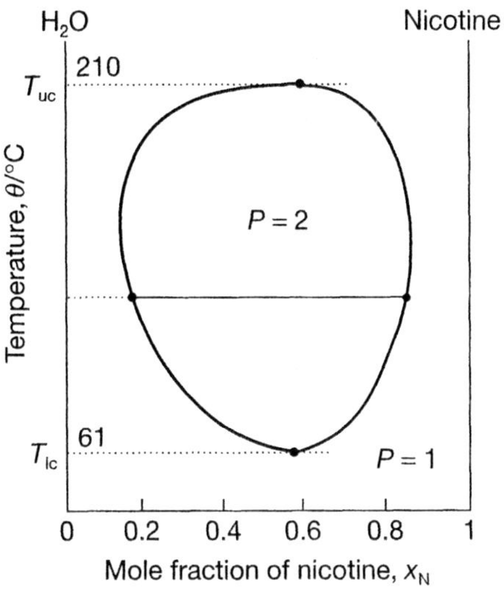

8.25 The temperature-composition diagram for water and nicotine, which has both upper and lower critical temperatures. Note the high temperatures for the liquid (especially the water): the diagram corresponds to a sample under pressure.

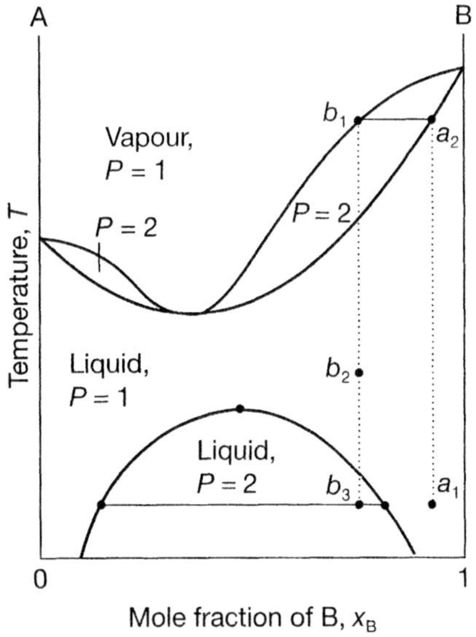

8.26 The temperature-composition diagram for a binary system in which the upper critical temperature is less than the boiling point at all compositions. The mixture forms a low-boiling azeotrope.

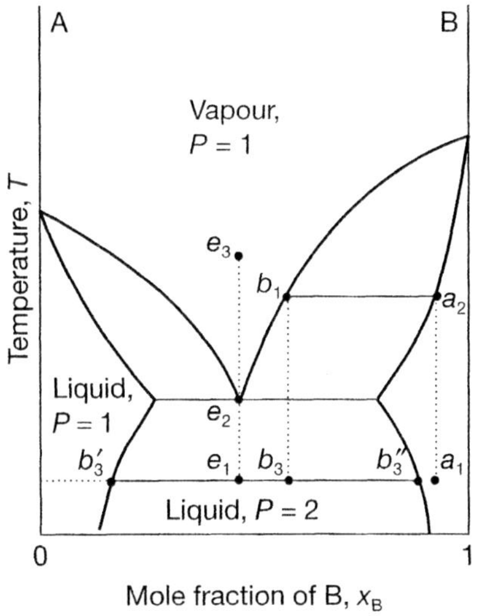

8.27 The temperature-composition diagram for a binary system in which boiling occurs before the two liquids are fully miscible.

example is nicotine and water, which are partially miscible between 6°C and 210°C (Fig. 8.25).

(c) The distillation of partially miscible liquids

Consider a pair of liquids that are partially miscible and form a low-boiling azeotrope. This combination is quite common because both properties reflect the tendency of the two kinds of molecule to avoid each other. There are two possibilities: one in which the liquids become fully miscible before they boil, the other in which boiling occurs before mixing is complete.

Figure 8.26 shows the phase diagram for two components that become fully miscible before they boil. Distillation of a mixture of composition a_1 leads to a vapour of composition b_1, which condenses to the completely miscible single-phase solution at b_2. Phase separation occurs only when this distillate is cooled to a point in the two-phase liquid region, such as b_3. This description applies only to the first drop of distillate. If distillation continues, the composition of the remaining liquid changes. In the end, when the whole sample has evaporated and condensed, the composition is back to a_1.

Figure 8.27 shows the second possibility, in which there is no upper critical solution temperature. The distillate obtained from a liquid initially of composition a_1 has composition b_3 and is a two-phase mixture. One phase has composition b_3' and the other has composition b_3''.

The behaviour of a system of composition represented by the isopleth e in Fig. 8.27 is interesting. A system at e_1 forms two phases, which persist (but with changing proportions) up to the boiling point at e_2. The vapour of this mixture has the same composition as the liquid (the liquid is an azeotrope). Similarly, condensing a vapour of composition e_3 gives a two-phase liquid of the same overall composition. At a fixed temperature, the mixture vaporizes and condenses like a single substance.

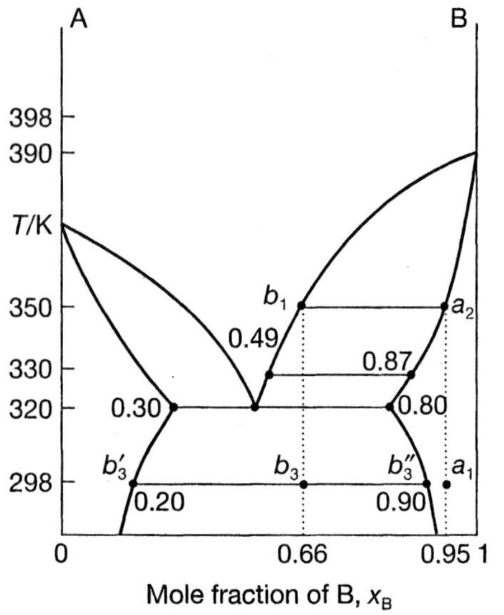

8.28 The points of the phase diagram in Fig. 8.27 that are discussed in Example 8.3.

Example 8.3 Interpreting a phase diagram

State the changes that occur when a mixture of composition $x_B = 0.95$ (a_1) in Fig. 8.28 is boiled and the vapour condensed.

Method The area in which the point lies gives the number of phases; the compositions of the phases are given by the points at the intersections of the horizontal tie line with the phase boundaries; the relative abundances are given by the lever rule (eqn 7).

Answer The initial point is in the one-phase region. When heated it boils at 350 K (a_2) giving a vapour of composition $x_B = 0.66$ (b_1). The liquid gets richer in B, and the last drop (of pure B) evaporates at 390 K. The boiling range of the liquid is therefore 350 to 390 K. If the initial vapour is drawn off, it has a composition $x_B = 0.66$. This composition would be maintained if the sample were very large, but for a finite sample it shifts to higher values and ultimately to $x_B = 0.95$. Cooling the distillate corresponds to moving down the $x_B = 0.66$ isopleth. At 330 K, for instance, the liquid phase has composition $x_B = 0.87$, the vapour $x_B = 0.49$; their relative proportions are 1:3. At 320 K the sample consists of three phases: the vapour and two liquids. One liquid phase has composition $x_B = 0.30$; the other has composition $x_B = 0.80$ in the ratio 0.62:1. Further cooling moves the system into the two-phase region, and at 298 K the compositions are 0.20 and 0.90 in the ratio 0.82:1. As further distillate boils over, the overall composition of the distillate becomes richer in B. When the last drop has been condensed the phase composition is the same as at the beginning.

Self-test 8.3 Repeat the discussion, beginning at the point $x_B = 0.4$, $T = 298$ K.

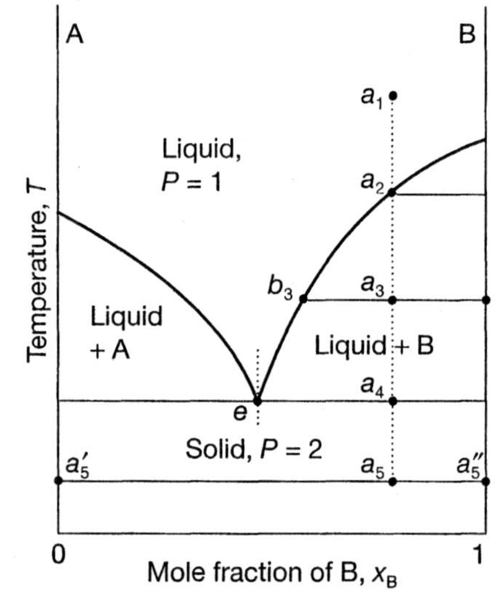

8.29 The temperature–composition phase diagram for two almost immiscible solids and their completely miscible liquids. Note the similarity to Fig. 8.27. The isopleth through e corresponds to the eutectic composition, the mixture with lowest melting point.

8.6 Liquid–solid phase diagrams

Knowledge of the temperature–composition diagrams for solid mixtures guides the design of important industrial processes, such as the manufacture of semiconductors and liquid crystal displays (Boxes 8.1 and 8.2). In this section, we shall consider systems where solid and liquid phases may both be present at temperatures below the boiling point.

Consider the two-component liquid of composition a_1 in Fig. 8.29. The changes that occur may be expressed as follows.

1 $a_1 \rightarrow a_2$. The system enters the two-phase region labelled 'Liquid + B'. Pure solid B begins to come out of solution and the remaining liquid becomes richer in A.

2 $a_2 \rightarrow a_3$. More of the solid forms, and the relative amounts of the solid and liquid (which are in equilibrium) are given by the lever rule. At this stage there are roughly equal amounts of each. The liquid phase is richer in A than before (its composition is given by b_3) because some B has been deposited.

3 $a_3 \rightarrow a_4$. At the end of this step, there is less liquid than at a_3, and its composition is given by e. This liquid now freezes to give a two-phase system of pure B and pure A.

Box 8.1 Liquid crystals

A *mesophase* is a phase intermediate between solid and liquid. Mesophases are of great importance in biology, for they occur as lipid bilayers and in vesicular systems. A mesophase may arise when molecules have highly anisotropic shapes, such as being long and thin, as in (**1**), or disk-like, as in (**2**). When the solid melts, some aspects of the long-range order characteristic of the solid may be retained, and the new phase may be a *liquid crystal*, a substance having liquid-like imperfect long-range order in at least one direction in space but positional or orientational order in at least one other direction. A *thermotropic* liquid crystal displays a transition to the liquid crystalline phase as the temperature is changed. A *lyotropic* liquid crystal is a solution that displays a transition to the liquid crystalline phase as the composition is changed.

One type of retained long-range order gives rise to a *smectic phase* (from the Greek word for soapy), in which the molecules align themselves in layers (see the illustration). Other materials, and some smectic liquid crystals at higher

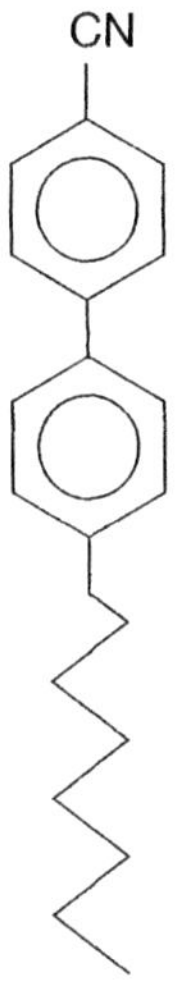

1

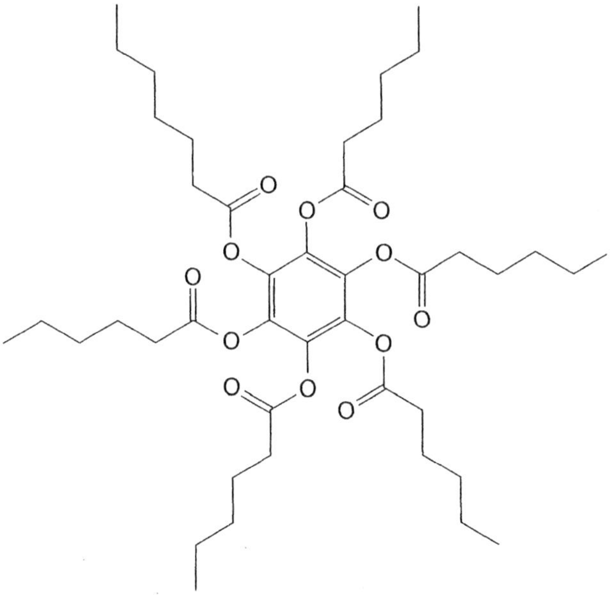

2

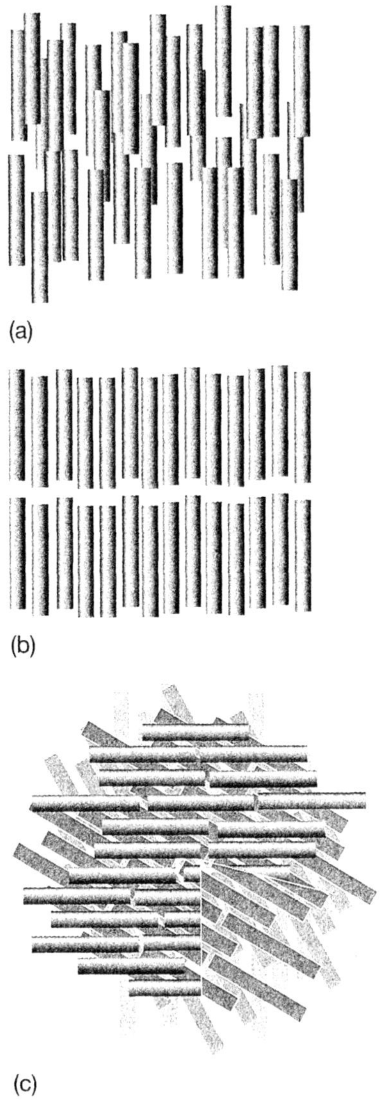

The arrangement of molecules in (a) the nematic phase, (b) the smectic phase, and (c) the cholesteric phase of liquid crystals. In the cholesteric phase, the stacking of layers continues to give a helical arrangement of molecules.

temperatures, lack the layered structure but retain a parallel alignment; this mesophase is called a *nematic phase* (from the Greek for thread, which refers to the observed defect structure of the phase). In the *cholesteric phase* (from the Greek for bile solid) the molecules lie in sheets at angles that change slightly between each sheet. That is, they form helical structures with a pitch that depends on the temperature. As a result, cholesteric liquid crystals diffract light and have colours that depend on the temperature. The strongly anisotropic optical properties of nematic liquid crystals, and their response to electric fields, form the basis of their use as data displays (LCDs).

In a 'twisted nematic' LCD, the liquid crystal is held between two flat plates about 10 μm apart. The inner surface of each plate is coated with a transparent conducting material, such as indium–tin oxide. The plates also have a surface that causes the liquid crystal to adopt a particular orientation at its interface and are typically set at 90° to each other but 270° in a 'supertwist' arrangement. The entire assembly is set between polarizing filters. The incident light passes through the outer polarizer; then its plane of polarization is rotated as it passes through the twisted nematic and, depending on the setting of the second polarizer, will pass through (if that is how the second polarizer is arranged). When a potential difference is applied across the cell, the helical arrangement is lost and the plane of the light is no longer rotated and will be blocked by the second polarizer.

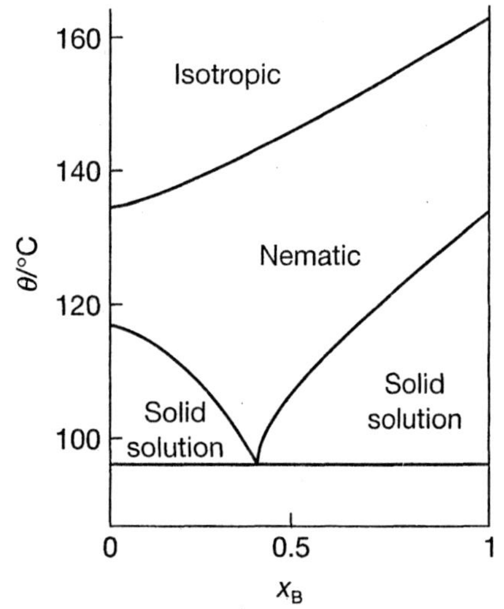

The phase diagram at 1 atm for a binary system of two liquid crystalline materials, 4,4.-dimethoxyazoxybenzene (A) and 4,4.-diethoxyazoxybenzene (B).

Although there are many liquid crystalline materials, some difficulty is often experienced in achieving a technologically useful temperature range for the existence of the mesophase. To overcome this difficulty, mixtures can be used. An example of the type of phase diagram that is then obtained is shown in the illustration. As can be seen, the mesophase exists over a wider range of temperatures than either liquid crystalline material alone.

Discussion questions

1. Disk-like molecules, such as (**2**), can form *discotic* liquid crystal mesophases that resemble a column, with the aromatic rings stacked one on top of the other and separated by very small distances (less than 0.5 nm). It has been proposed that such liquid crystals can act as 'molecular wires' for the conduction of electrons along the column. Explain the rationale for this proposal.

2. Some polymers can form liquid crystal mesophases with unusual physical properties. For example, liquid crystalline Kevlar (**3**) is strong enough to be the material of choice for bulletproof vests and is stable at temperatures up to 600 K. What molecular interactions contribute to the formation, thermal stability, and mechanical strength of liquid crystal mesophases in Kevlar?

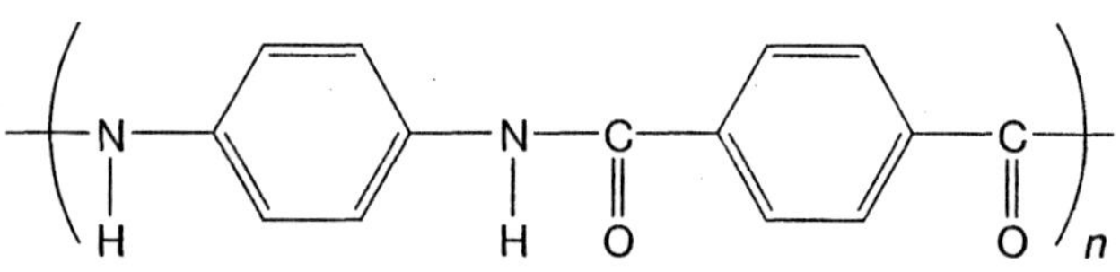

3

Further reading

P.J. Collings and M. Hird, *Introduction to liquid crystals: chemistry and physics*. Taylor & Francis, London (1997).

J.D. Lister and R. Shashidhar, Structure of liquid crystals. In *Encyclopedia of applied physics* (ed. G.L. Trigg), **8**, 515. VCH, New York (1994).

Box 8.2 Ultrapurity and controlled impurity

Advances in technology have called for materials of extreme purity. For example, semiconductor devices consist of almost perfectly pure silicon or germanium doped to a precisely controlled extent. For these materials to operate successfully, the impurity level must be kept down to less than 1 ppb (1 part in 10^9, which corresponds to a small grain of salt in 5 t of sugar).[6]

In the technique of *zone refining* the sample is in the form of a narrow cylinder. This cylinder is heated in a thin disk-like zone which is swept from one end of the sample to the other. The advancing liquid zone accumulates the impurities as it passes. In practice, a train of hot and cold zones are swept repeatedly from one end to the other as shown in the first illustration. The zone at the end of the sample is the impurity dump: when the heater has gone by, it cools to a dirty solid which can be discarded.

The technique makes use of the non-equilibrium properties of the system. It relies on the impurities being more soluble in the molten sample than in the solid, and sweeps them up by passing a molten zone repeatedly from one end to the other along a sample. The phase diagram in the second illustration gives some insight into the process. Consider a liquid (this represents the molten zone) on the isopleth through a_1, and let it cool without the entire sample coming to overall equilibrium. If the temperature falls to a_2 a solid of composition b_2 is deposited and the remaining liquid (the zone where the heater has moved on) is at a_2'. Cooling that liquid down an isopleth passing through a_2' deposits solid of composition b_3 and leaves liquid at a_3'. The process continues until the last drop of liquid to solidify is heavily contaminated with B. There is plenty of everyday evidence that impure liquids freeze in this way. For example, an ice cube is clear near the surface but misty in the core: the water used to make ice normally contains dissolved air; freezing proceeds from the outside, and air is accumulated in the retreating liquid phase. It cannot escape from the interior of the cube, and so when that freezes it occludes the air in a mist of tiny bubbles.

A modification of zone refining is *zone levelling*. It is used to introduce controlled amounts of impurity (for example, of indium into germanium). A sample rich in the required dopant is put at the head of the main sample, and made molten. The zone is then dragged repeatedly in alternate directions through the sample, where it deposits a uniform distribution of the impurity.

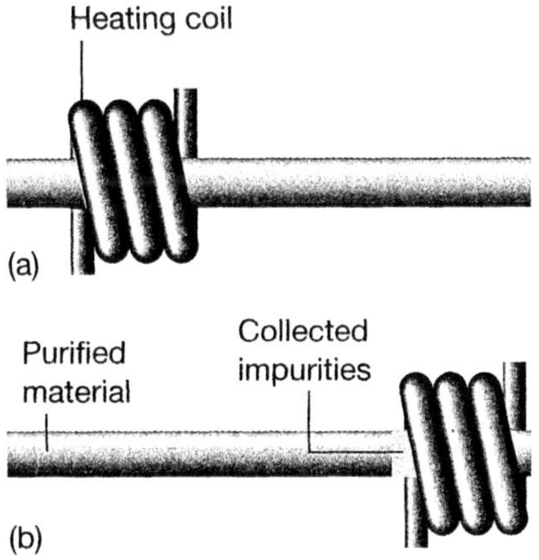

The procedure for zone refining. (a) Initially, impurities are distributed uniformly along the sample. (b) After a molten zone is passed along the rod, the impurities are more concentrated at the right. In practice, a series of molten zones are passed along the rod from left to right.

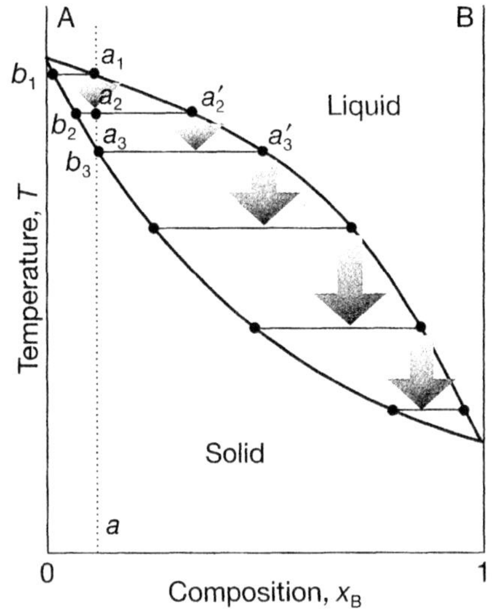

A binary temperature–composition diagram can be used to discuss zone refining, as explained in the text.

6. 1 t = 10^3 kg.

Discussion questions

1. Use a phase diagram like that in the second illustration to indicate how zone levelling may be described.

2. The technique of *float zoning*, which is similar to zone refining, has produced very pure samples of silicon for use in the semiconductor industry. Consult a textbook of materials science or metallurgy and prepare a discussion of the principles, advantages, and disadvantages of float zoning.

Further reading

H.-G. Lee, *Chemical thermodynamics for metals and materials*. Imperial College Press, London (1999).

D.R. Gaskell, *Introduction to the thermodynamics of materials*. Taylor & Francis, London (1995).

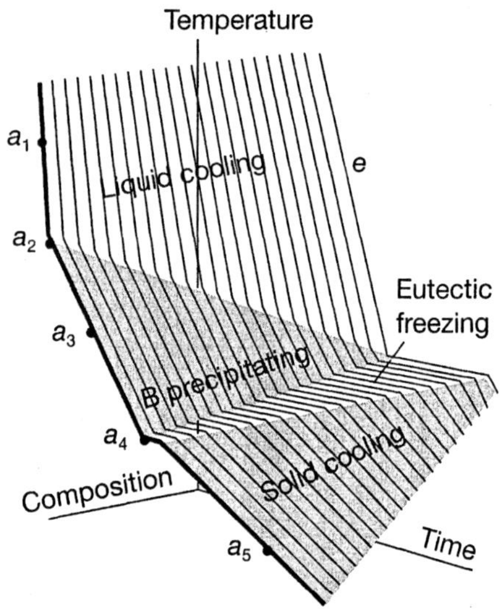

8.30 The cooling curves for the system shown in Fig. 8.29. For isopleth a, the rate of cooling slows at a_2 because solid B deposits from solution. There is a complete halt at a_4 while the eutectic solidifies. This halt is longest for the eutectic isopleth, e. The eutectic halt shortens again for compositions beyond e (richer in A). Cooling curves are used to construct the phase diagram.

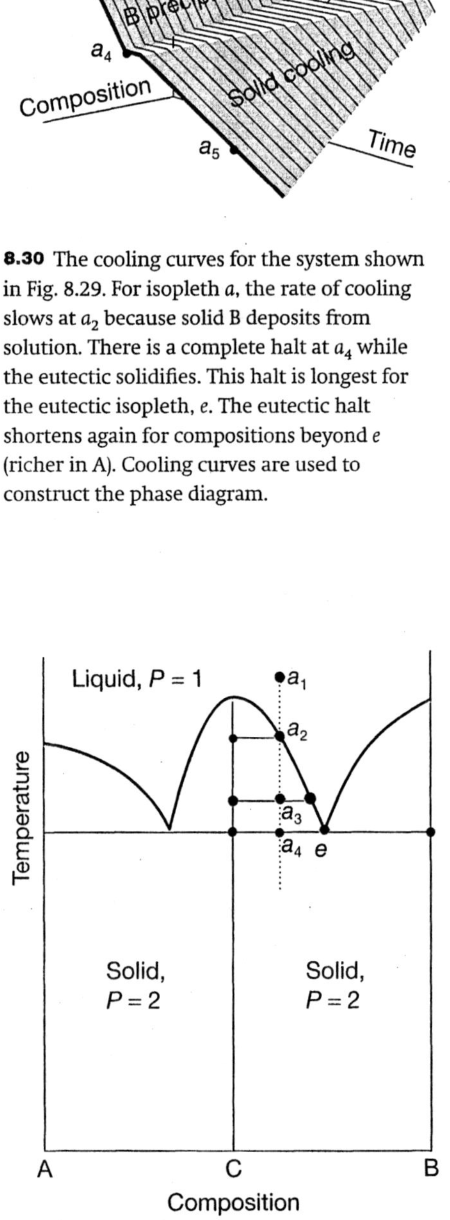

8.31 The phase diagram for a system in which A and B react to form a compound C = AB. This resembles two versions of Fig. 8.29 in each half of the diagram. The constituent C is a true compound, not just an equimolar mixture.

(a) Eutectics

The isopleth at e in Fig. 8.29 corresponds to the **eutectic** composition, the mixture with the lowest melting point.[7] A liquid with the eutectic composition freezes at a single temperature, without previously depositing solid A or B. A solid with the eutectic composition melts, without change of composition, at the lowest temperature of any mixture. Solutions of composition to the right of e deposit B as they cool, and solutions to the left deposit A: only the eutectic mixture (apart from pure A or pure B) solidifies at a single definite temperature ($F' = 0$ when $C = 2$ and $P = 3$) without gradually unloading one or other of the components from the liquid.

One technologically important eutectic is solder, which has mass composition of about 67 per cent tin and 33 per cent lead and melts at 183°C. The eutectic formed by 23 per cent NaCl and 77 per cent H_2O by mass melts at −21.1°C. When salt is added to ice under isothermal conditions (for example, when spread on an icy road) the mixture melts if the temperature is above −21.1°C (and the eutectic composition has been achieved). When salt is added to ice under adiabatic conditions (for example, when added to ice in a vacuum flask) the ice melts, but in doing so it absorbs heat from the rest of the mixture. The temperature of the system falls and, if enough salt is added, cooling continues down to the eutectic temperature. Eutectic formation occurs in the great majority of binary alloy systems, and is of great importance for the microstructure of solid materials. Although a eutectic solid is a two-phase system, it crystallizes out in a nearly homogeneous mixture of microcrystals. The two microcrystalline phases can be distinguished by microscopy and structural techniques such as X-ray diffraction.

Thermal analysis is a very useful practical way of detecting eutectics. We can see how it is used by considering the rate of cooling down the isopleth through a_1 in Fig. 8.29. The liquid cools steadily until it reaches a_2, when B begins to be deposited (Fig. 8.30). Cooling is now slower because the solidification of B is exothermic and retards the cooling. When the remaining liquid reaches the eutectic composition, the temperature remains constant ($F' = 0$) until the whole sample has solidified: this region of constant temperature is the eutectic halt. If the liquid has the eutectic composition e initially, the liquid cools steadily down to the freezing temperature of the eutectic, when there is a long **eutectic halt** as the entire sample solidifies (like the freezing of a pure liquid).

Monitoring the cooling curves at different overall compositions gives a clear indication of the structure of the phase diagram. The solid–liquid boundary is given by the points at which the rate of cooling changes. The longest eutectic halt gives the location of the eutectic composition and its melting temperature.

(b) Reacting systems

Many binary mixtures react to produce compounds, and technologically important examples of this behaviour include the III/V semiconductors, such as the gallium arsenide system, which forms the compound GaAs. Although three constituents are present, there are only two components because GaAs is formed from the reaction Ga + As ⇌ GaAs. We shall illustrate some of the principles involved with a system that forms a compound C that also forms eutectic mixtures with the species A and B (Fig. 8.31).

A system prepared by mixing an excess of B with A consists of C and unreacted B. This is a binary B, C system, which we suppose forms a eutectic. The principal change from the eutectic phase diagram in Fig. 8.29 is that the whole of the phase diagram is squeezed into the range of compositions lying between equal amounts of A and B ($x_B = 0.5$, marked C in Fig. 8.31) and pure B. The interpretation of the

7. The name comes from the Greek words for 'easily melted'.

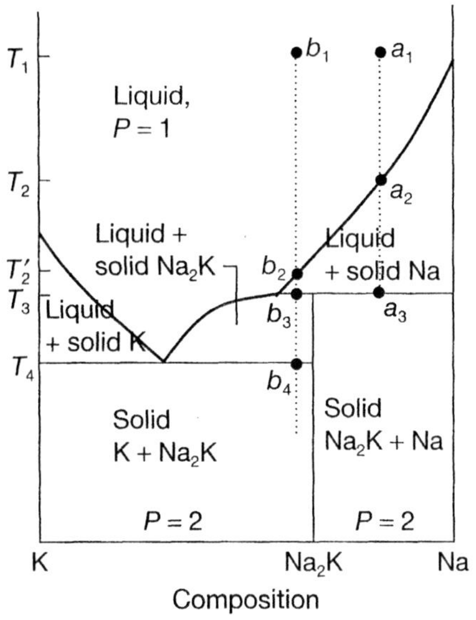

8.32 The phase diagram for an actual system (sodium and potassium) like that shown in Fig. 8.31, but with two differences. One is that the compound is Na_2K, corresponding to A_2B and not AB as in that illustration. The second is that the compound exists only as the solid, not as the liquid. The transformation of the compound at its melting point is an example of incongruent melting.

information in the diagram is obtained in the same way as for Fig. 8.29. The solid deposited on cooling along the isopleth a is the compound C. At temperatures below a_4 there are two solid phases, one consisting of C and the other of B.

(c) Incongruent melting

In some cases the compound C is not stable as a liquid. An example is the alloy Na_2K, which survives only as a solid (Fig. 8.32). Consider what happens as a liquid at a_1 is cooled:

1 $a_1 \rightarrow a_2$. Some solid Na is deposited, and the remaining liquid is richer in K.

2 $a_2 \rightarrow$ just below a_3. The sample is now entirely solid, and consists of solid Na and solid Na_2K.

Now consider the isopleth through b_1:

1 $b_1 \rightarrow b_2$. No obvious change occurs until the phase boundary is reached at b_2 when solid Na begins to deposit.

2 $b_2 \rightarrow b_3$. Solid Na deposits, but at b_3 a reaction occurs to form Na_2K: this compound is formed by the K atoms diffusing into the solid Na.

At this stage the liquid Na/K mixture is in equilibrium with a little solid Na_2K, but there is still no liquid compound.

3 $b_3 \rightarrow b_4$. As cooling continues, the amount of solid compound increases until at b_4 the liquid reaches its eutectic composition. It then solidifies to give a two-phase solid consisting of solid K and solid Na_2K.

If the solid is reheated, the sequence of events is reversed. No liquid Na_2K forms at any stage because it is too unstable to exist as a liquid. This behaviour is an example of **incongruent melting**, in which a compound melts into its components and does not itself form a liquid phase.

Checklist of key ideas

Phases, components, and degrees of freedom

8.1 Definitions

☐ phase, a state of matter that is uniform throughout, not only in chemical composition but also in physical state

☐ constituent, a chemical species (an ion or a molecule)

☐ component, a chemically independent constituent of a system

☐ variance, the number of intensive variables that can be changed independently without disturbing the number of phases in equilibrium

☐ degree of freedom, another name for variance

☐ bivariant system, a system with two degrees of freedom

8.2 The phase rule

☐ phase rule, $F = C - P + 2$

☐ application of the phase rule to one-component systems

☐ thermal analysis, a technique for detecting phase transitions that takes advantage of the effect of the enthalpy change during a first-order transition

☐ differential scanning calorimetry, a technique used for thermal analysis

☐ diamond anvil cell, a device used for high-pressure studies

Two-component systems

8.3 Vapour pressure diagrams

☐ vapour pressure of a mixture, $p = p_B^* + (p_A^* - p_B^*)x_A$

☐ composition of the vapour, $y_A = x_A p_A^*/\{p_B^* + (p_A^* - p_B^*)x_A\}$, $y_B = 1 - y_A$

☐ total vapour pressure of a mixture, $p = p_A^* p_B^*/\{p_A^* + (p_B^* - p_A^*)y_A\}$

☐ overall composition and the interpretation of phase diagrams

☐ isopleth, a line of constant composition

☐ tie line, a line joining two points representing phases in equilibrium

☐ lever rule, $n_\alpha l_\alpha = n_\beta l_\beta$

8.4 Temperature–composition diagrams

☐ temperature–composition diagram, a phase diagram in which the boundaries show the composition of the phases that are in equilibrium at various temperatures

☐ simple distillation, a process in which vapour is withdrawn and condensed

☐ fractional distillation, a process in which the boiling and condensation cycle is repeated successively

☐ theoretical plate, an effective vaporization and condensation step that is required to achieve a condensate of given composition from a given distillate

☐ azeotrope, a mixture that boils without change of composition

☐ the behaviour of immiscible liquids

8.5 Liquid–liquid phase diagrams

☐ partially miscible liquids, liquids that do not mix in all proportions at all temperatures

☐ phase separation and the interpretation of a liquid–liquid phase diagram

☐ upper critical solution temperature, the highest temperature at which phase separation occurs in a binary liquid mixture

☐ lower critical solution temperature, the temperature below which the components of a binary mixture mix in all proportions and above which they form two phases

☐ the interpretation of distillation of partially miscible liquids in terms of phase diagrams

8.6 Liquid–solid phase diagrams

☐ the interpretation of liquid–solid phase diagrams

☐ eutectic, the mixture with the lowest melting point; a liquid with the eutectic composition freezes at a single temperature

☐ eutectic halt, a delay in cooling while the eutectic freezes

☐ the effect of reactions on the appearance of phase diagrams

☐ incongruent melting, when a compound melts into its components and does not itself form a liquid phase

Further reading

Articles of general interest

J.S. Alper, The Gibbs phase rule revisited: interrelationships between components and phases. *J. Chem. Educ.* **76**, 1567 (1999).

R. Battino, The critical point and the number of degrees of freedom. *J. Chem. Educ.* **68**, 276 (1991).

W.B. Daniels, High-pressure techniques. In *Encyclopedia of applied physics* (ed. G.L. Trigg), **7**, 495. VCH, New York (1995).

A. Jayaraman, The diamond-anvil high-pressure cell. *Scientific American* **250** (4), 42 (1984).

N.K. Kildahl, Journey around a phase diagram. *J. Chem. Educ.* **71**, 1052 (1994).

J. Kochansky, Liquid systems with more than two immiscible phases. *J. Chem. Educ.* **68**, 655 (1991).

A.D. Pelton, Phase equilibria. In *Encyclopedia of applied physics* (ed. G.L. Trigg), **13**, 297. VCH, New York (1995).

R.M. Rosenberg, Description of regions in two-component phase diagrams. *J. Chem. Educ.* **76**, 2223 (1999).

R.J. Stead and K. Stead, Phase diagrams for ternary liquid systems. *J. Chem. Educ.* **67**, 385 (1990).

M. Zhao, Z. Wang, and L. Xiao, Determining the number of independent components by Brinkley's method. *J. Chem. Educ.* **69**, 539 (1992).

J. Zhou, W.S.W. Ho, and N.N. Li, Separation processes. In *Encyclopedia of applied physics* (ed. G.L. Trigg), **17**, 599. VCH, New York (1996).

Texts and sources of data and information

A. Alper, *Phase diagrams*, Vols. 1, 2, and 3. Academic Press, New York (1970).

W.D. Callister, Jr., *Materials science and engineering, an introduction*. Wiley, New York (2000).

M. Hillert, *Phase equilibria, phase diagrams and phase transformations: a thermodynamic basis*. Cambridge University Press (1998).

A. Reisman, *Phase equilibria*. Academic Press, New York (1970).

S.I. Sandler (ed.), *Models for thermodynamic and phase equilibria*. Marcel Dekker, New York (1994).

S.I. Sandler, *Chemical and engineering thermodynamics*. Wiley, New York (1998).

M.A. White, *Properties of materials*. Oxford University Press, Oxford (1999).

J. Wisniak, *Phase diagrams: A literature source book*. Elsevier, Amsterdam (1981–86).

Exercises

Discussion questions

8.1 (a) Define the following terms: phase, constituent, component, and degree of freedom.

8.1 (b) What factors determine the number of theoretical plates required to achieve a desired degree of separation in fractional distillation?

8.2 (a) Draw phase diagrams for the following types of systems. Label the regions and intersections of the diagrams, stating what materials (possibly compounds or azeotropes) are present and whether they are solid, liquid, or gas. (a) One-component, pressure–temperature diagram; liquid density greater than that of solid. (b) Two-component, temperature–composition, solid–liquid diagram; one compound AB formed that melts congruently, negligible solid–solid solubility.

8.2 (b) Draw phase diagrams for the following types of systems. Label the regions and intersections of the diagrams, stating what materials (possibly compounds or azeotropes) are present and whether they are solid, liquid, or gas. (a) Two-component, temperature-composition, solid-liquid diagram; one compound of formula AB_2 that melts incongruently; negligible solid-solid solubility. (b) Two-component, constant temperature-composition, liquid-vapour diagram; formation of an azeotrope at $x_B = 0.333$, complete miscibility.

8.3 (a) Label the regions of the phase diagram in Fig. 8.33. State what substances (if compounds, give their formulas) exist in each region. Label each substance in each region as solid, liquid, or gas.

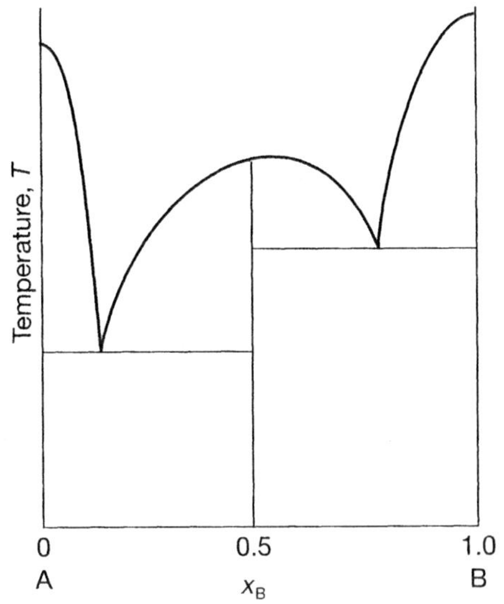

Fig. 8.33

8.3 (b) Label the regions of the phase diagram in Fig. 8.34. State what substances (if compounds, give their formulas) exist in each region. Label each substance in each region as solid, liquid, or gas.

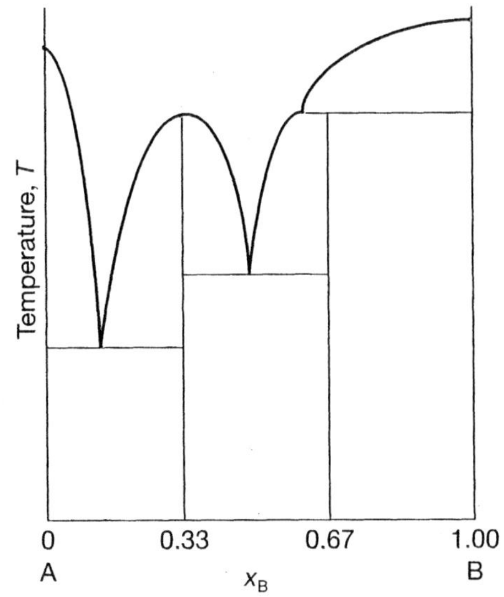

Fig. 8.34

Numerical exercises

8.4 (a) At 90°C, the vapour pressure of methylbenzene is 400 Torr and that of 1,2-dimethylbenzene is 150 Torr. What is the composition of a liquid mixture that boils at 90°C when the pressure is 0.50 atm? What is the composition of the vapour produced?

8.4 (b) At 90°C, the vapour pressure of 1,2-dimethylbenzene is 20 kPa and that of 1,3-dimethylbenzene is 18 kPa. What is the composition of a liquid mixture that boils at 90°C when the pressure is 19 kPa? What is the composition of the vapour produced?

8.5 (a) The vapour pressure of pure liquid A at 300 K is 575 Torr and that of pure liquid B is 390 Torr. These two compounds form ideal liquid and gaseous mixtures. Consider the equilibrium composition of a mixture in which the mole fraction of A in the vapour is 0.350. Calculate the total pressure of the vapour and the composition of the liquid mixture.

8.5 (b) The vapour pressure of pure liquid A at 293 K is 68.8 kPa and that of pure liquid B is 82.1 kPa. These two compounds form ideal liquid and gaseous mixtures. Consider the equilibrium composition of a mixture in which the mole fraction of A in the vapour is 0.612. Calculate the total pressure of the vapour and the composition of the liquid mixture.

8.6 (a) It is found that the boiling point of a binary solution of A and B with $x_A = 0.6589$ is 88°C. At this temperature the vapour pressures of pure A and B are 957.0 Torr and 379.5 Torr, respectively. (a) Is this solution ideal? (b) What is the initial composition of the vapour above the solution?

8.6 (b) It is found that the boiling point of a binary solution of A and B with $x_A = 0.4217$ is 96°C. At this temperature the vapour pressures of pure A and B are 110.1 kPa and 94.93 kPa, respectively. (a) Is this solution ideal? (b) What is the initial composition of the vapour above the solution?

8.7 (a) Dibromoethene (DE, $p^*_{DE} = 172$ Torr at 358 K) and dibromopropene (DP, $p^*_{DP} = 128$ Torr at 358 K) form a nearly ideal solution. If $z_{DE} = 0.60$, what is (a) p_{total} when the system is all liquid, (b) the composition of the vapour when the system is still almost all liquid?

8.7 (b) Benzene and toluene form nearly ideal solutions. At 20°C the vapour pressures of pure benzene and toluene are 74 Torr and 22 Torr, respectively. The solution is boiled by reducing the external pressure below the vapour pressure. Calculate (a) the pressure when boiling begins, (b) the composition of each component in the vapour, and (c) the vapour pressure when only a few drops of liquid remain. Assume that the rate of vaporization is low enough for the temperature to remain constant at 20°C.

8.8 (a) The following temperature/composition data were obtained for a mixture of octane (O) and methylbenzene (M) at 760 Torr, where x is the mole fraction in the liquid and y the mole fraction in the vapour at equilibrium.

θ/°C	110.9	112.0	114.0	115.8	117.3	119.0	121.1	123.0
x_M	0.908	0.795	0.615	0.527	0.408	0.300	0.203	0.097
y_M	0.923	0.836	0.698	0.624	0.527	0.410	0.297	0.164

The boiling points are 110.6°C and 125.6°C for M and O, respectively. Plot the temperature/composition diagram for the mixture. What is the composition of the vapour in equilibrium with the liquid of composition (a) $x_M = 0.250$ and (b) $x_O = 0.250$?

8.8 (b) The following temperature/composition data were obtained for a mixture of two liquids A and B at 1.00 atm, where x is the mole fraction in the liquid and y the mole fraction in the vapour at equilibrium.

θ/°C	125	130	135	140	145	150
x_A	0.91	0.65	0.45	0.30	0.18	0.098
y_A	0.99	0.91	0.77	0.61	0.45	0.25

The boiling points are 124°C for A and 155°C for B. Plot the temperature/composition diagram for the mixture. What is the composition of the vapour in equilibrium with the liquid of composition (a) $x_A = 0.50$ and (b) $x_B = 0.33$?

8.9 (a) State the number of components in the following systems. (a) NaH_2PO_4 in water at equilibrium with water vapour but disregarding the fact that the salt is ionized. (b) The same, but taking into account the ionization of the salt.

8.9 (b) State the number of components for a system in which $AlCl_3$ is dissolved in water, noting that hydrolysis and precipitation of $Al(OH)_3$ occur.

8.10 (a) Blue $CuSO_4 \cdot 5H_2O$ crystals release their water of hydration when heated. How many phases and components are present in an otherwise empty heated container?

8.10 (b) Ammonium chloride, NH_4Cl, decomposes when it is heated. (a) How many components and phases are present when the salt is heated in an otherwise empty container? (b) Now suppose that additional ammonia is also present. How many components and phases are present?

8.11 (a) A saturated solution of Na_2SO_4, with excess of the solid, is present at equilibrium with its vapour in a closed vessel. (a) How many phases and components are present? (b) What is the variance (the number of degrees of freedom) of the system? Identify the independent variables.

8.11 (b) Suppose that the solution referred to in Exercise 8.11a is not saturated. (a) How many phases and components are present? (b) What is the variance (the number of degrees of freedom) of the system? Identify the independent variables.

8.12 (a) Methylethyl ether (A) and diborane, B_2H_6 (B), form a compound that melts congruently at 133 K. The system exhibits two eutectics, one at 25 mole per cent B and 123 K and a second at 90 mole per cent B and 104 K. The melting points of pure A and B are 131 K and 110 K, respectively. Sketch the phase diagram for this system. Assume negligible solid–solid solubility.

8.12 (b) Sketch the phase diagram of the system NH_3/N_2H_4 given that the two substances do not form a compound with each other, that NH_3 freezes at −78°C and N_2H_4 freezes at +2°C, and that a eutectic is formed when the mole fraction of N_2H_4 is 0.07 and that the eutectic melts at −80°C.

8.13 (a) Figure 8.35 shows the phase diagram for two partially miscible liquids, which can be taken to be that for water (A) and 2-methyl-1-propanol (B). Describe what will be observed when a mixture of composition $x_B = 0.8$ is heated, at each stage giving the number, composition, and relative amounts of the phases present.

8.13 (b) Figure 8.36 is the phase diagram for silver and tin. Label the regions, and describe what will be observed when liquids of compositions a and b are cooled to 200 K.

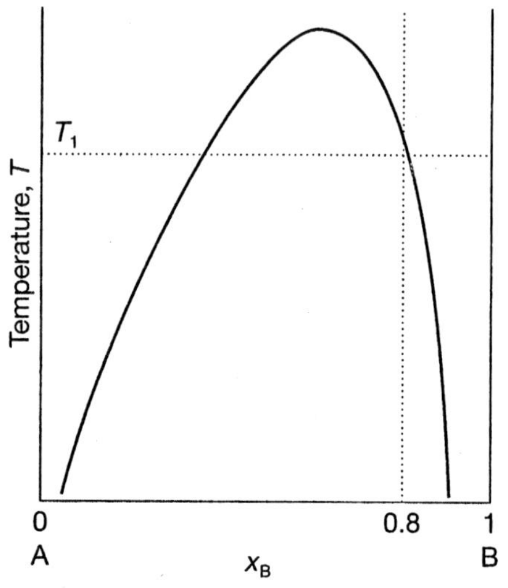

Fig. 8.35

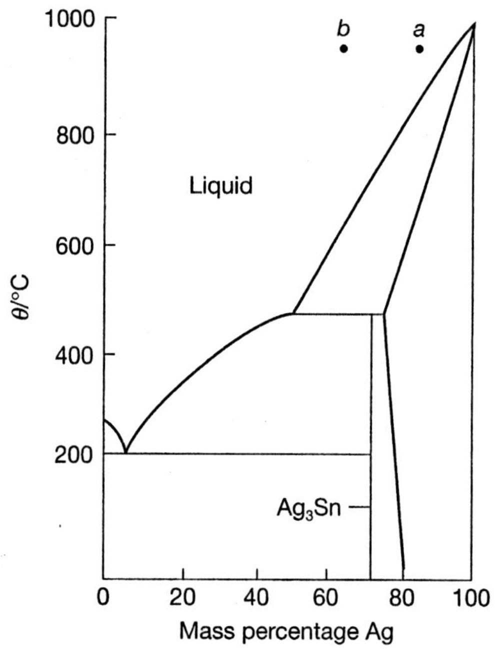

Fig. 8.36

8.14 (a) Indicate on the phase diagram in Fig. 8.37 the feature that denotes incongruent melting. What is the composition of the eutectic mixture and at what temperature does it melt?

8.14 (b) Indicate on the phase diagram in Fig. 8.38 the feature that denotes incongruent melting. What is the composition of the eutectic mixture and at what temperature does it melt?

8.15 (a) Sketch the cooling curves for the isopleths a and b in Fig. 8.37.

8.15 (b) Sketch the cooling curves for the isopleths a and b in Fig. 8.38.

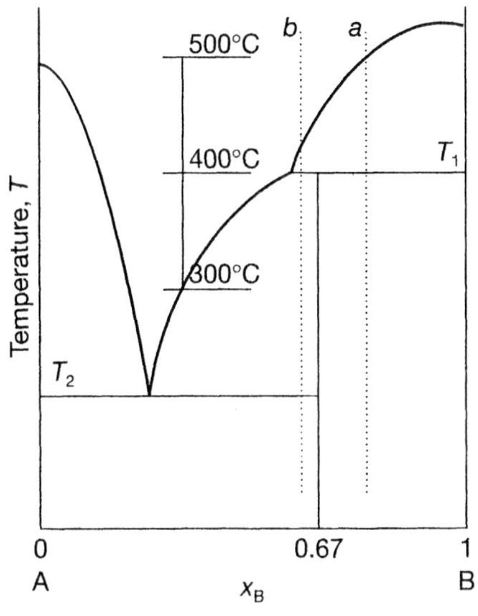

Fig. 8.37

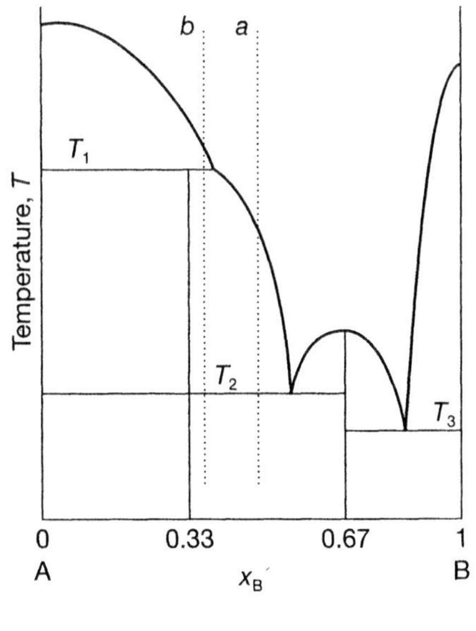

Fig. 8.38

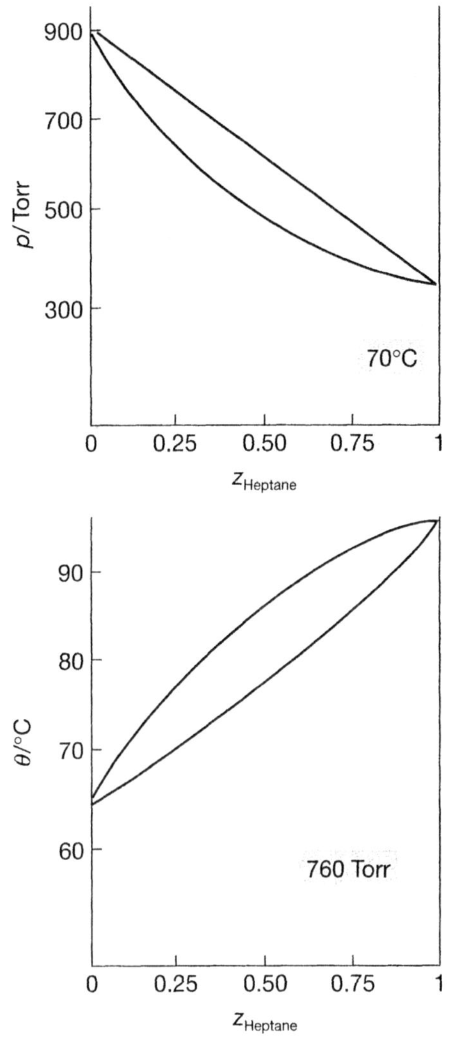

Fig. 8.39

8.16 (a) Use the phase diagram in Fig. 8.36 to state (a) the solubility of Ag in Sn at 800°C, (b) the solubility of Ag_3Sn in Ag at 460°C, (c) the solubility of Ag_3Sn in Ag at 300°C.

8.16 (b) Use the phase diagram in Fig. 8.37 to state (a) the solubility of B in A at 500°C, (b) the solubility of AB_2 in A at 390°C, (c) the solubility of AB_2 in B at 300°C.

8.17 (a) Figure 8.39 shows the experimentally determined phase diagrams for the nearly ideal solution of hexane and heptane. (a) Label the regions of the diagrams as to which phases are present. (b) For a solution containing 1 mol each of hexane and heptane, estimate the vapour pressure at 70°C when vaporization on reduction of the external pressure just begins. (c) What is the vapour pressure of the solution at 70°C when just one drop of liquid remains? (d) Estimate from the figures the mole fraction of hexane in the liquid and vapour phases for the conditions of part b. (e) What are the mole fractions for the conditions of part c? (f) At 85°C and 760 Torr, what are the amounts of substance in the liquid and vapour phases when $z_{heptane} = 0.40$?

8.17 (b) Uranium tetrafluoride and zirconium tetrafluoride melt at 1035°C and 912°C, respectively. They form a continuous series of solid solutions with a minimum melting temperature of 765°C and composition $x(ZrF_4) = 0.77$. At 900°C, the liquid solution of composition $x(ZrF_4) = 0.28$ is in equilibrium with a solid solution of composition $x(ZrF_4) = 0.14$. At 850°C the two compositions are 0.87 and 0.90, respectively. Sketch the phase diagram for this system and state what is observed when a liquid of composition $x(ZrF_4) = 0.40$ is cooled slowly from 900°C to 500°C.

8.18 (a) Methane (melting point 91 K) and tetrafluoromethane (melting point 89 K) do not form solid solutions with each other, and as liquids they are only partially miscible. The upper critical temperature of the liquid mixture is 94 K at $x(CF_4) = 0.43$ and the eutectic temperature is 84 K at $x(CF_4) = 0.88$. At 86 K, the phase in equilibrium with the tetrafluoromethane-rich solution changes from solid methane to a methane-rich liquid. At that temperature, the two liquid solutions that are in mutual equilibrium have the compositions $x(CF_4) = 0.10$ and $x(CF_4) = 0.80$. Sketch the phase diagram.

8.18 (b) Describe the phase changes that take place when a liquid mixture of 4.0 mol B_2H_6 (melting point 131 K) and 1.0 mol CH_3OCH_3 (melting point 135 K) is cooled from 140 K to 90 K. These substances form a compound $(CH_3)_2OB_2H_6$ that melts congruently at 133 K. The system exhibits one eutectic at $x(B_2H_6) = 0.25$ and 123 K and another at $x(B_2H_6) = 0.90$ and 104 K.

8.19 (a) Refer to the information in Exercise 8.18b and sketch the cooling curves for liquid mixtures in which $x(B_2H_6)$ is (a) 0.10, (b) 0.30, (c) 0.50, (d) 0.80, and (e) 0.95.

8.19 (b) Refer to the information in Exercise 8.18a and sketch the cooling curves for liquid mixtures in which $x(CF_4)$ is (a) 0.10, (b) 0.30, (c) 0.50, (d) 0.80, and (e) 0.95.

8.20 (a) Hexane and perfluorohexane show partial miscibility below 22.70°C. The critical concentration at the upper critical temperature is $x = 0.355$, where x is the mole fraction of C_6F_{14}. At 22.0°C the two solutions in equilibrium have $x = 0.24$ and $x = 0.48$, respectively, and at 21.5°C the mole fractions are 0.22 and 0.51. Sketch the phase diagram. Describe the phase changes that occur when perfluorohexane is added to a fixed amount of hexane at (a) 23°C, (b) 22°C.

8.20 (b) Two liquids, A and B, show partial miscibility below 52.4°C. The critical concentration at the upper critical temperature is $x = 0.459$, where x is the mole fraction of A. At 40.0°C the two solutions in equilibrium have $x = 0.22$ and $x = 0.60$, respectively, and at 42.5°C the mole fractions are 0.24 and 0.48. Sketch the phase diagram. Describe the phase changes that occur when B is added to a fixed amount of A at (a) 48°C, (b) 52.4°C.

Problems†

Numerical problems

8.1‡ 1-Butanol and chlorobenzene form a minimum-boiling azeotropic system. The mole fraction of 1-butanol in the liquid (x) and vapour (y) phases at 1.000 atm is given below for a variety of boiling temperatures (H. Artigas, C. Lafuente, P. Cea, F.M. Royo, and J.S. Urieta, *J. Chem. Eng. Data* **42**, 132 (1997)).

T/K	396.57	393.94	391.60	390.15	389.03	388.66	388.57
x	0.1065	0.1700	0.2646	0.3687	0.5017	0.6091	0.7171
y	0.2859	0.3691	0.4505	0.5138	0.5840	0.6409	0.7070

Pure chlorobenzene boils at 404.86 K. (a) Construct the chlorobenzene-rich portion of the phase diagram from the data. (b) Estimate the temperature at which a solution whose mole fraction of 1-butanol is 0.300 begins to boil. (c) State the compositions and relative proportions of the two phases present after a solution initially 0.300 1-butanol is heated to 393.94 K.

8.2‡ An *et al.* investigated the liquid–liquid coexistence curve of N,N-dimethylacetamide and heptane (X. An, H. Zhao, F. Fuguo, and W. Shen, *J. Chem. Thermodynamics* **28**, 1221 (1996)). Mole fractions of N,N-dimethylacetamide in the upper (x_1) and lower (x_2) phases of a two-phase region are given below as a function of temperature:

T/K	309.820	309.422	309.031	308.006	306.686
x_1	0.473	0.400	0.371	0.326	0.293
x_2	0.529	0.601	0.625	0.657	0.690

T/K	304.553	301.803	299.097	296.000	294.534
x_1	0.255	0.218	0.193	0.168	0.157
x_2	0.724	0.758	0.783	0.804	0.814

(a) Plot the phase diagram. (b) State the proportions and compositions of the two phases that form from mixing 0.750 mol of N,N-dimethylacetamide with 0.250 mol of heptane at 296.0 K. To what temperature must the mixture be heated to form a single-phase mixture?

8.3‡ The following data have been obtained for the liquid–vapour equilibrium compositions of mixtures of nitrogen and oxygen at 100 kPa.

T/K	77.3	78	80	82	84	86	88	90.2
$x(O_2)$	0	10	34	54	70	82	92	100
$y(O_2)$	0	2	11	22	35	52	73	100
$p^*(O_2)$/Torr	154	171	225	294	377	479	601	760

Plot the data on a temperature–composition diagram and determine the extent to which it fits the predictions for an ideal solution by calculating the activity coefficients of O_2 at each composition.

8.4 Phosphorus and sulfur form a series of binary compounds. The best characterized are P_4S_3, P_4S_7, and P_4S_{10}, all of which melt congruently. Assuming that only these three binary compounds of the two elements exist, (a) draw schematically only the P/S phase diagram. Label each region of the diagram with the substance that exists in that region and indicate its phase. Label the horizontal axis as x_S and give the numerical values of x_S that correspond to the compounds. The melting point of pure phosphorus is 44°C and that of pure sulfur is 119°C. (b) Draw, schematically, the cooling curve for a mixture of composition $x_S = 0.28$. Assume that a eutectic occurs at $x_S = 0.2$ and negligible solid–solid solubility.

8.5 The table below gives the break and halt temperatures found in the cooling curves of two metals A and B. Construct a phase diagram consistent with the data of these curves. Label the regions of the diagram, stating what phases and substances are present. Give the probable formulas of any compounds that form.

† Problems denoted with the symbol ‡ were supplied by Charles Trapp, Carmen Giunta, and Marshall Cady.

$100x_B$	$\theta_{break}/°C$	$\theta_{halt,1}/°C$	$\theta_{halt,2}/°C$
0		1100	
10.0	1060	700	
20.0	1000	700	
30.0	940	700	400
40.0	850	700	400
50.0	750	700	400
60.0	670	400	
70.0	550	400	
80.0		400	
90.0	450	400	
100.0		500	

8.6 Consider the phase diagram in Fig. 8.40, which represents a solid–liquid equilibrium. Label all regions of the diagram according to the chemical species that exist in that region and their phases. Indicate the number of species and phases present at the points labelled *b*, *d*, *e*, *f*, *g*, and *k*. Sketch cooling curves for compositions $x_B = 0.16, 0.23, 0.57, 0.67$, and 0.84.

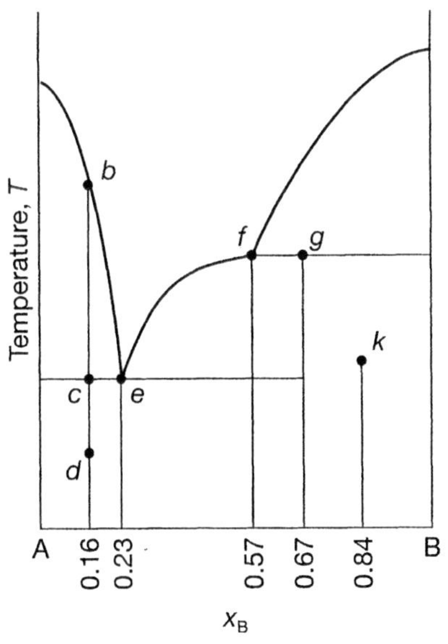

Fig. 8.40

8.7 Solutions of 3-methylphenylamine (MP) were made up in glycerol and then warmed from room temperature. The mixture became turbid at θ_1 and then cleared at θ_2. Plot the phase diagram using the data below, and find the upper and lower critical temperatures.

Mass%	18	20	40	60	80	85
$\theta_1/°C$	48	18	8	10	19	25
$\theta_2/°C$	53	90	120	118	83	53

Mass% denotes the mass percentage composition of MP. State what happens as MP is added dropwise to glycerol at 60°C. State the number of phases present at each composition and their relative amounts.

8.8 Sketch the phase diagram for the Mg/Cu system using the following information: $\theta_f(Mg) = 648°C$, $\theta_f(Cu) = 1085°C$; two intermetallic compounds are formed with $\theta_f(MgCu_2) = 800°C$ and $\theta_f(Mg_2Cu) = 580°C$; eutectics of mass percentage Mg composition and melting points 10 per cent (690°C), 33 per cent (560°C), and 65 per cent (380°C). A sample of Mg/Cu alloy containing 25 per cent Mg by mass was prepared in a crucible heated to 800°C in an inert atmosphere. Describe what will be observed if the melt is cooled slowly to room temperature. Specify the composition and relative abundances of the phases and sketch the cooling curve.

8.9‡ Figure 8.41 shows $\Delta_{mix}G(x_{Pb}, T)$ for a mixture of copper and lead. (a) What does the graph reveal about the miscibility of copper and lead and the spontaneity of solution formation? What is the variance (F) at (i) 1500 K, (ii) 1100 K? (b) Suppose that at 1500 K a mixture of composition (i) $x_{Pb} = 0.1$, (ii) $x_{Pb} = 0.7$, is slowly cooled to 1100 K. What is the equilibrium composition of the final mixture? Include an estimate of the relative amounts of each phase. (c) What is the solubility of (i) lead in copper, (ii) copper in lead at 1100 K?

8.10‡ The temperature–composition diagram for the Ca/Si binary system is shown in Fig. 8.42. (a) Identify eutectics, congruent

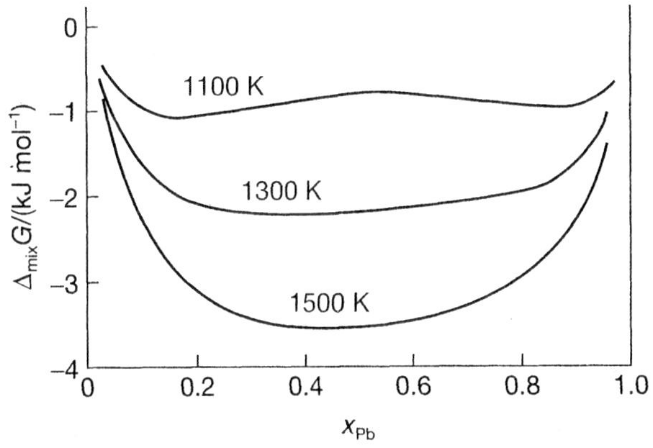

Fig. 8.41

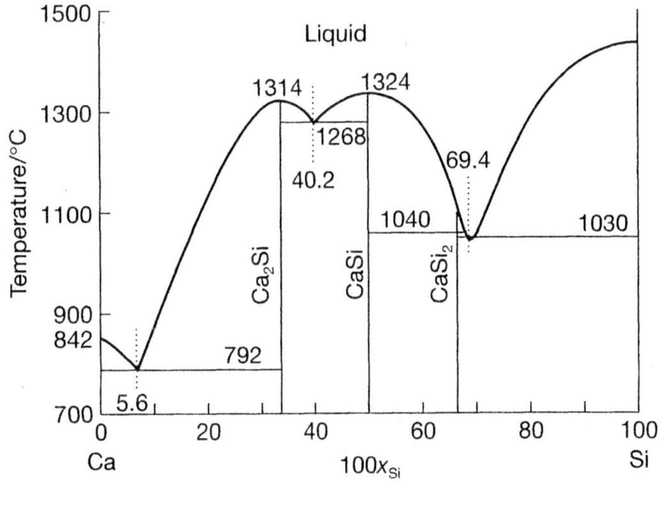

Fig. 8.42

melting compounds, and incongruent melting compounds. (b) If a 20 per cent by atom composition melt of silicon at 1500°C is cooled to 1000°C, what phases (and phase composition) would be at equilibrium? Estimate the relative amounts of each phase. (c) Describe the equilibrium phases observed when an 80 per cent by atom composition Si melt is cooled to 1030°C. What phases, and relative amounts, would be at equilibrium at a temperature (i) slightly higher than 1030°C, (ii) slightly lower than 1030°C? Draw a graph of the mole percentages of both Si(s) and $CaSi_2$(s) as a function of mole percentage of melt that is freezing at 1030°C.

8.11 Iron(II) chloride (melting point 677°C) and potassium chloride (melting point 776°C) form the compounds $KFeCl_3$ and K_2FeCl_4 at elevated temperatures. $KFeCl_3$ melts congruently at 380°C and K_2FeCl_4 melts incongruently at 399°C. Eutectics are formed with compositions $x = 0.38$ (melting point 351°C) and $x = 0.54$ (melting point 393°C), where x is the mole fraction of $FeCl_2$. The KCl solubility curve intersects the K_2FeCl_4 curve at $x = 0.34$. Sketch the phase diagram. State the phases that are in equilibrium when a mixture of composition $x = 0.36$ is cooled from 400°C to 300°C.

Theoretical problems

8.12 Show that two phases are in thermal equilibrium only if their temperatures are the same.

8.13 Show that two phases are in mechanical equilibrium only if their pressures are equal.

Applications: to chemical engineering

8.14 The compound *p*-azoxyanisole forms a liquid crystal. 5.0 g of the solid was placed in a tube, which was then evacuated and sealed. Use the phase rule to prove that the solid will melt at a definite temperature and that the liquid crystal phase will make a transition to a normal liquid phase at a definite temperature.

8.15 Magnesium oxide and nickel oxide withstand high temperatures. However, they do melt when the temperature is high enough and the behaviour of mixtures of the two is of considerable interest to the ceramics industry. Draw the temperature–composition diagram for the system using the data below, where x is the mole fraction of MgO in the solid and y its mole fraction in the liquid.

θ/°C	1960	2200	2400	2600	2800
x	0	0.35	0.60	0.83	1.00
y	0	0.18	0.38	0.65	1.00

State (a) the melting point of a mixture with $x = 0.30$, (b) the composition and proportion of the phases present when a solid of composition $x = 0.30$ is heated to 2200°C, (c) the temperature at which a liquid of composition $y = 0.70$ will begin to solidify.

8.16 The bismuth-cadmium phase diagram is of interest in metallurgy, and its general form can be estimated from expressions for the depression of the freezing point. Construct the diagram using the following data: $T_f(\text{Bi}) = 544.5$ K, $T_f(\text{Cd}) = 594$ K, $\Delta_{fus}H(\text{Bi}) = 10.88$ kJ mol^{-1}, $\Delta_{fus}H(\text{Cd}) = 6.07$ kJ mol^{-1}. The metals are mutually insoluble as solids. Use the phase diagram to state what would be observed when a liquid of composition $x(\text{Bi}) = 0.70$ is cooled slowly from 550 K. What are the relative abundances of the liquid and solid at (a) 460 K and (b) 350 K? Sketch the cooling curve for the mixture.

8.17‡ Carbon dioxide at high pressure is used to separate various compounds in citrus oil. The mole fraction of CO_2 in the liquid (x) and vapour (y) at 323.2 K is given below for a variety of pressures (Y. Iwai, T. Morotomi, K. Sakamoto, Y. Koga, and Y. Arai, *J. Chem. Eng. Data* **41**, 951 (1996)).

p/MPa	3.94	6.02	7.97	8.94	9.27
x	0.2873	0.4541	0.6650	0.7744	0.8338
y	0.9982	0.9980	0.9973	0.9958	0.9922

(a) Plot the portion of the phase diagram represented by these data. (b) State the compositions and relative proportions of the two phases present after an equimolar gas mixture is compressed to 6.02 MPa at 323.2 K.

8.18‡ Large quantities of silicon are made by reducing SiO_2 with carbon in arc furnaces at elevated temperatures. Many chemical species may be present in these high-temperature SiO_2/graphite mixtures, including C(s), SiO_2(s), SiO_2(l), Si(s), Si(l), SiC(s), SiO(g), and CO(g). The table below contains information about the temperature dependence of $G(T) - H_{SER} = a + bT$ for each of these species, and several others, in the temperature range 1500–2500 K and at 1 bar, where $G(T) - H_{SER}$ is the Gibbs energy at temperature T relative to the enthalpies of the pure constituent elements in their reference states at the conventional temperature. Use these data alone to answer the following questions. Assume perfect gas behaviour and ideal solutions when necessary.

	a/(kJ mol^{-1})	b/(kJ K^{-1} mol^{-1})
graphite	34.38	−0.040 54
Si(s)	44.281	−0.066 71
SiO_2(l)	−757.039	−0.184 56
Si(g)	484.667	−0.208 08
SiO(g)	−41.075	−0.281 23
O_2(g)	57.858	−0.268 56
SiO_2(s)	−818.448	−0.152 81
SiC(s)	2.521	−0.098 38
Si(l)	96.980	−0.097 43
SiO_2(g)	−209.039	−0.342 03
CO(g)	−54.581	−0.263 12
CO_2(g)	−302.873	−0.320 21

Calculate the normal melting point of silica and silicon. Make graphs of the vapour pressures of each in the temperature range 1500–2500 K and at a pressure of 1 bar. *Hint.* Use mathematical software or a spreadsheet. For a continuation of this story, see Problem 9.26.

8.19‡ The compound *tert*-amyl methyl ether (TAME) has potential use as a gasoline antiknock agent. The Redlich–Kister empirical equation for the excess Gibbs energy of mixing of methanol (1) and TAME (2) is:

$$G^E/RT = x_1 x_2 \sum_{n=0}^{4} A_n(1 - 2x_1)^n$$

where $A_n = 1.4170, -0.052\,45, 0.1599, 0.1061, -0.091\,59$ for $n = 0, 1, \ldots, 4$. The activity coefficients are given by

$$\ln \gamma_1 = x_2^2 \sum_{n=0}^{4} A_n f_n(x_1) \qquad \ln \gamma_2 = x_1^2 \sum_{n=0}^{4} (-1)^n A_n f_n(x_2)$$

where $f_n(x) = \{1 - 2(n+1)x\}(1 - 2x)^{n-1}$. The vapour pressure and molar volume of pure TAME are 6.09 kPa and 131.78 cm^3 mol^{-1}, respectively, at 288.15 K and those of methanol are 9.86 kPa and 40.254 $cm^3\ mol^{-1}$. (a) Draw a graph of ln γ against x_1. Do the graphs reflect correct values at infinite dilutions of either TAME or methanol? (b) Draw a graph of $G^E(x_1)$. What does this graph reveal about the spontaneity of mixing? Does the curve have the shape expected for a regular solution? (c) Draw a graph of total vapour pressure, $p = p_1 + p_2$, and also show the curves for $p(x_1)$ and $p(y_1)$, where y is the mole fraction in the vapour. What is the vapour pressure and composition of the azeotrope? What is the equilibrium pressure when $x_1 = 0.2$? (d) Draw a graph of total vapour and partial vapour pressures. The curves would be $p(x_1)$, $p_1(x_1)$, and $p_2(x_1)$. What do these curves reveal about deviations from ideality? Calculate the Henry's law constant for both methanol and TAME. (e) Estimate the vapour pressure of pure TAME at 288.15 K when subjected to an applied pressure of 2.0 bar. *Hint.* Use mathematical software or a spreadsheet.

9 Chemical equilibrium

This chapter develops the concept of chemical potential and shows how it is used to account for the equilibrium composition of chemical reactions. The equilibrium composition corresponds to a minimum in the Gibbs energy plotted against the extent of reaction, and by locating this minimum we establish the relation between the equilibrium constant and the standard Gibbs energy of reaction. The thermodynamic formulation of equilibrium enables us to establish the quantitative effects of changes in pressure, temperature, and pH.

Chemical reactions move towards a dynamic equilibrium in which both reactants and products are present but have no further tendency to undergo net change. In some cases, the concentration of products in the equilibrium mixture is so much greater than the concentration of unchanged reactants that for all practical purposes the reaction is 'complete'. However, in many important cases the equilibrium mixture has significant concentrations of both reactants and products. In this chapter we see how to use thermodynamics to predict the equilibrium composition under any reaction conditions.

Spontaneous chemical reactions

We have seen that the direction of spontaneous change at constant temperature and pressure is towards lower values of the Gibbs energy, *G*. The idea is entirely general, and in this chapter we apply it to the discussion of reactions.

9.1 The Gibbs energy minimum

We locate the equilibrium composition of a reaction mixture by calculating the Gibbs energy of the reaction mixture and identifying the composition that corresponds to minimum *G*.

(a) The reaction Gibbs energy

Consider the equilibrium $A \rightleftharpoons B$. Even though this reaction looks trivial, there are many examples of it, such as the isomerization of pentane to 2-methylbutane and the conversion of L-alanine to D-alanine. Suppose an infinitesimal amount $d\xi$ of A turns into B, then the change in the amount of A present is $dn_A = -d\xi$ and the change in the amount of B present is $dn_B = +d\xi$. The quantity ξ is called the **extent of reaction**; it has the dimensions of amount of substance and is reported in moles.

When the extent of reaction changes by a finite amount $\Delta\xi$, the amount of A present changes from $n_{A,0}$ to $n_{A,0} - \Delta\xi$ and the amount of B changes from $n_{B,0}$ to $n_{B,0} + \Delta\xi$. So, if initially 2.0 mol A is present and we wait until $\Delta\xi = +1.5$ mol, the amount of A remaining will be 0.5 mol.

The **reaction Gibbs energy**, $\Delta_r G$, is defined as the slope of the graph of the Gibbs energy plotted against the extent of reaction:

$$\Delta_r G = \left(\frac{\partial G}{\partial \xi}\right)_{p,T} \quad [9.1]$$

Although Δ normally signifies a *difference* in values, here Δ_r signifies a *derivative*, the slope of G with respect to ξ. However, to see that there is a close relationship with the normal usage, suppose the reaction advances by $d\xi$. The corresponding change in Gibbs energy is

$$dG = \mu_A\, dn_A + \mu_B\, dn_B = -\mu_A\, d\xi + \mu_B\, d\xi = (\mu_B - \mu_A)d\xi$$

This equation can be reorganized into

$$\left(\frac{\partial G}{\partial \xi}\right)_{p,T} = \mu_B - \mu_A$$

That is,

$$\Delta_r G = \mu_B - \mu_A \quad (9.2)$$

We see that $\Delta_r G$ can also be interpreted as the difference between the chemical potentials of the reactants and products *at the composition of the reaction mixture*.

Because chemical potential varies with composition, the slope of the plot of Gibbs energy against extent of reaction changes as the reaction proceeds. Moreover, because the reaction runs in the direction of decreasing G (that is, down the slope of G versus ξ), we see from eqn 2 that the reaction $A \rightarrow B$ is spontaneous when $\mu_A > \mu_B$, whereas the reverse reaction is spontaneous when $\mu_B > \mu_A$. The slope is zero, and the reaction is spontaneous in neither direction, when

$$\Delta_r G = 0 \quad (9.3)$$

This condition occurs when $\mu_B = \mu_A$ (Fig. 9.1). It follows that, if we can find the composition of the reaction mixture that ensures $\mu_B = \mu_A$, then we can identify the composition of the reaction mixture at equilibrium.

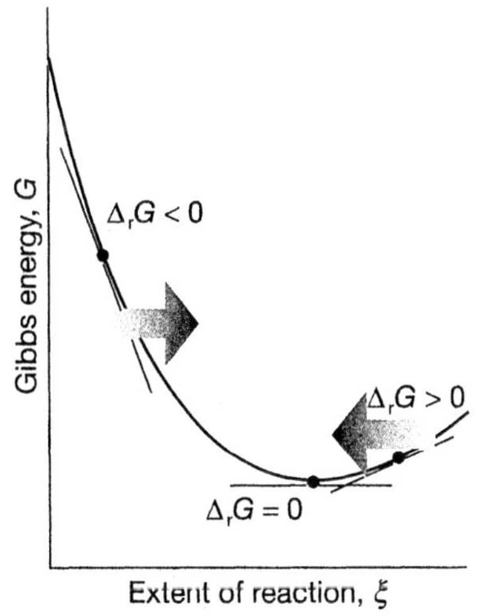

9.1 As the reaction advances (represented by motion from left to right along the horizontal axis) the slope of the Gibbs energy changes. Equilibrium corresponds to zero slope, at the foot of the valley.

(b) Exergonic and endergonic reactions

We can express the spontaneity of a reaction at constant temperature and pressure in terms of the reaction Gibbs energy:

If $\Delta_r G < 0$, the forward reaction is spontaneous.
If $\Delta_r G > 0$, the reverse reaction is spontaneous.
If $\Delta_r G = 0$, the reaction is at equilibrium.

A reaction for which $\Delta_r G < 0$ is called **exergonic** (from the Greek words for work-producing). The name signifies that, because the process is spontaneous, it can be used to drive another process, such as another reaction, or used to do non-expansion work. For example, the oxidation of carbohydrates can drive a number of biological processes, such as muscle contraction, synthesis of proteins, and brain activity (Box 9.2). A reaction for which $\Delta_r G > 0$ is called **endergonic** (signifying work-consuming). The reaction can be made to occur only by doing work on it, such as electrolysing water to reverse its spontaneous formation reaction. Reactions at equilibrium are spontaneous in neither direction: they are neither exergonic nor endergonic.

Box 9.1 Biological energy conversion: the synthesis of proteins and the oxidation of glucose

In biological cells, the energy released by the oxidation of foods (Box 2.2) is stored in adenosine triphosphate (ATP, **1**). The essence of ATP's action is its ability to lose its terminal phosphate group by hydrolysis and to form adenosine diphosphate (ADP):

$$\mathrm{ATP(aq)} + \mathrm{H_2O(l)} \longrightarrow \mathrm{ADP(aq)} + \mathrm{P_i(aq)} + \mathrm{H_3O^+(aq)}$$

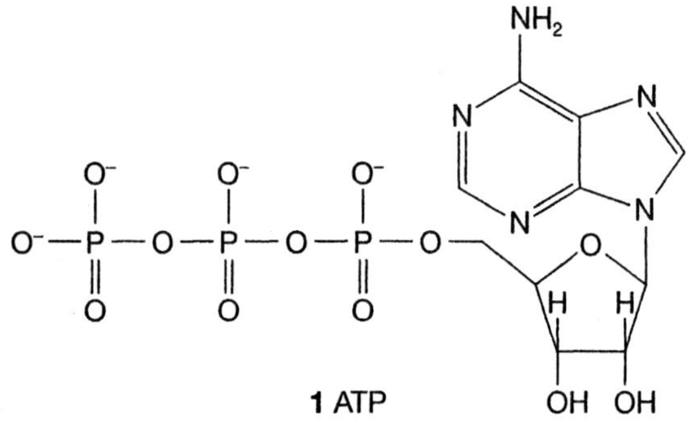

1 ATP

P_i denotes an inorganic phosphate group, such as $H_2PO_4^-$. The biological standard values for ATP hydrolysis at 37°C (310 K, blood temperature) are $\Delta_r G^{\oplus} = -30$ kJ mol^{-1}, $\Delta_r H^{\oplus} = -20$ kJ mol^{-1}, and $\Delta_r S^{\oplus} = +34$ J K^{-1} mol^{-1}. The hydrolysis is therefore exergonic ($\Delta_r G^{\oplus} < 0$) under these conditions and 30 kJ mol^{-1} is available for driving other reactions. Moreover, because the reaction entropy is large, the reaction Gibbs energy is sensitive to temperature. In view of its exergonicity the ADP–phosphate bond has been called a 'high-energy phosphate bond'. The name is intended to signify a high tendency to undergo reaction, and should not be confused with 'strong' bond. In fact, even in the biological sense it is not of very 'high energy'. The action of ATP depends on it being intermediate in activity. Thus ATP acts as a phosphate donor to a number of acceptors (for instance, glucose), but is recharged by more powerful phosphate donors in a number of biochemical processes.

Here we use the oxidation of glucose to CO_2 and H_2O by O_2 as an example of how the breakdown of foods leads to the formation of ATP in the cell. The process begins with *glycolysis*, the partial oxidation of glucose by nicotinamide adenine dinucleotide (NAD$^+$, **2**) to pyruvate ion, $CH_3COCO_2^-$, continues with the *citric acid cycle*, which oxidizes pyruvate to CO_2, and ends with *oxidative phosphorylation*, which reduces O_2 to H_2O. Glycolysis is the main source of energy during *anaerobic metabolism*, a form of metabolism in which inhaled O_2 does not play a role. The citric acid cycle and oxidative phosphorylation are the main mechanisms for the extraction of energy from carbohydrates during *aerobic metabolism*, a form of metabolism in which inhaled O_2 does play a role.

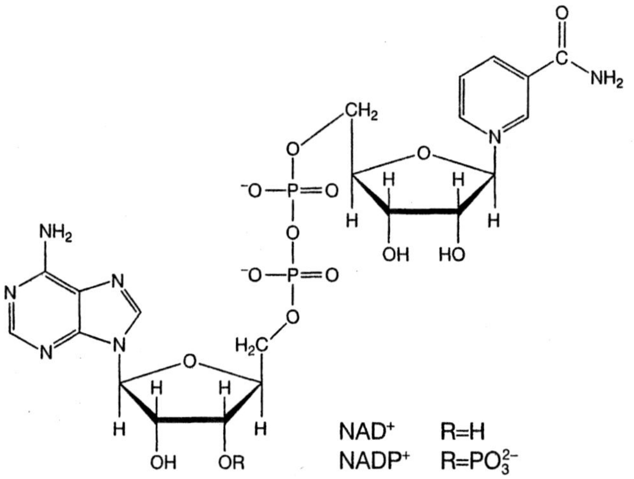

2

Glycolysis occurs in the *cytosol*, the aqueous material encapsulated by the cell membrane, and consists of 10 enzyme-catalysed reactions. At blood temperature, $\Delta_r G^{\oplus} = -147$ kJ mol^{-1} for the oxidation of glucose by NAD$^+$ to pyruvate ions. The oxidation of one glucose molecule is coupled to the conversion of two ADP molecules to two ATP molecules, so the net reaction of glycolysis is:

$$\mathrm{C_6H_{12}O_6(aq)} + 2\,\mathrm{NAD^+(aq)} + 2\,\mathrm{ADP(aq)} + 2\,\mathrm{P_i(aq)} + 2\,\mathrm{H_2O(l)} \longrightarrow 2\,\mathrm{CH_3COCO_2^-(aq)} + 2\,\mathrm{NADH(aq)} + 2\,\mathrm{ATP(aq)} + 2\,\mathrm{H_3O^+(aq)}$$

The standard reaction Gibbs energy is $(-147) - 2(-30)$ kJ mol^{-1} $= -87$ kJ mol^{-1}. The reaction is exergonic, and therefore spontaneous: the oxidation of glucose is used to 'recharge' the ATP. In cells that are deprived of O_2 pyruvate ion is reduced to lactate ion, $CH_3C(OH)CO_2^-$, by NADH.[5] Very strenuous exercise, such as bicycle racing, can decrease sharply the concentration of O_2 in muscle cells and the condition known as muscle fatigue results from increased concentrations of lactate ion.

The standard Gibbs energy of combustion of glucose is -2880 kJ mol^{-1}, so terminating its oxidation at pyruvate is a poor use of resources. In the presence of O_2, pyruvate is

5. In yeast, the terminal products are ethanol and CO_2.

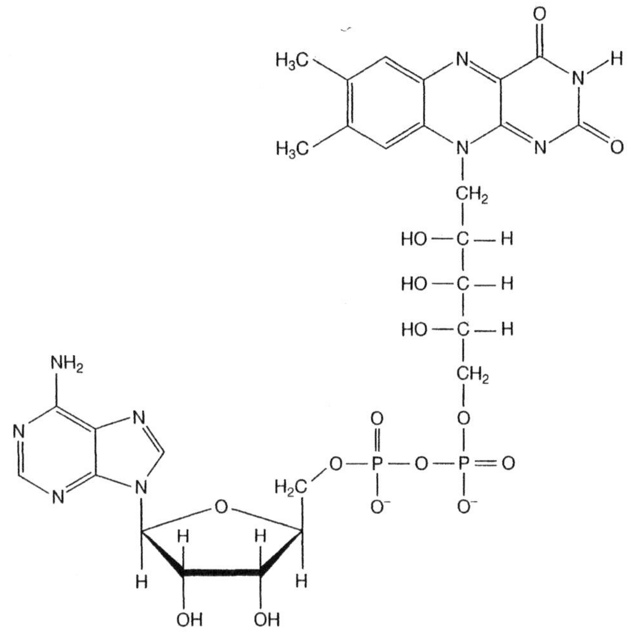

3

oxidized further during the citric acid cycle and oxidative phosphorylation, which occur in a special compartment of the cell called the *mitochondrion*. The citric acid cycle requires eight enzymes that drive the synthesis of ATP with the oxidation of pyruvate by NAD^+ and flavin adenine dinucleotide (FAD, **3**):

$$2\,CH_3COCO_2^-(aq) + 8\,NAD^+(aq) + 2\,FAD(aq) + 2\,ADP(aq) + 2\,P_i(aq) + 8\,H_2O(l) \longrightarrow 6\,CO_2(g) + 8\,NADH(aq) + 4\,H_3O^+(aq) + 2\,FADH_2(aq) + 2\,ATP(aq)$$

The NADH and $FADH_2$ go on to reduce O_2 during oxidative phosphorylation (Box 10.1), which also produces ATP. The citric acid cycle and oxidative phosphorylation generate as many as 38 ATP molecules for each glucose molecule consumed. Each mole of ATP extracts 30 kJ from the 2880 kJ supplied by 1 mol $C_6H_{12}O_6$ (180 g of glucose), so 1140 kJ is stored for later use. Therefore, aerobic oxidation of glucose is much more efficient than glycolysis.

In the cell, each ATP molecule can be used to drive an endergonic reaction for which $\Delta_r G^\ominus$ does not exceed +30 kJ mol^{-1}. For example, the biosynthesis of sucrose from glucose and fructose can be driven by plant enzymes because the reaction is endergonic to the extent $\Delta_r G^\ominus$ = +23 kJ mol^{-1}. The biosynthesis of proteins is strongly endergonic, not only on account of the enthalpy change but also on account of the large decrease in entropy that occurs when many amino acids are assembled into a precisely determined sequence. For instance, the formation of a peptide link is endergonic, with $\Delta_r G^\ominus$ = +17 kJ mol^{-1}, but the biosynthesis occurs indirectly and is equivalent to the consumption of three ATP molecules for each link. In a moderately small protein like myoglobin, with about 150 peptide links, the construction alone requires 450 ATP molecules, and therefore about 12 mol of glucose molecules for 1 mol of protein molecules.

Discussion questions

1. Here we investigate the molecular basis for the exergonicity of the ATP hydrolysis reaction. (a) The Gibbs energy of ATP hydrolysis becomes more negative with increasing pH (in the range $5 < pH < 9$) and this observation has been used to support the hypothesis that electrostatic repulsion between adjacent phosphate groups is a factor that controls the exergonicity of ATP hydrolysis. Provide a rationale for the hypothesis and discuss how the pH dependence of $\Delta_r G$ for ATP hydrolysis supports it. Do these electrostatic effects contribute to the $\Delta_r H$ or $\Delta_r S$ terms that determine the exergonicity of the reaction? (b) The average P—O bond length in the phosphate ion is 154 pm, which is short relative to the value of 173 pm expected for a P—O single bond. This observation has been used to support the hypothesis that electron delocalization in the phosphate ion, one of the hydrolysis products, and the absence of electron delocalization in the triphosphate group of the reactant are factors that control the exergonicity of ATP hydrolysis. Provide a rationale for the hypothesis and discuss how the bond length data support it. Does electron delocalization contribute to the $\Delta_r H$ or $\Delta_r S$ terms that determine the exergonicity of the reaction? (c) The standard reaction entropy for the hydrolysis of polyphosphates is positive; this is thought to be one of the factors that controls the exergonicity of ATP hydrolysis. Why would an increase in entropy accompany the hydrolysis of a triphosphate group into a disphosphate and a phosphate group?

2. At blood temperature, $\Delta_r G^\ominus$ = −218 kJ mol^{-1} and $\Delta_r H^\ominus$ = −120 kJ mol^{-1} for the production of lactate ion during anaerobic glycolysis. Provide a molecular interpretation for the observation that the reaction is more exergonic than it is exothermic.

Further reading

C.K. Matthews, K.E. van Holde, and K.G. Ahern, *Biochemistry*. Addison Wesley Longman, San Francisco (2000).

D.A. Harris, *Bioenergetics at a glance*. Blackwell Science, Oxford (1995).

9.2 The description of equilibrium

With the background established, we are now ready to see how to apply thermodynamics to the description of chemical equilibrium.

(a) Perfect gas equilibria

When A and B are perfect gases we can use eqn 7.14 ($\mu = \mu^\ominus + RT \ln p$, with p interpreted as $p/p^\ominus$) to write

$$\begin{aligned}\Delta_r G &= \mu_B - \mu_A = (\mu_B^\ominus + RT \ln p_B) - (\mu_A^\ominus + RT \ln p_A) \\ &= \Delta_r G^\ominus + RT \ln \frac{p_B}{p_A}\end{aligned} \quad (9.4)^\circ$$

If we denote the ratio of partial pressures by Q, we obtain

$$\Delta_r G = \Delta_r G^\ominus + RT \ln Q \qquad Q = \frac{p_B}{p_A} \quad (9.5)^\circ$$

The ratio Q is an example of a **reaction quotient**. It ranges from 0 (pure A) to infinity (pure B). The **standard reaction Gibbs energy**, $\Delta_r G^\ominus$, is defined (like the standard reaction enthalpy) as the difference in the standard molar Gibbs energies of the reactants and products. For our reaction

$$\Delta_r G^\ominus = G_{B,m}^\ominus - G_{A,m}^\ominus = \mu_B^\ominus - \mu_A^\ominus \quad (9.6)$$

In Section 4.6 we saw that the difference in standard molar Gibbs energies of the products and reactants is equal to the difference in their standard Gibbs energies of formation, so in practice we calculate $\Delta_r G^\ominus$ from

$$\Delta_r G^\ominus = \Delta_f G^\ominus(B) - \Delta_f G^\ominus(A) \quad (9.7)$$

At equilibrium $\Delta_r G = 0$. The ratio of partial pressures at equilibrium is denoted K, and eqn 5 becomes

$$0 = \Delta_r G^\ominus + RT \ln K$$

This expression rearranges to

$$RT \ln K = -\Delta_r G^\ominus \qquad K = \left(\frac{p_B}{p_A}\right)_{\text{equilibrium}} \quad (9.8)^\circ$$

This relation is a special case of one of the most important equations in chemical thermodynamics: it is the link between tables of thermodynamic data, such as those in the *Data section* at the end of this volume, and the chemically important equilibrium constant, K.

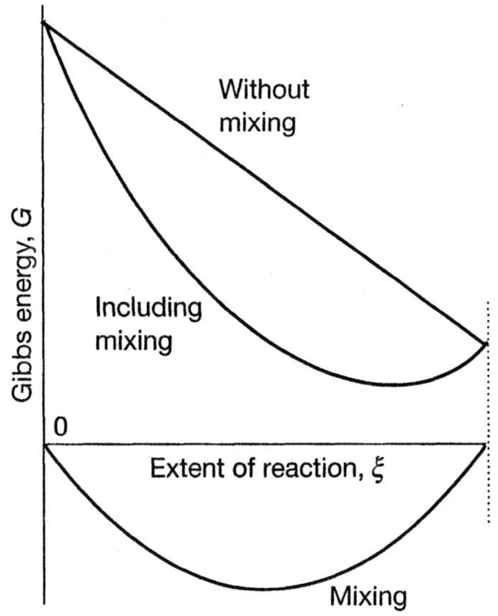

9.2 If the mixing of reactants and products is ignored, then the Gibbs energy changes linearly from its initial value (pure reactants) to its final value (pure products) and the slope of the line is $\Delta_r G^\ominus$. However, as products are produced, there is a further contribution to the Gibbs energy arising from their mixing (lowest curve). The sum of the two contributions has a minimum. That minimum corresponds to the equilibrium composition of the system.

Molecular interpretation 9.1

In molecular terms, the minimum in the Gibbs energy, which corresponds to $\Delta_r G = 0$, stems from the Gibbs energy of mixing of the two gases. Hence, an important contribution to the position of chemical equilibrium is the mixing of the products with the reactants as the products are formed.

Consider a hypothetical reaction in which A molecules change into B molecules without mingling together. The Gibbs energy of the system changes from $G^\ominus(A)$ to $G^\ominus(B)$ in proportion to the amount of B that had been formed, and the slope of the plot of G against the extent of reaction is a constant and equal to $\Delta_r G^\ominus$ at all stages of the reaction (Fig. 9.2). There is no minimum in the graph (except at pure B).

However, in fact, the newly produced B molecules mix with the surviving A molecules. We have seen that the contribution of a mixing process to the change in Gibbs energy (eqn 7.18) is

$$\Delta_{mix}G = nRT(x_A \ln x_A + x_B \ln x_B)$$

This expression makes a U-shaped contribution to the total change in Gibbs energy. As can be seen from the illustration, there is now a minimum in the Gibbs energy, and its position corresponds to the equilibrium composition of the reaction mixture.

We see from eqn 8 that, when $\Delta_r G^\ominus > 0$, the equilibrium constant $K < 1$. Therefore, at equilibrium the partial pressure of A exceeds that of B, which means that the reactant A is favoured in the equilibrium. When $\Delta_r G^\ominus < 0$, the equilibrium constant $K > 1$, so at equilibrium the partial pressure of B exceeds that of A. Now the product B is favoured in the equilibrium.

(b) The general case of a reaction

We can easily extend the argument that led to eqn 8 to a general reaction. First, we need to generalize the concept of extent of reaction.

Consider the reaction $2\,A + B \rightarrow 3\,C + D$. A more sophisticated way of expressing the chemical equation is to write it in the symbolic form

$$0 = 3\,C + D - 2\,A - B$$

by subtracting the reactants from both sides (and replacing the arrow by an equals sign). This equation has the form

$$0 = \sum_J \nu_J J \tag{9.9}$$

where J denotes the substances and the ν_J are the corresponding **stoichiometric numbers** in the chemical equation.[1] In our example, these numbers have the values

$$\nu_A = -2 \qquad \nu_B = -1 \qquad \nu_C = +3 \qquad \nu_D = +1$$

A stoichiometric number is positive for products and negative for reactants. Then we define ξ so that if it changes by $\Delta\xi$, then the change in the amount of any species J is $\nu_J\Delta\xi$.

Illustration 9.1

To express the equation

$$N_2(g) + 3\,H_2(g) \longrightarrow 2\,NH_3(g) \tag{9.10}$$

in the notation of eqn 9, we rearrange it to

$$0 = 2\,NH_3(g) - N_2(g) - 3\,H_2(g)$$

1. Stoichiometric *numbers* may be positive or negative; stoichiometric *coefficients* are always positive. Few, however, make the distinction between the two types of quantity.

and then identify the stoichiometric numbers as $\nu_{N_2} = -1$, $\nu_{H_2} = -3$, and $\nu_{NH_3} = +2$. Therefore, if initially there is 10 mol N_2 present, then when the extent of reaction changes from $\xi = 0$ to $\xi = 1$ mol, implying that $\Delta\xi = +1$ mol, the amount of N_2 changes from 10 mol to 9 mol. All the N_2 has been consumed when $\xi = 10$ mol. When $\Delta\xi = +1$ mol, the amount of H_2 changes by $-3 \times (1\ \text{mol}) = -3$ mol and the amount of NH_3 changes by $+2 \times (1\ \text{mol}) = +2$ mol.

The reaction Gibbs energy, $\Delta_r G$, is defined in the same way as before, eqn 1. In the *Justification* below, we show that the Gibbs energy of reaction can always be written

$$\Delta_r G = \Delta_r G^\ominus + RT \ln Q \tag{9.11}$$

with the standard reaction Gibbs energy calculated from

$$\Delta_r G^\ominus = \sum_{\text{Products}} \nu \Delta_f G^\ominus - \sum_{\text{Reactants}} \nu \Delta_f G^\ominus \tag{9.12}$$

or, more formally,

$$\Delta_r G^\ominus = \sum_J \nu_J \Delta_f G^\ominus(J) \tag{9.13}$$

The reaction quotient, Q, has the form

$$Q = \frac{\text{activities of products}}{\text{activities of reactants}} \tag{9.14}$$

with each species raised to the power given by its stoichiometric coefficient. More formally, to write the general expression for Q we introduce the symbol Π to denote the product of what follows it (just as Σ denotes the sum), and define Q as

$$Q = \prod_J a_J^{\nu_J} \qquad [9.15]$$

Because reactants have negative stoichiometric numbers, they automatically appear as the denominator when the product is written out explicitly. Recall from Table 7.3 that, for pure solids and liquids, the activity is 1, so such substances make no contribution to Q even though they may appear in the chemical equation.

Illustration 9.2

Consider the reaction

$$2\,A + 3\,B \rightarrow C + 2\,D$$

in which case $\nu_A = -2$, $\nu_B = -3$, $\nu_C = +1$, and $\nu_D = +2$. The reaction quotient is then

$$Q = \frac{a_C a_D^2}{a_A^2 a_B^3}$$

Justification 9.1

Consider the reaction with stoichiometric numbers ν_J. When the reaction advances by $d\xi$, the amounts of reactants and products change by $dn_J = \nu_J \, d\xi$. The resulting infinitesimal change in the Gibbs energy at constant temperature and pressure is

$$dG = \sum_J \mu_J \, dn_J = \left(\sum_J \nu_J \mu_J \right) d\xi \tag{9.16}$$

It follows that

$$\Delta_r G = \left(\frac{\partial G}{\partial \xi} \right)_{p,T} = \sum_J \nu_J \mu_J \tag{9.17}$$

To make further progress, we note that the chemical potential of a species J is related to its activity by

$$\mu_J = \mu_J^{\ominus} + RT \ln a_J$$

When this expression is substituted into eqn 17 we obtain

We use $a \ln x = \ln x^a$ and $\ln x + \ln y + \cdots = \ln xy \cdots$.

$$\Delta_r G = \overbrace{\sum_J \nu_J \mu_J^{\ominus}}^{\Delta_r G^{\ominus}} + RT \sum_J \nu_J \ln a_J = \Delta_r G^{\ominus} + RT \sum_J \ln a_J^{\nu_J}$$

$$= \Delta_r G^{\ominus} + RT \ln \overbrace{\prod_J a_J^{\nu_J}}^{Q} = \Delta_r G^{\ominus} + RT \ln Q$$

with Q given by eqn 15.

Now we conclude the argument based on eqn 11. At equilibrium, the slope of G is zero: $\Delta_r G = 0$. The activities then have their equilibrium values and we can write

$$K = \left(\prod_J a_J^{\nu_J} \right)_{\text{equilibrium}} \tag*{[9.18]}$$

This expression has the same form as Q, eqn 15, but is evaluated using equilibrium activities.[2] From now on, we shall not write the 'equilibrium' subscript explicitly, and will rely on the context to make it clear that for K we use equilibrium values and for Q we use the values at the specified stage of the reaction.

An equilibrium constant K expressed in terms of activities (or fugacities) is called a **thermodynamic equilibrium constant**. Note that, because activities are dimensionless numbers, the thermodynamic equilibrium constant is also dimensionless. In elementary applications, the activities that occur in eqn 18 are often replaced by the numerical values of molalities or molar concentrations, and fugacities are replaced by partial pressures. In either case, the resulting expressions are only approximations. The approximation is particularly severe for electrolyte solutions, for in them activity coefficients differ from 1 even in very dilute solutions.

At this point we set $\Delta_r G = 0$ in eqn 11 and replace Q by K. We immediately obtain

$$RT \ln K = -\Delta_r G^{\ominus} \tag{9.19}$$

2. For the special case treated in *Illustration* 9.2, $K = \left(\dfrac{a_C a_D^2}{a_A^2 a_B^3} \right)_{\text{equilibrium}}$.

This is an exact and highly important thermodynamic relation, for it enables us to predict the equilibrium constant of any reaction from tables of thermodynamic data, and hence to predict the equilibrium composition of the reaction mixture.[3]

Example 9.1 Calculating an equilibrium constant

Calculate the equilibrium constant for the ammonia synthesis reaction, eqn 10, at 298 K and show how K is related to the partial pressures of the species at equilibrium when the overall pressure is low enough for the gases to be treated as perfect.

Method Calculate the standard reaction Gibbs energy from eqn 12 and convert it to the value of the equilibrium constant by using eqn 19. The expression for the equilibrium constant is obtained from eqn 19 (or via eqn 14) and, because the gases are taken to be perfect, we replace each fugacity by a partial pressure.

Answer The standard Gibbs energy of the reaction is

$$\begin{aligned}\Delta_r G^{\ominus} &= 2\Delta_f G^{\ominus}(\mathrm{NH_3, g}) - \{\Delta_f G^{\ominus}(\mathrm{N_2, g}) + 3\Delta_f G^{\ominus}(\mathrm{H_2, g})\} \\ &= 2\Delta_f G^{\ominus}(\mathrm{NH_3, g}) = 2 \times (-16.5\ \mathrm{kJ\ mol^{-1}})\end{aligned}$$

Then,

$$\ln K = -\frac{2 \times (-16.5 \times 10^3\ \mathrm{J\,mol^{-1}})}{(8.3145\ \mathrm{J\,K^{-1}\,mol^{-1}}) \times (298\ \mathrm{K})} = \frac{2 \times 16.5 \times 10^3}{8.3145 \times 298}$$

Hence, $K = 6.1 \times 10^5$. This result is thermodynamically exact. The thermodynamic equilibrium constant for the reaction is

$$K = \frac{a_{\mathrm{NH_3}}^2}{a_{\mathrm{N_2}} a_{\mathrm{H_2}}^3} = \frac{(f_{\mathrm{NH_3}}/p^{\ominus})^2}{(f_{\mathrm{N_2}}/p^{\ominus})(f_{\mathrm{H_2}}/p^{\ominus})^3} = \frac{f_{\mathrm{NH_3}}^2 p^{\ominus 2}}{f_{\mathrm{N_2}} f_{\mathrm{H_2}}^3}$$

and this ratio has exactly the value we have just calculated. At low overall pressures fugacities can be replaced by partial pressures and an approximate form of the equilibrium constant is

$$K = \frac{p_{\mathrm{NH_3}}^2 p^{\ominus 2}}{p_{\mathrm{N_2}} p_{\mathrm{H_2}}^3}$$

Self-test 9.1 Evaluate the equilibrium constant for $\mathrm{N_2O_4(g)} \rightleftharpoons 2\,\mathrm{NO_2(g)}$ at 298 K. [$K = 0.15$]

Example 9.2 Estimating the degree of dissociation at equilibrium

The standard Gibbs energy of reaction for the decomposition $\mathrm{H_2O(g)} \rightarrow \mathrm{H_2(g)} + \frac{1}{2}\mathrm{O_2(g)}$ is $+118.08\ \mathrm{kJ\ mol^{-1}}$ at 2300 K. What is the degree of dissociation of H_2O at 2300 K and 1.00 bar?

Method The equilibrium constant is obtained from the standard Gibbs energy of reaction by using eqn 19, so the task is to relate the degree of dissociation, α,

3. In Chapter 20 we shall see that the right-hand side may be expressed in terms of spectroscopic data; so this expression also provides a link between spectroscopy and equilibrium composition.

to K and then to find its numerical value. Proceed by expressing the equilibrium compositions in terms of α, and solve for α in terms of K. Because the standard Gibbs energy of reaction is large and positive, we can anticipate that K will be small, and hence that $\alpha \ll 1$, which opens the way to making approximations to obtain its numerical value.

Answer The equilibrium constant is obtained from eqn 19 in the form

$$\ln K = -\frac{\Delta_r G^{\ominus}}{RT} = -\frac{(+118.08 \times 10^3\ \text{J mol}^{-1})}{(8.3145\ \text{J K}^{-1}\ \text{mol}^{-1}) \times (2300\ \text{K})} = -\frac{118.08 \times 10^3}{8.3145 \times 2300}$$

It follows that $K = 2.08 \times 10^{-3}$. The equilibrium composition can be expressed in terms of α by drawing up the following table:

	H_2O	H_2	O_2
Initial amount	n	0	0
Change to reach equilibrium	$-\alpha n$	$+\alpha n$	$+\frac{1}{2}\alpha n$
Amount at equilibrium	$(1-\alpha)n$	αn	$\frac{1}{2}\alpha n$
Mole fraction	$\frac{1-\alpha}{1+\frac{1}{2}\alpha}$	$\frac{\alpha}{1+\frac{1}{2}\alpha}$	$\frac{\frac{1}{2}\alpha}{1+\frac{1}{2}\alpha}$
Partial pressure	$\frac{(1-\alpha)p}{1+\frac{1}{2}\alpha}$	$\frac{\alpha p}{1+\frac{1}{2}\alpha}$	$\frac{\frac{1}{2}\alpha p}{1+\frac{1}{2}\alpha}$

The equilibrium constant is therefore

$$K = \frac{p_{H_2} p_{O_2}^{1/2}}{p_{H_2O}} = \frac{\alpha^{3/2} p^{1/2}}{(1-\alpha)(2+\alpha)^{1/2}}$$

In this expression, we have written p in place of $p/p^{\ominus}$, to keep the notation simple. Now make the approximation that $\alpha \ll 1$, and hence obtain

$$K \approx \frac{\alpha^{3/2} p^{1/2}}{2^{1/2}}$$

Under the stated condition, $p = 1.00$ bar (that is, $p/p^{\ominus} = 1.00$), so

$$\alpha \approx (2^{1/2}K)^{2/3} = 0.0205$$

That is, about 2 per cent of the water has decomposed.

Comment Always check that the approximation is consistent with the final answer. In this case $\alpha \ll 1$ in accord with the original assumption.

The web site contains links to online tools for the estimation of equilibrium constants of gas-phase reactions.

Self-test 9.2 Given that the standard Gibbs energy of reaction at 2000 K is $+135.2$ kJ mol^{-1} for the same reaction, suppose that steam at 200 kPa is passed through a furnace tube at that temperature. Calculate the mole fraction of O_2 present in the output gas stream. [0.002 21]

(c) The relation between equilibrium constants

The only remaining problem is to express the thermodynamic equilibrium constant in terms of the mole fractions, x_J, or molalities, b_J, of the species. To do so, we need

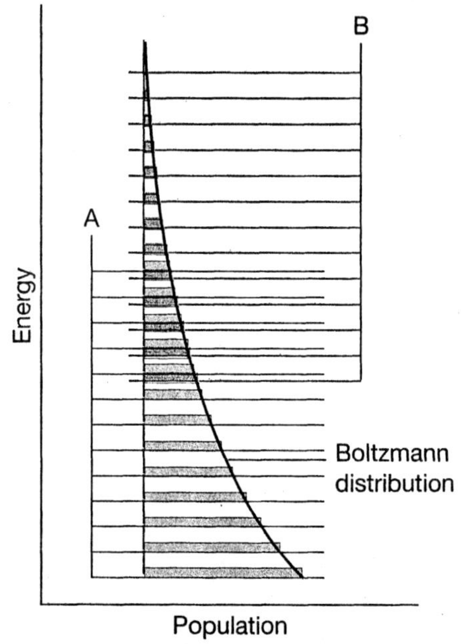

9.3 The Boltzmann distribution of populations over the energy levels of two species A and B with similar densities of energy levels; the reaction $A \rightarrow B$ is endothermic in this example. The bulk of the population is associated with the species A, so that species is dominant at equilibrium.

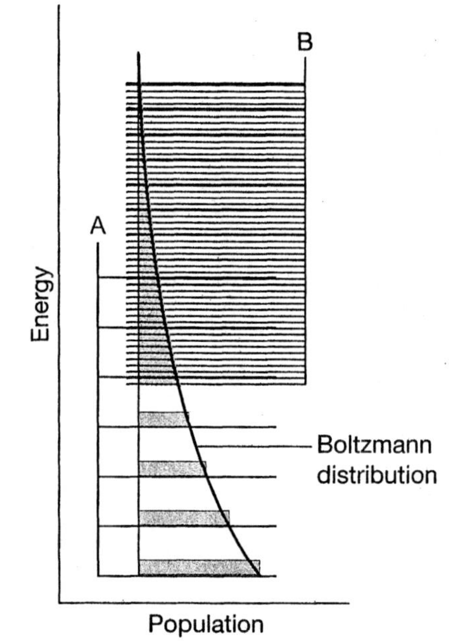

9.4 Even though the reaction $A \rightarrow B$ is endothermic, the density of energy levels in B is so much greater than that in A that the population associated with B is greater than that associated with A, so B is dominant at equilibrium.

to know the activity coefficients, and then to use $a_J = \gamma_J x_J$ or $a_J = \gamma_J b_J/b^\ominus$ (recalling that the activity coefficients depend on the choice). For example, in the latter case, for an equilibrium of the form $A + B \rightleftharpoons C + D$, where all four species are solutes, we write

$$K = \frac{a_C a_D}{a_A a_B} = \frac{\gamma_C \gamma_D}{\gamma_A \gamma_B} \times \frac{b_C b_D}{b_A b_B} = K_\gamma K_b \qquad (9.20)$$

The activity coefficients must be evaluated at the equilibrium composition of the mixture, which may involve a complicated calculation, because the latter is known only if the equilibrium composition is already known. In elementary applications, and to begin the iterative calculation of the concentrations in a real example, the assumption is often made that the activity coefficients are all so close to unity that $K_\gamma = 1$. Then we obtain the result widely used in elementary chemistry that $K \approx K_b$, and equilibria are discussed in terms of molalities (or molar concentrations) themselves. In Chapter 10 we shall see a way of making better estimates of activity coefficients for equilibria involving ions.

Molecular interpretation 9.2

We can obtain a deeper insight into the origin and significance of the equilibrium constant by considering the Boltzmann distribution of molecules over the available states of a system composed of reactants and products (recall *Molecular interpretation* 2.1). When atoms can exchange partners, as in a reaction, the available states of the system include arrangements in which the atoms are present in the form of reactants and in the form of products: these arrangements have their characteristic sets of energy levels, but the Boltzmann distribution does not distinguish between their identities, only their energies. The atoms distribute themselves over both sets of energy levels in accord with the Boltzmann distribution (Fig. 9.3). At a given temperature, there will be a specific distribution of populations, and hence a specific composition of the reaction mixture.

It can be appreciated from the illustration that, if the reactants and products both have similar arrays of molecular energy levels, then the dominant species in a reaction mixture at equilibrium will be the species with the lower set of energy levels. However, the fact that the Gibbs energy occurs in the expression is a signal that entropy plays a role as well as energy. Its role can be appreciated by referring to Fig. 9.4, We see that, although the B energy levels lie higher than the A energy levels, in this instance they are much more closely spaced. As a result, their total population may be considerable and B could even dominate in the reaction mixture at equilibrium. Closely spaced energy levels correlate with a high entropy, so in this case we see that entropy effects dominate adverse energy effects. This competition is mirrored in eqn 19, as can be seen most clearly by using $\Delta_r G^\ominus = \Delta_r H^\ominus - T\Delta_r S^\ominus$ and writing it in the form

$$K = e^{-\Delta_r H^\ominus/RT} e^{\Delta_r S^\ominus/R} \qquad (9.21)$$

Note that a positive reaction enthalpy results in a lowering of the equilibrium constant (that is, an endothermic reaction can be expected to have an equilibrium composition that favours the reactants). However, if there is positive reaction entropy, then the equilibrium composition may favour products, despite the endothermic character of the reaction.

(d) Equilibria in biological systems

We saw in Section 7.7d that for biological systems it is appropriate to adopt the biological standard state, in which pH = 7. It follows from eqn 7.58 that the relation between the thermodynamic and biological standard Gibbs energies of reaction for a reaction of the form

$$A + \nu H^+(aq) \rightarrow P \qquad (9.22)$$

is

$$\Delta_r G^{\oplus} = \Delta_r G^{\ominus} + 7\nu RT \ln 10 \qquad (9.23)$$

Note that there is no difference between the two standard values if hydrogen ions are not involved in the reaction ($\nu = 0$).

Illustration 9.3

Consider the reaction

$$NADH(aq) + H^+(aq) \longrightarrow NAD^+(aq) + H_2(g)$$

at 37°C, for which $\Delta_r G^{\ominus} = -21.8$ kJ mol^{-1}. NADH is the reduced form of nicotinamide adenine dinucleotide and NAD^+ is its oxidized form; the molecules play an important role in the later stages of the respiratory process (Box 10.2). It follows that, because $\nu = 1$ and 7 ln 10 = 16.1,

$$\begin{aligned}\Delta_r G^{\oplus} &= -21.8 \text{ kJ mol}^{-1} + 16.1 \times (8.3145 \times 10^{-3} \text{ kJ K}^{-1} \text{ mol}^{-1}) \times (310 \text{ K}) \\ &= +19.7 \text{ kJ mol}^{-1}\end{aligned}$$

Note that the biological standard value is opposite in sign (in this example) to the thermodynamic standard value.

Self-test 9.3 For a particular reaction of the form $A \rightarrow B + 2 H^+$ in aqueous solution, it was found that $\Delta_r G^{\ominus} = +20$ kJ mol^{-1} at 28°C. Estimate the value of $\Delta_r G^{\oplus}$. [−61 kJ mol^{-1}]

The response of equilibria to the conditions

Equilibria respond to changes in pressure, temperature, and concentrations of reactants and products. The equilibrium constant for a reaction is not affected by the presence of a catalyst or an enzyme (a biological catalyst). Catalysts increase the rate at which equilibrium is attained but do not affect its position (Sections 26.5 and 26.6). However, it is important to note that in industry reactions rarely reach equilibrium, partly on account of the rates at which reactants mix. As we shall see in Sections 26.7–10, catalysts can have some unexpected effects and may change the composition of the reaction mixture under non-equilibrium conditions.

9.3 How equilibria respond to pressure

The equilibrium constant depends on the value of $\Delta_r G^\ominus$, which is defined at a single, standard pressure. The value of $\Delta_r G^\ominus$, and hence of K, is therefore independent of the pressure at which the equilibrium is actually established. Formally, we may express this independence as

$$\left(\frac{\partial K}{\partial p}\right)_T = 0 \qquad (9.24)$$

The conclusion that K is independent of pressure does not necessarily mean that the equilibrium composition is independent of the pressure. However, before considering the consequences of pressure, we need to distinguish between the two ways in which pressure may be applied. The pressure within a reaction vessel can be increased by injecting an inert gas into it. However, so long as the gases are perfect, this addition of gas leaves all the partial pressures of the reacting gases unchanged: the partial pressure of a perfect gas is the pressure it would exert if it were alone in the container, so the presence of another gas has no effect. Put another way, the addition of an inert gas leaves the molar concentrations of the original gases unchanged, as they continue to occupy the same volume. It follows that pressurization by the addition of an inert gas has no effect on the equilibrium composition of the system (provided the gases are perfect). Alternatively, the pressure of the system may be increased by confining the gases to a smaller volume (that is, by compression). Now the partial pressures are changed. Put another way, their molar concentrations are modified because the volume the gases occupy is reduced.

We need to consider the role of compression and understand how changes in partial pressures can be consistent with the general result expressed in eqn 24 that the equilibrium constant itself is independent of the pressure. We shall see that compression can adjust the individual partial pressures of the reactants and products in such a way that, although each one changes, their ratio (as it appears in the equilibrium constant) remains the same. Consider, for instance, the perfect gas equilibrium $A \rightleftharpoons 2\,B$, for which the equilibrium constant is

$$K = \frac{p_B^2}{p_A p^\ominus}$$

The right-hand side of this expression remains constant only if an increase in p_A cancels an increase in the *square* of p_B. This relatively steep increase of p_A compared to p_B will occur if the equilibrium composition shifts in favour of A at the expense of B. Then the number of A molecules will increase as the volume of the container is decreased and its partial pressure will rise more rapidly than can be ascribed to a simple change in volume alone (Fig. 9.5).

The increase in the number of A molecules and the corresponding decrease in the number of B molecules brought about by compression is a special case of a principle proposed by the French chemist Henri Le Chatelier.[4] **Le Chatelier's principle** states that:

A system at equilibrium, when subjected to a disturbance, responds in a way that tends to minimize the effect of the disturbance

The principle implies that, if a system at equilibrium is compressed, then the reaction will adjust so as to minimize the increase in pressure. This it can do by reducing the number of particles in the gas phase, which implies a shift $A \leftarrow 2\,B$.

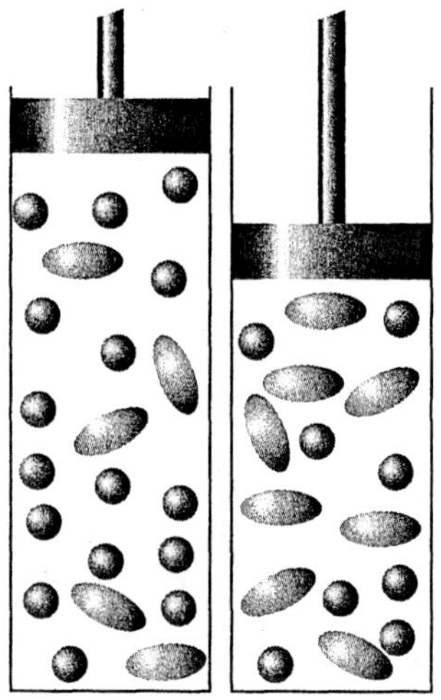

9.5 When a reaction at equilibrium is compressed (from *a* to *b*), the reaction responds by reducing the number of molecules in the gas phase (in this case by producing the dimers represented by the ellipses).

4. Le Chatelier also invented oxyacetylene welding.

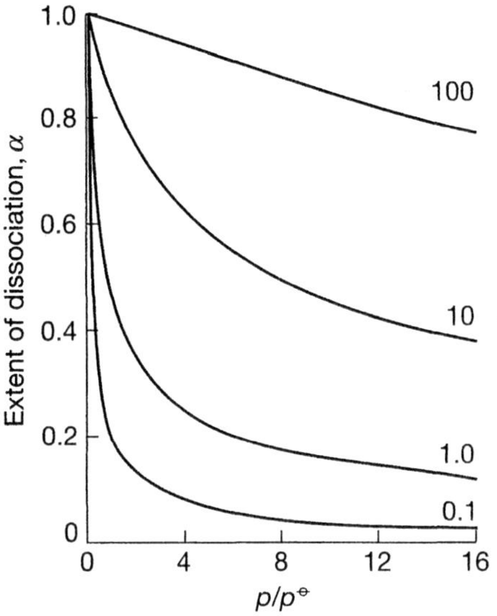

9.6 The pressure dependence of the degree of dissociation, α, at equilibrium for an $A(g) \rightleftharpoons 2\,B(g)$ reaction for different values of the equilibrium constant K. The value $\alpha = 0$ corresponds to pure A; $\alpha = 1$ corresponds to pure B.

The quantitative treatment of the effect of compression set out in the *Justification* below leads to the conclusion that the extent of dissociation, α, of A into 2 B is

$$\alpha = \left(\frac{1}{1 + 4p/K}\right)^{1/2} \qquad (9.25)$$

where p is to be understood as $p/p^{\ominus}$. This formula shows that, even though K is independent of pressure, the amounts of A and B do depend on pressure (Fig. 9.6). It also shows that, as p is increased, α decreases, in accord with Le Chatelier's principle.

Justification 9.2

Suppose that there is an amount n of A present initially (and no B). At equilibrium the amount of A is $(1 - \alpha)n$ and the amount of B is $2\alpha n$. It follows that the mole fractions present at equilibrium are

$$x_A = \frac{(1-\alpha)n}{(1-\alpha)n + 2\alpha n} = \frac{1-\alpha}{1+\alpha} \qquad x_B = \frac{2\alpha}{1+\alpha}$$

The equilibrium constant for the reaction (with p understood as $p/p^{\ominus}$) is

$$K = \frac{p_B^2}{p_A} = \frac{x_B^2 p^2}{x_A p} = \frac{4\alpha^2 p}{1-\alpha^2}$$

This expression rearranges into eqn 25.

Illustration 9.4

To predict the effect of an increase in pressure on the composition of the ammonia synthesis at equilibrium, eqn 10, we note that the number of gas molecules decreases (from 4 to 2). So, Le Chatelier's principle predicts that an increase in pressure will favour the product. The equilibrium constant is

$$K = \frac{p_{NH_3}^2}{p_{N_2} p_{H_2}^3} = \frac{x_{NH_3}^2 p^2}{x_{N_2} x_{H_2}^3 p^4} = \frac{K_x}{p^2}$$

where K_x is the part of the equilibrium constant expression that contains the equilibrium mole fractions of reactants and products. Therefore, doubling the pressure must increase K_x by a factor of 4 to preserve the value of K.

Self-test 9.4 Predict the effect of a 10-fold pressure increase on the equilibrium composition of the reaction $3\,N_2(g) + H_2(g) \rightarrow 2\,N_3H(g)$.

[100-fold increase in K_x]

9.4 The response of equilibria to temperature

Le Chatelier's principle predicts that a system at equilibrium will tend to shift in the endothermic direction if the temperature is raised, for then energy is absorbed as heat and the rise in temperature is opposed. Conversely, an equilibrium can be expected to shift in the exothermic direction if the temperature is lowered, for then energy is released and the reduction in temperature is opposed. These conclusions can be summarized as follows:

Exothermic reactions: increased temperature favours the reactants.

Endothermic reactions: increased temperature favours the products.

We shall now justify these remarks and see how to express the changes quantitatively.

(a) The van 't Hoff equation

The van 't Hoff equation, which is derived in the *Justification* below, is an expression for the slope of a plot of the equilibrium constant (specifically, ln K) as a function of temperature. It may be expressed in either of two ways:

$$\text{(a)}\ \frac{\mathrm{d}\ln K}{\mathrm{d}T}=\frac{\Delta_r H^{\ominus}}{RT^2} \qquad \text{(b)}\ \frac{\mathrm{d}\ln K}{\mathrm{d}(1/T)}=-\frac{\Delta_r H^{\ominus}}{R} \tag{9.26}$$

Justification 9.3

From eqn 19, we know that

$$\ln K=-\frac{\Delta_r G^{\ominus}}{RT}$$

Differentiation of ln K with respect to temperature then gives

$$\frac{\mathrm{d}\ln K}{\mathrm{d}T}=-\frac{1}{R}\frac{\mathrm{d}(\Delta_r G^{\ominus}/T)}{\mathrm{d}T}$$

The differentials are complete because K and $\Delta_r G^{\ominus}$ depend only on temperature, not on pressure. To develop this equation we use the Gibbs–Helmholtz equation (eqn 5.13) in the form

$$\frac{\mathrm{d}(\Delta_r G^{\ominus}/T)}{\mathrm{d}T}=-\frac{\Delta_r H^{\ominus}}{T^2}$$

where $\Delta_r H^{\ominus}$ is the standard reaction enthalpy at the temperature T. Combining the two equations gives the van 't Hoff equation, eqn 26a. The second form of the equation is obtained by noting that

$$\frac{\mathrm{d}(1/T)}{\mathrm{d}T}=-\frac{1}{T^2}, \qquad \text{so} \quad \mathrm{d}T=-T^2\,\mathrm{d}(1/T)$$

Equation 26a shows that $\mathrm{d}\ln K/\mathrm{d}T<0$ (and therefore that $\mathrm{d}K/\mathrm{d}T<0$) for a reaction that is exothermic under standard conditions ($\Delta_r H^{\ominus}<0$). A negative slope means that ln K, and therefore K itself, decreases as the temperature rises. Therefore, as asserted above, in the case of an exothermic reaction the equilibrium shifts away from products. The opposite occurs in the case of endothermic reactions.

Some insight into the thermodynamic basis of this behaviour comes from the expression $\Delta_r G=\Delta_r H-T\Delta_r S$ written in the form $-\Delta_r G/T=-\Delta_r H/T+\Delta_r S$. When the reaction is exothermic, $-\Delta_r H/T$ corresponds to a positive change of entropy of the surroundings and favours the formation of products. When the temperature is raised, $-\Delta_r H/T$ decreases, and the increasing entropy of the surroundings has a less important role. As a result, the equilibrium lies less to the right. When the reaction is endothermic, the principal factor is the increasing entropy of the reaction system. The importance of the unfavourable change of entropy of the surroundings is reduced if the temperature is raised (because then $\Delta_r H/T$ is smaller), and the reaction is able to shift towards products.

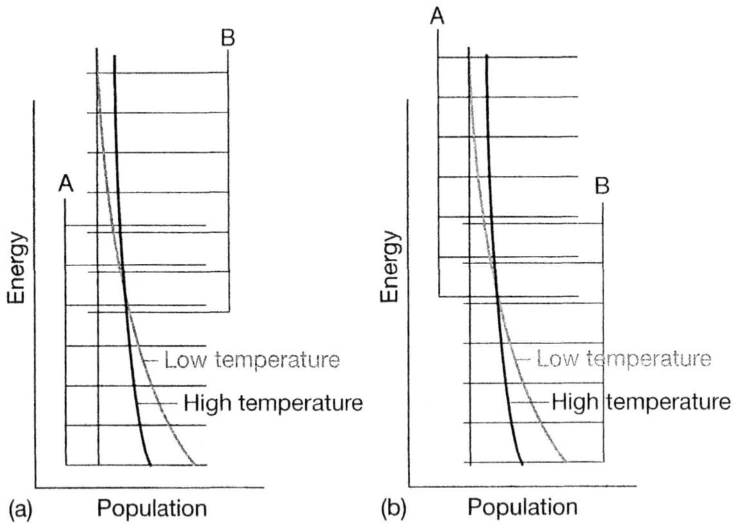

9.7 The effect of temperature on a chemical equilibrium can be interpreted in terms of the change in the Boltzmann distribution with temperature and the effect of that change in the population of the species. (a) In an endothermic reaction, the population of B increases at the expense of A as the temperature is raised. (b) In an exothermic reaction, the opposite happens.

Molecular interpretation 9.3

The typical arrangement of energy levels for an endothermic reaction is shown in Fig. 9.7a. When the temperature is increased, the Boltzmann distribution adjusts and the populations change as shown. The change corresponds to an increased population of the higher energy states at the expense of the population of the lower energy states. We see that the states that arise from the B molecules become more populated at the expense of the A molecules. Therefore, the total population of B states increases, and B becomes more abundant in the equilibrium mixture. Conversely, if the reaction is exothermic (Fig. 9.7b), then an increase in temperature increases the population of the A states (which start at higher energy) at the expense of the B states, so the reactants become more abundant.

Example 9.3 Measuring a reaction enthalpy

The data below show the temperature variation of the equilibrium constant of the reaction

$$Ag_2CO_3(s) \rightleftharpoons Ag_2O(s) + CO_2(g)$$

Calculate the standard reaction enthalpy of the decomposition.

T/K	350	400	450	500
K	3.98×10^{-4}	1.41×10^{-2}	1.86×10^{-1}	1.48

Method It follows from eqn 26b that, provided the reaction enthalpy can be assumed to be independent of temperature, a plot of $-\ln K$ against $1/T$ should be a straight line of slope $\Delta_r H^\ominus/R$.

Answer We draw up the following table:

T/K	350	400	450	500
$(10^3\ \mathrm{K})/T$	2.86	2.50	2.22	2.00
$-\ln K$	7.83	4.26	1.68	−0.39

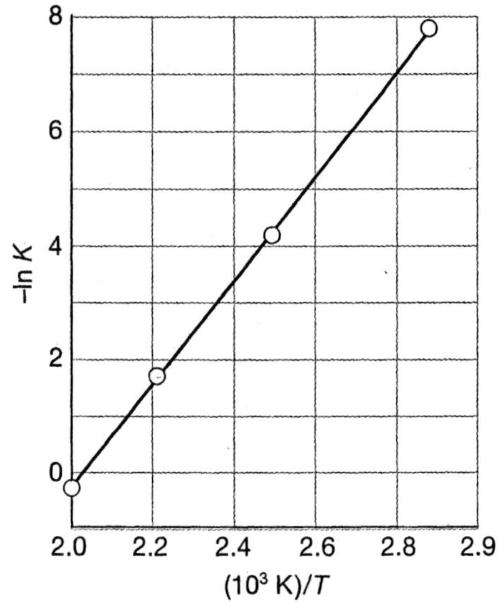

9.8 When $-\ln K$ is plotted against $1/T$, a straight line is expected with slope equal to $\Delta_r H^\ominus/R$. This is a noncalorimetric method for the measurement of reaction enthalpies.

These points are plotted in Fig. 9.8. The slope of the graph is $+9.6 \times 10^3$, so

$$\Delta_r H^\ominus = (+9.6 \times 10^3\ \text{K}) \times R = +80\ \text{kJ mol}^{-1}$$

Comment This is a noncalorimetric method of determining $\Delta_r H^\ominus$. A drawback is that the reaction enthalpy is actually temperature-dependent, so the plot is not expected to be perfectly linear. However, the temperature dependence is weak in many cases, so the plot is reasonably straight. In practice, the method is not very accurate, but it is often the only method available.

Self-test 9.5 The equilibrium constant of the reaction $2\ SO_2(g) + O_2(g) \rightleftharpoons 2\ SO_3(g)$ is 4.0×10^{24} at 300 K, 2.5×10^{10} at 500 K, and 3.0×10^4 at 700 K. Estimate the reaction enthalpy at 500 K. [-200 kJ mol^{-1}]

(b) The value of *K* at different temperatures

To find the value of the equilibrium constant at a temperature T_2 in terms of its value K_1 at another temperature T_1, we integrate eqn 26b between these two temperatures:

$$\ln K_2 - \ln K_1 = -\frac{1}{R}\int_{1/T_1}^{1/T_2} \Delta_r H^\ominus\, d(1/T) \qquad (9.27)$$

If we suppose that $\Delta_r H^\ominus$ varies only slightly with temperature over the temperature range of interest, we may take it outside the integral. It follows that

$$\ln K_2 - \ln K_1 = -\frac{\Delta_r H^\ominus}{R}\left(\frac{1}{T_2} - \frac{1}{T_1}\right) \qquad (9.28)$$

Illustration 9.5

To estimate the equilibrium constant for the synthesis of ammonia at 500 K from its value at 298 K (6.1×10^5 for the reaction as written in eqn 10) we use the standard reaction enthalpy, which can be obtained from Table 2.6 in the *Data section* by using $\Delta_r H^\ominus = 2\Delta_f H^\ominus(NH_3, g)$, and assume that its value is constant over the range of temperatures. Then, with $\Delta_r H^\ominus = -92.2$ kJ mol^{-1}, from eqn 28 we find

$$\ln K_2 = \ln(6.1 \times 10^5) - \frac{(-92.2 \times 10^3\ \text{J mol}^{-1})}{8.3145\ \text{J K}^{-1}\ \text{mol}^{-1}}\left(\frac{1}{500\ \text{K}} - \frac{1}{298\ \text{K}}\right) = -1.71$$

It follows that $K_2 = 0.18$.

Self-test 9.6 The equilibrium constant for $N_2O_4(g) \rightleftharpoons 2\ NO_2(g)$ was calculated in Self-test 9.1. Estimate its value at 100°C. [15]

Knowledge of the temperature dependence of the equilibrium constant for a reaction can be useful in the design of laboratory and industrial processes. For example, synthetic chemists can improve the yield of a reaction by changing the temperature of the reaction mixture. Also, reduction of a metal oxide with carbon or carbon monoxide results in the extraction of the metal when the process is carried out at a temperature for which the equilibrium constant for the reaction is much greater than one (Box 9.2).

Box 9.2 The extraction of metals from their oxides

Metals can be obtained from their oxides by reduction with carbon or carbon monoxide if any of the equilibria

$$MO(s) + C(s) \rightleftharpoons M(s) + CO(g)$$
$$MO(s) + \tfrac{1}{2}C(s) \rightleftharpoons M(s) + \tfrac{1}{2}CO_2(g)$$
$$MO(s) + CO(g) \rightleftharpoons M(s) + CO_2(g)$$

lie to the right (i.e. have $K > 1$). As we shall see, these equilibria can be discussed in terms of the thermodynamic functions for the reactions

(i) $M(s) + \tfrac{1}{2}O_2(g) \longrightarrow MO(s)$
(ii) $\tfrac{1}{2}C(s) + \tfrac{1}{2}O_2(g) \longrightarrow \tfrac{1}{2}CO_2(g)$
(iii) $C(s) + \tfrac{1}{2}O_2(g) \longrightarrow CO(g)$
(iv) $CO(g) + \tfrac{1}{2}O_2(g) \longrightarrow CO_2(g)$

The temperature dependences of the standard Gibbs energies of reactions (i)–(iv) depend on the reaction entropy through $d\Delta_rG^\ominus/dT = -\Delta_rS^\ominus$. Because in reaction (iii) there is a net increase in the amount of gas, the standard reaction entropy is large and positive; therefore, its $\Delta_rG^\ominus$ decreases sharply with increasing temperature. In reaction (iv), there is a similar net decrease in the amount of gas, so $\Delta_rG^\ominus$ increases sharply with increasing temperature. In reaction (ii), the amount of gas is constant, so the entropy change is small and $\Delta_rG^\ominus$ changes only slightly with temperature. These remarks are summarized in the illustration, which is called an *Ellingham diagram*. Note that $\Delta_rG^\ominus$ decreases upwards!

At room temperature, $\Delta_rG^\ominus$ is dominated by the contribution of the reaction enthalpy ($T\Delta_rS^\ominus$ being relatively small), so the order of increasing $\Delta_rG^\ominus$ is the same as the order of increasing $\Delta_rH^\ominus$ (Al_2O_3 is most exothermic; Ag_2O is least). The standard reaction entropy is similar for all metals because in each case gaseous oxygen is eliminated and a compact, solid oxide is formed. As a result, the temperature dependence of the standard Gibbs energy of oxidation should be similar for all metals, as is shown by the similar slopes of the lines in the diagram. The kinks at high temperatures correspond to the evaporation of the metals; less pronounced kinks occur at the melting temperatures of the metals and the oxides.

Successful reduction of the oxide depends on the outcome of the competition of the carbon for the oxygen bound to the metal. The standard Gibbs energies for the reductions can be expressed in terms of the standard Gibbs energies for the reactions above:

$$MO(s) + C(s) \longrightarrow M(s) + CO(g) \qquad \Delta_rG^\ominus = \Delta_rG^\ominus(iii) - \Delta_rG^\ominus(i)$$
$$MO(s) + \tfrac{1}{2}C(s) \longrightarrow M(s) + \tfrac{1}{2}CO_2(g) \qquad \Delta_rG^\ominus = \Delta_rG^\ominus(ii) - \Delta_rG^\ominus(i)$$
$$MO(s) + CO(g) \longrightarrow M(s) + CO_2(g) \qquad \Delta_rG^\ominus = \Delta_rG^\ominus(iv) - \Delta_rG^\ominus(i)$$

The equilibrium lies to the right if $\Delta_rG^\ominus < 0$. This is the case when the line for reaction (i) lies below (is more positive than) the line for one of the reactions (ii) to (iv).

The spontaneity of a reduction at any temperature can be predicted simply by looking at the diagram: a metal oxide is reduced by any carbon reaction lying above it, because the overall reaction then has $\Delta_rG^\ominus < 0$. For example, CuO can be reduced to Cu at any temperature above room temperature. Even in the absence of carbon, Ag_2O decomposes when heated above 200°C because then the standard Gibbs energy for reaction (i) becomes positive (and the reverse reaction is then spontaneous). On the other hand, Al_2O_3 is not reduced by carbon until the temperature has been raised to above 2000°C.

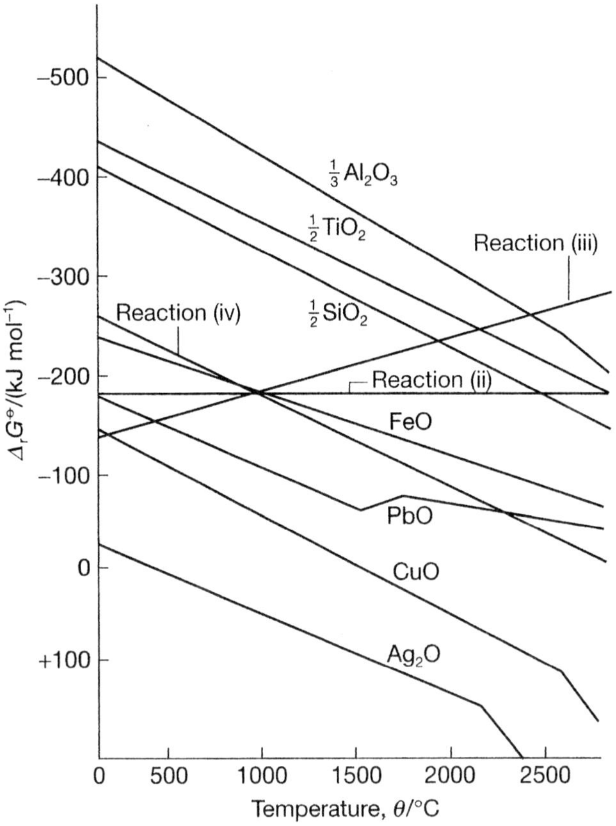

An Ellingham diagram for the discussion of metal ore reduction. Note that $\Delta_rG^\ominus$ is most negative at the top of the diagram.

Discussion questions

1. Use the Ellingham diagram in the illustration to identify the lowest temperature at which zinc oxide can be reduced to zinc metal by carbon.

2. How may an Ellingham diagram be used to decide whether one metal may be used to reduce the oxide of another metal?

Further reading

H.-G. Lee, *Chemical thermodynamics for metals and materials*. Imperial College Press, London (1999).

D.L. Johnson and G.B. Stracher (eds.), *Thermodynamic loop applications in materials systems*. Minerals, Metals, and Materials Society, Warrendale (1995).

9.5 The response of equilibria to pH

One of the most important examples of chemical equilibrium is the one that exists when acids and bases are present in solution. The proton transfer equilibria characteristic of solutions of Brønsted acids and bases[6] are established very rapidly (within a few nanoseconds) and are highly responsive to changes in the pH, where

$$\mathrm{pH} = -\log a_{\mathrm{H_3O^+}} \qquad [9.29]$$

and H_3O^+ is the hydronium ion, a representation of the state of the proton in aqueous solution. At low concentrations, the activity of hydronium ions is approximately equal to the numerical value of their molality or molar concentration, so a determination of pH is an indication of hydronium ion concentration. However, many thermodynamic observables depend on pH itself, and there is no need to make this approximation and interpretation.

(a) Acid–base equilibria in water

We need three equilibrium constants to discuss the effect of pH on equilibria involving the transfer of protons between species. One is the acidity constant, K_a:

$$\mathrm{HA(aq)} + \mathrm{H_2O(l)} \rightleftharpoons \mathrm{H_3O^+(aq)} + \mathrm{A^-(aq)} \qquad K_a = \frac{a_{\mathrm{H_3O^+}}a_{\mathrm{A^-}}}{a_{\mathrm{HA}}} \qquad (9.30)$$

It is common to report values of K_a in terms of its negative common logarithm, pK_a:

$$pK_a = -\log K_a \qquad [9.31]$$

The second equilibrium constant we require is the basicity constant, K_b, for a base B in water:

$$\mathrm{B(aq)} + \mathrm{H_2O(l)} \rightleftharpoons \mathrm{HB^+(aq)} + \mathrm{OH^-(aq)} \qquad K_b = \frac{a_{\mathrm{HB^+}}a_{\mathrm{OH^-}}}{a_{\mathrm{B}}} \qquad [9.32]$$

The third equilibrium to consider is the **autoprotolysis** of water, the equilibrium

$$2\,\mathrm{H_2O(l)} \rightleftharpoons \mathrm{H_3O^+(aq)} + \mathrm{OH^-(aq)} \qquad K_w = a_{\mathrm{H_3O^+}}a_{\mathrm{OH^-}} \qquad [9.33]$$

At 25°C, $K_w = 1.008 \times 10^{-14}$ ($pK_w = 14.00$), showing that only a few of the water molecules are ionized. If we introduce $\mathrm{pOH} = -\log a_{\mathrm{OH^-}}$ by analogy with pH, then it follows that

$$pK_w = \mathrm{pH} + \mathrm{pOH} \qquad (9.34)$$

As may be verified by multiplying the expressions for K_a and K_b together, the acidity constant of the conjugate acid HB^+ is related to the basicity constant of the base B by

$$K_aK_b = K_w \qquad (9.35)$$

The same relation holds between the acidity constant of an acid HA and the basicity constant of its conjugate base A^-. In each case, in elementary calculations it is common to replace activities by the numerical values of molar concentrations.

(b) The pH of acids and bases

The distinction between weak and strong acids and bases is an illustration of the different types of behaviour shown in Fig. 9.9: strong acids and bases are those for which the minimum Gibbs energy of the solution lies close to the products; weak

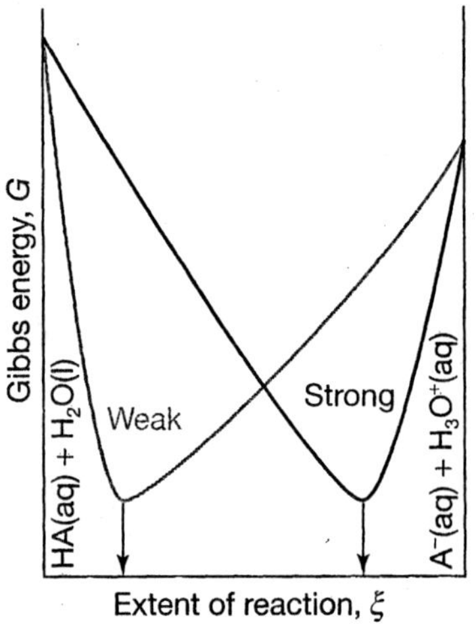

9.9 The Gibbs energy for a solution of a weak acid has a minimum that lies close to HA, and little deprotonation occurs at equilibrium. For a strong acid, the minimum lies close to products and deprotonation is almost complete.

6. It will be familiar from introductory chemistry that a Brønsted *acid* is a proton donor and a Brønsted *base* is a proton acceptor.

acids and bases are those for which the minimum lies close to the reactants. However, because the extent of proton transfer is so small (for typical solutions), an approximation is that the molar concentration of HA or B is unchanged from its nominal value. Moreover, because the molar concentrations of the species produced by the proton transfer are equal (to a good approximation),[7] the expression for K_a in eqn 30 simplifies to[8]

$$K_a \approx \frac{[H_3O^+]^2}{[HA]} \quad \text{or} \quad [H_3O^+] \approx (K_a[HA])^{1/2}$$

It then follows, by taking the negative logarithms of both sides, that

$$pH \approx \tfrac{1}{2}pK_a - \tfrac{1}{2}\log[HA] \qquad \{9.36\}$$

A similar expression can be derived from eqn 32 for the pH of a solution of a weak base: eqn 32 is used to determine pOH and then eqn 34 is used to convert pOH to pH. The result is

$$pH = pK_w - \tfrac{1}{2}pK_b + \tfrac{1}{2}\log[B] \qquad \{9.37\}$$

(c) Acid–base titrations

We shall suppose that we are titrating a volume V_A of a solution of a weak acid of nominal molar concentration A_0 (the analyte) with a solution of a strong base MOH of molar concentration B (the titrant). At the stoichiometric point of the titration,[9] enough titrant has been added for the amount of weak acid it supplies to match the amount of strong base in the analyte.

Our aim is to calculate the pH at any stage of the titration. The result can be calculated 'exactly' (see below) but it is more instructive to see how to make approximations at each stage. The 'exact' pH curve is shown in Fig. 9.10 together with the approximate formulas derived below.

When a volume V_B of the titrant has been added to the analyte the total volume of the analyte is $V = V_A + V_B$. The approximations we shall make are based on the fact that the acid is weak, and therefore that HA is more abundant than any A^- ions in the solution. Furthermore, when HA is present, it provides hydronium ions that greatly outnumber any that come from the autoprotolysis of water. Finally, when excess base is present, the OH^- ions it provides dominate any that come from the water autoprotolysis.

At the start of the titration, the analyte is a solution of a weak acid, so we can use eqn 36 to estimate its pH:

$$pH = \tfrac{1}{2}pK_a - \tfrac{1}{2}\log A_0 \qquad \{9.38\}$$

After the addition of some base (but before the stoichiometric point is reached), the concentration of A^- ions stems almost entirely from the salt that is present as a result of the reaction

$$HA(aq) + OH^-(aq) \longrightarrow A^-(aq) + H_2O(l)$$

We may neglect the relatively small contribution of A^- ions from the weak acid that remains after the reaction. Therefore, $[A^-] = S$, where S is the concentration of the salt that is present after the reaction between HA and the titrant. The amount of HA molecules that remains is the original amount, A_0V_A, less the amount of HA

7. That is, in an acid solution $[H_3O^+] \approx [A^-]$ and in a basic solution $[OH^-] \approx [HB^+]$.
8. As before, the notation {...} for the equation number reminds us that units (of [J]) have been discarded.
9. The stoichiometric point of a titration is still widely called the 'equivalence point'.

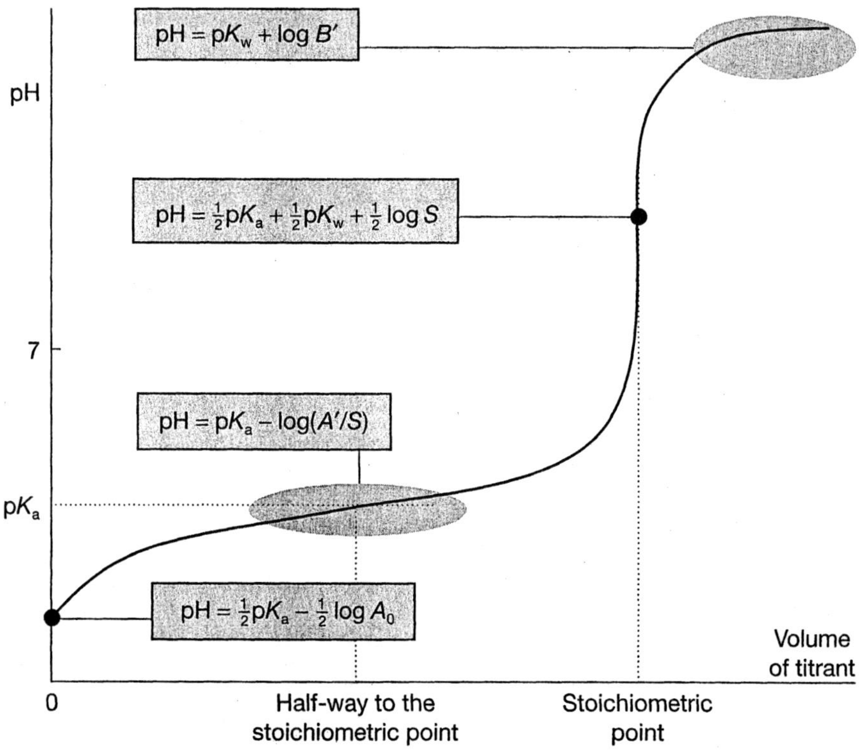

9.10 A summary of the regions of the pH curve of the titration of a weak acid with a strong base, and the equations used in different regions.

molecules that has been converted to salt, so the molar concentration of HA is $A' = (A_0V_A/V) - S$. This calculation ignores the small additional loss of HA as a result of its deprotonation in solution. Hence[10]

$$K_a = \frac{a_{H_3O^+}a_{A^-}}{a_{HA}} \approx \frac{a_{H_3O^+}S}{A'} \tag{9.39}$$

It follows that

$$pH = pK_a - \log\frac{A'}{S} \tag{9.40}$$

This expression is called the **Henderson–Hasselbalch equation**. The general form of this equation, once we recognize that A' is the molar concentration of acid in the solution and S is the molar concentration of base, is

$$pH = pK_a - \log\frac{[\text{acid}]}{[\text{base}]} \tag{9.41}$$

When the molar concentrations of acid and salt (acting as the base) are equal,

$$pH = pK_a \tag{9.42}$$

Hence the pK_a of the acid can be measured directly from the pH of the mixture. In practice this is done by recording the pH during a titration and then examining the record for the pH half-way to the stoichiometric point (as indicated in Fig. 9.10).

At the stoichiometric point, the H_3O^+ ions in the solution stem from the influence of the OH^- ions on the autoprotolysis equilibrium, and the OH^- ions are produced by

10. We are making the doubtful approximation that the activity coefficient of the A^- ions is close to 1.

the proton transfer equilibrium from H_2O to A^-. Because only a small amount of HA is formed in this way, the concentration of A^- ions is almost exactly that of the salt, and we can write $[A^-] \approx S$. The number of OH^- ions that arise from the proton transfer equilibrium greatly outnumber those produced by the water autoprotolysis, so we can set $[HA] \approx [OH^-]$:

$$K_b = \frac{a_{HA}a_{OH^-}}{a_{A^-}} \approx \frac{[OH^-]^2}{S}$$

Then, because $K_aK_b = K_w$, we can conclude that

$$[OH^-] \approx \left(\frac{SK_w}{K_a}\right)^{1/2} \qquad \{9.43\}$$

Therefore, after taking negative logarithms of both sides and using $pH = pK_w - pOH$,

$$pH = \tfrac{1}{2}pK_a + \tfrac{1}{2}pK_w + \tfrac{1}{2}\log S \qquad \{9.44\}$$

The corresponding expression for the pH at the stoichiometric point of a titration of a weak base with a strong acid is

$$pH = \tfrac{1}{2}pK_a - \tfrac{1}{2}\log S \qquad \{9.45\}$$

Illustration 9.6

The stoichiometric point of a titration of 25.00 mL of 0.100 M HClO(aq) with 0.100 M NaOH(aq) occurs when the molar concentration of NaClO is 0.050 mol L^{-1} (because the volume of the solution has increased from 25.00 mL to 50.00 mL). Therefore, the pH is

$$pH = \tfrac{1}{2} \times 7.43 + \tfrac{1}{2} \times 14.00 + \tfrac{1}{2}\log 0.050 = 10.1$$

Self-test 9.7 Estimate the pH at the stoichiometric point of a titration of 25.00 mL of 0.200 M NH_3(aq) with 0.300 M HCl(aq). [5.1]

When so much strong base has been added that the titration has been carried well past the stoichiometric point, the pH is determined by the excess base present. Then, $[H_3O^+] \approx K_w/[OH^-]$, so, if we write the molar concentration of excess base as B', this expression can be written

$$pH = pK_w + \log B' \qquad \{9.46\}$$

In this expression, as in all the preceding ones, the molar concentrations must take into account the change of volume that occurs as the titrant is added to the analyte.

(d) The 'exact' form of the pH curve

All the foregoing has supposed that one particular species dominates the pH of the mixture and we have invoked approximations to estimate the pH of the solution. In fact, no such approximations are necessary, for it is quite easy to set up and solve the 'exact' equation for the pH of the solution. We refer to 'exact' rather than exact, because we continue to ignore activity coefficients and treat all activities as molar concentrations. As we show in the following *Justification*, at any stage of the titration

$$v = \frac{K_aA_0[H_3O^+] + (K_w - [H_3O^+]^2)([H_3O^+] + K_a)}{([H_3O^+] + K_a)([H_3O^+]^2 + B_0[H_3O^+] - K_w)} \qquad v = \frac{V_B}{V_A} \qquad (9.47)$$

Justification 9.4

The five conditions we use are:

(a) Electrical neutrality: $[H_3O^+] + [M^+] = [A^-] + [OH^-]$, M^+ from added base

(b) Conservation of A groups: $[A^-] + [HA] = \dfrac{A_0V_A}{V_A + V_B}$

where V_A is the (fixed) initial volume of acid and V_B is the volume of titrant (base) added.

(c) Concentration of M^+ groups: $[M^+] = \dfrac{B_0V_B}{V_A + V_B}$

(d) Deprotonation equilibrium of HA: $[HA]K_a = [H_3O^+][A^-]$

(e) Autoprotolysis equilibrium: $K_w = [H_3O^+][OH^-]$

First, we express condition (b) in terms of $[H_3O^+]$ and $[A^-]$ by using condition (d) to eliminate [HA]:

$$[A^-] = \frac{A_0K_aV_A}{(V_A + V_B)([H_3O^+] + K_a)}$$

Next, we use this relation and condition (c), and at the same time we use condition (e) and (a) to eliminate $[OH^-]$:

$$[H_3O^+] + \frac{B_0V_B}{V_A + V_B} = \frac{K_aA_0V_A}{(V_A + V_B)([H_3O^+] + K_a)} + \frac{K_w}{[H_3O^+]}$$

Now it is just a matter of algebra: we multiply through by $(V_A + V_B)[H_3O^+]$, and then divide through by V_A. The result, after a small rearrangement of terms and writing $v = V_B/V_A$ is eqn 47.

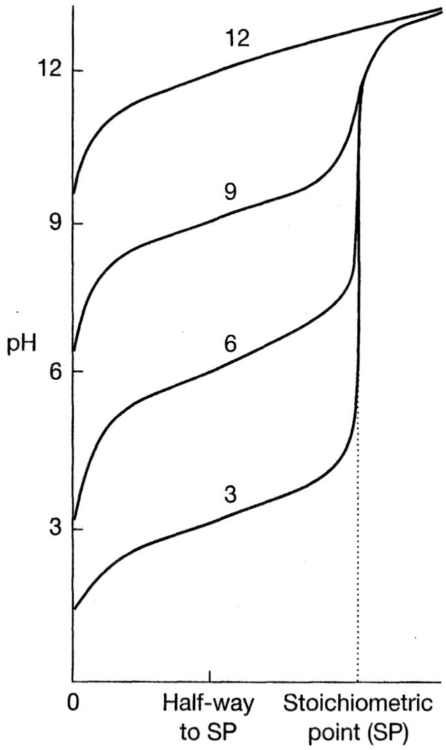

9.11 The 'exact' pH curves for the titration of a weak acid and strong base for different values of the pK_a of the acid.

Equation 47 is a cubic equation in the hydronium ion concentration. However, modern mathematical software makes such expressions easy to plot, and a plot of pH against volume of titrant added is shown in Fig. 9.11 for a variety of values of K_a. We see that the curves have exactly the form expected in the regions treated by approximation.

(e) Using equilibria to stabilize pH: buffers

The slow variation of the pH in the vicinity of $S = A'$, when the molar concentrations of the salt and acid are equal, is the basis of **buffer action**, the ability of a solution to oppose changes in pH when small amounts of strong acids and bases are added to the solution. The physical basis of buffer action is that the existence of an abundant supply of A^- ions (because a salt is present) can remove any H_3O^+ ions brought by additional strong acid; moreover, the numerous HA molecules can supply H_3O^+ ions to react with any strong base that is added.

Example 9.4 Estimating the pH of a buffer solution

Estimate the pH of an aqueous buffer solution that contains 0.200 mol L^{-1} KH_2PO_4 and 0.100 mol L^{-1} K_2HPO_4.

Table 9.1* Acidity constants in water at 298 K

	pK_{a1}	pK_{a2}	pK_{a3}
Acetic acid, CH_3COOH	4.76		
Ammonium ion, NH_4^+	9.25		
Carbonic acid, H_2CO_3	6.35	10.33	
Hydrocyanic acid	9.31		
Phosphoric acid, H_3PO_4	2.16	7.21	12.32

* More values are given in the *Data section*.

Method The pH of a solution of a weak acid and its salt can be estimated from the Henderson–Hasselbalch equation. To do so, we must identify the acid HA and its conjugate base A^-.

Answer In this example, the acid is the anion $H_2PO_4^-$ and its conjugate base is the anion HPO_4^{2-}:

$$H_2PO_4^-(aq) + H_2O(l) \rightleftharpoons H_3O^+(aq) + HPO_4^{2-}(aq)$$

The acidity constant we require is therefore pK_a for $H_2PO_4^-$ (the pK_{a2} of H_3PO_4) which from Table 9.1 is 7.21. Then, with $A' = 0.200$ mol L^{-1} and $S = 0.100$ mol L^{-1}, eqn 40 gives the pH of the solution as

$$pH = 7.21 - \log\frac{0.200}{0.100} = 6.91$$

Hence, the solution should buffer close to pH = 7.

Self-test 9.8 Calculate the pH of an aqueous buffer solution which contains 0.100 mol L^{-1} NH_3 and 0.200 mol L^{-1} NH_4Cl. [8.95; more realistically: 9]

(f) Using equilibria to detect changes in pH: indicators

The rapid change of pH near the stoichiometric point in a titration is the basis of indicator detection and of the more modern electronic detection techniques. An **acid–base indicator** is normally some large, water-soluble, weakly acidic organic molecule that can exist as acid (HIn) or conjugate base (In^-) forms that differ in colour. The two forms are in equilibrium in solution:

$$HIn(aq) + H_2O(l) \rightleftharpoons In^-(aq) + H_3O^+(aq)$$

If we make the usual assumption that the solution is so dilute that the activity of water is 1, then the equilibrium is described by the constant

$$K_{In} = \frac{a_{H_3O^+}a_{In^-}}{a_{HIn}} \quad [9.48]$$

We find the ratio of acid and base forms at a given pH by rearranging this expression to

$$\log\frac{[HIn]}{[In^-]} \approx pK_{In} - pH \quad (9.49)$$

Therefore, when $pH < pK_{In}$, the indicator is predominantly in its acidic form and has the corresponding colour; when the $pH > pK_{In}$, the indicator is mainly in its basic form. The **end point** is the pH of the solution when both forms are present in equal abundance, which occurs when $pH = pK_{In}$.

At the stoichiometric point of an acid–base titration, the pH changes sharply through several units, and if the pH passes through pK_{In} there is a pronounced colour change. With a well-chosen indicator, the end point coincides with the stoichiometric point of the titration.

Care must be taken to use an indicator that changes colour at the pH appropriate to the type of titration. Thus, in a weak acid–strong base titration, the stoichiometric point lies at the pH given by eqn 44, so an indicator that changes at that pH must be selected. Broadly speaking, an indicator with $pK_{In} > 7$ is required because the stoichiometric point lies at $pH > 7$. Similarly, in a strong acid–weak base titration, an indicator changing near the pH given by eqn 45 should be used. The stoichiometric points of such titrations lie at $pH < 7$, so an indicator with $pK_{In} < 7$ is required.

Checklist of key ideas

Spontaneous chemical reactions

9.1 The Gibbs energy minimum

- ☐ Extent of reaction (ξ): when the extent of reaction changes by a finite amount $\Delta\xi$, the amount of A present changes from $n_{A,0}$ to $n_{A,0} - \Delta\xi$
- ☐ reaction Gibbs energy, the slope of the graph of the Gibbs energy plotted against the extent of reaction: $\Delta_r G = (\partial G/\partial \xi)_{p,T}$
- ☐ relation of $\Delta_r G$ to the chemical potentials of the species, $\Delta_r G = \mu_B - \mu_A$
- ☐ condition of equilibrium, $\Delta_r G = 0$
- ☐ exergonic reaction, a reaction for which $\Delta_r G < 0$; such a reaction can be used to drive another process
- ☐ endergonic reaction, a reaction for which $\Delta_r G > 0$

9.2 The description of equilibrium

- ☐ general statement of a reaction, $0 = \sum_J \nu_J J$
- ☐ reaction quotient, $Q = \prod_J a_J^{\nu_J}$
- ☐ standard reaction Gibbs energy, $\Delta_r G^{\ominus}$
- ☐ equilibrium constant (K) in terms of $\Delta_r G^{\ominus}$, $\Delta_r G^{\ominus} = -RT \ln K$
- ☐ general expression for $\Delta_r G$ at an arbitrary stage of the reaction, $\Delta_r G = \Delta_r G^{\ominus} + RT \ln Q$
- ☐ $\Delta_r G^{\ominus}$ in terms of Gibbs energies of formation, $\Delta_r G^{\ominus} = \sum_{\text{Products}} \nu \Delta_f G^{\ominus} - \sum_{\text{Reactants}} \nu \Delta_f G^{\ominus} = \sum_J \nu_J \Delta_f G^{\ominus}(J)$
- ☐ thermodynamic equilibrium constant, an equilibrium constant K expressed in terms of activities (or fugacities): $K = (\prod_J a_J^{\nu_J})_{\text{equilibrium}}$
- ☐ relations between equilibrium constants, $K = K_\gamma K_b$
- ☐ equilibria in biological systems, $\Delta_r G^{\oplus} = \Delta_r G^{\ominus} + 7\nu RT \ln 10$

The response of equilibria to the conditions

- ☐ null effect of a catalyst

9.3 How equilibria respond to pressure

- ☐ response of K to changes in pressure, $(\partial K/\partial p)_T = 0$
- ☐ effect of pressure on composition
- ☐ Le Chatelier's principle: A system at equilibrium, when subjected to a disturbance, responds in a way that tends to minimize the effect of the disturbance

9.4 The response of equilibria to temperature

- ☐ Le Chatelier's principle for the effect of temperature: increased temperature favours the reactants in exothermic reactions and the products in endothermic reactions
- ☐ van 't Hoff equation for the effect of temperature on K, $d \ln K/dT = \Delta_r H^{\ominus}/RT^2$
- ☐ the equilibrium constant at one temperature in terms of its value at another temperature, $\ln K_2 - \ln K_1 = -(\Delta_r H^{\ominus}/R)(1/T_2 - 1/T_1)$

9.5 The response of equilibria to pH

- ☐ pH of a solution, $\text{pH} = -\log a_{H_3O^+}$
- ☐ acidity constant, $K_a = a_{H_3O^+}a_{A^-}/a_{HA}$
- ☐ the significance of $\text{p}K_a = -\log K_a$
- ☐ basicity constant, $K_b = a_{HB^+}a_{OH^-}/a_B$
- ☐ the relation between K_a and K_b, $K_aK_b = K_w$
- ☐ autoprotolysis constant, $K_w = a_{H_3O^+}a_{OH^-}$
- ☐ the relation between pH and pOH, $\text{pH} + \text{pOH} = \text{p}K_w$
- ☐ pH of a solution of a weak acid, $\text{pH} = \frac{1}{2}\text{p}K_a - \frac{1}{2}\log[\text{HA}]$
- ☐ changes in pH in the course of a titration, Fig. 9.10
- ☐ pH of a solution of a weak base, $\text{pH} = \text{p}K_w - \frac{1}{2}\text{p}K_b + \frac{1}{2}\log[\text{B}]$
- ☐ Henderson–Hasselbalch equation, $\text{pH} = \text{p}K_a - \log([\text{acid}]/[\text{base}])$
- ☐ the pH at the stoichiometric point of a titration of a weak acid with a strong base, $\text{pH} = \frac{1}{2}\text{p}K_a + \frac{1}{2}\text{p}K_w + \frac{1}{2}\log S$
- ☐ the pH after the stoichiometric point
- ☐ buffer action, the ability of a solution to oppose changes in pH when small amounts of strong acids and bases are added to the solution
- ☐ acid–base indicator, a water-soluble, weakly acidic organic molecule that can exist as acid (HIn) or conjugate base (In^-) forms that differ in colour
- ☐ indicator constant, $K_{In} = a_{H_3O^+}a_{In^-}/a_{HIn}$
- ☐ selection of indicators, $\text{p}K_{In} \approx$ pH of stoichiometric point

Further reading

Articles of general interest

R.T. Allsop and N.H. George, Le Chatelier—a redundant principle? *Educ. in Chem.* **21**, 54 (1984).

K. Anderson, Practical calculation of the equilibrium constant and the enthalpy of reaction at different temperatures. *J. Chem. Educ.* **71**, 474 (1994).

S. Chaston, Thermodynamic diagrams. *J. Chem. Educ.* **76**, 216 (1999).

N.C. Craig, The chemists' delta. *J. Chem. Educ.* **64**, 668 (1987).

R. deLevie, Explicit expression of the general form of the titration curve in terms of concentration: Writing a single closed-form expression for the titration curve for a variety of titrations without using approximations or segmentation. *J. Chem. Educ.* **70**, 209 (1993).

M. Deumiè, B. Boulil, and O. Henri-Rousseau, On the minimum of the Gibbs free energy involved in chemical equilibrium: Some further comments on the stability of the system. *J. Chem. Educ.* **64**, 201 (1987).

A. Dumon, A. Lichanot, and E. Poquet, Describing chemical transformation: From the extent of reaction ξ to the reaction advancement ratio χ. *J. Chem. Educ.* **70**, 29 (1993).

J. Gold and V. Gold, Le Chatelier's principle and the laws of van 't Hoff. *Educ. in Chem.* **22**, 82 (1985).

H.R. Kemp, The effect of temperature and pressure on equilibria: A derivation of the van 't Hoff rules. *J. Chem. Educ.* **64**, 482 (1987).

S.R. Logan, Entropy of mixing and homogeneous equilibria. *Educ. in Chem.* **25**, 44 (1988).

J.J. MacDonald, Equilibrium, free energy, and entropy: Rates and differences. *J. Chem. Educ.* **67**, 380 (1990).

J.J. MacDonald, Equilibria and $\Delta G^{\ominus}$. *J. Chem. Educ.* **67**, 745 (1990).

R.M. Rosenberg, I.M. Klotz, Spontaneity and the equilibrium constant: advantages of the Planck function. *J. Chem. Educ.* **76**, 1448 (1999).

G. Schmitz, The uncertainty of pH. *J. Chem. Educ.* **71**, 117 (1994).

M.J. Schultz, Why equilibrium? Understanding the role of entropy of mixing. *J. Chem. Educ.* **76**, 1391 (1999).

R.S. Treptow, Free energy versus extent of reaction: understanding the difference between ΔG and $\partial G/\partial \xi$. *J. Chem. Educ.* **73**, 51 (1996).

H. Xijun and Y. Xiuping, Influences of temperature and pressure on chemical equilibrium in non-ideal systems. *J. Chem. Educ.* **68**, 295 (1991).

Texts and sources of data and information

M.J. Blandamer, *Chemical equilibria in solution.* Ellis Horwood/Prentice Hall, Hemel Hempstead (1992).

W.A. Cramer and D.A. Knaff, *Energy transduction in biological membranes, a textbook of bioenergetics.* Springer Verlag, New York (1990).

W.E. Dasent, *Inorganic energetics.* Cambridge University Press (1982).

K. Denbigh, *The principles of chemical equilibrium, with applications in chemistry and chemical engineering.* Cambridge University Press (1981).

R.N. Goldberg and Y.B. Tewari, Thermodynamics of enzyme-catalyzed reactions. *J. Phys. Chem. Ref. Data.* Part 1: **22**, 515 (1993). Part 2: **23**, 547 (1994). Part 3: **23**, 1035 (1994). Part 4: **24**, 1669 (1995). Part 5: **24**, 1765 (1995).

F.M. Harold, *The vital force: a study of bioenergetics.* W.H. Freeman and Co., New York (1986).

R. Holub and P. Vónka, *The chemical equilibrium of gaseous systems.* Reidel, Boston (1976).

D.A. Johnson, *Some thermodynamic aspects of inorganic chemistry.* Cambridge University Press (1982).

W.R. Smith and R.W. Missen, *Chemical reaction equilibrium analysis.* Wiley, New York (1982).

Exercises

Oct 19

Discussion questions

9.1 (a) Explain how the mixing of reactants and products affects the position of chemical equilibrium.

9.1 (b) Suggest how the thermodynamic equilibrium constant may respond differently to changes in pressure and temperature from the equilibrium constant expressed in terms of partial pressures.

9.2 (a) Explain Le Chatelier's principle in terms of thermodynamic quantities.

9.2 (b) Explain the physical basis of the van 't Hoff equation for the temperature dependence of K.

9.3 (a) Describe the changes in pH that take place during the titration of a weak acid with a strong base.

9.3 (b) Describe the changes in pH that take place during the titration of a weak base with a strong acid.

9.4 (a) What approximations are made in the derivation of the Henderson–Hasselbalch equation?

9.4 (b) Describe the basis of buffer action and indicator detection.

Numerical exercises

9.5 (a) The equilibrium constant for the isomerization of *cis*-2-butene to *trans*-2-butene is $K = 2.07$ at 400 K. Calculate the standard reaction Gibbs energy.

9.5 (b) The equilibrium constant for the dissociation of Br_2 at 1600 K is $K = 0.255$. Calculate the standard reaction Gibbs energy.

9.6 (a) The standard reaction Gibbs energy of the isomerization of *cis*-2-pentene to *trans*-2-pentene at 400 K is -3.67 kJ mol^{-1}. Calculate the equilibrium constant of the isomerization.

9.6 (b) The standard reaction Gibbs energy of the decomposition of $CaCO_3$ to CaO and CO_2 at 1173 K is $+0.178$ kJ mol^{-1}. Calculate the equilibrium constant of the isomerization.

9.7 (a) At 2257 K and 1.00 atm total pressure, water is 1.77 per cent decomposed at equilibrium by way of the reaction $2\,H_2O(g) \rightleftharpoons 2\,H_2(g) + O_2(g)$. Calculate (a) K, (b) $\Delta_r G^{\ominus}$, and (c) $\Delta_r G$ at this temperature.

9.7 (b) For the equilibrium, $N_2O_4(g) \rightleftharpoons 2\,NO_2(g)$, the degree of dissociation, α_e, at 298 K is 0.201 at 1.00 bar total pressure. Calculate (a) $\Delta_r G$, (2) K, and (3) $\Delta_r G^{\ominus}$ at 298 K.

9.8 (a) Dinitrogen tetroxide is 18.46 per cent dissociated at 25°C and 1.00 bar in the equilibrium $N_2O_4(g) \rightleftharpoons 2\,NO_2(g)$. Calculate (a) K at 25°C, (b) $\Delta_r G^{\ominus}$, (c) K at 100°C given that $\Delta_r H^{\ominus} = +57.2$ kJ mol^{-1} over the temperature range.

9.2b **9.8 (b)** Molecular bromine is 24 per cent dissociated at 1600 K and 1.00 bar in the equilibrium $Br_2(g) \rightleftharpoons 2\,Br(g)$. Calculate (a) K at 25°C, (b) $\Delta_r G^{\ominus}$, (c) K at 2000°C given that $\Delta_r H^{\ominus} = +112$ kJ mol^{-1} over the temperature range.

9.9 (a) From information in the *Data section*, calculate the standard Gibbs energy and the equilibrium constant at (a) 298 K and (b) 400 K for the reaction $PbO(s) + CO(g) \rightleftharpoons Pb(s) + CO_2(g)$. Assume that the reaction enthalpy is independent of temperature.

9.9 (b) From information in the *Data section*, calculate the standard Gibbs energy and the equilibrium constant at (a) 25°C and (b) 50°C for the reaction $CH_4(g) + 3\,Cl_2(g) \rightleftharpoons CHCl_3(l) + 3\,HCl(g)$. Assume that the reaction enthalpy is independent of temperature.

9.10 (a) In the gas-phase reaction $2\,A + B \rightleftharpoons 3\,C + 2\,D$, it was found that, when 1.00 mol A, 2.00 mol B, and 1.00 mol D were mixed and allowed to come to equilibrium at 25°C, the resulting mixture contained 0.90 mol C at a total pressure of 1.00 bar. Calculate (a) the mole fractions of each species at equilibrium, (b) K_x, (c) K, and (d) $\Delta_r G^\ominus$.

9.10 (b) In the gas-phase reaction $A + B \rightleftharpoons C + 2\,D$, it was found that, when 2.00 mol A, 1.00 mol B, and 3.00 mol D were mixed and allowed to come to equilibrium at 25°C, the resulting mixture contained 0.79 mol C at a total pressure of 1.00 bar. Calculate (a) the mole fractions of each species at equilibrium, (b) K_x, (c) K, and (d) $\Delta_r G^\ominus$.

9.11 (a) The standard reaction enthalpy of $Zn(s) + H_2O(g) \rightarrow ZnO(s) + H_2(g)$ is approximately constant at +224 kJ mol^{-1} from 920 K up to 1280 K. The standard reaction Gibbs energy is +33 kJ mol^{-1} at 1280 K. Estimate the temperature at which the equilibrium constant becomes greater than 1.

9.11 (b) The standard enthalpy of a certain reaction is approximately constant at +125 kJ mol^{-1} from 800 K up to 1500 K. The standard reaction Gibbs energy is +22 kJ mol^{-1} at 1120 K. Estimate the temperature at which the equilibrium constant becomes greater than 1.

9.12 (a) The equilibrium constant of the reaction $2\,C_3H_6(g) \rightleftharpoons C_2H_4(g) + C_4H_8(g)$ is found to fit the expression

$$\ln K = -1.04 - \frac{1088}{T/K} + \frac{1.51 \times 10^5}{(T/K)^2}$$

between 300 K and 600 K. Calculate the standard reaction enthalpy and standard reaction entropy at 400 K.

9.12 (b) The equilibrium constant of a reaction is found to fit the expression

$$\ln K = -2.04 - \frac{1176}{T/K} + \frac{2.1 \times 10^7}{(T/K)^3}$$

between 400 K and 500 K. Calculate the standard reaction enthalpy and standard reaction entropy at 450 K.

9.13 (a) The standard reaction Gibbs energy of the isomerization of borneol ($C_{10}H_{17}OH$) to isoborneol in the gas phase at 503 K is +9.4 kJ mol^{-1}. Calculate the reaction Gibbs energy in a mixture consisting of 0.15 mol of borneol and 0.30 mol of isoborneol when the total pressure is 600 Torr.

9.13 (b) The equilibrium pressure of H_2 over solid uranium and uranium hydride, UH_3, at 500 K is 1.04 Torr. Calculate the standard Gibbs energy of formation of $UH_3(s)$ at 500 K.

9.14 (a) Calculate the percentage change in the equilibrium constant K_x of the reaction $H_2CO(g) \rightleftharpoons CO(g) + H_2(g)$ when the total pressure is increased from 1.0 bar to 2.0 bar at constant temperature.

9.14 (b) Calculate the percentage change in the equilibrium constant K_x of the reaction $CH_3OH(g) + NOCl(g) \rightleftharpoons HCl(g) + CH_3NO_2(g)$ when the total pressure is increased from 1.0 bar to 2.0 bar at constant temperature.

9.15 (a) The equilibrium constant for the gas-phase isomerization of borneol ($C_{10}H_{17}OH$) to isoborneol at 503 K is 0.106. A mixture consisting of 7.50 g of borneol and 14.0 g of isoborneol in a container of volume 5.0 L is heated to 503 K and allowed to come to equilibrium. Calculate the mole fractions of the two substances at equilibrium.

9.15 (b) The equilibrium constant for the reaction $N_2(g) + O_2(g) \rightleftharpoons 2\,NO(g)$ is 1.69×10^{-3} at 2300 K. A mixture consisting of 5.0 g of nitrogen and 2.0 g of oxygen in a container of volume 1.0 L is heated to 2300 K and allowed to come to equilibrium. Calculate the mole fraction of NO at equilibrium.

9.16 (a) Use the data in Table 2.6 of the *Data section* to decide which of the following reactions have $K > 1$ at 298 K.

(a) $HCl(g) + NH_3(g) \rightleftharpoons NH_4Cl(s)$

(b) $2\,Al_2O_3(s) + 3\,Si(s) \rightleftharpoons 3\,SiO_2(s) + 4\,Al(s)$

(c) $Fe(s) + H_2S(g) \rightleftharpoons FeS(s) + H_2(g)$

9.16 (b) Use the data in Table 2.6 of the *Data section* to decide which of the following reactions have $K > 1$ at 298 K.

(a) $FeS_2(s) + 2\,H_2(g) \rightleftharpoons Fe(s) + 2\,H_2S(g)$

(b) $2\,H_2O_2(l) + H_2S(g) \rightleftharpoons H_2SO_4(l) + 2\,H_2(g)$

9.17 (a) Which of the equilibria in Exercise 9.16a are favoured (in the sense of K increasing) by a rise in temperature at constant pressure?

9.17 (b) Which of the equilibria in Exercise 9.16b are favoured (in the sense of K increasing) by a reduction in temperature at constant pressure?

9.18 (a) What is the standard enthalpy of a reaction for which the equilibrium constant is (a) doubled, (b) halved when the temperature is increased by 10 K at 298 K?

9.18 (b) What is the standard enthalpy of a reaction for which the equilibrium constant is (a) doubled, (b) halved when the temperature is increased by 15 K at 310 K?

9.19 (a) The standard Gibbs energy of formation of $NH_3(g)$ is −16.5 kJ mol^{-1} at 298 K. What is the reaction Gibbs energy when the partial pressure of the N_2, H_2, and NH_3 (treated as perfect gases) are 3.0 bar, 1.0 bar, and 4.0 bar, respectively? What is the spontaneous direction of the reaction in this case?

9.19 (b) The dissociation vapour pressure of NH_4Cl at 427°C is 608 kPa but at 459°C it has risen to 1115 kPa. Calculate (a) the equilibrium constant, (b) the standard reaction Gibbs energy, (c) the standard enthalpy, (d) the standard entropy of dissociation, all at 427°C. Assume that the vapour behaves as a perfect gas and that $\Delta H^\ominus$ and $\Delta S^\ominus$ are independent of temperature in the range given.

9.20 (a) Estimate the temperature at which $CaCO_3$(calcite) decomposes.

9.20 (b) Estimate the temperature at which $CuSO_4 \cdot 5H_2O$ undergoes dehydration.

9.21 (a) At the half-way point in the titration of a weak acid with a strong base the pH was measured as 5.40. (a) What is the acidity constant and the pK_a of the acid? (b) What is the pH of the solution that is 0.015 M in the acid?

9.21 (b) At the half-way point in the titration of a weak acid with a strong base the pH was measured as 4.82. (a) What is the acidity constant and the pK_a of the acid? (b) What is the pH of the solution that is 0.025 M in the acid?

9.22 (a) Calculate the pH of (a) 0.10 M $NH_4Cl(aq)$, (b) 0.10 M $NaCH_3CO_2(aq)$, (c) 0.100 M $CH_3COOH(aq)$.

9.22 (b) Calculate the pH of (a) 0.10 M $NaHCO_2(aq)$, (b) 0.20 M $NaC_6H_5CO_2(aq)$, (c) 0.150 M $HCN(aq)$.

9.23 (a) Calculate the pH at the stoichiometric point of the titration of 25.00 mL of 0.100 M lactic acid with 0.150 M NaOH(aq).

9.23 (b) Calculate the pH at the stoichiometric point of the titration of 25.00 mL of 0.100 M chlorous acid with 0.175 M NaOH(aq). The pK_a of chlorous acid is 1.96.

9.24 (a) Sketch the pH curve of a solution containing 0.10 M $NaCH_3CO_2$(aq) and a variable amount of acetic acid.

9.24 (b) Sketch the pH curve of a solution containing 0.15 M $NaC_6H_5CO_2$(aq) and a variable amount of benzoic acid.

9.25 (a) From the information in Table 9.1, select suitable buffers for (a) pH = 2.2 and (b) pH = 7.0.

9.25 (b) From the information in Table 9.1, select suitable buffers for (a) pH = 4.6 and (b) pH = 10.8.

Problems†

Numerical problems

9.1 The equilibrium constant for the reaction, $I_2(s) + Br_2(g) \rightleftharpoons 2\,IBr(g)$ is 0.164 at 25°C. (a) Calculate $\Delta_r G^\ominus$ for this reaction. (b) Bromine gas is introduced into a container with excess solid iodine. The pressure and temperature are held at 0.164 atm and 25°C, respectively. Find the partial pressure of IBr(g) at equilibrium. Assume that all the bromine is in the liquid form and that the vapour pressure of iodine is negligible. (c) In fact, solid iodine has a measurable vapour pressure at 25°C. In this case, how would the calculation have to be modified?

9.2 Consider the decomposition of methane, $CH_4(g)$, into the elements $H_2(g)$ and C(s, graphite). (a) Given that $\Delta_f H^\ominus(CH_4, g) = -74.85$ kJ mol^{-1} and that $\Delta_f S^\ominus(CH_4, g) = -80.67$ J K^{-1} mol^{-1} at 298 K, calculate the value of the equilibrium constant at 298 K. (b) Assuming that $\Delta_f H^\ominus$ is independent of temperature, calculate K at 50°C. (c) Calculate the degree of dissociation, α_e, of methane at 25°C and a total pressure of 0.010 bar. (d) Without doing any numerical calculations, explain how the degree of dissociation for this reaction will change as the pressure and temperature are varied.

9.3 The equilibrium pressure of H_2 over U(s) and UH_3(s) between 450 K and 715 K fits the expression

$$\ln(p/\text{Pa}) = 69.32 - \frac{1.464 \times 10^4}{T/\text{K}} - 5.65\ln(T/\text{K})$$

Find an expression for the standard enthalpy of formation of UH_3(s) and from it calculate $\Delta_r C_p^\ominus$.

9.4 The degree of dissociation, α_e, of CO_2(g) into CO(g) and O_2(g) at high temperatures was found to vary with temperature as follows:

T/K	1395	1443	1498
$\alpha_e/10^{-4}$	1.44	2.50	4.71

Assuming $\Delta_r H^\ominus$ to be constant over this temperature range, calculate K, $\Delta_r G^\ominus$, $\Delta_r H^\ominus$, and $\Delta_r S^\ominus$. Make any justifiable approximations.

9.5 The standard reaction enthalpy for the decomposition of $CaCl_2 \cdot NH_3$(s) into $CaCl_2$(s) and NH_3(g) is nearly constant at +78 kJ mol^{-1} between 350 K and 470 K. The equilibrium pressure of NH_3 in the presence of $CaCl_2 \cdot NH_3$ is 12.8 Torr at 400 K. Find an expression for the temperature dependence of $\Delta_r G^\ominus$ in the same range.

9.6 Calculate the equilibrium constant of the reaction $CO(g) + H_2(g) \rightleftharpoons H_2CO(g)$ given that, for the production of liquid formaldehyde, $\Delta_r G^\ominus = +28.95$ kJ mol^{-1} at 298 K and that the vapour pressure of formaldehyde is 1500 Torr at that temperature.

9.7 Acetic acid was evaporated in container of volume 21.45 cm^3 at 437 K and at an external pressure of 764.3 Torr, and the container was then sealed. The mass of acid present in the sealed container was 0.0519 g. The experiment was repeated with the same container but at 471 K, and it was found that 0.0380 g of acetic acid was present. Calculate the equilibrium constant for the dimerization of the acid in the vapour and the enthalpy of vaporization.

9.8 A sealed container was filled with 0.300 mol H_2(g), 0.400 mol I_2(g), and 0.200 mol HI(g) at 870 K and total pressure 1.00 bar. Calculate the amounts of the components in the mixture at equilibrium given that $K = 870$ for the reaction $H_2(g) + I_2(g) \rightleftharpoons 2\,HI(g)$.

9.9 For the gas-phase reaction $2\,A + B \rightarrow 3\,C + 2\,D$, it was found that, when 1.00 mol A, 2.00 mol B, and 1.00 mol D were mixed and came to equilibrium at 1.00 bar, the resulting mixture contained 0.90 mol C. Calculate the equilibrium constant.

9.10 The dissociation of I_2 can be monitored by measuring the total pressure, and three sets of results are as follows:

T/K	973	1073	1173
$100p$/atm	6.244	7.500	9.181
$10^4 n_I$	2.4709	2.4555	2.4366

where n_I is the amount of I atoms per mole of I_2 molecules in the mixture, which occupied 342.68 cm^3. Calculate the equilibrium constants of the dissociation and the standard enthalpy of dissociation at the mean temperature.

9.11‡ Alberty *et al.* report thermochemical properties of polycyclic aromatic hydrocarbons, many estimated by group additivity methods (R.A. Alberty, M.B. Chung, and A.K. Reif, *J. Phys. Chem. Ref. Data* **18**, 77 (1989)). Some of their data are shown below for three isomeric pyrenes of formula $C_{20}H_{12}$ at 1000 K.

Species	$\Delta_f H^\ominus$/(kJ mol^{-1})	$S_m^\ominus$/(J K^{-1} mol^{-1})
perylene	+253.2	987.9
benzo(e)pyrene	+253.2	993.7
benzo(a)pyrene	+262.4	999.4

What are the equilibrium mole fractions of these three isomers at 1000 K?

9.12‡ Thorn *et al.* recently carried out a study of Cl_2O(g) by photoelectron ionization (R.P. Thorn, L.J. Stief, S.-C. Kuo, and

† Problems denoted with the symbol ‡ were supplied by Charles Trapp, Carmen Giunta, and Marshall Cady.

R.B. Klemm, *J. Phys. Chem.* **100**, 14 178 (1996)). From their measurements, they report $\Delta_f H^\ominus(Cl_2O) = +77.2$ kJ mol^{-1}. They combined this measurement with literature data on the reaction $Cl_2O\,(g) + H_2O(g) \rightarrow 2\,HOCl(g)$, for which $K = 8.2 \times 10^{-2}$ and $\Delta_r S^\ominus = +16.38$ J K^{-1} mol^{-1}, and with readily available thermodynamic data on water vapour to report a value for $\Delta_f H^\ominus(HOCl)$. Calculate that value. All quantities refer to 298 K.

9.13‡ The 1980s saw reports of $\Delta_f H^\ominus(SiH_2)$ ranging from 243 to 289 kJ mol^{-1}. For example, the lower value appeared was cited in the review article by R. Walsh (*Acc. Chem. Res.* **14**, 246 (1981)); Walsh now leans toward the upper end of the range (H.M. Frey, R. Walsh, and I.M. Watts, *J. Chem. Soc., Chem. Commun.* 1189 (1986)). The higher value was reported in S.-K. Shin and J.L. Beauchamp, *J. Phys. Chem.* **90**, 1507 (1986). If the standard enthalpy of formation is uncertain by this amount, by what factor is the equilibrium constant for the formation of SiH_2 from its elements uncertain at (a) 298 K, (b) 700 K?

Theoretical problems

9.14 Show that, if K_p increases with pressure, then K_ϕ must decrease, where $K = K_p K_\phi$, ϕ denoting the fugacity coefficient.

9.15 Express the equilibrium constant of a gas-phase reaction $A + 3\,B \rightleftharpoons 2\,C$ in terms of the equilibrium value of the extent of reaction, ξ, given that initially A and B were present in stoichiometric proportions. Find an expression for ξ as a function of the total pressure, p, of the reaction mixture and sketch a graph of the expression obtained.

9.16 When light passes through a cell of length l containing an absorbing gas at a pressure p, the absorption is proportional to pl. Consider the equilibrium $2\,NO_2 \rightleftharpoons N_2O_4$, with NO_2 the absorbing species. Show that, when two cells of lengths l_1 and l_2 are used, and the pressures needed to obtain equal absorptions are p_1 and p_2, respectively, then the equilibrium constant is given by

$$K = \frac{(p_1\rho^2 - p_2)^2}{\rho(\rho - 1)(p_2 - p_1\rho)p^\ominus}$$

with $\rho = l_1/l_2$. The following data were obtained (R.J. Nordstrum and W.H. Chan, *J. Phys. Chem.* **80**, 847 (1976)):

Absorbance	p_1/Torr	p_2/Torr
0.05	1.00	5.47
0.10	2.10	12.00
0.15	3.15	18.65

with $l_1 = 395$ mm and $l_2 = 75$ mm. Determine the equilibrium constant of the reaction.

9.17 Find an expression for the standard reaction Gibbs energy at a temperature T' in terms of its value at another temperature T and the coefficients a, b, and c in the expression for the molar heat capacity listed in Table 2.2. Evaluate the standard Gibbs energy of formation of $H_2O(l)$ at 372 K from its value at 298 K.

9.18 Derive an 'exact' expression for $v = V_B/V_A$ for the titration of an aqueous solution of a weak base with an aqueous solution of a strong acid. *Hint.* See *Justification 9.4*.

Applications: to biology, environmental science, and chemical engineering

9.19 Hydrogen and carbon monoxide have been investigated for use in fuel cells, so their solubilities in molten salts are of interest. Their solubilities in a molten $NaNO_3/KNO_3$ mixture were examined (E. Desimoni and P.G. Zambonin, *J. Chem. Soc. Faraday Trans. 1*, 2014 (1973)) with the following results:

$$\log s_{H_2} = -5.39 - \frac{768}{T/K} \qquad \log s_{CO} = -5.98 - \frac{980}{T/K}$$

where s is the solubility in mol cm^{-3} bar^{-1}. Calculate the standard molar enthalpies of solution of the two gases at 570 K.

9.20 Under biochemical standard conditions, aerobic respiration produces approximately 38 molecules of ATP per molecule of glucose that is completely oxidized. (a) What is the percentage efficiency of aerobic respiration under biochemical standard conditions? (b) The following conditions are more likely to be observed in a living cell: $p_{CO_2} = 5.3 \times 10^{-2}$ atm, $p_{O_2} = 0.132$ atm, [glucose] $= 5.6 \times 10^{-2}$ mol L^{-1}, [ATP] = [ADP] = [P_i] $= 1.0 \times 10^{-4}$ mol L^{-1}, pH = 7.4, $T = 310$ K. Assuming that activities can be replaced by the numerical values of molar concentrations, calculate the efficiency of aerobic respiration under these physiological conditions. (c) A typical diesel engine operates between $T_c = 873$ K and $T_h = 1923$ K with an efficiency that is approximately 75 per cent of the theoretical limit of $(1 - T_c/T_h)$ (see Section 4.2). Compare the efficiency of a typical diesel engine with that of aerobic respiration under typical physiological conditions (see part b). Why is biological energy conversion more or less efficient than energy conversion in a diesel engine?

9.21 It has been observed in living cells that the anion ATP^{4-} forms a 1:1 complex with the Mg^{2+} ion. The cation and anion form two coordinate covalent bonds: one that involves a negatively charged oxygen belonging to the terminal phosphate group of ATP^{4-} and another that involves a negatively charged oxygen belonging to the phosphate group adjacent to the terminal phosphate group of ATP^{4-}. Is hydrolysis of $MgATP^{2-}$ more or less exergonic than hydrolysis of ATP^{4-} under identical intracellular conditions? Explain your answer. *Hint.* See Discussion question 1 of Box 9.1.

9.22‡ The dimerization of ClO in the Antarctic winter stratosphere is believed to play an important part in that region's severe seasonal depletion of ozone. The following equilibrium constants are based on measurements by R.A. Cox and G.D. Hayman (*Nature* **332**, 796 (1988)) on the reaction $2\,ClO(g) \rightarrow (ClO)_2(g)$.

T/K	233	248	258	268	273
K	4.13×10^8	5.00×10^7	1.45×10^7	5.37×10^6	3.20×10^6

T/K	280	288	295	303
K	9.62×10^5	4.28×10^5	1.67×10^5	7.02×10^4

(a) Derive the values of $\Delta_r H^\ominus$ and $\Delta_r S^\ominus$ for this reaction. (b) Compute the standard enthalpy of formation and the standard molar entropy of $(ClO)_2$ given $\Delta_f H^\ominus(ClO) = +101.8$ kJ mol^{-1} and $S_m^\ominus(ClO) = 226.6$ J K^{-1} mol^{-1} (*CRC Handbook*, 2000).

9.23‡ Acid rain is an environmental concern in many parts of the world. In assessing the acidity of rainfall, it is important to have an idea of the acidity of natural rainwater. Assuming that natural rainwater (that is, uncontaminated with nitric or sulfuric acids) is in equilibrium with 3.6×10^{-4} atm CO_2 (Henry's law constant 1.25×10^6 Torr), what is the pH of natural rainwater? What would the pH of natural rainwater have been in pre-industrial times, when the partial pressure of CO_2 was about 2.8×10^{-4} atm? State your approximations explicitly.

9.24‡ Nitric acid hydrates have received much attention as possible catalysts for heterogeneous reactions that bring about the Antarctic ozone hole. Worsnop *et al.* investigated the thermodynamic stability of these hydrates under conditions typical of the polar winter stratosphere (D.R. Worsnop, L.E. Fox, M.S. Zahniser, and S.C. Wofsy, *Science* **259**, 71 (1993)). Standard reaction Gibbs energies can be computed for the following reactions at 190 K from their data:

(i) $H_2O(g) \longrightarrow H_2O(s)$ $\Delta_r G^{\ominus} = -23.6\ \text{kJ mol}^{-1}$

(ii) $H_2O(g) + HNO_3(g) \longrightarrow HNO_3 \cdot H_2O(s)$ $\Delta_r G^{\ominus} = -57.2\ \text{kJ mol}^{-1}$

(iii) $2\,H_2O(g) + HNO_3(g) \longrightarrow HNO_3 \cdot 2H_2O(s)$ $\Delta_r G^{\ominus} = -85.6\ \text{kJ mol}^{-1}$

(iv) $3\,H_2O(g) + HNO_3(g) \longrightarrow HNO_3 \cdot 3H_2O(s)$ $\Delta_r G^{\ominus} = -112.8\ \text{kJ mol}^{-1}$

Which solid is thermodynamically most stable at 190 K if $p_{H_2O} = 1.3 \times 10^{-7}$ bar and $p_{HNO_3} = 4.1 \times 10^{-10}$ bar? (*Hint.* Try computing $\Delta_r G^{\ominus}$ for each reaction under the prevailing conditions; if more than one solid forms spontaneously, examine $\Delta_r G^{\ominus}$ for the conversion of one solid to another.)

9.25‡ The ocean's ability to dissolve carbon is an important sink when modelling the carbon cycle. (a) Draw a graph of α_J versus pH for carbon dioxide solutions in the pH range 0–14 where the fraction α_J of species J is defined as

$$\alpha_J = \frac{[J]}{[\text{total carbon}]} = \frac{[J]}{[CO_2] + [H_2CO_3] + [H_2CO_3^-] + [CO_3^{2-}]}$$

Assume that the solutions are ideal. (b) What are the major chemical species at the physiological pH of 7.4 and equilibrium? (c) What is the maximum mass of ocean carbon dissolved as CO_2, H_2CO_3, HCO_3^-, CO_3^{2-} if the ocean has an average pH of 8, an average temperature of 298 K, and a volume of $1.37 \times 10^{18}\ \text{m}^3$? The partial pressure of atmospheric carbon dioxide is 3.3×10^{-4} atm. *Hint.* Use mathematical software.

9.26‡ Refer to Problem 8.18. (a) Draw an Ellingham diagram for the smelting reduction of silica with graphite in the temperature range 1500–2000 K and at 1 bar. What does the diagram reveal about the evolution of carbon dioxide during smelting? (b) Determine the equilibrium purity of silicon made by reducing silica with graphite at 2000 K and 1 bar. Assume that graphite is soluble in the silicon but that SiO_2 and SiC are not. Also determine the equilibrium partial pressures of SiO(g) and CO(g). Equilibria to consider include $SiO_2(l) + C(\text{melt}) \rightleftharpoons SiO(g) + CO(g)$, $SiO_2(l) + 2\,C(\text{melt}) \rightleftharpoons Si(l) + 2\,CO(g)$, and $SiO_2(l) + 3\,C(\text{melt}) \rightleftharpoons SiC(s) + 2\,CO(g)$. *Hint.* Use mathematical software.

9.27‡ Suppose that an iron catalyst at a particular manufacturing plant produces ammonia in the most cost-effective manner at 450°C when the pressure is such that $\Delta_r G$ for the reaction $\frac{1}{2}N_2(g) + \frac{2}{3}H_2(g) \rightarrow NH_3(g)$ is equal to $-500\ \text{J mol}^{-1}$. (a) What pressure is needed? (b) Now suppose that a new catalyst is developed that is most cost-effective at 400°C when the pressure gives the same value of $\Delta_r G$. What pressure is needed when the new catalyst is used? What are the advantages of the new catalyst? Assume that (i) all gases are perfect gases or that (ii) all gases are van der Waals gases. Isotherms of $\Delta_r G(T, p)$ in the pressure range 100 atm $\le p \le$ 400 atm are needed to derive the answer. (c) Do the isotherms you plotted confirm Le Chatelier's principle concerning the response of equilibrium changes in temperature and pressure?

10 Equilibrium electrochemistry

The principles of thermodynamics established in the preceding chapters can be applied to solutions of electrolytes. One important extension of the previous material, however, is the need to take into account activity coefficients, for they differ significantly from 1 on account of the strong ionic interactions in electrolyte solutions. These coefficients are best treated as empirical quantities, but it is possible to estimate them in very dilute solutions. The bulk of this chapter is concerned with the description of the thermodynamic properties of reactions that take place in electrochemical cells, in which, as the reaction proceeds, it drives electrons through an external circuit. Thermodynamic arguments can be used to derive an expression for the electric potential of such cells and the potential can be related to their composition. There are two major topics developed in this connection. One is the definition and tabulation of standard potentials; the second is the use of these standard potentials to predict the equilibrium constants and other thermodynamic properties of chemical reactions.

Although we can discuss the thermodynamic properties of electrolyte solutions in terms of chemical potentials and activities in much the same way as solutions of non-electrolytes, electrolyte solutions have a number of distinctive features. One is the presence of strong interactions between ions in solution, which means that deviations from ideality are marked even in dilute systems. Therefore, we must equip ourselves with means of dealing with activity coefficients that differ significantly from 1. A second feature is that, because many reactions of ions involve the transfer of electrons, they can be studied (and utilized) by allowing them to take place in an electrochemical cell. Measurements like those described in this chapter provide data that are very useful for discussing the characteristics of electrolyte solutions and of ionic equilibria in solution.

The thermodynamic properties of ions in solution

Many of the concepts described in previous chapters carry over without change into the discussion of electrolyte solutions. In this section we focus on departures from ideality.

10.1 Thermodynamic functions of formation

The standard enthalpy and Gibbs energy of a reaction involving ions in solution are expressed in terms of standard enthalpies and Gibbs energies of formation listed in Table 2.6 in the *Data section*. These properties are used in exactly the same way as those for neutral compounds.

(a) Standard functions of formation of ions

The values of $\Delta_f H^\ominus$ and $\Delta_f G^\ominus$ refer to the formation of solutions of ions from the reference states of the parent elements. A major problem, however, is that solutions of cations cannot be prepared without their accompanying anions. Therefore, although the standard enthalpy of an overall reaction such as

$$Ag(s) + \tfrac{1}{2}Cl_2(g) \longrightarrow Ag^+(aq) + Cl^-(aq)$$

for which

$$\Delta_r H^\ominus = \Delta_f H^\ominus(Ag^+, aq) + \Delta_f H^\ominus(Cl^-, aq)$$

is meaningful and measurable (and found to be -61.58 kJ mol^{-1}), the enthalpies of the individual formation reactions are not measurable.

This problem is solved by defining one ion, conventionally the hydrogen ion, to have zero standard enthalpy and Gibbs energy of formation at all temperatures:

$$\Delta_f H^\ominus(H^+, aq) = 0 \qquad \Delta_f G^\ominus(H^+, aq) = 0 \qquad [10.1]$$

In essence, this definition adjusts the actual values of the enthalpies and Gibbs energies of formation of ions by a fixed amount, which is chosen so that the standard value for one of them, $H^+(aq)$, has the value zero. Then for the reaction

$$\tfrac{1}{2}H_2(g) + \tfrac{1}{2}Cl_2(g) \longrightarrow H^+(aq) + Cl^-(aq) \qquad \Delta_r G^\ominus = -131.23 \text{ kJ mol}^{-1}$$

we can write

$$\Delta_r G^\ominus = \Delta_f G^\ominus(H^+, aq) + \Delta_f G^\ominus(Cl^-, aq)$$
$$= \Delta_f G^\ominus(Cl^-, aq)$$

and hence identify $\Delta_f G^\ominus(Cl^-, aq)$ as -131.23 kJ mol^{-1}. All the Gibbs energies and enthalpies of formation of ions in Tables 10.1 and 2.6 were calculated in the same way.

Table 10.1* Standard thermodynamic functions of formation of ions in aqueous solution at 298 K

Ion	$\Delta_f H^\ominus$/(kJ mol^{-1})	$\Delta_f G^\ominus$/(kJ mol^{-1})
Cl^-	−167.2	−131.2
Cu^{2+}	+64.8	+65.5
H^+	0	0
K^+	−252.4	−283.3
Na^+	−240.1	−261.9
PO_4^{3-}	−1277.0	−1019.0

* More values are given in the *Data section*; see Table 2.6.

Illustration 10.1

With the value of $\Delta_f G^\ominus(Cl^-, aq)$ established, we can find the value of $\Delta_f G^\ominus(Ag^+, aq)$ from

$$Ag(s) + \tfrac{1}{2}Cl_2(g) \longrightarrow Ag^+(aq) + Cl^-(aq) \qquad \Delta_r G^\ominus = -54.12 \text{ kJ mol}^{-1}$$

which leads to $\Delta_f G^\ominus(Ag^+, aq) = +77.11$ kJ mol^{-1}.

Self-test 10.1 The standard enthalpy of formation of $AgNO_3(aq)$ is -99.4 kJ mol^{-1} at 298 K. Calculate the standard enthalpy of formation of the nitrate ion in water by using the standard enthalpy of formation of $Ag^+(aq)$ in Table 2.6. [-205.0 kJ mol^{-1}]

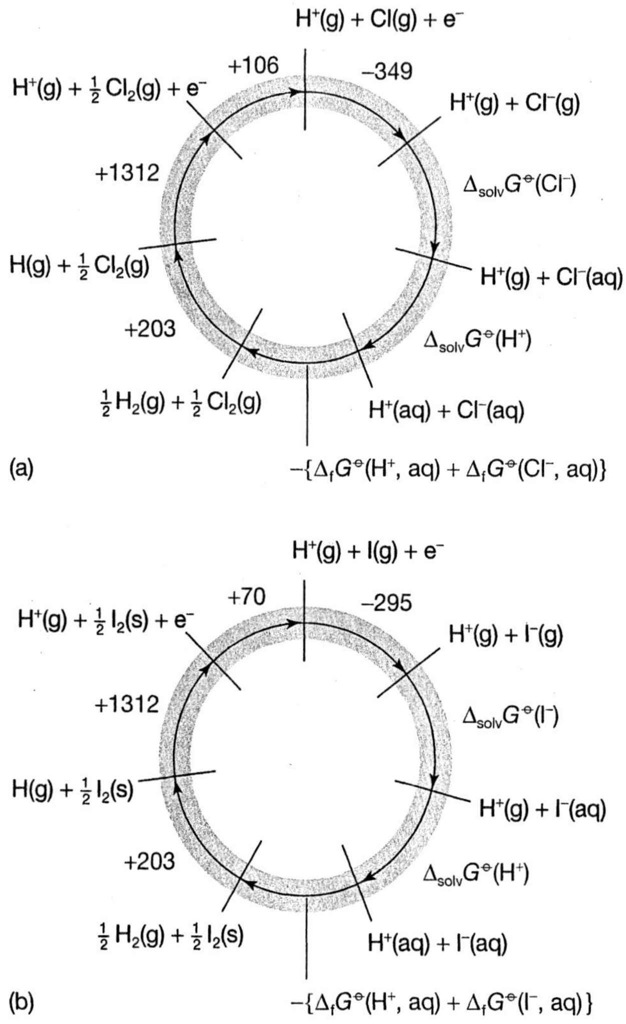

10.1 The thermodynamic cycles for the discussion of the Gibbs energies of solvation (hydration) and formation of (a) chloride ions, (b) iodide ions in aqueous solution. The sum of the changes in Gibbs energies around the cycle sum to zero because *G* is a state function.

(b) Contributions to the Gibbs energy of formation

The factors responsible for the magnitude of the Gibbs energy of formation of an ion in solution can be identified by analysing it in terms of a thermodynamic cycle. As an illustration, we consider the standard Gibbs energy of formation of Cl^- in water, which is −131 kJ mol^{-1}. We do so by treating the formation reaction

$$\tfrac{1}{2}H_2(g) + \tfrac{1}{2}X_2(g) \longrightarrow H^+(aq) + X^-(aq)$$

as the outcome of the sequence of steps shown in Fig. 10.1 (with values taken from the *Data section*, particularly Tables 2.6, 13.4, and 13.5).[1] The sum of the Gibbs energies for all the steps around a closed cycle is zero, so

$$\Delta_f G^\ominus(Cl^-, aq) = 1272\ \text{kJ mol}^{-1} + \Delta_{solv}G^\ominus(H^+) + \Delta_{solv}G^\ominus(Cl^-)$$

1. The standard Gibbs energies of formation of the gas-phase ions are unknown. We have therefore used ionization energies and electron affinities (Section 13.4f), and have assumed that any differences from the Gibbs energies arising from conversion to enthalpy and the inclusion of entropies to obtain Gibbs energies in the formation of H^+ are cancelled by the corresponding terms in the electron gain of X. The conclusions from the cycles are therefore only approximate.

Table 10.2* Relative permittivities (dielectric constants) at 298 K

	ε_r
Ammonia	16.9 22.4 (at −33°C)
Benzene	2.274
Ethanol	24.30
Water	78.54

* More values are given in the *Data section*.

An important point to note is that the value of $\Delta_f G^\ominus$ of an ion X is not determined by the properties of X alone but includes contributions from the dissociation, ionization, and hydration of hydrogen.

Gibbs energies of solvation of individual ions may be estimated from an equation derived by Max Born, who identified $\Delta_{solv}G^\ominus$ with the electrical work of transferring an ion from a vacuum into the solvent treated as a continuous dielectric of relative permittivity ε_r (Table 10.2). The resulting **Born equation**, which is derived in the *Justification* below, is

$$\Delta_{solv}G^\ominus = -\frac{z_i^2 e^2 N_A}{8\pi\varepsilon_0 r_i}\left(1 - \frac{1}{\varepsilon_r}\right) \tag{10.2}$$

where z_i is the charge number of the ion and r_i its radius (N_A is Avogadro's constant).[2] Note that $\Delta_{solv}G^\ominus < 0$, and that $\Delta_{solv}G^\ominus$ is strongly negative for small, highly charged ions in media of high relative permittivity. For water at 25°C,

$$\Delta_{solv}G^\ominus = -\frac{z_i^2}{(r_i/\text{pm})} \times (6.86 \times 10^4 \text{ kJ mol}^{-1}) \tag{10.3}$$

Justification 10.1

The electrical concepts required in this derivation are reviewed in *Further information 2*. The strategy of the calculation is to identify the Gibbs energy of solvation with the work of transferring an ion from a vacuum into the solvent. That work is calculated by taking the difference of the work of charging an ion when it is in the solution and the work of charging the same ion when it is in a vacuum. We model an ion as a sphere of radius r_i immersed in a medium of permittivity ε. When the charge of the sphere is q, the electric potential, ϕ, at its surface is

$$\phi = \frac{q}{4\pi\varepsilon r_i}$$

The work of bringing up a charge dq to the sphere is $\phi\, dq$. Therefore, the total work of charging the sphere from 0 to $z_i e$ is

$$w = \int_0^{z_i e} \phi\, dq = \frac{1}{4\pi\varepsilon r_i}\int_0^{z_i e} q\, dq = \frac{z_i^2 e^2}{8\pi\varepsilon r_i}$$

This electrical work of charging, when multiplied by Avogadro's constant, is the molar Gibbs energy for charging the ions.

The work of charging an ion in a vacuum is obtained by setting $\varepsilon = \varepsilon_0$, the vacuum permittivity. The corresponding value for charging the ion in a medium is obtained by setting $\varepsilon = \varepsilon_r\varepsilon_0$, where ε_r is the relative permittivity of the medium. It follows that the change in molar Gibbs energy that accompanies the transfer of ions from a vacuum to a solvent is the difference of these two quantities:

$$\Delta_{solv}G^\ominus = \frac{z_i^2 e^2 N_A}{8\pi\varepsilon r_i} - \frac{z_i^2 e^2 N_A}{8\pi\varepsilon_0 r_i}$$

which can easily be rearranged into eqn 2.

2. Ionic radii are given in Table 23.3.

Illustration 10.2

To see how closely the Born equation reproduces the experimental data, we calculate the difference in the values of $\Delta_f G^{\ominus}$ for Cl^- and I^- in water, for which $\varepsilon_r = 78.54$ at 25°C, given their radii as 181 pm and 220 pm (Table 23.3), respectively, is

$$\Delta_{solv}G^{\ominus}(Cl^-) - \Delta_{solv}G^{\ominus}(I^-) = -\left(\frac{1}{181} - \frac{1}{220}\right) \times (6.86 \times 10^4 \text{ kJ mol}^{-1})$$

$$= -67 \text{ kJ mol}^{-1}$$

This estimated difference is in good agreement with the experimental difference, which is -61 kJ mol^{-1}.

Self-test 10.2 Estimate the value of $\Delta_{solv}G^{\ominus}(Cl^-, aq) - \Delta_{solv}G^{\ominus}(Br^-, aq)$ from experimental data and from the Born equation.

[-27 kJ mol^{-1} experimental; -29 kJ mol^{-1} calculated]

(c) Standard entropies of ions in solution

Although the partial molar entropy of the solute in an electrolyte solution can be measured, there is no experimental way of ascribing a part of that entropy to the cations and a part to the anions. Therefore, yet again we are forced to define the partial molar entropy of one species and set up a table of values for other ions on that basis. The entropies of ions in solution are reported on a scale in which the standard entropy of the H^+ ions in water is taken as zero at all temperatures:

$$S^{\ominus}(H^+, aq) = 0 \qquad [10.4]$$

The values based on this choice are listed in Table 10.3 and more fully in Table 2.6 in the *Data section*.

Because the entropies of ions in water are values relative to the hydrogen ion in water, they may be either positive or negative. A positive entropy means that an ion has a higher partial molar entropy than H^+ in water and a negative entropy means that the ion has a lower partial molar entropy than H^+ in water. For instance, the entropy of Cl^-(aq) is +57 J K^{-1} mol^{-1} and that of Mg^{2+}(aq) is -128 J K^{-1} mol^{-1}. Partial molar ion entropies vary as expected on the basis that they are related to the degree to which the ions order the water molecules around them in the solution. Small, highly charged ions induce local structure in the surrounding water, and the disorder of the solution is decreased more than in the case of large, singly charged ions. The absolute, Third-law standard partial molar entropy of the proton in water can be estimated by proposing a model of the structure it induces, and there is some agreement on the value -21 J K^{-1} mol^{-1}. The negative value indicates that the proton induces order in the solvent.

Table 10.3* Standard entropies of ions in aqueous solution at 298 K

	$S_m^{\ominus}$/(J K^{-1} mol^{-1})	
Cl^-	+56.5	
Cu^{2+}	−99.6	
H^+	0	by definition
K^+	−102.5	
Na^+	−59.0	
PO_4^{3-}	−221.8	

* More values are given in the *Data section*; see Table 2.6.

10.2 Ion activities

Interactions between ions are so strong that the approximation of replacing activities by molalities is valid only in very dilute solutions (less than 10^{-3} mol kg^{-1} in total ion concentration) and in precise work activities themselves must be used.

(a) The definition of activity

We saw in Section 7.7c that the chemical potential of a solute in a real solution is related to its activity a by

$$\mu = \mu^{\ominus} + RT \ln a \tag{10.5}$$

where the standard state is a hypothetical solution with molality $b^{\ominus} = 1$ mol kg^{-1} in which the ions are behaving ideally. The activity is related to the molality,[3] b, by

$$a = \gamma \frac{b}{b^{\ominus}} \tag{10.6}$$

where the activity coefficient, γ, depends on the composition, molality, and temperature of the solution. As the solution approaches ideality (in the sense of obeying Henry's law) at low molalities, the activity coefficient tends towards 1:

$$\gamma \to 1 \quad \text{and} \quad a \to b/b^{\ominus} \quad \text{as} \quad b \to 0 \tag{10.7}$$

Because all deviations from ideality are carried in the activity coefficient, the chemical potential can be written[4]

$$\mu = \mu^{\ominus} + RT \ln b + RT \ln \gamma = \mu^{\text{ideal}} + RT \ln \gamma \qquad \{10.8\}$$

where μ^{ideal} is the chemical potential of the ideal-dilute solution of the same molality.

(b) Mean activity coefficients

If the chemical potential of a univalent cation M^+ is denoted μ_+ and that of a univalent anion X^- is denoted μ_-, the total molar Gibbs energy of the ions in the electrically neutral solution is the sum of these partial molar quantities. The molar Gibbs energy of an *ideal* solution is

$$G_{\text{m}}^{\text{ideal}} = \mu_+^{\text{ideal}} + \mu_-^{\text{ideal}} \qquad (10.9)^{\circ}$$

However, for a *real* solution of M^+ and X^- of the same molality,

$$G_{\text{m}} = \mu_+ + \mu_- = \mu_+^{\text{ideal}} + \mu_-^{\text{ideal}} + RT \ln \gamma_+ + RT \ln \gamma_- = G_{\text{m}}^{\text{ideal}} + RT \ln \gamma_+\gamma_- \tag{10.10}$$

All the deviations from ideality are contained in the last term.

There is no experimental way of separating the product $\gamma_+\gamma_-$ into contributions from the cations and the anions. The best we can do experimentally is to assign responsibility for the nonideality equally to both kinds of ion. Therefore, for a 1,1-electrolyte, we introduce the **mean activity coefficient** as the geometric mean of the individual coefficients:

The geometric mean of x and y is $(xy)^{1/2}$.

$$\gamma_{\pm} = (\gamma_+\gamma_-)^{1/2} \qquad [10.11]$$

and express the individual chemical potentials of the ions as

$$\mu_+ = \mu_+^{\text{ideal}} + RT \ln \gamma_{\pm} \qquad \mu_- = \mu_-^{\text{ideal}} + RT \ln \gamma_{\pm} \tag{10.12}$$

The sum of these two chemical potentials is the same as before, eqn 10, but now the nonideality is shared equally.

We can generalize this approach to the case of a compound M_pX_q that dissolves to give a solution of p cations and q anions from each formula unit. The molar Gibbs energy of the ions is the sum of their partial molar Gibbs energies:

$$G_{\text{m}} = p\mu_+ + q\mu_- = G_{\text{m}}^{\text{ideal}} + pRT \ln \gamma_+ + qRT \ln \gamma_- \tag{10.13}$$

3. The IUPAC recommendation for the symbol is either m or b; whereas the former is more commonly used, b has the advantage that it is less likely to be confused with mass, m.
4. In accord with the convention adopted in earlier chapters, the notation has been simplified by interpreting b as $b/b^{\ominus}$ and labelling the equation { ... }.

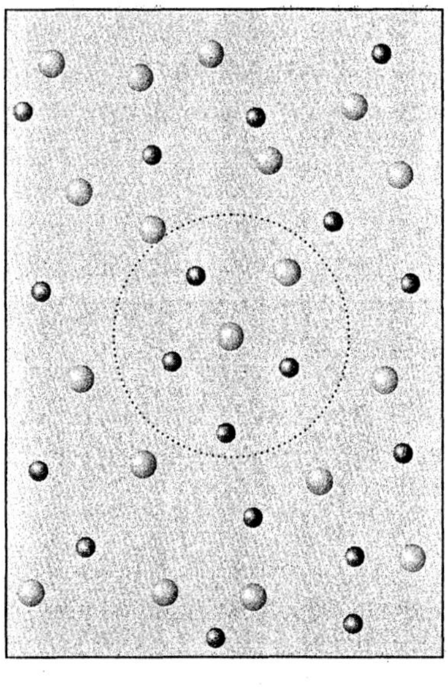

10.2 The picture underlying the Debye–Hückel theory is of a tendency for anions to be found around cations, and of cations to be found around anions (one such local clustering region is shown by the circle). The ions are in ceaseless motion, and the diagram represents a snapshot of their motion. The solutions to which the theory applies are far less concentrated than shown here.

Table 10.4 Ionic strength and molality, $I = k \times b/b^\ominus$

k	X^-	X^{2-}	X^{3-}	X^{4-}
M^+	1	3	6	10
M^{2+}	3	4	15	12
M^{3+}	6	15	9	42
M^{4+}	10	12	42	16

For example, the ionic strength of an M_2X_3 solution of molality b, which is understood to give M^{3+} and X^{2-} ions in solution, is $15b/b^\ominus$.

If we introduce the mean activity coefficient

$$\gamma_\pm = (\gamma_+^p \gamma_-^q)^{1/s} \qquad s = p + q \tag*{[10.14]}$$

and write the chemical potential of each ion as

$$\mu_i = \mu_i^{\text{ideal}} + RT \ln \gamma_\pm \tag{10.15}$$

we get the same expression as in eqn 13 for G_m when we write

$$G = p\mu_+ + q\mu_- \tag{10.16}$$

However, both types of ion now share equal responsibility for the nonideality.

(c) The Debye–Hückel limiting law

The long range and strength of the Coulombic interaction between ions means that it is likely to be primarily responsible for the departures from ideality in ionic solutions and to dominate all the other contributions to nonideality. This domination is the basis of the **Debye–Hückel theory** of ionic solutions, which was devised by Peter Debye and Erich Hückel in 1923. We give here a qualitative account of the theory and its principal conclusions. The calculation itself, which is a profound example of how a seemingly intractable problem can be formulated and then resolved by drawing on physical insight, is outlined in the *Justification* below.

Oppositely charged ions attract one another. As a result, anions are more likely to be found near cations in solution, and vice versa (Fig. 10.2). Overall the solution is electrically neutral, but near any given ion there is an excess of counter ions (ions of opposite charge). Averaged over time, counter ions are more likely to be found near any given ion. This time-averaged, spherical haze around the central ion, in which counter ions outnumber ions of the same charge as the central ion, has a net charge equal in magnitude but opposite in sign to that on the central ion, and is called its **ionic atmosphere**. The energy, and therefore the chemical potential, of any given central ion is lowered as a result of its electrostatic interaction with its ionic atmosphere. This lowering of energy appears as the difference between the molar Gibbs energy G_m and the ideal value G_m^{ideal} of the solute, and hence can be identified with $RT \ln \gamma_\pm$. The stabilization of ions by their interaction with their ionic atmospheres is part of the explanation why chemists commonly use dilute solutions, in which the stabilization is less important, to achieve precipitation of ions from electrolyte solutions.

The model leads to the result that at very low concentrations the activity coefficient can be calculated from the **Debye–Hückel limiting law**

$$\log \gamma_\pm = -|z_+z_-|AI^{1/2} \tag{10.17}$$

where $A = 0.509$ for an aqueous solution at 25°C and I is the dimensionless **ionic strength** of the solution:

$$I = \tfrac{1}{2}\sum_i z_i^2(b_i/b^\ominus) \tag*{[10.18]}$$

In this expression z_i is the charge number of an ion i (positive for cations and negative for anions) and b_i is its molality. The ionic strength occurs widely wherever ionic solutions are discussed, as we shall see. The sum extends over all the ions present in the solution. For solutions consisting of two types of ion at molalities b_+ and b_-,

$$I = \tfrac{1}{2}(b_+z_+^2 + b_-z_-^2)/b^\ominus \tag{10.19}$$

The ionic strength emphasizes the charges of the ions because the charge numbers occur as their squares. Table 10.4 summarizes the relation of ionic strength and molality in an easily usable form.

Illustration 10.3

The mean activity coefficient of 5.0×10^{-3} mol kg^{-1} KCl(aq) at 25°C is calculated by writing

$$I = \tfrac{1}{2}(b_+ + b_-)/b^\ominus = b/b^\ominus$$

where b is the molality of the solution (and $b_+ = b_- = b$). Then, from eqn 17,

$$\log \gamma_\pm = -0.509 \times (5.0 \times 10^{-3})^{1/2} = -0.036$$

Hence, $\gamma_\pm = 0.92$. The experimental value is 0.927.

Self-test 10.3 Calculate the ionic strength and the mean activity coefficient of 1.00 mmol kg^{-1} $CaCl_2$(aq) at 25°C. [3.00 mmol kg^{-1}, 0.880]

Justification 10.2

Imagine a solution in which all the ions have their actual positions, but in which their Coulombic interactions have been turned off. The difference in molar Gibbs energy between the ideal and real solutions is equal to w_e, the electrical work of charging the system in this arrangement. Therefore, for a salt M_pX_q,

$$\ln \gamma_\pm = \frac{w_e}{sRT} \qquad s = p + q \tag{10.20}$$

It follows that we must first find the final distribution of the ions and then the work of charging them in that distribution.

The Coulomb potential at a distance r from an isolated ion of charge z_ie in a medium of permittivity ε is

$$\phi_i = \frac{Z_i}{r} \qquad Z_i = \frac{z_ie}{4\pi\varepsilon} \tag{10.21}$$

The ionic atmosphere causes the potential to decay with distance more sharply than this expression implies. Such shielding is a familiar problem in electrostatics, and its effect is taken into account by replacing the Coulomb potential by the **shielded Coulomb potential**, an expression of the form

$$\phi_i = \frac{Z_i}{r}e^{-r/r_D} \tag{10.22}$$

where r_D is called the **Debye length**. When r_D is large, the shielded potential is virtually the same as the unshielded potential. When it is small, the shielded potential is much smaller than the unshielded potential, even for short distances (Fig. 10.3).

To calculate r_D, we need to know how the **charge density**, ρ_i, of the ionic atmosphere, the charge in a small region divided by the volume of the region, varies with distance from the ion. This step draws on another standard result of electrostatics, in which charge density and potential are related by **Poisson's equation**:

$$\nabla^2\phi = -\frac{\rho}{\varepsilon} \tag{10.23}$$

Because we are considering only a spherical ionic atmosphere, we can use a simplified form of this equation in which the charge density varies only with distance from the central ion:

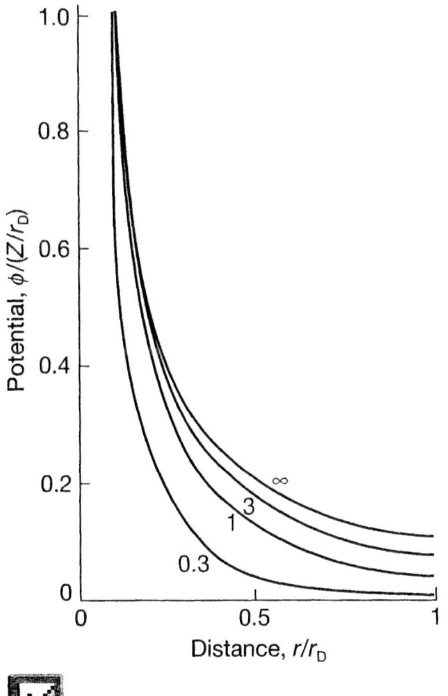

10.3 The variation of the shielded Coulomb potential with distance for different values of the Debye length, r_D/a. The smaller the Debye length, the more sharply the potential decays to zero. In each case, a is an arbitrary unit of length.

$$\frac{1}{r^2}\frac{\mathrm{d}}{\mathrm{d}r}\left(r^2\frac{\mathrm{d}\phi_i}{\mathrm{d}r}\right) = -\frac{\rho_i}{\varepsilon}$$

Substitution of the expression for the shielded potential, eqn 22, results in

$$r_D^2 = -\frac{\varepsilon\phi_i}{\rho_i} \tag{10.24}$$

To solve this equation we need to relate ρ_i and ϕ_i.

For the next step we draw on the fact that the energy of an ion depends on its closeness to the central ion, and then use the Boltzmann distribution (*Molecular interpretation* 2.1) to work out the probability that an ion will be found at each distance. The energy of an ion of charge z_je at a distance where it experiences the potential ϕ_i of the central ion i relative to its energy when it is far away in the bulk solution is its charge times the potential:

$$E = z_j e\phi_i$$

Therefore, according to the Boltzmann distribution, the ratio of the molar concentration, c_j, of ions at a distance r and the molar concentration in the bulk, c_j°, where the energy is zero, is:

$$\frac{c_j}{c_j^\circ} = \mathrm{e}^{-E/kT}$$

The charge density, ρ_i, at a distance r from the ion i is the molar concentration of each type of ion multiplied by the charge per mole of ions, z_ieN_A. The quantity eN_A, the magnitude of the charge per mole of electrons, is the Faraday constant, F. It follows that

$$\rho_i = c_+z_+F + c_-z_-F = c_+^\circ z_+F\,\mathrm{e}^{-z_+e\phi_i/kT} + c_-^\circ z_-F\,\mathrm{e}^{-z_-e\phi_i/kT} \tag{10.25a}$$

At this stage we need to simplify the expression to avoid the awkward exponential terms. Because the average electrostatic interaction energy is small compared with kT we may write eqn 25a as

$$\rho_i = (c_+^\circ z_+ + c_-^\circ z_-)F - (c_+^\circ z_+^2 + c_-^\circ z_-^2)\frac{F^2\phi_i}{RT} + \cdots \tag{10.25b}$$

(To obtain this expression, we have replaced an e by F/N_A and recognized that $N_Ak = R$.) The first term in the expansion is zero because it is the charge density in the bulk, uniform solution, and the solution is electrically neutral. The unwritten terms are assumed to be too small to be significant. The one remaining term can be expressed in terms of the ionic strength, eqn 18, by noting that in the dilute aqueous solutions we are considering there is little difference between molality and molar concentration, and $c \approx b\rho$, where ρ is the mass density of the solvent

$$c_+^\circ z_+^2 + c_-^\circ z_-^2 \approx (b_+^\circ z_+^2 + b_-^\circ z_-^2)\rho = 2Ib^\ominus\rho$$

With these approximations, eqn 25 becomes

$$\rho_i = -\frac{2\rho F^2 Ib^\ominus\phi_i}{RT}$$

We can now solve eqn 24 for r_D:

$$r_D = \left(\frac{\varepsilon RT}{2\rho F^2 Ib^\ominus}\right)^{1/2} \tag{10.26}$$

To calculate the activity coefficient we need to find the electrical work of charging the central ion when it is surrounded by its atmosphere. To do so, we need to know

Table 10.5* Mean activity coefficients in water at 298 K

b/(mol kg^{-1})	KCl	$CaCl_2$
0.001	0.966	0.888
0.01	0.902	0.732
0.1	0.770	0.524
1.0	0.607	0.725

* More values are given in the *Data section*.

the potential at the ion due to its atmosphere, ϕ_{atmos}. This potential is the difference between the total potential, given by eqn 22, and the potential due to the central ion itself:

$$\phi_{atmos} = \phi - \phi_{central\ ion} = Z_i\left(\frac{e^{-r/r_D}}{r} - \frac{1}{r}\right)$$

The potential at the central ion (at $r = 0$) is obtained by taking the limit of this expression as $r \to 0$, and is

$$\phi_{atmos}(0) = -\frac{Z_i}{r_D}$$

This expression shows us that the potential of the ionic atmosphere is equivalent to the potential arising from a single charge of equal magnitude but opposite sign to that of the central ion and located at a distance r_D from the ion. If the charge of the central ion were q and not z_ie, then the potential due to its atmosphere would be

$$\phi_{atmos}(0) = -\frac{q}{4\pi\varepsilon r_D}$$

The work of adding a charge dq to a region where the electrical potential is $\phi_{atmos}(0)$ is

$$dw_e = \phi_{atmos}(0)\,dq$$

Therefore, the total molar work of fully charging the ions is

$$w_e = N_A\int_0^{z_ie}\phi_{atmos}(0)\,dq = -\frac{N_A}{4\pi\varepsilon r_D}\int_0^{z_ie} q\,dq$$

$$= -\frac{N_Az_i^2e^2}{8\pi\varepsilon r_D} = -\frac{z_i^2F^2}{8\pi\varepsilon N_A r_D}$$

It follows from eqn 20 that the mean activity coefficient of the ions is

$$\ln\gamma_\pm = \frac{pw_{e,+} + qw_{e,-}}{sRT} = -\frac{(pz_+^2 + qz_-^2)F^2}{8\pi\varepsilon s N_A RTr_D}$$

However, for neutrality $pz_+ + qz_- = 0$;[5] therefore

$$\ln\gamma_\pm = -\frac{|z_+z_-|F^2}{8\pi\varepsilon N_A RTr_D}$$

When we replace r_D by using eqn 26, and convert to common logarithms, this expression turns into eqn 17 with

$$A = \frac{F^3}{4\pi N_A \ln 10}\left(\frac{\rho b^\ominus}{2\varepsilon^3R^3T^3}\right)^{1/2} \qquad (10.27)$$

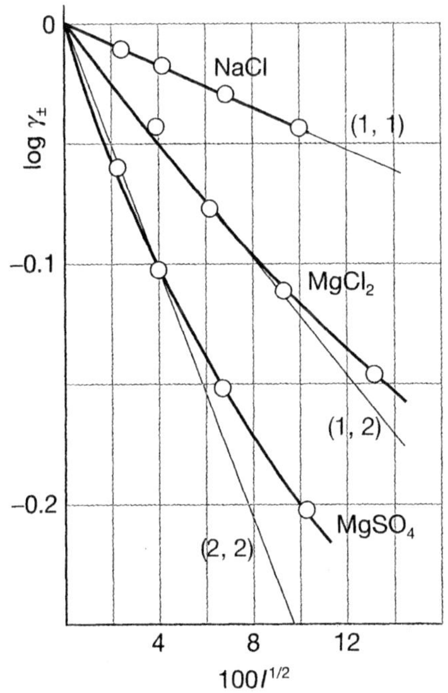

10.4 An experimental test of the Debye–Hückel limiting law. Although there are marked deviations for moderate ionic strengths, the limiting slopes as $I \to 0$ are in good agreement with the theory, so the law can be used for extrapolating data to very low molalities.

The name 'limiting law' is applied to eqn 17 because ionic solutions of moderate molalities may have activity coefficients that differ from the values given by this expression, yet all solutions are expected to conform as $b \to 0$. Table 10.5 lists some experimental values of activity coefficients for salts of various valence types. Figure 10.4 shows some of these values plotted against $I^{1/2}$, and compares them with

5. For this step, multiply $pz_+ + qz_- = 0$ by p and also, separately, by q; add the two expressions and rearrange the result by using $p + q = s$ and $z_+z_- = -|z_+z_-|$.

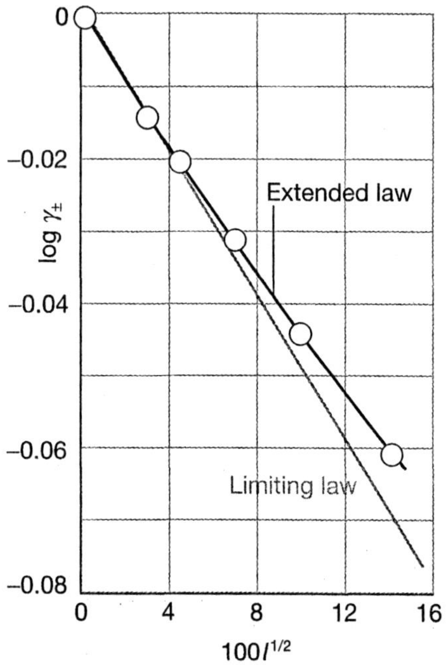

10.5 The extended Debye-Hückel law gives agreement with experiment over a wider range of molalities (as shown here for a 1,1-electrolyte), but it fails at higher molalities.

the theoretical straight lines calculated from eqn 17. The agreement at very low molalities (less than about 1 mmol kg^{-1}, depending on charge type) is impressive, and convincing evidence in support of the model. Nevertheless, the departures from the theoretical curves above these molalities are large, and show that the approximations are valid only at very low concentrations.

(d) The extended Debye–Hückel law

When the ionic strength of the solution is too high for the limiting law to be valid, the activity coefficient may be estimated from the **extended Debye-Hückel law**:

$$\ln \gamma_\pm = -\frac{A|z_+z_-|I^{1/2}}{1 + BI^{1/2}} \qquad (10.28)$$

where B is another dimensionless constant. Although B can be interpreted as a measure of the closest approach of the ions, it is best regarded as an adjustable empirical parameter. A curve drawn in this way is shown in Fig. 10.5. It is clear that eqn 28 accounts for some activity coefficients over a moderate range of dilute solutions (up to about 0.1 mol kg^{-1}); nevertheless it remains very poor near 1 mol kg^{-1}.

Current theories of activity coefficients for ionic solutes take an indirect route. They set up a theory for the dependence of the activity coefficient of the solvent on the concentration of the solute, and then use the Gibbs-Duhem equation (eqn 7.12) to estimate the activity coefficient of the solute. The results are reasonably reliable for solutions with molalities greater than about 0.1 mol kg^{-1} and are valuable for the discussion of mixed salt solutions, such as sea water.

Electrochemical cells

An **electrochemical cell** consists of two **electrodes**, or metallic conductors, in contact with an **electrolyte**, an ionic conductor (which may be a solution, a liquid, or a solid). An electrode and its electrolyte comprise an **electrode compartment**. The two electrodes may share the same compartment. The various kinds of electrode are summarized in Table 10.6 and illustrated in Fig. 10.6. Any 'inert metal' shown as part of the specification is present to act as a source or sink of electrons, but takes no other part in the reaction other than acting as a catalyst for it. If the electrolytes are different, the two compartments may be joined by a **salt bridge**, which is a tube containing a concentrated electrolyte solution in agar jelly that completes the electrical circuit and enables the cell to function.

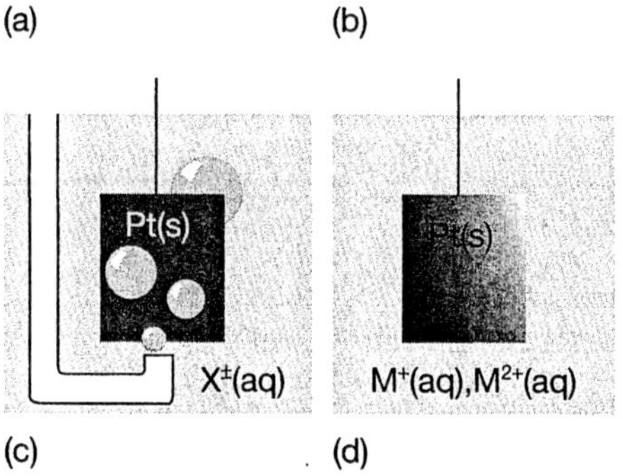

10.6 Typical electrode types: (a) metal–metal ion, (b) metal–insoluble salt, (c) gas, (d) redox electrodes.

Table 10.6 Varieties of electrode

Electrode type	Designation	Redox couple	Half-reaction
Metal/metal-ion	$M(s)\|M^+(aq)$	M^+/M	$M^+(aq) + e^- \longrightarrow M(s)$
Gas electrode	$Pt(s)\|X_2(g)\|X^+(aq)$	X^+/X_2	$X^+(aq) + e^- \longrightarrow \frac{1}{2}X_2(g)$
	$Pt(s)\|X_2(g)\|X^-(aq)$	X_2/X^-	$\frac{1}{2}X_2(g) + e^- \longrightarrow X^-(aq)$
Metal/insoluble-salt	$M(s)\|MX(s)\|X^-(aq)$	$MX/M,X^-$	$MX(s) + e^- \longrightarrow M(s) + X^-(aq)$
Redox	$Pt(s)\|M^+(aq),M^{2+}(aq)$	M^{2+}/M^+	$M^{2+}(aq) + e^- \longrightarrow M^+(aq)$

A **galvanic cell** is an electrochemical cell that produces electricity as a result of the spontaneous reaction occurring inside it. An **electrolytic cell** is an electrochemical cell in which a non-spontaneous reaction is driven by an external source of current.

10.3 Half-reactions and electrodes

It will be familiar from introductory chemistry courses that **oxidation** is the removal of electrons from a species and **reduction** is the addition of electrons to a species. A **redox reaction** is a reaction in which there is a transfer of electrons from one species to another. The electron transfer may be accompanied by other events, such as atom or ion transfer, but the net effect is electron transfer and hence a change in oxidation number of an element. The **reducing agent** (or 'reductant') is the electron donor; the **oxidizing agent** (or 'oxidant') is the electron acceptor.

(a) Half-reactions

Any redox reaction may be expressed as the difference of two reduction **half-reactions**, which are conceptual reactions showing the gain of electrons. For example, the reduction of Cu^{2+} ions by zinc can be expressed as the difference of the following two half-reactions:

$$Cu^{2+}(aq) + 2\,e^- \longrightarrow Cu(s) \qquad Zn^{2+}(aq) + 2\,e^- \longrightarrow Zn(s)$$

The difference of the two (copper − zinc) is

$$Cu^{2+}(aq) + Zn(s) \longrightarrow Cu(s) + Zn^{2+}(aq) \tag{10.29}$$

Even reactions that are not redox reactions may often be expressed as the difference of two reduction half-reactions.

Example 10.1 Expressing a reaction in terms of half-reactions

Express the dissolution of silver chloride in water as the difference of two reduction half-reactions.

Method First, write the overall chemical equation. Then select one of the reactants, and write a half-reaction in which the reactant is reduced to one of the products. Next, subtract that half-reaction from the overall reaction to identify the second half-reaction. In practice, it is easier to reverse the half-reaction and add it to the overall reaction. Write the resulting half-reaction as a reduction (by reversing it).

Answer The chemical equation of the overall reaction is

$$AgCl(s) \longrightarrow Ag^+(aq) + Cl^-(aq)$$

We select as one half-reaction the reduction of AgCl (more precisely, the reduction of the Ag(I) in AgCl to Ag(0)):

$$AgCl(s) + e^- \longrightarrow Ag(s) + Cl^-(aq)$$

Reversing this half-reaction gives

$$Ag(s) + Cl^-(aq) \longrightarrow AgCl(s) + e^-$$

and adding it to the overall reaction gives

$$AgCl(s) + Ag(s) + Cl^-(aq) \longrightarrow Ag^+(aq) + Cl^-(aq) + AgCl(s) + e^-$$

Now cancel species that appear on both sides of the arrow:

$Ag(s) \longrightarrow Ag^+(aq) + e^-$

Finally, reverse this half-reaction to obtain a reduction half-reaction:

$Ag^+(aq) + e^- \longrightarrow Ag(s)$

Comment There is no net change of oxidation number in dissolution of AgCl, so it is not a redox reaction.

Self-test 10.4 Express the formation of H_2O from H_2 and O_2 in acidic solution (a redox reaction) as the difference of two reduction half-reactions.
[$4\,H^+(aq) + 4\,e^- \rightarrow 2\,H_2(g)$, $O_2(g) + 4\,H^+(aq) + 4\,e^- \rightarrow 2\,H_2O(l)$]

The reduced and oxidized species in a half-reaction form a **redox couple**. Thus, the redox couples mentioned so far are Cu^{2+}/Cu and Zn^{2+}/Zn. In general we write a couple as Ox/Red and the corresponding reduction half-reaction as

$$Ox + \nu e^- \longrightarrow Red \qquad (10.30)$$

We shall often find it useful to express the composition of an electrode compartment in terms of the reaction quotient, Q, for the half-reaction. This quotient is defined like the reaction quotient for the overall reaction, but the electrons are ignored. Thus, for the copper half-reaction $Cu^{2+}(aq) + 2\,e^- \rightarrow Cu(s)$, we write

$$Q = \frac{1}{a_{Cu^{2+}}}$$

We have used the fact that the pure metal (the standard state of the element) has unit activity (recall Table 7.3).

Illustration 10.4

The reaction quotient for the reduction of O_2 to H_2O in acid solution,

$O_2(g) + 4\,H^+(aq) + 4\,e^- \longrightarrow 2\,H_2O(l)$

is

$$Q = \frac{a_{H_2O}^2}{a_{H^+}^4 (f_{O_2}/p^{\ominus})} \approx \frac{p^{\ominus}}{a_{H^+}^4 p_{O_2}}$$

The approximations used in the second step are that the activity of water is 1 (because the solution is dilute) and the oxygen behaves as a perfect gas.

Self-test 10.5 Write the half-reaction and the reaction quotient for a chlorine gas electrode. [$Cl_2(g) + 2\,e^- \rightarrow 2\,Cl^-(aq)$, $Q = a_{Cl^-}^2 p^{\ominus}/p_{Cl_2}$]

(b) Reactions at electrodes

The reduction and oxidation processes responsible for the overall reaction in a cell are separated in space: oxidation takes place in one electrode compartment and reduction takes place in the other compartment. As the reaction proceeds, the electrons released in the oxidation

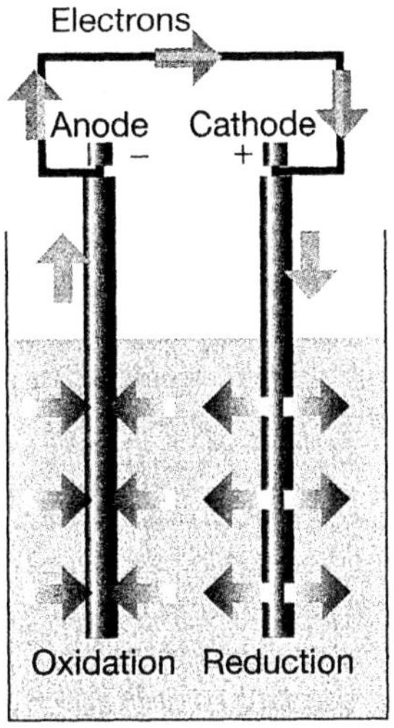

10.7 When a spontaneous reaction takes place in a galvanic cell, electrons are deposited in one electrode (the site of oxidation, the anode) and collected from another (the site of reduction, the cathode), and so there is a net flow of current which can be used to do work. Note that the + sign of the cathode can be interpreted as indicating the electrode at which electrons enter the cell, and the – sign of the anode is where the electrons leave the cell.

$$\mathrm{Red_1} \longrightarrow \mathrm{Ox_1} + \nu\,\mathrm{e^-}$$

at one electrode travel through the external circuit and re-enter the cell through the other electrode. There they bring about reduction:

$$\mathrm{Ox_2} + \nu\,\mathrm{e^-} \longrightarrow \mathrm{Red_2}$$

The electrode at which oxidation occurs is called the **anode**; the electrode at which reduction occurs is called the **cathode**.

In a galvanic cell, the cathode has a higher potential than the anode: the species undergoing reduction, Ox_2, withdraws electrons from its electrode (the cathode, Fig. 10.7), so leaving a relative positive charge on it (corresponding to a high potential). At the anode, oxidation results in the transfer of electrons to the electrode, so giving it a relative negative charge (corresponding to a low potential). In an electrolytic cell, the anode is still the location of oxidation (by definition), but now electrons must be withdrawn from the species in that compartment because oxidation does not occur spontaneously, and at the cathode there must be a supply of electrons to drive the reduction. Therefore, in an electrolytic cell the anode must be made relatively positive to the cathode (Fig. 10.8).

10.4 Varieties of cells

The simplest type of cell has a single electrolyte common to both electrodes (as in Fig. 10.7). In some cases it is necessary to immerse the electrodes in different electrolytes, as in the 'Daniell cell' in which the redox couple at one electrode is Cu^{2+}/Cu and at the other is Zn^{2+}/Zn (Fig. 10.9). In an **electrolyte concentration cell**, the electrode compartments are identical except for the concentrations of the electrolytes. In an **electrode concentration cell** the electrodes themselves have different

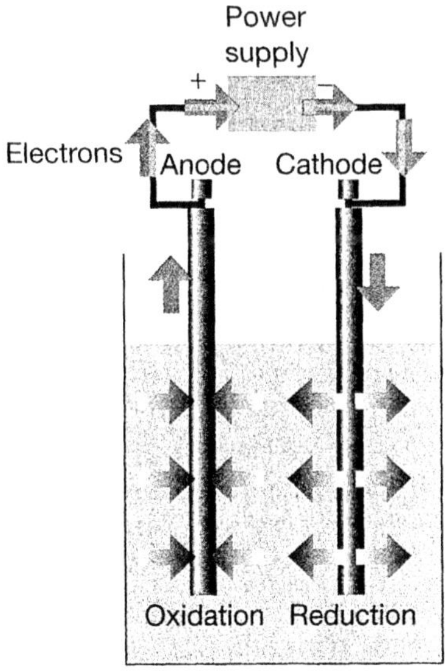

10.8 In an electrolytic cell, electrons are forced through the circuit by an external source. Although the cathode is still the site of reduction, it is now the negative electrode whereas the anode, the site of oxidation, is positive.

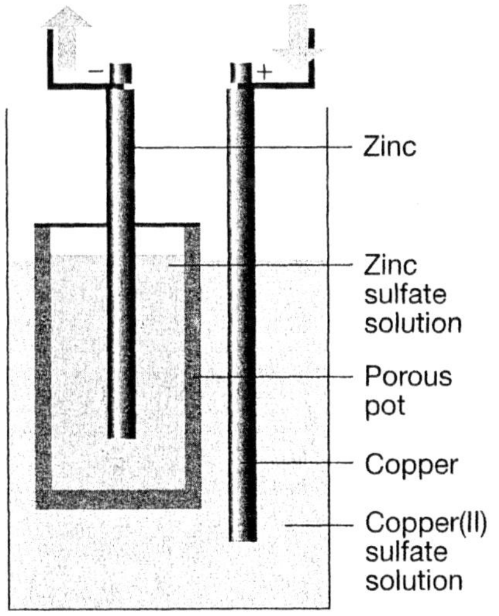

10.9 One version of the Daniell cell. The copper electrode is the cathode and the zinc electrode is the anode. Electrons leave the cell from the zinc and enter it again through the copper electrode.

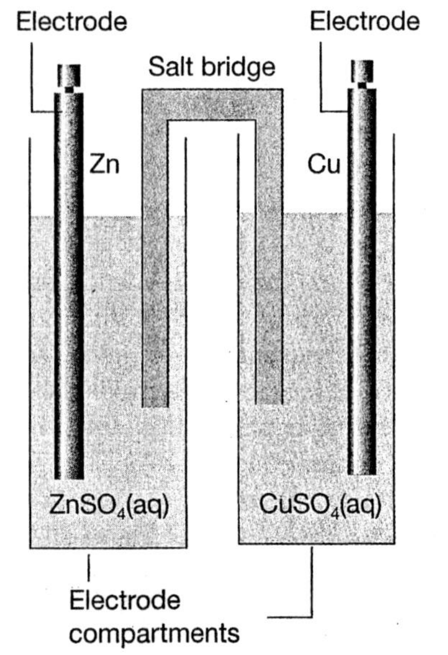

10.10 The salt bridge, essentially an inverted U-tube full of concentrated salt solution in a jelly, has two opposing liquid junction potentials that almost cancel.

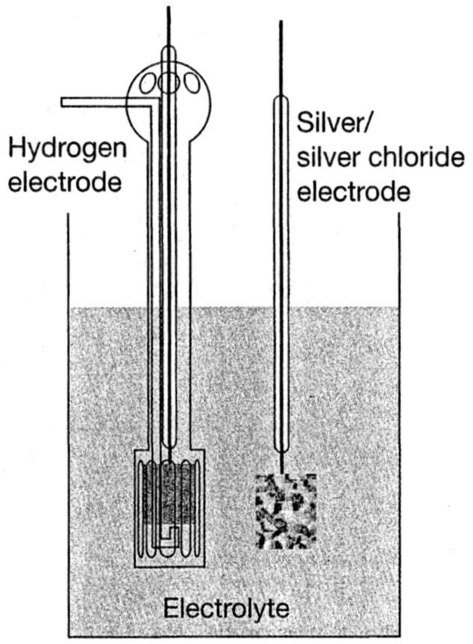

10.11 A typical cell for measuring a standard potential consists of a hydrogen electrode (on the left) and the electrode for the couple of interest (on the right).

concentrations, either because they are gas electrodes operating at different pressures or because they are amalgams (solutions in mercury) with different concentrations.

(a) Liquid junction potentials

In a cell with two different electrolyte solutions in contact, as in the Daniell cell, there is an additional source of potential difference across the interface of the two electrolytes. This potential is called the **liquid junction potential**, E_{lj}. Another example of a junction potential is that between different concentrations of hydrochloric acid. At the junction, the mobile H^+ ions diffuse into the more dilute solution. The bulkier Cl^- ions follow, but initially do so more slowly, which results in a potential difference at the junction. The potential then settles down to a value such that, after that brief initial period, the ions diffuse at the same rates. Electrolyte concentration cells always have a liquid junction; electrode concentration cells do not.

The contribution of the liquid junction to the potential can be reduced (to about 1 to 2 mV) by joining the electrolyte compartments through a salt bridge (Fig. 10.10). The reason for the success of the salt bridge is that the liquid junction potentials at either end are largely independent of the concentrations of the two dilute solutions, and so nearly cancel.

(b) Notation

In the notation for cells, phase boundaries are denoted by a vertical bar. For example, the cell in Fig. 10.11 is denoted by the 'cell diagram'

$$\mathrm{Pt(s)|H_2(g)|HCl(aq)|AgCl(s)|Ag(s)}$$

A liquid junction is denoted by ⁝, so the cell in Fig. 10.9, is denoted

$$\mathrm{Zn(s)|ZnSO_4(aq)} \ \vdots \ \mathrm{CuSO_4(aq)|Cu(s)}$$

A double vertical line, ||, denotes an interface for which it is assumed that the junction potential has been eliminated. Thus the cell in Fig. 10.10 is denoted

$$\mathrm{Zn(s)|ZnSO_4(aq)||CuSO_4(aq)|Cu(s)}$$

An electrolyte concentration cell in which the liquid junction potential is assumed to be eliminated is denoted

$$\mathrm{Pt(s)|H_2(g)|HCl(aq,}\ b_1\mathrm{)||HCl(aq,}\ b_2\mathrm{)|H_2(g)|Pt(s)}$$

(c) The cell reaction

The current produced by a galvanic cell arises from the spontaneous chemical reaction taking place inside it. The **cell reaction** is the reaction in the cell written on the assumption that the right-hand electrode is the cathode, and hence that the spontaneous reaction is one in which reduction is taking place in the right-hand compartment. Later we see how to predict if the right-hand electrode is in fact the cathode; if it is, then the cell reaction is spontaneous as written. If the left-hand electrode turns out to be the cathode, then the reverse of the corresponding cell reaction is spontaneous.

To write the cell reaction corresponding to a cell diagram, we first write the right-hand half-reaction as a reduction (because we have assumed that to be spontaneous). Then we subtract from it the left-hand reduction half-reaction (for, by implication, that electrode is the site of oxidation). Thus, in the cell

$$\mathrm{Zn(s)|ZnSO_4(aq)||CuSO_4(aq)|Cu(s)}$$

the two electrodes and their reduction half-reactions are

Right-hand electrode: $Cu^{2+}(aq) + 2\,e^- \longrightarrow Cu(s)$

Left-hand electrode: $Zn^{2+}(aq) + 2\,e^- \longrightarrow Zn(s)$

Hence, the overall cell reaction is the difference:

$$Cu^{2+}(aq) + Zn(s) \longrightarrow Cu(s) + Zn^{2+}(aq)$$

(d) The cell potential

A cell in which the overall cell reaction has not reached chemical equilibrium can do electrical work as the reaction drives electrons through an external circuit. The work that a given transfer of electrons can accomplish depends on the potential difference between the two electrodes. This potential difference is called the **cell potential** and is measured in volts, V. When the cell potential is large, a given number of electrons travelling between the electrodes can do a large amount of electrical work. When the cell potential is small, the same number of electrons can do only a small amount of work. A cell in which the overall reaction is at equilibrium can do no work, and then the cell potential is zero.

According to the discussion in Section 4.5e, we know that the maximum electrical work that a system (the cell) can do is given by the value of ΔG, and in particular that for a spontaneous process (in which both ΔG and w are negative) at constant temperature and pressure

$$w_{e,max} = \Delta G \tag{10.31}$$

Here we use w_e to denote electrical work, a specific kind of non-expansion work. It follows that, to draw thermodynamic conclusions from measurements of the work a cell can do, we must ensure that the cell is operating reversibly. Only then is it producing maximum work and only then can eqn 31 be used to relate that work to ΔG. Moreover, we saw in Section 9.1a that the reaction Gibbs energy is actually a property relating to a specified composition of the reaction mixture. Therefore, to measure $\Delta_r G$ we must ensure that the cell is operating reversibly at a specific, constant composition. Both these conditions are achieved by measuring the cell potential when it is balanced by an exactly opposing source of potential so that the cell reaction occurs reversibly, the composition is constant, and no current flows: in effect, the cell reaction is poised for change, but not actually changing. The resulting potential difference is called the **electromotive force** (emf), E, of the cell.

(e) The relation between E and $\Delta_r G$

The relation between the reaction Gibbs energy and the emf of the cell is

$$-\nu FE = \Delta_r G \tag{10.32}$$

where F is the **Faraday constant**, $F = eN_A$, the magnitude of the charge per mole of electrons. This equation, which is derived in the *Justification* below, is the key connection between electrical measurements on the one hand and thermodynamic properties on the other. It will be the basis of all that follows.

Justification 10.3

We consider the change in G when the cell reaction advances by an infinitesimal amount $d\xi$ at some composition. We saw in *Justification* 9.1 that, at constant temperature and pressure, G changes by

$$dG = \sum_J \mu_J \, dn_J = \sum_J \nu_J \mu_J \, d\xi$$

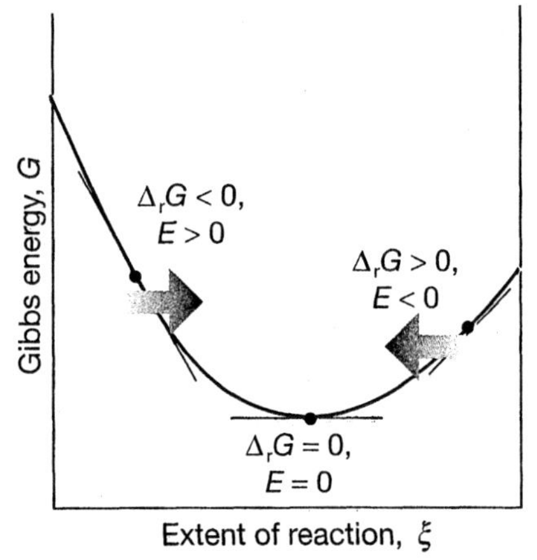

10.12 As explained in Chapter 9, a spontaneous reaction occurs in the direction of decreasing Gibbs energy. When expressed in terms of a cell potential, the spontaneous direction of change can be expressed in terms of the cell emf, E. The reaction is spontaneous as written (from left to right on the illustration) when $E > 0$. The reverse reaction is spontaneous when $E < 0$. When the cell reaction is at equilibrium, the cell potential is zero.

The reaction Gibbs energy, $\Delta_r G$, at the specified composition is

$$\Delta_r G = \left(\frac{\partial G}{\partial \xi}\right)_{p,T} = \sum_J \nu_J \mu_J$$

So we can write

$$dG = \Delta_r G \, d\xi$$

The maximum non-expansion (electrical) work that the reaction can do as it advances by $d\xi$ at constant temperature and pressure is therefore

$$dw_e = \Delta_r G \, d\xi$$

This work is infinitesimal, and the composition of the system is virtually constant when it occurs.

Suppose that the reaction advances by $d\xi$; then $\nu \, d\xi$ electrons must travel from the anode to the cathode. The total charge transported between the electrodes when this change occurs is $-\nu e N_A \, d\xi$ (because $\nu \, d\xi$ is the amount of electrons and the charge per mole of electrons is $-eN_A$). Hence, the total charge transported is $-\nu F \, d\xi$ because $eN_A = F$.

The work done when an infinitesimal charge $-\nu F \, d\xi$ travels from the anode to the cathode is equal to the product of the charge and the potential difference E (see Table 2.1 and *Further information 2*):

$$dw_e = -\nu F E \, d\xi$$

When we equate this relation to the one above, the advancement $d\xi$ cancels, and we obtain eqn 32.

It follows from eqn 32 that, by knowing the reaction Gibbs energy at a specified composition, we can state the cell emf at that composition. Note that a negative reaction Gibbs energy, corresponding to a spontaneous cell reaction, corresponds to a positive cell emf. Another way of looking at the content of eqn 32 is that it shows that the driving power of a cell (that is, its emf) is proportional to the slope of the Gibbs energy with respect to the extent of reaction. It is plausible that a reaction that is far from equilibrium (when the slope is steep) has a strong tendency to drive electrons through an external circuit (Fig. 10.12). When the slope is close to zero (when the cell reaction is close to equilibrium), the emf is small.

Illustration 10.5

To estimate the emf to expect for a typical cell we set $\Delta_r G \approx -100 \text{ kJ mol}^{-1}$ and $\nu = 1$; we find

$$E \approx -\frac{(-100 \times 10^3 \text{ J mol}^{-1})}{1 \times (9.6 \times 10^4 \text{ C mol}^{-1})} \approx 1 \text{ V}$$

We have used 1 J = 1 C V.

(f) The Nernst equation

We can go on to relate the emf to the activities of the participants in the cell reaction. We know from eqn 9.11 that the reaction Gibbs energy is related to the composition of the reaction mixture by

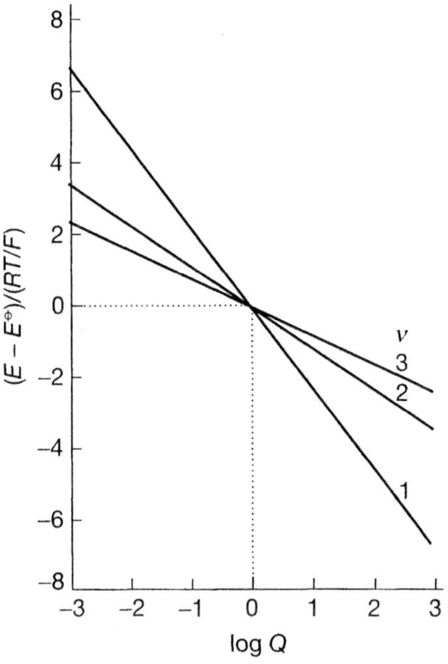

10.13 The variation of cell emf with the value of the reaction quotient for the cell reaction for different values of ν (the number of electrons transferred). At 298 K, $RT/F = 25.69$ mV, so the vertical scale refers to multiples of this value.

$$\Delta_r G = \Delta_r G^{\ominus} + RT \ln Q$$

where Q is the reaction quotient. It follows, on division of both sides by $-\nu F$, that

$$E = -\frac{\Delta_r G^{\ominus}}{\nu F} - \frac{RT}{\nu F}\ln Q$$

The first term on the right is written

$$E^{\ominus} = -\frac{\Delta_r G^{\ominus}}{\nu F} \qquad [10.33]$$

and called the **standard emf** of the cell. That is, the standard emf is the standard reaction Gibbs energy expressed as a potential (in volts). It follows that

$$E = E^{\ominus} - \frac{RT}{\nu F}\ln Q \qquad (10.34)$$

This equation for the emf in terms of the composition is called the **Nernst equation**; the dependence of cell potential on composition that it predicts is summarized in Fig. 10.13.

We see from eqn 34 that the standard emf (which will shortly move to centre stage of the exposition) can be interpreted as the emf when all the reactants and products are in their standard states, for then all activities are 1, so $Q = 1$ and $\ln Q = 0$. However, the fact that the standard emf is merely a disguised form of the standard reaction Gibbs energy should always be kept in mind and underlies all its applications.

Illustration 10.6

Because $RT/F = 25.7$ mV at 25°C, a practical form of the Nernst equation is

$$E = E^{\ominus} - \frac{25.7\text{ mV}}{\nu}\ln Q$$

It then follows that, for a reaction in which $\nu = 1$, if Q is increased by a factor of 10, then the emf decreases by 59.2 mV.

(g) Concentration cells

We can use the Nernst equation to derive an expression for the emf of an electrolyte concentration cell. Consider the cell

$$M|M^+(aq, L)||M^+(aq, R)|M$$

where the solutions L and R have different molalities. The cell reaction is

$$M^+(aq, R) \longrightarrow M^+(aq, L) \qquad Q = \frac{a_L}{a_R} \qquad \nu = 1$$

The standard emf of the cell is zero, because a cell cannot drive a current through a circuit when the two electrode compartments are identical (and, specifically, $\Delta_r G^{\ominus} = 0$ for the cell reaction). Therefore, the emf when the compartments have different concentrations is

$$E = -\frac{RT}{F}\ln\frac{a_L}{a_R} \approx -\frac{RT}{F}\ln\frac{b_L}{b_R} \qquad (10.35)$$

If R is the more concentrated solution, $E > 0$. Physically, the positive potential arises because positive ions tend to be reduced, so withdrawing electrons from the electrode. This process is dominant in the more concentrated right-hand electrode compartment.

One important example of a system that resembles this description is the biological cell membrane, which is more permeable to K^+ ions than to either Na^+ or Cl^- ions. In the enzyme (Na^+–K^+)-ATPase, the hydrolysis of ATP drives the pumping of ions across the membrane. For example, the concentration of K^+ inside an inactive nerve cell is about 20 times that on the outside, whereas the concentration of Na^+ outside the cell is about 10 times that on the inside. The difference in concentrations of ions results in a transmembrane potential difference of about −62 mV, with the negative sign denoting that the inside has a negative potential. This potential difference is also called the **resting potential** of the cell membrane.

The transmembrane potential difference plays a particularly interesting role in the transmission of nerve impulses. Upon receiving an impulse, which is called an **action potential**, a site in the nerve cell membrane becomes transiently permeable to Na^+ and the transmembrane potential difference changes. To propagate along a nerve cell, the action potential must change the transmembrane potential difference by at least 20 mV, to values that are less negative than −40 mV. Propagation occurs when an action potential in one site of the membrane triggers an action potential in an adjacent site, with sites behind the moving action potential returning to the resting potential.

(h) Cells at equilibrium

A special case of the Nernst equation has great importance in electrochemistry. Suppose the reaction has reached equilibrium; then $Q = K$, where K is the equilibrium constant of the cell reaction. However, a chemical reaction at equilibrium cannot do work, and hence it generates zero potential difference between the electrodes of a galvanic cell. Therefore, setting $E = 0$ and $Q = K$ in the Nernst equation gives

$$\ln K = \frac{\nu F E^{\ominus}}{RT} \tag{10.36}$$

This very important equation lets us predict equilibrium constants from measured standard cell potentials. However, before we use it extensively, we need to establish a further result.

Illustration 10.7

Because the standard potential of the Daniell cell is +1.10 V, the equilibrium constant for the cell reaction $Cu^{2+}(aq) + Zn(s) \rightarrow Cu(s) + Zn^{2+}(aq)$, for which $\nu = 2$, is $K = 1.5 \times 10^{37}$ at 298.15 K. We conclude that the displacement of copper by zinc goes virtually to completion.

10.5 Standard potentials

A galvanic cell is a combination of two electrodes, and each one can be considered as making a characteristic contribution to the overall cell potential. Although it is not possible to measure the contribution of a single electrode, we can define the

Table 10.7* Standard potentials at 298 K

Couple	$E^\ominus$/V
$Ce^{4+}(aq) + e^- \longrightarrow Ce^{3+}(aq)$	+1.61
$Cu^{2+}(aq) + 2e^- \longrightarrow Cu(s)$	+0.34
$AgCl(s) + e^- \longrightarrow Ag(s) + Cl^-(aq)$	+0.22
$2H^+(aq) + 2e^- \longrightarrow H_2(g)$	0
$Zn^{2+}(aq) + 2e^- \longrightarrow Zn(s)$	−0.76
$Na^+(aq) + e^- \longrightarrow Na(s)$	−2.71

* More values are given in the *Data section*.

potential of one of the electrodes as zero and then assign values to others on that basis. The specially selected electrode is the **standard hydrogen electrode** (SHE):

$$Pt(s)|H_2(g)|H^+(aq) \qquad E^\ominus = 0 \qquad [10.37]$$

at all temperatures. The **standard potential**, $E^\ominus$, of another couple is then assigned by constructing a cell in which it is the right-hand electrode and the standard hydrogen electrode is the left-hand electrode. For example, the standard potential of the Ag^+/Ag couple is the standard emf of the following cell:[6]

$$Pt(s)|H_2(g)|H^+(aq)\,||\,Ag^+(aq)|Ag(s) \qquad E^\ominus(Ag^+/Ag) = E^\ominus = +0.80\text{ V}$$

Likewise, the standard potential of the $AgCl/Ag,Cl^-$ couple is the standard emf of the following cell:

$$Pt(s)|H_2(g)|H^+(aq)\,||\,Cl^-(aq)|AgCl(s)|Ag(s) \qquad E^\ominus(AgCl/Ag,Cl^-) = E^\ominus = +0.22\text{ V}$$

Although a standard potential is written as though it refers to a half-reaction such as

$$AgCl(s) + e^- \longrightarrow Ag(s) + Cl^-(aq) \qquad E^\ominus(AgCl/Ag,Cl^-) = +0.22\text{ V}$$

it should be understood that these equations are only shorthand for writing

$$AgCl(s) + \tfrac{1}{2}H_2(g) \longrightarrow Ag(s) + H^+(aq) + Cl^-(aq) \qquad E^\ominus = +0.22\text{ V}$$

and that the standard potential is determined by properties of the hydrogen electrode as well as the couple to which the potential refers. Table 10.7 lists standard potentials at 298 K.

An important feature of standard emf of cells and standard potentials of electrodes is that they are unchanged if the chemical equation for the cell reaction or a half-reaction is multiplied by a numerical factor. A numerical factor increases the value of the standard Gibbs energy for the reaction. However, it also increases the number of electrons transferred by the same factor, and by eqn 33 the value of $E^\ominus$ remains unchanged.

(a) The cell emf in terms of individual standard potentials

The standard emf of a cell formed from any two electrodes is calculated by taking the difference of their standard potentials. This rule follows from the fact that a cell such as

$$Ag(s)|Ag^+(aq)\,||\,Cl^-(aq)|AgCl(s)|Ag(s)$$

is equivalent to two cells joined back-to-back:

$$Ag(s)|Ag^+(aq)\,||\,H^+(aq)|H_2(g)|Pt(s)\text{—}Pt(s)|H_2(g)|H^+(aq)\,||\,Cl^-(aq)|AgCl(s)|Ag(s)$$

The overall potential of this composite cell, and therefore of the cell of interest, is

$$E^\ominus = E^\ominus(AgCl/Ag,Cl^-) - E^\ominus(Ag^+/Ag) = -0.58\text{ V}$$

The standard potentials in Table 10.7 can all be used in the same way, and the standard emf of the cell is obtained by subtracting the standard potential of the left-hand electrode from the standard potential of the right-hand electrode:

$$E^\ominus = E^\ominus(\text{right}) - E^\ominus(\text{left}) \qquad (10.38)$$

Because $\Delta G^\ominus = -\nu FE^\ominus$, it then follows that, if the result gives $E^\ominus > 0$, then the corresponding cell reaction has $K > 1$.

6. Here and from now on, all values refer to 298 K.

Illustration 10.8

To calculate the equilibrium constant for the disproportionation $2\,Cu^{+}(aq) \rightarrow Cu(s) + Cu^{2+}(aq)$ at 298.15 K we need

Right-hand electrode: $Cu(s)|Cu^{+}(aq)$ $\quad Cu^{+}(aq) + e^{-} \longrightarrow Cu(aq)$ $\quad E^{\ominus} = +0.52\text{ V}$

Left-hand electrode: $Pt(s)|Cu^{2+}(aq),Cu^{+}(aq)$ $\quad Cu^{2+}(aq) + e^{-} \longrightarrow Cu^{+}(s)$ $\quad E^{\ominus} = +0.16\text{ V}$

The standard emf of the cell is therefore

$$E^{\ominus} = +0.52\text{ V} - 0.16\text{ V} = +0.36\text{ V}$$

Then, because $\nu = 1$,

$$\ln K = \frac{0.36\text{ V}}{0.025\,693\text{ V}} = \frac{0.36}{0.025\,693}$$

Hence, $K = 1.2 \times 10^{6}$.

Example 10.2 Calculating an equilibrium constant

Evaluate the solubility constant of silver chloride (the equilibrium constant for the dissolution of AgCl(s)) and its solubility from cell potential data at 298.15 K.

Method We need to find an electrode combination that reproduces the solubility equilibrium, and then identify the solubility constant with the equilibrium constant of the cell reaction. Calculate the standard emf of the cell from the standard potentials and use the result in eqn 36. For the solubility, S, which is the concentration of solute at equilibrium, express the solubility product in terms of S and solve the resulting equation for S.

Answer We saw in Example 10.1 that the dissolution can be expressed as the difference of the following half-reactions:

$$AgCl(s) + e^{-} \longrightarrow Ag(s) + Cl^{-}(aq) \qquad E^{\ominus} = +0.22\text{ V}$$

$$Ag^{+}(aq) + e^{-} \longrightarrow Ag(s) \qquad E^{\ominus} = +0.80\text{ V}$$

The standard emf is therefore −0.58 V. Then, because $\nu = 1$,

$$\ln K = \frac{\nu E^{\ominus}}{RT/F} = \frac{1 \times (-0.58\text{ V})}{2.5693 \times 10^{-2}\text{ V}} = -\frac{0.58}{2.5693 \times 10^{-2}}$$

Therefore, $K = 1.6 \times 10^{-10}$. Because $K = a_{Ag^{+}}a_{Cl^{-}} \approx S^{2}$, it follows that $S = 1.3 \times 10^{-5}\text{ mol kg}^{-1}$.

Self-test 10.6 Calculate the solubility constant and the solubility of mercury(I) chloride at 298.15 K. *Hint.* The mercury(I) ion is the diatomic species Hg_2^{2+}. [2.6×10^{-18}, $8.7 \times 10^{-7}\text{ mol kg}^{-1}$]

(b) The measurement of standard potentials

The procedure for measuring a standard potential can be illustrated by considering a specific case, the silver chloride electrode. The measurement is made on the 'Harned cell':

$$\mathrm{Pt(s)|H_2(g)|HCl(aq)|AgCl(s)|Ag(s)} \qquad \tfrac{1}{2}\mathrm{H_2(g) + AgCl(s) \longrightarrow HCl(aq) + Ag(s)}$$

for which the Nernst equation is

$$E = E^{\ominus}(\mathrm{AgCl/Ag,Cl^-}) - \frac{RT}{F}\ln\frac{a_{\mathrm{H^+}}a_{\mathrm{Cl^-}}}{(f_{\mathrm{H_2}}/p^{\ominus})^{1/2}}$$

We shall set $f = p^{\ominus}$ from now on, and for simplicity write the standard potential as $E^{\ominus}$. The activities can be expressed in terms of the molality b and the mean activity coefficient, $\gamma_{\pm}$, through eqn 11:

$$E = E^{\ominus} - \frac{RT}{F}\ln b^2 - \frac{RT}{F}\ln \gamma_{\pm}^2$$

This expression rearranges to

$$E + \frac{2RT}{F}\ln b = E^{\ominus} - \frac{2RT}{F}\ln \gamma_{\pm} \tag{10.39}$$

From the Debye–Hückel limiting law for a 1,1-electrolyte, we know that $\ln \gamma_{\pm} \propto -b^{1/2}$ (the natural logarithm used here is proportional to the common logarithm that appears in eqn 17). Therefore, with the constant of proportionality in this relation written as $(F/2RT)C$:

$\ln x = \ln 10 \log x \approx 2.303 \log x.$

$$E + \frac{2RT}{F}\ln b = E^{\ominus} + Cb^{1/2}$$

The expression on the left is evaluated at a range of molalities, plotted against $b^{1/2}$, and extrapolated to $b = 0$.[7] The intercept at $b^{1/2} = 0$ is the value of $E^{\ominus}$ for the silver/silver-chloride electrode.

Example 10.3 Determining the standard emf of a cell

The emf of the cell $\mathrm{Pt(s)|H_2(g,}\ p^{\ominus})|\mathrm{HCl(aq,}\ b)|\mathrm{AgCl(s)|Ag(s)}$ at 25°C has the following values:

$b/(10^{-3}\,b^{\ominus})$	3.215	5.619	9.138	25.63
E/V	0.520 53	0.492 57	0.468 60	0.418 24

Determine the standard emf of the cell.

Method Proceed as described above, but adjusted to the cell reaction. Start by writing the Nernst equation for the cell, and then express the activities that occur in Q in terms of the mean activity coefficient. The latter can be written as proportional to $b^{1/2}$ by using the Debye–Hückel limiting law. However, there is no need to write all the constants because the standard emf is obtained by extrapolation, as explained in the text.

Answer The ionic equation for the cell reaction is

$$\tfrac{1}{2}\mathrm{H_2(g) + AgCl(s) \longrightarrow Ag(s) + H^+(aq) + Cl^-(aq)} \qquad Q = a_{\mathrm{H^+}}a_{\mathrm{Cl^-}} \qquad \nu = 1$$

because the solid is at unit activity and the gas is at 1 bar. The Nernst equation is therefore

7. In precise work, the $b^{1/2}$ term is brought to the left, and a higher order correction term from the extended Debye–Hückel law is used on the right.

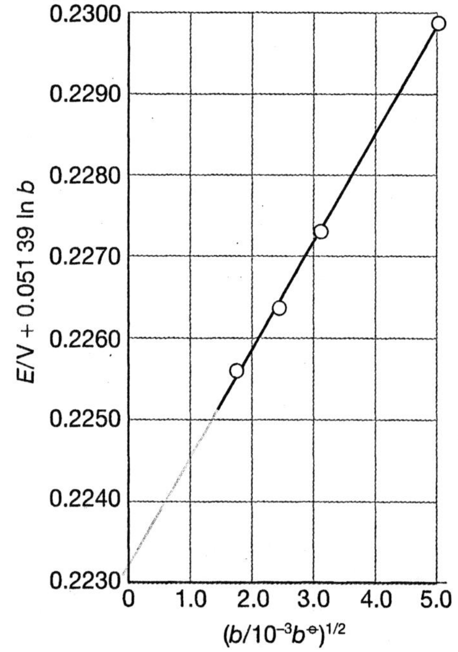

10.14 The plot, and the extrapolation used for the experimental measurement of a standard cell emf. The intercept at $b^{1/2} = 0$ is $E^{\ominus}$.

$$E = E^{\ominus} - \frac{RT}{F}\ln a_{H^+}a_{Cl^-}$$

The activities are related to the molality, b, of HCl by

$$a_{H^+}a_{Cl^-} = \gamma_{\pm}^2 b_{H^+}b_{Cl^-} = \gamma_{\pm}^2 b^2$$

because $b_{H^+} = b_{Cl^-} = b$ for a fully deprotonated acid. Only a little work is then needed to convert the Nernst equation into

$$E + \frac{2RT}{F}\ln b = E^{\ominus} + Cb^{1/2}$$

where C is a collection of constants that come from the limiting law. We now draw up the following table, using $2RT/F = 0.051\,39$ V:

$b/(10^{-3}\,b^{\ominus})$	3.215	5.619	9.138	25.63
$(b/(10^{-3}\,b^{\ominus}))^{1/2}$	1.793	2.370	3.023	5.063
E/V	0.520 53	0.492 57	0.468 60	0.418 24
E/V + 0.051 39 ln b	0.2256	0.2263	0.2273	0.2299

The data are plotted in Fig. 10.14; as can be seen, they extrapolate to $E^{\ominus} = 0.2232$ V.

Self-test 10.7 The data below are for the cell Pt(s)|H_2(g, $p^{\ominus}$)|HBr(aq, b)|AgBr(s)|Ag(s) at 25°C. Determine the standard emf of the cell.

$b/(10^{-4}\,b^{\ominus})$	4.042	8.444	37.19
E/V	0.473 81	0.436 36	0.361 73

[0.071 V]

(c) The measurement of activity coefficients

Once the standard potential of an electrode in a cell is known, we can use it to determine the activities of the ions with respect to which it is reversible by measuring the cell emf with the ions at the concentration of interest. For example, the mean activity coefficient of the ions in hydrochloric acid of molality b is obtained from eqn 39 in the form

$$\ln \gamma_{\pm} = \frac{E^{\ominus} - E}{2RT/F} - \ln b \qquad \{10.40\}$$

once E has been measured.

Applications of standard potentials

Cell emfs are a convenient source of data on the Gibbs energies, enthalpies, and entropies of reactions. In practice the standard values of these quantities are the ones normally determined.

10.6 The electrochemical series

We have seen that for two redox couples Ox_1/Red_1 and Ox_2/Red_2, and the cell

$$Red_1,Ox_1 \,||\, Red_2,Ox_2 \qquad E^{\ominus} = E_2^{\ominus} - E_1^{\ominus} \qquad (10.41a)$$

that the cell reaction

$$\mathrm{Red_1 + Ox_2 \longrightarrow Ox_1 + Red_2} \qquad (10.41b)$$

is spontaneous as written if $E^\ominus > 0$, and therefore if $E_2^\ominus > E_1^\ominus$. Because in the cell reaction Red_1 reduces Ox_2, we can conclude that

Red_1 has a thermodynamic tendency to reduce Ox_2 if $E_1^\ominus < E_2^\ominus$

More briefly: low reduces high. The reactions of the electron transport chains of respiration and photosynthesis are good applications of this principle (Box 10.1).

Box 10.1 Biological energy conversion: driving the synthesis of ATP with electron transfer reactions

Many important biochemical processes involve electron transfer reactions. Here, we examine the electrochemical reactions associated with photosynthesis and the aerobic oxidation of glucose. These processes are related by the reactions

$$\mathrm{C_6H_{12}O_6(s) + 6\,O_2(g) \underset{\text{photosynthesis}}{\overset{\text{glycolysis and the citric acid cycle}}{\rightleftarrows}} 6\,CO_2(g) + 6\,H_2O(l)}$$

The exergonic oxidation of one $C_6H_{12}O_6$ molecule requires the transfer of 24 electrons to six O_2 molecules. The half-reactions for the oxidation of glucose and the reduction of O_2 are

$$\mathrm{C_6H_{12}O_6(s) + 6\,H_2O(l) \longrightarrow 6\,CO_2(g) + 24\,H^+(aq) + 24\,e^-}$$
$$\mathrm{6\,O_2(g) + 24\,H^+(aq) + 24\,e^- \longrightarrow 12\,H_2O(l)}$$

The electrons do not flow directly from glucose to O_2. In the cell, glucose is oxidized to CO_2 by NAD^+ and FAD during glycolysis and the citric acid cycle (Box 9.1):

$$\mathrm{C_6H_{12}O_6(s) + 10\,NAD^+ + 2\,FAD + 4\,ADP + 4\,P_i^- + 2\,H_2O \longrightarrow 6\,CO_2 + 10\,NADH + 2\,FADH_2 + 4\,ATP + 6\,H^+}$$

In the mitochondrial electron transport chain, electrons from the powerful reducing agents NADH and $FADH_2$ pass through four membrane-bound protein complexes and two mobile electron carriers before reducing O_2 to H_2O. We shall see that the electron transfer reactions drive the synthesis of ATP at three of the membrane protein complexes.

In complex I (NADH:ubiquinone oxidoreductase), NADH is oxidized by ubiquinone (UQ, **1**) in a two-electron reaction:

$$\mathrm{H^+ + NADH + UQ \xrightarrow{\text{complex I}} NAD^+ + UQH_2}$$
$$E^\oplus = +0.42\ \mathrm{V}, \quad \Delta_r G^\oplus = -81\ \mathrm{kJ\,mol^{-1}}$$

Additional UQ molecules are reduced by $FADH_2$ in complex II (succinate dehydrogenase):

$$\mathrm{FADH_2 + UQ \xrightarrow{\text{complex II}} FAD + UQH_2}$$
$$E^\oplus = +0.015\ \mathrm{V}, \quad \Delta_r G^\oplus = -2.9\ \mathrm{kJ\,mol^{-1}}$$

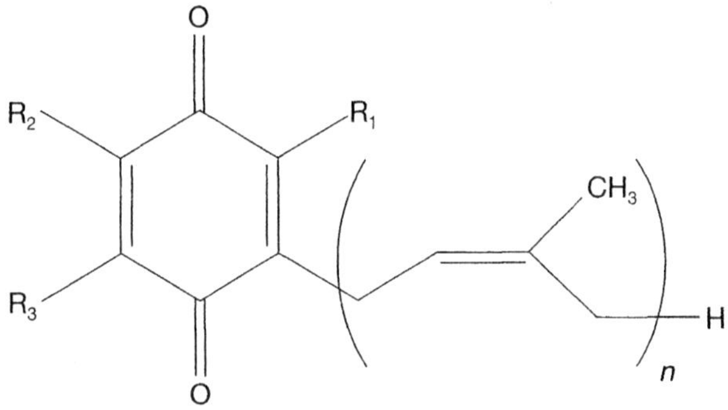

Ubiquinone: $R_1 = CH_3$, $R_2 = R_3 = OCH_3$
Plastoquinone: $R_1 = H$, $R_2 = R_3 = CH_3$

1

Reduced ubiquinone migrates to complex III (ubiquinol: cytrochrome *c* oxidoreductase), which catalyses the reduction of the protein cytochrome *c* (Cyt *c*). Cytochrome *c* contains the haem *c* group (**2**), the central iron ion of which can exist in oxidation states +3 and +2. The net reaction catalysed by complex III is

$$\mathrm{UQH_2 + 2\,Fe^{3+}(Cyt\ \mathit{c}) \xrightarrow{\text{complex III}} UQ + 2\,Fe^{2+}(Cyt\ \mathit{c}) + 2\,H^+}$$
$$E^\oplus = +0.15\ \mathrm{V}, \quad \Delta_r G^\oplus = -30\ \mathrm{kJ\,mol^{-1}}$$

Reduced cytochrome *c* carries electrons from complex III to complex IV (ferrocytochrome *c* oxidase), where O_2 is reduced to H_2O:

$$\mathrm{2\,Fe^{2+}(Cyt\ \mathit{c}) + 2\,H^+ + \tfrac{1}{2}\,O_2 \xrightarrow{\text{complex IV}} 2\,Fe^{3+}(Cyt\ \mathit{c}) + H_2O}$$
$$E^\oplus = +0.815\ \mathrm{V}, \quad \Delta_r G^\oplus = -109\ \mathrm{kJ\,mol^{-1}}$$

The reactions that occur in complexes I, III, and IV are sufficiently exergonic to drive the synthesis of ATP in the process called *oxidative phosphorylation*:

$$\mathrm{ADP + P_i + H^+ \longrightarrow ATP} \qquad \Delta_r G^\oplus = +30\ \mathrm{kJ\,mol^{-1}}$$

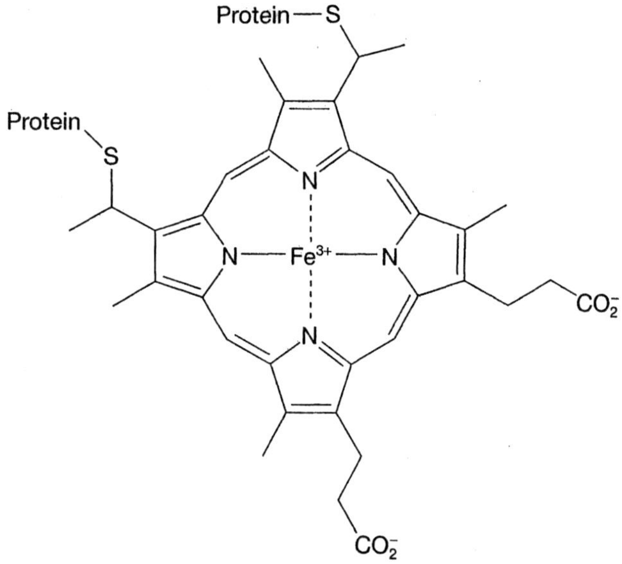

2

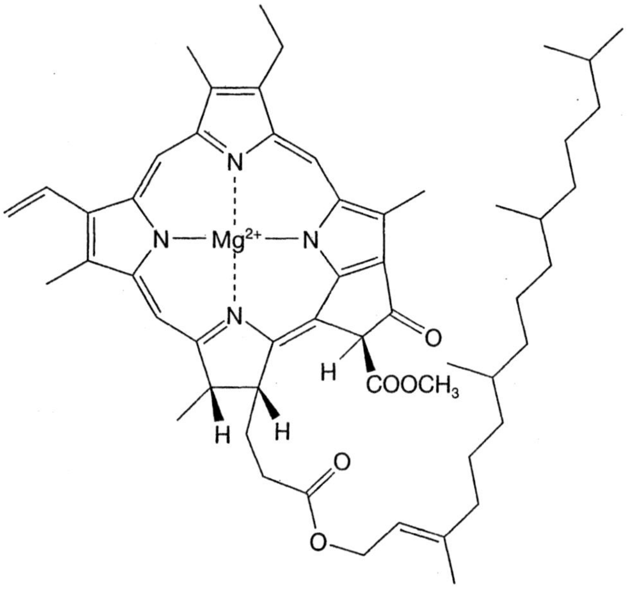

3

Experiments show that 11 molecules of ATP are made for every three molecules of NADH and one molecule of $FADH_2$ that are oxidized by the mitochondrial electron transport chain. The ATP is then hydrolysed on demand to perform useful biochemical work throughout the cell.

In plant photosynthesis, solar energy drives the endergonic reduction of CO_2 to glucose, with concomitant oxidation of water to O_2 ($\Delta_r G^\ominus = +2880$ kJ mol^{-1}). The process takes place in the *chloroplast*, a special organelle of the plant cell. Electrons flow from reductant to oxidant via a series of electrochemical reactions that are coupled to the synthesis of ATP. First, the leaf absorbs solar energy and transfers it to membrane protein complexes known as photosystem I and photosystem II.[8] The absorption of energy from light decreases the reduction potential of special dimers of chlorophyll *a* molecules (**3**) known as P700 (in photosystem I) and P680 (in photosystem II). In their high-energy or excited states, P680 and P700 initiate electron transfer reactions that culminate in the oxidation of water to O_2 and the reduction of $NADP^+$ to NADPH:[9]

$$2\,NADP^+ + 2\,H_2O \xrightarrow{\text{light}} O_2 + 2\,NADPH + 2\,H^+$$

It is clear that energy from light is required to drive this reaction because, in the dark, $E^\ominus = -1.135$ V and $\Delta_r G^\ominus = +438.0$ kJ mol^{-1}.

Working together, photosystem I and the enzyme ferredoxin:$NADP^+$ oxidoreductase catalyse the light-induced oxidation of $NADP^+$ to NADPH. The electrons required for this process come initially from P700 in its excited state. The resulting $P700^+$ is then reduced by the mobile carrier plastocyanin (Pc), a protein in which the bound copper ion can exist in oxidation states +2 and +1. The net reaction is

$$NADP^+ + 2\,Cu^+(Pc) + H^+ \xrightarrow{\text{light, photosystem I}} NADPH + 2\,Cu^{2+}(Pc)$$

Oxidized plastocyanin accepts electrons from reduced plastoquinone (PQ, **1**). The process is catalysed by the cytochrome b_6f complex, a membrane protein complex that resembles complex III of mitochondria:

$$PQH_2 + 2\,Cu^{2+}(Pc) \xrightarrow{\text{cyt } b_6f \text{ complex}} PQ + 2\,H^+ + 2\,Cu^+(Pc)$$
$$E^\ominus = +0.370\text{ V}, \quad \Delta_r G^\ominus = -71.4\text{ kJ mol}^{-1}$$

This reaction is sufficiently exergonic to drive the synthesis of ATP in the process known as *photophosphorylation*.

Plastoquinone is reduced by water in a process catalysed by light and photosystem II. The electrons required for the reduction of plastoquinone come initially from P680 in its excited state. The resulting $P680^+$ is then reduced ultimately by water. The net reaction is

$$H_2O + PQ \xrightarrow{\text{light, photosystem II}} \tfrac{1}{2}O_2 + PQH_2$$

In this way, plant photosynthesis uses an abundant source of electrons (water) and of energy (the Sun) to drive the endergonic reduction of $NADP^+$, with concomitant synthesis of

8. See Section 26.11 and Box 26.2 for details of the energy transfer process.
9. $NADP^+$ is a form of NAD^+ with a phosphate group covalently attached to the carbohydrate portion of the molecule. See Box 9.1.

ATP. Experiments show that, for each molecule of NADPH formed in the chloroplast of green plants, one molecule of ATP is synthesized.

The ATP and NADPH molecules formed by the light-induced electron transfer reactions of plant photosynthesis participate directly in the reduction of CO_2 to glucose in the chloroplast:

$$6\,CO_2 + 12\,NADPH + 12\,ATP + 12\,H^+ \longrightarrow C_6H_{12}O_6 + 12\,NADP^+ + 12\,ADP + 12\,P_i^- + 6\,H_2O$$

In summary, electrochemical reactions mediated by membrane protein complexes harness energy in the form of ATP. Plant photosynthesis uses solar energy to transfer electrons from a poor reductant (water) to carbon dioxide. In the process, high-energy molecules (carbohydrates, such as glucose) are synthesized in the cell. Animals feed on the carbohydrates derived from photosynthesis. During aerobic metabolism, the O_2 released by photosynthesis as a waste product is used to oxidize carbohydrates to CO_2. This drives biological processes, such as biosynthesis, muscle contraction, cell division, and nerve conduction. Hence, the sustenance of life on Earth depends on a tightly regulated carbon–oxygen cycle that is driven by solar energy.

Discussion questions

1. In anaerobic bacteria, the source of carbon may be a molecule other than glucose and the final electron acceptor is some molecule other than O_2. Could a bacterium evolve to utilize the ethanol/nitrate pair instead of the glucose/O_2 pair as a source of metabolic energy?

2. The photosynthetic oxidation of water to O_2 occurs in an enzyme that contains four manganese ions, each of which can exist in oxidation states ranging from +2 to +4. The electrochemical production of one molecule of O_2 requires the oxidation of two molecules of water by a total of four electrons. However, the excited state of P680 can donate only one electron at a time to plastoquinone. Explain how electron transfer mediated by P680 can lead to the formation of a molecule of O_2 in photosystem II. *Hint.* See V.A. Szalai and G.W. Brudvig, How plants produce dioxygen. *American Scientist* **86**, 542 (1998).

Further reading

C.K. Matthews, K.E. van Holde, and K.G. Ahern, *Biochemistry*. Addison Wesley Longman, San Francisco (2000).

M. Saraste, Oxidative phosphorylation at the *fin de siècle*. *Science* **283**, 1488 (1999).

Table 10.8* The electrochemical series of the metals

Least strongly reducing
Gold
Platinum
Silver
Mercury
Copper
(Hydrogen)
Lead
Tin
Nickel
Iron
Zinc
Chromium
Aluminium
Magnesium
Sodium
Calcium
Potassium
Most strongly reducing

* The complete series can be inferred from Table 10.7.

Illustration 10.9

Because $E^\ominus(Zn^{2+}, Zn) = -0.76\ V < E^\ominus(Cu^{2+}, Cu) = +0.34\ V$, zinc has a thermodynamic tendency to reduce Cu^{2+} ions in aqueous solution.

Table 10.8 shows a part of the **electrochemical series**, the metallic elements (and hydrogen) arranged in the order of their reducing power as measured by their standard potentials in aqueous solution. A metal low in the series (with a lower standard potential) can reduce the ions of metals with higher standard potentials. This conclusion is qualitative. The quantitative value of K is obtained by doing the calculations we have described previously. For example, to determine whether zinc can displace magnesium from aqueous solutions at 298 K, we note that zinc lies above magnesium in the electrochemical series, so zinc cannot reduce magnesium ions in aqueous solution. Zinc can reduce hydrogen ions, because hydrogen lies higher in the series. However, even for reactions that are thermodynamically favourable, there may be kinetic factors that result in very slow rates of reaction.

10.7 The measurement of pH and pK_a

The half-reaction at a hydrogen electrode is

$$H^+(aq) + e^- \longrightarrow \tfrac{1}{2}H_2(g) \qquad Q = \frac{(f_{H_2}/p^\ominus)^{1/2}}{a_{H^+}} \qquad \nu = 1$$

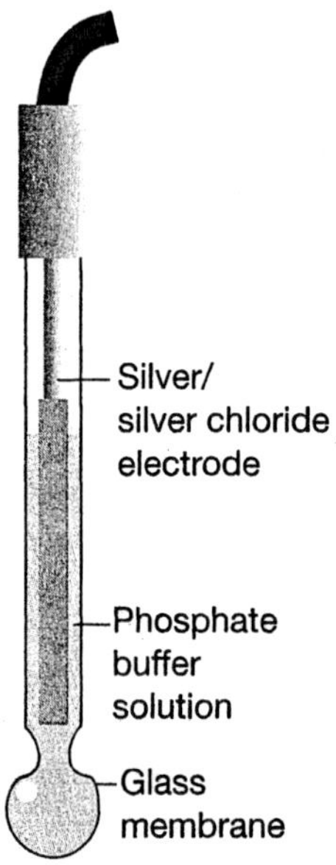

10.15 The glass electrode. It is commonly used in conjunction with a calomel electrode that makes contact with the test solution through a salt bridge.

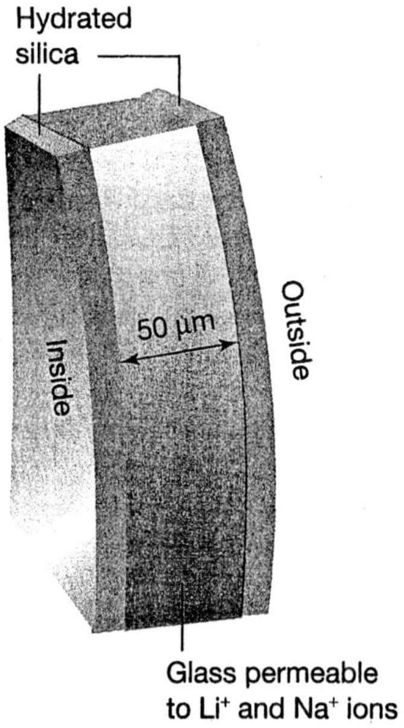

10.16 A section through the wall of a glass electrode.

If we let $f_{H_2} = p^\ominus$ and $E^\ominus(H^+/H_2) = 0$, the electrode potential is

$$E(H^+/H_2) = \frac{RT}{F}\ln a_{H^+} = -\frac{RT\ln 10}{F}\,\text{pH} \qquad (10.42a)$$

This expression makes sense physically. Increasing the pH above 0 (lowering the hydrogen ion activity below 1) decreases the tendency of the positive ions to become reduced at the electrode, so we should expect its potential to become negative. At 25°C, when $RT/F = 25.69$ mV, this relation becomes

$$E(H^+/H_2) = -59.16\ \text{mV} \times \text{pH} \qquad (10.42b)$$

Each unit increase in pH decreases the electrode potential by 59 mV.

The measurement of the pH of a solution is simple in principle, for it is based on the measurement of the potential of a hydrogen electrode immersed in the solution. The left-hand electrode of the cell is typically a saturated calomel (Hg_2Cl_2(s)) reference electrode with potential E(cal); the right-hand electrode is the hydrogen electrode with potential given by eqn 42. The pH of the cell is therefore

$$\text{pH} = \frac{E + E(\text{cal})}{(-59.16\ V)} \qquad (10.43)$$

The practical definition of the pH of a solution X is

$$\text{pH} = \text{pH(S)} - \frac{FE}{RT\ln 10} \qquad [10.44]$$

where E is the emf of the cell

$$\text{Pt(s)}|\text{H}_2\text{(g)}|\text{S(aq)}||3.5\ \text{M KCl(aq)}||\text{X(aq)}|\text{H}_2\text{(g)}|\text{Pt(s)}$$

and S is a solution of standard pH. The currently recommended primary standards include a saturated aqueous solution of potassium hydrogen tartrate, which has pH = 3.557 at 25°C, and 0.0100 mol kg^{-1} disodium tetraborate, which has pH = 9.180 at that temperature.

In practice, indirect methods are much more convenient and the hydrogen electrode is replaced by the **glass electrode** (Fig. 10.15). This electrode is sensitive to hydrogen ion activity, and has a potential proportional to pH. It is filled with a phosphate buffer containing Cl^- ions, and conveniently has $E = 0$ when the external medium is at pH = 7. The glass electrode is much more convenient to handle than the gas electrode itself and is calibrated using solutions of known pH.

The responsiveness of a glass electrode to the hydronium ion activity is a result of complex processes at the interface between the glass membrane and the solutions on either side of it. The membrane itself is permeable to Na^+ and Li^+ ions but not to H^+ ions. Therefore, the potential difference across the glass membrane must arise by a mechanism different from that responsible for biological transmembrane potentials. A clue to the mechanism comes from a detailed inspection of the glass membrane, for each face is coated with a thin layer of hydrated silica (Fig. 10.16). The hydrogen ions in the test solution modify this layer to an extent that depends on their activity in the solution, and the charge modification of the outside layer is transmitted to the inner layer by the Na^+ and Li^+ ions in the glass. The hydronium ion activity gives rise to a membrane potential by this indirect mechanism.

The electrochemical determination of pH opens up a route to the electrochemical determination of pK_a, for we saw in Section 9.5c that the pK_a of an acid is equal to the pH of a solution containing equal amounts of the acid and its conjugate base.

Electrodes can be constructed with potentials that are sensitive to the concentrations of ions other than the hydronium ion. Some examples of these **ion-selective electrodes** are discussed in Box 10.2.

Box 10.2 Species-selective electrodes

A glass electrode is an example of an *ion-selective electrode*, an electrode that generates a potential in response to the presence of a solution of specific ions. Electrodes sensitive to hydrogen ions, and hence to pH, are typically glasses based on lithium silicate doped with heavy-metal oxides. The glass can be responsive to Na^+, K^+, and NH_4^+ ions by being doped with Al_2O_3 and B_2O_3.

A suitably adapted glass electrode can be used to detect the presence of certain gases. A simple form of a gas-sensing electrode consists of a glass electrode contained in an outer sleeve filled with an aqueous solution and separated from the test solution by a membrane that is permeable to gas. When a gas such as sulfur dioxide or ammonia diffuses into the aqueous solution, it modifies its pH, which in turn affects the potential of the glass electrode. The presence of an enzyme that converts a compound, such as urea or an amino acid, into ammonia, which then affects the pH, can be used to detect these organic compounds.

Somewhat more sophisticated devices are used as ion-selective electrodes that give potentials according to the presence of specific ions present in a test solution. In one arrangement, a porous lipophilic (hydrocarbon-attracting) membrane is attached to a small reservoir of a hydrophobic (water-repelling) liquid, such as dioctylphenylphosphonate, that saturates it (see the illustration). The liquid contains a chelating agent, such as $(RO)_2PO_2^-$ with R a C_8 to C_{18} chain, that acts as a kind of solubilizing agent for the ions with which it can form a complex. The chelated ions are able to migrate through the lipophilic membrane, and hence give rise to a transmembrane potential, which is detected by a silver/silver chloride electrode in the interior of the assembly. Electrodes of this construction can be designed to be sensitive to a variety of ionic species, including calcium, zinc, iron, lead, and copper ions.

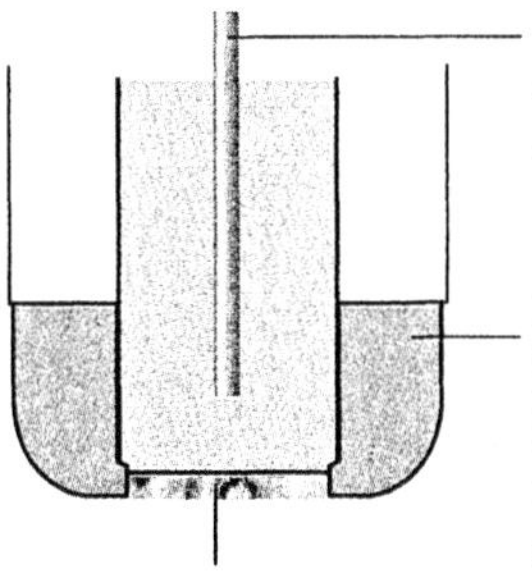

The structure of an ion-selective electrode. Chelated ions are able to migrate through the lipophilic membrane.

In theory, the transmembrane potential should be determined entirely by differences in the activity of the species that the electrode was designed to detect. In practice, a small potential difference, called the *asymmetry potential*, is observed even when the activity of the test species is the same on both sides of the membrane.[10] Furthermore, all species-selective electrodes are sensitive to more than one species. For example, a Na^+ selective electrode also responds, albeit less effectively, to the activity of K^+ ions in the test solution. As a result of these effects, the potential of an electrode sensitive to species X^+ that is also susceptible to interference by species Y^+ is given by a modified form of the Nernst equation:

$$E = E_{ap} + \beta\frac{RT}{F}\ln(a_{X^+} + k_{X,Y}a_{Y^+})$$

where E_{ap} is the asymmetry potential, β is an experimental parameter that captures deviations from the Nernst equation, and $k_{X,Y}$ is the *selectivity coefficient* of the electrode and is related to the response of the electrode to the interfering species Y^+. A value of $\beta = 1$ indicates that the electrode responds to the activity of ions in solution in a way that is consistent with the Nernst equation and, in practice, most species-selective electrodes of high quality have $\beta \approx 1$. The selectivity coefficient, and hence interference effects, can be minimized when designing and manufacturing a species-selective electrode. For precise work, it is necessary to calibrate the response of the electrode by measuring E_{ap}, β, and $k_{X,Y}$ before performing experiments on solutions of unknown concentration of X^+.

Discussion questions

1. Suggest a calibration procedure for a Na^+-selective electrode, in which the values of E_{ap}, β, and $k_{X,Y}$ are measured experimentally. Assume that K^+ is the only interfering species.

2. Chemists and biochemists are often interested in measuring the concentration of an ion in solution, rather than its activity. The strategy for correlating directly the value of an electrode potential to a value of ion concentration consists of performing all measurements in solutions of constant and high ionic strength. Explain why this is a useful strategy.

Further reading

D.C. Harris, *Quantitative chemical analysis*. W.H. Freeman & Co., New York (1999).

J. Koryta, *Ions, electrodes, and membranes*. Wiley, New York (1992).

10. The asymmetry potential is due to the fact that it is not possible to manufacture a membrane material that has the same structure and the same chemical properties throughout.

10.8 Thermodynamic functions

The standard emf of a cell is related to the standard reaction Gibbs energy through eqn 33 ($\Delta_r G^\ominus = -\nu F E^\ominus$). Therefore, by measuring $E^\ominus$ we can obtain this important thermodynamic quantity. Its value can then be used to calculate the Gibbs energy of formation of ions using the convention explained in Section 10.1a.

Illustration 10.10

The cell reaction taking place in

$$\mathrm{Pt(s)|H_2|H^+(aq)||Ag^+(aq)|Ag(s)} \qquad E^\ominus = +0.7996\ \mathrm{V}$$

is

$$\mathrm{Ag^+(aq)} + \tfrac{1}{2}\mathrm{H_2(g)} \longrightarrow \mathrm{H^+(aq) + Ag(s)} \qquad \Delta_r G^\ominus = -\Delta_f G^\ominus(\mathrm{Ag^+, aq})$$

Therefore, with $\nu = 1$, we find

$$\Delta_f G^\ominus(\mathrm{Ag^+, aq}) = -(-FE^\ominus) = +77.15\ \mathrm{kJ\ mol^{-1}}$$

which is in close agreement with the value in Table 2.6 of the *Data section*.

Example 10.4 Evaluating a standard potential from two others

Given that the standard potentials of the Cu^{2+}/Cu and Cu^+/Cu couples are +0.340 V and +0.522 V, respectively, evaluate $E^\ominus(Cu^{2+}, Cu^+)$.

Method First, we note that reaction Gibbs energies may be added (as in a Hess's law analysis of reaction enthalpies). Therefore, we should convert the $E^\ominus$ values to $\Delta G^\ominus$ values by using eqn 33, add them appropriately, and then convert the overall $\Delta G^\ominus$ to the required $E^\ominus$ by using eqn 33 again. This roundabout procedure is necessary because, as we shall see, although the factor F cancels, the factor ν in general does not.

Answer The electrode reactions are as follows:

$$\text{(a)}\ \mathrm{Cu^{2+}(aq) + 2\,e^- \longrightarrow Cu(s)} \qquad E^\ominus = +0.340\ \mathrm{V}, \qquad \text{so}\ \ \Delta_r G^\ominus = -2(0.340\ \mathrm{V})F$$

$$\text{(b)}\ \mathrm{Cu^{+}(aq) + e^- \longrightarrow Cu(s)} \qquad E^\ominus = +0.522\ \mathrm{V}, \qquad \text{so}\ \ \Delta_r G^\ominus = -(0.522\ \mathrm{V})F$$

The required reaction is

$$\text{(c)}\ \mathrm{Cu^{2+}(aq) + e^- \longrightarrow Cu^+(aq)} \qquad E^\ominus = -\Delta_r G^\ominus/F$$

Because (c) = (a) – (b), the standard Gibbs energy of reaction (c) is

$$\Delta_r G^\ominus = \Delta_r G^\ominus(\mathrm{a}) - \Delta_r G^\ominus(\mathrm{b}) = (-0.158\ \mathrm{V}) \times F$$

Therefore, $E^\ominus = +0.158$ V.

Comment Note that we cannot combine the $E^\ominus$ values directly and we must always work via $\Delta G^\ominus$. Alternatively, note that the generalization of the calculation illustrated above is

$$\nu_c E^\ominus(\mathrm{c}) = \nu_a E^\ominus(\mathrm{a}) - \nu_b E^\ominus(\mathrm{b})$$

Self-test 10.8 Calculate the standard potential of the Fe^{3+}/Fe couple from the values for the Fe^{3+}/Fe^{2+} and Fe^{2+}/Fe couples. [–0.037 V]

The temperature coefficient of the standard cell emf, $dE^\ominus/dT$, gives the standard entropy of the cell reaction. This conclusion follows from the thermodynamic relation $(\partial G/\partial T)_p = -S$ and eqn 33, which combine to give[11]

$$\frac{dE^\ominus}{dT} = \frac{\Delta_r S^\ominus}{\nu F} \qquad (10.45)$$

Hence we have an electrochemical technique for obtaining standard reaction entropies and through them the entropies of ions in solution.

Finally, we can combine the results obtained so far and use them to obtain the standard reaction enthalpy:

$$\Delta_r H^\ominus = \Delta_r G^\ominus + T\Delta_r S^\ominus = -\nu F\left(E^\ominus - T\frac{dE^\ominus}{dT}\right) \qquad (10.46)$$

This expression provides a noncalorimetric method for measuring $\Delta_r H^\ominus$ and, through the convention $\Delta_f H^\ominus(H^+, aq) = 0$, the standard enthalpies of formation of ions in solution. Thus, electrical measurements can be used to calculate all the thermodynamic properties with which this chapter began.

Example 10.5 **Using the temperature coefficient of the cell potential**

The standard emf of the cell Pt(s)|H_2(g)|HBr(aq)|AgBr(s)|Ag(s) was measured over a range of temperatures, and the data were fitted to the following polynomial:

$$E^\ominus/V = 0.071\,31 - 4.99\times10^{-4}(T/K - 298) - 3.45\times10^{-6}(T/K - 298)^2$$

Evaluate the standard reaction Gibbs energy, enthalpy, and entropy at 298 K.

Method The standard Gibbs energy of reaction is obtained by using eqn 33 after evaluating $E^\ominus$ at 298 K and by using 1 V C = 1 J. The standard entropy of reaction is obtained by using eqn 45, which involves differentiating the polynomial with respect to T and then setting $T = 298$ K. The reaction enthalpy is obtained by combining the values of the standard Gibbs energy and entropy.

Answer At $T = 298$ K, $E^\ominus = +0.071\,31$ V, so

$$\begin{aligned}\Delta_r G^\ominus &= -\nu F E^\ominus = -(1)\times(9.6485\times10^4\ \text{C mol}^{-1})\times(+0.071\,31\ \text{V})\\ &= -6.880\times10^3\ \text{V C mol}^{-1} = -6.880\ \text{kJ mol}^{-1}\end{aligned}$$

The temperature coefficient of the cell potential is

$$\frac{dE^\ominus}{dT} = -4.99\times10^{-4}\ \text{V K}^{-1} - 2(3.45\times10^{-6})(T/K - 298)\ \text{V K}^{-1}$$

At $T = 298$ K this expression evaluates to

$$\frac{dE}{dT} = -4.99\times10^{-4}\ \text{V K}^{-1}$$

So, from eqn 45, the reaction entropy is

$$\begin{aligned}\Delta_r S^\ominus &= 1\times(9.6485\times10^4\ \text{C mol}^{-1})\times(-4.99\times10^{-4}\ \text{V K}^{-1})\\ &= -48.2\ \text{J K}^{-1}\ \text{mol}^{-1}\end{aligned}$$

11. The derivative is complete because $E^\ominus$, like $\Delta_r G^\ominus$, is independent of the pressure.

It then follows that

$$\Delta_r H^\ominus = \Delta_r G^\ominus + T\Delta_r S^\ominus = -6.880\ \text{kJ mol}^{-1} + (298\ \text{K}) \times (-0.0482\ \text{kJ K}^{-1}\ \text{mol}^{-1})$$
$$= -21.2\ \text{kJ mol}^{-1}$$

Comment One difficulty with this procedure lies in the accurate measurement of small temperature coefficients of cell potential. Nevertheless, it is another example of the striking ability of thermodynamics to relate the apparently unrelated, in this case to relate electrical measurements to thermal properties.

Self-test 10.9 Predict the standard potential of the Harned cell at 303 K from tables of thermodynamic data. [+0.2222 V]

Checklist of key ideas

The thermodynamic properties of ions in solution

10.1 Thermodynamic functions of formation

- ☐ standard enthalpy of formation of ions in solution, the standard enthalpy change accompanying the formation of solutions of ions from the reference states of the parent elements
- ☐ standard enthalpy and Gibbs energy of formation of ions, based on $\Delta_f H^\ominus(H^+, aq) = 0$ and $\Delta_f G^\ominus(H^+, aq) = 0$
- ☐ contributions to $\Delta_f G^\ominus$, the use of a thermodynamic cycle
- ☐ Born equation, $\Delta_{solv} G^\ominus = -(z_i^2 e^2 N_A / 8\pi\varepsilon_0 r_i)(1 - 1/\varepsilon_r)$
- ☐ standard (partial molar) entropies of ions in water, based on $S^\ominus(H^+, aq) = 0$

10.2 Ion activities

- ☐ the activity of an ion in solution, $\mu = \mu^\ominus + RT \ln a$
- ☐ the activity coefficient, $a = \gamma b/b^\ominus$
- ☐ the chemical potential of an ion, $\mu = \mu^{ideal} + RT \ln \gamma$
- ☐ mean activity coefficient, the geometric mean of the individual coefficients: $\gamma_\pm = (\gamma_+^p \gamma_-^q)^{1/(p+q)}$
- ☐ total Gibbs energy of ions in solution, $G = p\mu_+ + q\mu_-$
- ☐ Debye–Hückel theory, a theory of activity coefficients of electrolyte solutions based on the assumption that coulombic interactions between ions are dominant
- ☐ ionic atmosphere, time-averaged, spherical haze of charge around the central ion, in which counter ions outnumber ions of the same charge as the central ion
- ☐ Debye–Hückel limiting law, $\log \gamma_\pm = -|z_+ z_-| A I^{1/2}$
- ☐ shielded Coulomb potential, $\phi_i = (z_i e/4\pi\varepsilon_0 r)e^{-r/r_D}$
- ☐ Debye length, r_D, the distance in which the shielded Coulomb potential falls to 1/e of its unshielded value; a measure of the thickness of the ionic atmosphere
- ☐ charge density, ρ_i, the charge in a small region divided by the volume of the region
- ☐ Poisson's equation, an equation for calculating the potential arising from a given charge distribution, $\nabla^2\phi = -\rho/\varepsilon$
- ☐ ionic strength, $I = \frac{1}{2}\sum_i z_i^2(b_i/b^\ominus)$
- ☐ extended Debye–Hückel law, $\ln \gamma_\pm = -|z_+ z_-| A I^{1/2}/(1 + BI^{1/2})$

Electrochemical cells

- ☐ electrochemical cell, an arrangement in which a chemical reaction does electrical work or in which electrical work is used to bring about a chemical reaction
- ☐ electrode, a metallic conductor that makes contact with an electrolyte
- ☐ electrolyte, an ionic conductor
- ☐ electrode compartment, an electrode and its electrolyte
- ☐ salt bridge, a concentrated electrolyte solution in agar jelly that completes the electrical circuit between two electrode compartments
- ☐ galvanic cell, an electrochemical cell that produces electricity as a result of the spontaneous reaction occurring inside it.
- ☐ electrolytic cell, an electrochemical cell in which a non-spontaneous reaction is driven by an external source of current

10.3 Half-reactions and electrodes

- ☐ oxidation, the removal of electrons from a species
- ☐ reduction, the addition of electrons to a species
- ☐ redox reaction, a reaction in which there is a transfer of electrons from one species to another
- ☐ reducing agent (or 'reductant'), an electron donor in a redox reaction
- ☐ oxidizing agent (or 'oxidant'), an electron acceptor in a redox reaction
- ☐ half-reaction, a conceptual reaction showing the gain of electrons
- ☐ redox couple, the reduced and oxidized species in a half-reaction
- ☐ reaction quotient for a half-reaction, as for the reaction quotient for the overall reaction, but the electrons are ignored

- ☐ anode, the electrode at which oxidation occurs
- ☐ cathode, the electrode at which reduction occurs

10.4 Varieties of cells

- ☐ electrolyte concentration cell, a galvanic cell in which the electrode compartments are identical except for the concentrations of the electrolytes
- ☐ electrode concentration cell, a galvanic cell in which the electrodes themselves have different concentrations
- ☐ liquid junction potential, the potential difference across the interface of the two electrolytes
- ☐ cell notation: | phase boundary; ⋮ liquid junction; || an interface at which the junction potential has been eliminated
- ☐ the convention for writing the cell reaction, the reaction in the cell written on the assumption that the right-hand electrode is the cathode
- ☐ cell potential, the potential difference between the two electrodes of a galvanic cell
- ☐ electromotive force (emf), the cell potential when it is balanced by an exactly opposing source of potential so that the cell reaction occurs reversibly, the composition is constant, and no current flows
- ☐ the cell potential in terms of the reaction Gibbs energy, $-\nu FE = \Delta_r G$
- ☐ Faraday constant, $F = eN_A$
- ☐ standard emf, the standard reaction Gibbs energy expressed as a potential: $E^{\ominus} = -\Delta_r G^{\ominus}/\nu F$
- ☐ Nernst equation, the equation for the emf of a cell in terms of the composition: $E = E^{\ominus} - (RT/\nu F)\ln Q$
- ☐ transmembrane potential, $E = -(RT/F)\ln(b_L/b_R)$
- ☐ resting potential, the potential difference across the membrane of a neuron at rest
- ☐ action potential, the potential difference across the membrane of a neuron when a signal is propagating
- ☐ equilibrium constant for the cell reaction, $\ln K = \nu FE^{\ominus}/RT$

10.5 Standard potentials

- ☐ standard hydrogen electrode (SHE), $Pt(s)|H_2(g)|H^+(aq)$, $E^{\ominus} = 0$ at all temperatures
- ☐ standard potential of a couple ($E^{\ominus}$), the standard emf of a cell in which a couple forms the right-hand electrode and the standard hydrogen electrode is the left-hand electrode
- ☐ the standard emf in terms of the standard potentials of the electrodes, the difference of electrode potentials: $E^{\ominus} = E^{\ominus}(\text{right}) - E^{\ominus}(\text{left})$
- ☐ the graphical determination of the standard potential
- ☐ the electrochemical determination of activity coefficients

Applications of standard potentials

10.6 The electrochemical series

- ☐ the interpretation of the relative values of standard potentials, Red_1 has a thermodynamic tendency to reduce Ox_2 if $E_1^{\ominus} < E_2^{\ominus}$
- ☐ electrochemical series, the metallic elements (and hydrogen) arranged in the order of their reducing power as measured by their standard potentials in aqueous solution

10.7 The measurement of pH and pK_a

- ☐ the measurement of pH and pK_a, $\text{pH} = \text{pH(S)} - FE/(RT\ln 10)$
- ☐ glass electrode, an electrode sensitive to hydrogen ion activity
- ☐ ion-selective electrode, an electrode sensitive to the concentrations of ions other than the hydronium ion

10.8 Thermodynamic functions

- ☐ the temperature coefficient of cell potential, $dE^{\ominus}/dT = \Delta_r S^{\ominus}/\nu F$
- ☐ the determination of standard reaction entropy and enthalpy, $\Delta_r S^{\ominus} = \nu F\,dE^{\ominus}/dT$, $\Delta_r H^{\ominus} = -\nu F(E^{\ominus} - T\,dE^{\ominus}/dT)$

Further reading

Articles of general interest

A.K. Covington, R.G. Bates, and D.A. Durst, Definition of pH scales, standard reference values, and related terminology. *Pure. Appl. Chem.* **57**, 531 (1985).

R.L. DeKock, Tendency of reaction, electrochemistry, and units. *J. Chem. Educ.* **73**, 955 (1996).

A.-S. Feiner and A.J. McEvoy, The Nernst equation. *J. Chem. Educ.* **71**, 493 (1994).

J.J. MacDonald, Cathodes, terminals, and signs. *Educ. in Chem.* **25**, 52 (1988).

Y. Marcus, Ionic radii in aqueous solutions. *Chem. Rev.* **88**, 1475 (1988).

P. Millet, Electric potential distribution in an electrochemical cell. *J. Chem. Educ.* **73**, 956 (1996).

P.J. Morgan and E. Gileadi, Alleviating the common confusion caused by polarity in electrochemistry. *J. Chem. Educ.* **66**, 912 (1989).

Texts and sources of data and information

R.R. Adžić and E.B. Yeager, Electrochemistry. In *Encyclopedia of applied physics* (ed. G.L. Trigg), **5**, 223. VCH, New York (1993).

M.S. Antelman, *The encyclopedia of chemical electrode potentials.* Plenum, New York (1982).

A.J. Bard (ed.), *Encyclopedia of electrochemistry of the elements.* Marcel Dekker (1973–1986).

A.J. Bard and L.R. Faulkner, *Electrochemical methods.* Wiley, New York (2000).

A.J. Bard, R. Parsons, and J. Jordan (ed.), *Standard potentials in aqueous solution.* Marcel Dekker, New York (1985).

J.O'M. Bockris, B.E. Conway, *et al.* (ed.), *Comprehensive treatise on electrochemistry.* 1–10. Plenum, New York (1980–85).

C.M.A. Brett and A.M.O. Brett, *Electroanalysis.* Oxford University Press (1998).

R.G. Compton and G.H.W. Sanders, *Electrode potentials.* Oxford University Press (1998).

W.A. Cramer and D.A. Knaff, *Energy transduction in biological membranes, a textbook of bioenergetics*. Springer-Verlag, New York (1990).

D.R. Crow, *Principles and applications of electrochemistry*. Blackie, London (1994).

J. Goodisman, *Electrochemistry: Theoretical foundations*. Wiley, New York (1987).

C.H. Hamann, A. Hamnett, and W. Vielstich, *Electrochemistry*. Wiley-VCH, Weinheim (1998).

D.A. Harris, *Bioenergetics at a glance*. Blackwell Science, Oxford (1995).

V.M.M. Lobo, *Handbook of electrolyte solutions, Parts A and B*. Elsevier, Amsterdam (1990).

D. Pletcher and F.C. Walsh, *Industrial electrochemistry*. Chapman and Hall, London (1993).

P. Rieger, *Electrochemistry*. Chapman and Hall, London (1993).

Exercises

Discussion questions

10.1 (a) Discuss the factors responsible for the magnitude of the Gibbs energy of formation of an ion in solution.

10.1 (b) Describe the general features of the Debye–Hückel theory of electrolyte solutions.

10.2 (a) Distinguish between galvanic and electrolytic cells.

10.2 (b) Distinguish between cell potential and electromotive force.

10.3 (a) Describe a method for the determination of the standard potential of a cell.

10.3 (b) Describe a method for the determination of the pH of an aqueous solution.

Numerical exercises

10.4 (a) Calculate $\Delta_r H^\ominus$ for the reaction $Zn(s) + CuSO_4(aq) \rightarrow ZnSO_4(aq) + Cu(s)$ from the information in Table 2.6 in the *Data section*.

10.4 (b) Calculate $\Delta_r H^\ominus$ for the reaction $NaCl(aq) + AgNO_3(aq) \rightarrow AgCl(s) + NaNO_3(aq)$ from the information in Table 2.6 in the *Data section*.

10.5 (a) Calculate the molar solubility of mercury(II) chloride at 25°C from standard Gibbs energies of formation.

10.5 (b) Calculate the molar solubility of lead(II) sulfide at 25°C where the standard Gibbs energy of formation is −98.7 kJ mol^{-1}.

10.6 (a) Estimate the standard Gibbs energy of formation of $F^-(aq)$ from the value for $Cl^-(aq)$, taking the radius of F^- as 131 pm.

10.6 (b) Estimate the standard Gibbs energy of formation of $NO_3^-(aq)$ from the value for $Cl^-(aq)$, taking the radius of the nitrate ion as 189 pm (its thermochemical radius).

10.7 (a) Relate the ionic strengths of (a) KCl, (b) $FeCl_3$, and (c) $CuSO_4$ solutions to their molalities, *b*.

10.7 (b) Relate the ionic strengths of (a) $MgCl_2$, (b) $Al_2(SO_4)_3$, and (c) $Fe_2(SO_4)_3$ solutions to their molalities, *b*.

10.8 (a) Calculate the ionic strength of a solution that is 0.10 mol kg^{-1} in KCl(aq) and 0.20 mol kg^{-1} in $CuSO_4$(aq).

10.8 (b) Calculate the ionic strength of a solution that is 0.040 mol kg^{-1} in $K_3[Fe(CN)_6]$(aq), 0.030 mol kg^{-1} in KCl(aq), and 0.050 mol kg^{-1} in NaBr(aq).

10.9 (a) Calculate the masses of (a) $Ca(NO_3)_2$ and, separately, (b) NaCl to add to a 0.150 mol kg^{-1} solution of KNO_3(aq) containing 500 g of solvent to raise its ionic strength to 0.250.

10.9 (b) Calculate the masses of (a) KNO_3 and, separately, (b) $Ba(NO_3)_2$ to add to a 0.110 mol kg^{-1} solution of KNO_3(aq) containing 500 g of solvent to raise its ionic strength to 1.00.

10.10 (a) What molality of $CuSO_4$ has the same ionic strength as 1.00 mol kg^{-1} KCl(aq)?

10.10 (b) What molality of $Al_2(SO_4)_3$ has the same ionic strength as 0.500 mol kg^{-1} $Ca(NO_3)_2$(aq)?

10.11 (a) Express the mean activity coefficient of the ions in a solution of $CaCl_2$ in terms of the activity coefficients of the individual ions.

10.11 (b) Express the mean activity coefficient of the ions in a solution of $Al_2(SO_4)_3$ in terms of the activity coefficients of the individual ions.

10.12 (a) Estimate the mean ionic activity coefficient and activity of a solution that is 0.010 mol kg^{-1} $CaCl_2$(aq) and 0.030 mol kg^{-1} NaF(aq).

10.12 (b) Estimate the mean ionic activity coefficient and activity of a solution that is 0.020 mol kg^{-1} NaCl(aq) and 0.035 mol kg^{-1} $Ca(NO_3)_2$(aq).

10.13 (a) The mean activity coefficient in a 0.500 mol kg^{-1} $LaCl_3$(aq) solution is 0.303 at 25°C. What is the percentage error in the value predicted by the Debye–Hückel limiting law?

10.13 (b) The mean activity coefficient in a 0.100 mol kg^{-1} $CaCl_2$(aq) solution is 0.524 at 25°C. What is the percentage error in the value predicted by the Debye–Hückel limiting law?

10.14 (a) The mean activity coefficients of HBr in three dilute aqueous solutions at 25°C are 0.930 (at 5.0 mmol kg^{-1}), 0.907 (at 10.0 mmol kg^{-1}), and 0.879 (at 20.0 mmol kg^{-1}). Estimate the value of *B* in the extended Debye–Hückel law.

10.14 (b) The mean activity coefficients of KCl in three dilute aqueous solutions at 25°C are 0.927 (at 5.0 mmol kg^{-1}), 0.902 (at 10.0 mmol kg^{-1}), and 0.816 (at 50.0 mmol kg^{-1}). Estimate the value of *B* in the extended Debye–Hückel law.

10.15 (a) For CaF_2, $K_s = 3.9 \times 10^{-11}$ at 25°C and the standard Gibbs energy of formation of CaF_2(s) is −1167 kJ mol^{-1}. Calculate the standard Gibbs energy of formation of CaF_2(aq).

10.15 (b) For PbI_2, $K_s = 1.4 \times 10^{-8}$ at 25°C and the standard Gibbs energy of formation of $PbI_2(s)$ is −173.64 kJ mol^{-1}. Calculate the standard Gibbs energy of formation of $PbI_2(aq)$.

10.16 (a) Consider a hydrogen electrode in aqueous HBr solution at 25°C operating at 1.15 atm. Calculate the change in the electrode potential when the molality of the acid is changed from 5.0 mmol kg^{-1} to 20.0 mol kg^{-1}. Activity coefficients are given in Exercise 10.14a.

10.16 (b) Consider a hydrogen electrode in aqueous HCl solution at 25°C operating at 105 kPa. Calculate the change in the electrode potential when the molality of the acid is changed from 5.0 mmol kg^{-1} to 50 mol kg^{-1}. Activity coefficients are given in Table 10.5.

10.17 (a) Devise a cell in which the cell reaction is $Mn(s) + Cl_2(g) \rightarrow MnCl_2(aq)$. Give the half-reactions for the electrodes and from the standard cell emf of 2.54 V deduce the standard potential of the Mn^{2+}/Mn couple.

10.17 (b) Devise a cell in which the cell reaction is $Cd(s) + Ni(OH)_3(s) \rightarrow Cd(OH)_2(s) + Ni(OH)_2(s)$. Give the half-reactions for the electrodes.

10.18 (a) Write the cell reactions and electrode half-reactions for the following cells:

(a) $Zn|ZnSO_4(aq) \| AgNO_3(aq)|Ag$

(b) $Cd|CdCl_2(aq) \| HNO_3(aq)|H_2(g)|Pt$

(c) $Pt|K_3[(CN)_6](aq),K_4[(CN)_6](aq) \| CrCl_3(aq)|Cr$

10.18 (b) Write the cell reactions and electrode half-reactions for the following cells:

(a) $Pt|Cl_2(g)|HCl(aq) \| K_2CrO_4(aq)|Ag_2CrO_4(s)|Ag$

(b) $Pt|Fe^{3+}(aq),Fe^{2+}(aq) \| Sn^{4+}(aq),Sn^{2+}(aq)|Pt$

(c) $Cu|Cu^{2+}(aq) \| Mn^{2+}(aq),H^+(aq)|MnO_2(s)|Pt$

10.19 (a) Devise cells in which the following are the reactions:

(a) $Zn(s) + CuSO_4(aq) \longrightarrow ZnSO_4(aq) + Cu(s)$

(b) $2\,AgCl(s) + H_2(g) \longrightarrow 2\,HCl(aq) + 2\,Ag(s)$

(c) $2\,H_2(g) + O_2(g) \longrightarrow 2\,H_2O(l)$

10.19 (b) Devise cells in which the following are the reactions:

(a) $2\,Na(s) + 2\,H_2O(l) \longrightarrow 2\,NaOH(aq) + H_2(g)$

(b) $H_2(g) + I_2(g) \longrightarrow 2\,HI(aq)$

(c) $H_3O^+(aq) + OH^-(aq) \longrightarrow H_2O(l)$

10.20 (a) Use standard potentials to calculate the standard emfs of the cells in Exercise 10.18a.

10.20 (b) Use standard potentials to calculate the standard emfs of the cells in Exercise 10.18b.

10.21 (a) Use standard potentials to calculate the standard emfs of the cells in Exercise 10.19a.

10.21 (b) Use standard potentials to calculate the standard emfs of the cells in Exercise 10.19b.

10.22 (a) Calculate the standard Gibbs energies at 25°C of the following reactions from the standard potential data in Table 10.7:

(a) $2\,Na(s) + 2\,H_2O(l) \longrightarrow 2\,NaOH(aq) + H_2(g)$

(b) $2\,K(s) + 2\,H_2O(l) \longrightarrow 2\,KOH(aq) + H_2(g)$

10.22 (b) Calculate the standard Gibbs energies at 25°C of the following reactions from the standard potential data in Table 10.7:

(a) $K_2S_2O_8(aq) + 2\,KI(aq) \longrightarrow I_2(s) + 2\,K_2SO_4(aq)$

(b) $Pb(s) + Zn(NO_3)_2(aq) \longrightarrow Pb(NO_3)_2(aq) + Zn(s)$

10.23 (a) The standard reaction Gibbs energy for $K_2CrO_4(aq) + 2\,Ag(s) + 2\,FeCl_3(aq) \rightarrow Ag_2CrO_4(s) + 2\,FeCl_2(aq) + 2\,KCl(aq)$ is −62.5 kJ mol^{-1} at 298 K. (a) Calculate the standard emf of the corresponding galvanic cell and (b) the standard potential of the $Ag_2CrO_4/Ag,CrO_4^{2-}$ couple.

10.23 (b) Two half-cell reactions may be combined in such a way as to form (a) a new half-cell reaction or (b) a complete cell reaction. Illustrate both (a) and (b) by using the half-cell reactions listed below and calculate $E^\ominus$ for both the new half-cell and complete cell reaction.

(i) $2\,H_2O(l) + 2\,e^- \longrightarrow H_2(g) + 2\,OH^-(aq)$ $E_1^\ominus = -0.828$ V

(ii) $Ag^+(aq) + e^- \longrightarrow Ag(s)$ $E_2^\ominus = +0.799$ V

10.24 (a) Calculate the standard potential of the couple $Ag_2S,H_2O/Ag,S^{2-},O_2,H^+$ from the following data:

$Ag_2S(s) + 2\,e^- \longrightarrow 2\,Ag(s) + S^{2-}(aq)$ $E^\ominus = -0.69$ V

$O_2(g) + 4\,H^+(aq) + 4\,e^- \longrightarrow 2\,H_2O(l)$ $E^\ominus = +1.23$ V

10.24 (b) Consider the cell $Pt|H_2(g, p^\ominus)|HCl(aq)|AgCl(s)|Ag$, for which the cell reaction is $2\,AgCl(s) + H_2(g) \rightarrow 2\,Ag(s) + 2\,HCl(aq)$. At 25°C and a molality of HCl of 0.010 mol kg^{-1}, $E = +0.4658$ V. (a) Write the Nernst equation for the cell reaction, (b) Calculate $\Delta_r G$ for the cell reaction. (c) Assuming that the Debye–Hückel limiting law holds at this concentration, calculate $E^\ominus$(AgCl, Ag).

10.25 (a) Use the Debye–Hückel limiting law and the Nernst equation to estimate the emf of the cell $Ag|AgBr(s)|KBr(aq, 0.050\ mol\ kg^{-1}) \| Cd(NO_3)_2(aq, 0.010\ mol\ kg^{-1})|Cd$ at 25°C.

10.25 (b) Use the information in Tables 10.7 to calculate the standard emf of the cell $Ag|AgNO_3(aq) \| Fe(NO_3)_2(aq)|Fe$ and the standard Gibbs energy and enthalpy of the cell reaction at 25°C. Estimate the value of $\Delta_r G^\ominus$ at 35°C.

10.26 (a) Calculate the equilibrium constants of the following reactions at 25°C from standard potential data:

(a) $Sn(s) + Sn^{4+}(aq) \rightleftharpoons 2\,Sn^{2+}(aq)$

(b) $Sn(s) + 2\,AgCl(s) \rightleftharpoons SnCl_2(aq) + 2\,Ag(s)$

10.26 (b) Calculate the equilibrium constants of the following reactions at 25°C from standard potential data:

(a) $Sn(s) + CuSO_4(aq) \rightleftharpoons Cu(s) + SnSO_4(aq)$

(b) $Cu^{2+}(aq) + Cu(s) \rightleftharpoons 2\,Cu^+(aq)$

10.27 (a) Use the standard emfs of the couples Au^+/Au(+1.69 V), Au^{3+}/Au(+1.40 V), and Fe^{3+}/Fe^{2+}(+0.77 V) to calculate $E^\ominus$ and the equilibrium constant for the reaction $2\,Fe^{2+}(aq) + Au^{3+}(aq) \rightleftharpoons 2\,Fe^{3+}(aq) + Au^+(aq)$.

10.27 (b) Determine the standard emf of a cell in which the reaction is $Co^{3+}(aq) + 3\,Cl^-(aq) + 3\,Ag(s) \rightarrow 3\,AgCl(s) + Co(s)$ from the standard potentials of the couples $Ag/AgCl,Cl^-$ (+0.22 V), Co^{3+}/Co^{2+} (+1.81 V), and Co^{2+}/Co (−0.28 V).

10.28 (a) The solubilities of AgCl and $BaSO_4$ in water are 1.34×10^{-5} mol kg^{-1} and 9.51×10^{-4} mol kg^{-1}, respectively, at 25°C. Calculate their solubility constants. Is there any significant difference when activity coefficients are ignored?

10.28 (b) The solubilities of AgI and Bi_2S_3 in water are 1.2×10^{-8} mol kg^{-1} and 1.6×10^{-20} mol kg^{-1}, respectively, at 25°C. Calculate their solubility constants. Is there any significant difference when activity coefficients are ignored?

10.29 (a) Derive an expression for the potential of an electrode for which the half-reaction is the reduction of $Cr_2O_7^{2-}$ ions to Cr^{3+} ions in acidic solution.

10.29 (b) Derive an expression for the potential of an electrode for which the half-reaction is the reduction of MnO_4^- ions to Mn^{2+} ions in acidic solution.

10.30 (a) The emf of the cell Pt|H_2(g)|HCl(aq)|AgCl(s)|Ag is +0.322 V at 25°C. What is the pH of the electrolyte solution?

10.30 (b) The emf of the cell Pt|H_2(g)|HI(aq)|AgI(s)|Ag is 1.00 V at 25°C. What is the pH of the electrolyte solution?

10.31 (a) The solubility of AgBr is 2.6 μmol kg^{-1} at 25°C. What is the emf of the cell Ag|AgBr(aq)|AgBr(s)|Ag at that temperature?

10.31 (b) The solubility of AgI is 12 nmol kg^{-1} at 25°C. What is the emf of the cell Ag|AgI(aq)|AgI(s)|Ag at that temperature?

10.32 (a) The emf of the cell Ag|AgI(s)|AgI(aq)|Ag is +0.9509 V at 25°C. Calculate (a) the solubility of AgI and (b) its solubility constant.

10.32 (b) The emf of the cell Bi|Bi_2S_3(s)|Bi_2S_3(aq)|Bi is +0.96 V at 25°C. Calculate (a) the solubility of Bi_2S_3 and (b) its solubility constant.

Problems†

Numerical problems

10.1 Devise a cell in which the overall reaction is Pb(s) + Hg_2SO_4(s) → $PbSO_4$(s) + 2 Hg(l). What is its potential when the electrolyte is saturated with both salts at 25°C? The solubility constants of Hg_2SO_4 and $PbSO_4$ are 6.6×10^{-7} and 1.6×10^{-8}, respectively.

10.2 Given that $\Delta_r G^\ominus = -212.7$ kJ mol^{-1} for the reaction in the Daniell cell at 25°C, and $b(CuSO_4) = 1.0 \times 10^{-3}$ mol kg^{-1} and $b(ZnSO_4) = 3.0 \times 10^{-3}$ mol kg^{-1}, calculate (a) the ionic strengths of the solutions, (b) the mean ionic activity coefficients in the compartments, (c) the reaction quotient, (d) the standard cell emf, and (e) the cell emf. (Take $\gamma_+ = \gamma_- = \gamma_\pm$ in the respective compartments.)

10.3 Although the hydrogen electrode may be conceptually the simplest electrode and is the basis for our reference state of electrical potential in electrochemical systems, it is cumbersome to use. Therefore, several substitutes for it have been devised. One of these alternatives is the quinhydrone electrode (quinhydrone, Q·QH_2, is a complex of quinone, $C_6H_4O_2$ = Q, and hydroquinone, $C_6H_4O_2H_2$ = QH_2). The electrode half-reaction is Q(aq) + 2 H^+(aq) + 2 e^- → QH_2(aq), $E^\ominus = +0.6994$ V. If the cell Hg|Hg_2Cl_2(s)|HCl(aq)|Q·QH_2|Au is prepared, and the measured cell potential is +0.190 V, what is the pH of the HCl solution? Assume that the Debye–Hückel limiting law is applicable.

10.4 Consider the cell, Zn(s)|$ZnCl_2$(0.0050 mol kg^{-1})|Hg_2Cl_2(s) Hg(l), for which the cell reaction is Hg_2Cl_2(s) + Zn(s) → 2 Hg(l) + 2 Cl^-(aq) + Zn^{2+}(aq). Given that $E^\ominus(Zn^{2+}, Zn) = -0.7628$ V, $E^\ominus(Hg_2Cl_2, Hg) = +0.2676$ V, and that the emf is +1.2272 V, (a) write the Nernst equation for the cell. Determine (b) the standard emf, (c) $\Delta_r G$, $\Delta_r G^\ominus$, and K for the cell reaction, (d) the mean ionic activity and activity coefficient of $ZnCl_2$ from the measured emf, and (e) the mean ionic activity coefficient of $ZnCl_2$ from the Debye–Hückel limiting law. (f) Given that $(\partial E/\partial T)_p = -4.52 \times 10^{-4}$ V K^{-1}, calculate ΔS and ΔH.

10.5 The emf of the cell Pt|H_2(g, $p^\ominus$)|HCl(aq, b)|Hg_2Cl_2(s)|Hg(l) has been measured with high precision (G.J. Hills and D.J.G. Ives, *J. Chem. Soc.*, 311 (1951)) with the following results at 25°C:

b/(mmol kg^{-1})	1.6077	3.0769	5.0403	7.6938	10.9474
E/V	0.600 80	0.568 25	0.543 66	0.522 67	0.505 32

Determine the standard emf of the cell and the mean activity coefficient of HCl at these molalities. (Make a least-squares fit of the data to the best straight line.)

10.6 Careful measurements of the emf of the cell Pt|H_2(g, $p^\ominus$)|NaOH(aq, 0.0100 mol kg^{-1}), NaCl(aq, 0.011 25 mol kg^{-1})|AgCl(s)|Ag have been reported (C.P. Bezboruah, M.F.G.F.C. Camoes, A.K. Covington, and J.V. Dobson, *J. Chem. Soc. Faraday Trans. I*, **69**, 949 (1973)). Among the data is the following information:

θ/°C	20.0	25.0	30.0
E/V	1.047 74	1.048 64	1.049 42

Calculate pK_w at these temperatures and the standard enthalpy and entropy of the autoprotolysis of water at 25.0°C.

10.7 Measurements of the emf of cells of the type Ag|AgX(s)MX(b_1)|M_xHg|MX(b_2)|AgX(s)|Ag, where M_xHg denotes an amalgam and the electrolyte is an alkali metal halide dissolved in ethylene glycol, have been reported (U. Sen, *J. Chem. Soc. Faraday Trans. I* **69**, 2006 (1973)) and some values for LiCl are given below. Estimate the activity coefficient at the concentration marked * and then use this value to calculate activity coefficients from the measured cell potential at the other concentrations. Base your answer on the following version of the extended Debye–Hückel law:

$$\log \gamma = -\frac{AI^{1/2}}{1 - BI^{1/2}} + kI$$

with $A = 1.461$, $B = 1.70$, $k = 0.20$, and $I = b/b^\ominus$. For $b_2 = 0.091\,41$ mol kg^{-1}:

b_1/(mol kg^{-1})	0.0555	0.091 41*	0.1652	0.2171	1.040	1.350
E/V	−0.0220	0.0000	0.0263	0.0379	0.1156	0.1336

10.8 Suppose the extended Debye–Hückel law for a 1,1-electrolyte is written in the simplified form $\log \gamma_\pm = -0.509I^{1/2} + kI$, where k is a constant and $I = b/b^\ominus$. Show that a plot of y against I, where

$$y = E + 0.1183 \log I - 0.0602I^{1/2}$$

should give a straight line with intercept $E^\ominus$ and slope $-0.1183k$. Apply the technique to the following data (at 25°C) on the cell Pt|H_2(g, $p^\ominus$)|HCl(aq, b)|AgCl(s)|Ag:

b/(mmol kg^{-1})	123.8	25.63	9.138	5.619	3.215
E/mV	341.99	418.24	468.60	492.57	520.53

† Problems denoted with the symbol ‡ were supplied by Charles Trapp, Carmen Giunta, and Marshall Cady.

(a) Find the standard emf and the standard potential of the $AgCl/Ag,Cl^-$ couple. (b) The emf was measured as 352.4 mV when $b = 100.0\ mol\ kg^{-1}$. What is the pH and the mean ionic activity coefficient?

10.9 The mean activity coefficients for aqueous solutions of NaCl at 25°C are given below. Confirm that they support the Debye–Hückel limiting law and that an improved fit is obtained with the extended law.

$b/(mmol\ kg^{-1})$	1.0	2.0	5.0	10.0	20.0
$\gamma_\pm$	0.9649	0.9519	0.9275	0.9024	0.8712

10.10 The standard potential of the $AgCl/Ag,Cl^-$ couple has been measured very carefully over a range of temperature (R.G. Bates and V.E. Bowers, *J. Res. Nat. Bur. Stand.* **53**, 283 (1954)) and the results were found to fit the expression

$$E^\ominus/V = 0.236\ 59 - 4.8564 \times 10^{-4}(\theta/°C) - 3.4205 \times 10^{-6}(\theta/°C)^2 + 5.869 \times 10^{-9}(\theta/°C)^3$$

Calculate the standard Gibbs energy and enthalpy of formation of $Cl^-(aq)$ and its entropy at 298 K.

10.11‡ The list of standard potentials compiled by S.G. Bratsch (*J. Phys. Chem. Ref. Data* **18**, 1 (1989)) shows that europium has the least negative reduction potential for the M^{3+}/M couple in strong acid (pH = 0.000) of all the lanthanides, $E^\ominus(Eu^{3+}/Eu) = -1.991$ V, while the next easiest to reduce is ytterbium, with $E^\ominus(Yb^{3+}/Yb) = -2.19$ V. (a) Describe the criteria for a reducing agent which would deposit Eu metal from an acidic solution of M^{3+} lanthanide ions while leaving the other lanthanides in solution under standard conditions (unit activities and fugacities; pH = 0). Is there such a reagent in Table 10.7? (b) Now suppose the conditions are not standard. What is the maximum value of the ratio $a_{Yb^{3+}}/a_{Eu^{3+}}$ such that Eu still deposits spontaneously (just barely) but Yb does not?

10.12‡ (a) Derive a general relation for $(\partial E/\partial p)_{T,n}$ for electrochemical cells employing reactants in any state of matter. (b) E. Cohen and K. Piepenbroek (*Z. physik. Chem.* **167A**, 365 (1933)) calculated the change in volume for the reaction $TlCl(s) + CNS^-(aq) \rightarrow TlCNS(s) + Cl^-$ at 30°C from density data and obtained $\Delta_r V = -2.666 \pm 0.080\ cm^3\ mol^{-1}$. They also measured the emf of the cell Tl(Hg)|TlCNS(s)|KCNS:KCl|TlCl|Tl(Hg) at pressures up to 1500 atm. Their results are given in the following table:

p/atm	1.00	250	500	750	1000	1250	1500
E/mV	8.56	9.27	9.98	10.69	11.39	12.11	12.82

From this information, obtain $(\partial E/\partial p)_{T,n}$ at 30°C and compare to the value obtained from $\Delta_r V$. (c) Fit the data to a polynomial for E against p. How constant is $(\partial E/\partial p)_{T,n}$? (d) From the polynomial, estimate an effective isothermal compressibility for the cell as a whole.

10.13‡ Because the hydrogen electrode is the fundamental reference electrode, it is important to study the effect of pressure on cells containing it. W.R. Hainsworth, H.J. Rowley, and D.A. MacInnes (*J. Am. Chem. Soc.* **46**, 1437 (1924)) investigated the influence of pressure on the hydrogen/calomel cell, $Pt|H_2(g, p)|HCl(0.1\ M)|Hg_2Cl_2|Hg(l)$, up to pressures of 1000 atm at 25°C with the following results, where $\Delta E = E(p) - E(1\ atm)$.

p/atm	1.00	10	38	51	108	210	380	430
ΔE/mV	0	29.5	47	51	61	69	79	82

p/atm	560	720	900	1020
ΔE/mV	86	91	95	98

(a) Derive an expression for ΔE using the perfect gas law. Compare the expression with the experimental results. (b) Fit the values of ΔE and $(\partial E/\partial p)_{T,n}$ to a polynomial expression in p. (c) Obtain a theoretical equation for these quantities using both the virial form of the van der Waals equation and the empirical virial equation $Z = 1 + 0.000\ 537(p/atm) + 3.5 \times 10^{-8}(p/atm)^2$ used by Hainsworth *et al.* Plot the experimental and theoretical data for this cell on the same graph. How well do they fit? (d) Calculate the fugacity of H_2 from the cell potential data.

10.14 Use the data below to confirm that the Debye–Hückel limiting law correctly predicts the limiting values of the mean activity coefficient of acetic acid by demonstrating that pK_a', where $K_a = K_a'K_\gamma$, plotted against $(\alpha b)^{1/2}$, where b is the molality of the acid and α its degree of deprotonation, should be a straight line.

$b/(mmol\ kg^{-1})$	0.0280	0.1114	0.2184	1.0283	2.414	5.9115
α	0.5393	0.3277	0.2477	0.1238	0.0829	0.0540

10.15‡ J.W. Larson (*J. Chem. Eng. Data* **40**, 1276 (1995)) studied the thermochemistry of vanadium(V) ions in aqueous solution. At neutral pH, the dominant species are $H_2VO_4^-$ and $V_4O_{12}^{4-}$. The equilibrium constant for the reaction $4\ H_2VO_4^-(aq) \rightarrow V_4O_{12}^{4-}(aq) + 4\ H_2O(l)$ was determined to be 106.4 at 50°C. How is vanadium distributed between these two species at 50°C and unit ionic strength if the total vanadium concentration is $0.10\ mmol\ kg^{-1}$? Take the ionic strength to be 1.000 and use the following extended Debye–Hückel equation to estimate activity coefficients:

$$\log \gamma = -\frac{Az^2I^{1/2}}{1 + I^{1/2}}$$

where $A = 0.5373$ at 50°C.

10.16‡ The table below summarizes the emf observed for the cell $Pd|H_2(g, 1\ bar)|BH(aq, b), B(aq, b)|AgCl(s)|Ag$. Each measurement is made at equimolar concentrations of 2-aminopyridinium chloride (BH) and 2-aminopyridine (B). The data are for 25°C and it is found that $E^\ominus = 0.222\ 51$ V. Use the data to determine pK_a for the acid at 25°C and the mean activity coefficient ($\gamma_\pm$) of BH as a function of molality (b) and ionic strength (I). Use the extended Debye–Hückel equation for the mean activity coefficient in the form:

$$-\log \gamma_\pm = \frac{AI^{1/2}}{1 + BI^{1/2}} - kb$$

where $A = 0.5091$ and B and k are parameters that depend upon the ions. Draw a graph of the mean activity coefficient with $b = 0.04\ mol\ kg^{-1}$ and $0 \le I \le 0.1$.

$b/(mol\ kg^{-1})$	0.01	0.02	0.03	0.04	0.05
E(25°C)/V	0.744 52	0.728 53	0.719 28	0.713 14	0.708 09

$b/(mol\ kg^{-1})$	0.06	0.07	0.08	0.09	0.10
E(25°C)/V	0.703 80	0.700 59	0.697 90	0.695 71	0.693 38

Hint. Use mathematical software or a spreadsheet.

10.17‡ The table below summarizes $pK_a(T)$ for BH of Problem 10.16; add your value of pK_a(25°C) calculated in Problem 10.16 to those of the table. Determine the coefficients a_0, a_1, and a_2 such that the regression equation $pK_a(T) = a_0 + a_1/(T/K) + a_2 \ln(T/K)$ gives the best least-squares fit of the data. Draw graphs of $\Delta_r G^\ominus(T)$, $\Delta_r H^\ominus(T)$, and $\Delta_r S^\ominus(T)$ for the proton transfer reaction.

θ/°C	5	10	15	20	25	30	35	40
pK_a	7.177	7.060	6.949	6.841	?	6.640	6.543	6.449

10.18 The Sb|Sb_2O_3(s)|OH^-(aq) electrode is reversible with respect to OH^- ions. Derive an expression for its potential in terms of (a) the pOH and (b) the pH of the solution. By how much does the potential change when the molality of NaOH(aq) in the electrode compartment is increased from 0.010 mol kg^{-1} to 0.050 mol kg^{-1} at 25°C? Use the Debye–Hückel limiting law to estimate any activity coefficients required.

10.19 Superheavy elements are now of considerable interest, particularly because signs of stability are starting to emerge with element 114, which has recently been made. Shortly before it was (falsely) believed that the first superheavy element had been discovered, an attempt was made to predict the chemical properties of ununpentium (Uup, element 115, O.L. Keller, C.W. Nestor, and B. Fricke, *J. Phys. Chem.* **78**, 1945 (1974)). In one part of the paper the standard enthalpy and entropy of the reaction $Uup^+(aq) + \frac{1}{2}H_2(g) \rightarrow Uup(s) + H^+(aq)$ were estimated from the following data: $\Delta_{sub}H^{\ominus}(Uup) = +1.5$ eV, $I(Uup) = 5.52$ eV, $\Delta_{hyd}H^{\ominus}(Uup^+, aq) = -3.22$ eV, $S^{\ominus}(Uup^+, aq) = +1.34$ meV K^{-1}, $S^{\ominus}(Uup, s) = 0.69$ meV K^{-1}. Estimate the expected standard potential of the Uup^+/Uup couple.

Theoretical problems

10.20 Show that the solubility, S, of a sparingly soluble 1:1 salt is related to its solubility constant by $S = K_s^{1/2}e^{1.172S^{1/2}}$.

10.21 Suppose that a sparingly soluble salt MX has solubility constant K_s and solubility S. Show that in an ideal solution that is of concentration C of a freely soluble salt NX, the solubility of MX is changed to

$$S' = \tfrac{1}{2}(C^2 + 4K_s^2)^{1/2} - \tfrac{1}{2}C$$

and that $S' \approx K_s/C$ when K_s is small (in a sense to be specified).

10.22 Show that, if the ionic strength of a solution of the sparingly soluble salt MX and the freely soluble salt NX is dominated by the concentration C of the latter, and if it is valid to use the Debye–Hückel limiting law, the solubility S' in the mixed solution is given by

$$S' = \frac{K_s e^{4.606AC^{1/2}}}{C}$$

when K_s is small (in a sense to be specified).

10.23 Show that the freezing-point depression of a real solution in which the solvent of molar mass M has activity a_A obeys

$$\frac{d\ln a_A}{d(\Delta T)} = -\frac{M}{K_f}$$

and use the Gibbs–Duhem equation to show that

$$\frac{d\ln a_B}{d(\Delta T)} = -\frac{1}{b_B K_f}$$

where a_B is the solute activity and b_B is its molality. Use the Debye–Hückel limiting law to show that the osmotic coefficient (ϕ, Problem 7.21) is given by $\phi = 1 - \frac{1}{3}A'I$ with $A' = 2.303A$ and $I = b/b^{\ominus}$.

Applications: to biology and chemical engineering

10.24 The addition of a small amount of a salt, such as $(NH_4)_2SO_4$, to a solution containing a charged protein increases the protein's solubility in water. This observation is called the *salting-in effect*. However, the addition of large amounts of salt can decrease the solubility of the protein to such an extent that the protein precipitates from solution. This observation is called the *salting-out effect* and is used widely by biochemists to isolate and purify proteins. Consider the equilibrium $PX_\nu(s) \rightleftharpoons P^{\nu+}(aq) + \nu X^-(aq)$, where $P^{\nu+}$ is a polycationic protein of charge $+\nu$ and X^- is its counter ion. Use Le Chatelier's principle and the physical principles behind the Debye–Hückel limiting law to provide a molecular interpretation for the salting-in and salting-out effects.

10.25 The standard potentials of proteins are not commonly measured by the methods described in this chapter because proteins often lose their native structure and function when they react on the surfaces of electrodes. In an alternative method, the oxidized protein is allowed to react with an appropriate electron donor in solution. The standard potential of the protein is then determined from the Nernst equation, the equilibrium concentrations of all species in solution, and the known standard potential of the electron donor. We shall illustrate this method with the protein cytochrome *c*. The one-electron reaction between cytochrome *c*, cyt, and 2,6-dichloroindophenol, D, can be followed spectrophotometrically because each of the four species in solution has a distinct colour, or absorption spectrum. We write the reaction as $cyt_{ox} + D_{red} \rightleftharpoons cyt_{red} + D_{ox}$, where the subscripts 'ox' and 'red' refer to oxidized and reduced states, respectively. (a) Consider $E^{\ominus}_{cyt}$ and $E^{\ominus}_{D}$ to be the standard potentials of cytochrome *c* and D, respectively. Show that, at equilibrium('eq'), a plot of $\ln([D_{ox}]_{eq}/[D_{red}]_{eq})$ versus $\ln([cyt_{ox}]_{eq}/[cyt_{red}]_{eq})$ is linear with slope of one and *y*-intercept $F(E^{\ominus}_{cyt} - E^{\ominus}_{D})/RT$, where equilibrium activities are replaced by the numerical values of equilibrium molar concentrations. (b) The following data were obtained for the reaction between oxidized cytochrome *c* and reduced D in a pH 6.5 buffer at 298 K. The ratios $[D_{ox}]_{eq}/[D_{red}]_{eq}$ and $[cyt_{ox}]_{eq}/[cyt_{red}]_{eq}$ were adjusted by titrating a solution containing oxidized cytochrome *c* and reduced D with a solution of sodium ascorbate, which is a strong reductant. From the data and the standard potential of D of 0.237 V, determine the standard potential cytochrome *c* at pH 6.5 and 298K.

$[D_{ox}]_{eq}/[D_{red}]_{eq}$	0.002 79	0.008 43	0.0257	0.0497
$[cyt_{ox}]_{eq}/[cyt_{red}]_{eq}$	0.0106	0.0230	0.0894	0.197
$[D_{ox}]_{eq}/[D_{red}]_{eq}$	0.0748	0.238	0.534	
$[cyt_{ox}]_{eq}/[cyt_{red}]_{eq}$	0.335	0.809	1.39	

10.26 Sodium fluoride is routinely added to public water supplies because it is known that fluoride ion can prevent tooth decay. In a fluoride-selective electrode used in the analysis of water samples a crystal of LaF_3 doped with Eu^{2+}, denoted as Eu^{2+}:LaF_3, provides a semipermeable barrier between the test solution and the solution inside the electrode (the filling solution), which contains 0.1 mol kg^{-1} NaF(aq) and 0.1 mol kg^{-1} NaCl(aq). A silver–silver chloride electrode immersed in the filling solution is connected to a potentiometer and the emf of the cell can be measured against an appropriate reference electrode. It follows that the half-cell for a fluoride-selective electrode is represented by

$$Ag(s)|AgCl(s)|NaCl(aq, b_1), NaF(aq, b_1)|Eu^{2+}:LaF_3(s)|F^-(aq, b_2)$$

where b_1 and b_2 are the molalities of fluoride ion in the filling and test solutions, respectively. (a) Derive an expression for the emf of this half-cell. (b) The fluoride-selective electrode just described is not sensitive to HF(aq). Hydroxide ion is the only interfering species, with $k_{F^-,OH^-} = 0.1$. Use this information and the fact that K_a of HF is 3.5×10^{-4} at 298 K to specify a range of pH values in which the electrode responds accurately to the activity of F^- in the test solution at 298 K.

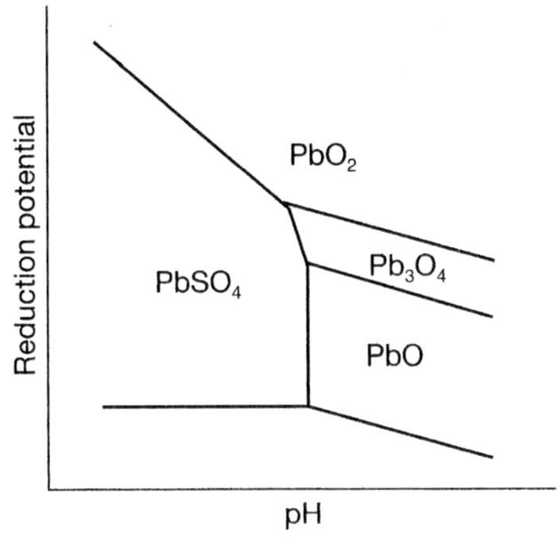

Fig. 10.17

10.27‡ A cell of the lead–acid battery may be represented as Pb(s)|$PbSO_4$(s)|H_2SO_4(aq)|PbO_2(s)|$PbSO_4$(s)|Pb(s) where the aqueous solution is about 35 per cent by mass sulfuric acid. The sulfuric acid molality, b, is related to the solution density, d, at 25°C by the empirical equation:

$$b(\mathrm{H_2SO_4})/(\mathrm{mol\ kg^{-1}}) = a(d - d_{25})/(\mathrm{g\ cm^{-3}}) + c(d - d_{25})^2/(\mathrm{g\ cm^{-3}})^2$$

where d_{25} is the density of pure water at 25°C, $a = 14.523$, and $c = 25.031$. (a) Draw graphs of molality, mass percentage, and molar concentration functions of density for densities in the range 1.0–1.4 g cm^{-3}. (b) Write balanced equations for the half-reactions and the cell reaction for the lead–acid battery. Determine $E^{\ominus}$, $\Delta_r G^{\ominus}$, $\Delta_r S^{\ominus}$, and $\Delta_r H^{\ominus}$ for the cell reaction at 25°C. Estimate $E^{\ominus}$(15°C). Calculate the cell emf at 25°C when $Q = 6.0 \times 10^{-5}$. (c) Figure 10.17 shows a simplified 'Pourbaix diagram' for the redox chemistry of $PbO_2/PbSO_4$/Pb systems. The horizontal and slanted lines of the diagram are equilibrium half-cell potentials when all chemical species have unit activity. Vertical lines are non-redox equilibria. Prove that the redox potential lines at 25°C are given by $E = E^{\ominus} - (0.592\ \mathrm{V})\,(\nu_H/\nu)\,\mathrm{pH}$, where ν_H is the number of hydrogen ions used in the half- reaction. Calculate $E(PbO_2/PbSO_4)$ at pH values of 5 and 8. What is $E(PbSO_4/Pb)$ at any pH?

10.28 Fuel cells provide electrical power for spacecraft (as in the NASA space shuttles) and also show promise as power sources for automobiles. A fuel cell develops an electric potential from the chemical reaction between reagents supplied from an outside source. What is the emf of a cell fuelled by (a) hydrogen and oxygen, (b) the combustion of butane at 1.0 bar and 298 K?

Part 3

Change

Part 3 considers the processes by which change occurs. We prepare the ground for a discussion of the rates of reactions by considering the motion of molecules in gases and in liquids. Then we establish the precise meaning of reaction rate, and see how the overall rate, and the complex behaviour of some reactions, may be expressed in terms of elementary steps and the atomic events that take place when molecules meet. Characteristic physical and chemical events take place at surfaces, including catalysis, and we see how to describe them. A special type of surface is that of an electrode, and we shall see how to describe and understand the rate at which electrons are transferred between an electrode and species in solution.

24 Molecules in motion

One of the simplest types of molecular motion to describe is the random motion of molecules of a perfect gas. We see that the kinetic theory can be used to account for the pressure of a gas and the rates at which molecules and energy migrate through gases. Molecular mobility is particularly important in liquids. Another simple kind of motion is the largely uniform motion of ions in solution in the presence of an electric field. Molecular and ionic motion have common features and, by considering them from a more general viewpoint, we derive expressions that govern the migration of properties through matter. One of the most useful consequences of this general approach is the formulation of the diffusion equation, which is an equation that shows how matter and energy spread through media of various kinds. Finally, we build a simple model for all types of molecular motion, in which the molecules migrate in a series of small steps, and see that it accounts for many of the properties of migrating molecules in both gases and condensed phases.

The general approach we describe in this chapter provides techniques for discussing the motion of all kinds of particles in all kinds of fluids. We set the scene by considering a simple type of motion, that of molecules in a perfect gas, and go on to see that molecular motion in liquids shows a number of similarities. We shall concentrate on the **transport properties** of a substance, its ability to transfer matter, energy, or some other property from one place to another. Four examples of transport properties are

Diffusion, the migration of matter down a concentration gradient.

Thermal conduction, the migration of energy down a temperature gradient.

Electric conduction, the migration of electric charge along an electrical potential gradient.

Viscosity, the migration of linear momentum down a velocity gradient.

It is convenient to include in the discussion **effusion**, the emergence of a gas from a container through a small hole.

Molecular motion in gases

Here we present the kinetic model of a perfect gas as a starting point for the discussion of its transport properties. In the **kinetic model** of gases we assume that the only contribution to the energy of the gas is from the kinetic energies of the

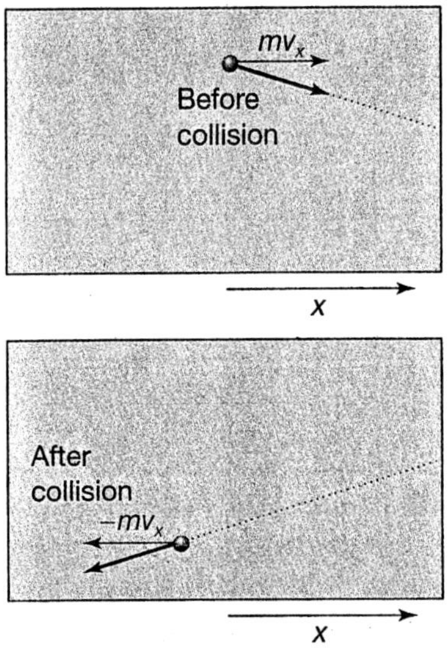

24.1 The pressure of a gas arises from the impact of its molecules on the walls. In an elastic collision of a molecule with a wall perpendicular to the x-axis, the x-component of velocity is reversed but the y- and z-components are unchanged.

molecules. The kinetic model is one of the most remarkable—and arguably most beautiful—models in physical chemistry for, from a set of very slender assumptions, powerful quantitative conclusions can be deduced.

24.1 The kinetic model of gases

The kinetic model is based on three assumptions:

1 The gas consists of molecules of mass m in ceaseless random motion.

2 The size of the molecules is negligible, in the sense that their diameters are much smaller than the average distance travelled between collisions.

3 The molecules interact only through brief, infrequent, and elastic collisions.

An **elastic collision** is a collision in which the total translational kinetic energy of the molecules is conserved.

(a) Pressure and molecular speeds

From the very economical assumptions of the kinetic model, we show in the following *Justification* that the pressure and volume of the gas are related by

$$pV = \tfrac{1}{3}nMc^2 \qquad (24.1)°$$

where $M = mN_A$, the molar mass of the molecules, and c is the **root mean square speed** of the molecules, the square root of the mean of the squares of the speeds, v, of the molecules:

$$c = \langle v^2 \rangle^{1/2} \qquad [24.2]$$

Justification 24.1

Consider the arrangement in Fig. 24.1. When a particle of mass m that is travelling with a component of velocity v_x parallel to the x-axis collides with the wall on the right and is reflected, its linear momentum (the product of its mass and its velocity) changes from mv_x before the collision to $-mv_x$ after the collision (when it is travelling in the opposite direction). The x-component of momentum therefore changes by $2mv_x$ on each collision (the y- and z-components are unchanged). Many molecules collide with the wall in an interval Δt, and the total change of momentum is the product of the change in momentum of each molecule multiplied by the number of molecules that reach the wall during the interval.

Because a molecule with velocity component v_x can travel a distance $v_x\Delta t$ along the x-axis in an interval Δt, all the molecules within a distance $v_x\Delta t$ of the wall will strike it if they are travelling towards it (Fig. 24.2). It follows that, if the wall has area A, then all the particles in a volume $A \times v_x\Delta t$ will reach the wall (if they are travelling towards it). The number density of particles is nN_A/V, where n is the total amount of molecules in the container of volume V and N_A is Avogadro's constant, so the number of molecules in the volume $Av_x\Delta t$ is $(nN_A/V) \times Av_x\Delta t$.

At any instant, half the particles are moving to the right and half are moving to the left. Therefore, the average number of collisions with the wall during the interval Δt is $\tfrac{1}{2}nN_AAv_x\Delta t/V$. The total momentum change in that interval is the product of this number and the change $2mv_x$:

$$\text{Momentum change} = \frac{nN_AAv_x\Delta t}{2V} \times 2mv_x = \frac{nmAN_Av_x^2\Delta t}{V} = \frac{nMAv_x^2\Delta t}{V}$$

where $M = mN_A$.

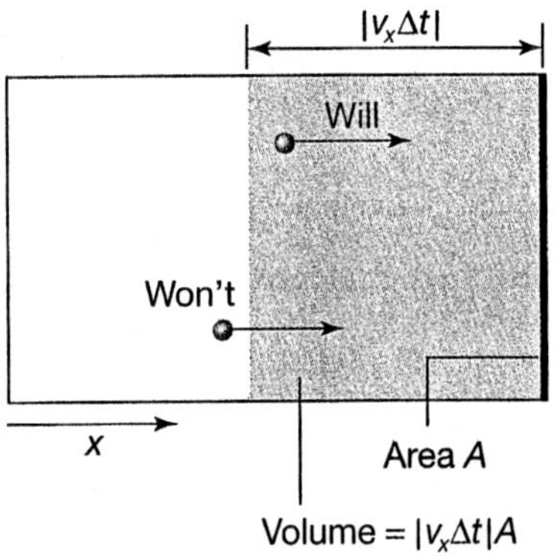

24.2 A molecule will reach the wall on the right within an interval Δt if it is within a distance $v_x\Delta t$ of the wall and travelling to the right.

Next, to find the force, we calculate the rate of change of momentum, which is this change of momentum divided by the interval Δt during which it occurs:

$$\text{Rate of change of momentum} = \frac{nMAv_x^2}{V}$$

This rate of change of momentum is equal to the force (by Newton's second law of motion). It follows that the pressure, the force divided by the area, is

$$\text{Pressure} = \frac{nMv_x^2}{V}$$

Not all the molecules travel with the same velocity, so the detected pressure, p, is the average (denoted $\langle \ldots \rangle$) of the quantity just calculated:

$$p = \frac{nM\langle v_x^2 \rangle}{V}$$

This expression already resembles the perfect gas equation of state.

Because the molecules are moving randomly and there is no net flow in a particular direction, the average speed along x is the same as that in the y- and z-directions. It follows that

$$c = (\langle v_x^2 \rangle + \langle v_y^2 \rangle + \langle v_z^2 \rangle)^{1/2} = (3\langle v_x^2 \rangle)^{1/2}$$

implying that $\langle v_x^2 \rangle = \frac{1}{3}c^2$. Equation 1 now follows immediately.

Equation 1 is one of the key results of the kinetic model. We see that, if the root mean square speed of the molecules depends only on the temperature, then at constant temperature

$$pV = \text{constant}$$

which is the content of Boyle's law. Moreover, for eqn 1 to be the equation of state of a perfect gas, its right-hand side must be equal to nRT. It follows that the root mean square speed of the molecules in a gas at a temperature T must be

$$c = \left(\frac{3RT}{M}\right)^{1/2} \qquad (24.3)^\circ$$

We can conclude that *the root mean square speed of the molecules of a gas is proportional to the square root of the temperature and inversely proportional to the square root of the molar mass*. That is, the higher the temperature, the higher the root mean square speed of the molecules, and at a given temperature, heavy molecules travel more slowly than light molecules. Sound waves are pressure waves, and for them to propagate the molecules of the gas must move to form regions of high and low pressure. Therefore, we should expect the root mean square speeds of molecules to be comparable to the speed of sound in air (340 m s^{-1}). The root mean square speed of N_2 molecules, for instance, is found from eqn 3 to be 515 m s^{-1}.

Equation 3 is an expression for the mean square speed of molecules. However, in an actual gas the speeds of individual molecules span a wide range, and the collisions in the gas continually redistribute the speeds among the molecules. Before a collision, a molecule may be travelling rapidly, but after a collision it may be accelerated to a very high speed, only to be slowed again by the next collision. The fraction of molecules that have speeds in the range v to $v + dv$ is proportional to the width of the range, and is written $f(v)\,dv$, where $f(v)$ is called the **distribution of speeds**.

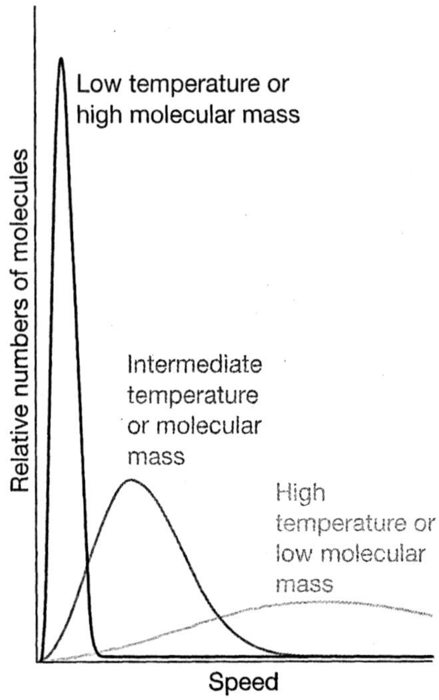

24.3 The distribution of molecular speeds with temperature and molar mass. Note that the most probable speed (corresponding to the peak of the distribution) increases with temperature and with decreasing molar mass, and simultaneously the distribution becomes broader.

The precise form of f for molecules of a gas at a temperature T was derived by J.C. Maxwell, and is

$$f(v) = 4\pi\left(\frac{M}{2\pi RT}\right)^{3/2} v^2 e^{-Mv^2/2RT} \qquad (24.4)$$

This expression is called the **Maxwell distribution of speeds**, and is derived in the following *Justification*. Figure 24.3 summarizes its main features. We see that the range of speeds broadens as the temperature increases. Lighter molecules also have a broader distribution of speeds than heavy molecules. To use eqn 4 to calculate the fraction of molecules in a given narrow range of speeds, Δv, we evaluate $f(v)$ at the speed of interest, then multiply it by the width of the range of speeds of interest; that is, we form $f(v)\Delta v$. To use the distribution to calculate the fraction in a range of speeds that is too wide to be treated as infinitesimal, we evaluate the integral:

$$\text{Fraction in the range } v_1 \text{ to } v_2 = \int_{v_1}^{v_2} f(v)\,dv \qquad (24.5)$$

This integral is the area under the graph of f as a function of v and, except in special cases, has to be evaluated numerically by using mathematical software (Fig. 24.4).

Justification 24.2

The Boltzmann distribution is a key result of physical chemistry and is treated fully in Section 19.1. It implies that the fraction of molecules with velocity components v_x, v_y, v_z is proportional to an exponential function of their kinetic energy, which is

$$E = \tfrac{1}{2}mv_x^2 + \tfrac{1}{2}mv_y^2 + \tfrac{1}{2}mv_z^2$$

Therefore, we can write

$$f = Ke^{-E/kT} = Ke^{-(\frac{1}{2}mv_x^2+\frac{1}{2}mv_y^2+\frac{1}{2}mv_z^2)/kT} = Ke^{-mv_x^2/2kT}e^{-mv_y^2/2kT}e^{-mv_z^2/2kT}$$

where K is a constant of proportionality and $f\,dv_x\,dv_y\,dv_z$ is the fraction of molecules in the velocity range v_x to $v_x + dv_x$, v_y to $v_y + dv_y$, and v_z to $v_z + dv_z$. We see that the fraction factorizes into three factors, one for each axis, and we can write $f = f(v_x)f(v_y)f(v_z)$ with

$$f(v_x) = K^{1/3}e^{-mv_x^2/kT}$$

For any a, x, y, z, ... we may write $a^{x+y+z+\cdots} = a^x a^y a^z \ldots$.

and likewise for the two other directions.

To determine the constant K, we note that a molecule must have a velocity somewhere in the range $-\infty < v_x < \infty$, so

$$\int_{-\infty}^{\infty} f(v_x)\,dv_x = 1$$

Substitution of the expression for $f(v_x)$ then gives

$$1 = K^{1/3}\int_{-\infty}^{\infty} e^{-mv_x^2/2kT}\,dv_x = K^{1/3}\left(\frac{2\pi kT}{m}\right)^{1/2}$$

Therefore, $K = (m/2\pi kT)^{3/2} = (M/2\pi RT)^{3/2}$, where M is the molar mass of the molecules. At this stage we know that

$$f(v_x) = \left(\frac{M}{2\pi RT}\right)^{1/2} e^{-Mv_x^2/2RT} \qquad (24.6)$$

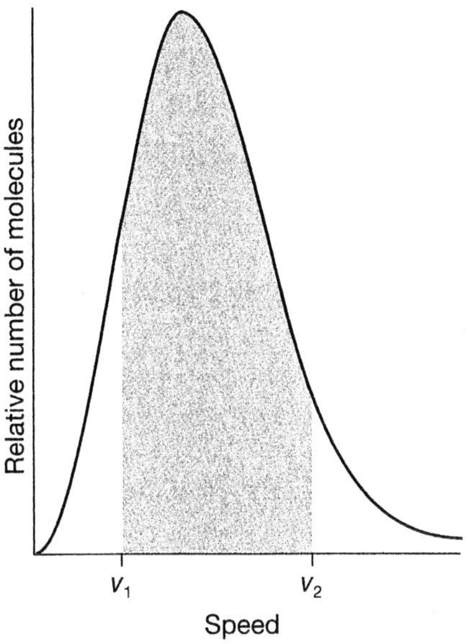

24.4 To calculate the probability that a molecule will have a speed in the range v_1 to v_2, we integrate the distribution between those two limits; the integral is equal to the area of the curve between the limits, as shown shaded here.

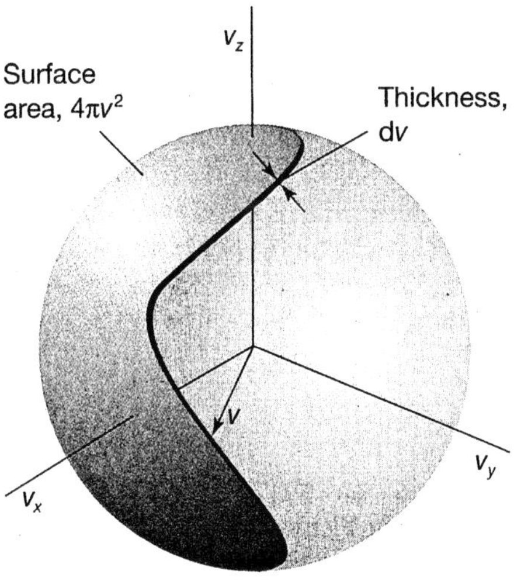

24.5 To evaluate the probability that a molecule has a speed in the range v to $v + dv$, we evaluate the total probability that the molecule will have a speed that is anywhere in a spherical shell of radius $v = (v_x^2 + v_y^2 + v_z^2)^{1/2}$ and thickness dv by summing the probabilities that it is in a volume element $dv_x\, dv_y\, dv_z$ at a distance v from the origin.

The probability that a molecule has a velocity in the range v_x to $v_x + dv_x$, v_y to $v_y + dv_y$, v_z to $v_z + dv_z$ is

$$f(v_x)f(v_y)f(v_z)\,dv_x\,dv_y\,dv_z = \left(\frac{M}{2\pi RT}\right)^{3/2} e^{-Mv^2/2RT}\,dv_x\,dv_y\,dv_z$$

where $v^2 = v_x^2 + v_y^2 + v_z^2$. The probability $f(v)\,dv$ that the molecules have a speed in the range v to $v + dv$ regardless of direction is the sum of the probabilities that the velocity lies in any of the volume elements $dv_x\,dv_y\,dv_z$ forming a spherical shell of radius v and thickness dv (Fig. 24.5). The sum of the volume elements on the right-hand side of the last equation is the volume of this shell, $4\pi v^2\,dv$. Therefore,

$$f(v) = 4\pi\left(\frac{M}{2\pi RT}\right)^{3/2} v^2 e^{-Mv^2/2RT}$$

as given in eqn 4.

Example 24.1 Calculating the mean speed of molecules in a gas

What is the mean speed, $\bar{c}$, of N_2 molecules in air at 25°C?

Method We are asked to calculate the mean speed, not the root mean square speed. A mean speed is calculated by multiplying each speed by the fraction of molecules that have that speed, and then adding all the products together. When the speed varies over a continuous range, the sum is replaced by an integral. To employ this approach here, we note that the fraction of molecules with a speed in the range v to $v + dv$ is $f(v)\,dv$, so the product of this fraction and the speed is $vf(v)\,dv$. The mean speed, $\bar{c}$, is obtained by evaluating the integral

$$\bar{c} = \int_0^\infty vf(v)\,dv$$

with f given in eqn 4.

Answer The integral required is

$$\bar{c} = 4\pi\left(\frac{M}{2\pi RT}\right)^{3/2}\int_0^\infty v^3 e^{-Mv^2/2RT}\,dv$$

$$= 4\pi\left(\frac{M}{2\pi RT}\right)^{3/2} \times \frac{1}{2}\left(\frac{2RT}{M}\right)^2 = \left(\frac{8RT}{\pi M}\right)^{1/2}$$

Substitution of the data then gives

$$\bar{c} = \left(\frac{8 \times (8.3141\ \mathrm{J\,K^{-1}\,mol^{-1}}) \times (298\ \mathrm{K})}{\pi \times (28.02 \times 10^{-3}\ \mathrm{kg\,mol^{-1}})}\right)^{1/2} = 475\ \mathrm{m\,s^{-1}}$$

where we have used $1\ \mathrm{J} = 1\ \mathrm{kg\,m^2\,s^{-2}}$. To evaluate the integral we have used the standard result from tables of integrals (or software) that

$$\int_0^\infty x^3 e^{-ax^2}\,dx = \frac{1}{2a^2}$$

Self-test 24.1 Evaluate the root mean square speed of the molecules by integration. You will need the integral

$$\int_0^\infty x^4 e^{-ax^2}\,dx = \frac{3}{8}\left(\frac{\pi}{a^5}\right)^{1/2}$$

[$c = (3RT/M)^{1/2}$, 515 m s^{-1}]

As shown in Example 24.1, we can use the Maxwell distribution to evaluate the **mean speed**, $\bar{c}$, of the molecules in a gas:

$$\bar{c} = \left(\frac{8RT}{\pi M}\right)^{1/2} \tag{24.7}$$

To find the location of this peak, differentiate f with respect to v and look for the value of v at which the derivative is zero (other than at $v = 0$ and $v = \infty$).

We can identify the **most probable speed**, c^*, from the location of the peak of the distribution:

$$c^* = \left(\frac{2RT}{M}\right)^{1/2} \tag{24.8}$$

Figure 24.6 summarizes these results.

The **relative mean speed**, $\bar{c}_{\rm rel}$, the mean speed with which one molecule approaches another, can also be calculated from the distribution:

$$\bar{c}_{\rm rel} = 2^{1/2}\bar{c} \tag{24.9}$$

This result is much harder to derive, but the diagram in Fig. 24.7 should help to show that it is plausible. The last result can also be generalized to the relative mean speed of two dissimilar molecules of masses $m_{\rm A}$ and $m_{\rm B}$:

$$\bar{c}_{\rm rel} = \left(\frac{8kT}{\pi\mu}\right)^{1/2} \qquad \mu = \frac{m_{\rm A}m_{\rm B}}{m_{\rm A}+m_{\rm B}} \tag{24.10}$$

Note that the molecular masses (not the molar masses) and Boltzmann's constant, $k = R/N_{\rm A}$, appear in this expression; the quantity μ is called the **reduced mass** of the molecules. Equation 10 turns into eqn 7 when the molecules are identical (that is, $m_{\rm A} = m_{\rm B} = m$, so $\mu = \frac{1}{2}m$).

The Maxwell distribution has been verified experimentally. For example, molecular speeds can be measured directly with a velocity selector (Fig. 24.8). The spinning cylinder has channels that permit the passage of only those molecules moving through them at the appropriate speed, and the number of molecules can be determined by collecting them at a detector.

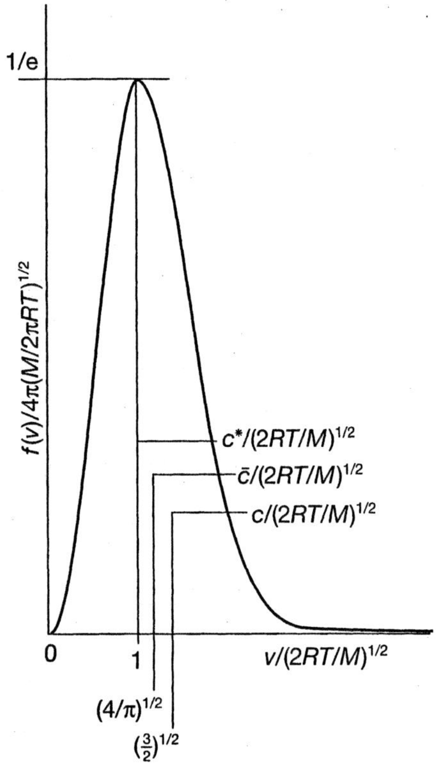

24.6 A summary of the conclusions that can be deduced from the Maxwell distribution for molecules of molar mass M at a temperature T: c^* is the most probable speed, $\bar{c}$ is the mean speed, and c is the root mean square speed.

(b) The collision frequency

A qualitative picture of the events taking place in a gas was first described in Section 1.1. The kinetic model enables us to make that picture more quantitative. In particular, it enables us to calculate the frequency with which molecular collisions occur and the distance a molecule travels on average between collisions.

We count a 'hit' whenever the centres of two molecules come within a distance d of each other, where d, the **collision diameter**, is of the order of the actual diameters of the molecules (for impenetrable hard spheres d is the diameter). As we show in the following *Justification*, we can use kinetic theory to deduce that the **collision frequency**, z, the number of collisions made by one molecule divided by the time interval during which the collisions are counted, when there are N molecules in a volume V is

$$z = \sigma\bar{c}_{\rm rel}\mathcal{N} \tag{24.11}^\circ$$

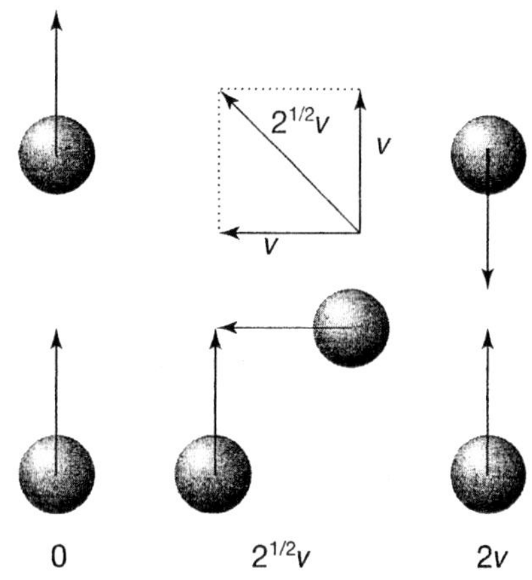

24.7 A simplified version of the argument to show that the mean relative speed of molecules in a gas is related to the mean speed. When the molecules are moving in the same direction, the mean relative speed is zero; it is $2v$ when the molecules are approaching each other. A typical mean direction of approach is from the side, and the mean speed of approach is then $2^{1/2}v$. The last direction of approach is the most characteristic, so the mean speed of approach can be expected to be about $2^{1/2}v$. This value is confirmed by more detailed calculation.

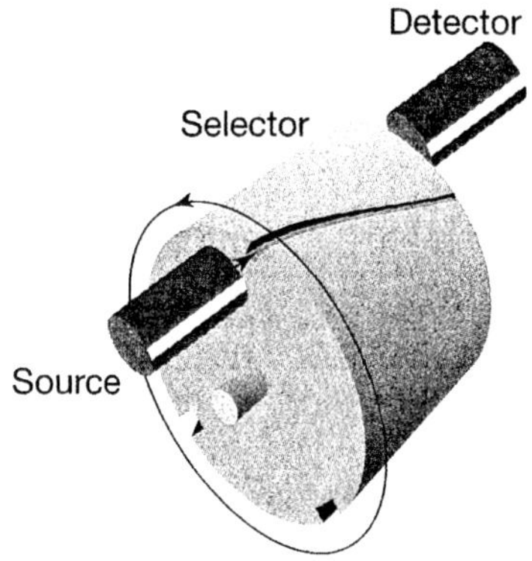

24.8 A velocity selector. The molecules are produced in the source (which may be an oven with a small hole in one wall), and travel in a beam towards the rotating channels. Only if the speed of a molecule is such as to carry it along the channel that rotates into its path will it reach the detector. Thus, the number of slow molecules can be counted by rotating the discs slowly, and the number of fast molecules counted by rotating the discs rapidly.

Table 24.1* Collision cross-sections

	σ/nm^2
Benzene, C_6H_6	0.88
Carbon dioxide, CO_2	0.52
Helium, He	0.21
Nitrogen, N_2	0.43

* More values are given in the *Data section*.

with $\mathcal{N} = N/V$ and $\bar{c}_{rel}$ given in eqn 10. The area $\sigma = \pi d^2$ is called the **collision cross-section** of the molecules. Some typical collision cross-sections are given in Table 24.1 (they are obtained by the techniques described in Section 21.7). In terms of the pressure,

$$z = \frac{\sigma \bar{c}_{rel} p}{kT} \qquad (24.12)^{\circ}$$

Justification 24.3

We consider the positions of all the molecules except one to be frozen. Then we note what happens as one mobile molecule travels through the gas with a mean relative speed $\bar{c}_{rel}$ for a time Δt. In doing so it sweeps out a 'collision tube' of cross-sectional area $\sigma = \pi d^2$ and length $\bar{c}_{rel}\Delta t$, and therefore of volume $\sigma \bar{c}_{rel}\Delta t$ (Fig. 24.9). The number

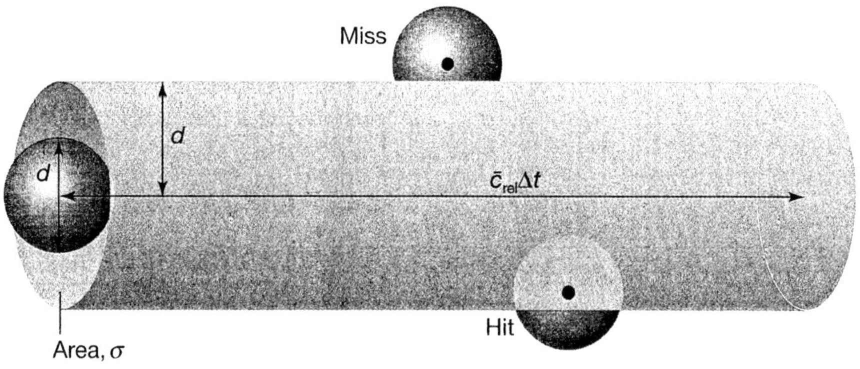

24.9 In an interval Δt, a molecule of diameter d sweeps out a tube of radius d and length $\bar{c}_{rel}\Delta t$. As it does so it encounters other molecules with centres that lie within the tube, and each such encounter counts as one collision. In reality, the tube is not straight, but changes direction at each collision. Nevertheless, the volume swept out is the same, and this straightened version of the tube can be used as a basis of the calculation.

of stationary molecules with centres inside the collision tube is given by the volume of the tube multiplied by the number density $\mathcal{N} = N/V$, and is $\mathcal{N}\sigma\bar{c}_{rel}\Delta t$. The number of hits scored in the interval Δt is equal to this number, so the number of collisions divided by the time interval, is $\mathcal{N}\sigma\bar{c}_{rel}$. The expression in terms of the pressure of the gas is obtained by using the perfect gas equation to write

$$\frac{N}{V} = \frac{nN_A}{V} = \frac{pN_A}{RT} = \frac{p}{kT}$$

Equation 11 shows that, at constant volume, the collision frequency increases with increasing temperature. The reason for this increase is that the mean relative speed increases with temperature. Equation 12 shows that, at constant temperature, the collision frequency is proportional to the pressure. Such a proportionality is plausible, for the greater the pressure, the greater the number density of molecules in the sample, and the rate at which they encounter one another is greater even though their average speed remains the same. For an N_2 molecule in a sample at 1 atm and 25°C, $z \approx 5 \times 10^9\ s^{-1}$, so a given molecule collides about 5×10^9 times each second. We are beginning to appreciate the timescale of events in gases.

(c) The mean free path

Once we have the collision frequency, we can calculate the **mean free path**, λ (lambda), the average distance a molecule travels between collisions. If a molecule collides with a frequency z, it spends a time $1/z$ in free flight between collisions, and therefore travels a distance $(1/z)\bar{c}$. Therefore, the mean free path is

$$\lambda = \frac{\bar{c}}{z} \qquad (24.13)$$

Substitution of the expression for z in eqn 12 gives

$$\lambda = \frac{kT}{2^{1/2}\sigma p} \qquad (24.14)$$

Doubling the pressure reduces the mean free path by half. A typical mean free path in nitrogen gas at 1 atm is 70 nm, or about 10^3 molecular diameters. Although the temperature appears in eqn 14, in a sample of constant volume, the pressure is proportional to T, so T/p remains constant when the temperature is increased. Therefore, the mean free path is independent of the temperature in a sample of gas in a container of fixed volume. The distance between collisions is determined by the number of molecules present in the given volume, not by the speed at which they travel.

In summary, a typical gas (N_2 or O_2) at 1 atm and 25°C can be thought of as a collection of molecules travelling with a mean speed of about 500 m s^{-1}. Each molecule makes a collision within about 1 ns, and between collisions it travels about 10^3 molecular diameters. The kinetic model of gases is valid (and the gas behaves nearly perfectly) if the diameter of the molecules is much smaller than the mean free path ($d \ll \lambda$), for then the molecules spend most of their time far from one another (Box 24.1).

24.2 Collisions with walls and surfaces

The key result for accounting for transport in the gas phase is the rate at which molecules strike an area (which may be an imaginary area embedded in the gas, or part of a real wall). The **collision flux**, Z_W, is the number of collisions with the area

Box 24.1 The Sun as a ball of perfect gas

The kinetic theory of gases is valid when the size of the particles is negligible compared with their mean free path. It may seem absurd, therefore, to expect the kinetic theory and, as a consequence, the perfect gas law, to be applicable to the dense matter of stellar interiors. In the Sun, for instance, the mean density is about 1.50 times that of liquid water at its centre and comparable to that of water about half way to its surface. However, we have to realize that the state of matter is that of a *plasma*, in which the electrons have been stripped from the atoms of hydrogen and helium that make up the bulk of the matter of stars. As a result, the particles making up the plasma have diameters comparable to those of nuclei, or about 10 fm. Therefore, a mean free path of only 0.1 pm satisfies the criterion for the validity of the kinetic theory and the perfect gas law. We can therefore use $pV = nRT$ as the equation of state for the stellar interior.

As for any perfect gas, the pressure in the interior of the Sun is related to the mass density, $\rho = m/V$, by $p = \rho RT/M$. Atoms are stripped of their electrons in the interior of stars so, if we suppose that the interior consists of ionized hydrogen atoms, the mean molar mass is one-half the molar mass of hydrogen, or 0.5 g mol^{-1} (the mean of the molar mass of H^+ and e^-, the latter being virtually 0). Half way to the centre of the Sun, the temperature is 3.6 MK and the mass density is 1.20 g cm^{-3} (slightly denser than water); so the pressure there works out as 7.2×10^{13} Pa, or about 720 million atmospheres.

We can combine this result with the expression for the pressure from kinetic theory ($p = \frac{1}{3}NMc^2/V$). Because the total kinetic energy of the particles is $E_K = \frac{1}{2}Nmc^2$, we can write $p = \frac{2}{3}E_K/V$. That is, the pressure of the plasma is related to the *kinetic energy density*, $\rho_K = E_K/V$, the kinetic energy of the molecules in a region divided by the volume of the region, by $p = \frac{2}{3}\rho_K$. It follows that the kinetic energy density half way to the centre of the Sun is about 0.11 GJ cm^{-3}. In contrast, on a warm day (25°C) on Earth, the (translational) kinetic energy density of our atmosphere is only 0.15 J cm^{-3}.

Discussion questions

1. The principal components of the atmosphere of the Earth are diatomic molecules, which can rotate as well as translate. Given that the translational kinetic energy density of the atmosphere is 0.15 J cm^{-3}, what is the total kinetic energy density, including rotation?

2.[1] In the *standard model* of stellar structure (I. Nicholson, *The sun*; Rand McNally, New York (1982)), the interior of the Sun is thought to consist of 36 per cent H and 64 per cent He by mass, at a density of 158 g cm^{-3} very close to the centre. Both atoms are completely ionized. The approximate dimensions of the nuclei can be calculated from the formula $r_{\text{nucleus}} = 1.4 \times 10^{-15} A^{1/3}$ m, where A is the mass number. The size of the free electron, $r_e \approx 10^{-18}$ m, is negligible compared to the size of the nuclei. (a) Calculate the excluded volume in 1.0 cm^3 of the stellar interior and on that basis decide upon the applicability of the perfect gas law to this system. (b) The standard model suggests that the pressure in the stellar interior is 2.5×10^{11} atm. Calculate the temperature of the Sun's interior based on the perfect gas model. The generally accepted standard model value is 16 MK. (c) Would a van der Waals type of equation (with $a = 0$) give a better value for T?

Further reading

D.B. Clark, The ideal gas law at the center of the sum. *J. Chem. Educ.* **66**, 826 (1989).

C.R. Cowley, *An introduction to cosmochemistry*. Cambridge University Press (1995).

1. Problem supplied by Carmen Giunta and Charles Trapp.

in a given time interval divided by the area and the duration of the interval. The **collision frequency**, the number of hits per second, is obtained by multiplication of the collision flux by the area of interest. We show in the *Justification* below that the collision flux is

$$Z_W = \frac{p}{(2\pi mkT)^{1/2}} \qquad (24.15)°$$

When $p = 100$ kPa (1.00 bar) and $T = 300$ K, $Z_W \approx 3 \times 10^{23}$ cm^{-2} s^{-1}.

Justification 24.4

Consider a wall of area A perpendicular to the x-axis (as in Fig. 24.2). If a molecule has $v_x > 0$ (that is, it is travelling in the direction of positive x), then it will strike the wall within an interval Δt if it lies within a distance $v_x\Delta t$ of the wall. Therefore, all molecules in the volume $Av_x\Delta t$, and with positive x-component of velocities, will strike the wall in the interval Δt. The total number of collisions in this interval is therefore the volume $Av_x\Delta t$ multiplied by the number density, $\mathcal{N}$, of molecules. However, to take account of the presence of a range of velocities in the sample, we must sum the result over all the positive values of v_x weighted by the probability distribution of velocities (eqn 6):

$$\text{Number of collisions} = \mathcal{N}A\Delta t \int_0^\infty v_x f(v_x)\,\mathrm{d}x$$

The collision flux is the number of collisions divided by A and Δt, so

$$Z_W = \mathcal{N}\int_0^\infty v_x f(v_x)\,\mathrm{d}x$$

Then, using the velocity distribution in eqn 6,

$$\int_0^\infty v_x f(v_x)\,\mathrm{d}v_x = \left(\frac{m}{2\pi kT}\right)^{1/2} \int_0^\infty v_x \mathrm{e}^{-mv_x^2/2kT}\,\mathrm{d}v_x$$

$$= \left(\frac{kT}{2\pi m}\right)^{1/2}$$

Therefore,

$$Z_W = \mathcal{N}\left(\frac{kT}{2\pi m}\right)^{1/2} = \tfrac{1}{4}\bar{c}\mathcal{N} \qquad (24.16)^\circ$$

Substitution of $\mathcal{N} = nN_A/V = p/kT$ gives eqn 15.

24.3 The rate of effusion

The essential empirical observations on effusion are summarized by **Graham's law of effusion**, which states that the rate of effusion is inversely proportional to the square root of the molar mass. The basis of this result is that, as remarked above, the mean speed of molecules is inversely proportional to $M^{1/2}$, so the rate at which they strike the area of the hole is also inversely proportional to $M^{1/2}$. However, by using the expression for the rate of collisions, we can obtain a more detailed expression for the rate of effusion and hence use effusion data more effectively.

When a gas at a pressure p and temperature T is separated from a vacuum by a small hole, the rate of escape of its molecules is equal to the rate at which they strike the area of the hole (which is given by eqn 15). Therefore, for a hole of area A_0,

$$\text{Rate of effusion} = Z_W A_0 = \frac{pA_0}{(2\pi mkT)^{1/2}} = \frac{pA_0N_A}{(2\pi MRT)^{1/2}} \qquad (24.17)^\circ$$

(In the last step we have used $R = N_Ak$ and $M = mN_A$.) This rate is inversely proportional to $M^{1/2}$, in accord with Graham's law.

Differential equations are reviewed in *Further information 1.*

Example 24.2 Deducing the time dependence of the pressure inside an effusion oven

Derive an expression that shows how the pressure of a gas inside an effusion oven (a heated chamber with a small hole in one wall) varies with time if the oven is not replenished as the gas escapes.

Method The rate of effusion is proportional to the pressure of the gas in the container so, as gas effuses and the pressure falls, the rate of effusion will decrease. To find the explicit expression, set up a differential equation relating dp/dt to p, and then integrate it. The rate of effusion, as given by eqn 17, is the number of molecules that leave the container in a given interval divided by the duration of the interval. The first step is to relate the rate of change of pressure to the rate of change of number of molecules by using the perfect gas law in the form $pV = NkT$.

Answer The rate of change of pressure of a gas in a container at constant pressure and temperature is related to the rate of change of the number of molecules present by

$$\frac{dp}{dt} = \frac{kT}{V}\frac{dN}{dt}$$

The rate of change of the number of molecules is equal to the collision frequency with the hole, and that in turn is equal to the collision flux multiplied by the area of the hole:

$$\frac{dN}{dt} = -Z_W A_0 = -\frac{pA_0}{(2\pi mkT)^{1/2}}$$

Substitution of this expression into the one above gives

$$\frac{dp}{dt} = -\left(\frac{kT}{2\pi m}\right)^{1/2}\frac{pA_0}{V}$$

This expression integrates to

$$p = p_0 e^{-t/\tau} \qquad \tau = \left(\frac{2\pi m}{kT}\right)^{1/2}\frac{V}{A_0}$$

Comment The pressure falls exponentially towards zero; the decrease is faster the higher the temperature, the bigger the hole, and the lower the mass of the molecules.

Self-test 24.2 Show that $t_{1/2}$, the time required for the pressure to decrease to half its initial value, is independent of the initial pressure. [$t_{1/2} = \tau \ln 2$]

Equation 17 is the basis of the **Knudsen method** for the determination of the vapour pressures of liquids and solids, particularly of substances with very low vapour pressures. Thus, if the vapour pressure of a sample is p, and it is enclosed in a cavity with a small hole, then the rate of loss of mass from the container is proportional to p.

Example 24.3 Calculating the vapour pressure from a mass loss

Caesium (m.p. 29°C, b.p. 686°C) was introduced into a container and heated to 500°C. When a hole of diameter 0.50 mm was opened in the container for 100 s, a mass loss of 385 mg was measured. Calculate the vapour pressure of liquid caesium at 500 K.

Method The pressure of vapour is constant inside the container despite the effusion of atoms because the hot liquid metal replenishes the vapour. The rate of effusion is therefore constant, and given by eqn 17. To express the rate in terms of mass, multiply the number of atoms that escape by the mass of each atom.

Answer The mass loss Δm in an interval Δt is related to the collision flux by

$$\Delta m = Z_W A_0 m \Delta t$$

where A_0 is the area of the hole and m is the mass of one atom. It follows that

$$Z_W = \frac{\Delta m}{A_0 m \Delta t}$$

Because Z_W is related to the pressure by eqn 15, we can write

$$p = \left(\frac{2\pi RT}{M}\right)^{1/2} \frac{\Delta m}{A_0 \Delta t}$$

Because $M = 132.9\ \text{g mol}^{-1}$, substitution of the data gives $p = 11$ kPa (using $1\ \text{Pa} = 1\ \text{N m}^{-2} = 1\ \text{J m}^{-3}$), or 83 Torr.

Self-test 24.3 How long would it take 1.0 g of Cs atoms to effuse out of the oven under the same conditions? [260 s]

24.4 Transport properties of a perfect gas

Transport properties are commonly expressed in terms of a number of 'phenomenological' equations, or equations that are empirical summaries of experimental observations. These phenomenological equations apply to all kinds of properties and media. In the following sections, we introduce the equations for the general case and then show how to calculate the parameters that appear in these phenomenological expressions.

(a) The phenomenological equations

The rate of migration of a property is measured by its **flux**, J, the quantity of that property passing through a given area in a given time interval divided by the area and the duration of the interval. If matter is flowing (as in diffusion), we speak of a **matter flux** of so many molecules per square metre per second; if the property is energy (as in thermal conduction), then we speak of the **energy flux** and express it in joules per square metre per second, and so on.

Experimental observations on transport properties show that the flux of a property is usually proportional to the first derivative of some other related property. For example, the flux of matter diffusing parallel to the z-axis of a container is found to be proportional to the first derivative of the concentration:

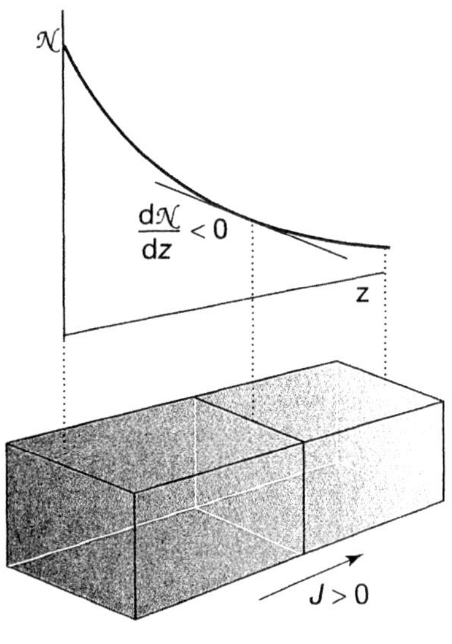

24.10 The flux of particles down a concentration gradient. Fick's first law states that the flux of matter (the number of particles passing through an imaginary window in a given interval divided by the area of the window and the length of the interval) is proportional to the density gradient at that point.

$$J(\text{matter}) \propto \frac{d\mathcal{N}}{dz} \qquad (24.18)$$

where $\mathcal{N}$ is the number density of particles with units number per metre cubed (m^{-3}). The SI units of J are number per metre squared per second ($m^{-2}\ s^{-1}$). The proportionality of the flux of matter to the concentration gradient is sometimes called **Fick's first law of diffusion**: the law implies that, if the concentration varies steeply with position, then diffusion will be fast. There is no net flux if the concentration is uniform ($d\mathcal{N}/dz = 0$). Similarly, the rate of thermal conduction (the flux of the energy associated with thermal motion) is found to be proportional to the temperature gradient:

$$J(\text{energy}) \propto \frac{dT}{dz} \qquad (24.19)$$

The SI units of this flux are joules per metre squared per second ($J\ m^{-2}\ s^{-1}$).

A positive value of J signifies a flux towards positive z; a negative value of J signifies a flux towards negative z. Because matter flows down a concentration gradient, from high concentration to low concentration, J is positive if $d\mathcal{N}/dz$ is negative (Fig. 24.10). Therefore, the coefficient of proportionality in eqn 19 must be negative, and we write it $-D$:

$$J(\text{matter}) = -D\frac{d\mathcal{N}}{dz} \qquad (24.20)$$

The constant D is called the **diffusion coefficient**; its SI units are metre squared per second ($m^2\ s^{-1}$). Energy migrates down a temperature gradient, and the same reasoning leads to

$$J(\text{energy}) = -\kappa\frac{dT}{dz} \qquad (24.21)$$

where κ is the **coefficient of thermal conductivity**. The SI units of κ are joules per kelvin per metre per second ($J\ K^{-1}\ m^{-1}\ s^{-1}$). Some experimental values are given in Table 24.2.

To see the connection between the flux of momentum and the viscosity, consider a fluid in a state of **Newtonian flow**, which can be imagined as occurring by a series of layers moving past one another (Fig. 24.11). The layer next to the wall of the vessel is stationary, and the velocity of successive layers varies linearly with distance, z, from the wall. Molecules ceaselessly move between the layers and bring with them the x-component of linear momentum they possessed in their original layer. A layer is retarded by molecules arriving from a more slowly moving layer because they have a low momentum in the x-direction. A layer is accelerated by molecules arriving from a more rapidly moving layer. We interpret the net retarding effect as the fluid's viscosity.

Because the retarding effect depends on the transfer of the x-component of linear momentum into the layer of interest, the viscosity depends on the flux of this x-component in the z-direction. The flux of the x-component of momentum is proportional to dv_x/dz because there is no net flux when all the layers move at the same velocity. We can therefore write

$$J(x\text{-component of momentum}) = -\eta\frac{dv_x}{dz} \qquad (24.22)$$

The constant of proportionality, η, is the **coefficient of viscosity** (or simply 'the viscosity'). Its units are kilogram per metre per second ($kg\ m^{-1}\ s^{-1}$). Viscosities are often reported in poise (P), where $1\ P = 10^{-1}\ kg\ m^{-1}\ s^{-1}$. Some experimental values are given in Table 24.2.

Table 24.2* Transport properties of gases at 1 atm

	$\kappa/(J\ K^{-1}\ m^{-1}\ s^{-1})$	$\eta/\mu P$†	
	273 K	273 K	293 K
Ar	0.0163	210	223
CO_2	0.0145	136	147
He	0.1442	187	196
N_2	0.0240	166	176

* More values are given in the *Data section*.
† $\mu P = 10^{-7}\ kg\ m^{-1}\ s^{-1}$.

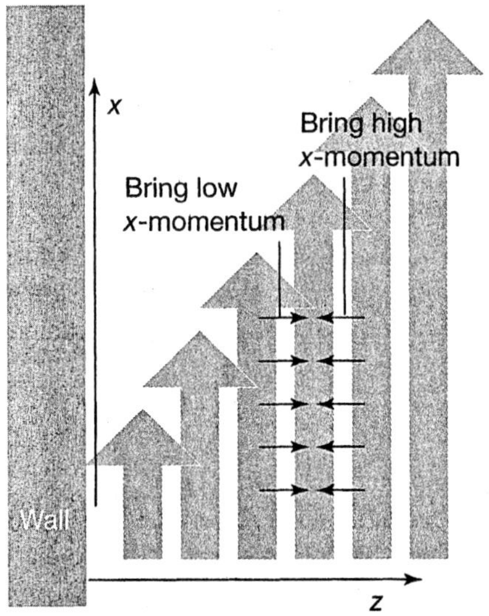

24.11 The viscosity of a fluid arises from the transport of linear momentum. In this illustration the fluid is undergoing laminar flow, and particles bring their initial momentum when they enter a new layer. If they arrive with high x-component of momentum they accelerate the layer; if they arrive with low x-component of momentum they retard the layer.

Table 24.3 Transport properties of perfect gases

Property	Transported quantity	Simple kinetic theory	Units
Diffusion	Matter	$D = \frac{1}{3}\lambda\bar{c}$	$m^2\ s^{-1}$
Thermal conductivity	Energy	$\kappa = \frac{1}{3}\lambda\bar{c}C_{V,m}[A] = \dfrac{\bar{c}C_{V,m}}{(3\sqrt{2})\sigma N_A}$	$J\ K^{-1}\ m^{-1}\ s^{-1}$
Viscosity	Momentum	$\eta = \frac{1}{3}\lambda\bar{c}m\mathcal{N} = \dfrac{m\bar{c}}{(3\sqrt{2})\sigma}$	$kg\ m^{-1}\ s^{-1}$

(b) Diffusion

We shall now see how the kinetic theory is used to justify Fick's law, and deduce the values of the transport coefficients of a perfect gas. As shown in the *Justification* below and summarized in Table 24.3, the kinetic theory leads to the result that, for a perfect gas,

$$D = \tfrac{1}{3}\lambda\bar{c} \qquad (24.23)^{\circ}$$

The mean free path, λ, decreases as the pressure is increased (Section 1.3b), so D decreases with increasing pressure and, as a result, the gas molecules diffuse more slowly. The mean speed, $\bar{c}$, increases with the temperature (Section 24.1a), so D also increases with temperature. As a result, molecules in a hot sample diffuse more quickly than those in a cool sample (for a given concentration gradient). Because the mean free path increases when the collision cross-section of the molecules decreases, the diffusion coefficient is greater for small molecules than for large molecules.

Justification 24.5

Consider the arrangement depicted in Fig. 24.12. On average, the molecules passing through the area A at $z = 0$ have travelled about one mean free path λ since their last collision. Therefore, the number density where they originated is $\mathcal{N}(z)$ evaluated at $z = -\lambda$. This number density is approximately

$$\mathcal{N}(-\lambda) = \mathcal{N}(0) - \lambda\left(\frac{d\mathcal{N}}{dz}\right)_0 \qquad (24.24)$$

This relation, and others like it that follow, is based on the Taylor expansion of a function, $f(x) = f(0) + (df/dx)_0 x + \cdots$, truncated after the second term. The subscript 0 indicates that the slope df/dx should be evaluated at $x = 0$.

The average number of impacts on the imaginary window of area A_0 during an interval Δt is $Z_W A_0 \Delta t$, with $Z_W = \frac{1}{4}\mathcal{N}\bar{c}$ (eqn 16). Therefore, the flux from left to right, $J(L \to R)$, arising from the supply of molecules on the left, is

$$J(L \to R) = \frac{\frac{1}{4}A_0\mathcal{N}(-\lambda)\bar{c}\Delta t}{A_0\Delta t} = \tfrac{1}{4}\mathcal{N}(-\lambda)\bar{c} \qquad (24.25a)$$

There is also a flux of molecules from right to left. On average, the molecules making the journey have originated from $z = +\lambda$ where the number density is $\mathcal{N}(\lambda)$. Therefore,

$$J(L \leftarrow R) = -\tfrac{1}{4}\mathcal{N}(\lambda)\bar{c} \qquad (24.25b)$$

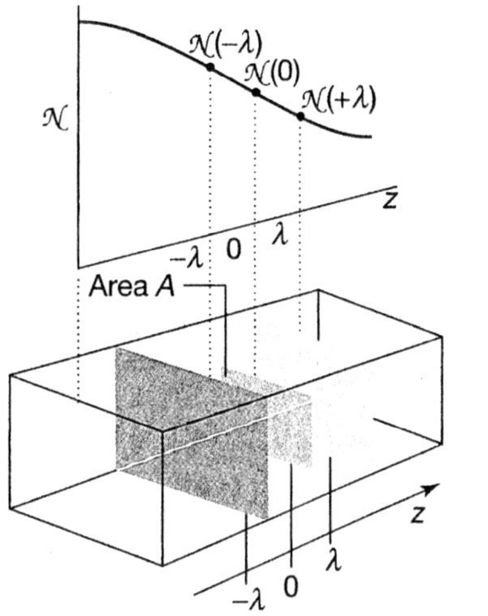

24.12 The calculation of the rate of diffusion of a gas considers the net flux of molecules through a plane of area A as a result of arrivals from on average a distance λ away in each direction, where λ is the mean free path.

The average number density at $z = +\lambda$ is approximately

$$\mathcal{N}(\lambda) = \mathcal{N}(0) + \lambda\left(\frac{d\mathcal{N}}{dz}\right)_0 \tag{24.26}$$

The net flux is

$$\begin{aligned} J_z &= J(L \rightarrow R) + J(L \leftarrow R) \\ &= \tfrac{1}{4}\bar{c}\left\{\left[\mathcal{N}(0) - \lambda\left(\frac{d\mathcal{N}}{dz}\right)_0\right] - \left[\mathcal{N}(0) + \lambda\left(\frac{d\mathcal{N}}{dz}\right)_0\right]\right\} \\ &= -\tfrac{1}{2}\bar{c}\lambda\left(\frac{d\mathcal{N}}{dz}\right)_0 \end{aligned} \tag{24.27}$$

This equation shows that the flux is proportional to the first derivative of the concentration, in agreement with Fick's law.

At this stage it looks as though we can pick out a value of the diffusion coefficient by comparing eqns 20 and 27, so obtaining $D = \frac{1}{2}\lambda\bar{c}$. It must be remembered, however, that the calculation is quite crude, and is little more than an assessment of the order of magnitude of D. One aspect that has not been taken into account is illustrated in Fig. 24.13, which shows that, although a molecule may have begun its journey very close to the window, it could have a long flight before it gets there. Because the path is long, the molecule is likely to collide before reaching the window, so it ought to be added to the graveyard of other molecules that have collided. To take this effect into account involves a lot of work, but the end result is the appearance of a factor of $\frac{2}{3}$ representing the lower flux. The modification results in eqn 23.

(c) Thermal conduction

According to the kinetic theory of gases, and as shown in the *Justification* below, the thermal conductivity of a perfect gas A having molar concentration [A] is given by the expression

$$\kappa = \tfrac{1}{3}\lambda\bar{c}C_{V,m}[\mathrm{A}] \tag{24.28°}$$

where $C_{V,m}$ is the molar heat capacity at constant volume.

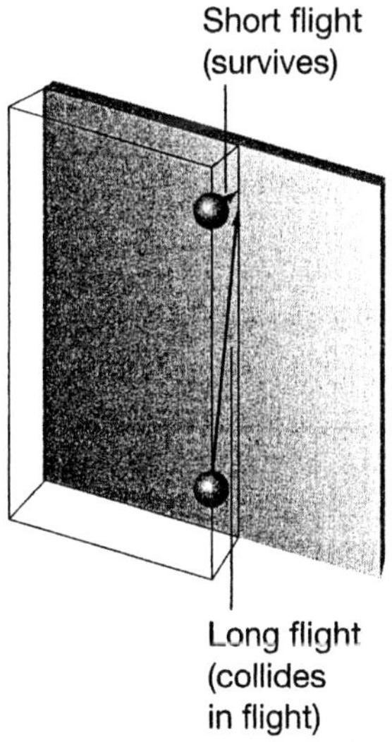

24.13 One approximation ignored in the simple treatment is that some particles might make a long flight to the plane even though they are only a short perpendicular distance away, and therefore they have a higher chance of colliding during their journey.

Justification 24.6

According to the equipartition theorem (Section 20.3), each molecule carries an average energy $\varepsilon = \nu kT$, where ν is a number of the order of 1. For monatomic particles, $\nu = \frac{3}{2}$. When one molecule passes through the imaginary window, it transports that energy on average. We suppose that the number density is uniform but that the temperature is not. On average, molecules arrive from the left after travelling a mean free path from their last collision in a hotter region, and therefore with a higher energy. Molecules also arrive from the right after travelling a mean free path from a cooler region. The two opposing energy fluxes are therefore

$$\begin{aligned} J(L \rightarrow R) &= \tfrac{1}{4}\bar{c}\mathcal{N}\varepsilon(-\lambda) \qquad \varepsilon(-\lambda) = \nu k\left\{T - \lambda\left(\frac{dT}{dz}\right)_0\right\} \\ J(L \leftarrow R) &= -\tfrac{1}{4}\bar{c}\mathcal{N}\varepsilon(\lambda) \qquad \varepsilon(\lambda) = \nu k\left\{T + \lambda\left(\frac{dT}{dz}\right)_0\right\} \end{aligned} \tag{24.29}$$

and the net flux is

$$J_z = J(\mathrm{L}\rightarrow\mathrm{R}) + J(\mathrm{L}\leftarrow\mathrm{R}) = -\tfrac{1}{2}\nu k\lambda\bar{c}\mathcal{N}\left(\frac{\mathrm{d}T}{\mathrm{d}z}\right)_0 \quad (24.30)$$

As before, we multiply by $\frac{2}{3}$ to take long flight paths into account, and so arrive at

$$J_z = -\tfrac{1}{3}\nu k\lambda\bar{c}\mathcal{N}\left(\frac{\mathrm{d}T}{\mathrm{d}z}\right)_0 \quad (24.31)$$

The energy flux is proportional to the temperature gradient, as we wanted to show. Comparison of this equation with eqn 21 shows that

$$\kappa = \tfrac{1}{3}\nu k\lambda\bar{c}\mathcal{N} \quad (24.32)$$

Equation 28 then follows from $C_{V,\mathrm{m}} = \nu kN_\mathrm{A}$ for a perfect gas, where [A] is the molar concentration of A. For this step, we use $\mathcal{N} = N/V = nN_\mathrm{A}/V = N_\mathrm{A}[\mathrm{A}]$.

Because λ is inversely proportional to the pressure, and hence inversely proportional to the molar concentration of the gas, it follows from eqn 28 that the thermal conductivity is independent of the pressure. The physical reason for this independence is that the thermal conductivity can be expected to be large when many molecules are available to transport the energy, but the presence of so many molecules limits their mean free path and they cannot carry the energy over a great distance. These two effects balance. The thermal conductivity is indeed found experimentally to be independent of the pressure, except when the pressure is very low, when $\kappa \propto p$. At low pressures λ exceeds the dimensions of the apparatus, and the distance over which the energy is transported is determined by the size of the container and not by the other molecules present. The flux is still proportional to the number of carriers, but the length of the journey no longer depends on λ, so $\kappa \propto [\mathrm{A}]$, which implies that $\kappa \propto p$.

(d) The viscosity of a perfect gas

We have seen that viscosity is related to the flux of momentum. As shown in the *Justification* below, the expression obtained from the kinetic theory of gases is

$$\eta = \tfrac{1}{3}M\lambda\bar{c}[\mathrm{A}] \quad (24.33)^\circ$$

where [A] is the molar concentration of the gas molecules and M is their molar mass.

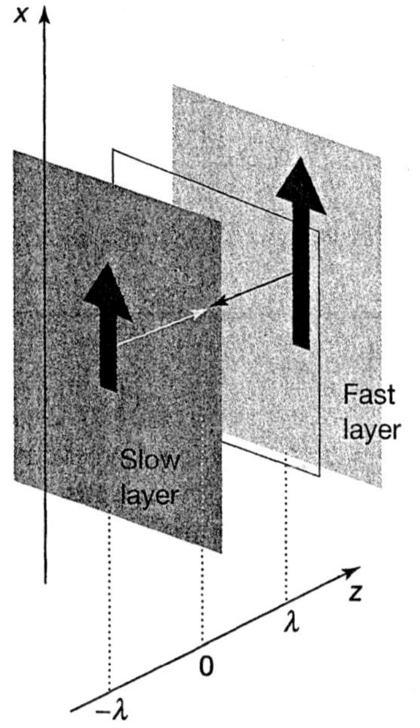

24.14 The calculation of the viscosity of a gas examines the net x-component of momentum brought to a plane from faster and slower layers on average a mean free path away in each direction.

Justification 24.7

Molecules travelling from the right in Fig. 24.14 (from a fast layer to a slower one) transport a momentum $mv_x(\lambda)$ to their new layer at $z = 0$; those travelling from the left transport $mv_x(-\lambda)$ to it. If it is assumed that the density is uniform, the collision flux is $\frac{1}{4}\mathcal{N}\bar{c}$. Those arriving from the right on average carry a momentum

$$mv_x(\lambda) = mv_x(0) + m\lambda\left(\frac{\mathrm{d}v_x}{\mathrm{d}z}\right)_0$$

Those arriving from the left bring a momentum

$$mv_x(-\lambda) = mv_x(0) - m\lambda\left(\frac{\mathrm{d}v_x}{\mathrm{d}z}\right)_0$$

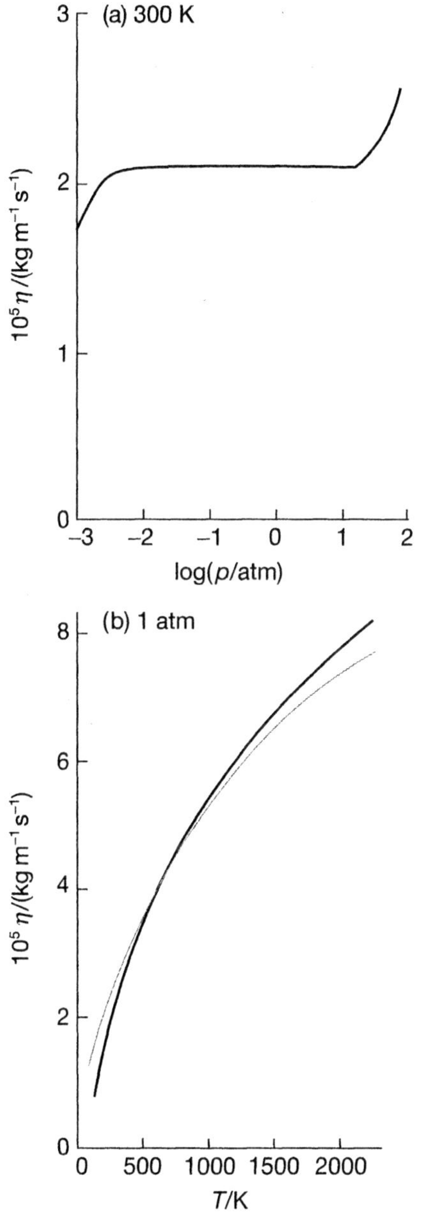

24.15 The experimental results for (a) the pressure dependence of the viscosity of argon, and (b) its temperature dependence. The broken line in the latter is the calculated value. Fitting the observed and calculating curves is one way of determining the collision cross-section.

The net flux of x-momentum in the z-direction is therefore

$$J = \tfrac{1}{4}\mathcal{N}\bar{c}\left\{\left[mv_x(0) - m\lambda\left(\frac{dv_x}{dz}\right)_0\right] - \left[mv_x(0) - m\lambda\left(\frac{dv_x}{dz}\right)_0\right]\right\}$$
$$= -\tfrac{1}{2}\mathcal{N}m\lambda\bar{c}\left(\frac{dv_x}{dz}\right)_0$$

The flux is proportional to the velocity gradient, as we wished to show. Comparison of this expression with eqn 22, and multiplication by $\frac{2}{3}$ in the normal way, leads to

$$\eta = \tfrac{1}{3}\mathcal{N}m\lambda\bar{c} \qquad (24.34)$$

which can easily be converted into eqn 33.

The viscosity is independent of the pressure: $\lambda \propto 1/p$ and $[A] \propto p$, implying that $\eta \propto \bar{c}$, independent of p. The physical reason is the same as for the thermal conductivity: more molecules are available to transport the momentum, but they carry it less far on account of the decrease in mean free path. An unexpected result is that, because $\bar{c} \propto T^{1/2}$, the viscosity coefficient is proportional to $T^{1/2}$. That is, the viscosity of a gas *increases* with temperature. This conclusion is explained when we remember that at high temperatures the molecules travel more quickly, so the flux of momentum is greater.[2]

There are two main techniques for measuring viscosities of gases. One technique depends on the rate of damping of the torsional oscillations of a disk hanging in the gas. The half-life of the decay of the oscillation depends on the viscosity and the design of the apparatus, and the apparatus needs to be calibrated. The other method is based on **Poiseuille's formula** for the rate of flow of a fluid through a tube of radius r:

$$\frac{dV}{dt} = \frac{(p_1^2 - p_2^2)\pi r^4}{16l\eta p_0} \qquad (24.35)$$

where V is the volume flowing, p_1 and p_2 are the pressures at each end of the tube of length l, and p_0 is the pressure at which the volume is measured.

Such measurements confirm that the viscosities of gases are independent of pressure over a wide range. For instance, the results for argon from 10^{-3} atm to 10^2 atm are shown in Fig. 24.15, and we see that η is constant from about 0.01 atm to 20 atm. The measurements also confirm (to a lesser extent) the $T^{1/2}$ dependence. The dotted line in the illustration shows the calculated values using $\sigma = 22 \times 10^{-20}\ \text{m}^2$, implying a collision diameter of 260 pm, in contrast to the van der Waals diameter of 335 pm obtained from the density of the solid. The agreement is not too bad, considering the simplicity of the model, especially the neglect of intermolecular forces.

Illustration 24.1

In a Poiseuille flow experiment to measure the viscosity of air at 298 K, the sample was allowed to flow through a tube of length 100 cm and internal diameter 1.00 mm. The high-pressure end was at 765 Torr and the low-pressure end was at 760 Torr. The volume was measured at the latter pressure.

2. As we shall see in Section 24.5, the viscosity of a liquid *decreases* with increase in temperature because intermolecular interactions must be overcome.

In 100 s, 90.2 cm^3 of air passed through the tube. The viscosity of air at 298 K is found by reorganizing the Poiseuille formula, eqn 35, into

$$\eta = \frac{(p_1^2 - p_2^2)\pi r^4}{16lp_0(\mathrm{d}V/\mathrm{d}t)}$$

and substituting the data (after converting the pressures to pascals by using 1 Torr = 133.3 Pa):

$$\eta = \frac{\{(765 \times 133.3\ \mathrm{Pa})^2 - (760 \times 133.3\ \mathrm{Pa})^2\} \times \pi \times (5.00 \times 10^{-4}\ \mathrm{m})^4}{16 \times (1.00 \times 10^{-1}\ \mathrm{m}) \times (760 \times 133.3\ \mathrm{Pa}) \times \left(\dfrac{9.02 \times 10^{-5}\ \mathrm{m}^3}{100\ \mathrm{s}}\right)}$$

$$= 1.82 \times 10^{-4}\ \mathrm{kg\ m^{-1}\ s^{-1}}$$

where we have used 1 Pa = 1 kg m^{-1} s^{-2}. The kinetic theory expression gives $\eta = 1.4 \times 10^{-5}$ kg m^{-1} s^{-1}, so the agreement is quite good. Viscosities are commonly expressed in centipoise (cP) or (for gases) micropoise (μP), the conversion being 1 cP = 10^{-3} kg m^{-1} s^{-1}; the viscosity of air at 20°C is about 180 μP.

Self-test 24.4 What volume would be collected if the pressure gradient were doubled, other conditions remaining constant? [181 cm^3]

Molecular motion in liquids

We outlined what is currently known about the structure of simple liquids in Section 21.8. Here we consider a particularly simple type of motion through a liquid, that of an ion, and see that the information that motion provides can be used to infer the behaviour of uncharged species too.

24.5 Experimental results

The motion of molecules in liquids can be studied experimentally by a variety of methods. Relaxation time measurements in NMR and ESR (Section 18.5b) can be interpreted in terms of the mobilities of the molecules, and have been used to show that big molecules in viscous fluids typically rotate in a series of small (about 5°) steps, whereas small molecules in nonviscous fluids typically jump through about 1 radian (57°) in each step. Another important technique is **inelastic neutron scattering**, in which the energy neutrons collect or discard as they pass through a sample is interpreted in terms of the motion of its particles. The same technique is used to examine the internal dynamics of macromolecules.

More mundane than these experiments are viscosity measurements (Table 24.4). For a molecule to move in a liquid, it must acquire at least a minimum energy to escape from its neighbours. The probability that a molecule has at least an energy E_a is proportional to $e^{-E_a/RT}$, so the mobility of the molecules in the liquid should follow this type of temperature dependence. Because the coefficient of viscosity, η, is inversely proportional to the mobility of the particles, we should expect that

$$\eta \propto e^{E_a/RT} \qquad (24.36)$$

(Note the positive sign of the exponent.) This expression implies that the viscosity should decrease sharply with increasing temperature. Such a variation is found

Table 24.4* Viscosities of liquids at 298 K, $\eta/(10^{-3}$ kg m^{-1} s$^{-1})$

Benzene	0.601
Mercury	1.55
Pentane	0.224
Water†	0.891

* More values are given in the *Data section*.
† The viscosity of water corresponds to 0.891 cP.

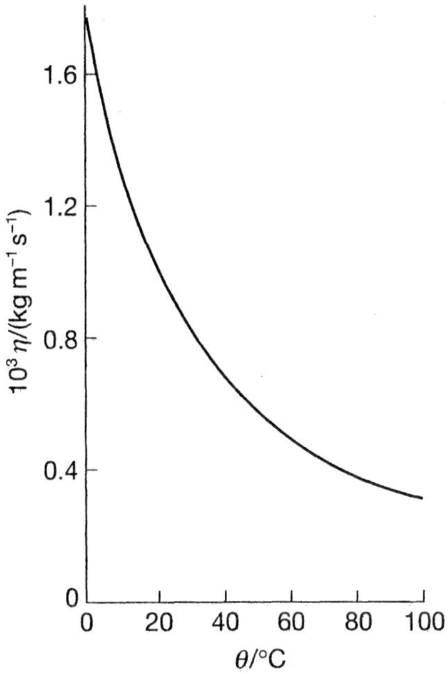

24.16 The experimental temperature dependence of the viscosity of water. As the temperature is increased, more molecules are able to escape from the potential wells provided by their neighbours, and so the liquid becomes more fluid. A plot of ln η against $1/T$ is a straight line (over a small range) with positive slope.

experimentally, at least over reasonably small temperature ranges (Fig. 24.16). The activation energy typical of viscosity is comparable to the mean potential energy of intermolecular interactions.

One problem with the interpretation of viscosity measurements is that the change in density of the liquid as it is heated makes a pronounced contribution to the temperature variation of the viscosity. Thus, the temperature dependence of viscosity at constant volume, when the density is constant, is much less than that at constant pressure. The intermolecular interactions between the molecules of the liquid govern the magnitude of E_a, but the problem of calculating it is immensely difficult and still largely unsolved. At low temperatures, the viscosity of water decreases as the pressure is increased. This behaviour is consistent with the rupture of hydrogen bonds.

24.6 The conductivities of electrolyte solutions

Further insight into the nature of molecular motion can be obtained by studying the motion of ions in solution, for ions can be dragged through the solvent by the application of a potential difference between two electrodes immersed in the sample. By studying the transport of charge through electrolyte solutions it is possible to build up a picture of the events that occur in them and, in some cases, to extrapolate the conclusions to species that have zero charge; that is, to neutral molecules.

(a) Conductance and conductivity

The fundamental measurement used to study the motion of ions is that of the electrical resistance, R, of the solution. The **conductance**, G, of a solution is the inverse of its resistance R: $G = 1/R$. As resistance is expressed in ohms, Ω, the conductance of a sample is expressed in Ω^{-1}. The reciprocal ohm used to be called the mho, but its official designation is now the siemens, S, and $1\text{ S} = 1\,\Omega^{-1} = 1\text{ C V}^{-1}\text{ s}^{-1}$. The conductance of a sample decreases with its length l and increases with its cross-sectional area A. We therefore write

$$G = \frac{\kappa A}{l} \tag{24.37}$$

where κ is the **conductivity**. With the conductance in siemens and the dimensions in metres, it follows that the SI units of κ are siemens per metre (S m^{-1}).

The conductivity of a solution depends on the number of ions present, and it is normal to introduce the **molar conductivity**, Λ_m, which is defined as

$$\Lambda_m = \frac{\kappa}{c} \qquad [24.38]$$

where c is the molar concentration of the added electrolyte. The SI unit of molar conductivity is siemens metre-squared per mole (S m^2 mol^{-1}), and typical values are about 10 mS m^2 mol^{-1} (where 1 mS = 10^{-3} S).

The molar conductivity is found to vary with the concentration. One reason for this variation is that the number of ions in the solution might not be proportional to the concentration of the electrolyte. For instance, the concentration of ions in a solution of a weak acid depends on the concentration of the acid in a complicated way, and doubling the concentration of the acid added does not double the number of ions. Secondly, because ions interact strongly with one another, the conductivity of a solution is not exactly proportional to the number of ions present.

The concentration dependence of molar conductivities indicates that there are two classes of electrolyte. The characteristic of a **strong electrolyte** is that its molar conductivity depends only slightly on the molar concentration (and in general decreases slightly as the concentration is increased, Fig. 24.17). The characteristic

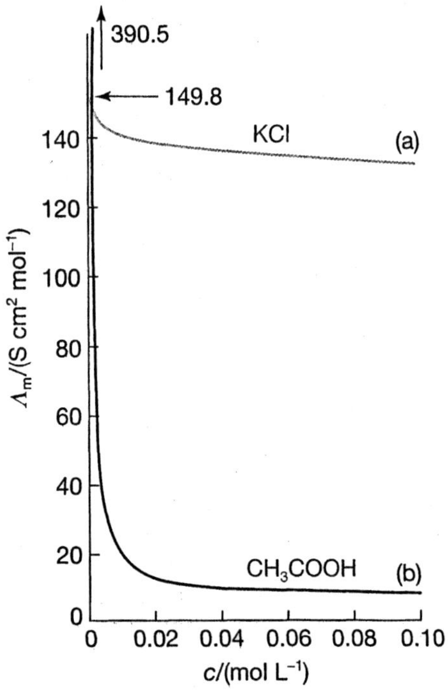

24.17 The concentration dependence of the molar conductivities of (a) a typical strong electrolyte (aqueous potassium chloride) and (b) a typical weak electrolyte (aqueous acetic acid).

of a **weak electrolyte** is that its molar conductivity is normal at concentrations close to zero, but falls sharply to low values as the concentration increases. The classification depends on the solvent employed as well as the solute: lithium chloride, for example, is a strong electrolyte in water but a weak electrolyte in propanone.

(b) Strong electrolytes

Strong electrolytes are substances that are virtually fully ionized in solution, and include ionic solids and strong acids. As a result of their complete ionization, the concentration of ions in solution is proportional to the concentration of strong electrolyte added.

In an extensive series of measurements during the nineteenth century, Friedrich Kohlrausch showed that, at low concentrations, the molar conductivities of strong electrolytes vary linearly with the square root of the concentration:

$$\Lambda_m = \Lambda_m^\circ - \mathcal{K}c^{1/2} \tag{24.39}$$

This variation is called **Kohlrausch's law**. The constant Λ_m° is the **limiting molar conductivity**, the molar conductivity in the limit of zero concentration (when the ions are effectively infinitely far apart and do not interact with one another). The constant $\mathcal{K}$ is found to depend more on the stoichiometry of the electrolyte (that is, whether it is of the form MA, or M_2A, etc.) than on its specific identity. In due course we shall see that the $c^{1/2}$ dependence arises from interactions between ions: when charge is conducted ionically, ions of one charge are moving past the ions of interest and retard its progress.

Kohlrausch was also able to show that Λ_m° can be expressed as the sum of contributions from its individual ions. If the limiting molar conductivity of the cations is denoted λ_+ and that of the anions λ_-, then his **law of the independent migration of ions** states that

$$\Lambda_m^\circ = \nu_+\lambda_+ + \nu_-\lambda_- \tag{24.40}^\circ$$

where ν_+ and ν_- are the numbers of cations and anions per formula unit of electrolyte (for example, $\nu_+ = \nu_- = 1$ for HCl, NaCl, and $CuSO_4$, but $\nu_+ = 1$, $\nu_- = 2$ for $MgCl_2$). This simple result, which can be understood on the grounds that the ions migrate independently in the limit of zero concentration, lets us predict the limiting molar conductivity of any strong electrolyte from the data in Table 24.5.

Illustration 24.2

The limiting molar conductivity of $BaCl_2$ in water at 298 K is

$$\Lambda_m^\circ = (12.72 + 2 \times 7.63)\ \text{mS m}^2\ \text{mol}^{-1} = 27.98\ \text{mS m}^2\ \text{mol}^{-1}$$

Table 24.5* Limiting ionic conductivities in water at 298 K, λ/(mS m^2 mol^{-1})

H^+	34.96	OH^-	19.91
Na^+	5.01	Cl^-	7.63
K^+	7.35	Br^-	7.81
Zn^{2+}	10.56	SO_4^{2-}	16.00

* More values are given in the *Data section*.

(c) Weak electrolytes

Weak electrolytes are not fully ionized in solution. They include weak Brønsted acids and bases, such as CH_3COOH and NH_3. The marked concentration dependence of their molar conductivities arises from the displacement of the equilibrium

$$\text{HA(aq)} + \text{H}_2\text{O(l)} \rightleftharpoons \text{H}_3\text{O}^+\text{(aq)} + \text{A}^-\text{(aq)} \qquad K_a = \frac{a_{\text{H}_3\text{O}^+}a_{\text{A}^-}}{a_{\text{HA}}} \tag{24.41}$$

towards products at low molar concentrations.

The conductivity depends on the number of ions in the solution, and therefore on the **degree of ionization**, α, of the electrolyte; when referring to weak acids, we speak instead of the **degree of deprotonation**. It is defined so that, for the acid HA at a molar concentration c, at equilibrium

$$[H_3O^+] = \alpha c \qquad [A^-] = \alpha c \qquad [HA] = (1 - \alpha)c \tag{24.42}$$

If we ignore activity coefficients, the acidity constant, K_a, is approximately

$$K_a = \frac{\alpha^2 c}{1 - \alpha} \tag*{(24.43)°}$$

from which it follows that

$$\alpha = \frac{K_a}{2c}\left\{\left(1 + \frac{4c}{K_a}\right)^{1/2} - 1\right\} \tag*{(24.44)°}$$

The acid is fully deprotonated at infinite dilution, and its molar conductivity is then Λ_m°. Because only a fraction α is actually present as ions in the actual solution, the measured molar conductivity Λ_m is given by

$$\Lambda_m = \alpha \Lambda_m^\circ \tag*{(24.45)°}$$

with α given by eqn 44.

Illustration 24.3

The molar conductivity of 0.0100 M $CH_3COOH(aq)$ at 298 K is $\Lambda_m = 1.65\ \mathrm{mS\ m^2\ mol^{-1}}$. The degree of deprotonation, α, is calculated from eqn 45 with $\Lambda_m^\circ = 39.05\ \mathrm{mS\ m^2\ mol^{-1}}$ (Table 24.5). It follows that $\alpha = 0.0423$. The acidity constant, K_a, is calculated by substitution of α into eqn 43, which gives $K_a = 1.9 \times 10^{-5}$.

Self-test 24.5 The molar conductivity of 0.0250 M HCOOH(aq) is $4.61\ \mathrm{mS\ m^2\ mol^{-1}}$. Determine the p$K_a$ of the acid. [3.44]

Once we know K_a, we can use eqns 44 and 45 to predict the concentration dependence of the molar conductivity. The result agrees quite well with the experimental curve in Fig. 24.17. More usefully, we can use the concentration dependence of Λ_m in measurements of the limiting molar conductance. First, we rearrange eqn 43 into

$$\frac{1}{\alpha} = 1 + \frac{\alpha c}{K_a} \tag*{(24.46)°}$$

Then, by using eqn 45, we obtain **Ostwald's dilution law**:

$$\frac{1}{\Lambda_m} = \frac{1}{\Lambda_m^\circ} + \frac{\Lambda_m c}{K_a(\Lambda_m^\circ)^2} \tag*{(24.47)°}$$

This equation implies that, if $1/\Lambda_m$ is plotted against $c\Lambda_m$, then the intercept at $c = 0$ will be $1/\Lambda_m^\circ$ (Fig. 24.18).

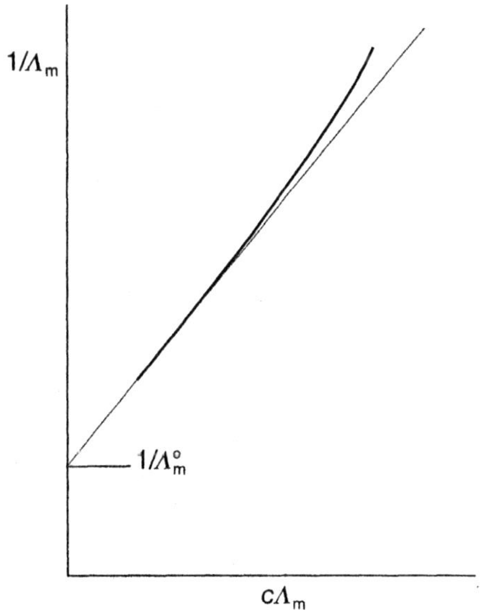

24.18 The graph used to determine the limiting value of the molar conductivity of a solution by extrapolation to zero concentration.

24.7 The mobilities of ions

To interpret conductivity measurements we need to know why ions move at different rates, why they have different molar conductivities, and why the molar conductivities of strong electrolytes decrease with the square root of the molar concentration. The central idea in this section is that, although the motion of an ion remains largely random, the presence of an electric field biases its motion, and the ion undergoes net migration through the solution.

(a) The drift speed

When the potential difference between two electrodes a distance l apart is $\Delta\phi$, the ions in the solution between them experience a uniform electric field of magnitude

$$\mathcal{E} = \frac{\Delta\phi}{l} \tag{24.48}$$

In such a field, an ion of charge[3] ze experiences a force of magnitude

$$\mathcal{F} = ze\mathcal{E} = \frac{ze\Delta\phi}{l} \tag{24.49}$$

A cation responds to the application of the field by accelerating towards the negative electrode and an anion responds by accelerating towards the positive electrode. However, this acceleration is short lived. As the ion moves through the solvent it experiences a frictional retarding force, $\mathcal{F}_{\text{fric}}$, proportional to its speed. If we assume that the Stokes formula (eqn 22.34) for a sphere of radius a and speed s applies even on a microscopic scale (and independent evidence from magnetic resonance suggests that it often gives at least the right order of magnitude), then we can write this retarding force as

$$\mathcal{F}_{\text{fric}} = fs \qquad f = 6\pi\eta a \tag{24.50}$$

The two forces act in opposite directions, and the ions quickly reach a terminal speed, the **drift speed**, when the accelerating force is balanced by the viscous drag. The net force is zero when

$$s = \frac{ze\mathcal{E}}{f} \tag{24.51}$$

It follows that the drift speed of an ion is proportional to the strength of the applied field. We write

$$s = u\mathcal{E} \tag*{[24.52]}$$

where u is called the **mobility** of the ion (Table 24.6). Comparison of eqns 51 and 52 and use of eqn 50 shows that

$$u = \frac{ze}{f} = \frac{ze}{6\pi\eta a} \tag{24.53}$$

Table 24.6* Ionic mobilities in water at 298 K, $u/(10^{-8}\ \text{m}^2\ \text{s}^{-1}\ \text{V}^{-1})$

H^+	36.23	OH^-	20.64
Na^+	5.19	Cl^-	7.91
K^+	7.62	Br^-	8.09
Zn^{2+}	5.47	SO_4^{2-}	8.29

* More values are given in the *Data section*.

Illustration 24.4

For an order of magnitude estimate we can take $z = 1$ and a the radius of an ion such as Cs^+ (which might be typical of a smaller ion plus its hydration sphere), which is 170 pm. For the viscosity, we use $\eta = 1.0$ cP (1.0×10^{-3} kg m^{-1} s^{-1}, Table 24.4). Then $u \approx 5 \times 10^{-8}$ m^2 V^{-1} s^{-1}. This value means that, when there is a potential difference of 1 V across a solution of length 1 cm (so $\mathcal{E} = 100$ V m^{-1}), the drift speed is typically about 5 μm s^{-1}. That speed might seem slow, but not when expressed on a molecular scale, for it corresponds to an ion passing about 10^4 solvent molecules per second.

3. In this chapter we disregard the sign of the charge number and so avoid notational complications.

Because the drift speed governs the rate at which charge is transported, we might expect the conductivity to decrease with increasing solution viscosity and ion size. Experiments confirm these predictions for bulky ions (such as R_4N^+ and RCO_2^-) but not for small ions. For example, the molar conductivities of the alkali metal ions increase from Li^+ to Cs^+ (Table 24.6) even though the ionic radii increase. The paradox is resolved when we realize that the radius a in the Stokes formula is the **hydrodynamic radius** (or 'Stokes radius') of the ion, its effective radius in the solution taking into account all the H_2O molecules it carries in its hydration sphere. Small ions give rise to stronger electric fields than large ones,[4] so small ions are more extensively solvated than big ions. Thus, an ion of small ionic radius may have a large hydrodynamic radius because it drags many solvent molecules through the solution as it migrates. The hydrating H_2O molecules are often very labile, however, and NMR and isotope studies have shown that the exchange between the coordination sphere of the ion and the bulk solvent is very rapid.

The proton, although it is very small, has a very high molar conductivity (Table 24.6)! Proton and ^{17}O-NMR show that the times characteristic of protons hopping from one molecule to the next are about 1.5 ps, which is comparable to the time that inelastic neutron scattering shows it takes a water molecule to reorientate through about 1 rad (1 to 2 ps). According to the **Grotthuss mechanism**,[5] there is an effective motion of a proton that involves the rearrangement of bonds in a group of water molecules. However, the actual mechanism is still highly contentious. Attention now focuses on the $H_9O_4^+$ unit, in which the nearly trigonal planar H_3O^+ ion[6] is linked to three strongly solvating H_2O molecules. This cluster of atoms is itself hydrated, but the hydrogen bonds in the secondary sphere are weaker than in the primary sphere. It is envisaged that the rate-determining step is the cleavage of one of the weaker hydrogen bonds of this secondary sphere (Fig. 24.19a). After this bond cleavage has taken place, and the released molecule has rotated through a few degrees (a process that takes about 1 ps), there is a rapid adjustment of bond lengths and angles in the remaining cluster, to form an $H_5O_2^+$ cation of structure $H_2O \cdots H^+ \cdots OH_2$ (Fig. 24.19b). Shortly after this reorganization has occurred, a new $H_9O_4^+$ cluster forms as other molecules rotate into a position where they can become members of a secondary hydration sphere, but now the positive charge is located one molecule to the right of its initial location (Fig. 24.19c). According to this model, there is no coordinated motion of a proton along a chain of molecules, simply a very rapid hopping between neighbouring sites, with a low activation energy. The model is consistent with the observation that the molar conductivity of protons increases as the pressure is raised, for increasing pressure ruptures the hydrogen bonds in water. The mobility of NH_4^+ in liquid ammonia is also anomalous and presumably occurs by an analogous mechanism.

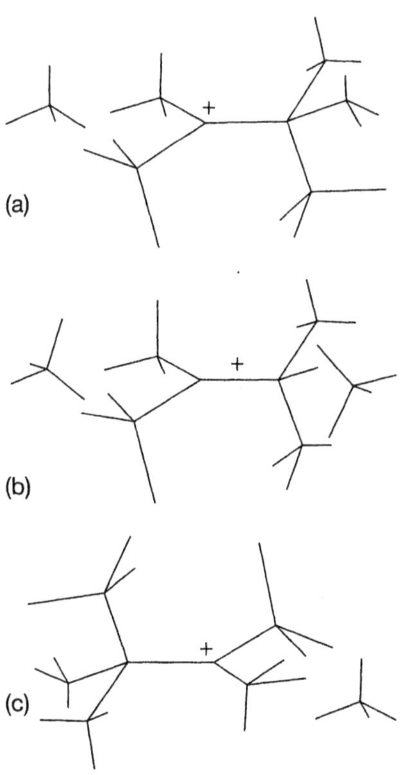

24.19 The mechanism of conduction by water as proposed by N. Agmon (*Chem. Phys. Letts.*, **244**, 456 (1995)). Proton transfer between neighbouring molecules occurs when one molecule rotates into such a position that an O—H···O hydrogen bond can flip into being an O···H—O hydrogen bond. See text for a description of the steps.

(b) Mobility and conductivity

Ionic mobilities provide a link between measurable and theoretical quantities. As a first step we establish in the *Justification* below the following relation between an ion's mobility and its molar conductivity:

$$\lambda = zuF \qquad (24.54)^\circ$$

where F is the Faraday constant ($F = N_A e$).

4. The electric field at the surface of a sphere of radius r is proportional to ze/r^2, so the smaller the radius the stronger the field.

5. The name of the mechanism is an allusion to an early view advanced by von Grotthuss, in which it was supposed that chains of dipoles were responsible for the transport of charge in water.

6. In the gas phase, H_3O^+ is trigonal pyramidal.

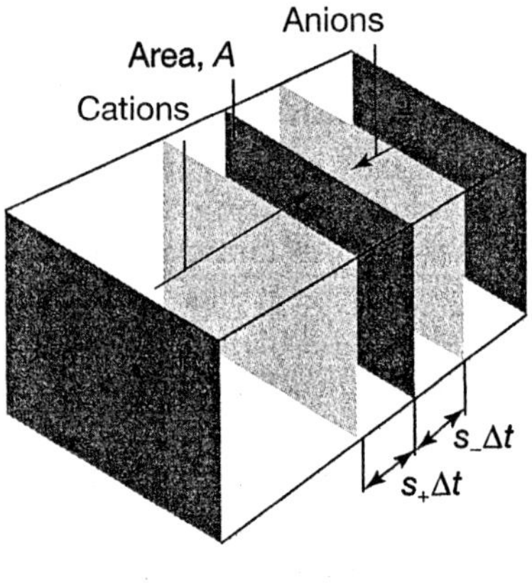

24.20 In the calculation of the current, all the cations within a distance $s_+\Delta t$ (that is, those in the volume $s_+A\Delta t$) will pass through the area A. The anions in the corresponding volume the other side of the window will also contribute to the current similarly.

Justification 24.8

To keep the calculation simple, we ignore signs in the following, and concentrate on the magnitudes of quantities: the direction of ion flux can always be decided by common sense.

Consider a solution of a fully dissociated strong electrolyte at a molar concentration c. Let each formula unit give rise to ν_+ cations of charge z_+e and ν_- anions of charge z_-e. The molar concentration of each type of ion is therefore νc (with $\nu = \nu_+$ or ν_-), and the number density of each type is νcN_A. The number of ions of one kind that pass through an imaginary window of area A during an interval Δt is equal to the number within the distance $s\Delta t$ (Fig. 24.20), and therefore to the number in the volume $s\Delta tA$. (The same sort of argument was used in Section 24.1 in the discussion of the pressure of a gas.) The number of ions of that kind in this volume is equal to $s\Delta tA\nu cN_A$. The flux through the window (the number of this type of ion passing through the window divided by the area of the window and the duration of the interval) is therefore

$$J(\text{ions}) = \frac{s\Delta tA\nu cN_A}{A\Delta t} = s\nu cN_A$$

Each ion carries a charge ze, so the flux of charge is

$$J(\text{charge}) = zs\nu ceN_A = zs\nu cF$$

Because $s = u\mathcal{E}$, the flux is

$$J(\text{charge}) = zu\nu cF\mathcal{E}$$

The current, I, through the window due to the ions we are considering is the charge flux times the area:

$$I = JA = zu\nu cF\mathcal{E}A$$

Because the electric field is the potential gradient, $\Delta\phi/l$, we can write

$$I = \frac{zu\nu cFA\Delta\phi}{l} \tag{24.55}$$

Current and potential difference are related by Ohm's law, that

$$I = \frac{\Delta\phi}{R} = G\Delta\phi = \frac{\kappa A\Delta\phi}{l}$$

where we have used eqn 37 ($G = \kappa A/l$). Note that the proportionality of current to potential difference ($I \propto \Delta\phi$) is another example of a phenomenological flux equation like those introduced in Section 24.4a. Comparison of the last two expressions gives $\kappa = zu\nu cF$. Division by the molar concentration of ions, νc, then results in eqn 54.

Equation 54 applies to the cations and to the anions. Therefore, for the solution itself in the limit of zero concentration (when there are no interionic interactions),

$$\Lambda_m^\circ = (z_+u_+\nu_+ + z_-u_-\nu_-)F \tag{24.56°}$$

For a symmetrical z:z electrolyte (for example, $CuSO_4$ with $z = 2$), this equation simplifies to

$$\Lambda_m^\circ = z(u_+ + u_-)F \tag{24.57°}$$

Illustration 24.5

Earlier, we estimated the typical ionic mobility as 5×10^{-8} m^2 V^{-1} s^{-1}; so, with $z = 1$ for both the cation and anion, we can estimate that a typical limiting molar conductivity should be about 10 mS m^2 mol^{-1}, in accord with experiment. The experimental value for KCl, for instance, is 15 mS m^2 mol^{-1}.

(c) Transport numbers

The **transport number**, $t_\pm$, is defined as the fraction of total current carried by the ions of a specified type. For a solution of two kinds of ion, the transport numbers of the cations (t_+) and anions (t_-) are

$$t_\pm = \frac{I_\pm}{I} \qquad [24.58]$$

where $I_\pm$ is the current carried by the cation (I_+) or anion (I_-) and I is the total current through the solution. Because the total current is the sum of the cation and anion currents, it follows that

$$t_+ + t_- = 1 \qquad (24.59)$$

The **limiting transport number**, $t_\pm^\circ$, is defined in the same way but for the limit of zero concentration of the electrolyte solution. We shall consider only these limiting values from now on, for that avoids the problem of ionic interactions.

The current that can be ascribed to each type of ion is related to the mobility of the ion by eqn 55. Hence the relation between $t_\pm^\circ$ and $u_\pm$ is

$$t_\pm^\circ = \frac{z_\pm \nu_\pm u_\pm}{z_+\nu_+u_+ + z_-\nu_-u_-} \qquad (24.60)^\circ$$

However, because $z_+\nu_+ = z_-\nu_-$ for any electrolyte, eqn 60 simplifies to

$$t_\pm^\circ = \frac{u_\pm}{u_+ + u_-} \qquad (24.61)^\circ$$

Moreover, because the ionic conductivities are related to the mobilities by eqn 54, it also follows from eqn 60 that

$$t_\pm^\circ = \frac{\nu_\pm\lambda_\pm}{\nu_+\lambda_+ + \nu_-\lambda_-} = \frac{\nu_\pm\lambda_\pm}{\Lambda_m^\circ} \qquad (24.62)^\circ$$

and hence, for each type of ion,

$$\nu_\pm\lambda_\pm = t_\pm^\circ \Lambda_m^\circ \qquad (24.63)^\circ$$

Consequently, because there are independent ways of measuring transport numbers of ions, we can determine the individual ionic conductivities and (through eqn 54) the ionic mobilities.

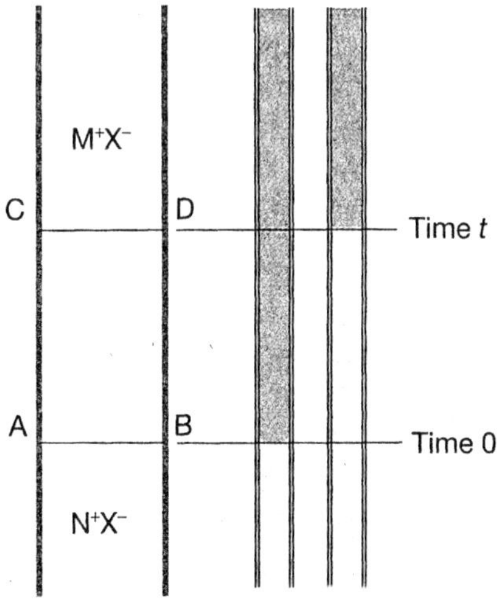

24.21 In the moving boundary method for the measurement of transport numbers the distance moved by the boundary is observed as a current is passed. All the M ions in the volume between AB and CD must have passed through CD if the boundary moves from AB to CD.

(d) The measurement of transport numbers

One of the most accurate methods for measuring transport numbers is the **moving boundary method**, in which the motion of a boundary between two ionic solutions having a common ion is observed as a current flows.

Let MX be the salt of interest and NX a salt giving a denser solution. The solution of NX is called the **indicator solution**; it occupies the lower part of a vertical tube of cross-sectional area A (Fig. 24.21). The MX solution, which is called the **leading solution**, occupies the upper part of the tube. There is a sharp boundary between the two solutions. The indicator solution must be denser than the leading solution, and the mobility of the M ions must be greater than that of the N ions.[7] Thus, if any M ions diffuse into the lower solution, they will be pulled upwards more rapidly than the N ions around them, and the boundary will reform. The interpretation of the experiment makes use of the relation (see the *Justification* below) between the distance, l, moved by the boundary in the time Δt for which a current I is passed for a period Δt:

$$t_+ = \frac{z_+ clAF}{I\Delta t} \qquad (24.64)$$

Hence, by measuring the distance moved, the transport number and hence the conductivity and mobility of the ions can be determined.

Justification 24.9

When a current I is passed for a time Δt, the boundary moves from AB to CD, so all the M ions in the volume between AB and CD must have passed through CD. That number is $clAN_A$, so the charge that the M ions transfer through the plane is $z_+ clAeN_A$. However, the *total* charge transferred when a current I flows for an interval Δt is $I\Delta t$. Therefore, the fraction due to the motion of the M ions, which is their transport number, is given by eqn 64.

In the **Hittorf method**, an electrolytic cell is divided into three compartments and a charge $I\Delta t$ is passed. An amount $I\Delta t/z_+F$ of cations is discharged at the cathode but an amount $t_+ I\Delta t/z_+F$ of cations migrates into the cathode compartment. The net change in the amount of cations in that compartment is therefore

$$\text{Net change} = (t_+ - 1)\frac{I\Delta t}{z_+F} = -t_-\frac{I\Delta t}{z_+F} \qquad (24.65)$$

Hence, by measuring the change of composition in the cathode compartment, the anion transport number t_- can be deduced. Likewise, the change in composition of the anode compartment is $-t_+ I\Delta t/z_-F$, which gives the cation transport number, t_+.

Transport numbers may also be measured by using galvanic cells. In particular, the measurement is made on a **cell with transference**, which is a galvanic cell with a liquid boundary across which ions may pass from one electrode compartment to the other. An example is the cell

$$\mathrm{Ag(s)|AgCl(s)|HCl}(m_1)\mathrm{|HCl}(m_2)\mathrm{|AgCl(s)|Ag(s)}$$

for which the emf is E_t and the electrodes are reversible with respect to anions (Cl^-). The corresponding cell without transference is

$$\mathrm{Ag(s)|AgCl(s)|HCl}(m_1)\mathrm{|H_2(g)|Pt(s)|H_2(g)|HCl}(m_2)\mathrm{|AgCl(s)|Ag(s)}$$

and its emf is E. We show in the *Justification* below that the two emfs are related by

$$E_t = 2t_+E \qquad (24.66)$$

Therefore, comparison of the two emfs gives the transport number of the *counter ion* of the ion with respect to which the electrodes are reversible (in this case, the transport number of H^+).

7. One procedure is to add bromothymol blue indicator to a slightly alkaline solution of the ion of interest and to use a cadmium electrode at the lower end of the vertical tube. The electrode produces Cd^{2+} ions, which are slow moving and slightly acidic (the hydrated ion is a Brønsted acid), and the boundary is revealed by the colour change of the indicator.

Justification 24.10

The argument is similar to that used to analyse the Hittorf method. Consider the consequences of passing 1 mol of electrons through the cell with transport specified above. In the right-hand electrode compartment, 1 mol Cl^- is formed but t_- mol Cl^- migrates out of it across the junction. The net change is $(1 - t_-)$ mol $= t_+$ mol. At the same time, t_+ mol H^+ migrates into the compartment. In the left-hand electrode compartment 1 mol Cl^- is removed from solution (to form 1 mol AgCl), but t_- mol Cl^- flows in across the junction. The net change is therefore $(-1 + t_-)$ mol $= -t_+$ mol Cl^-. At the same time, t_+ mol H^+ flows out. The reaction Gibbs energy is therefore

$$\Delta_r G = t_+\{\mu(Cl^-, m_2) - \mu(Cl^-, m_1) + \mu(H^+, m_2) - \mu(H^+, m_1)\} = 2t_+ RT \ln \frac{a_2}{a_1}$$

Because $\Delta_r G = -FE$, it follows that

$$E_t = -\frac{2t_+ RT}{F} \ln \frac{a_2}{a_1}$$

For the same cell without transference, the Nernst equation gives

$$E_t = -\frac{RT}{F} \ln \frac{a_2}{a_1}$$

and the ratio of the two cell emfs is $2t_+$.

24.8 Conductivities and ion–ion interactions

The remaining problem is to account for the $c^{1/2}$ dependence of the Kohlrausch law (eqn 39). In Section 10.2c we saw something similar: the activity coefficients of ions at low concentrations also depend on $c^{1/2}$ and depend on their charge type rather than their specific identities. That $c^{1/2}$ dependence was explained in terms of the properties of the ionic atmosphere around each ion, and we can suspect that the same explanation applies here too.

To accommodate the effect of motion, we need to modify the picture of an ionic atmosphere as a spherical haze of charge. Because the ions forming the atmosphere do not adjust to the moving ion immediately, the atmosphere is incompletely formed in front of the moving ion and incompletely decayed behind the ion (Fig. 24.22). The overall effect is the displacement of the centre of charge of the atmosphere a short distance behind the moving ion. Because the two charges are opposite, the result is a retardation of the moving ion. This reduction of the ions' mobility is called the **relaxation effect**. A confirmation of the picture is obtained by observing the conductivities of ions at high frequencies, which are greater than at low frequencies: the atmosphere does not have time to follow the rapidly changing direction of motion of the ion, and the effect of the field averages to zero.

The ionic atmosphere has another effect on the motion of the ions. We have seen that the moving ion experiences a viscous drag. When the ionic atmosphere is present this drag is enhanced because the ionic atmosphere moves in an opposite direction to the central ion. The enhanced viscous drag, which is called the **electrophoretic effect**, reduces the mobility of the ions, and hence also reduces their conductivities.

The quantitative formulation of these effects is far from simple, but the **Debye–Hückel–Onsager theory** is an attempt to obtain quantitative expressions at about

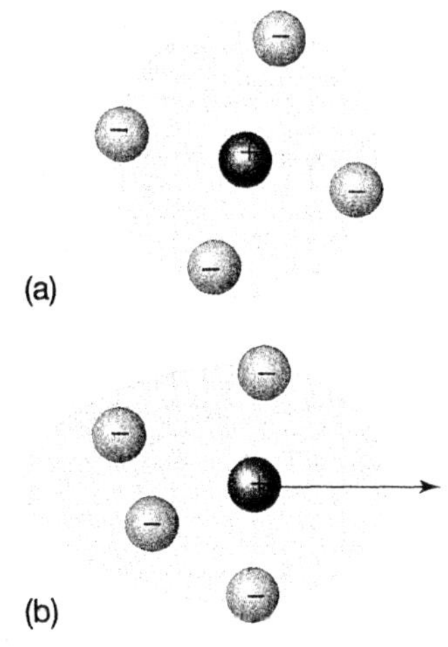

24.22 (a) In the absence of an applied field, the ionic atmosphere is spherically symmetric, but (b) when a field is present it is distorted and the centres of negative and positive charge no longer coincide. The attraction between the opposite charges retards the motion of the central ion.

Table 24.7* Debye–Hückel–Onsager coefficients for (1,1)-electrolytes at 298 K

Solvent	$A/(\text{mS m}^2\ \text{mol}^{-1}/(\text{mol L}^{-1})^{1/2})$	$B/(\text{mol L}^{-1})^{-1/2}$
Methanol	15.61	0.923
Propanone	32.8	1.63
Water	6.02	0.229

* More values are given in the *Data section*.

the same level of sophistication as the Debye–Hückel theory itself. The theory leads to a Kohlrausch-like expression in which

$$\mathcal{K} = A + B\Lambda_m^\circ \tag{24.67}$$

with

$$A = \frac{z^2eF^2}{3\pi\eta}\left(\frac{2}{\varepsilon RT}\right)^{1/2} \qquad B = \frac{qz^3eF}{24\pi\varepsilon RT}\left(\frac{2}{\varepsilon RT}\right)^{1/2} \tag{24.68}$$

where ε is the electric permittivity of the solvent (Section 22.1e) and $q = 0.586$ for a 1,1-electrolyte (Table 24.7). The slopes of the conductivity curves are predicted to depend on the charge type of the electrolyte, in accord with the Kohlrausch law, and some comparisons between theory and experiment are shown in Fig. 24.23. The agreement is quite good at very low molar concentrations (less than about 10^{-3} M, depending on the charge type).

Molecular dynamics calculations can also shed light on electric conduction. The key equation is a special case of a **Green–Kubo relation**, which expresses a transport property in terms of the fluctuations in microscopic properties of a system. The electrical conductivity is related to the fluctuations in the instantaneous electric current, j, in the sample that arises from variations in the velocities of the ions:

$$\kappa = \frac{1}{kTV}\int_0^\infty \langle j(0)j(t)\rangle\,dt \qquad j = \sum_i^N z_i e v_i \tag{24.69}$$

where v_i is the velocity of the ion i at a given instant and the angular brackets denote an average over the sample. If the ions are very mobile, there will be large fluctuations in the instantaneous currents in the sample, and the conductivity of the medium will be high. If the ions are locked into position, as in an ionic solid, there will be no instantaneous currents, and the ionic conductivity will be zero. The velocities of the ions are calculated explicitly in a molecular dynamics simulation, so the **correlation function**, the quantity $\langle j(0)j(t)\rangle$, can be evaluated reasonably simply by taking the average of a large number of such calculations with different initial conditions.

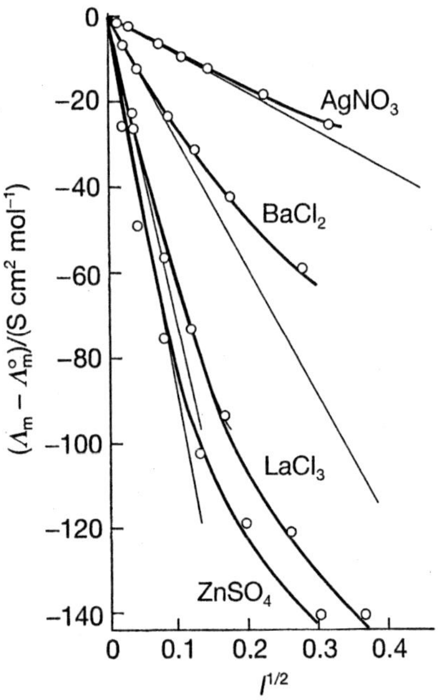

24.23 The dependence of molar conductivities on the square root of the ionic strength, and comparison (dotted lines) with the dependence predicted by the Debye–Hückel–Onsager theory.

Diffusion

We are now in a position to extend the discussion of ionic motion to cover the migration of neutral molecules and of ions in the absence of an applied electric field. We shall do this by expressing ion motion in a more general way than hitherto, and will then discover that the same equations apply even when the charge on the particles is zero.

24.9 The thermodynamic view

We saw in Part 1 that, at constant temperature and pressure, the maximum non-expansion work that can be done per mole when a substance moves from a location where its chemical potential is μ to a location where its chemical potential is $\mu + d\mu$ is $dw = d\mu$. In a system in which the chemical potential depends on the position x,

$$dw = d\mu = \left(\frac{\partial \mu}{\partial x}\right)_{p,T} dx \tag{24.70}$$

We also saw in Chapter 2 (Table 2.1) that in general work can always be expressed in terms of an opposing force (which here we write $\mathcal{F}$), and that

$$dw = -\mathcal{F}\,dx \tag{24.71}$$

By comparing these two expressions, we see that the slope of the chemical potential can be interpreted as an effective force per mole of molecules. We write this **thermodynamic force** as

$$\mathcal{F} = -\left(\frac{\partial \mu}{\partial x}\right)_{p,T} \qquad [24.72]$$

There is not necessarily a real force pushing the particles down the slope of the chemical potential. As we shall see, the force may represent the spontaneous tendency of the molecules to disperse as a consequence of the Second Law and the hunt for maximum entropy.

(a) The thermodynamic force of a concentration gradient

In a solution in which the activity of the solute is a, the chemical potential is

$$\mu = \mu^{\ominus} + RT \ln a$$

If the solution is not uniform the activity depends on the position and we can write

$$\mathcal{F} = -RT\left(\frac{\partial \ln a}{\partial x}\right)_{p,T} \tag{24.73}$$

A useful expression involving derivatives of logarithms is $d \ln y/dx = (1/y)(dy/dx)$.

If the solution is ideal, a may be replaced by the molar concentration c, and then

$$\mathcal{F} = -\frac{RT}{c}\left(\frac{\partial c}{\partial x}\right)_{p,T} \qquad (24.74)^{\circ}$$

Example 24.4 Calculating the thermodynamic force

Suppose the concentration of a solute decays exponentially along the length of a container. Calculate the thermodynamic force on the solute at 25°C given that the concentration falls to half its value in 10 cm.

Method According to eqn 74, the thermodynamic force is calculated by differentiating the concentration with respect to distance. Therefore, write an expression for the variation of the concentration with distance, and then differentiate it.

Answer The concentration varies with position as

$$c = c_0 e^{-x/\lambda}$$

where λ is the decay constant. Therefore,

$$\frac{dc}{dx} = -\frac{c}{\lambda}$$

Equation 74 then implies that

$$\mathcal{F} = \frac{RT}{\lambda}$$

We know that the concentration falls to $\frac{1}{2}c_0$ at $x = 10$ cm, so we can find λ from $\frac{1}{2} = e^{-(10\,\text{cm})/\lambda}$. That is, $\lambda = (10\text{ cm}/\ln 2)$. It follows that

$$\mathcal{F} = (8.314\,51\text{ J K}^{-1}\text{ mol}^{-1}) \times (298\text{ K}) \times \ln 2/(1.0 \times 10^{-1}\text{ m}) = 17\text{ kN mol}^{-1}$$

where we have used 1 J = 1 N m.

Self-test 24.6 Calculate the thermodynamic force on the molecules of molar mass M in a vertical tube in a gravitational field on the surface of the Earth, and evaluate $\mathcal{F}$ for molecules of molar mass 100 g mol^{-1}. Comment on its magnitude relative to that just calculated.

[$\mathcal{F} = -Mg = -0.98$ N mol^{-1}; the force arising from the concentration gradient greatly dominates that arising from the gravitational gradient]

(b) Fick's first law of diffusion

In Section 24.4a we saw that Fick's first law of diffusion (that the particle flux is proportional to the concentration gradient) could be deduced from the kinetic theory of gases. We shall now show that it can be deduced more generally and that it applies to the diffusion of species in condensed phases too.

We suppose that the flux of diffusing particles is motion in response to a thermodynamic force arising from a concentration gradient. The particles reach a steady drift speed, s, when the thermodynamic force, $\mathcal{F}$, is matched by the viscous drag. This drift speed is proportional to the thermodynamic force, and we write $s \propto \mathcal{F}$. However, the particle flux, J, is proportional to the drift speed, and the thermodynamic force is proportional to the concentration gradient, dc/dx. The chain of proportionalities ($J \propto s$, $s \propto \mathcal{F}$, and $\mathcal{F} \propto dc/dx$) implies that $J \propto dc/dx$, which is the content of Fick's law.

(c) The Einstein relation

Fick's law can be written[8]

$$J = -D\frac{dc}{dx} \tag{24.75}$$

In this expression, D is the diffusion coefficient and dc/dx is the slope of the molar concentration. The flux is related to the drift speed by

$$J = sc \tag{24.76}$$

This relation follows from the argument that we have used several times before. Thus, all particles within a distance $s\Delta t$, and therefore in a volume $s\Delta tA$, can pass through a window of area A in an interval Δt. Hence, the amount of substance that can pass through the window in that interval is $s\Delta tAc$. Therefore,

$$sc = -D\frac{dc}{dx}$$

If now we express dc/dx in terms of $\mathcal{F}$ by using eqn 74, we find

$$s = -\frac{D}{c}\frac{dc}{dx} = \frac{D\mathcal{F}}{RT} \tag{24.77}$$

8. This expression is derived from eqn 20 by dividing both sides by the Avogadro constant, which converts numbers into amounts (numbers of moles).

Therefore, once we know the effective force and the diffusion coefficient, D, we can calculate the drift speed of the particles (and vice versa) whatever the origin of the force.

There is one case where we already know the drift speed and the effective force acting on a particle: an ion in solution has a drift speed $s = u\mathcal{E}$ when it experiences a force $ez\mathcal{E}$ from an electric field of strength $\mathcal{E}$ (so $\mathcal{F} = N_A ez\mathcal{E} = zF\mathcal{E}$). Therefore, substituting these known values into eqn 77 gives

$$u\mathcal{E} = \frac{zF\mathcal{E}D}{RT}$$

and hence

$$u = \frac{zFD}{RT} \qquad (24.78)$$

This equation rearranges into the very important result known as the **Einstein relation** between the diffusion coefficient and the ionic mobility:

$$D = \frac{uRT}{zF} \qquad (24.79)^{\circ}$$

On inserting the typical value $u = 5 \times 10^{-8}\ \text{m}^2\ \text{s}^{-1}\ \text{V}^{-1}$, we find $D \approx 1 \times 10^{-9}\ \text{m}^2\ \text{s}^{-1}$ at 25°C as a typical value of the diffusion coefficient of an ion in water.

(d) The Nernst–Einstein equation

The Einstein relation provides a link between the molar conductivity of an electrolyte and the diffusion coefficients of its ions. First, by using eqns 54 ($\lambda = zuF$) and 78 we write

$$\lambda = zuF = \frac{z^2DF^2}{RT} \qquad (24.80)^{\circ}$$

for each type of ion. Then, from $\Lambda_m^{\circ} = \nu_+\lambda_+ + \nu_-\lambda_-$, the limiting molar conductivity is

$$\Lambda_m = (\nu_+z_+^2D_+ + \nu_-z_-^2D_-)\frac{F^2}{RT} \qquad (24.81)^{\circ}$$

which is the **Nernst–Einstein equation**. An application of this equation is to the determination of ionic diffusion coefficients from conductivity measurements; another is to the prediction of conductivities using models of ionic diffusion (see below).

(e) The Stokes–Einstein equation

Equations 53 ($u = ez/f$) and 76 relate the mobility of an ion to the frictional force and to the diffusion coefficient, respectively. We can combine the two expressions using $eR/F = k$ into the **Stokes–Einstein equation**:

$$D = \frac{kT}{f} \qquad (24.82)$$

If the frictional force is described by Stokes's law, then we also obtain a relation between the diffusion coefficient and the viscosity of the medium:

$$D = \frac{kT}{6\pi\eta a} \qquad (24.83)$$

An important feature of eqn 82 (and of its special case, eqn 83) is that it makes no reference to the charge of the diffusing species. Therefore, the equation also applies

Table 24.8* Diffusion coefficients at 298 K, $D/(10^{-9}\,\mathrm{m^2\,s^{-1}})$

H^+ in water	9.31
I_2 in hexane	4.05
Na^+ in water	1.33
Sucrose in water	0.522

* More values are given in the *Data section*.

in the limit of vanishingly small charge; that is, it also applies to neutral molecules. Consequently, we may use viscosity measurements to estimate the diffusion coefficients for electrically neutral molecules in solution (Table 24.8). It must not be forgotten, however, that both equations depend on the assumption that the viscous drag is proportional to the speed.

Example 24.5 Interpreting the mobility of an ion

Use the experimental value of the mobility to evaluate the diffusion coefficient, the limiting molar conductivity, and the hydrodynamic radius of a sulfate ion in aqueous solution.

Method The starting point is the mobility of the ion, which is given in Table 24.6. The diffusion coefficient can then be determined from the Einstein relation, eqn 79. The ionic conductivity is related to the mobility by eqn 54. To estimate the hydrodynamic radius, a, of the ion, use the Stokes–Einstein relation to find f and the Stokes law to relate f to a.

Answer From Table 24.6, the mobility of SO_4^{2-} is $8.29 \times 10^{-8}\,\mathrm{m^2\,s^{-1}\,V^{-1}}$. It follows from eqn 79 that

$$D = \frac{uRT}{zF} = 1.1 \times 10^{-9}\,\mathrm{m^2\,s^{-1}}$$

From eqn 54 it follows that

$$\lambda = zuF = 16\,\mathrm{mS\,m^2\,mol^{-1}}$$

Finally, from $f = 6\pi\eta a$ using 0.891 cP (or $8.91 \times 10^{-4}\,\mathrm{kg\,m^{-1}\,s^{-1}}$) for the viscosity of water (Table 24.4):

$$a = \frac{kT}{6\pi\eta D} = 220\,\mathrm{pm}$$

Comment The bond length in SO_4^{2-} is 144 pm, so the radius calculated here is plausible and consistent with a small degree of solvation.

Self-test 24.7 Repeat the calculation for the NH_4^+ ion.

[$1.96 \times 10^{-9}\,\mathrm{m^2\,s^{-1}}$, $7.4\,\mathrm{mS\,m^2\,mol^{-1}}$, 125 pm]

Experimental support for the relations derived above comes from conductivity measurements. In particular, **Walden's rule** is the empirical observation that the product $\eta\Lambda_m$ is very approximately constant for the same ions in different solvents (but there are numerous exceptions). Because $\Lambda_m \propto D$, and we have just seen that $D \propto 1/\eta$, we do indeed predict that $\Lambda_m \propto 1/\eta$, as Walden's rule implies. The usefulness of the rule, however, is muddied by the role of solvation: different solvents solvate the same ions to different extents, so both the hydrodynamic radius and the viscosity change with the solvent.

24.10 The diffusion equation

We now turn to the discussion of time-dependent diffusion processes, where we are interested in the spreading of inhomogeneities with time. One example is the temperature of a metal bar that has been heated at one end: if the source of heat is removed, then the bar gradually settles down into a state of uniform temperature.

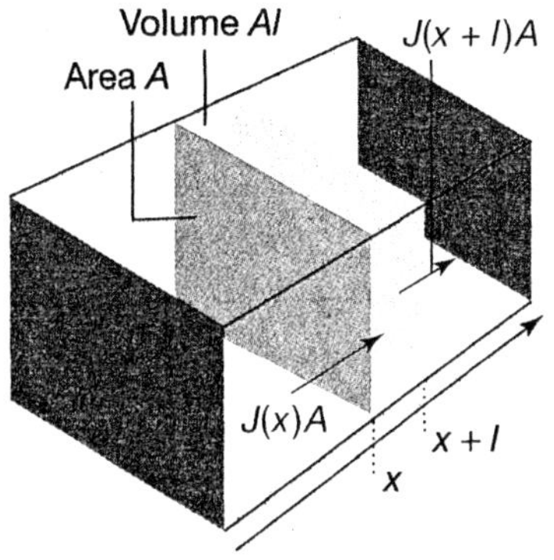

24.24 The net flux in a region is the difference between the flux entering from the region of high concentration (on the left) and the flux leaving to the region of low concentration (on the right).

When the source of heat is maintained and the bar can radiate, it settles down into a steady state of nonuniform temperature. Another example (and one more relevant to chemistry) is the concentration distribution in a solvent to which a solute is added. We shall focus on the description of the diffusion of particles, but similar arguments apply to the diffusion of physical properties, such as temperature. Our aim is to obtain an equation for the rate of change of the concentration of particles in an inhomogeneous region.

The central equation of this section is the **diffusion equation**,[9] which relates the rate of change of concentration at a point to the spatial variation of the concentration at that point:

$$\frac{\partial c}{\partial t} = D\frac{\partial^2 c}{\partial x^2} \tag{24.84}$$

We show in the following *Justification* that the diffusion equation follows from Fick's first law of diffusion.

Justification 24.11

Consider a thin slab of cross-sectional area A that extends from x to $x + l$ (Fig. 24.24). Let the concentration at x be c at the time t. The amount (number of moles) of particles that enter the slab in the infinitesimal interval dt is $JA\,dt$, so the rate of increase in molar concentration inside the slab (which has volume Al) on account of the flux from the left is

$$\frac{\partial c}{\partial t} = \frac{JA\,dt}{Al\,dt} = \frac{J}{l}$$

There is also an outflow through the right-hand window. The flux through that window is J', and the rate of change of concentration that results is

$$\frac{\partial c}{\partial t} = -\frac{J'A\,dt}{Al\,dt} = -\frac{J'}{l}$$

The net rate of change of concentration is therefore

$$\frac{\partial c}{\partial t} = \frac{J - J'}{l}$$

Each flux is proportional to the concentration gradient at the window. So, by using Fick's first law, we can write

$$J - J' = -D\frac{\partial c}{\partial x} + D\frac{\partial c'}{\partial x} = -D\frac{\partial c}{\partial x} + D\frac{\partial}{\partial x}\left\{c + \left(\frac{\partial c}{\partial x}\right)l\right\} = Dl\frac{\partial^2 c}{\partial x^2}$$

When this relation is substituted into the expression for the rate of change of concentration in the slab, we get eqn 84.

(a) The significance of the diffusion equation

The diffusion equation shows that the rate of change of concentration is proportional to the curvature (more precisely, to the second derivative) of the concentration with respect to distance. If the concentration changes sharply from point to point (if the distribution is highly wrinkled) then the concentration changes rapidly with time. If the curvature is zero, then the concentration is constant in time. If the

9. This equation is sometimes called 'Fick's second law of diffusion'.

concentration decreases linearly with distance, then the concentration at any point is constant because the inflow of particles is exactly balanced by the outflow.

The diffusion equation can be regarded as a mathematical formulation of the intuitive notion that there is a natural tendency for the wrinkles in a distribution to disappear. More succinctly: Nature abhors a wrinkle.

(b) Diffusion with convection

The transport of particles arising from the motion of a streaming fluid is called **convection**. If for the moment we ignore diffusion, then the flux of particles through an area A in an interval Δt when the fluid is flowing at a velocity v can be calculated in the way we have used several times before (by counting the particles within a distance $v\Delta t$), and is

$$J = \frac{cAv\Delta t}{A\Delta t} = cv \qquad (24.85)$$

This J is called the **convective flux**. The rate of change of concentration in a slab of thickness l and area A is, by the same argument as before,

$$\frac{\partial c}{\partial t} = \frac{J - J'}{l} = \left\{c - \left[c + \left(\frac{\partial c}{\partial x}\right)l\right]\right\}\frac{v}{l} = -v\frac{\partial c}{\partial x} \qquad (24.86)$$

(We have assumed that the velocity does not depend on the position.)

When both diffusion and convection occur, the total change of concentration in a region is the sum of the two effects, and the **generalized diffusion equation** is

$$\frac{\partial c}{\partial t} = D\frac{\partial^2 c}{\partial x^2} - v\frac{\partial c}{\partial x} \qquad (24.87)$$

A further refinement, which is important in chemistry, is the possibility that the concentrations of particles may change as a result of reaction. When reactions are included in eqn 87 (Section 27.2), we get a powerful differential equation for discussing the properties of reacting, diffusing, convecting systems and which is the basis of reactor design in chemical industry and of the utilization of resources in living cells.

(c) Solutions of the diffusion equation

The diffusion equation, eqn 84, is a second-order differential equation with respect to space and a first-order differential equation with respect to time. Therefore, we must specify two boundary conditions for the spatial dependence and a single initial condition for the time dependence.

As an illustration, consider a solvent in which the solute is initially coated on one surface of the container (for example, a layer of sugar on the bottom of a deep beaker of water). The single initial condition is that at $t = 0$ all N_0 particles are concentrated on the yz-plane (of area A) at $x = 0$. The two boundary conditions are derived from the requirements: (1) that the concentration must everywhere be finite and (2) that the total amount (number of moles) of particles present is n_0 (with $n_0 = N_0/N_A$) at all times. These requirements imply that the flux of particles is zero at the top and bottom surfaces of the system. Under these conditions it is found that

$$c(x, t) = \frac{n_0}{A(\pi Dt)^{1/2}} e^{-x^2/4Dt} \qquad (24.88)$$

as may be verified by direct substitution. Figure 24.25 shows the shape of the concentration distribution at various times, and it is clear that the concentration spreads and tends to uniformity.

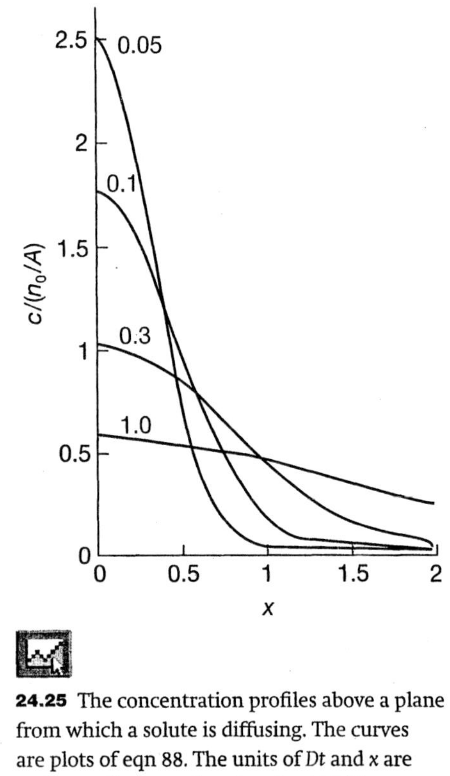

24.25 The concentration profiles above a plane from which a solute is diffusing. The curves are plots of eqn 88. The units of Dt and x are arbitrary, but are related so that Dt/x^2 is dimensionless. For example, if x is in metres, Dt would be in metres^2; so, for $D = 10^{-9}\ m^2\ s^{-1}$, $Dt = 0.1\ m^2$ corresponds to $t = 10^8$ s.

Another useful result is for a localized concentration of solute in a three-dimensional solvent (a sugar lump suspended in a large flask of water). The concentration of diffused solute is spherically symmetrical, and at a radius r is

$$c(r, t) = \frac{n_0}{8(\pi Dt)^{3/2}} e^{-r^2/4Dt} \quad (24.89)$$

Other chemically (and physically) interesting arrangements, such as transport of substances across biological membranes (Box 24.2), can be treated. In many cases the solutions are more cumbersome.

(d) The measurement of diffusion coefficients

The solutions of the diffusion equation are useful for experimental determinations of diffusion coefficients. In the **capillary technique**, a capillary tube, open at one end and containing a solution, is immersed in a well stirred larger quantity of solvent, and the change of concentration in the tube is monitored. The solute diffuses from the open end of the capillary at a rate that can be calculated by solving the diffusion equation with the appropriate boundary conditions, so D may be determined. In the **diaphragm technique**, the diffusion occurs through the capillary pores of a sintered glass diaphragm separating the well-stirred solution and solvent. The concentrations are monitored and then related to the solutions of the diffusion equation corresponding to this arrangement. Diffusion coefficients may also be measured by the dynamic light scattering technique described in Section 22.8 and by NMR.

Box 24.2 Transport across biological membranes

Controlled transport of molecules and ions across biological membranes is at the heart of a number of key cellular processes, such as the generation of action potentials in neurons (Section 10.4g), the transfer of glucose into red blood cells, and the synthesis of ATP by oxidative phosphorylation and photophosphorylation (Box 10.1). Here we examine in some detail the various ways in which non-electrolytes and electrolytes cross the lipid bilayer.

Suppose that a lipid bilayer provides a barrier that slows down the transfer of molecules or ions into or out of the cell. The thermodynamic tendency to transport a species A through the bilayer is partially determined by an activity gradient across the membrane, which results in a difference in molar Gibbs energy between the inside and the outside of the cell

$$\Delta G_m = G_{m,in} - G_{m,out} = RT \ln \frac{a_{in}}{a_{out}}$$

The equation implies that transport into the cell of either neutral or charged species is thermodynamically favourable if $a_{in} < a_{out}$ or, if we set the activity coefficients to unity, if $[A]_{in} < [A]_{out}$. If A is an ion, there is a second component of ΔG_m that arises from differences in Coulomb repulsions on each side of the bilayer and gives rise to a membrane potential difference $\Delta\phi = \phi_{in} - \phi_{out}$. The final expression for ΔG is then

$$\Delta G_m = RT \ln \frac{[A]_{in}}{[A]_{out}} + zF\Delta\phi$$

where z is the ion charge number and F is Faraday's constant. This equation implies that there is a tendency, called *passive transport*, for a species to move down concentration and membrane potential gradients. It is also possible to move a species against these gradients, but the process must now be driven by the exergonic hydrolysis of ATP. This process is called *active transport*.

Consider the passive transport of an uncharged species A across a lipid bilayer of thickness l. To simplify the problem, we will assume that the concentration of A is always maintained at $[A] = [A]_0$ on one surface of the membrane and at $[A] = 0$ on the other surface, perhaps by a perfect balance between the rate of the process that produces A on one side and the rate of another process that consumes A completely on the other side. This is one example of a steady-state assumption, which will be discussed in more detail in Section 25.7. Then $\partial[A]/\partial t = 0$ and the diffusion equation simplifies to

$$D\frac{d^2[A]}{dx^2} = 0$$

where D is the diffusion coefficient and the steady-state assumption makes partial derivatives unnecessary. We use the boundary conditions $[A](0) = [A]_0$ and $[A](l) = 0$ to solve this differential equation and the result, which may be verified by differentiation, is

$$[A](x) = [A]_0\left(1 - \frac{x}{l}\right)$$

which implies that the [A] decreases linearly inside the membrane. We now use Fick's first law to calculate the flux J of A through the membrane and the result is

$$J = D\frac{[A]_0}{l}$$

However, we need to modify this equation slightly to account for the fact that the concentration of A on the surface of a membrane is not always equal to the concentration of A measured in the bulk solution, which we assume to be aqueous. This difference arises from the significant difference in the solubility of A in an aqueous environment and in the solution-membrane interface. One way to deal with this problem is to define a *partition coefficient* κ as

$$\kappa = \frac{[A]_0}{[A]_s}$$

where $[A]_s$ is the concentration of A in the bulk aqueous solution. It follows that

$$J = \kappa D\frac{[A]_s}{l}$$

In spite of the assumptions that led to its final form, this equation describes adequately the passive transport of many non-electrolytes through membranes of blood cells.

In many cases the equation just derived underestimates the flux and the implication is that the membrane is more permeable than expected. However, because the permeability increases only for certain species and not others, this is evidence that transport can be mediated by carriers. One example is the transporter protein that carries glucose into cells.

A characteristic of a carrier C is that it binds to the transported species A and the dissociation of the AC complex is described by

$$AC \rightleftharpoons A + C \qquad K = \frac{[A][C]}{[AC]}$$

where we have used concentrations instead of activities. After writing $[C]_0 = [C] + [AC]$, where $[C]_0$ is the total concentration of carrier, it follows that

$$[AC] = \frac{[A][C]_0}{[A] + K}$$

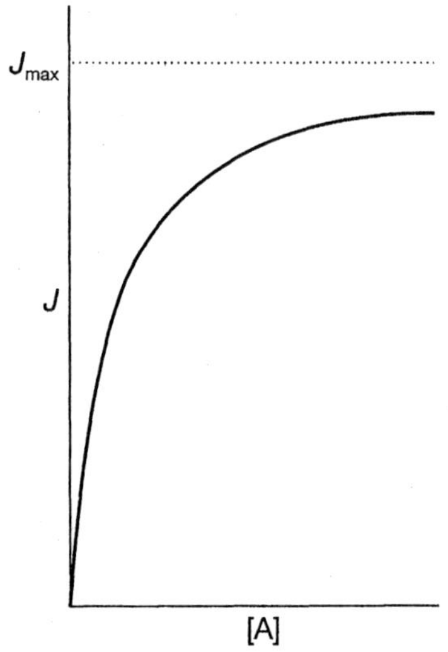

The flux of the species AC through a membrane varies with the concentration of the species A. The behaviour shown in the illustration and explained in the text is characteristic of mediated transport of A, with C as a carrier molecule.

We can now use the expression for J derived above to write an expression for the flux of the species AC through the membrane:

$$J = \frac{\kappa_{AC}D_{AC}[C]_0}{l}\frac{[A]}{[A] + K} = J_{max}\frac{[A]}{[A] + K}$$

where κ_{AC} and D_{AC} are the partition coefficient and diffusion coefficient of the species AC, respectively. We see from the first illustration that when $[A] \ll K$ the flux varies linearly with [A] and that the flux reaches a maximum value of $J_{max} = \kappa_{AC}D_{AC}[C]_0/l$ when $[A] \gg K$. This behaviour is characteristic of mediated transport.

The transport of ions through bilayers also needs to be mediated because charged species do not partition well into the hydrophobic environment of the membrane. There are two mechanisms for ion transport: mediation by a carrier molecule or transport through a *channel former*, a protein that creates a hydrophilic pore through which the ion can pass. An example of a channel former is the polypeptide gramicidin A, which increases the membrane permeability to cations such as H^+, K^+, and Na^+.

Ion channels are proteins that effect the movement of specific ions down a membrane potential gradient. They are highly selective, so there is a channel protein for Ca^{2+}, another for Cl^-, and so on. The opening of the gate may be triggered by potential differences between the two sides of the membrane or by the binding of an *effector* molecule to a specific receptor site on the channel. It is now possible to measure currents and conductances through single ion channel proteins; they are of the order of 10^{-12} A and 10^{-11} – 10^{-10} S, respectively.

Ions such as H^+, Na^+, K^+, and Ca^{2+} are often transported actively across membranes by integral proteins called *ion pumps*. Ion pumps are molecular machines that work by adopting conformations that are permeable to one ion but not others depending on the state of phosphorylation of the protein. Because protein phosphorylation requires dephosphorylation of ATP, the conformational change that opens or closes the pump is endergonic and requires the use of energy stored during metabolism. An example of an ion pump is the Na^+/K^+-ATPase already discussed in Section 10.4g in connection with neuronal action potentials.

Another important ion pump is the H^+-ATPase, which occurs in the inner membranes of the chloroplast and the mitochondrion (Box 10.1). The *chemiosmotic theory* proposed by Peter Mitchell explains how H^+-ATPases use the flux of protons through the membrane to power the phosphorylation of ADP to ATP. Some of the electron transfer reactions of respiration and photosynthesis also carry protons from one side of the membrane to the other, creating a proton gradient. We can estimate the Gibbs energy available for phosphorylation by writing

$$\Delta G_m = RT \ln \frac{[H^+]_{in}}{[H^+]_{out}} + F\Delta\phi$$

where we have used $z = +1$. After substituting $\Delta pH = pH_{in} - pH_{out} = -\log[H^+]_{in} + \log[H^+]_{out}$, it follows that

$$\Delta G_m = F\Delta\phi - 2.303RT\Delta pH$$

In the mitochondrion, $\Delta pH \approx -1.4$ and $\Delta\phi \approx 0.14$ V, so $\Delta G_m \approx +21.5$ kJ mol^{-1}. Because 31 kJ mol^{-1} is needed for phosphorylation, we conclude that at least 2 mol of protons (and probably much more) must flow through the membrane for the phosphorylation of 1 mol of ADP.

The coupling of the proton flow to the phosphorylation is itself of considerable interest and we present a model proposed by Paul Boyer.[10] Structural studies by John Walker show that the channel through which the protons flow is linked in tandem to a unit composed of six protein molecules arranged in pairs of α and β subunits to form three interlocked $\alpha\beta$ segments (see the second illustration). The conformations of the three pairs may be loose (L), tight (T), or open (O), and one of each type is present at each stage. A protein at the centre of the interlocked structure, the γ subunit shown as a white arrow, rotates and induces structural changes that cycle each of the three segments between L, T, and O conformations. At the start of a cycle, a T unit holds an ATP molecule. Then ADP and a P_i group migrate into the

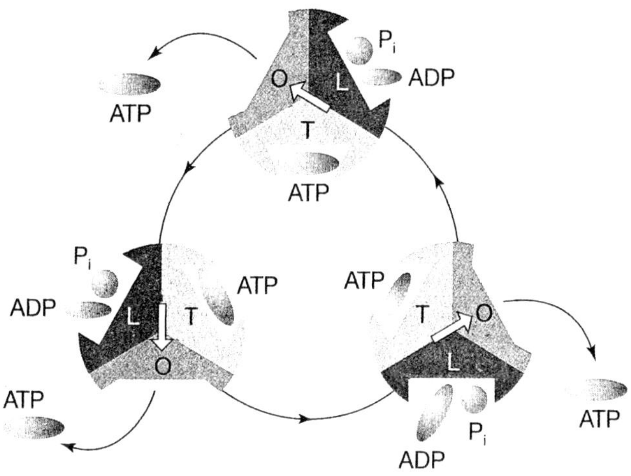

The mechanism of action of H^+-ATPase, a molecular motor that transports protons across the mitochondrial membrane and catalyses either the formation or hydrolysis of ATP.

L site and, as it closes into T, the earlier T site opens into O and releases its ATP. The ADP and P_i in the T site meanwhile condense into ATP, and the new L site is ready for the cycle to begin again. The proton flux drives the rotation of the γ subunit, and hence the conformational changes of the α/β segments, as well as providing the energy for the condensation reaction itself. Several key aspects of this mechanism have been confirmed experimentally. For example, the rotation of the γ subunit has been visualized directly by using single-molecule spectroscopy (Box 17.2).

Discussion questions

1. Provide a molecular interpretation for the observation that mediated transport leads to a maximum flux J_{max} when the concentration of the transported species becomes very large.

2. If the mitochondrial membrane potential were 70 mV, as is common for other membranes, how much ATP could be synthesized from the transport of 4 mol H^+, assuming the pH difference remains the same?

Further reading

C. Miller, Ion channels: doing hard chemistry with hard ions. *Curr. Op. Chem. Biol.* **4**, 148 (2000).

D. Stack, A.G.W. Leslie, and J.E. Walker, Molecular architecture of the rotary motor in ATP synthase. *Science* **286**, 1700 (1999).

10. Paul Boyer and John Walker received the Nobel Prize for chemistry in 1997 for their work on the mechanism of H^+-ATPase. They shared the award with Jens Skou, who was recognized for his work on the Na^+/K^+-ATPase. Peter Mitchell received the Nobel Prize for chemistry in 1978 for his chemiosmotic theory.

24.11 Diffusion probabilities

The solutions of the diffusion equation can be used to predict the concentration of particles (or the value of some other physical quantity, such as the temperature in a nonuniform system) at any location. We can also use them to calculate the net distance through which the particles diffuse in a given time.

Example 24.6 Calculating the net distance of diffusion

Calculate the net distance travelled on average by particles in a time t if they have a diffusion constant D.

Method We need to calculate the probability that a particle will be found at a certain distance from the origin, and then calculate the average by weighting each distance by that probability.

Answer The number of particles in a slab of thickness dx and area A at x, where the molar concentration is c, is $cAN_A\,dx$. The probability that any of the $N_0 = n_0N_A$ particles is in the slab is therefore $cAN_A\,dx/N_0$. If the particle is in the slab, it has travelled a distance x from the origin. Therefore, the mean distance travelled by all the particles is the sum of each x weighted by the probability of its occurrence:

$$\langle x\rangle = \int_0^{\infty} \frac{xcAN_A}{N_0}\,dx = \frac{1}{(\pi Dt)^{1/2}}\int_0^{\infty} xe^{-x^2/4Dt}\,dx = 2\left(\frac{Dt}{\pi}\right)^{1/2}$$

Comment The average distance of diffusion varies as the square root of the lapsed time. If we use the Stokes–Einstein relation for the diffusion coefficient, the mean distance travelled by particles of radius a in a solvent of viscosity η is

$$\langle x\rangle = \left(\frac{2kTt}{3\pi^2\eta a}\right)^{1/2}$$

Self-test 24.8 Derive an expression for the root mean square distance travelled by diffusing particles in a time t. $[\langle x^2\rangle^{1/2} = (2Dt)^{1/2}]$

As shown in Example 24.6, the average distance travelled by diffusing particle in a time t (in an arrangement like that illustrated in Fig. 24.25) is

$$\langle x\rangle = 2\left(\frac{Dt}{\pi}\right)^{1/2} \qquad (24.90)$$

and the root mean square distance travelled in the same time is

$$\langle x^2\rangle^{1/2} = (2Dt)^{1/2} \qquad (24.91)$$

The latter is a valuable measure of the spread of particles when they can diffuse in both directions from the origin (for then $\langle x\rangle = 0$ at all times). The root mean square distance travelled by particles with a typical diffusion coefficient ($D = 5\times10^{-10}$ m^2 s^{-1}) is illustrated in Fig. 24.26, which shows how long it takes for diffusion to increase the net distance travelled on average to about 1 cm in an unstirred solution. The graph shows that diffusion is a very slow process (which is why solutions are stirred, to encourage mixing by convection).

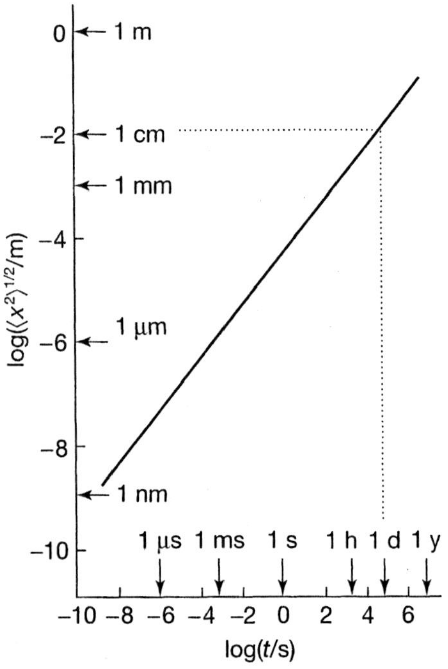

24.26 The root mean square distance covered by particles with $D = 5\times10^{-10}$ m^2 s^{-1}. Note the great slowness of diffusion.

24.12 The statistical view

An intuitive picture of diffusion is of the particles moving in a series of small steps and gradually migrating from their original positions. We shall explore this idea using a model in which the particles can jump through a distance λ in a time τ. The total distance travelled by a particle in a time t is therefore $t\lambda/\tau$. However, the particle will not necessarily be found at that distance from the origin. The direction of each step may be different, and the net distance travelled must take the changing directions into account.

If we simplify the discussion by allowing the particles to travel only along a straight line (the x-axis), and for each step (to the left or the right) to be through the same distance λ, then we obtain the **one-dimensional random walk.**[11] We show in the *Justification* below that the probability of a particle being at a distance x from the origin after a time t is

$$P = \left(\frac{2\tau}{\pi t}\right)^{1/2} e^{-x^2\tau/2t\lambda^2} \tag{24.92}$$

Justification 24.12

Consider a one-dimensional random walk in which each step is through a distance λ to the left or right. The net distance travelled after N steps is equal to the difference between the number of steps to the right (N_R) and to the left (N_L), and is $(N_R - N_L)\lambda$. We write $n = N_R - N_L$ and the total number of steps as $N = N_R + N_L$.

The number of ways of performing a walk with a given net distance of travel $n\lambda$ is the number of ways of making N_R steps to the right and N_L steps to the left, and is given by the binomial coefficient

$$W = \frac{N!}{N_L!N_R!} = \frac{N!}{\{\frac{1}{2}(N+n)\}!\{\frac{1}{2}(N-n)\}!}$$

The probability of the net distance walked being $n\lambda$ is

$$P = \frac{\text{number of paths with } N_R \text{ steps to the right}}{\text{total number of steps}}$$

$$= \frac{W}{2^N} = \frac{N!}{\{\frac{1}{2}(N+n)\}!\{\frac{1}{2}(N-n)\}!2^N}$$

The use of Stirling's approximation (Section 19.1a) in the form

$$\ln x! \approx \ln(2\pi)^{1/2} + (x + \tfrac{1}{2})\ln x - x$$

gives (after quite a lot of algebra)

$$\ln P = \ln\left(\frac{2}{\pi N}\right)^{1/2} - \tfrac{1}{2}(N+n+1)\ln\left(1+\frac{n}{N}\right) - \tfrac{1}{2}(N-n+1)\ln\left(1-\frac{n}{N}\right)$$

For small net distances ($n \ll N$) we can use the approximation $\ln(1 \pm x) \approx \pm x - \frac{1}{2}x^2$, and so obtain

$$\ln P \approx \ln\left(\frac{2}{\pi N}\right)^{1/2} - \frac{n^2}{2N}$$

11. The same model was used in the discussion of a one-dimensional random coil in Section 22.2a.

At this point, we note that the number of steps taken in a time t is $N = t/\tau$ and the net distance travelled from the origin is $x = n\lambda$. Substitution of these quantities into the equation just derived gives an expression that rearranges into eqn 92.

The differences of detail between eqns 88 and 92 arise from the fact that in the present calculation the particles can migrate in either direction from the origin. Moreover, they can be found only at discrete points separated by λ instead of being anywhere on a continuous line. The fact that the two expressions are so similar suggests that diffusion can indeed be interpreted as the outcome of a large number of steps in random directions.

We can now relate the coefficient D to the step length λ and the rate at which the jumps occur. Thus, by comparing the two exponents in eqn 88 and eqn 92 we can immediately write down the **Einstein–Smoluchowski equation**:

$$D = \frac{\lambda^2}{2\tau} \qquad (24.93)$$

Illustration 24.6

Suppose that a SO_4^{2-} ion jumps through its own diameter each time it makes a move in an aqueous solution; then, because $D = 1.1 \times 10^{-9}\ m^2\ s^{-1}$ and $a = 210$ pm (as deduced from mobility measurements), it follows from $\lambda = 2a$ that $\tau = 80$ ps. Because τ is the time for one jump, the ion makes 1×10^{10} jumps per second.

The Einstein–Smoluchowski equation is the central connection between the microscopic details of particle motion and the macroscopic parameters relating to diffusion (for example, the diffusion coefficient and, through the Stokes–Einstein relation, the viscosity). It also brings us back full circle to the properties of the perfect gas. For if we interpret λ/τ as $\bar{c}$, the mean speed of the molecules, and interpret λ as a mean free path, then we can recognize in the Einstein–Smoluchowski equation exactly the same expression as we obtained from the kinetic theory of gases in *Justification 24.5*. That is, the diffusion of a perfect gas is a random walk with an average step size equal to the mean free path.

Checklist of key ideas

Molecular motion in gases

24.1 The kinetic model of gases

- ☐ transport property, the ability of a substance to transfer matter, energy, or some other property from one place to another
- ☐ diffusion, the migration of matter down a concentration gradient
- ☐ thermal conduction, the migration of energy down a temperature gradient
- ☐ electric conduction, the migration of electric charge along an electrical potential gradient
- ☐ viscosity, the migration of linear momentum down a velocity gradient
- ☐ effusion, the emergence of a gas from a container through a small hole
- ☐ kinetic model, a model of a gas in which the only contribution to the energy is from the kinetic energies of the molecules
- ☐ elastic collision, a collision in which the total translational kinetic energy of the molecules is conserved
- ☐ pressure of a gas, $pV = \frac{1}{3}nMc^2$
- ☐ root mean square speed, the square root of the mean of the squares of the speeds: $c = \langle v^2 \rangle^{1/2} = (3RT/M)^{1/2}$
- ☐ distribution of speeds, the function $f(v)$ which, through $f(v)\,dv$, gives the fraction of molecules that have speeds in the range v to $v + dv$

- ☐ Maxwell distribution of speeds, $f(v) = 4\pi(M/2\pi RT)^{3/2}v^2 e^{-Mv^2/2RT}$
- ☐ mean speed, $\bar{c} = (8RT/\pi M)^{1/2}$
- ☐ most probable speed, $c^* = (2RT/M)^{1/2}$
- ☐ relative mean speed, $\bar{c}_{rel} = (8kT/\pi\mu)^{1/2}$
- ☐ reduced mass, $\mu = m_A m_B/(m_A + m_B)$
- ☐ collision diameter, the distance of approach corresponding to a collision
- ☐ collision frequency, the number of collisions made by a molecule in an interval divided by the length of the interval, $z = \sigma\bar{c}_{rel}\mathcal{N}$
- ☐ collision cross-section, $\sigma = \pi d^2$
- ☐ mean free path, the average distance a molecule travels between collisions: $\lambda = \bar{c}/z$

24.2 Collisions with walls and surfaces

- ☐ collision flux, Z_W, the number of collisions with an area in a given time interval divided by the area and the duration of the interval, $Z_W = p/(2\pi mkT)^{1/2}$
- ☐ collision frequency, the collision flux multiplied by the area of the region of interest

24.3 The rate of effusion

- ☐ Graham's law of effusion: the rate of effusion is inversely proportional to the square root of the molar mass
- ☐ rate of effusion: rate $= pA_0/(2\pi mkT)^{1/2}$
- ☐ Knudsen method, a method for the determination of the vapour pressures of liquids and solids

24.4 Transport properties of a perfect gas

- ☐ flux, the quantity of a property passing through a given area in a given time interval divided by the area and the duration of the interval
- ☐ matter flux, the flux of matter, $J(\text{matter}) \propto d\mathcal{N}/dz$
- ☐ energy flux, the flux of energy, $J(\text{energy}) \propto dT/dz$
- ☐ momentum flux, $J(\text{momentum}) \propto dv/dz$
- ☐ Fick's first law of diffusion: the flux of matter is proportional to the concentration gradient, $J(\text{matter}) = -D\, d\mathcal{N}/dz$
- ☐ diffusion coefficient, the coefficient in Fick's first law
- ☐ coefficient of thermal conductivity, the coefficient κ in $J(\text{energy}) = -\kappa\, dT/dz$
- ☐ Newtonian flow, flow that occurs by a series of layers moving past one another
- ☐ coefficient of viscosity, the coefficient η in $J(\text{momentum}) = -\eta\, dv_x/dz$
- ☐ diffusion coefficient for a perfect gas, $D = \frac{1}{3}\lambda\bar{c}$
- ☐ coefficient of thermal conductivity for a perfect gas, $\kappa = \frac{1}{3}\lambda\bar{c}C_{V,m}[\text{A}]$
- ☐ coefficient of viscosity of a perfect gas, $\eta = \frac{1}{3}M\lambda\bar{c}[\text{A}]$
- ☐ Poiseuille's formula for the rate of flow, $dV/dt = (p_1^2 - p_2^2)\pi r^4/16l\eta p_0$

Molecular motion in liquids

24.5 Experimental results

- ☐ inelastic neutron scattering, scattering in which neutrons collect or discard energy as they pass through a sample

24.6 The conductivities of electrolyte solutions

- ☐ conductance, G, the inverse of resistance
- ☐ conductivity, the constant κ in $G = \kappa A/l$
- ☐ molar conductivity, $\Lambda_m = \kappa/c$
- ☐ strong electrolyte, an electrolyte with a molar conductivity that varies only slightly with concentration
- ☐ weak electrolyte, an electrolyte with a molar conductivity that is normal at concentrations close to zero, but falls sharply to low values as the concentration increases
- ☐ Kohlrausch's law, for the concentration dependence of the molar conductivity of a strong electrolyte, $\Lambda_m = \Lambda_m^\circ - \mathcal{K}c^{1/2}$
- ☐ limiting molar conductivity, Λ_m°, the molar conductivity at zero concentration
- ☐ law of the independent migration of ions, $\Lambda_m^\circ = \nu_+\lambda_+ + \nu_-\lambda_-$
- ☐ degree of ionization (or deprotonation), the parameter α in $[\text{H}_3\text{O}^+] = \alpha c$, $[\text{A}^-] = \alpha c$, and $[\text{HA}] = (1 - \alpha)c$ for the deprotonation of a monoprotic acid
- ☐ Ostwald's dilution law, $1/\Lambda_m = 1/\Lambda_m^\circ + \Lambda_m c/K_a(\Lambda_m^\circ)^2$

24.7 The mobilities of ions

- ☐ drift speed, the terminal speed when an accelerating force is balanced by the viscous drag: $s = u\mathcal{E}$
- ☐ mobility of an ion, the coefficient u in the expression $s = u\mathcal{E}$; $u = ze/6\pi\eta a$
- ☐ hydrodynamic radius (Stokes radius), the effective radius of a particle in solution
- ☐ Grotthuss mechanism, a mechanism for the conduction of protons in solution
- ☐ ionic conductivity, the contribution of ions of one type to the molar conductivity: $\lambda = zuF$
- ☐ transport number, the fraction of total current carried by the ions of a specified type, $t_\pm = I_\pm/I$
- ☐ limiting transport number, the transport number at zero concentration, $t_\pm^\circ = u_\pm/(u_+ + u_-)$
- ☐ moving boundary method, a method for determining transport numbers based on the observation of the motion of a boundary between two ionic solutions having a common ion
- ☐ indicator solution, the solution that occupies the lower part of a vertical tube
- ☐ leading solution, the solution that occupies the upper part of a vertical tube
- ☐ Hittorf method, a method for determining transport numbers based on the determination of the changes in concentration when a current flows
- ☐ cell with transference, a cell with a liquid boundary

24.8 Conductivities and ion–ion interactions

- ☐ relaxation effect, the reduction of an ion's mobility due to distortion of the ionic atmosphere
- ☐ electrophoretic effect, the enhanced viscous drag due to the countercurrent of oppositely charged ions
- ☐ Debye–Hückel–Onsager theory, a theory of the concentration dependence of the molar conductivity of a strong electrolyte, $\mathcal{K} = A + B\Lambda_m^\circ$
- ☐ Green–Kubo relation, the relation between a transport property and fluctuations in microscopic properties of a system, $\kappa = \dfrac{1}{kTV}\displaystyle\int_0^\infty \langle j(0)j(t)\rangle\, dt$
- ☐ correlation function, the average of an observable compared at different times, $\langle j(0)j(t)\rangle$

Diffusion

24.9 The thermodynamic view

- ☐ thermodynamic force, $\mathcal{F} = -(\partial\mu/\partial x)_{p,T}$
- ☐ Einstein relation, $D = uRT/zF$
- ☐ Nernst–Einstein relation, $\Lambda_m = (\nu_+ z_+^2 D_+ + \nu_- z_-^2 D_-)(F^2/RT)$
- ☐ Stokes–Einstein equation, $D = kT/f$
- ☐ Walden's rule: The product $\eta\Lambda_m$ is very approximately constant for the same ions in different solvents

24.10 The diffusion equation

- ☐ diffusion equation, the relation between the rate of change of concentration at a point and the spatial variation of the concentration at that point: $\partial c/\partial t = D\partial^2 c/\partial x^2$

- ☐ convection, the transport of particles arising from the motion of a streaming fluid
- ☐ convective flux, $J = cv$
- ☐ generalized diffusion equation, the diffusion equation including convection: $\partial c/\partial t = D\partial^2 c/\partial x^2 - v\partial c/\partial x$
- ☐ capillary technique, a technique for determining the diffusion coefficient
- ☐ diaphragm technique, a technique for determining the diffusion coefficient

24.11 Diffusion probabilities

- ☐ average distance travelled in diffusion, $\langle x\rangle = 2(Dt/\pi)^{1/2}$
- ☐ root mean square distance travelled in diffusion, $\langle x^2\rangle^{1/2} = (2Dt)^{1/2}$

24.12 The statistical view

- ☐ one-dimensional random walk, a random walk along a straight line, $P = (2\tau/\pi t)^{1/2} e^{-x^2\tau/2t\lambda^2}$
- ☐ Einstein–Smoluchowski equation, $D = \lambda^2/2\tau$

Further reading

Articles of general interest

N. Agmon, The Grotthuss mechanism. *Chem. Phys. Letts.* **244**, 456 (1995).

P.J. Dunlop, B.J. Steel, and J.E. Lane, Experimental methods for studying diffusion in liquids, gases, and solids. In *Techniques of chemistry* (ed. A. Weissberger and B.W. Rossiter), **4**, 205. Wiley-Interscience, New York (1972).

B.L. Earl, Confusion in the expressions for transport coefficients. *J. Chem. Educ.* **66**, 147 (1989).

H.G. Herz, B.M. Braun, K.J. Müller, and R. Mauer, What is the physical significance of the pictures representing the Grotthuss H^+ conductance mechanism? *J. Chem. Educ.* **64**, 777 (1987).

T. Kenney. Graham's law: defining gas velocities. *J. Chem. Educ.* **67**, 871 (1990).

D.G. Leaist, Diffusion and ionic conduction in liquids. In *Encyclopedia of applied physics* (ed. G.L. Trigg), **5**, 661. VCH, New York (1993).

M.S.P. Sansom, I.H. Shrivastave, K.M. Ranatunga, and G.R. Smith, Simulations of ion channels: watching ions and water move. *Trends in Biol. Sci.* **25**, 368 (2000).

M. Spiro, Determination of transference numbers. In *Techniques of chemistry* (ed. A. Weissberger and B.W. Rossiter), **2A**, 205. Wiley-Interscience, New York (1971).

Texts and sources of data and information

A.J. Bard and L.R. Faulkner, *Electrochemical methods: fundamentals and applications*. Wiley, New York (2000).

R.B. Bird, W.E. Stewart, and E.N. Lightfoot, *Transport phenomena*. Wiley, New-York (1960).

J. Crank, *The mathematics of diffusion*. Clarendon Press, Oxford (1975).

C.E. Hecht, *Statistical mechanics and kinetic theory*. Dover, New York (1998).

J.O. Hirschfelder, *The molecular theory of gases and liquids*. Wiley, New York (1964).

W. Kauzmann, *Kinetic theory of gases*. Addison-Wesley, Reading (1966).

G.A. Krestov, N.P. Novosyolov, and I.S. Perelygin, *Ionic solvation*. Ellis Horwood/ Prentice Hall, Hemel Hempstead (1993).

D.R. Lide (ed.), *CRC Handbook of Chemistry and Physics*, Sections 5 and 6. CRC Press, Boca Raton (2000).

J.N. Murrell and A.D. Jenkins, *Properties of liquids and solutions*. Wiley-Interscience, New York (1994).

NIST thermodynamic and transport properties of pure fluids database, NIST standard reference database 12, National Institute of Standards and Testing, Gaithersburg (1995). This material is on CD-rom.

J.S. Rowlinson and F.L. Swinton, *Liquids and liquid mixtures*. Butterworths, London (1982).

D. Tabor, *Gases, liquids, and solids*. Cambridge University Press (1991).

K.E. van Holde, W.C. Johnson, and P.S. Ho, *Principles of physical biochemistry*. Prentice Hall, Upper Saddle River (1998).

A.J. Walton, *Three phases of matter*. Oxford University Press (1983).

Exercises

Discussion questions

24.1 (a) Use the kinetic theory to justify the following observations: (a) the rate of a reaction in the gas phase depends on the energy with which two molecules collide, which in turn depends on their speeds; (b) in the Earth's atmosphere, light gases, such as H_2 and He, are rare but heavier gases, such as O_2, CO_2, and N_2, are abundant.

24.1 (b) Provide a molecular interpretation for each of the following processes: diffusion, thermal conduction, electric conduction, and viscosity.

24.2 (a) Provide a molecular interpretation for the observation that the viscosity of a gas increases with temperature, whereas the viscosity of a liquid decreases with increasing temperature.

24.2 (b) Discuss the mechanism of proton conduction in liquid water.

24.3 (a) State the limits to the generality of the following expressions: (a) $J = -D(\mathrm{d}c/\mathrm{d}x)$, (b) $D = kT/f$, and (c) $D = kT/6\pi\eta a$.

24.3 (b) Discuss how nuclear magnetic resonance spectroscopy, inelastic neutron scattering, and dynamic light scattering may be used to measure the mobility of molecules in liquids.

Numerical exercises

24.4 (a) Determine the ratios of (a) the mean speeds, (b) the mean kinetic energies of H_2 molecules and Hg atoms at 20°C.

24.4 (b) Determine the ratios of (a) the mean speeds, (b) the mean kinetic energies of He atoms and Hg atoms at 25°C.

24.5 (a) A 1.0 L glass bulb contains 1.0×10^{23} H_2 molecules. If the pressure exerted by the gas is 100 kPa, what are (a) the temperature of the gas, (b) the root mean square speeds of the molecules. (c) Would the temperature be different if they were O_2 molecules?

24.5 (b) The best laboratory vacuum pump can generate a vacuum of about 1 nTorr. At 25°C and assuming that air consists of N_2 molecules with a collision diameter of 395 pm, calculate (a) the mean speed of the molecules, (b) the mean free path, (c) the collision frequency in the gas.

24.6 (a) At what pressure does the mean free path of argon at 25°C become comparable to the size of a 1 L vessel that contains it? Take $\sigma = 0.36$ nm^2.

24.6 (b) At what pressure does the mean free path of argon at 25°C become comparable to the diameters of the atoms themselves?

24.7 (a) At an altitude of 20 km the temperature is 217 K and the pressure 0.050 atm. What is the mean free path of N_2 molecules? ($\sigma = 0.43$ nm^2).

24.7 (b) At an altitude of 15 km the temperature is 217 K and the pressure 12.1 kPa. What is the mean free path of N_2 molecules? ($\sigma = 0.43$ nm^2).

24.8 (a) How many collisions does a single Ar atom make in 1.0 s when the temperature is 25°C and the pressure is (a) 10 atm, (b) 1.0 atm, (c) 1.0 μatm?

24.8 (b) How many collisions per second does an N_2 molecule make at an altitude of 15 km? (See Exercise 24.7b for data.)

24.9 (a) Calculate the mean free path of molecules in air using $\sigma = 0.43$ nm^2 at 25°C and (a) 10 atm, (b) 1.0 atm, (c) 1.0×10^{-6} atm.

24.9 (b) Calculate the mean free path of carbon dioxide molecules in air using $\sigma = 0.52$ nm^2 at 25°C and (a) 15 atm, (b) 1.0 bar, (c) 1.0 Torr.

24.10 (a) Use the Maxwell distribution of speeds to estimate the fraction of N_2 molecules at 500 K that have speeds in the range 290 to 300 m s^{-1}.

24.10 (b) Use the Maxwell distribution of speeds to estimate the fraction of CO_2 molecules at 300 K that have speeds in the range 200 to 250 m s^{-1}.

24.11 (a) A solid surface with dimensions 2.5 mm × 3.0 mm is exposed to argon gas at 90 Pa and 500 K. How many collisions do the Ar atoms make with this surface in 15 s?

24.11 (b) A solid surface with dimensions 3.5 mm × 4.0 cm is exposed to helium gas at 111 Pa and 1500 K. How many collisions do the He atoms make with this surface in 10 s?

24.12 (a) An effusion cell has a circular hole of diameter 2.50 mm. If the molar mass of the solid in the cell is 260 g mol^{-1} and its vapour pressure is 0.835 Pa at 400 K, by how much will the mass of the solid decrease in a period of 2.00 h?

24.12 (b) An effusion cell has a circular hole of diameter 3.00 mm. If the molar mass of the solid in the cell is 300 g mol^{-1} and its vapour pressure is 0.224 Pa at 450 K, by how much will the mass of the solid decrease in a period of 24.00 h?

24.13 (a) Calculate the flux of energy arising from a temperature gradient of 2.5 K m^{-1} in a sample of argon in which the mean temperature is 273 K.

24.13 (b) Calculate the flux of energy arising from a temperature gradient of 3.5 K m^{-1} in a sample of hydrogen in which the mean temperature is 260 K.

24.14 (a) Use the experimental value of the thermal conductivity of neon (Table 24.2) to estimate the collision cross-section of Ne atoms at 273 K.

24.14 (b) Use the experimental value of the thermal conductivity of nitrogen (Table 24.2) to estimate the collision cross-section of N_2 molecules at 298 K.

24.15 (a) In a double-glazed window, the panes of glass are separated by 5.0 cm. What is the rate of transfer of heat by conduction from the warm room (25°C) to the cold exterior (−10°C) through a window of area 1.0 m^2? What power of heater is required to make good the loss of heat?

24.15 (b) Two sheets of copper of area 1.50 m^2 are separated by 10.0 cm. What is the rate of transfer of heat by conduction from the warm sheet (50°C) to the cold sheet (−10°C). What is the rate of loss of heat?

24.16 (a) A manometer was connected to a bulb containing carbon dioxide under slight pressure. The gas was allowed to escape through a small pinhole, and the time for the manometer reading to drop from 75 cm to 50 cm was 52 s. When the experiment was repeated using nitrogen (for which $M = 28.02$ g mol^{-1}) the same fall took place in 42 s. Calculate the molar mass of carbon dioxide.

24.16 (b) A manometer was connected to a bulb containing nitrogen under slight pressure. The gas was allowed to escape through a small pinhole, and the time for the manometer reading to drop from 65.1 cm to 42.1 cm was 18.5 s. When the experiment was repeated using a fluorocarbon gas, the same fall took place in 82.3 s. Calculate the molar mass of the fluorocarbon.

24.17 (a) A space vehicle of internal volume 3.0 m^3 is struck by a meteor and a hole of radius 0.10 mm is formed. If the oxygen pressure within the vehicle is initially 80 kPa and its temperature 298 K, how long will the pressure take to fall to 70 kPa?

24.17 (b) A container of internal volume 22.0 m^3 was punctured, and a hole of radius 0.050 mm was formed. If the nitrogen pressure within the vehicle is initially 122 kPa and its temperature 293 K, how long will the pressure take to fall to 105 kPa?

24.18 (a) Use the experimental value of the coefficient of viscosity for neon (Table 24.2) to estimate the collision cross-section of Ne atoms at 273 K.

24.18 (b) Use the experimental value of the coefficient of viscosity for nitrogen (Table 24.2) to estimate the collision cross-section of the molecules at 273 K.

24.19 (a) Calculate the inlet pressure required to maintain a flow rate of 9.5×10^5 L h^{-1} of nitrogen at 293 K flowing through a pipe of length 8.50 m and diameter 1.00 cm. The pressure of gas as it leaves the tube is 1.00 bar. The volume of the gas is measured at that pressure.

24.19 (b) Calculate the inlet pressure required to maintain a flow rate of 8.70 $cm^3\ s^{-1}$ of nitrogen at 300 K flowing through a pipe of length 10.5 m and diameter 15 mm. The pressure of gas as it leaves the tube is 1.00 bar. The volume of the gas is measured at that pressure.

24.20 (a) Calculate the viscosity of air at (a) 273 K, (b) 298 K, (c) 1000 K. Take $\sigma \approx 0.40\ nm^2$. (The experimental values are 173 μP at 273 K, 182 μP at 20°C, and 394 μP at 600°C.)

24.20 (b) Calculate the viscosity of benzene vapour at (a) 273 K, (b) 298 K, (c) 1000 K. Take $\sigma \approx 0.88\ nm^2$.

24.21 (a) Calculate the thermal conductivities of (a) argon, (b) helium at 300 K and 1.0 mbar. Each gas is confined in a cubic vessel of side 10 cm, one wall being at 310 K and the one opposite at 295 K. What is the rate of flow of energy as heat from one wall to the other in each case?

24.21 (b) Calculate the thermal conductivities of (a) neon, (b) nitrogen at 300 K and 15 mbar. Each gas is confined in a cubic vessel of side 15 cm, one wall being at 305 K and the one opposite at 295 K. What is the rate of flow of energy as heat from one wall to the other in each case?

24.22 (a) The viscosity of carbon dioxide was measured by comparing its rate of flow through a long narrow tube (using Poiseuille's formula) with that of argon. For the same pressure differential, the same volume of carbon dioxide passed through the tube in 55 s as argon in 83 s. The viscosity of argon at 25°C is 208 μP; what is the viscosity of carbon dioxide? Estimate the molecular diameter of carbon dioxide.

24.22 (b) The viscosity of a chlorofluorocarbon (CFC) was measured by comparing its rate of flow through a long narrow tube (using Poiseuille's formula) with that of argon. For the same pressure differential, the same volume of the CFC passed through the tube in 72.0 s as argon in 18.0 s. The viscosity of argon at 25°C is 208 μP; what is the viscosity of the CFC? Estimate the molecular diameter of the CFC. Take $M = 200\ g\ mol^{-1}$.

24.23 (a) Calculate the thermal conductivity of argon ($C_{V,m} = 12.5\ J\ K^{-1}\ mol^{-1}$, $\sigma = 0.36\ nm^2$) at room temperature (20°C).

24.23 (b) Calculate the thermal conductivity of nitrogen ($C_{V,m} = 20.8\ J\ K^{-1}\ mol^{-1}$, $\sigma = 0.43\ nm^2$) at room temperature (20°C).

24.24 (a) Calculate the diffusion constant of argon at 25°C and (a) 1.00 Pa, (b) 100 kPa, (c) 10.0 MPa. If a pressure gradient of 0.10 atm cm^{-1} is established in a pipe, what is the flow of gas due to diffusion?

24.24 (b) Calculate the diffusion constant of nitrogen at 25°C and (a) 10.0 Pa, (b) 100 kPa, (c) 15.0 MPa. If a pressure gradient of 0.20 bar m^{-1} is established in a pipe, what is the flow of gas due to diffusion?

24.25 (a) The mobility of a chloride ion in aqueous solution at 25°C is $7.91 \times 10^{-8}\ m^2\ s^{-1}\ V^{-1}$. Calculate the molar ionic conductivity.

24.25 (b) The mobility of an acetate ion in aqueous solution at 25°C is $4.24 \times 10^{-8}\ m^2\ s^{-1}\ V^{-1}$. Calculate the molar ionic conductivity.

24.26 (a) The mobility of a Rb^+ ion in aqueous solution is $7.92 \times 10^{-8}\ m^2\ s^{-1}\ V^{-1}$ at 25°C. The potential difference between two electrodes placed in the solution is 35.0 V. If the electrodes are 8.00 mm apart, what is the drift speed of the Rb^+ ion?

24.26 (b) The mobility of a Li^+ ion in aqueous solution is $4.01 \times 10^{-8}\ m^2\ s^{-1}\ V^{-1}$ at 25°C. The potential difference between two electrodes placed in the solution is 12.0 V. If the electrodes are 1.00 cm apart, what is the drift speed of the ion?

24.27 (a) What fraction of the total current is carried by Li^+ when current flows through an aqueous solution of LiBr at 25°C?

24.27 (b) What fraction of the total current is carried by Cl^- when current flows through an aqueous solution of NaCl at 25°C?

24.28 (a) The limiting molar conductivities of KCl, KNO_3, and $AgNO_3$ are 14.99 mS $m^2\ mol^{-1}$, 14.50 mS $m^2\ mol^{-1}$, and 13.34 mS $m^2\ mol^{-1}$, respectively (all at 25°C). What is the limiting molar conductivity of AgCl at this temperature?

24.28 (b) The limiting molar conductivities of NaI, $NaCH_3CO_2$, and $Mg(CH_3CO_2)_2$ are 12.69 mS $m^2\ mol^{-1}$, 9.10 mS $m^2\ mol^{-1}$, and 18.78 mS $m^2\ mol^{-1}$, respectively (all at 25°C). What is the limiting molar conductivity of MgI_2 at this temperature?

24.29 (a) At 25°C the molar ionic conductivities of Li^+, Na^+, and K^+ are 3.87 mS $m^2\ mol^{-1}$, 5.01 mS $m^2\ mol^{-1}$, and 7.35 mS $m^2\ mol^{-1}$, respectively. What are their mobilities?

24.29 (b) At 25°C the molar ionic conductivities of F^-, Cl^-, and Br^- are 5.54 mS $m^2\ mol^{-1}$, 7.635 mS $m^2\ mol^{-1}$, and 7.81 mS $m^2\ mol^{-1}$, respectively. What are their mobilities?

24.30 (a) The mobility of a NO_3^- ion in aqueous solution at 25°C is $7.40 \times 10^{-8}\ m^2\ s^{-1}\ V^{-1}$. Calculate its diffusion coefficient in water at 25°C.

24.30 (b) The mobility of a $CH_3CO_2^-$ ion in aqueous solution at 25°C is $4.24 \times 10^{-8}\ m^2\ s^{-1}\ V^{-1}$. Calculate its diffusion coefficient in water at 25°C.

24.31 (a) The diffusion coefficient of CCl_4 in heptane at 25°C is $3.17 \times 10^{-9}\ m^2\ s^{-1}$. Estimate the time required for a CCl_4 molecule to have a root mean square displacement of 5.0 mm.

24.31 (b) The diffusion coefficient of I_2 in hexane at 25°C is $4.05 \times 10^{-9}\ m^2\ s^{-1}$. Estimate the time required for an iodine molecule to have a root mean square displacement of 1.0 cm.

24.32 (a) Estimate the effective radius of a sucrose molecule in water at 25°C given that its diffusion coefficient is $5.2 \times 10^{-10}\ m^2\ s^{-1}$ and that the viscosity of water is 1.00 cP.

24.32 (b) Estimate the effective radius of a glycine molecule in water at 25°C given that its diffusion coefficient is $1.055 \times 10^{-9}\ m^2\ s^{-1}$ and that the viscosity of water is 1.00 cP.

24.33 (a) The diffusion coefficient for molecular iodine in benzene is $2.13 \times 10^{-9}\ m^2\ s^{-1}$. How long does a molecule take to jump through about one molecular diameter (approximately the fundamental jump length for translational motion)?

24.33 (b) The diffusion coefficient for CCl_4 in heptane is $3.17 \times 10^{-9}\ m^2\ s^{-1}$. How long does a molecule take to jump through about one molecular diameter (approximately the fundamental jump length for translational motion)?

24.34 (a) What are the root mean square distances travelled by an iodine molecule in benzene and by a sucrose molecule in water at 25°C in 1.0 s?

24.34 (b) About how long, on average, does it take for the molecules in Exercise 24.34a to drift to a point (a) 1.0 mm, (b) 1.0 cm from their starting points?

Problems†

Numerical problems

24.1 Instead of the arrangement in Fig. 24.8, the speed of molecules can also be measured with a rotating slotted-disc apparatus, which consists of five coaxial 5.0 cm diameter disks separated by 1.0 cm, the slots in their rims being displaced by 2.0° between neighbours. The relative intensities, I, of the detected beam of Kr atoms for two different temperatures and at a series of rotation rates were as follows:

ν/Hz	20	40	80	100	120
I(40 K)	0.846	0.513	0.069	0.015	0.002
I(100 K)	0.592	0.485	0.217	0.119	0.057

Find the distributions of molecular velocities, $f(v_x)$, at these temperatures, and check that they conform to the theoretical prediction for a one-dimensional system.

24.2 Cars were timed by police radar as they passed in both directions below a bridge. Their velocities (kilometres per hour, numbers of cars in parentheses) to the east and west were as follows: 80 E (40), 85 E (62), 90 E (53), 95 E (12), 100 E (2); 80 W (38), 85 W (59), 90 W (60), 100 W (2). What are (a) the mean velocity, (b) the mean speed, (c) the root mean square speed?

24.3 A population consists of people of the following heights (in metres, numbers of individuals in brackets): 1.80 (1), 1.82 (2), 1.84 (4), 1.86 (7), 1.88 (10), 1.90 (15), 1.92 (9), 1.94 (4), 1.96 (0), 1.98 (1). What are (a) the mean height, (b) the root mean square height of the population?

24.4 Calculate the ratio of the thermal conductivities of gaseous hydrogen at 300 K to gaseous hydrogen at 10 K. Be circumspect, and think about the modes of motion that are thermally active at the two temperatures.

24.5 In the Knudsen method for the determination of vapour pressure, a weighed amount of a sample is heated inside a container, in the wall of which there is a small hole. The mass loss over a period of time can be related to the vapour pressure at the temperature of the experiment. If Δm is the mass lost in an interval Δt through a circular hole of radius R, find an expression relating the vapour pressure, p, to Δm and Δt. A Knudsen cell was used to determine the vapour pressure of germanium at 1000°C. During an interval of 7200 s the mass loss through a hole of radius 0.50 mm amounted to 43 μg. What is the vapour pressure of germanium at 1000°C? Assume the gas to be monatomic.

24.6 The nuclide ^{244}Bk (berkelium) decays by producing α particles, which capture electrons and form He atoms. Its half-life is 4.4 h. A sample of mass 1.0 mg was placed in a container of volume 1.0 cm^3 that was impermeable to α radiation, but there was also a hole of radius 2.0 μm in the wall. What is the pressure of helium at 298 K, inside the container after (a) 1.0 h, (b) 10 h?

24.7 An atomic beam is designed to function with (a) cadmium, (b) mercury. The source is an oven maintained at 380 K, there being a small slit of dimensions 1.0 cm × 1.0 cm × 10^{-3} cm. The vapour pressure of cadmium is 0.13 Pa and that of mercury is 152 kPa at this temperature. What is the atomic current (the number of atoms per unit time) in the beams?

24.8 Conductivities are often measured by comparing the resistance of a cell filled with the sample to its resistance when filled with some standard solution, such as aqueous potassium chloride. The conductivity of water is 76 mS m^{-1} at 25°C and the conductivity of 0.100 mol L^{-1} KCl(aq) is 1.1639 S m^{-1}. A cell had a resistance of 33.21 Ω when filled with 0.100 mol L^{-1} KCl(aq) and 300.0 Ω when filled with 0.100 mol L^{-1} CH_3COOH. What is the molar conductivity of acetic acid at that concentration and temperature?

24.9 The resistances of a series of aqueous NaCl solutions, formed by successive dilution of a sample, were measured in a cell with cell constant (the constant C in the relation $\kappa = C/R$) equal to 0.2063 cm^{-1}. The following values were found:

c/(mol L^{-1})	0.000 50	0.0010	0.0050	0.010	0.020	0.050
R/Ω	3314	1669	342.1	174.1	89.08	37.14

Verify that the molar conductivity follows the Kohlrausch law and find the limiting molar conductivity. Determine the coefficient $\mathcal{K}$. Use the value of $\mathcal{K}$ (which should depend only on the nature, not the identity of the ions) and the information that $\lambda(Na^+) = 5.01$ mS m^2 mol^{-1} and $\lambda(I^-) = 7.68$ mS m^2 mol^{-1} to predict (a) the molar conductivity, (b) the conductivity, (c) the resistance it would show in the cell, of 0.010 mol L^{-1} NaI(aq) at 25°C.

24.10 After correction for the water conductivity, the conductivity of a saturated aqueous solution of AgCl at 25°C was found to be 0.1887 mS m^{-1}. What is the solubility of silver chloride at this temperature?

24.11 What are the drift speeds of Li^+, Na^+, and K^+ in water when a potential difference of 10 V is applied across a 1.00-cm conductivity cell? How long would it take an ion to move from one electrode to the other? In conductivity measurements it is normal to use alternating current: what are the displacements of the ions in (a) centimetres, (b) solvent diameters, about 300 pm, during a half cycle of 1.0 kHz applied potential?

24.12 The mobilities of H^+ and Cl^- at 25°C in water are 3.623×10^{-7} m^2 s^{-1} V^{-1} and 7.91×10^{-8} m^2 s^{-1} V^{-1}, respectively. What proportion of the current is carried by the protons in 1.0×10^{-3} M HCl(aq)? What fraction do they carry when the NaCl is added to the acid so that the solution is 1.0 mol L^{-1} in the salt? Note how concentration as well as mobility governs the transport of current.

24.13 In a moving boundary experiment on KCl the apparatus consisted of a tube of internal diameter 4.146 mm, and it contained aqueous KCl at a concentration of 0.021 mol L^{-1}. A steady current of 18.2 mA was passed, and the boundary advanced as follows:

Δt/s	200	400	600	800	1000
x/mm	64	128	192	254	318

Find the transport number of K^+, its mobility, and its ionic conductivity.

24.14 The proton possesses abnormal mobility in water, but does it behave normally in liquid ammonia? To investigate this question, a moving-boundary technique was used to determine the transport

† Problems denoted with the symbol ‡ were supplied by Charles Trapp, Carmen Giunta, and Marshall Cady.

number of NH_4^+ in liquid ammonia (the analogue of H_3O^+ in liquid water) at −40°C (J. Baldwin, J. Evans, and J.B. Gill, *J. Chem. Soc. A*, 3389 (1971)). A steady current of 5.000 mA was passed for 2500 s, during which time the boundary formed between mercury(II) iodide and ammonium iodide solutions in ammonia moved 286.9 mm in a 0.013 65 mol kg^{-1} solution and 92.03 mm in a 0.042 55 mol kg^{-1} solution. Calculate the transport number of NH_4^+ at these concentrations, and comment on the mobility of the proton in liquid ammonia. The bore of the tube is 4.146 mm and the density of liquid ammonia is 0.682 g cm^{-3}.

24.15 A dilute solution of potassium permanganate in water at 25°C was prepared. The solution was in a horizontal tube of length 10 cm, and at first there was a linear gradation of intensity of the purple solution from the left (where the concentration was 0.100 mol L^{-1}) to the right (where the concentration was 0.050 mol L^{-1}). What are the magnitude and sign of the thermodynamic force acting on the solute (a) close to the left face of the container, (b) in the middle, (c) close to the right face? Give the force per mole and force per molecule in each case.

24.16 Estimate the diffusion coefficients and the effective hydrodynamic radii of the alkali metal cations in water from their mobilities at 25°C. Estimate the approximate number of water molecules that are dragged along by the cations. Ionic radii are given in Table 21.3.

24.17 Nuclear magnetic resonance can be used to determine the mobility of molecules in liquids. A set of measurements on methane in carbon tetrachloride showed that its diffusion coefficient is 2.05×10^{-9} m^2 s^{-1} at 0°C and 2.89×10^{-9} m^2 s^{-1} at 25°C. Deduce what information you can about the mobility of methane in carbon tetrachloride.

24.18 A concentrated sucrose solution is poured into a cylinder of diameter 5.0 cm. The solution consisted of 10 g of sugar in 5.0 cm^3 of water. A further 1.0 L of water is then poured very carefully on top of the layer, without disturbing the layer. Ignore gravitational effects, and pay attention only to diffusional processes. Find the concentration at 5.0 cm above the lower layer after a lapse of (a) 10 s, (b) 1.0 y.

24.19 In a series of observations on the displacement of rubber latex spheres of radius 0.212 μm, the mean square displacements after selected time intervals were on average as follows:

t/s	30	60	90	120
$10^{12}\langle x^2\rangle$/m^2	88.2	113.5	128	144

These results were originally used to find the value of the Avogadro constant, but there are now better ways of determining N_A, so the data can be used to find another quantity. Find the effective viscosity of water at the temperature of this experiment (25°C).

24.20‡ A.K. Srivastava, R.A. Samant, and S.D. Patankar (*J. Chem. Eng. Data* **41**, 431 (1996)) measured the conductance of several salts in a binary solvent mixture of water and a dipolar aprotic solvent 1,3-dioxolan-2-one (ethylene carbonate). They report the following conductances at 25°C in a solvent 80 per cent 1,3-dioxolan-2-one by mass:

NaI

c/(mmol L^{-1})	32.02	20.28	12.06	8.64	2.85	1.24	0.83
Λ_m/(S cm^2 mol^{-1})	50.26	51.99	54.01	55.75	57.99	58.44	58.67

KI

c/(mmol L^{-1})	17.68	10.88	87.19	2.67	1.28	0.83	0.19
Λ_m/(S cm^2 mol^{-1})	42.45	45.91	47.53	51.81	54.09	55.78	57.42

Calculate Λ_m° for NaI and KI in this solvent and λ°(Na) − λ°(K). Compare your results to the analogous quantities in aqueous solution using Table 24.5 in the *Data section*.

24.21‡ A. Fenghour, W.A. Wakeham, V. Vesovic, J.T.R. Watson, J. Millat, and E. Vogel (*J. Phys. Chem. Ref. Data* **24**, 1649 (1995)) have compiled an extensive table of viscosity coefficients for ammonia in the liquid and vapour phases. Deduce the effective molecular diameter of NH_3 based on each of the following vapour-phase viscosity coefficients: (a) $\eta = 9.08 \times 10^{-6}$ kg m^{-1} s^{-1} at 270 K and 1.00 bar; (b) $\eta = 1.749 \times 10^{-5}$ kg m^{-1} s^{-1} at 490 K and 10.0 bar.

24.22‡ G. Bakale, K. Lacmann, and W.F. Schmidt (*J. Phys. Chem.* **100**, 12 477 (1996)) measured the mobility of singly charged C_{60}^- ions in a variety of nonpolar solvents. In cyclohexane at 22°C, the mobility is 1.1 cm^2 V^{-1} s^{-1}. Estimate the effective radius of the C_{60}^- ion. The viscosity of the solvent is 0.93×10^{-3} kg m^{-1} s^{-1}. *Comment.* The researchers interpreted the substantial difference between the radius so determined and the van der Waals radius of neutral C_{60} in terms of a solvation layer around the ion.

Theoretical problems

24.23 Start from the Maxwell–Boltzmann distribution and derive an expression for the most probable speed of a gas of molecules at a temperature T. Go on to demonstrate the validity of the equipartition conclusion that the average translational kinetic energy of molecules free to move in three dimensions is $\frac{3}{2}kT$.

24.24 Consider molecules that are confined to move in a plane (a two-dimensional gas). Calculate the distribution of speeds and determine the mean speed of the molecules at a temperature T.

24.25 A specially constructed velocity-selector accepts a beam of molecules from an oven at a temperature T but blocks the passage of molecules with a speed greater than the mean. What is the mean speed of the emerging beam, relative to the initial value, treated as a one-dimensional problem?

24.26 What is the proportion of gas molecules having (a) more than, (b) less than the root mean square speed? (c) What are the proportions having speeds greater and smaller than the mean speed?

24.27 Calculate the fractions of molecules in a gas that have a speed in a range Δv at the speed nc^* relative to those in the same range at c^* itself. This calculation can be used to estimate the fraction of very energetic molecules (which is important for reactions). Evaluate the ratio for $n = 3$ and $n = 4$.

24.28 Show how the ratio of two transport numbers t' and t'' for two cations in a mixture depends on their concentrations c' and c'', and their mobilities u' and u''.

24.29 Confirm that eqn 88 is a solution of the diffusion equation with the correct initial value.

24.30 The diffusion equation is valid when many elementary steps are taken in the time interval of interest; but the random walk calculation lets us discuss distributions for short times as well as for long. Use eqn 92 to calculate the probability of being six paces from the origin (that is, at $x = 6\lambda$) after (a) four, (b) six, (c) twelve steps.

24.31 Use mathematical software to calculate P in a one-dimensional random walk, and evaluate the probability of being at $x = n\lambda$ for $n = 6, 10, 14, \ldots, 60$. Compare the numerical value with the analytical value in the limit of a large number of steps. At what value of n is the discrepancy no more than 0.1 per cent?

24.32‡ A dilute solution of a weak (1,1)-electrolyte contains both neutral ion pairs and ions in equilibrium ($AB \rightleftharpoons A^+ + B^-$). Prove that molar conductivities are related to the degree of ionization by the equations:

$$\frac{1}{\Lambda_m} = \frac{1}{\Lambda_m(\alpha)} + \frac{(1-\alpha)\Lambda_m^\circ}{\alpha^2 \Lambda_m(\alpha)^2}$$

where Λ_m° is the molar conductivity at infinite dilution.

Applications: to astrophysics and biology

24.33 Calculate the escape velocity (the minimum initial velocity that will take an object to infinity) from the surface of a planet of radius R. What is the value for (a) the Earth, $R = 6.37 \times 10^6$ m, $g = 9.81$ m s^{-2}, (b) Mars, $R = 3.38 \times 10^6$ m, $m_{Mars}/m_{Earth} = 0.108$. At what temperatures do H_2, He, and O_2 molecules have mean speeds equal to their escape speeds? What proportion of the molecules have enough speed to escape when the temperature is (a) 240 K, (b) 1500 K? Calculations of this kind are very important in considering the composition of planetary atmospheres.

24.34‡ Interstellar space is quite different from the gaseous environments we commonly encounter on Earth. For instance, a typical density of the medium is about 1 atom cm^{-3} and that atom is typically H; the effective temperature due to stellar background radiation is about 10 000 K. Estimate the diffusion coefficient and thermal conductivity of H under these conditions. *Comment*. Energy is in fact transferred much more effectively by radiation.

24.35 Enrico Fermi, the great Italian scientist, was a master at making good approximate calculations based on little or no actual data. Hence, such calculations are often called 'Fermi calculations'. Do a Fermi calculation on how long it would take for a gaseous air-borne cold virus of molar mass 100 kg mol^{-1} to travel the distance between two conversing people 1.0 m apart by diffusion in still air.

24.36 The diffusion coefficient of a particular kind of tRNA molecule is $D = 1.0 \times 10^{-11}$ m^2 s^{-1} in the medium of a cell interior. How long does it take molecules produced in the cell nucleus to reach the walls of the cell at a distance 1.0 μm, corresponding to the radius of the cell?

24.37‡ In this problem, we examine a model for the transport of oxygen from air in the lungs to blood. First, show that, for the initial and boundary conditions $c(x, t) = c(x, 0) = c_0$, $(0 < x < \infty)$ and $c(0, t) = c_s$, $(0 \le t \le \infty)$ where c_0 and c_s are constants, the concentration, $c(x, t)$, of a species is given by

$$c(x,t) = c_0 + (c_s - c_0)\{1 - \mathrm{erf}(\xi)\} \qquad \xi(x,t) = \frac{x}{(4Dt)^{1/2}}$$

where erf(ξ) is the error function (*Justification 12.3*) and the concentration $c(x, t)$ evolves by diffusion from the *yz*-plane of constant concentration, such as might occur if a condensed phase is absorbing a species from a gas phase. Now draw graphs of concentration profiles at several different times of your choice for the diffusion of oxygen into water at 298 K (when $D = 2.10 \times 10^{-9}$ m^2 s^{-1}) on a spatial scale comparable to passage of oxygen from lungs through alveoli into the blood. Use $c_0 = 0$ and set c_s equal to the solubility of oxygen in water. *Hint*. Use mathematical software.

25

The rates of chemical reactions

This chapter is the first of a sequence that explores the rates of chemical reactions. The chapter begins with a discussion of the definition of reaction rate and outlines the techniques for its measurement. The results of such measurements show that reaction rates depend on the concentration of reactants (and products) in characteristic ways that can be expressed in terms of differential equations known as rate laws. The solutions of these equations are used to predict the concentrations of species at any time after the start of the reaction. The form of the rate law also provides insight into the series of elementary steps by which a reaction takes place. The key task in this connection is the construction of a rate law from a proposed mechanism and its comparison with experiment. Simple elementary steps have simple rate laws, and these rate laws can be combined together by invoking one or more approximations. These approximations include the concept of the rate-determining stage of a reaction, the steady-state concentration of a reaction intermediate, and the existence of a pre-equilibrium.

This chapter introduces the principles of **chemical kinetics**, the study of reaction rates, by showing how the rates of reactions may be measured and interpreted. The remaining chapters of this part of the text then develop this material in more detail and apply it to more complicated or more specialized cases. The rate of a chemical reaction might depend on variables under our control, such as the pressure, the temperature, and the presence of a catalyst, and we may be able to optimize the rate by the appropriate choice of conditions. The study of reaction rates also leads to an understanding of the **mechanisms** of reactions, their analysis into a sequence of elementary steps.

Empirical chemical kinetics

The first steps in the kinetic analysis of reactions are to establish the stoichiometry of the reaction and identify any side reactions. The basic data of chemical kinetics are then the concentrations of the reactants and products at different times after a reaction has been initiated. The rates of most chemical reactions are sensitive to the temperature, so in conventional experiments the temperature of the reaction mixture must be held constant throughout the course of the reaction. This requirement puts severe demands on the design of an experiment. Gas-phase reactions, for instance, are often carried out in a vessel held in contact with a substantial block of metal. Liquid-phase reactions, including flow reactions, must be carried out in an efficient thermostat. Special efforts have to be made to study reactions at low temperatures, as in the study of the kinds of reactions that take place in interstellar clouds. Thus, supersonic expansion of the reaction gas can be used to attain temperatures as low as 10 K. For work in the liquid phase and the solid phase, very low temperatures are often reached by flowing cold liquid or cold gas around the reaction vessel. Alternatively, the entire reaction vessel is immersed in a thermally insulated container filled with a cryogenic liquid, such as liquid helium (for work at around 4 K) or liquid nitrogen (for work at around 77 K). Non-isothermal conditions are sometimes employed. For instance, the shelf-life of an expensive pharmaceutical may be explored by slowly raising the temperature of a single sample.

25.1 Experimental techniques

The method used to monitor concentrations depends on the species involved and the rapidity with which their concentrations change. Many reactions reach equilibrium over periods of minutes or hours, and several techniques may then be used to follow the changing concentrations.

(a) Monitoring the progress of a reaction

A reaction in which at least one component is a gas might result in an overall change in pressure in a system of constant volume, so its progress may be followed by recording the variation of pressure with time.

Example 25.1 Monitoring the variation in pressure

Predict how the total pressure varies during the gas-phase decomposition $2\,N_2O_5(g) \rightarrow 4\,NO_2(g) + O_2(g)$.

Method The total pressure (at constant volume and temperature and assuming perfect gas behaviour) is proportional to the number of gas-phase molecules. Therefore, because each mole of N_2O_5 gives rise to $\frac{5}{2}$ mol of gas molecules, we can expect the pressure to rise to $\frac{5}{2}$ times its initial value. To confirm this conclusion, express the progress of the reaction in terms of the fraction, α, of N_2O_5 molecules that have reacted.

Answer Let the initial pressure be p_0 and the initial amount of N_2O_5 molecules present be n. When a fraction α of the N_2O_5 molecules has decomposed, the amounts of the components in the reaction mixture are:

	N_2O_5	NO_2	O_2	Total
Amount:	$n(1-\alpha)$	$2\alpha n$	$\frac{1}{2}\alpha n$	$n(1+\frac{3}{2}\alpha)$

When $\alpha = 0$ the pressure is p_0, so at any stage the total pressure is

$$p = (1 + \tfrac{3}{2}\alpha)p_0$$

When the reaction is complete, the pressure will have risen to $\frac{5}{2}$ times its initial value.

Self-test 25.1 Repeat the calculation for $2\,NOBr(g) \rightarrow 2\,NO(g) + Br_2(g)$.
[$p = (1 + \frac{1}{2}\alpha)p_0$]

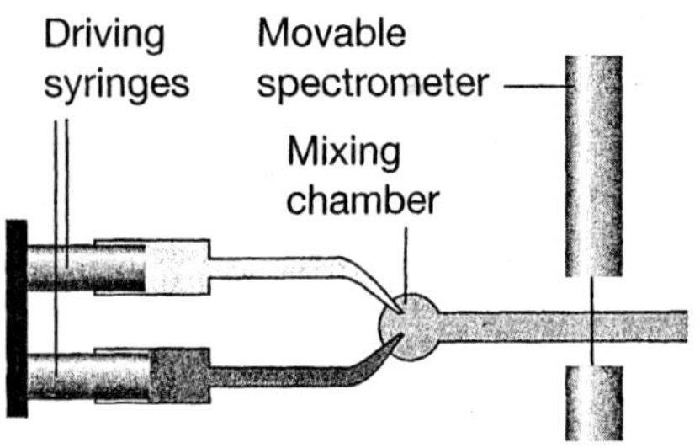

25.1 The arrangement used in the flow technique for studying reaction rates. The reactants are injected into the mixing chamber at a steady rate. The location of the spectrometer corresponds to different times after initiation.

Spectrophotometry, the measurement of absorption of radiation in a particular spectral region, is widely applicable, and is especially useful when one substance in the reaction mixture has a strong characteristic absorption in a conveniently accessible region of the electromagnetic spectrum. For example, the progress of the reaction

$$H_2(g) + Br_2(g) \longrightarrow 2\,HBr(g)$$

can be followed by measuring the absorption of visible light by bromine. A reaction that changes the number or type of ions present in a solution may be followed by monitoring the electrical conductivity of the solution. The replacement of neutral molecules by ionic products can result in dramatic changes in the conductivity, as in the reaction

$$(CH_3)_3CCl(aq) + H_2O(l) \longrightarrow (CH_3)_3COH(aq) + H^+(aq) + Cl^-(aq)$$

If hydrogen ions are produced or consumed, the reaction may be followed by monitoring the pH of the solution.

Other methods of determining composition include emission spectroscopy, mass spectrometry, gas chromatography, nuclear magnetic resonance, and electron spin resonance (for reactions involving radicals or paramagnetic *d*-metal ions).

(b) Application of the techniques

In a **real-time analysis** the composition of the system is analysed while the reaction is in progress. Either a small sample is withdrawn or the bulk solution is monitored. In the **flow method** the reactants are mixed as they flow together in a chamber (Fig. 25.1). The reaction continues as the thoroughly mixed solutions flow through the outlet tube, and observation of the composition at different positions along the tube is equivalent to the observation of the reaction mixture at different times after mixing. The disadvantage of conventional flow techniques is that a large volume of reactant solution is necessary. This makes the study of fast reactions particularly difficult because to spread the reaction over a length of tube the flow must be rapid. This disadvantage is avoided by the **stopped-flow technique**, in which the reagents are mixed very quickly in a small chamber fitted with a syringe instead of an outlet tube (Fig. 25.2). The flow ceases when the plunger of the syringe reaches a stop, and the reaction continues in the mixed solutions. Observations, usually using spectroscopic techniques such as ultraviolet–visible absorption, circular dichroism, and fluorescence emission, are made on the sample as a function of time. The technique allows for the study of reactions that occur on the millisecond to second timescale. The suitability of the stopped-flow method to the study of small samples means that it is appropriate for many biochemical reactions, and it has been widely used to study the kinetics of protein folding and enzyme action (Box 25.1).

Very fast reactions can be studied by **flash photolysis**, in which the sample is exposed to a brief flash of light that initiates the reaction, and then the contents of the reaction chamber are monitored. Most work is now done with lasers with photolysis pulse widths that range from femtoseconds to nanoseconds (Section 17.5). The apparatus used for flash photolysis studies is based on the experimental design for time-resolved spectroscopy (Section 17.7e). Reactions occurring on a picosecond or femtosecond timescale may be probed by using electronic absorption or emission, infrared absorption, or Raman scattering. The spectra are recorded at a series of times following laser excitation. The laser pulse can initiate the reaction by forming a reactive species, such as an excited electronic state of a molecule, a radical, or an ion. We discuss examples of excited state reactions in Chapter 26. An example of radical generation is the light-induced dissociation of $Cl_2(g)$ to yield Cl atoms that react with HBr to make HCl and Br according to the following sequence:

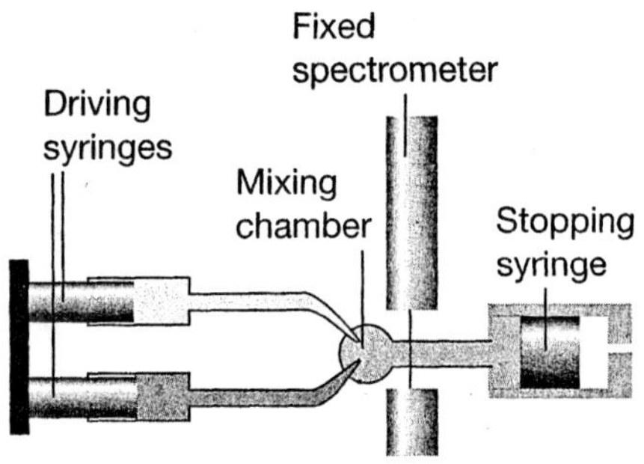

25.2 In the stopped-flow technique the reagents are driven quickly into the mixing chamber by the driving syringes and then the time dependence of the concentrations is monitored.

Box 25.1 Kinetics of the helix–coil transition in polypeptides

We saw in Box 20.1 that a simple statistical model accounts for the thermodynamic aspects of the helix–coil transition in polypeptides. The unfolding of a helix begins somewhere in the middle of the chain with a nucleation step, which is less favourable than the remaining helix to coil conversions, and continues in a cooperative fashion, in which the polymer becomes increasingly more susceptible to structural changes as more conversions occur. Here we examine the kinetics of the helix–coil transition, focusing primarily on experimental strategies and some recent results.

Earlier work on folding and unfolding of small polypeptides and large proteins relied primarily on rapid mixing and stopped-flow techniques (Section 25.1). In a typical stopped-flow experiment, a sample of the protein with a high concentration of a chemical denaturant, such as urea or guanidinium hydrochloride, is mixed with a solution containing a much lower concentration of the same denaturant. Upon entering the mixing chamber, the denaturant is diluted and the protein refolds. Unfolding is observed by mixing a sample of folded protein with a solution containing a high concentration of denaturant. These experiments are ideal for sorting out events in the millisecond timescale, such as the formation of contacts between helical segments in a large protein. However, the available data also indicate that, in a number of proteins, a significant portion of the folding process occurs in less than 1 ms, a time range not accessible by the stopped-flow technique. More recent temperature-jump and flash photolysis experiments have uncovered faster events. For example, at ambient temperature the formation of a loop between helical or sheet segments may be as fast as 1 μs and the formation of tightly packed cores with significant tertiary structure occurs in the 10–100 μs range. Among the fastest events are the formation and denaturation of helices and sheets from fully unfolded or folded peptide chains, respectively. Here, we examine how the laser-induced temperature-jump technique has been used in the study of the helix–coil transition.

The laser-induced temperature-jump technique takes advantage of the fact that proteins unfold, or melt, at high temperatures and each protein has a characteristic melting temperature (Section 22.5). Proteins also lose their native structures at very low temperatures, a process known as *cold denaturation*, and re-fold when the temperature in increased but kept significantly below the melting temperature. Hence, a temperature-jump experiment can be configured to monitor either folding or unfolding of a polypeptide, depending on the initial and final temperatures of the sample. The challenge of using melting or cold denaturation as the basis of kinetic measurements lies in increasing the temperature of the sample very quickly so fast relaxation proccess can be monitored. A number of clever strategies have been employed. In one example, a pulsed laser excites dissolved dye molecules that decay largely by internal conversion, or heat transfer to the solution. Another variation makes use of direct excitation of overtones of the O—H or O—D stretching modes of H_2O or D_2O, respectively, with a pulsed infrared laser. The latter strategy leads to temperature jumps in a small irradiated volume of about 20 K in less than 100 ps. Relaxation of the sample can then be probed by a variety of spectroscopic techniques, including absorption, emission, or Raman scattering. For example, the infrared absorption spectrum of a polypeptide is sensitive to polypeptide conformation, as the N—H stretching vibrations in the range 1630–1670 cm^{-1} are significantly different in the helix and coil forms.

Much of the kinetic work on the helix–coil transition has been conducted in small synthetic polypeptides rich in alanine, an amino acid that is known to stabilize helical structures. Both experimental results and statistical mechanical calculations suggest that the mechanism of unfolding consists of at least two steps: a very fast step in which amino acids at either end of a helical segment undergo transitions to coil regions and a slower rate-determining step that corresponds to the cooperative melting of the rest of the chain and loss of helical content. Using *h* and *c* to denote an amino acid residue belonging to a helical and coil region, respectively, the mechanism may be summarized as follows:

hhhh ... ⟶ *chhh* ...	very fast
chhh ... ⟶ *cccc* ...	rate-determining step

The rate-determining step is thought to account for the relaxation time of 160 ns measured with a laser-induced temperature jump between 282.5 K and 300.6 K in an alanine-rich polypeptide containing 21 amino acids. It is thought that the limitation on the rate of the helix–coil transition in this peptide arises from an activation energy barrier of 1.7 kJ mol^{-1} associated with nucleation events of the form ... *hhhh* ... → ... *hhch* ... in the middle of the chain. Therefore, nucleation is not only thermodynamically unfavourable but also kinetically slow. Models that use concepts of statistical thermodynamics also suggest that a *hhhh* ... → *chhh* ... transition at either end of a helical segment has a significantly lower activation energy on account of the converting amino acid not being flanked by *h* regions.

The time constant for the helix–coil transition has also been measured in proteins. In apomyoglobin (myoglobin lacking the haem co-factor), the unfolding of the helices appears to have a relaxation time of about 50 ns, even shorter than in synthetic peptides. It is difficult to interpret these results because we do not yet know how the amino acid sequence or interactions between helices in a folded protein affect the helix–coil relaxation time.

Discussion questions

1. Propose a set of experiments in which analysis of the line-shapes of NMR transitions (Section 18.8) is used to monitor fast events in protein folding or unfolding. What are the advantages and disadvantages of this NMR method over methods that use electronic or vibrational spectroscopy?

2. Consider a mechanism for the helix–coil transition in which nucleation occurs in the middle of the chain:

$$hhhh\ldots \rightleftharpoons hchh\ldots$$
$$hchh\ldots \rightleftharpoons cccc\ldots$$

We saw above that this type of nucleation is relatively slow, so neither step may be rate-determining. (a) Set up the rate equations for this alternate mechanism. (b) Apply the steady-state approximation and show that, under these circumstances, the mechanism is equivalent to $hhhh\ldots \rightleftharpoons cccc\ldots$ (c) Use your knowledge of experimental techniques and your results from parts (a) and (b) to support or refute the following statement: It is very difficult to obtain experimental evidence for intermediates in protein unfolding by performing simple rate measurements and one must resort to special time-resolved or trapping techniques to detect intermediates directly.

Further reading

R.H. Callender, R.B. Dyer, R. Gilmanshin, and W.H. Woodruff, Fast events in protein folding: the time evolution of primary processes. *Ann. Rev. Phys. Chem.* **49**, 173 (1998).

U. Mayor, C.M. Johnson, V. Daggett, and A.R. Fersht, Protein folding and unfolding in microseconds to nanoseconds by experiment and simulation. *Proc. Natl. Acad. Sci USA* **97**, 13518 (2000).

$$Cl_2 + h\nu \longrightarrow Cl + Cl$$
$$Cl + HBr \longrightarrow HCl^* + Br$$
$$HCl^* + M \longrightarrow HCl + M$$

Here HCl^* denotes a thermally excited HCl molecule and M is a body (an unreactive molecule or the wall of the container) that removes the excess energy stored in HCl by the reaction between Cl and HBr.

In contrast to real-time analysis, **quenching methods** are based on stopping, or quenching, the reaction after it has been allowed to proceed for a certain time. In this way the composition is analysed at leisure and reaction intermediates may be trapped. These methods are suitable only for reactions that are slow enough for there to be little reaction during the time it takes to quench the mixture, so that reaction timescales on the order of hundreds of milliseconds can be studied with ease. In the **chemical quench flow method**, the reactants are mixed in much the same way as in the flow method but the reaction is quenched by another reagent, such as solution of acid or base, after the mixture has travelled along a fixed length of the outlet tube. Different reaction times can be selected by varying the flow rate along the outlet tube. An advantage of the chemical quench flow method over the stopped-flow method is that spectroscopic fingerprints are not needed in order to measure the concentration of reactants and products. Once the reaction has been quenched, the solution may be examined by rather 'slow' techniques, such as gel electrophoresis, mass spectrometry, and chromatography. In the **freeze quench method**, the reaction is quenched by cooling the mixture within milliseconds and the concentrations of reactants, intermediates, and products are measured spectroscopically.

25.2 The rates of reactions

Reaction rates depend on the composition and the temperature of the reaction mixture. The next few sections look at these observations in more detail.

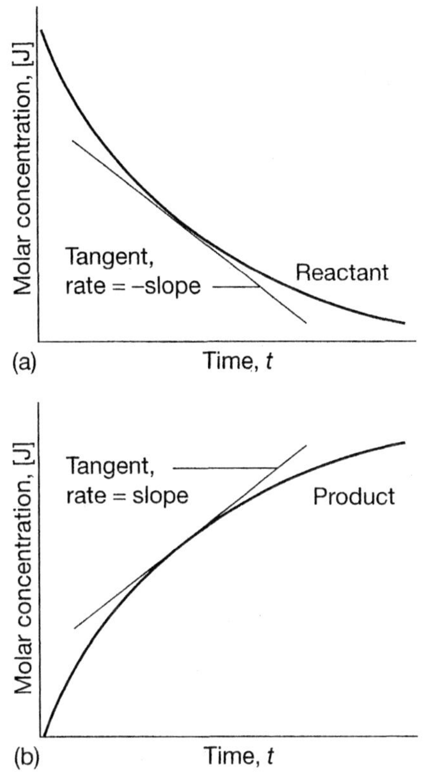

25.3 (a) The definition of (instantaneous) rate as the slope of the tangent drawn to the curve showing the variation of concentration with time. (b) For negative slopes, the sign is changed when reporting the rate, so all reaction rates are positive.

(a) The definition of rate

Consider a reaction of the form A + 2 B → 3 C + D, in which at some instant the molar concentration of a participant J is [J]. The instantaneous **rate of consumption** of one of the reactants at a given time is –d[R]/d*t*, where R is A or B. This rate is a positive quantity (Fig. 25.3). The **rate of formation** of one of the products (C or D, which we denote P) is d[P]/d*t* (note the difference in sign). This rate is also positive.

It follows from the stoichiometry for the reaction A + 2 B → 3 C + D that

$$\frac{d[D]}{dt} = \frac{1}{3}\frac{d[C]}{dt} = -\frac{d[A]}{dt} = -\frac{1}{2}\frac{d[B]}{dt}$$

so there are several rates connected with the reaction. The problem of having several possibly different rates to describe the same reaction is avoided by defining the unique **rate of reaction**, v, as the rate of change of the extent of reaction, ξ (the quantity introduced in Section 9.2b):

$$v = \frac{d\xi}{dt} \qquad [25.1a]$$

Because $\nu_J\xi = n_J - n_{J,0}$, the change in the extent of reaction is related to the change in the amount of each substance J by $\nu_J\, d\xi = dn_J$, so

$$v = \frac{1}{\nu_J}\frac{dn_J}{dt} \qquad (25.1b)$$

(Remember that ν_J is negative for reactants and positive for products.) For a homogeneous reaction we divide by the (constant) volume of the system and write

$$v = \frac{1}{\nu_J}\frac{d[J]}{dt} \qquad (25.1c)$$

For a heterogeneous reaction, we divide instead by the surface area of the species and write

$$v = \frac{1}{\nu_J}\frac{d\sigma_J}{dt} \qquad (25.1d)$$

where σ_J is the surface density of J. In each case there is now a single rate for the entire reaction (for the chemical equation as written). With molar concentrations in moles per litre and time in seconds, reaction rates of homogeneous reactions are reported in moles per litre per second (mol L^{-1} s^{-1}) or related units. For heterogeneous reactions, rates are expressed in moles per square metre per second (mol m^{-2} s^{-1}) or related units.

Illustration 25.1

If the rate of formation of NO in the reaction 2 NOBr(g) → 2 NO(g) + Br_2(g) is reported as 0.16 mmol L^{-1} s^{-1}, we use $\nu_{NO} = +2$ to report that $v = 0.080$ mmol L^{-1} s^{-1}. Because $\nu_{NOBr} = -2$ it follows that d[NOBr]/d*t* = –0.16 mmol L^{-1} s^{-1}. The rate of consumption of NOBr is therefore 0.16 mmol L^{-1} s^{-1}.

Self-test 25.2 The rate of change of molar concentration of CH_3 radicals in the reaction 2 CH_3(g) → CH_3CH_3(g) was reported as d[CH_3]/d*t* = –1.2 mol L^{-1} s^{-1} under particular conditions. What is (a) the rate of reaction and (b) the rate of formation of CH_3CH_3? [(a) 0.60 mol L^{-1} s^{-1}, (b) 0.60 mol L^{-1} s^{-1}]

(b) Rate laws and rate constants

The rate of reaction is often found to be proportional to the concentrations of the reactants raised to a power. For example, the rate of a reaction may be found to be proportional to the molar concentrations of two reactants A and B, so we write

$$v = k[\mathrm{A}][\mathrm{B}] \tag{25.2}$$

with each concentration raised to the first power. The coefficient k is called the **rate constant** for the reaction. The rate constant is independent of the concentrations but depends on the temperature. An experimentally determined equation of this kind is called the **rate law** of the reaction. More formally, a rate law is an equation that expresses the rate of reaction as a function of the concentrations of all the species present in the overall chemical equation for the reaction at some time:

$$v = f([\mathrm{A}],[\mathrm{B}], \ldots) \qquad [25.3\mathrm{a}]$$

For homogeneous gas-phase reactions, it is often more convenient to express the rate law in terms of partial pressures, which are related to molar concentrations by $p_\mathrm{J} = RT[\mathrm{A}]$. In this case, we write

$$v = f(p_\mathrm{A}, p_\mathrm{B}, \ldots) \qquad [25.3\mathrm{b}]$$

The rate law of a reaction is determined experimentally, and cannot in general be inferred from the chemical equation for the reaction. The reaction of hydrogen and bromine, for example, has a very simple stoichiometry, $\mathrm{H_2(g) + Br_2(g) \rightarrow 2\,HBr(g)}$, but its rate law is complicated:

$$v = \frac{k[\mathrm{H_2}][\mathrm{Br_2}]^{3/2}}{[\mathrm{Br_2}] + k'[\mathrm{HBr}]} \tag{25.4}$$

In certain cases the rate law does reflect the stoichiometry of the reaction; but that is either a coincidence or reflects a feature of the underlying reaction mechanism (see later).

A practical application of a rate law is that, once we know the law and the value of the rate constant, we can predict the rate of reaction from the composition of the mixture. Moreover, as we shall see later, by knowing the rate law, we can go on to predict the composition of the reaction mixture at a later stage of the reaction. Moreover, a rate law is a guide to the mechanism of the reaction, for any proposed mechanism must be consistent with the observed rate law.

(c) Reaction order

Many reactions are found to have rate laws of the form

$$v = k[\mathrm{A}]^a[\mathrm{B}]^b \ldots \tag{25.5}$$

The power to which the concentration of a species (a product or a reactant) is raised in a rate law of this kind is the **order** of the reaction with respect to that species. A reaction with the rate law in eqn 2 is **first-order** in A and first-order in B. The **overall order** of a reaction with a rate law like that in eqn 5 is the sum of the individual orders, $a + b + \cdots$. The rate law in eqn 2 is therefore second-order overall.

A reaction need not have an integral order, and many gas-phase reactions do not. For example, a reaction having the rate law

$$v = k[\mathrm{A}]^{1/2}[\mathrm{B}] \tag{25.6}$$

is half-order in A, first-order in B, and three-halves-order overall. Some reactions obey a **zero-order rate law**, and therefore have a rate that is independent of the concentration of the reactant (so long as some is present). Thus, the catalytic decomposition of phosphine (PH_3) on hot tungsten at high pressures has the rate law

$$v = k \tag{25.7}$$

The PH_3 decomposes at a constant rate until it has almost entirely disappeared. Only heterogeneous reactions can have rate laws that are zero-order overall.

When a rate law is not of the form in eqn 5, the reaction does not have an overall order and may not even have definite orders with respect to each participant. Thus, although eqn 4 shows that the reaction of hydrogen and bromine is first-order in H_2, the reaction has an indefinite order with respect to both Br_2 and HBr and has no overall order.

These remarks point to three problems. First, we must see how to identify the rate law and obtain the rate constant from the experimental data. We concentrate on this aspect in this chapter. Second, we must see how to construct reaction mechanisms that are consistent with the rate law. We shall introduce the techniques of doing so in this chapter and develop them further in Chapter 26. Third, we must account for the values of the rate constants and explain their temperature dependence. We shall see a little of what is involved in this chapter, but leave the details until Chapter 27.

(d) The determination of the rate law

The determination of a rate law is simplified by the **isolation method** in which the concentrations of all the reactants except one are in large excess. If B is in large excess, for example, then to a good approximation its concentration is constant throughout the reaction. Although the true rate law might be $v = k[A][B]$, we can approximate [B] by $[B]_0$ and write

$$v = k'[A] \qquad k' = k[B]_0 \tag{25.8}$$

which has the form of a first-order rate law. Because the true rate law has been forced into first-order form by assuming that the concentration of B is constant, it is called a **pseudofirst-order rate law**. The dependence of the rate on the concentration of each of the reactants may be found by isolating them in turn (by having all the other substances present in large excess), and so constructing a picture of the overall rate law.

In the **method of initial rates**, which is often used in conjunction with the isolation method, the rate is measured at the beginning of the reaction for several different initial concentrations of reactants. We shall suppose that the rate law for a reaction with A isolated is $v = k[A]^a$; then its initial rate, v_0, is given by the initial values of the concentration of A, and we write $v_0 = k[A]_0^a$. Taking logarithms gives:

$$\log v_0 = \log k + a \log[A]_0 \tag{25.9}$$

For a series of initial concentrations, a plot of the logarithms of the initial rates against the logarithms of the initial concentrations of A should be a straight line with slope a.

Example 25.2 Using the method of initial rates

The recombination of iodine atoms in the gas phase in the presence of argon was investigated and the order of the reaction was determined by the method of initial rates. The initial rates of reaction of $2\,I(g) + Ar(g) \rightarrow I_2(g) + Ar(g)$ were as follows:

$[I]_0/(10^{-5}$ mol $L^{-1})$		1.0	2.0	4.0	6.0
$v_0/($mol L^{-1} $s^{-1})$	(a)	8.70×10^{-4}	3.48×10^{-3}	1.39×10^{-2}	3.13×10^{-2}
	(b)	4.35×10^{-3}	1.74×10^{-2}	6.96×10^{-2}	1.57×10^{-1}
	(c)	8.69×10^{-3}	3.47×10^{-2}	1.38×10^{-1}	3.13×10^{-1}

The Ar concentrations are (a) 1.0 mmol L^{-1}, (b) 5.0 mmol L^{-1}, and (c) 10 mmol L^{-1}. Determine the orders of reaction with respect to the I and Ar atom concentrations and the rate constant.

Method Plot the logarithm of the initial rate, log v_0, against log$[I]_0$ for a given concentration of Ar, and, separately, against log$[Ar]_0$ for a given concentration of I. The slopes of the two lines are the orders of reaction with respect to I and Ar, respectively. The intercepts with the vertical axis at log[X] = 0 give log k.

Answer The plots are shown in Fig. 25.4. The slopes are 2 and 1 respectively, so the (initial) rate law is

$$v_0 = k[I]_0^2[Ar]_0$$

This rate law signifies that the reaction is second-order in [I], first-order in [Ar], and third-order overall. The intercept corresponds to $k = 9\times10^9$ mol^{-2} L^2 s^{-1}.

Comment The units of k come automatically from the calculation, and are always such as to convert the product of concentrations to concentration per unit time (for example, mol L^{-1} s^{-1}).

Self-test 25.3 The initial rate of a reaction depended on concentration of a substance J as follows:

$[J]_0/($mmol $L^{-1})$	5.0	8.2	17	30
$v_0/(10^{-7}$ mol L^{-1} $s^{-1})$	3.6	9.6	41	130

Determine the order of the reaction with respect to J and calculate the rate constant. [2, 1.4×10^{-2} L mol^{-1} s^{-1}]

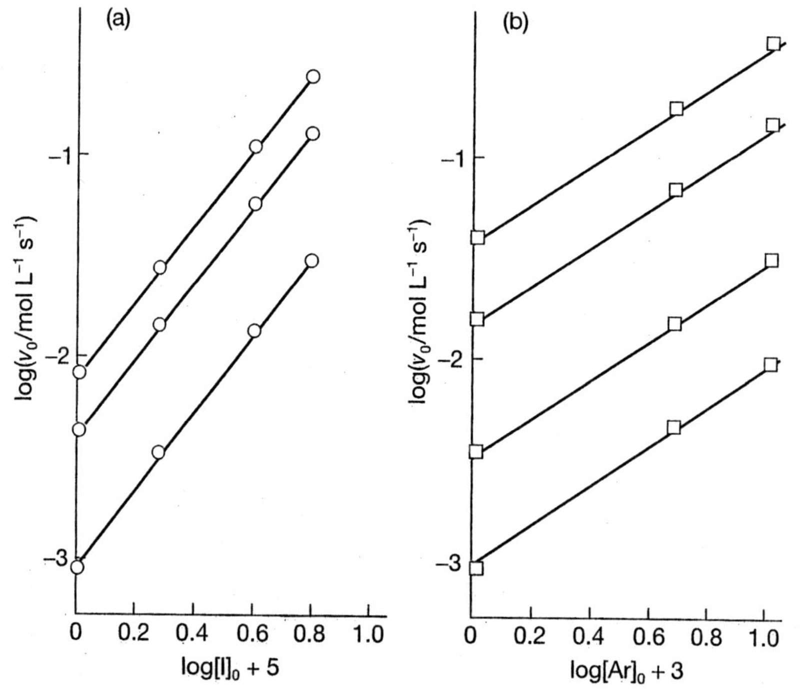

25.4 The plot of log v_0 against (a) log$[I]_0$ for a given $[Ar]_0$, and (b) log$[Ar]_0$ for a given $[I]_0$.

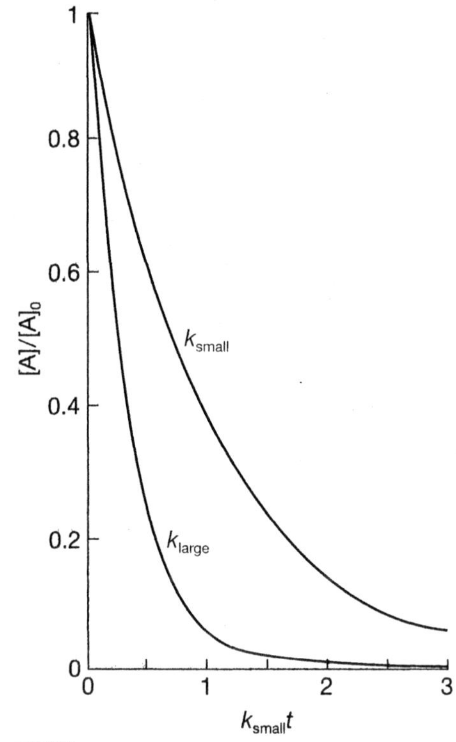

25.5 The exponential decay of the reactant in a first-order reaction. The larger the rate constant, the more rapid the decay: here $k_{large} = 3k_{small}$.

The method of initial rates might not reveal the full rate law, for the products may participate in the reaction and affect the rate. For example, products participate in the synthesis of HBr, because eqn 4 shows that the full rate law depends on the concentration of HBr. To avoid this difficulty, the rate law should be fitted to the data throughout the reaction. The fitting may be done, in simple cases at least, by using a proposed rate law to predict the concentration of any component at any time, and comparing it with the data. A law should also be tested by observing whether the addition of products or, for gas-phase reactions, a change in the surface-to-volume ratio in the reaction chamber affects the rate.

25.3 Integrated rate laws

Because rate laws are differential equations, we must integrate them if we want to find the concentrations as a function of time. Even the most complex rate laws may be integrated numerically. However, in a number of simple cases analytical solutions are easily obtained, and prove to be very useful. We shall examine a few of these simple cases here.

(a) First-order reactions

As shown in the *Justification* below, the first-order rate law for the consumption of a reactant A

$$\frac{d[A]}{dt} = -k[A] \tag{25.10a}$$

has the solution

$$\ln\left(\frac{[A]}{[A]_0}\right) = -kt \qquad [A] = [A]_0 e^{-kt} \tag{25.10b}$$

These two equations are versions of an **integrated rate law**, the integrated form of the rate law.

Justification 25.1

Equation 10a rearranges to

$$\frac{d[A]}{[A]} = -k\,dt$$

which can be integrated directly because k is a constant independent of t. Initially (at $t = 0$), the concentration of A is $[A]_0$, and at a later time t it is $[A]$, so we make these values the limits of the integrals and write

$$\int_{[A]_0}^{[A]} \frac{d[A]}{[A]} = -k\int_0^t dt$$

Because the integral of $1/x$ is $\ln x$, eqn 10b is obtained immediately.

The web site contains links to databases of rate constants of chemical reactions.

Equation 10b shows that, if $\ln([A]/[A]_0)$ is plotted against t, then a first-order reaction will give a straight line of slope $-k$. Some rate constants determined in this way are given in Table 25.1. The second expression in eqn 10b shows that in a first-order reaction the reactant concentration decreases exponentially with time with a rate determined by k (Fig. 25.5).

Table 25.1* Kinetic data for first-order reactions

Reaction	Phase	θ/°C	k/s^{-1}	$t_{1/2}$
$2\,N_2O_5 \longrightarrow 4\,NO_2 + O_2$	g	25	3.38×10^{-5}	5.70 h
$2\,N_2O_5 \longrightarrow 4\,NO_2 + O_2$	Br_2(l)	25	4.27×10^{-5}	4.51 h
$C_2H_6 \longrightarrow 2\,CH_3$	g	700	5.36×10^{-4}	21.6 min

* More values are given in the *Data section*.

Example 25.3 Analysing a first-order reaction

The variation in the partial pressure of azomethane with time was followed at 600 K, with the results given below. Confirm that the decomposition

$$CH_3N_2CH_3(g) \longrightarrow CH_3CH_3(g) + N_2(g)$$

is first-order in azomethane, and find the rate constant at 600 K.

t/s	0	1000	2000	3000	4000
p/(10^{-2} Torr)	8.20	5.72	3.99	2.78	1.94

Method As indicated in the text, to confirm that a reaction is first-order, plot ln ([A]/[A]$_0$) against time and expect a straight line. Because the partial pressure of a gas is proportional to its concentration, an equivalent procedure is to plot $\ln(p/p_0)$ against t. If a straight line is obtained, its slope can be identified with $-k$.

Answer We draw up the following table:

t/s	0	1000	2000	3000	4000
$\ln(p/p_0)$	0	−0.360	−0.720	−1.082	−1.441

Figure 25.6 shows the plot of $\ln(p/p_0)$ against t. The plot is straight, confirming a first-order reaction, and its slope is $-3.6 \times 10^{-4}\,s^{-1}$. Therefore, $k = 3.6 \times 10^{-4}\,s^{-1}$.

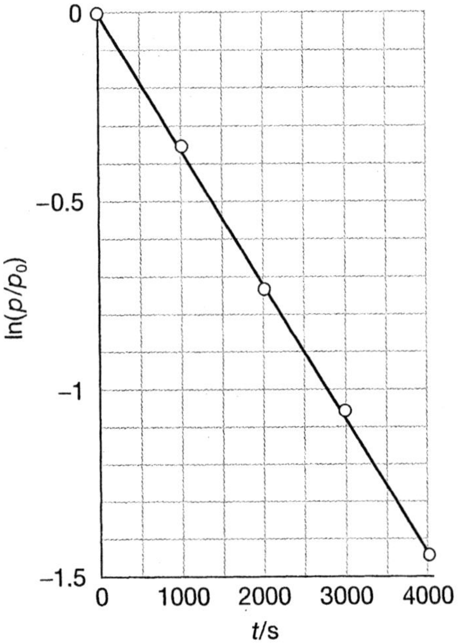

25.6 The determination of the rate constant of a first-order reaction: a straight line is obtained when ln [A] (or, as here, ln p) is plotted against t; the slope gives k.

Self-test 25.4 In a particular experiment, it was found that the concentration of N_2O_5 in liquid bromine varied with time as follows:

t/s	0	200	400	600	1000
[N_2O_5]/(mol L^{-1})	0.110	0.073	0.048	0.032	0.014

Confirm that the reaction is first-order in N_2O_5 and determine the rate constant. [$k = 2.1 \times 10^{-3}\,s^{-1}$]

(b) Half-lives and time constants

A useful indication of the rate of a first-order chemical reaction is the **half-life**, $t_{1/2}$, of a substance, the time taken for the concentration of a reactant to fall to half its initial value. The time for [A] to decrease from [A]$_0$ to $\frac{1}{2}$[A]$_0$ in a first-order reaction is given by eqn 10b as

$$kt_{1/2} = -\ln\left(\frac{\frac{1}{2}[A]_0}{[A]_0}\right) = -\ln\tfrac{1}{2} = \ln 2$$

Hence

$$t_{1/2} = \frac{\ln 2}{k} \tag{25.11}$$

($\ln 2 = 0.693$.) The main point to note about this result is that, for a first-order reaction, the half-life of a reactant is independent of its initial concentration. Hence, if the concentration of A at some *arbitrary* stage of the reaction is [A], then it will have fallen to $\frac{1}{2}$[A] after a further interval of $(\ln 2)/k$. Some half-lives are given in Table 25.1.

Another indication of the rate of a first-order reaction is the **time constant**, τ, the time required for the concentration of a reactant to fall to $1/e$ of its initial value. From eqn 10b it follows that

$$k\tau = -\ln\left(\frac{[\mathrm{A}]_0/\mathrm{e}}{[\mathrm{A}]_0}\right) = -\ln\frac{1}{\mathrm{e}} = 1$$

Hence, the time constant is the reciprocal of the rate constant:

$$\tau = \frac{1}{k} \tag{25.12}$$

(c) Second-order reactions

We show in the *Justification* below that the integrated form of the second-order rate law

$$\frac{\mathrm{d}[\mathrm{A}]}{\mathrm{d}t} = -k[\mathrm{A}]^2 \tag{25.13a}$$

is

$$\frac{1}{[\mathrm{A}]} - \frac{1}{[\mathrm{A}]_0} = kt \qquad [\mathrm{A}] = \frac{[\mathrm{A}]_0}{1 + kt[\mathrm{A}]_0} \tag{25.13b}$$

Justification 25.2

To integrate eqn 13a we rearrange it into

$$\frac{\mathrm{d}[\mathrm{A}]}{[\mathrm{A}]^2} = -k\,\mathrm{d}t$$

The concentration of A is $[\mathrm{A}_0]$ at $t = 0$ and [A] at a general time t later. Therefore,

$$-\int_{[\mathrm{A}]_0}^{[\mathrm{A}]} \frac{\mathrm{d}[\mathrm{A}]}{[\mathrm{A}]^2} = k\int_0^t \mathrm{d}t$$

Because the integral of $1/x^2$ is $-1/x$, we obtain eqn 13b by substitution of the limits.

The first expression in eqn 13b shows that to test for a second-order reaction we should plot $1/[\mathrm{A}]$ against t and expect a straight line. The slope of the graph is k. Some rate constants determined in this way are given in Table 25.2. The second expression lets us predict the concentration of A at any time after the start of the reaction. It shows that the concentration of A approaches zero more slowly than in a first-order reaction with the same initial rate (Fig. 25.7).

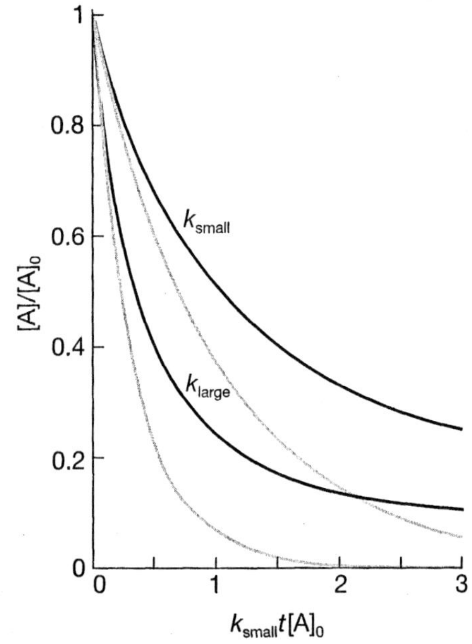

25.7 The variation with time of the concentration of a reactant in a second-order reaction. The pale line is the corresponding decay in a first-order reaction with the same initial rate. For this illustration, $k_{large} = 3k_{small}$.

Table 25.2* Kinetic data for second-order reactions

Reaction	Phase	θ/°C	k/(L mol^{-1} s^{-1})
$2\,NOBr \longrightarrow 2\,NO + Br_2$	g	10	0.80
$2\,I \longrightarrow I_2$	g	23	7×10^9
$CH_3Cl + CH_3O^-$	$CH_3OH(l)$	20	2.29×10^{-6}

* More values are given in the *Data section*.

It follows from eqn 13b by substituting $t = t_{1/2}$ and $[A] = \frac{1}{2}[A]_0$ that the half-life of a species A that is consumed in a second-order reaction is

$$t_{1/2} = \frac{1}{k[A]_0} \tag{25.14}$$

Therefore, unlike a first-order reaction, the half-life of a substance in a second-order reaction varies with the initial concentration. A practical consequence of this dependence is that species that decay by second-order reactions (which includes some environmentally harmful substances) may persist in low concentrations for long periods because their half-lives are long when their concentrations are low.

Another type of second-order reaction is one that is first-order in each of two reactants A and B:

$$\frac{d[A]}{dt} = -k[A][B] \tag{25.15}$$

Such a rate law cannot be integrated until we know how the concentration of B is related to that of A. For example, if the reaction is A + B → P, where P denotes products, and the initial concentrations are $[A]_0$ and $[B]_0$, then it is shown in the *Justification* below that, at a time t after the start of the reaction, the concentrations satisfy the relation

$$\ln\left(\frac{[B]/[B]_0}{[A]/[A]_0}\right) = ([B]_0 - [A]_0)kt \tag{25.16}$$

Therefore, a plot of the expression on the left against t should be a straight line from which k can be obtained. Note that, if $[A]_0 = [B]_0$, then the solutions are those already given in eqn 13b (but this solution cannot be found simply by setting $[A]_0 = [B]_0$ in eqn 16).

Justification 25.3

It follows from the reaction stoichiometry that, when the concentration of A has fallen to $[A]_0 - x$, the concentration of B will have fallen to $[B]_0 - x$ (because each A that disappears entails the disappearance of one B). It follows that

$$\frac{d[A]}{dt} = -k([A]_0 - x)([B]_0 - x)$$

Then, because $d[A]/dt = -dx/dt$, the rate law is

$$\frac{dx}{dt} = k([A]_0 - x)([B]_0 - x)$$

To solve an integral of the form $\int \frac{1}{(a-x)(b-x)}\mathrm{d}x$, where a and b are constants, we use the method of partial fractions. First we write

$$\frac{1}{(a-x)(b-x)} = \frac{1}{b-a}\left(\frac{1}{a-x} - \frac{1}{b-x}\right)$$

and integrate the expression on the right. It follows that

$$\int \frac{\mathrm{d}x}{(a-x)(b-x)} = \frac{1}{b-a}\left[\int \frac{\mathrm{d}x}{a-x} - \int \frac{\mathrm{d}x}{b-x}\right] = \frac{1}{b-a}\left(\ln\frac{1}{a-x} - \ln\frac{1}{b-x}\right) + \text{constant}$$

The initial condition is that $x = 0$ when $t = 0$; so the integration required is

$$\int_0^x \frac{\mathrm{d}x}{([\mathrm{A}]_0 - x)([\mathrm{B}]_0 - x)} = k\int_0^t \mathrm{d}t$$

The integral on the right is simply kt. The integral on the left is evaluated by using the method of partial fractions and by using $[\mathrm{A}] = [\mathrm{A}]_0$ and $[\mathrm{B}] = [\mathrm{B}]_0$ at $t = 0$ to give:

$$\int_0^x \frac{\mathrm{d}x}{([\mathrm{A}]_0 - x)([\mathrm{B}]_0 - x)} = \frac{1}{[\mathrm{B}]_0 - [\mathrm{A}]_0}\left\{\ln\left(\frac{[\mathrm{A}]_0}{[\mathrm{A}]_0 - x}\right) - \ln\left(\frac{[\mathrm{B}]_0}{[\mathrm{B}]_0 - x}\right)\right\}$$

This expression can be simplified and rearranged into eqn 16 by combining the two logarithms and noting that $[\mathrm{A}] = [\mathrm{A}]_0 - x$ and $[\mathrm{B}] = [\mathrm{B}]_0 - x$.

Similar calculations may be carried out to find the integrated rate laws for other orders, and some are listed in Table 25.3.

Table 25.3 Integrated rate laws

Order	Reaction	Rate law*	$t_{1/2}$
0	$\mathrm{A} \rightarrow \mathrm{P}$	$v = k$	$\frac{[\mathrm{A}]_0}{2k}$
		$kt = x$ for $0 \le x \le [\mathrm{A}]_0$	
1	$\mathrm{A} \rightarrow \mathrm{P}$	$v = k[\mathrm{A}]$	$\frac{\ln 2}{k}$
		$kt = \ln\frac{[\mathrm{A}]_0}{[\mathrm{A}]_0 - x}$	
2	$\mathrm{A} \rightarrow \mathrm{P}$	$v = k[\mathrm{A}]^2$	$\frac{1}{k[\mathrm{A}]_0}$
		$kt = \frac{x}{[\mathrm{A}]_0([\mathrm{A}]_0 - x)}$	
	$\mathrm{A} + \mathrm{B} \rightarrow \mathrm{P}$	$v = k[\mathrm{A}][\mathrm{B}]$	
		$kt = \frac{1}{[\mathrm{B}]_0 - [\mathrm{A}]_0}\ln\frac{[\mathrm{A}]_0([\mathrm{B}]_0 - x)}{([\mathrm{A}]_0 - x)[\mathrm{B}]_0}$	
	$\mathrm{A} + 2\mathrm{B} \rightarrow \mathrm{P}$	$v = k[\mathrm{A}][\mathrm{B}]$	
		$kt = \frac{1}{[\mathrm{B}]_0 - 2[\mathrm{A}]_0}\ln\frac{[\mathrm{A}]_0([\mathrm{B}]_0 - 2x)}{([\mathrm{A}]_0 - x)[\mathrm{B}]_0}$	
	$\mathrm{A} \rightarrow \mathrm{P}$ with autocatalysis		
		$v = k[\mathrm{A}][\mathrm{P}]$	
		$kt = \frac{1}{[\mathrm{A}]_0 + [\mathrm{P}]_0}\ln\frac{[\mathrm{A}]_0([\mathrm{P}]_0 + x)}{([\mathrm{A}]_0 - x)[\mathrm{P}]_0}$	
3	$\mathrm{A} + 2\mathrm{B} \rightarrow \mathrm{P}$	$v = k[\mathrm{A}][\mathrm{B}]^2$	
		$kt = \frac{2x}{(2[\mathrm{A}]_0 - [\mathrm{B}]_0)([\mathrm{B}]_0 - 2x)[\mathrm{B}]_0} + \frac{1}{(2[\mathrm{A}]_0 - [\mathrm{B}]_0)^2}\ln\frac{[\mathrm{A}]_0([\mathrm{B}]_0 - 2x)}{([\mathrm{A}]_0 - x)[\mathrm{B}]_0}$	
$n \ge 2$	$\mathrm{A} \rightarrow \mathrm{P}$	$v = k[\mathrm{A}]^n$	$\frac{2^{n-1} - 1}{(n-1)k[\mathrm{A}]_0^{n-1}}$
		$kt = \frac{1}{n-1}\left\{\frac{1}{([\mathrm{A}]_0 - x)^{n-1}} - \frac{1}{[\mathrm{A}]_0^{n-1}}\right\}$	

* $x = [\mathrm{P}]$, and $v = \mathrm{d}x/\mathrm{d}t$.

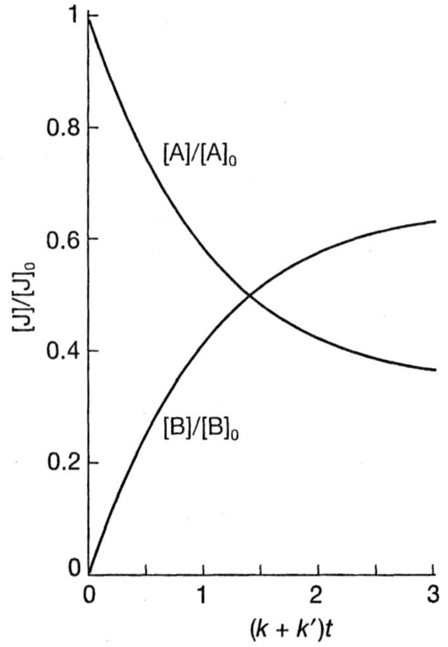

25.8 The approach of concentrations to their equilibrium values as predicted by eqn 20 for a reaction $A \rightleftharpoons B$ that is first-order in each direction, and for which $k = 2k'$.

25.4 Reactions approaching equilibrium

Because all the laws considered so far disregard the possibility that the reverse reaction is important, none of them describes the overall rate when the reaction is close to equilibrium. At that stage the products may be so abundant that the reverse reaction must be taken into account. In practice, however, most kinetic studies are made on reactions that are far from equilibrium, and the reverse reactions are unimportant.

(a) First-order reactions close to equilibrium

We can explore the variation of the composition with time close to chemical equilibrium by considering a reaction in which A forms B and both forward and reverse reactions are first-order (as in some isomerizations). The scheme we consider is

$$\begin{array}{ll} A \longrightarrow B & v = k[A] \\ B \longrightarrow A & v = k'[B] \end{array} \tag{25.17}$$

The concentration of A is reduced by the forward reaction (at a rate $k[A]$) but it is increased by the reverse reaction (at a rate $k'[B]$). The net rate of change is therefore

$$\frac{d[A]}{dt} = -k[A] + k'[B] \tag{25.18}$$

If the initial concentration of A is $[A]_0$, and no B is present initially, then at all times $[A] + [B] = [A]_0$. Therefore,

$$\frac{d[A]}{dt} = -k[A] + k'([A]_0 - [A]) = -(k + k')[A] + k'[A]_0 \tag{25.19}$$

The solution of this first-order differential equation (as may be checked by differentiation) is

$$[A] = \frac{k' + ke^{-(k+k')t}}{k' + k}[A]_0 \tag{25.20}$$

Figure 25.8 shows the time dependence predicted by this equation.

As $t \to \infty$, the concentrations reach their equilibrium values, which are given by eqn 20 as:

$$[A]_{eq} = \frac{k'[A]_0}{k + k'} \qquad [B]_{eq} = [A]_0 - [A]_\infty = \frac{k[A]_0}{k + k'} \tag{25.21}$$

It follows that the equilibrium constant of the reaction is[1]

$$K = \frac{[B]_{eq}}{[A]_{eq}} = \frac{k}{k'} \tag{25.22}$$

Exactly the same conclusion can be reached—more simply, in fact—by noting that, at equilibrium, the forward and reverse rates must be the same, so

$$k[A]_{eq} = k'[B]_{eq} \tag{25.23}$$

This relation rearranges into eqn 22.

Equation 22 is very important, because it relates the thermodynamic quantity, the equilibrium constant, to quantities relating to rates. The practical importance of

1. This expression is only approximate because thermodynamic equilibrium constants are expressed in terms of activities, not concentrations.

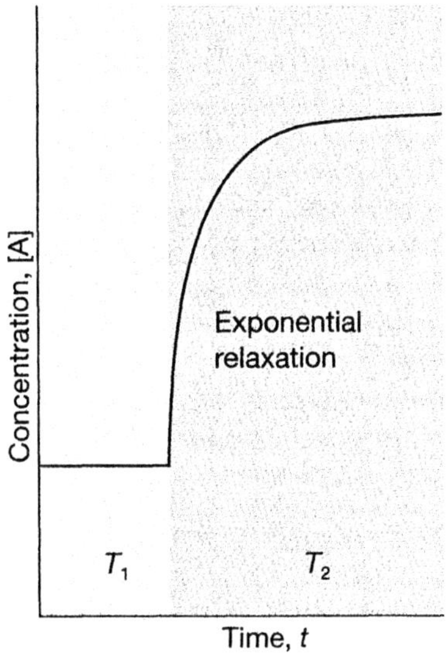

25.9 The relaxation to the new equilibrium composition when a reaction initially at equilibrium at a temperature T_1 is subjected to a sudden change of temperature, which takes it to T_2.

eqn 22 is that, if one of the rate constants can be measured, then the other may be obtained if the equilibrium constant is known.

For a more general reaction, the overall equilibrium constant can be expressed in terms of the rate constants for all the intermediate stages of the reaction mechanism:

$$K = \frac{k_a}{k'_a} \times \frac{k_b}{k'_b} \times \cdots$$

where the ks are the rate constants for the individual steps and the k's are for the corresponding reverse steps.

(b) Relaxation methods

The term **relaxation** denotes the return of a system to equilibrium. It is used in chemical kinetics to indicate that an externally applied influence has shifted the equilibrium position of a reaction, normally suddenly, and that the reaction is adjusting to the equilibrium composition characteristic of the new conditions (Fig. 25.9). We shall consider the response of reaction rates to a **temperature jump**, a sudden change in temperature. We know from Section 9.3a that the equilibrium composition of a reaction depends on the temperature (provided $\Delta_r H^\ominus$ is nonzero), so a shift in temperature acts as a perturbation on the system. One way of achieving a temperature jump is to discharge a capacitor through a sample made conducting by the addition of ions, but laser or microwave discharges can also be used. Temperature jumps of between 5 and 10 K can be achieved in about 1 μs with electrical discharges. The high energy output of pulsed lasers (Section 17.5) is sufficient to generate temperature jumps of between 10 and 30 K within nanoseconds in aqueous samples. Some equilibria are also sensitive to pressure, and **pressure-jump techniques** may then also be used.

When a sudden temperature increase is applied to a simple $A \rightleftharpoons B$ equilibrium that is first-order in each direction, we show in the *Justification* below that the composition relaxes exponentially to the new equilibrium composition:

$$x = x_0 e^{-t/\tau} \qquad \frac{1}{\tau} = k_a + k_b \tag{25.24}$$

where x is the departure from equilibrium at the new temperature and x_0 is the departure from equilibrium immediately after the temperature jump.

Justification 25.4

In the following analysis, we need to keep track of the fact that rate constants depend on temperature. At the initial temperature, when the rate constants are k'_a and k'_b, the net rate of change of [A] is

$$\frac{d[A]}{dt} = -k'_a[A] + k'_b[B]$$

At equilibrium under these conditions, $d[A]/dt = 0$, and the concentrations are $[A]'_{eq}$ and $[B]'_{eq}$. Therefore,

$$k'_a[A]'_{eq} = k'_b[B]'_{eq}$$

When the temperature is increased suddenly, the rate constants change to k_a and k_b, but the concentrations of A and B remain for an instant at their old equilibrium values. As the system is no longer at equilibrium, it readjusts to the new equilibrium concentrations, which are now given by

$$k_a[A]_{eq} = k_b[B]_{eq}$$

and it does so at a rate that depends on the new rate constants.

We write the deviation of [A] from its new equilibrium value as x, so $[A] = x + [A]_{eq}$ and $[B] = [B]_{eq} - x$. The concentration of A then changes as follows:

$$\frac{d[A]}{dt} = -k_a(x + [A]_{eq}) + k_b(-x + [B]_{eq}) = -(k_a + k_b)x$$

because the two terms involving the equilibrium concentrations cancel. Because $d[A]/dt = dx/dt$, this equation is a first-order differential equation with the solution given in eqn 24.

Equation 24 shows that the concentrations of A and B relax into the new equilibrium at a rate determined by the sum of the two new rate constants. Because the equilibrium constant under the new conditions is $K \approx k_a/k_b$, its value may be combined with the relaxation time measurement to find the individual k_a and k_b.

Example 25.4 Analysing a temperature-jump experiment

The $H_2O(l) \rightarrow H^+(aq) + OH^-(aq)$ reaction relaxes to equilibrium with a relaxation time of 37 μs at 298 K and pH ≈ 7, and $pK_w = 14.01$. Given that the forward reaction is first-order and the reverse is second-order overall, calculate the rate constants for the forward and reverse reactions.

Method We need to derive an expression for the relaxation time, τ (the time constant for return to equilibrium), in terms of k_1 (forward, first-order reaction) and k_2 (reverse, second-order reaction). We can proceed as above, but it will be necessary to make the assumption that the deviation from equilibrium (x) is so small that terms in x^2 can be neglected. Relate k_1 and k_2 through the equilibrium constant, but be careful with units because K_w is dimensionless.

Answer The forward rate at the final temperature is $k_1[H_2O]$ and the reverse rate is $k_2[H^+][OH^-]$. The net rate of ionization of H_2O is

$$\frac{d[H_2O]}{dt} = -k_1[H_2O] + k_2[H^+][OH^-]$$

We write $[H_2O] = [H_2O]_{eq} + x$, $[H^+] = [H^+]_{eq} - x$, and $[OH^-] = [OH^-]_{eq} - x$, and obtain

$$\frac{dx}{dt} = -\{k_1 + k_2([H^+]_{eq} + [OH^-]_{eq})\}x - k_1[H_2O]_{eq} + k_2[H^+]_{eq}[OH^-]_{eq} + k_2x^2$$

$$\approx -\{k_1 + k_2([H^+]_{eq} + [OH^-]_{eq})\}x$$

where we have neglected the term in x^2 and used the equilibrium condition to eliminate the terms that are independent of x. It follows that

$$\frac{1}{\tau} = k_1 + k_2([H^+]_{eq} + [OH^-]_{eq})$$

The equilibrium condition is

$$k_1[H_2O]_{eq} = k_2[H^+]_{eq}[OH^-]_{eq}$$

From this expression it follows that

$$\frac{k_1}{k_2} = \frac{[H^+]_{eq}[OH^-]_{eq}}{[H_2O]_{eq}} = \frac{K_w\,(\text{mol L}^{-1})^2}{[H_2O]_{eq}} = \frac{K_w}{55.6}\ \text{mol L}^{-1}$$

because the molar concentration of pure water is 55.6 mol L^{-1}. If we write $K = K_w/55.6 = 1.8 \times 10^{-16}$, we obtain

$$\frac{1}{\tau} = k_2\{(K\ \text{mol L}^{-1}) + [H^+]_{eq} + [OH^-]_{eq}\} = k_2\{K + K_w^{1/2} + K_w^{1/2}\}\text{mol L}^{-1}$$

$$= (2.0 \times 10^{-7}) \times k_2\ \text{mol L}^{-1}$$

Hence,

$$k_2 = \frac{1}{(3.7 \times 10^{-5}\ \text{s}) \times (2.0 \times 10^{-7}\ \text{mol L}^{-1})} = 1.4 \times 10^{11}\ \text{L mol}^{-1}\ \text{s}^{-1}$$

It follows that

$$k_1 = k_2 K\ \text{mol L}^{-1} = 2.4 \times 10^{-5}\ \text{s}^{-1}$$

Comment Notice how we keep track of units: K and K_w are dimensionless; k_2 is expressed in L mol^{-1} s^{-1} and k_1 is expressed in s^{-1}. The reaction is faster in ice, where $k_2 = 8.6 \times 10^{12}$ L mol^{-1} s^{-1}.

Self-test 25.5 Derive an expression for the relaxation time of a concentration when the reaction $A + B \rightleftharpoons C + D$ is second-order in both directions. $[1/\tau = k([A] + [B])_{eq} + k'([C] + [D])_{eq}]$

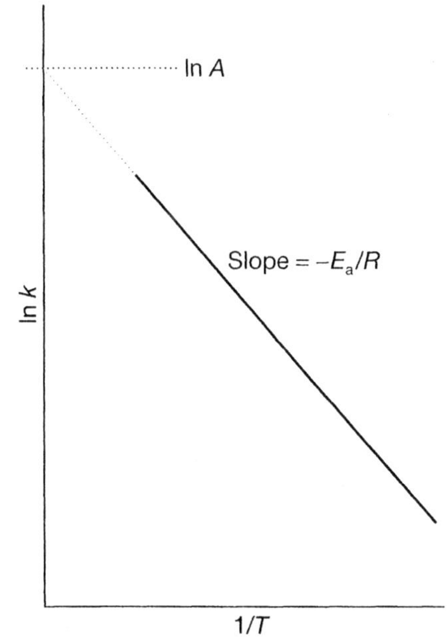

25.10 A plot of $\ln k$ against $1/T$ is a straight line when the reaction follows the behaviour described by the Arrhenius equation (eqn 25). The slope gives $-E_a/R$ and the intercept at $1/T = 0$ gives $\ln A$.

25.5 The temperature dependence of reaction rates

The rate constants of most reactions increase as the temperature is raised. Many reactions in solution fall somewhere in the range spanned by the hydrolysis of methyl ethanoate (where the rate constant at 35°C is 1.82 times that at 25°C) and the hydrolysis of sucrose (where the factor is 4.13).

(a) The Arrhenius parameters

It is found experimentally for many reactions that a plot of $\ln k$ against $1/T$ gives a straight line. This behaviour is normally expressed mathematically by introducing two parameters, one representing the intercept and the other the slope of the straight line, and writing the **Arrhenius equation**

$$\ln k = \ln A - \frac{E_a}{RT} \tag{25.25}$$

The parameter A, which corresponds to the intercept of the line at $1/T = 0$ (at infinite temperature, Fig. 25.10), is called the **pre-exponential factor** or the **frequency factor**. The parameter E_a, which is obtained from the slope of the line ($-E_a/R$), is called the **activation energy**. Collectively, the two quantities are called the **Arrhenius parameters** (Table 25.4).

Table 25.4* Arrhenius parameters

(1) First-order reactions	A/s^{-1}	$E_a/(kJ\ mol^{-1})$
$CH_3NC \longrightarrow CH_3CN$	3.98×10^{13}	160
$2\ N_2O_5 \longrightarrow 4\ NO_2 + O_2$	4.94×10^{13}	103.4
(2) Second-order reactions	$A/(L\ mol^{-1}\ s^{-1})$	$E_a/(kJ\ mol^{-1})$
$OH + H_2 \longrightarrow H_2O + H$	8.0×10^{10}	42
$NaC_2H_5O + CH_3I$ in ethanol	2.42×10^{11}	81.6

* More values are given in the *Data section*.

Example 25.5 Determining the Arrhenius parameters

The rate of the second-order decomposition of acetaldehyde (ethanal, CH_3CHO) was measured over the temperature range 700–1000 K, and the rate constants are reported below. Find E_a and A.

T/K	700	730	760	790	810	840	910	1000
$k/(L\ mol^{-1}\ s^{-1})$	0.011	0.035	0.105	0.343	0.789	2.17	20.0	145

Method According to eqn 25, the data can be analysed by plotting $\ln(k/L\ mol^{-1}\ s^{-1})$ against $1/(T/K)$ and getting a straight line. The slope of this line is $(-E_a/R)/K$ and the intercept at $1/T = 0$ is $\ln A$.

Answer We draw up the following table:

$(10^3\ K)/T$	1.43	1.37	1.32	1.27	1.23	1.19	1.10	1.00
$\ln(k/L\ mol^{-1}\ s^{-1})$	−4.51	−3.35	−2.25	−1.07	−0.24	0.77	3.00	4.98

Now plot $\ln k$ against $1/T$ (Fig. 25.11). The least-squares best fit of the line is with slope -2.27×10^4 and intercept 27.7. Therefore,

$$E_a = (2.27 \times 10^4\ K) \times (8.3145\ J\ K^{-1}\ mol^{-1}) = 189\ kJ\ mol^{-1}$$

$$A = e^{27.7}\ L\ mol^{-1}\ s^{-1} = 1.1 \times 10^{12}\ L\ mol^{-1}\ s^{-1}$$

Comment Note that A has the same units as k. The slopes and intercepts of graphs are always dimensionless, and care must be taken to relate the numerical value to the physical quantity by noting how the data have been plotted. In practice, A is obtained from one of the mid-range data values rather than using a lengthy extrapolation.

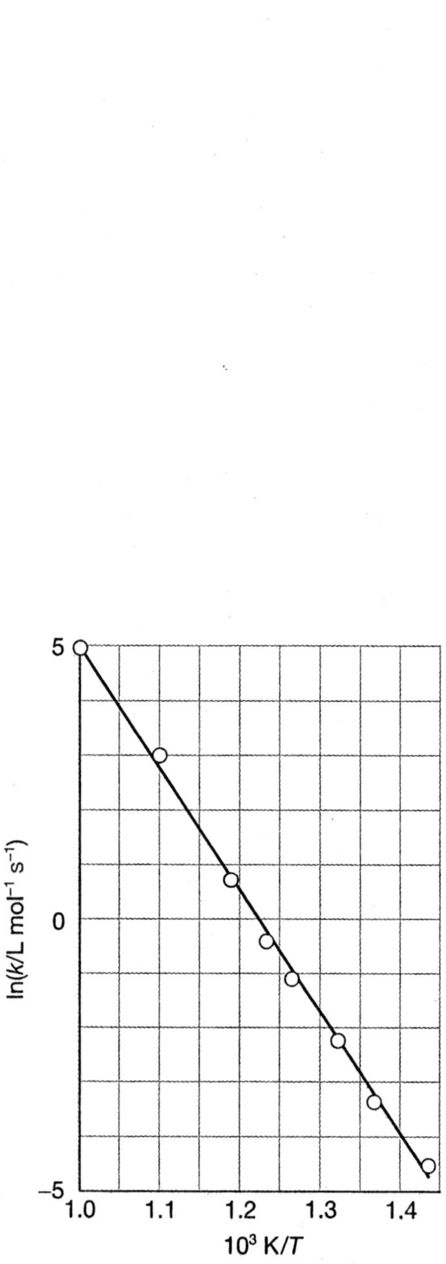

25.11 The Arrhenius plot using the data in Example 25.5.

Self-test 25.6 Determine A and E_a from the following data:

T/K	300	350	400	450	500
$k/(L\ mol^{-1}\ s^{-1})$	7.9×10^6	3.0×10^7	7.9×10^7	1.7×10^8	3.2×10^8

[8×10^{10} L mol^{-1} s^{-1}, 23 kJ mol^{-1}]

The fact that E_a is given by the slope of the plot of $\ln k$ against $1/T$ means that, the higher the activation energy, the stronger the temperature dependence of the rate constant (that is, the steeper the slope). *A high activation energy signifies that the rate constant depends strongly on temperature.* If a reaction has zero activation energy, its rate is independent of temperature. In some cases the activation energy is negative, which indicates that the rate decreases as the temperature is raised. We shall see that such behaviour is a signal that the reaction has a complex mechanism.

The temperature dependence of some reactions is not Arrhenius-like, in the sense that a straight line is not obtained when $\ln k$ is plotted against $1/T$. However, it is still possible to define an activation energy as

$$E_a = RT^2\left(\frac{d\ln k}{dT}\right) \qquad [25.26]$$

This definition reduces to the earlier one (as the slope of a straight line) for a temperature-independent activation energy. However, the definition in eqn 26 is more general than eqn 25, because it allows E_a to be obtained from the slope (at the temperature of interest) of a plot of $\ln k$ against $1/T$ even if the Arrhenius plot is not a straight line. Non-Arrhenius behaviour is sometimes a sign that quantum mechanical tunnelling is playing a significant role in the reaction (Section 25.7e).

(b) The interpretation of the parameters

For the present chapter we shall regard the Arrhenius parameters as purely empirical quantities that enable us to discuss the variation of rate constants with temperature. We begin by writing eqn 25 as

$$k = Ae^{-E_a/RT} \qquad (25.27)$$

To interpret E_a it is useful to consider how the potential energy changes in the course of a chemical reaction that begins with a collision between molecules of A and molecules of B (Fig. 25.12). As the reaction proceeds, A and B come into contact, distort, and begin to exchange or discard atoms. The **reaction coordinate** is the collection of motions, such as changes in interatomic distances and bond angles, that are directly involved in the formation of products from reactants. The potential energy rises to a maximum and the cluster of atoms that corresponds to the region close to the maximum is called the **activated complex**. After the maximum, the potential energy falls as the atoms rearrange in the cluster, and it reaches a value characteristic of the products. The climax of the reaction is at the peak of the potential energy, which corresponds to the activation energy E_a. Here two reactant molecules have come to such a degree of closeness and distortion that a small further distortion will send them in the direction of products. This crucial configuration is called the **transition state** of the reaction. Although some molecules entering the transition state might revert to reactants, if they pass through this configuration then it is inevitable that products will emerge from the encounter.[2]

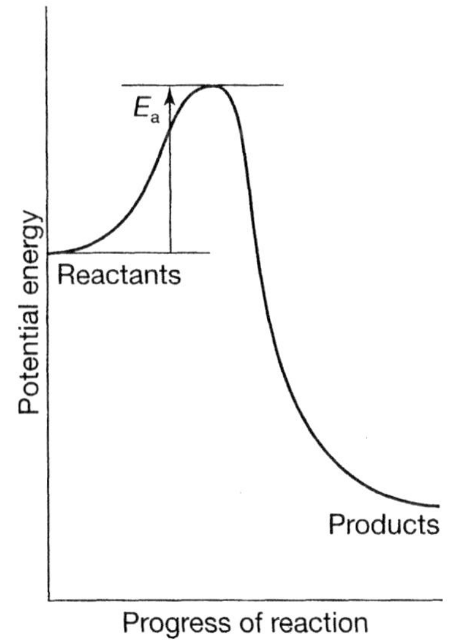

25.12 A potential energy profile for an exothermic reaction. The height of the barrier between the reactants and products is the activation energy of the reaction.

We also conclude from the preceding discussion that *the activation energy is the minimum kinetic energy that reactants must have in order to form products*. For example, in a gas-phase reaction there are numerous collisions each second, but only a tiny proportion are sufficiently energetic to lead to reaction. The fraction of collisions with a kinetic energy in excess of an energy E_a is given by the Boltzmann distribution as $e^{-E_a/RT}$. Hence, the exponential factor in eqn 27 can be interpreted as the fraction of collisions that have enough kinetic energy to lead to reaction.

The pre-exponential factor is a measure of the rate at which collisions occur irrespective of their energy. Hence, the product of A and the exponential factor, $e^{-E_a/RT}$, gives the rate of *successful* collisions. We shall develop these remarks in Chapter 27 and see that they have their analogues for reactions that take place in liquids.

2. The terms activated complex and transition state or transition structure are often used as synonyms; however, we shall preserve a distinction.

Accounting for the rate laws

We now move on to the second stage of the analysis of kinetic data, their explanation in terms of a postulated reaction mechanism.

25.6 Elementary reactions

Most reactions occur in a sequence of steps called **elementary reactions**, each of which involves only a small number of molecules or ions. A typical elementary reaction is

$$H + Br_2 \longrightarrow HBr + Br$$

(We do not specify the phase of the species in the chemical equation for an elementary reaction.) This equation signifies that an H atom attacks a Br_2 molecule to produce an HBr molecule and a Br atom. The **molecularity** of an elementary reaction is the number of molecules coming together to react in an elementary reaction. In a **unimolecular reaction**, a single molecule shakes itself apart or its atoms into a new arrangement, as in the isomerization of cyclopropane (**1**) to propene (**2**). In a **bimolecular reaction**, a pair of molecules collide and exchange energy, atoms, or groups of atoms, or undergo some other kind of change. It is most important to distinguish molecularity from order: reaction order is an empirical quantity, and obtained from the experimental rate law; molecularity refers to an elementary reaction proposed as an individual step in a mechanism.

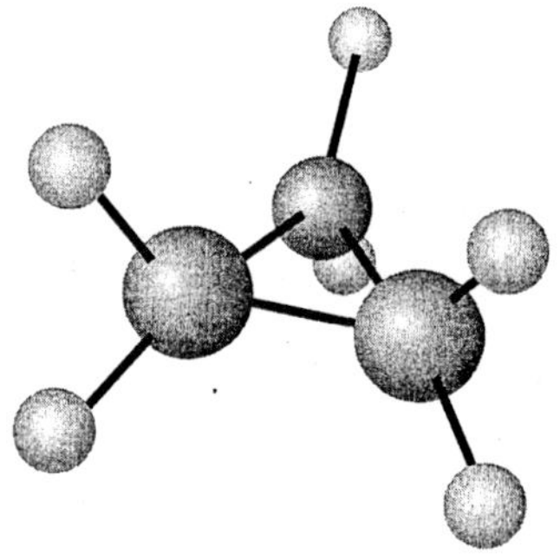

1 Cyclopropane

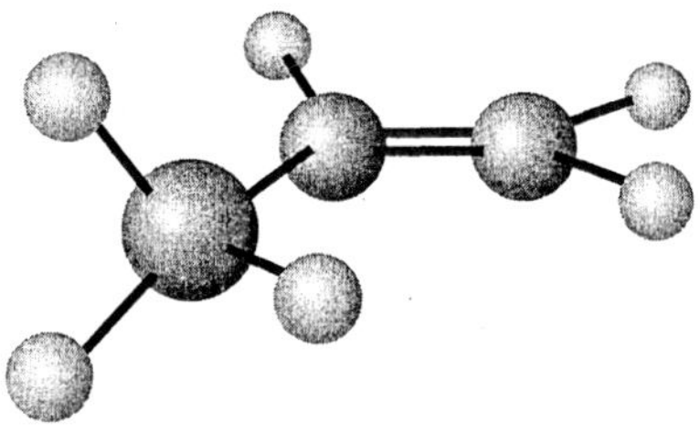

2 Propene

The rate law of a unimolecular elementary reaction is first-order in the reactant:

$$A \longrightarrow P \qquad \frac{d[A]}{dt} = -k[A] \tag{25.28}$$

where P denotes products (several different species may be formed). A unimolecular reaction is first-order because the number of A molecules that decay in a short interval is proportional to the number available to decay. (Ten times as many decay in the same interval when there are initially 1000 A molecules as when there are only 100 present.) Therefore, the rate of decomposition of A is proportional to its molar concentration.

An elementary bimolecular reaction has a second-order rate law:

$$A + B \longrightarrow P \qquad \frac{d[A]}{dt} = -k[A][B] \tag{25.29}$$

A bimolecular reaction is second-order because its rate is proportional to the rate at which the reactant species meet, which in turn is proportional to their concentrations. Therefore, if we have evidence that a reaction is a single-step, bimolecular process, we can write down the rate law (and then go on to test it). Bimolecular elementary reactions are believed to account for many homogeneous reactions, such as the dimerizations of alkenes and dienes and reactions such as

$$CH_3I(alc) + CH_3CH_2O^-(alc) \longrightarrow CH_3OCH_2CH_3(alc) + I^-(alc)$$

(where 'alc' signifies alcohol solution). There is evidence that the mechanism of this reaction is a single elementary step

$$CH_3I + CH_3CH_2O^- \longrightarrow CH_3OCH_2CH_3 + I^-$$

This mechanism is consistent with the observed rate law

$$v = k[CH_3I][CH_3CH_2O^-] \tag{25.30}$$

We shall see below how to combine a series of simple steps together into a mechanism and how to arrive at the corresponding rate law. For the present we emphasize that, *if the reaction is an elementary bimolecular process, then it has second-order kinetics, but if the kinetics are second-order, then the reaction might be complex.* The postulated mechanism can be explored only by detailed detective work on the system, and by investigating whether side products or intermediates appear during the course of the reaction. Detailed analysis of this kind was one of the ways, for example, in which the reaction $H_2(g) + I_2(g) \rightarrow 2\,HI(g)$ was shown to proceed by a complex reaction. For many years the reaction had been accepted on good, but insufficiently meticulous evidence as a fine example of a simple bimolecular reaction in which atoms exchanged partners during a collision.

25.7 Consecutive elementary reactions

Some reactions proceed through the formation of an intermediate (I), as in the consecutive unimolecular reactions

$$A \xrightarrow{k_a} I \xrightarrow{k_b} P$$

An example is the decay of a radioactive family, such as

$$^{239}U \xrightarrow{2.35\,\text{min}} {}^{239}Np \xrightarrow{2.35\,\text{day}} {}^{239}Pu$$

(The times are half-lives.) We can discover the characteristics of this type of reaction by setting up the rate laws for the net rate of change of the concentration of each substance.

(a) The variation of concentrations with time

The rate of unimolecular decomposition of A is

$$\frac{d[A]}{dt} = -k_a[A] \tag{25.31}$$

and A is not replenished. The intermediate I is formed from A (at a rate $k_a[A]$) but decays to P (at a rate $k_b[I]$). The net rate of formation of I is therefore

$$\frac{d[I]}{dt} = k_a[A] - k_b[I] \tag{25.32}$$

The product P is formed by the unimolecular decay of I:

$$\frac{d[P]}{dt} = k_b[I] \tag{25.33}$$

We suppose that initially only A is present, and that its concentration is $[A]_0$.

The first of the rate laws, eqn 31, is an ordinary first-order decay, so we can write

$$[A] = [A]_0 e^{-k_a t} \tag{25.34}$$

When this equation is substituted into eqn 32, we obtain after rearrangement

$$\frac{d[I]}{dt} + k_b[I] = k_a[A]_0 e^{-k_a t} \tag{25.35}$$

This differential equation has a standard form and, after setting $[I]_0 = 0$, the solution is

$$[I] = \frac{k_a}{k_b - k_a}(e^{-k_a t} - e^{-k_b t})[A]_0 \tag{25.36}$$

The solution of a first-order differential equation with the form

$$\frac{dy}{dx} + yf(x) = g(x)$$

is given by

$$e^{\int f(x)\,dx} y = \int e^{\int f(x)\,dx} g(x)\,dx + \text{constant}$$

Equation 35 is a special case of this standard form, with $f(x) = \text{constant}$.

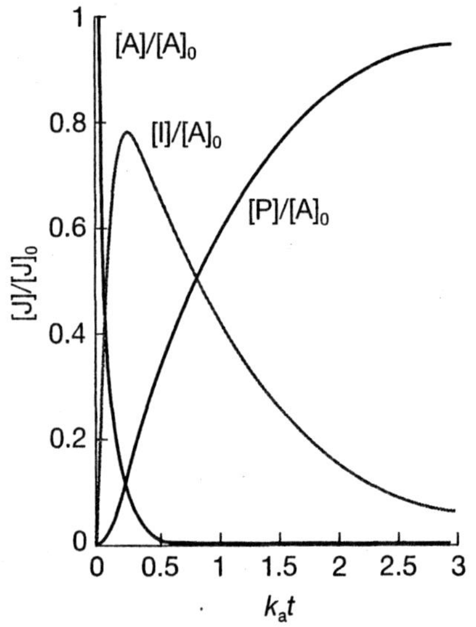

25.13 The concentrations of A, I, and P in the consecutive reaction scheme A → I → P. The curves are plots of eqns 34, 36, and 37 with $k_a = 10k_b$. If the intermediate I is in fact the desired product, it is important to be able to predict when its concentration is greatest; see Example 25.6.

At all times $[A] + [I] + [P] = [A]_0$, so it follows that

$$[P] = \left\{1 + \frac{k_a e^{-k_b t} - k_b e^{-k_a t}}{k_b - k_a}\right\}[A]_0 \tag{25.37}$$

The concentration of the intermediate I rises to a maximum, and then falls to zero (Fig. 25.13). The concentration of the product P rises from zero towards $[A]_0$.

Example 25.6 Analysing consecutive reactions

Suppose that in an industrial batch process a substance A produces the desired compound I that goes on to decay to a worthless product C, each step of the reaction being first-order. At what time will I be present in greatest concentration?

Method The time dependence of the concentration of I is given by eqn 36. We can find the time at which [I] passes through a maximum, t_{max}, by calculating d[I]/dt and setting the resulting rate equal to zero.

Answer It follows from eqn 36 that

$$\frac{d[I]}{dt} = -\frac{k_a[A]_0(k_a e^{-k_a t} - k_b e^{-k_b t})}{k_b - k_a}$$

This rate is equal to zero when

$$k_a e^{-k_a t} = k_b e^{-k_b t}$$

Therefore,

$$t_{max} = \frac{1}{k_a - k_b}\ln\frac{k_a}{k_b}$$

Comment For a given value of k_a, as k_b increases both the time at which [I] is a maximum and the yield of I decrease.

Self-test 25.7 Calculate the maximum concentration of I and justify the last remark. $[[I]_{max}/[A]_0 = (k_a/k_b)^c,\ c = k_b/(k_b - k_a)]$

(b) The rate-determining step

Suppose now that $k_b \gg k_a$; then whenever an I molecule is formed it decays rapidly into P. Because

$$e^{-k_b t} \ll e^{-k_a t} \qquad k_b - k_a \approx k_b$$

eqn 37 reduces to

$$[P] \approx (1 - e^{-k_a t})[A]_0 \tag{25.38}$$

which shows that the formation of the final product P depends on only the *smaller* of the two rate constants. That is, the rate of formation of P depends on the rate at which I is formed, not on the rate at which I changes into P. For this reason, the step A → I is called the 'rate-determining step' of the reaction. Its existence has been likened to building a six-lane highway up to a single-lane bridge: the traffic flow is governed by the rate of crossing the bridge. Similar remarks apply to more complicated reaction mechanisms, and in general the **rate-determining step** is the

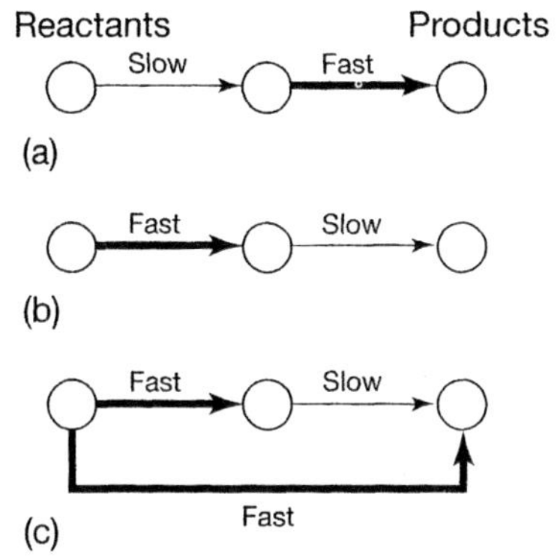

25.14 In these diagrams of reaction schemes, heavy arrows represent fast steps and light arrows represent slow steps. (a) The first step is rate-determining; (b) the second step is rate-determining; (c) although one step is slow, it is not rate-determining because there is a fast route that circumvents it.

slowest step in a mechanism and controls the overall rate of the reaction. However, the rate-determining step is not just the slowest step: it must be slow *and* be a crucial gateway for the formation of products. If a faster reaction can also lead to products, then the slowest step is irrelevant because the slow reaction can then be side-stepped (Fig. 25.14).

The rate law of a reaction that has a rate-determining step can often be written down almost by inspection. If the first step in a mechanism is rate-determining, then the rate of the overall reaction is equal to the rate of the first step because all subsequent steps are so fast that once the first intermediate is formed it results immediately in the formation of products. Figure 25.15 shows the reaction profile for a mechanism of this kind in which the slowest step is the one with the highest activation energy. Once over the initial barrier, the intermediates cascade into products. However, a rate-determining step may also stem from the low concentration of a crucial reactant and need not correspond to the step with highest activation barrier.

(c) The steady-state approximation

One feature of the calculation so far has probably not gone unnoticed: there is a considerable increase in mathematical complexity as soon as the reaction mechanism has more than a couple of steps. A reaction scheme involving many steps is nearly always unsolvable analytically, and alternative methods of solution are necessary. One approach is to integrate the rate laws numerically (see *Further information 1*). An alternative approach, which continues to be widely used because it leads to convenient expressions and more readily digestible results, is to make an approximation.

The **steady-state approximation** assumes that, after an initial **induction period**, an interval during which the concentrations of intermediates, I, rise from zero, and during the major part of the reaction, the rates of change of concentrations of all reaction intermediates are negligibly small (Fig. 25.16):

$$\frac{d[I]}{dt} \approx 0 \qquad (25.39)$$

This approximation greatly simplifies the discussion of reaction schemes. For example, when we apply the approximation to the consecutive first-order mechanism, we set $d[I]/dt = 0$ in eqn 32, which then becomes

$$k_a[A] - k_b[I] \approx 0$$

Then

$$[I] \approx (k_a/k_b)[A] \qquad (25.40)$$

On substituting this value of [I] into eqn 33, that equation becomes

$$\frac{d[P]}{dt} = k_b[I] \approx k_a[A] \qquad (25.41)$$

and we see that P is formed by a first-order decay of A, with a rate constant k_a, the rate constant of the slower, rate-determining, step. We can write down the solution of this equation at once by substituting the solution for [A], eqn 34, and integrating:

$$[P] = k_a[A]_0 \int_0^t e^{-k_a t}\,dt = (1 - e^{-k_a t})[A]_0 \qquad (25.42)$$

This is the same (approximate) result as before, eqn 38, but much more quickly obtained.

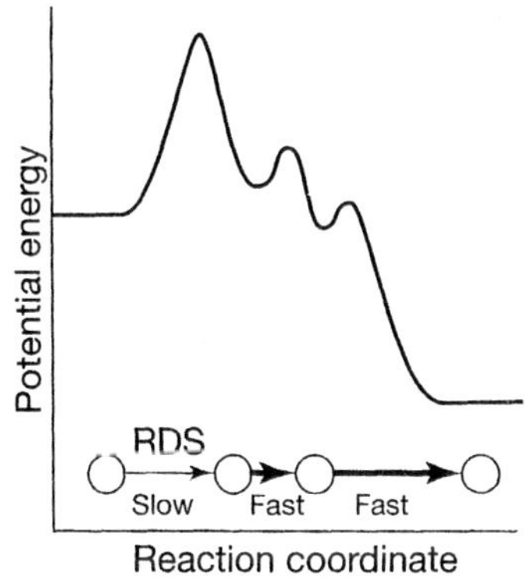

25.15 The reaction profile for a mechanism in which the first step (RDS) is rate-determining.

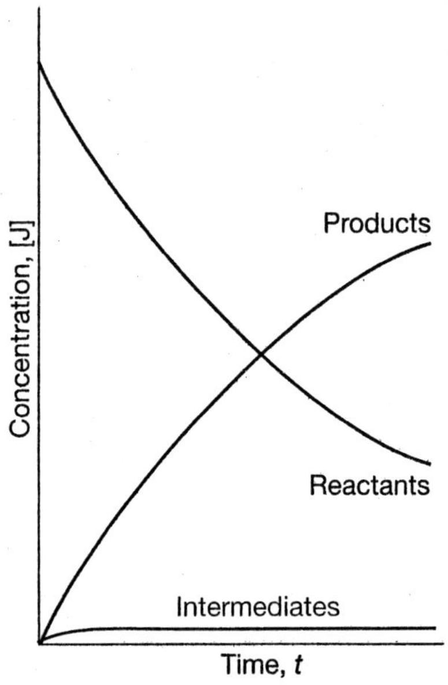

25.16 The basis of the steady-state approximation. It is supposed that the concentrations of intermediates remain small and hardly change during most of the course of the reaction.

Example 25.7 Using the steady-state approximation

Devise the rate law for the decomposition of N_2O_5,

$$2\,N_2O_5(g) \longrightarrow 4\,NO_2(g) + O_2(g)$$

on the basis of the following mechanism:

$$N_2O_5 \longrightarrow NO_2 + NO_3 \qquad k_a$$
$$NO_2 + NO_3 \longrightarrow N_2O_5 \qquad k'_a$$
$$NO_2 + NO_3 \longrightarrow NO_2 + O_2 + NO \qquad k_b$$
$$NO + N_2O_5 \longrightarrow NO_2 + NO_2 + NO_2 \qquad k_c$$

Method First identify the intermediates (species that occur in the reaction steps but do not appear in the overall reaction) and write expressions for their net rates of formation. Then, all net rates of change of the concentrations of intermediates are set equal to zero and the resulting equations are solved algebraically.

Answer The intermediates are NO and NO_3; the net rates of change of their concentrations are

$$\frac{d[NO]}{dt} = k_b[NO_2][NO_3] - k_c[NO][N_2O_5] \approx 0$$

$$\frac{d[NO_3]}{dt} = k_a[N_2O_5] - k'_a[NO_2][NO_3] - k_b[NO_2][NO_3] \approx 0$$

The net rate of change of concentration of N_2O_5 is

$$\frac{d[N_2O_5]}{dt} = -k_a[N_2O_5] + k'_a[NO_2][NO_3] - k_c[NO][N_2O_5]$$

and replacing the concentrations of the intermediates by using the equations above gives

$$\frac{d[N_2O_5]}{dt} = -\frac{2k_ak_b[N_2O_5]}{k'_a + k_b}$$

Comment The decomposition of N_2O_5 is problematic because its rate decreases more quickly than expected at low pressures. It is believed that this decrease is due to changes in the rate constants themselves (particularly k'_a).

Self-test 25.8 Derive the rate law for the decomposition of ozone in the reaction $2\,O_3(g) \rightarrow 3\,O_2(g)$ on the basis of the (incomplete) mechanism

$$O_3 \longrightarrow O_2 + O \qquad k_a$$
$$O_2 + O \longrightarrow O_3 \qquad k'_a$$
$$O + O_3 \longrightarrow O_2 + O_2 \qquad k_b$$

$[d[O_3]/dt = -2k_ak_b[O_3]^2/(k'_a[O_2] + k_b[O_3])]$

(d) Pre-equilibria

From a simple sequence of consecutive reactions we now turn to a slightly more complicated mechanism in which an intermediate I reaches an equilibrium with the reactants A and B:

$$A + B \rightleftharpoons I \longrightarrow P \qquad (25.43)$$

The rate constants are k_a and k_a' for the forward and reverse reactions of the equilibrium and k_b for the final step. This scheme involves a **pre-equilibrium**, in which an intermediate is in equilibrium with the reactants. A pre-equilibrium arises when the rates of formation of the intermediate and its decay back into reactants are much faster than its rate of formation of products; thus, the condition is possible when $k_a' \gg k_b$ but not when $k_b \gg k_a'$. Because we assume that A, B, and I are in equilibrium, we can write

$$K = \frac{[\mathrm{I}]}{[\mathrm{A}][\mathrm{B}]} \qquad K = \frac{k_a}{k_a'} \tag{25.44}$$

In writing these equations, we are presuming that the rate of reaction of I to form P is too slow to affect the maintenance of the pre-equilibrium (see the example below). The rate of formation of P may now be written:

$$\frac{d[\mathrm{P}]}{dt} = k_b[\mathrm{I}] = k_bK[\mathrm{A}][\mathrm{B}] \tag{25.45}$$

This rate law has the form of a second-order rate law with a composite rate constant:

$$\frac{d[\mathrm{P}]}{dt} = k[\mathrm{A}][\mathrm{B}] \qquad k = k_bK = \frac{k_ak_b}{k_a'} \tag{25.46}$$

Example 25.8 Analysing a pre-equilibrium

Repeat the pre-equilibrium calculation but without ignoring the fact that I is slowly leaking away as it forms P.

Method Begin by writing the net rates of change of the concentrations of the substances and then invoke the steady-state approximation for the intermediate I. Use the resulting expression to obtain the rate of change of the concentration of P.

Answer The net rates of change of P and I are

$$\frac{d[\mathrm{P}]}{dt} = k_b[\mathrm{I}]$$

$$\frac{d[\mathrm{I}]}{dt} = k_a[\mathrm{A}][\mathrm{B}] - k_a'[\mathrm{I}] - k_b[\mathrm{I}] \approx 0$$

The second equation solves to

$$[\mathrm{I}] \approx \frac{k_a[\mathrm{A}][\mathrm{B}]}{k_a' + k_b}$$

When we substitute this into the expression for the rate of formation of P, we obtain

$$\frac{d[\mathrm{P}]}{dt} \approx k[\mathrm{A}][\mathrm{B}] \qquad k = \frac{k_ak_b}{k_a' + k_b}$$

Comment This expression reduces to that in eqn 46 when the rate constant for the decay of I into products is much smaller than that for its decay into reactants, $k_b \ll k_a'$.

Self-test 25.9 Show that the pre-equilibrium mechanism in which $2\,\mathrm{A} \rightleftharpoons \mathrm{I}(K)$ followed by $\mathrm{I} + \mathrm{B} \rightarrow \mathrm{P}(k_b)$ results in an overall third-order reaction.

$[d[\mathrm{P}]/dt = k_bK[\mathrm{A}]^2[\mathrm{B}]]$

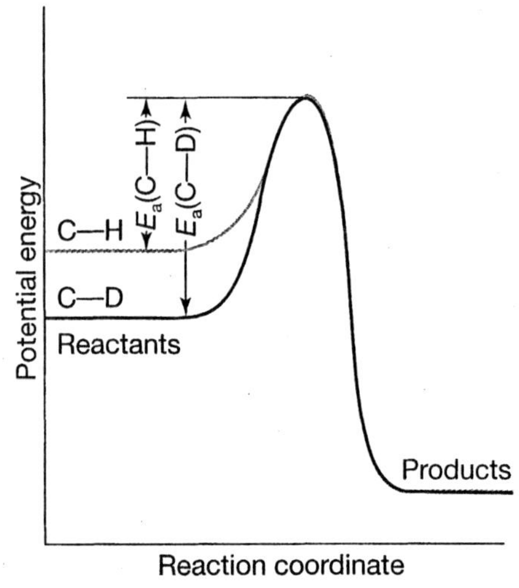

25.17 Changes in the reaction profile when a C—H bond undergoing cleavage is deuterated. In this illustration, the C—H and C—D bonds are modelled as simple harmonic oscillators. The only significant change is in the zero-point energy of the reactants, which is lower for C—D than for C—H. As a result, the activation energy is greater for C—D cleavage than for C—H cleavage.

(e) The kinetic isotope effect

The postulation of a plausible mechanism requires careful analysis of many experiments designed to determine the fate of atoms during the formation of products. Observation of the **kinetic isotope effect**, a decrease in the rate of a chemical reaction upon replacement of one atom in a reactant by a heavier isotope, facilitates the identification of bond-breaking events in the rate-determining step. A **primary kinetic isotope effect** is observed when the rate-determining step requires the scission of a bond involving the isotope. A **secondary kinetic isotope effect** is the reduction in reaction rate even though the bond involving the isotope is not broken to form product. In both cases, the effect arises from the increase in activation energy that accompanies the replacement of an atom by a heavier isotope on account of changes in the zero-point vibrational energies (Section 17.7d).

Here we consider the origin of the primary kinetic isotope effect in a reaction in which the rate-determining step is the scission of a C—H bond. The reaction coordinate corresponds to the stretching of the C—H bond and the potential energy profile is shown in Fig. 25.17. On deuteration, the dominant change is the reduction of the zero-point energy of the bond (because the deuterium atom is heavier). The whole reaction profile is not lowered, however, because the relevant vibration in the activated complex has a very low force constant, so there is little zero-point energy associated with the reaction coordinate in either isotopomeric form of the complex.

We assume that, to a good approximation, a change in the activation energy arises only from the change in zero-point energy of the stretching vibration, so

$$E_a(\text{C—D}) - E_a(\text{C—H}) = N_A\{\tfrac{1}{2}hc\tilde{\nu}(\text{C—H}) - \tfrac{1}{2}hc\tilde{\nu}(\text{C—D})\} \qquad (25.47)$$

where $\tilde{\nu}$ is the relevant vibrational wavenumber. From Section 16.9, we know that $\tilde{\nu}(\text{C—D}) = (\mu_{CH}/\mu_{CD})^{1/2}\tilde{\nu}(\text{C—H})$, where μ is the relevant effective mass. It follows that

$$E_a(\text{C—D}) - E_a(\text{C—H}) = \tfrac{1}{2}N_A hc\tilde{\nu}(\text{C—H})\left\{1 - \left(\frac{\mu_{CH}}{\mu_{CD}}\right)^{1/2}\right\} \qquad (25.48)$$

If we assume further that the pre-exponential factor does not change upon deuteration, then the rate constants for the two species should be in the ratio

$$\frac{k(\text{C—D})}{k(\text{C—H})} = e^{-\lambda} \qquad \text{with} \quad \lambda = \frac{hc\tilde{\nu}(\text{C—H})}{2kT}\left\{1 - \left(\frac{\mu_{CH}}{\mu_{CD}}\right)^{1/2}\right\} \qquad (25.49)$$

Note that $\lambda > 0$ because $\mu_{CD} > \mu_{CH}$ and that $k(\text{C—D})/k(\text{C—H})$ decreases with decreasing temperature.

Illustration 25.2

From infrared spectra, the fundamental vibrational wavenumber for stretching of a C—H bond is about 3000 cm^{-1}. From $\mu_{CH}/\mu_{CD} = 0.538$ and eqn 49, it follows that $k(\text{C—D})/k(\text{C—H}) = 0.145$ at 298 K. We predict that at room temperature C—H cleavage should be about seven times faster than C—D cleavage, other conditions being equal. However, experimental values of $k(\text{C—D})/k(\text{C—H})$ can differ significantly from those predicted by eqn 49 on account of the severity of the assumptions in the model.

Secondary kinetic isotope effects arise from differences in the zero-point energies between reactants and an activated complex with significantly different structure. When a change in the zero-point energy of a single vibration involving motion of a C—H group can account for the change in activation energy upon deuteration, we write:

$$E_a(\text{C—D}) - E_a(\text{C—H}) = N_A[\{\tfrac{1}{2}hc\tilde{\nu}^{\ddagger}(\text{C—H}) - \tfrac{1}{2}hc\tilde{\nu}(\text{C—H})\} - \{\tfrac{1}{2}hc\tilde{\nu}^{\ddagger}(\text{C—D}) - \tfrac{1}{2}hc\tilde{\nu}(\text{C—D})\}] \quad (25.50)$$

where $\tilde{\nu}^{\ddagger}$ and $\tilde{\nu}$ denote vibrational wavenumbers in the activated complex and reactant, respectively. With $\tilde{\nu}^{\ddagger}(\text{C—D}) = (\mu_{CH}/\mu_{CD})^{1/2}\tilde{\nu}^{\ddagger}(\text{C—H})$ and $\tilde{\nu}(\text{C—D}) = (\mu_{CH}/\mu_{CD})^{1/2}\tilde{\nu}(\text{C—H})$, it follows that

$$E_a(\text{C—D}) - E_a(\text{C—H}) = \tfrac{1}{2}N_A hc\{\tilde{\nu}^{\ddagger}(\text{C—H}) - \tilde{\nu}(\text{C—H})\}\left\{1 - \left(\frac{\mu_{CH}}{\mu_{CD}}\right)^{1/2}\right\} \quad (25.51)$$

and

$$\frac{k(\text{C—D})}{k(\text{C—H})} = e^{-\lambda} \quad \text{with} \quad \lambda = \frac{hc\{\tilde{\nu}^{\ddagger}(\text{C—H}) - \tilde{\nu}(\text{C—H})\}}{2kT}\left\{1 - \left(\frac{\mu_{CH}}{\mu_{CD}}\right)^{1/2}\right\} \quad (25.52)$$

Illustration 25.3

In the heterolytic dissociation

$$CHCl_3 \longrightarrow CHCl_2^+ + Cl^-$$

the activated complex resembles the product $CHCl_2^+$. From infrared spectra, the fundamental vibrational wavenumber for a bending motion involving the C—H group is about 800 cm^{-1} in $CHCl_3$ and about 1350 cm^{-1} in $CHCl_2^+$. Assuming that $\tilde{\nu}^{\ddagger}(\text{C—H}) = 1350$ cm^{-1} on account of the structural similarity between $CHCl_2^+$ and the activated complex, it follows from $\mu_{CH}/\mu_{CD} = 0.538$ and eqn 52 that $k(\text{C—D})/k(\text{C—H}) = 0.709$ at 298 K. We predict that at room temperature the dissociation of $CHCl_3$ should be about 40 per cent faster than dissociation of $CDCl_3$. Comparison with the result from Illustration 25.2 shows that the secondary kinetic isotope effect leads to higher values of $k(\text{C—D})/k(\text{C—H})$ than does the primary kinetic isotope effect. This conclusion is supported by a number of experimental observations.

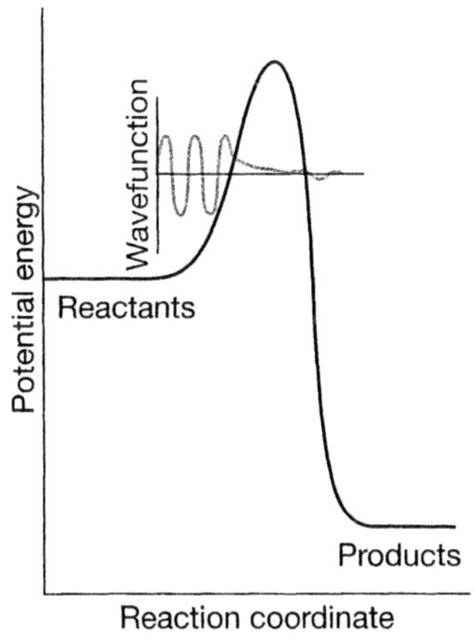

25.18 A proton can tunnel through the activation energy barrier that separates reactants from products, so the effective height of the barrier is reduced and the rate of the proton transfer reaction increases. The effect is represented by drawing the wavefunction of the proton near the barrier. Proton tunnelling is important only at low temperatures, when most of the reactants are trapped on the left of the barrier.

In some cases, substitution of deuterium for hydrogen results in values of $k(\text{C—D})/k(\text{C—H})$ that are too low to be accounted for by eqns 49 or 52, even when more complete models are used to predict ratios of rate constants. Such abnormal kinetic isotope effects are evidence for a path in which quantum mechanical tunnelling of hydrogen atoms takes place through the activation barrier (Fig. 25.18). We saw in Section 12.3 that the probability of tunnelling through a barrier decreases as the mass of the particle increases, so deuterium tunnels less efficiently through a barrier than hydrogen and its reactions are correspondingly slower. Quantum mechanical tunnelling can be the dominant process in reactions involving hydrogen atom or proton transfer when the temperature is so low that very few reactant molecules can overcome the activation energy barrier. We shall see in Chapter 29 that, because m_e is so small, tunnelling is also a very important contributor to the rates of electron transfer reactions.

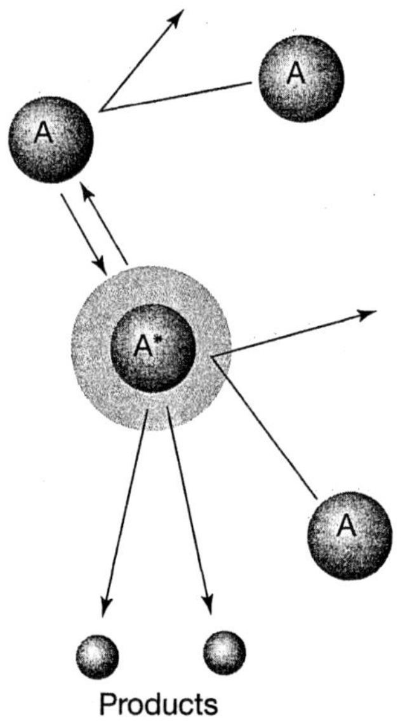

25.19 A representation of the Lindemann–Hinshelwood mechanism of unimolecular reactions. The species A is excited by collision with A, and the excited A molecule (A*) may either be deactivated by a collision with A or go on to decay by a unimolecular process to form products.

25.8 Unimolecular reactions

A number of gas-phase reactions follow first-order kinetics, as in the isomerization of cyclopropane mentioned earlier:

$$\textit{cyclo}\text{-}C_3H_6 \longrightarrow CH_3CH{=}CH_2 \qquad v = k[\textit{cyclo}\text{-}C_3H_6] \tag{25.53}$$

The problem with the interpretation of first-order rate laws is that presumably a molecule acquires enough energy to react as a result of its collisions with other molecules. However, collisions are simple bimolecular events, so how can they result in a first-order rate law? First-order gas-phase reactions are widely called 'unimolecular reactions' because they also involve an elementary unimolecular step in which the reactant molecule changes into the product. This term must be used with caution, though, because the overall mechanism has bimolecular as well as unimolecular steps.

(a) The Lindemann–Hinshelwood mechanism

The first successful explanation of unimolecular reactions was provided by Frederick Lindemann in 1921 and then elaborated by Cyril Hinshelwood. In the **Lindemann–Hinshelwood mechanism** it is supposed that a reactant molecule A becomes energetically excited by collision with another A molecule (Fig. 25.19):

$$A + A \longrightarrow A^* + A \qquad \frac{d[A^*]}{dt} = k_a[A]^2 \tag{25.54}$$

The energized molecule might lose its excess energy by collision with another molecule:

$$A + A^* \longrightarrow A + A \qquad \frac{d[A^*]}{dt} = -k_a'[A][A^*] \tag{25.55}$$

Alternatively, the excited molecule might shake itself apart and form products P. That is, it might undergo the unimolecular decay

$$A^* \longrightarrow P \qquad \frac{d[A^*]}{dt} = -k_b[A^*] \tag{25.56}$$

If the unimolecular step is slow enough to be the rate-determining step, the overall reaction will have first-order kinetics, as observed. This conclusion can be demonstrated explicitly by applying the steady-state approximation to the net rate of formation of A*:

$$\frac{d[A^*]}{dt} = k_a[A]^2 - k_a'[A][A^*] - k_b[A^*] \approx 0 \tag{25.57}$$

This equation solves to

$$[A^*] = \frac{k_a[A]^2}{k_b + k_a'[A]} \tag{25.58}$$

so the rate law for the formation of P is

$$\frac{d[P]}{dt} = k_b[A^*] = \frac{k_a k_b[A]^2}{k_b + k_a'[A]} \tag{25.59}$$

At this stage the rate law is not first-order. However, if the rate of deactivation by (A*,A) collisions is much greater than the rate of unimolecular decay, in the sense that

$$k_a'[A^*][A] \gg k_b[A^*] \qquad \text{or} \qquad k_a'[A] \gg k_b$$

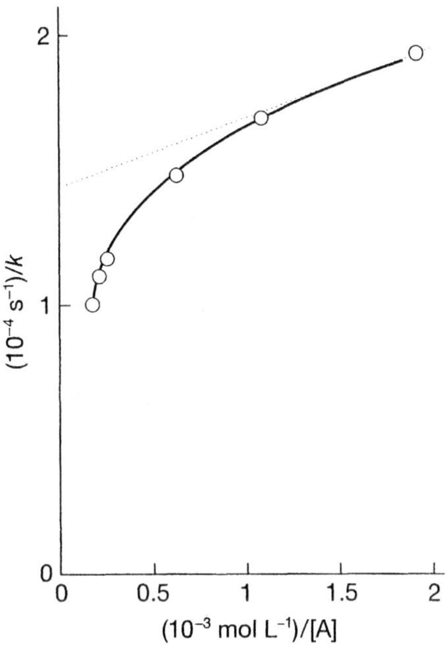

25.20 The pressure dependence of the unimolecular isomerization of *trans*-CHD=CHD showing a pronounced departure from the straight line predicted by eqn 63 based on the Lindemann–Hinshelwood mechanism.

then we can neglect k_b in the denominator of eqn 39 and obtain

$$\frac{d[P]}{dt} \approx k[A] \qquad k = k_a k_b / k_a' \tag{25.60}$$

Equation 60 is a first-order rate law, as we set out to show.

The Lindemann–Hinshelwood mechanism can be tested because it predicts that, as the concentration (and therefore the partial pressure) of A is reduced, the reaction should switch to overall second-order kinetics. Thus, when $k_a'[A] \ll k_b$, the rate law in eqn 59 is

$$\frac{d[P]}{dt} \approx k_a[A]^2 \tag{25.61}$$

The physical reason for the change of order is that at low pressures the rate-determining step is the bimolecular formation of A*. If we write the full rate law in eqn 59 as

$$\frac{d[P]}{dt} = k[A] \qquad k = \frac{k_a k_b [A]}{k_b + k_a'[A]} \tag{25.62}$$

then the expression for the effective rate-constant, k, can be rearranged to

$$\frac{1}{k} = \frac{k_a'}{k_a k_b} + \frac{1}{k_a[A]} \tag{25.63}$$

Hence, a test of the theory is to plot $1/k$ against $1/[A]$, and to expect a straight line.

Whereas the Lindemann–Hinshelwood mechanism agrees in general with the switch in order of unimolecular reactions, it does not agree in detail. Figure 25.20 shows a typical graph of $1/k$ against $1/[A]$. The graph has a pronounced curvature, corresponding to a larger value of k (a smaller value of $1/k$) at high pressures (low $1/[A]$) than would be expected by extrapolation of the reasonably linear low pressure (high $1/[A]$) data.

(b) The activation energy of a composite reaction

Although the rate of each step of a complex mechanism might increase with temperature and show Arrhenius behaviour, is that true of a composite reaction? To answer this question, we consider the high-pressure limit of the Lindemann–Hinshelwood mechanism as expressed in eqn 60. If each of the rate constants has an Arrhenius-like temperature dependence, we can use eqn 27 for each of them, and write

$$\begin{aligned} k = \frac{k_a k_b}{k_a'} &= \frac{(A_a e^{-E_a(a)/RT})(A_b e^{-E_a(b)/RT})}{(A_a' e^{-E_a'(a)/RT})} \\ &= \frac{A_a A_b}{A_a'} e^{-\{E_a(a)+E_a(b)-E_a'(a)\}/RT} \end{aligned} \tag{25.64}$$

That is, the composite rate constant k has an Arrhenius-like form with activation energy

$$E_a = E_a(a) + E_a(b) - E_a'(a) \tag{25.65}$$

Provided $E_a(a) + E_a(b) > E_a'(a)$, the activation energy is positive and the rate increases with temperature. However, it is conceivable that $E_a(a) + E_a(b) < E_a'(a)$ (Fig. 25.21), in which case the activation energy is negative and the rate will *decrease* as the temperature is raised. There is nothing remarkable about this behaviour: all it means is that the reverse reaction (corresponding to the deactivation of A*) is so sensitive to

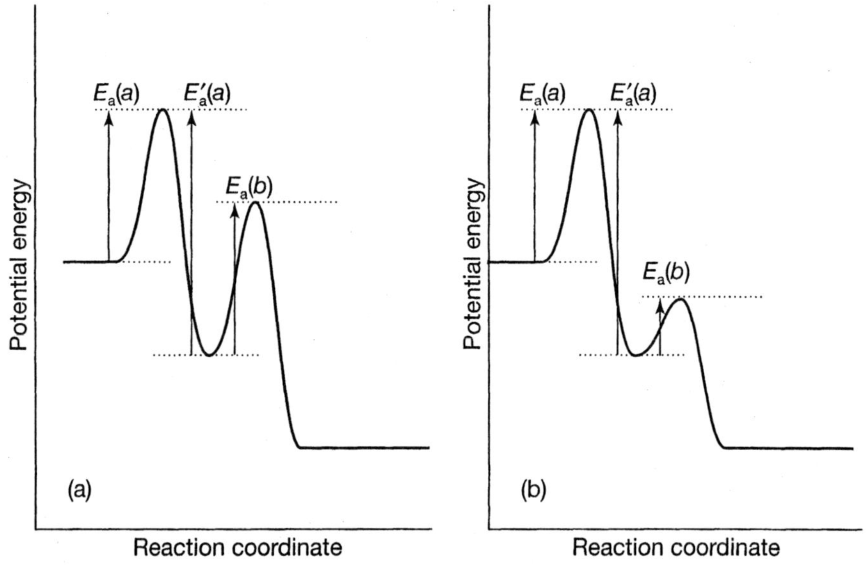

25.21 For a reaction with a pre-equilibrium, there are three activation energies to take into account, two referring to the reversible steps of the pre-equilibrium and one for the final step. The relative magnitudes of the activation energies determine whether the overall activation energy is (a) positive or (b) negative.

temperature that its rate increases sharply as the temperature is raised, and depletes the steady-state concentration of A*. The Lindemann–Hinshelwood mechanism is an unlikely candidate for this type of behaviour because the deactivation of A* has only a small activation energy, but there are reactions with analogous mechanisms in which a negative activation energy is observed.

When we examine the general rate law given in eqn 59, it is clear that the temperature dependence may be difficult to predict because each rate constant in the expression for k increases with temperature, and the outcome depends on whether the terms in the numerator dominate those in the denominator, or vice versa. The fact that so many reactions do show Arrhenius-like behaviour with positive activation energies suggests that their rate laws are in a 'simple' regime, like eqn 61 rather than eqn 59, and that the temperature dependence is dominated by the activation energy of the rate-determining stage.

Checklist of key ideas

- ☐ chemical kinetics, the study of reaction rates
- ☐ mechanism of reaction, the sequence of elementary steps involved in a reaction

Empirical chemical kinetics

25.1 Experimental techniques

- ☐ real-time analysis, a procedure in which the composition of a system is analysed while the reaction is in progress
- ☐ flow method, a procedure in which the composition of a system is analysed as the reactants flow into a mixing chamber
- ☐ stopped-flow technique, a procedure in which the reagents are mixed very quickly in a small chamber fitted with a syringe instead of an outlet tube
- ☐ flash photolysis, a procedure in which the reaction is initiated by a brief flash of light
- ☐ quenching methods, techniques based on stopping the reaction after it has been allowed to proceed for a certain time
- ☐ chemical quench flow method, a technique in which the reactants are mixed as in the flow method but the reaction is quenched by another reagent
- ☐ freeze quench method, a technique in which the reaction is quenched by cooling the mixture

25.2 The rates of reactions

- ☐ rate of consumption of a reactant R, $-d[R]/dt$
- ☐ rate of formation of a product P, $d[P]/dt$
- ☐ rate of reaction, $v = d\xi/dt$
- ☐ rate of homogeneous reaction, $v = (1/\nu_J)\, d[J]/dt$
- ☐ rate of heterogeneous reaction, $v = (1/\nu_J)\, d\sigma_J/dt$

☐ rate constant, the constant k in a rate law

☐ rate law, the rate as a function of concentration, $v = f([A], [B], \ldots)$

☐ reaction order, the power to which the concentration of a species is raised in a rate law

☐ first-order reaction, a reaction with a rate law of the form $v = k[A]$

☐ second-order reaction, a reaction with a rate law of the form $v = k[A]^2$

☐ overall order, the sum of the orders in a rate law of the form $v = k[A]^a[B]^b \ldots$

☐ zero-order rate law, a rate law of the form $v = k$

☐ isolation method, a procedure in which the concentrations of all the reactants except one are in large excess

☐ pseudofirst-order rate law, a rate law of the form $v = k'[A]$ with $k' = k[B]_0$ by maintaining B in large excess

☐ method of initial rates, a procedure in which the rate is measured at the beginning of the reaction for several different initial concentrations of reactants; $\log v_0 = \log k + a \log [A]_0$

25.3 Integrated rate laws

☐ integrated rate law, the integrated form of a rate law

☐ first-order integrated rate law, $\ln([A]/[A]_0) = -kt$, $[A] = [A]_0e^{-kt}$

☐ half life, $t_{1/2} = (\ln 2)/k$ (first-order reaction), $t_{1/2} = 1/(k[A]_0)$ (second-order reaction)

☐ time constant, the time required for the concentration of a reactant to fall to $1/e$ of its initial value

☐ second-order integrated rate law, $1/[A] - 1/[A]_0 = kt$, $[A] = [A]_0/(1 + kt[A]_0)$

25.4 Reactions approaching equilibrium

☐ approach to equilibrium, $[A] = \{(k' + ke^{-(k+k')t})/(k + k')\}[A]_0$

☐ equilibrium constant, $K = k/k'$

☐ relaxation, the return to equilibrium

☐ temperature jump, a procedure in which a sudden temperature rise is imposed and the system returns to equilibrium

☐ pressure-jump techniques, as for temperature jump, but with a sudden change in pressure

25.5 The temperature dependence of reaction rates

☐ Arrhenius equation, $\ln k = \ln A - E_a/RT$

☐ pre-exponential factor (frequency factor), the parameter A in the Arrhenius equation

☐ activation energy, the parameter E_a in the Arrhenius equation; the minimum kinetic energy for reaction during a molecular encounter

☐ Arrhenius parameters, the parameters A and E_a

☐ generalized activation energy, $E_a = RT^2(\mathrm{d} \ln k/\mathrm{d}T)$

☐ reaction coordinate, the collection of motions that are directly involved in the formation of products

☐ activated complex, the cluster of atoms that corresponds to the region close to the maximum potential energy along the reaction coordinate

☐ transition state, a configuration of atoms in the activated complex which, if attained, leads to products

Accounting for the rate laws

25.6 Elementary reactions

☐ elementary reaction, a step in a reaction mechanism

☐ molecularity, the number of molecules coming together to react in an elementary reaction

☐ unimolecular reaction, an elementary reaction involving a single reactant molecule

☐ bimolecular reaction, an elementary reaction involving the encounter of two reactant molecules

25.7 Consecutive elementary reactions

☐ consecutive first-order reactions, a sequence of first-order reactions; see Fig. 25.13

☐ rate-determining step, the step in a mechanism that controls the overall rate of the reaction; commonly but not necessarily the slowest step

☐ steady-state approximation, the rates of change of concentrations of all reaction intermediates are negligibly small: $\mathrm{d}[I]/\mathrm{d}t \approx 0$

☐ induction period, the initial stage of a reaction during which reaction intermediates are formed

☐ pre-equilibrium, a state in which an intermediate is in equilibrium with the reactants and which arises when the rates of formation of the intermediate and its decay back into reactants are much faster than its rate of formation of products

☐ kinetic isotope effect, the decrease in the rate of a chemical reaction upon replacement of one atom in a reactant by a heavier isotope

☐ primary kinetic isotope effect, the kinetic isotope effect observed when the rate-determining step requires the scission of a bond involving the isotope: $k(\mathrm{C{-}D})/k(\mathrm{C{-}H}) = e^{-\lambda}$, $\lambda = \{hc\tilde{\nu}(\mathrm{C{-}H})/2kT\}\{1 - (\mu_{CH}/\mu_{CD})^{1/2}\}$

☐ secondary kinetic isotope effect, the reduction in reaction rate even though the bond involving the isotope is not broken to form product: $k(\mathrm{C{-}D})/k(\mathrm{C{-}H}) = e^{-\lambda}$, $\lambda = \{hc[\tilde{\nu}^{\ddagger}(\mathrm{C{-}H}) - \tilde{\nu}(\mathrm{C{-}H})]/2kT\}\{1 - (\mu_{CH}/\mu_{CD})^{1/2}\}$

25.8 Unimolecular reactions

☐ Lindemann–Hinshelwood mechanism, a theory of 'unimolecular' reactions: $A + A \rightarrow A^* + A$, $A + A^* \rightarrow A + A$, $A^* \rightarrow P$; $\mathrm{d}[P]/\mathrm{d}t = k_ak_b[A]^2/(k_b + k_a'[A])$

☐ activation energies of composite reactions, $E_a = E_a(a) + E_a(b) - E_a'(a)$

Further reading

Articles of general interest

J. Andraos, How mathematics figures in chemistry: some examples. *J. Chem. Educ.* **76**, 258 (1999).

J. Andraos, A streamlined approach to solving simple and complex kinetic systems analytically. *J. Chem. Educ.* **76**, 1578 (1999).

M.N. Berberan-Santos and J.M.G. Martinho, Integration of kinetic rate equations by matrix methods. *J. Chem. Educ.* **67**, 375 (1990).

S. Bluestone and K.Y. Yan, A method to find the rate constants in chemical kinetics of a complex reaction. *J. Chem. Educ.* **72**, 884 (1995).

R.A.B. Bond, B.S. Martincigh, J.R. Mika, and R.H. Simoyi, The quasi-steady-state approximation: numerical validation. *J. Chem. Educ.* **75**, 1158 (1998).

R.W. Carr, Chemical kinetics. In *Encyclopedia of applied physics* (ed. G.L. Trigg), **3**, 345. VCH, New York (1992).

P.L. Corio, Chemical kinetics in constant-pressure systems. *J. Chem. Educ.* **76**, 1452 (1999).

T. Cvitaš, A new look at reaction rates. *J. Chem. Educ.* **76**, 1574 (1999).

J.G. Eberhardt and E. Levin, A simplified integration technique for reaction rate laws of integral order in several substances. *J. Chem. Educ.* **72**, 193 (1995).

G.I. Gellene, Application of kinetic approximations to the $A + B \rightleftharpoons C$ reaction system. *J. Chem. Educ.* **72**, 196 (1995).

J.M. Goodman, How do approximations affect the solutions to kinetic equations? *J. Chem. Educ.* **76**, 275 (1999).

E. Levin and J.G. Eberhart, Simplified rate-law integration for reactions that are first-order in each of the two reactants. *J. Chem. Educ.* **66**, 705 (1989).

H. Maskill, The extent of reaction and chemical kinetics. *Educ. in Chem.* **21**, 122 (1984).

J.C. Reeve, Some provocative opinions on the terminology of chemical kinetics. *J. Chem. Educ.* **68**, 728 (1991).

S.R. Logan, The meaning and significance of 'the activation energy' of a chemical reaction. *Educ. in Chem.* **23**, 148 (1986).

Texts and sources of data and information

C.H. Bamford, C.F. Tipper, and R.G. Compton (ed.), *Comprehensive chemical kinetics*, Vols. 1–38. Elsevier, Amsterdam (1969–2001).

K.A. Connors, *Chemical kinetics*. VCH, Weinheim (1990).

K.J. Laidler, *Chemical kinetics*. Harper and Row, New York (1987).

J.C. Lindon, G.E. Tranter, and J.L. Holmes (ed.), *Encyclopedia of spectroscopy and spectrometry*. Academic Press, San Diego (2000).

S.R. Logan, *Fundamentals of chemical kinetics*. Longman, Harlow (1996).

NDRL/NIST solution kinetics database, NIST standard reference database 40, National Institute of Standards and Technology, Gaithersburg (1994). For the URL, see the web site for this book.

NIST chemical kinetics database, NIST standard reference database 17, National Institute of Standards and Technology, Gaithersburg (1998). For the URL, see the web site for this book.

M.J. Pilling and P.W. Seakins, *Reaction kinetics*. Oxford University Press (1996).

J.I. Steinfeld, J.S. Francisco, and W.L. Hase, *Chemical kinetics and dynamics*. Prentice Hall, Englewood Cliffs (1998).

Exercises

Discussion questions

25.1 (a) Write a brief report on a recent research article in which at least one of the following techniques was used to study the kinetics of a chemical reaction: stopped-flow techniques, flash photolysis, chemical quench-flow methods, freeze quench methods, temperature-jump methods, or pressure-jump methods. Your report should be similar in content and size to one of the *Boxes* found throughout this text.

25.1 (b) Describe the main features, including advantages and disadvantages, of the following experimental methods for determining the rate law of a reaction: the isolation method, the method of initial rates, and fitting data to integrated rate law expressions.

25.2 (a) Distinguish between reaction order and molecularity.

25.2 (b) Assess the validity of the following statement: the rate-determining step is the slowest step in a reaction mechanism.

25.3 (a) Distinguish between a pre-equilibrium approximation and a steady-state approximation.

25.3 (b) Define the terms in and limit the generality of the expression $\ln k = \ln A - E_a/RT$.

25.4 (a) Distinguish between a primary and a secondary kinetic isotope effect. Discuss how kinetic isotope effects in general can provide insight into the mechanism of a reaction.

25.4 (b) Discuss the limitations of the generality of the expression $k = k_a k_b[A]/(k_b + k'_a[A])$ for the effective rate constant of a unimolecular reaction $A \rightarrow P$ with the following mechanism: $A + A \rightleftharpoons A^* + A$ (k_a, k'_a), $A^* \rightarrow P(k_b)$. Suggest an experimental procedure that may either support or refute the mechanism.

Numerical exercises

25.5 (a) The rate of the reaction $A + 2\,B \rightarrow 3\,C + D$ was reported as 1.0 mol L^{-1} s^{-1}. State the rates of formation and consumption of the participants.

25.5 (b) The rate of the reaction $A + 3\,B \rightarrow C + 2\,D$ was reported as 1.0 mol L^{-1} s^{-1}. State the rates of formation and consumption of the participants.

25.6 (a) The rate of formation of C in the reaction $2\,A + B \rightarrow 2\,C + 3\,D$ is 1.0 mol L^{-1} s^{-1}. State the reaction rate, and the rates of formation or consumption of A, B, and D.

25.6 (b) The rate of consumption of B in the reaction $A + 3\,B \rightarrow C + 2\,D$ is 1.0 mol L^{-1} s^{-1}. State the reaction rate, and the rates of formation or consumption of A, C, and D.

25.7 (a) The rate law for the reaction in Exercise 25.5a was found to be $v = k[A][B]$. What are the units of k? Express the rate law in terms of the rates of formation and consumption of (a) A, (b) C.

25.7 (b) The rate law for the reaction in Exercise 25.5b was found to be $v = k[A][B]^2$. What are the units of k? Express the rate law in terms of the rates of formation and consumption of (a) A, (b) C.

25.8 (a) The rate law for the reaction in Exercise 25.6a was reported as $d[C]/dt = k[A][B][C]$. Express the rate law in terms of the reaction rate; what are the units for k in each case?

25.8 (b) The rate law for the reaction in Exercise 25.6b was reported as $d[C]/dt = k[A][B][C]^{-1}$. Express the rate law in terms of the reaction rate; what are the units for k in each case?

25.9 (a) At 518°C, the rate of decomposition of a sample of gaseous acetaldehyde, initially at a pressure of 363 Torr, was 1.07 Torr s^{-1} when 5.0 per cent had reacted and 0.76 Torr s^{-1} when 20.0 per cent had reacted. Determine the order of the reaction.

25.9 (b) At 400 K, the rate of decomposition of a gaseous compound initially at a pressure of 12.6 kPa, was 9.71 Pa s^{-1} when 10.0 per cent had reacted and 7.67 Pa s^{-1} when 20.0 per cent had reacted. Determine the order of the reaction.

25.10 (a) At 518°C, the half-life for the decomposition of a sample of gaseous acetaldehyde (ethanal) initially at 363 Torr was 410 s. When the pressure was 169 Torr, the half-life was 880 s. Determine the order of the reaction.

25.10 (b) At 400 K, the half-life for the decomposition of a sample of a gaseous compound initially at 55.5 kPa was 340 s. When the pressure was 28.9 kPa, the half-life was 178 s. Determine the order of the reaction.

25.11 (a) The rate constant for the first-order decomposition of N_2O_5 in the reaction $2\,N_2O_5(g) \rightarrow 4\,NO_2(g) + O_2(g)$ is $k = 3.38 \times 10^{-5}\ s^{-1}$ at 25°C. What is the half-life of N_2O_5? What will be the pressure, initially 500 Torr, (a) 10 s, (b) 10 min after initiation of the reaction?

25.11 (b) The rate constant for the first-order decomposition of a compound A in the reaction $2\,A \rightarrow P$ is $k = 2.78 \times 10^{-7}\ s^{-1}$ at 25°C. What is the half-life of A? What will be the pressure, initially 32.1 kPa, (a) 10 s, (b) 10 min after initiation of the reaction?

25.12 (a) A second-order reaction of the type $A + B \rightarrow P$ was carried out in a solution that was initially 0.050 mol L^{-1} in A and 0.080 mol L^{-1} in B. After 1.0 h the concentration of A had fallen to 0.20 mol L^{-1}. (a) Calculate the rate constant. (b) What is the half-life of the reactants?

25.12 (b) A second-order reaction of the type $A + 2\,B \rightarrow P$ was carried out in a solution that was initially 0.075 mol L^{-1} in A and 0.030 mol L^{-1} in B. After 1.0 h the concentration of A had fallen to 0.045 mol L^{-1}. (a) Calculate the rate constant. (b) What is the half-life of the reactants?

25.13 (a) If the rate laws are expressed with (a) concentrations in moles per litre, (b) pressures in kilopascals, what are the units of the second-order and third-order rate constants?

25.13 (b) If the rate laws are expressed with (a) concentrations in molecules per metre cubed, (b) pressures in newtons per metre squared, what are the units of the second-order and third-order rate constants?

25.14 (a) The second-order rate constant for the reaction

$$CH_3COOC_2H_5(aq) + OH^-(aq) \longrightarrow CH_3CO_2^-(aq) + CH_3CH_2OH(aq)$$

is 0.11 L mol^{-1} s^{-1}. What is the concentration of ester after (a) 10 s, (b) 10 min when ethyl acetate is added to sodium hydroxide so that the initial concentrations are $[NaOH] = 0.050$ mol L^{-1} and $[CH_3COOC_2H_5] = 0.100$ mol L^{-1}?

25.14 (b) The second-order rate constant for the reaction $A + 2\,B \rightarrow C + D$ is 0.21 L mol^{-1} s^{-1}. What is the concentration of C after (a) 10 s, (b) 10 min when the reactants are mixed with initial concentrations of $[A] = 0.025$ mol L^{-1} and $[B] = 0.150$ mol L^{-1}?

25.15 (a) A reaction $2\,A \rightarrow P$ has a second-order rate law with $k = 3.50 \times 10^{-4}$ L mol^{-1} s^{-1}. Calculate the time required for the concentration of A to change from 0.260 mol L^{-1} to 0.011 mol L^{-1}.

25.15 (b) A reaction $2\,A \rightarrow P$ has a third-order rate law with $k = 3.50 \times 10^{-4}\ L^2\ mol^{-2}\ s^{-1}$. Calculate the time required for the concentration of A to change from 0.077 mol L^{-1} to 0.021 mol L^{-1}.

25.16 (a) The rate constant for the decomposition of a certain substance is 2.80×10^{-3} L mol^{-1} s^{-1} at 30°C and 1.38×10^{-2} L mol^{-1} s^{-1} at 50°C. Evaluate the Arrhenius parameters of the reaction.

25.16 (b) The rate constant for the decomposition of a certain substance is 1.70×10^{-2} L mol^{-1} s^{-1} at 24°C and 2.01×10^{-2} L mol^{-1} s^{-1} at 37°C. Evaluate the Arrhenius parameters of the reaction.

25.17 (a) The base-catalysed bromination of nitromethane-d_3 in water at room temperature (298 K) proceeds 4.3 times more slowly than the bromination of the undeuterated material. Account for this difference. Use $k_f(C{-}H) = 450\ N\ m^{-1}$.

25.17 (b) Predict the order of magnitude of the isotope effect on the relative rates of displacement of (a) ^1H and ^3H, (b) ^{16}O and ^{18}O. Will raising the temperature enhance the difference? Take $k_f(C{-}H) = 450\ N\ m^{-1}$, $k_f(C{-}O) = 1750\ N\ m^{-1}$.

25.18 (a) Show that $t_{1/2} \propto 1/[A]^{n-1}$ for a reaction that is nth-order in A.

25.18 (b) Deduce an expression for the time it takes for the concentration of a substance to fall to one-third its initial value in an nth-order reaction.

25.19 (a) The effective rate constant for a gaseous reaction that has a Lindemann–Hinshelwood mechanism is $2.50 \times 10^{-4}\ s^{-1}$ at 1.30 kPa and $2.10 \times 10^{-5}\ s^{-1}$ at 12 Pa. Calculate the rate constant for the activation step in the mechanism.

25.19 (b) The effective rate constant for a gaseous reaction that has a Lindemann–Hinshelwood mechanism is $1.7 \times 10^{-3}\ s^{-1}$ at 1.09 kPa and $2.2 \times 10^{-4}\ s^{-1}$ at 25 Pa. Calculate the rate constant for the activation step in the mechanism.

25.20 (a) The pK_a of NH_4^+ is 9.25 at 25°C. The rate constant at 25°C for the reaction of NH_4^+ and OH^- to form aqueous NH_3 is 4.0×10^{10} L mol^{-1} s^{-1}. Calculate the rate constant for proton transfer to NH_3. What relaxation time would be observed if a temperature jump were applied to a solution of 0.15 mol L^{-1} $NH_3(aq)$ at 25°C?

25.20 (b) The equilibrium $A \rightleftharpoons B + C$ at 25°C is subjected to a temperature jump that slightly increases the concentrations of B and C. The measured relaxation time is 3.0 μs. The equilibrium constant for the system is 2.0×10^{-16} at 25°C, and the equilibrium concentrations of B and C at 25°C are both 2.0×10^{-4} mol L^{-1}. Calculate the rate constants for steps (1) and (2).

Problems†

Numerical problems

25.1 The data below applies to the formation of urea from ammonium cyanate, $NH_4CNO \rightarrow NH_2CONH_2$. Initially 22.9 g of ammonium cyanate was dissolved in in enough water to prepare 1.00 L of solution. Determine the order of the reaction, the rate constant, and the mass of ammonium cyanate left after 300 min.

t/min	0	20.0	50.0	65.0	150
m(urea)/g	0	7.0	12.1	13.8	17.7

25.2 The data below apply to the reaction, $(CH_3)_3CBr + H_2O \rightarrow (CH_3)_3COH + HBr$. Determine the order of the reaction, the rate constant, and the molar concentration of $(CH_3)_3CBr$ after 43.8 h.

t/h	0	3.15	6.20	10.00	18.30	30.80
$[(CH_3)_3CBr]/(10^{-2}\ mol\ L^{-1})$	10.39	8.96	7.76	6.39	3.53	2.07

25.3 The thermal decomposition of an organic nitrile produced the following data;

$t/(10^3\ s)$	0	2.00	4.00	6.00	8.00	10.00	12.00	∞
[nitrile]/(mol L^{-1})	1.10	0.86	0.67	0.52	0.41	0.32	0.25	0

Determine the order of the reaction and the rate constant. Note that this reaction does not go to completion.

25.4 The following data have been obtained for the decomposition of $N_2O_5(g)$ at 67°C according to the reaction $2\ N_2O_5(g) \rightarrow 4\ NO_2(g) + O_2(g)$. Determine the order of the reaction, the rate constant, and the half-life. It is not necessary to obtain the result graphically; you may do a calculation using estimates of the rates of change of concentration.

t/min	0	1	2	3	4	5
$[N_2O_5]/(mol\ L^{-1})$	1.000	0.705	0.497	0.349	0.246	0.173

25.5 A first-order decomposition reaction is observed to have the following rate constants at the indicated temperatures. Estimate the activation energy.

$k/(10^{-3}\ s^{-1})$	2.46	45.1	576
θ/°C	0	20.0	40.0

25.6 The gas-phase decomposition of acetic acid at 1189 K proceeds by way of two parallel reactions:

$$(1)\ CH_3COOH \longrightarrow CH_4 + CO_2 \qquad k_1 = 3.74\ s^{-1}$$

$$(2)\ CH_3COOH \longrightarrow H_2C{=}C{=}O + H_2O \qquad k_2 = 4.65\ s^{-1}$$

What is the maximum percentage yield of the ketene CH_2CO obtainable at this temperature?

25.7 The composition of a liquid-phase reaction $2\ A \rightarrow B$ was followed by a spectrophotometric method with the following results:

t/min	0	10	20	30	40	∞
[B]/(mol L^{-1})	0	0.089	0.153	0.200	0.230	0.312

Determine the order of the reaction and its rate constant.

25.8 The ClO radical decays rapidly by way of the reaction $2\ ClO \rightarrow Cl_2 + O_2$. The following data have been obtained:

$t/(10^{-3}\ s)$	0.12	0.62	0.96	1.60	3.20	4.00	5.75
$[ClO]/(10^{-6}\ mol\ L^{-1})$	8.49	8.09	7.10	5.79	5.20	4.77	3.95

Determine the rate constant of the reaction and the average lifetime of a ClO radical.

25.9 Cyclopropane isomerizes into propene when heated to 500°C in the gas phase. The extent of conversion for various initial pressures has been followed by gas chromatography by allowing the reaction to proceed for a time with various initial pressures:

p_0/Torr	200	200	400	400	600	600
t/s	100	200	100	200	100	200
p/Torr	186	173	373	347	559	520

where p_0 is the initial pressure and p is the final pressure of cyclopropane. What are the order and rate constant for the reaction under these conditions?

25.10 The addition of hydrogen halides to alkenes has played a fundamental role in the investigation of organic reaction mechanisms. In one study (M.J. Haugh and D.R. Dalton, *J. Amer. Chem. Soc.* **97**, 5674 (1975)), high pressures of hydrogen chloride (up to 25 atm) and propene (up to 5 atm) were examined over a range of temperatures and the amount of 2-chloropropane formed was determined by NMR. Show that, if the reaction $A + B \rightarrow P$ proceeds for a short time δt, the concentration of product follows $[P]/[A] = k[A]^{m-1}[B]^n\delta t$ if the reaction is mth-order in A and nth-order in B. In a series of runs the ratio of [chloropropane] to [propene] was independent of [propene] but the ratio of [chloropropane] to [HCl] for constant amounts of propene depended on [HCl]. For $\delta t \approx 100$ h (which is short on the timescale of the reaction) the latter ratio rose from zero to 0.05, 0.03, 0.01 for $p(HCl) = 10$ atm, 7.5 atm, 5.0 atm, respectively. What are the orders of the reaction with respect to each reactant?

25.11 Use mathematical software or an electronic spreadsheet to examine the time dependence of [I] in the reaction mechanism $A \rightarrow I \rightarrow P\ (k_1, k_2)$. You may either integrate eqn 35 numerically (see *Further information 1*) or use eqn 36 directly. In all of the following calculations, use $[A]_0 = 1\ mol\ L^{-1}$ and a time range of 0 to 5 s. (a) Plot [I] against t for $k_1 = 10\ s^{-1}$ and $k_2 = 1\ s^{-1}$. (b) Increase the ratio k_2/k_1 steadily by decreasing the value of k_1 and examine the plot of [I] against t at each turn. What approximation about d[I]/dt becomes increasingly valid?

25.12 Show that the following mechanism can account for the rate law of the reaction in Problem 25.10:

$$HCl + HCl \rightleftharpoons (HCl)_2 \qquad K_1$$

$$HCl + CH_3CH{=}CH_2 \rightleftharpoons complex \qquad K_2$$

$$(HCl)_2 + complex \longrightarrow CH_3CHClCH_3 + HCl + HCl \qquad k\ (slow)$$

What further tests could you apply to verify this mechanism?

25.13 In the experiments described in Problems 25.10 and 25.12 an inverse temperature dependence of the reaction rate was observed, the overall rate of reaction at 70°C being roughly one-third that at

† Problems denoted with the symbol ‡ were supplied by Charles Trapp, Carmen Giunta, and Marshall Cady.

19°C. Estimate the apparent activation energy and the activation energy of the rate-determining step given that the enthalpies of the two equilibria are both of the order of -14 kJ mol^{-1}.

25.14 The second-order rate constants for the reaction of oxygen atoms with aromatic hydrocarbons have been measured (R. Atkinson and J.N. Pitts, *J. Phys. Chem.* **79**, 295 (1975)). In the reaction with benzene the rate constants are 1.44×10^7 L mol^{-1} s^{-1} at 300.3 K, 3.03×10^7 L mol^{-1} s^{-1} at 341.2 K, and 6.9×10^7 L mol^{-1} s^{-1} at 392.2 K. Find the pre-exponential factor and activation energy of the reaction.

25.15 In Problem 25.9 the isomerization of cyclopropane over a limited pressure range was examined. If the Lindemann mechanism of first-order reactions is to be tested we also need data at low pressures. These have been obtained (H.O. Pritchard, R.G. Sowden, and A.F. Trotman-Dickenson, *Proc. R. Soc.* **A 217**, 563 (1953)):

p/Torr	84.1	11.0	2.89	0.569	0.120	0.067
$10^4\,k_{\rm eff}$/s^{-1}	2.98	2.23	1.54	0.857	0.392	0.303

Test the Lindemann theory with these data.

25.16‡ P.W. Seakins, M.J. Pilling, L.T. Niiranen, D. Gutman, and L.N. Krasnoperov (*J. Phys. Chem.* **96**, 9847 (1992)) measured the forward and reverse rate constants for the gas-phase reaction $C_2H_5(g) + HBr(g) \rightarrow C_2H_6(g) + Br(g)$ and used their findings to compute thermodynamic parameters for C_2H_5. The reaction is bimolecular in both directions with Arrhenius parameters $A = 1.0 \times 10^9$ L mol^{-1} s^{-1}, $E_a = -4.2$ kJ mol^{-1} for the forward reaction and $k' = 1.4 \times 10^{11}$ L mol^{-1} s^{-1}, $E_a = 53.3$ kJ mol^{-1} for the reverse reaction. Compute $\Delta_f H^\ominus$, $S_m^\ominus$, and $\Delta_f G^\ominus$ of C_2H_5 at 298 K.

25.17‡ Here we consider the reaction $SiCl_3H(g) + Br_2(g) \rightarrow SiCl_3Br(g) + HBr(g)$. The temperature dependence of this reaction has been studied by measuring the degree of conversion, $\alpha(t)$, for initial mixtures of 4.0 volume percentage trichlorosilane and 10.0 volume percentage bromine in helium at 1 bar. The data at four different temperatures are as follows:

$\alpha(160°C, t) = 0.044$ at 6.07 min; 0.057 at 7.87 min; 0.076 at 10.00 min

$\alpha(180°C, t) = 0.137$ at 2.72 min; 0.188 at 3.91 min; 0.228 at 5.03 min; 0.273 at 6.67 min; 0.302 at 8.06 min

$\alpha(200°C, t) = 0.281$ at 1.39 min; 0.395 at 2.16 min; 0.516 at 3.36 min; 0.612 at 5.04 min

$\alpha(220°C, t) = 0.592$ at 0.67 min; 0.734 at 1.22 min; 0.866 at 2.16 min

For these reactants it is thought that the dissociation $Br_2(g) \rightarrow 2\,Br(g)$ is at equilibrium and that in the range 300–700 K, the equilibrium constant for the dissociation, can be approximated as $K = 10^{5.1}e^{-C/RT}$, where $C = 190.2$ kJ mol^{-1} and K is expressed in terms of concentrations in moles per centimetre cubed (mol cm^{-3}). No wall effects were observed. Reactions of HBr were discounted and no hexachlorodisilane was observed. (a) Assume a rate law of the form $d[SiCl_3Br]/dt = k[SiCl_3H]^a[Br_2]^b$ and determine the reaction orders a and b. (b) Determine the rate constant $k(T)$ for each of the four data sets. Evaluate the activation parameters. (c) Show that the proposed mechanism

$$SiCl_3H + \cdot Br \longrightarrow \cdot SiCl_3 + HBr$$
$$\cdot SiCl_3 + Br_2 \longrightarrow SiCl_3Br + \cdot Br$$

is consistent with the experimental rate law.

Theoretical problems

25.18 The reaction mechanism

$$A_2 \rightleftharpoons A + A \quad \text{(fast)}$$
$$A + B \longrightarrow P \quad \text{(slow)}$$

involves an intermediate A. Deduce the rate law for the reaction.

25.19 The equilibrium $A \rightleftharpoons B$ is first-order in both directions. Derive an expression for the concentration of A as a function of time when the initial molar concentrations of A and B are $[A]_0$ and $[B]_0$. What is the final composition of the system?

25.20 Derive an integrated expression for a second-order rate law $v = k[A][B]$ for a reaction of stoichiometry $2\,A + 3\,B \rightarrow P$.

25.21 Derive the integrated form of a third-order rate law $v = k[A]^2[B]$ in which the stoichiometry is $2\,A + B \rightarrow P$ and the reactants are initially present in (a) their stoichiometric proportions, (b) with B present initially in twice the amount.

25.22 Set up the rate equations for the reaction mechanism:

$$A \underset{}{\overset{k_a,\,k_a'}{\rightleftharpoons}} B \overset{k_b,\,k_b'}{\rightleftharpoons} C$$

Show that the mechanism is equivalent to

$$A \overset{k_{\rm eff},\,k_{\rm eff}'}{\rightleftharpoons} C$$

under specified circumstances.

25.23 Show that the ratio $t_{1/2}/t_{3/4}$, where $t_{1/2}$ is the half-life and $t_{3/4}$ is the time for the concentration of A to decrease to $\frac{3}{4}$ of its initial value (implying that $t_{3/4} < t_{1/2}$) can be written as a function of n alone, and can therefore be used as a rapid assessment of the order of a reaction.

25.24 Derive an equation for the steady-state rate of the sequence of reactions $A \rightleftharpoons B \rightleftharpoons C \rightleftharpoons D$, with [A] maintained at a fixed value and the product D removed as soon as it is formed.

25.25‡ For the second-order reaction $A + B \rightarrow$ Products, the rate of reaction, v, may be written

$$v = \frac{dx}{dt} = k([A]_0 - x)([B]_0 + x)$$

where x is the decrease in concentration of A or B as a result of reaction. Find an expression for the maximum rate and the conditions under which it applies. Draw a graph of v against x and, noting that v and x cannot be negative, identify the portion of the curve that corresponds to reality.

25.26‡ For the consecutive reaction $A \rightarrow B \rightarrow C$, Fig. 25.13 shows [B] plotted against time for $k_a = 10k_b$. For $[A]_0 = 1.0$ mol L^{-1} and $k_a = 1.0$ min^{-1}, plot [B] against t for $k_a/k_b = 5$, 1, and 0.5. For each case determine the time at which [B] reaches a maximum. For the only $A + B \rightarrow P$ reaction in Table 25.3, find an expression for x as a function of time.

Applications: to archaeology, biology, and environmental science

25.27 The half-life for the (first-order) radioactive decay of ^{14}C is 5730 y (it emits β rays with an energy of 0.16 MeV). An archaeological sample contained wood that had only 72 per cent of the ^{14}C found in living trees. What is its age?

25.28 One of the hazards of nuclear explosions is the generation of ^{90}Sr and its subsequent incorporation in place of calcium in bones. This nuclide emits β rays of energy 0.55 MeV, and has a half-life of 28.1 y. Suppose 1.00 μg was absorbed by a newly born child. How much will remain after (a) 18 y, (b) 70 y if none is lost metabolically?

25.29 Sucrose is readily hydrolysed to glucose and fructose in acidic solution. The hydrolysis is often monitored by measuring the angle of rotation of plane-polarized light passing through the solution. From the angle of rotation the concentration of sucrose can be determined. An experiment on the hydrolysis of sucrose in 0.50 M HCl(aq) produced the following data:

t/min	0	14	39	60	80
[sucrose]/(mol L^{-1})	0.316	0.300	0.274	0.256	0.238
t/min	110	140	170	210	
[sucrose]/(mol L^{-1})	0.211	0.190	0.170	0.146	

Determine the rate constant of the reaction and the average lifetime of a sucrose molecule.

25.30 Consider the following mechanism for renaturation of a double helix from its strands A and B:

$$\text{A} + \text{B} \rightleftharpoons \text{unstable helix} \qquad \text{(fast)}$$
$$\text{Unstable helix} \longrightarrow \text{stable double helix} \qquad \text{(slow)}$$

Derive the rate equation for the formation of the double helix and express the rate constant of the renaturation reaction in terms of the rate constants of the individual steps.

25.31‡ *Prebiotic reactions* are reactions that might have occurred under the conditions prevalent on the Earth before the first living creatures emerged and which can lead to analogues of molecules necessary for life as we now know it. To qualify, a reaction must proceed with favourable rates and equilibria. M.P. Robertson and S.I. Miller (*Science* **268**, 702(1995)) have studied the prebiotic synthesis of 5-substituted uracils, among them 5-hydroxymethyluracil (HMU). Amino acid analogues can be formed from HMU under prebiotic conditions by reaction with various nucleophiles, such as H_2S, HCN, indole, imidazole, etc. For the synthesis of HMU (the uracil analogue of serine) from uracil and formaldehyde (HCHO), the rate of addition is given by $\log\{k/(\text{L mol}^{-1}\,\text{s}^{-1})\} = 11.75 - 5488/(T/\text{K})$ (at pH = 7), and $\log K = -1.36 + 1794/(T/\text{K})$. For this reaction, calculate the rates and equilibrium constants over a range of temperatures corresponding to possible prebiotic conditions, such as 0–50°C, and plot them against temperature. Also, calculate the activation energy and the standard reaction Gibbs energy and enthalpy at 25°C. Prebiotic conditions are not likely to be standard conditions. Speculate about how the actual values of the reaction Gibbs energy and enthalpy might differ from the standard values. Do you expect that the reaction would still be favourable?

25.32‡ Methane is a by-product of a number of natural processes (such as digestion of cellulose in ruminant animals and anaerobic decomposition of organic waste matter) and industrial processes (such as food production and fossil fuel use). Reaction with the hydroxyl radical OH is the main path by which CH_4 is removed from the lower atmosphere. T. Gierczak, R.K. Talukdar, S.C. Herndon, G.L. Vaghjiani, and A.R. Ravishankara (*J. Phys. Chem.* **A 101**, 3125 (1997)) measured the rate constants for the elementary bimolecular gas-phase reaction of methane with the hydroxyl radical over a range of temperatures of importance to atmospheric chemistry. Deduce the Arrhenius parameters A and E_a from the following measurements:

T/K	295	295	223	218
k/(10^6 L mol^{-1} s^{-1})	3.70	3.55	0.494	0.452
T/K	213	206	200	195
k/(10^6 L mol^{-1} s^{-1})	0.379	0.295	0.241	0.217

25.33‡ As we saw in Problem 25.32, reaction with the hydroxyl radical OH is the main path by which CH_4, a by-product of many natural and industrial processes, is removed from the lower atmosphere. T. Gierczak, R.K. Talukdar, S.C. Herndon, G.L. Vaghjiani, and A.R. Ravishankara (*J. Phys. Chem.* **A 101**, 3125 (1997)) measured the rate constants for the bimolecular gas-phase reaction $CH_4(g) + OH(g) \rightarrow CH_3(g) + H_2O(g)$ and found $A = 1.13 \times 10^9$ L mol^{-1} s^{-1} and $E_a = 14.1$ kJ mol^{-1} for the Arrhenius parameters. (a) Estimate the rate of consumption of CH_4. Take the average OH concentration to be 1.5×10^{-15} mol L^{-1}, that of CH_4 to be 4.0×10^{-8} mol L^{-1}, and the temperature to be −10°C. (b) Estimate the global annual mass of CH_4 consumed by this reaction (which is slightly less than the amount introduced to the atmosphere) given an effective volume for the Earth's lower atmosphere of 4×10^{21} L.

25.34‡ T. Gierczak, R.K. Talukdar, S.C. Herndon, G.L. Vaghjiani, and A.R. Ravishankara (*J. Phys. Chem.* **A 101**, 3125 (1997)) measured the rate constants for the bimolecular gas-phase reaction of methane with the hydroxyl radical in several isotopic variations. From their data, the following Arrhenius parameters can be obtained:

	A/(L mol^{-1} s^{-1})	E_a/(kJ mol^{-1})
$CH_4 + OH \longrightarrow CH_3 + H_2O$	1.13×10^9	14.1
$CD_4 + OH \longrightarrow CD_3 + DOH$	6.0×10^8	17.5
$CH_4 + OD \longrightarrow CH_3 + DOH$	1.01×10^9	13.6

Compute the rate constants at 298 K, and interpret the kinetic isotope effects.

25.35‡ The oxidation of HSO_3^- by O_2 in aqueous solution is a reaction of importance to the processes of acid rain formation and flue gas desulfurization. R.E. Connick, Y.-X. Zhang, S. Lee, R. Adamic, and P. Chieng (*Inorg. Chem.* **34**, 4543 (1995)) report that the reaction $2\,HSO_3^- + O_2 \rightarrow 2\,SO_4^{2-} + 2\,H^+$ follows the rate law $v = k[HSO_3^-]^2[H^+]^2$. Given a pH of 5.6 and an oxygen molar concentration of 2.4×10^{-4} mol L^{-1} (both presumed constant), an initial HSO_3^- molar concentration of 5×10^{-5} mol L^{-1}, and a rate constant of 3.6×10^6 L^3 mol^{-3} s^{-1}, what is the initial rate of reaction? How long would it take for HSO_3^- to reach half its initial concentration?

26 The kinetics of complex reactions

This chapter extends the material introduced in Chapter 25 by showing how to deal with complex reaction mechanisms. First, we consider chain reactions and see that either complicated or simple rate laws can be obtained, depending on the conditions. Under certain circumstances, a chain reaction can become explosive, and we see some of the reasons for this behaviour. An important application of these more complicated techniques is to the kinetics of polymerization reactions. There are two major classes of polymerization process and the average molar mass of the product varies with time in distinctive ways. Second, we describe homogeneous catalysis and apply the associated concepts to enzyme-catalysed reactions. We also consider reactions in which the concentrations of the intermediates and products oscillate in time: although their rate laws may be well known, under certain circumstances the composition of a system may be unpredictable. Finally, we describe the principles of photochemistry.

Many reactions take place by mechanisms that involve several elementary steps. Some take place at a useful rate only after absorption of light or if a catalyst is present. Certain other reactions take place by mechanisms with positive or negative feedback, in which the products of the reaction influence the rate of formation of products. In this chapter we see how to develop the ideas introduced in Chapter 25 to deal with these special kinds of reactions.

Chain reactions

Many gas-phase reactions and liquid-phase polymerization reactions are **chain reactions**. In a chain reaction, a reaction intermediate produced in one step generates an intermediate in a subsequent step, then that intermediate generates another intermediate, and so on. The intermediates in a chain reaction are called **chain carriers**. In a **radical chain reaction** the chain carriers are radicals (species with unpaired electrons). Ions may also act as chain carriers. In nuclear fission the chain carriers are neutrons.

26.1 The rate laws of chain reactions

A chain reaction can have a simple rate law. As a first example, consider the **pyrolysis**, or thermal decomposition in the absence of air, of acetaldehyde (ethanal, CH_3CHO), which is found to be three-halves order in CH_3CHO:

$$CH_3CHO(g) \longrightarrow CH_4(g) + CO(g) \qquad v = k[CH_3CHO]^{3/2} \tag{26.1}$$

Some ethane is also detected. The **Rice–Herzfeld mechanism** for this reaction is as follows (the dot signifies an unpaired electron and marks a radical):

Initiation:	$CH_3CHO \longrightarrow \cdot CH_3 + \cdot CHO$	$v = k_i[CH_3CHO]$
Propagation:	$CH_3CHO + \cdot CH_3 \longrightarrow CH_4 + CH_3CO\cdot$	$v = k_p[CH_3CHO][\cdot CH_3]$
Propagation:	$CH_3CO\cdot \longrightarrow \cdot CH_3 + CO$	$v = k_p'[CH_3CO\cdot]$
Termination:	$\cdot CH_3 + \cdot CH_3 \longrightarrow CH_3CH_3$	$v = k_t[\cdot CH_3]^2$

The chain carriers $\cdot CH_3$ and $\cdot CHO$ are formed initially in the **initiation step**. To simplify the treatment, we shall ignore the subsequent reactions of $\cdot CHO$, except to note that it gives rise to the formation of CO and to the by-product H_2. The chain carrier $\cdot CH_3$ attacks other reactant molecules in the **propagation steps**, and each attack gives rise to a new carrier. Radicals combine and end the chain in the **termination step**.

To test the proposed mechanism we need to show that it leads to the observed rate law. According to the steady-state approximation (Section 25.7c), the net rate of change of the intermediates ($\cdot CH_3$ and $CH_3CO\cdot$) may be set equal to zero:

$$\frac{d[\cdot CH_3]}{dt} = k_i[CH_3CHO] - k_p[\cdot CH_3][CH_3CHO] + k_p'[CH_3CO\cdot] - 2k_t[\cdot CH_3]^2 = 0$$

$$\frac{d[CH_3CO\cdot]}{dt} = k_p[\cdot CH_3][CH_3CHO] - k_p'[CH_3CO\cdot] = 0$$

The sum of the two equations is

$$k_i[CH_3CHO] - 2k_t[\cdot CH_3]^2 = 0$$

which implies that the steady-state concentration of $\cdot CH_3$ radicals is

$$[\cdot CH_3] = \left(\frac{k_i}{2k_t}\right)^{1/2}[CH_3CHO]^{1/2} \tag{26.2}$$

It follows that the rate of formation of CH_4 is

$$\frac{d[CH_4]}{dt} = k_p[\cdot CH_3][CH_3CHO] = k_p\left(\frac{k_i}{2k_t}\right)^{1/2}[CH_3CHO]^{3/2} \tag{26.3}$$

which is in agreement with the three-halves order observed experimentally (eqn 1). However, this mechanism does not accommodate the formation of various known reaction by-products, such as propanone (CH_3COCH_3) and propanal (CH_3CH_2CHO).

In many cases, a chain reaction leads to a complicated rate law. An example is the **hydrogen–bromine reaction**:

$$H_2(g) + Br_2(g) \longrightarrow 2\,HBr(g) \qquad \frac{d[HBr]}{dt} = \frac{k[H_2][Br_2]^{3/2}}{[Br_2] + k'[HBr]} \tag{26.4}$$

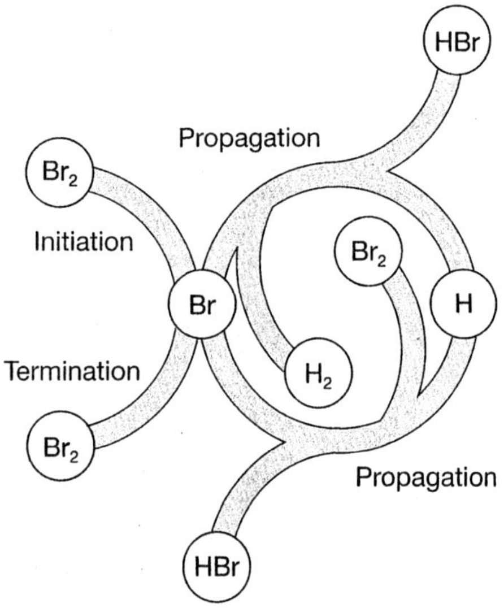

26.1 A schematic representation of the mechanism of the reaction between hydrogen and bromine. Note how the reactants and products are shown as arms to the circle, but the intermediates (H and Br) occur only within the circle. Similar diagrams are used to depict the action of catalysts.

The following mechanism has been proposed to account for this rate law (Fig. 26.1):

Initiation: $Br_2 + M \longrightarrow Br\cdot + Br\cdot + M \qquad v = k_i[Br_2][M]$

where M is either Br_2 or H_2. This step is an example of a **thermolysis**, a reaction initiated by heat or by vigorous intermolecular collisions.

Propagation:	$Br\cdot + H_2 \longrightarrow HBr + H\cdot$	$v = k_p[Br\cdot][H_2]$
	$H\cdot + Br_2 \longrightarrow HBr + Br\cdot$	$v = k_p'[H\cdot][Br_2]$
Retardation:	$H\cdot + HBr \longrightarrow H_2 + Br\cdot$	$v = k_r[H\cdot][HBr]$
Termination:	$Br\cdot + Br\cdot + M \longrightarrow Br_2 + M^*$	$v = k_t[Br\cdot]^2[M]$

A **retardation step** reduces the net rate of formation of product. In this case, the chain carrier H· attacks a molecule of HBr, the product. In the termination step, the third body M removes the energy of recombination. Other possible termination steps include the recombination of H atoms to form H_2 and combination of H and Br atoms. However, it turns out that only Br atom recombination is important. The net rate of formation of the product HBr is

$$\frac{d[HBr]}{dt} = k_p[Br\cdot][H_2] + k_p'[H\cdot][Br_2] - k_r[H\cdot][HBr]$$

We can now either analyse the rate equations numerically (see *Further information 1*) or look for approximate solutions and see if they agree with the empirical rate law. The following example illustrates the latter approach.

Example 26.1 Deriving the rate equation of a chain reaction

Derive the rate law for the formation of HBr according to the mechanism given above.

Method Make the steady-state approximation for the concentrations of any intermediates (H· and Br· in the present case) by setting the net rates of change of their concentrations equal to zero. Solve the resulting equations for the concentrations of the intermediates, and then use the resulting expressions in the equation for the net rate of formation of HBr.

Answer The net rates of formation of the two intermediates are

$$\frac{d[H\cdot]}{dt} = k_p[Br\cdot][H_2] - k_p'[H\cdot][Br_2] - k_r[H\cdot][HBr] = 0$$

$$\frac{d[Br\cdot]}{dt} = 2k_i[Br_2][M] - k_p[Br\cdot][H_2] + k_p'[H\cdot][Br_2] + k_r[H\cdot][HBr] - 2k_t[Br\cdot]^2[M] = 0$$

The steady-state concentrations of the intermediates are obtained by solving these two simultaneous equations and are

$$[Br\cdot] = \left(\frac{k_i}{k_t}\right)^{1/2}[Br_2]^{1/2} \qquad [H\cdot] = \frac{k_p(k_i/k_t)^{1/2}[H_2][Br_2]^{1/2}}{k_p'[Br_2] + k_r[HBr]}$$

Note that [M] has cancelled. When we substitute these concentrations into the expression for d[HBr]/dt, we obtain

$$\frac{d[HBr]}{dt} = \frac{2k_p(k_i/k_t)^{1/2}[H_2][Br_2]^{3/2}}{[Br_2] + (k_r/k_p')[HBr]}$$

This equation has the same form as the empirical rate law (eqn 4), so the two empirical rate constants can be identified as

$$k = 2k_p\left(\frac{k_i}{k_t}\right)^{1/2} \qquad k' = \frac{k_r}{k_p'}$$

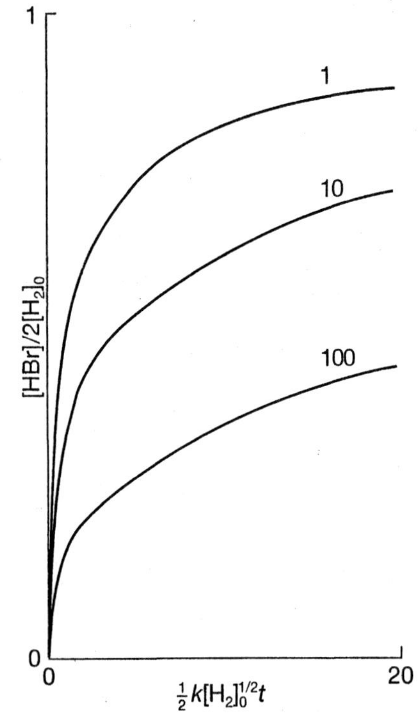

26.2 The numerical integration of the HBr rate law, Example 26.1, can be used to explore how the concentration of HBr changes with time. These runs began with stoichiometric proportions of H_2 and Br_2; the curves are labelled with the value of $2k' - 1$.

Comment The presence of HBr in the denominator of the rate law is a sign that it is acting as an inhibitor and reducing the rate of formation of product. Likewise, the presence of Br_2 stems from its role in the removal of the reactive ·H radicals from the chain. Numerical integration of the rate law with mathematical software shows the predicted time dependence of the concentration of HBr for this mechanism (Fig. 26.2).

Self-test 26.1 Deduce the rate law for the production of HBr when the initiation step is the photolysis, or light-induced decomposition, of Br_2 into two bromine atoms, Br·. Let the photolysis rate be $v = I_{abs}$, where I_{abs} is the intensity of absorbed radiation. [See eqn 39, below]

26.2 Explosions

A **thermal explosion** is a very rapid reaction arising from a rapid increase of reaction rate with increasing temperature. The temperature of the system rises if the energy released by an exothermic reaction cannot escape, and the reaction goes faster. The acceleration of the rate results in an even faster rise of temperature, so the reaction becomes catastrophically fast. A **chain-branching explosion** occurs when the number of chain centres grows exponentially.

An example of both types of explosion is the reaction between hydrogen and oxygen:

$$2\,H_2(g) + O_2(g) \longrightarrow 2\,H_2O(g)$$

Although the net reaction is very simple, the mechanism is very complex and has not yet been fully elucidated. A chain reaction is involved, and the chain carriers include H·, ·O·, and ·OH. Some steps involving H· are:

Initiation:	$H_2 \longrightarrow H\cdot + H\cdot$	$v = \text{constant } (v_{init})$
Propagation:	$H_2 + \cdot OH \longrightarrow \cdot H + H_2O$	$v = k_p[H_2][\cdot OH]$
Branching:	$\cdot O_2\cdot + \cdot H \longrightarrow \cdot O\cdot + \cdot OH$	$v = k_b[\cdot O_2\cdot][H\cdot]$
	$\cdot O\cdot + H_2 \longrightarrow \cdot OH + \cdot H$	$v = k_b'[\cdot O\cdot][H_2]$
Termination:	$H\cdot + \text{wall} \longrightarrow \tfrac{1}{2}H_2$	$v = k_t[H\cdot]$
	$H\cdot + O_2 + M \longrightarrow HO_2\cdot + M^*$	$v = k_t'[H\cdot][O_2][M]$

A **branching step** is an elementary reaction that produces more than one chain carrier. Recall that an O atom has the ground-state configuration $[He]2s^2 2p^4$, with two unpaired electrons.

The occurrence of an explosion depends on the temperature and pressure of the system, and the **explosion regions** for the reaction, the conditions under which explosion occurs, are shown in Fig. 26.3. At very low pressures the system is outside the explosion region and the mixture reacts smoothly. At these pressures the chain carriers produced in the branching steps can reach the walls of the container where they combine. Increasing the pressure along a vertical line in the illustration takes the system through the **first explosion limit** (provided the temperature is greater than about 730 K). The chain carriers react before reaching the walls and the branching reactions are explosively efficient. The reaction is smooth when the pressure is above the **second explosion limit.** The concentration of molecules in the gas is then so great that the radicals produced in the branching reactions recombine via termination steps.

In the last elementary reaction of the mechanism, the third body (M) can remove excess energy. At low pressures three-particle collisions are unimportant and recombination is much slower. At higher pressures, when three-particle collisions are important, the explosive propagation of the chain slows down because the branching steps are diverted into simple propagation steps (see the following example). When the pressure is increased to above the **third explosion limit**, the reaction rate increases so much that a thermal explosion occurs.

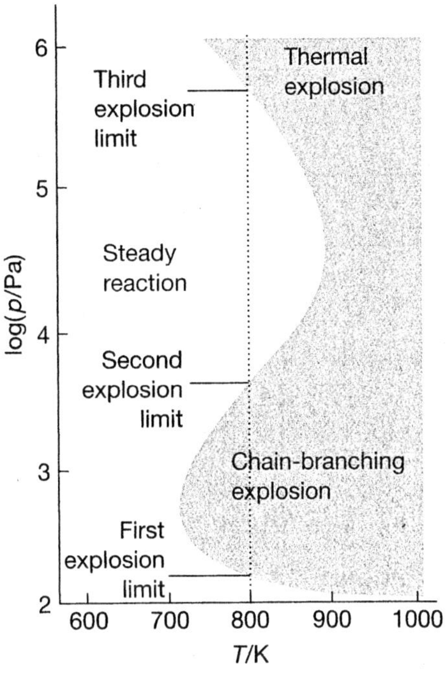

26.3 The explosion limits of the $H_2 + O_2$ reaction. In the explosive regions the reaction proceeds explosively when heated homogeneously.

Example 26.2 Examining the explosion behaviour of a chain reaction

For the reaction of hydrogen and oxygen described above, show that an explosion occurs when the rate of chain branching exceeds that of chain termination.

Method Identify the onset of explosion with the rapid increase in the concentration of radicals, and for simplicity identify that concentration with the concentration of H· atoms, which probably outnumber the highly reactive ·OH and ·O· radicals. Set up the corresponding rate laws for the reaction intermediates and then apply the steady-state approximation.

Answer The rate of formation of radicals, v_{rad}, is identified with d[H·]/dt; therefore we write

$$v_{rad} = v_{init} + k_p[\cdot OH][H_2] - k_b[H\cdot][O_2] + k_b'[\cdot O\cdot][H_2] - k_t[H\cdot] - k_t'[H\cdot][O_2][M]$$

Applying the steady-state approximation to ·OH and ·O· gives

$$\frac{d[\cdot OH]}{dt} = -k_p[\cdot OH][H_2] + k_b[H\cdot][O_2] + k_b'[\cdot O\cdot][H_2] = 0$$

$$\frac{d[\cdot O\cdot]}{dt} = k_b[H\cdot][O_2] - k_b'[\cdot O\cdot][H_2] = 0$$

The solutions of these two algebraic equations are

$$[\cdot O\cdot] = \frac{k_b[H\cdot][O_2]}{k_b'[H_2]} \qquad [\cdot OH] = \frac{2k_b[H\cdot][O_2]}{k_p[H_2]}$$

The rate of formation of radicals is therefore

$$v_{rad} = v_{init} + (2k_b[O_2] - k_t - k_t'[O_2][M])[H\cdot]$$

We write $k_{branch} = 2k_b[O_2]$, a measure of the rate of the more important chain-branching step, and $k_{term} = k_t + k_t'[O_2][M]$, a measure of the rate of chain termination. Then,

$$\frac{d[H\cdot]}{dt} = v_{init} + (k_{branch} - k_{term})[H\cdot]$$

There are two solutions. At low O_2 concentrations, termination dominates branching, so $k_{term} > k_{branch}$. Then,

$$[H\cdot] = \frac{v_{init}}{k_{term} - k_{branch}}(1 - e^{-(k_{term}-k_{branch})t})$$

As can be seen from Fig. 26.4a, in this regime there is steady combustion of hydrogen. At high O_2 concentrations, branching dominates termination, or $k_{branch} > k_{term}$. Then,

$$[H\cdot] = \frac{v_{init}}{k_{branch} - k_{term}}(e^{(k_{branch}-k_{term})t} - 1)$$

There is now an explosive increase in the concentration of radicals (Fig. 26.4b).

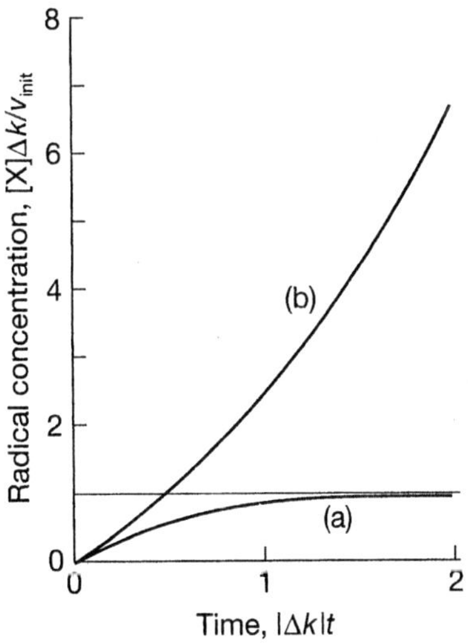

26.4 The concentration of radicals in the fuel-rich regime of the hydrogen–oxygen reaction (a) under steady combustion conditions, (b) in the explosive region. For this graph, $\Delta k = k_{branch} - k_{term}$.

Comment Although the steady-state approximation does not hold under explosive conditions, the calculation at least gives an indication of the basis for the transition from smooth combustion to explosion.

Self-test 26.2 Calculate the variation in radical composition when the rates of branching and termination are equal. $[[H\cdot] = v_{init}t]$

Not all explosions are due to chain reactions. Solid-state explosions, such as the explosion of ammonium nitrate or TNT (2,4,6-trinitrotoluene), for instance, are simply decompositions that occur very rapidly with the production of large amounts of gas-phase molecules.

Polymerization kinetics

In **stepwise polymerization** any two monomers present in the reaction mixture can link together at any time and growth of the polymer is not confined to chains that are already forming (Fig. 26.5). As a result, monomers are removed early in the reaction and, as we shall see, the average molar mass of the product grows with time. In **chain polymerization** an activated monomer, M, attacks another monomer, links to it, then that unit attacks another monomer, and so on. The monomer is used up as it becomes linked to the growing chains (Fig. 26.6). High polymers are formed rapidly and only the yield, not the average molar mass, of the polymer is increased by allowing long reaction times.

26.3 Stepwise polymerization

Stepwise polymerization commonly proceeds by a condensation reaction, in which a small molecule (typically H_2O) is eliminated in each step. Stepwise polymerization is the mechanism of production of polyamides, as in the formation of nylon-66:

$$\begin{aligned}H_2N(CH_2)_6NH_2 + HOOC(CH_2)_4COOH &\longrightarrow H_2N(CH_2)_6NHCO(CH_2)_4COOH + H_2O\\ &\longrightarrow H\text{—}[NH(CH_2)_6NHCO(CH_2)_4CO]_n\text{—}OH\end{aligned}$$

Polyesters and polyurethanes are formed similarly (the latter without elimination). A polyester, for example, can be regarded as the outcome of the stepwise condensation of a hydroxyacid HO—M—COOH. We shall consider the formation of a polyester from such a monomer, and measure its progress in terms of the concentration of the —COOH groups in the sample (which we denote A), for these groups gradually disappear as the condensation proceeds. Because the condensation reaction can occur between molecules containing any number of monomer units, chains of many different lengths can grow in the reaction mixture.

In the absence of a catalyst, we can expect the condensation to be overall second-order in the concentration of the —OH and —COOH (or A) groups, and write

$$\frac{d[A]}{dt} = -k[OH][A] \tag{26.5a}$$

However, because there is one —OH group for each —COOH group, this equation is the same as

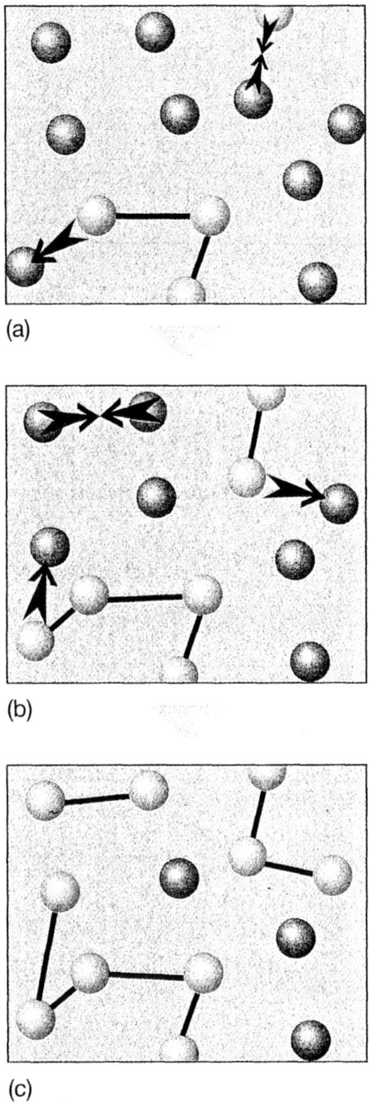

26.5 In stepwise polymerization, growth can start at any pair of monomers, and so new chains begin to form throughout the reaction.

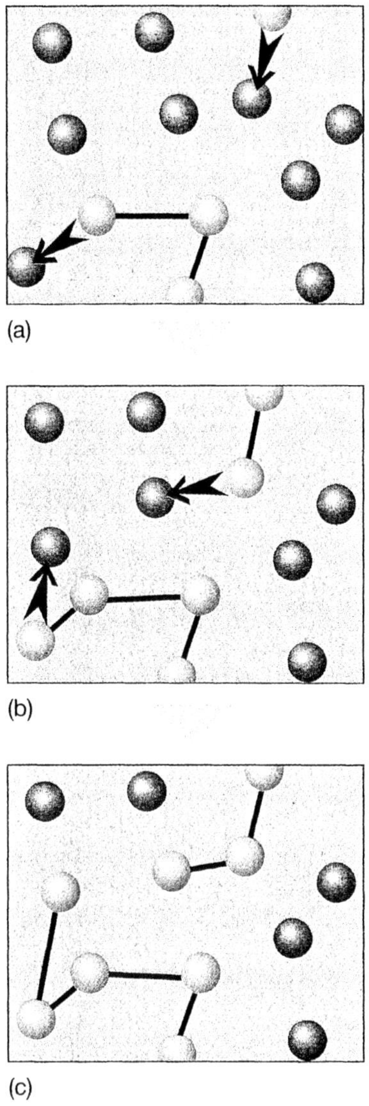

26.6 The process of chain polymerization. Chains grow as each chain acquires additional monomers.

$$\frac{d[A]}{dt} = -k[A]^2 \qquad (26.5b)$$

If we assume that the rate constant for the condensation is independent of the chain length, then k remains constant throughout the reaction. The solution of this rate law is given by eqn 25.13, and is

$$[A] = \frac{[A]_0}{1 + kt[A]_0} \qquad (26.6)$$

The fraction, p, of —COOH groups that have condensed at time t is, after application of eqn 6:

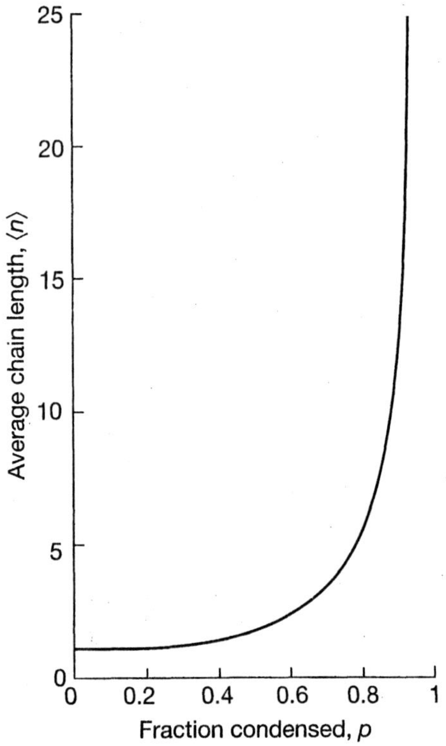

26.7 The average chain length of a polymer as a function of the fraction of reacted monomers, p. Note that p must be very close to 1 for the chains to be long.

$$p = \frac{[A]_0 - [A]}{[A]_0} = \frac{kt[A]_0}{1 + kt[A]_0} \qquad (26.7)$$

Next, we calculate the **degree of polymerization**, which is defined as the average number of monomer residues per polymer molecule. This quantity is the ratio of the initial concentration of A, $[A]_0$, to the concentration of end groups, [A], at the time of interest, because there is one –A group per polymer molecule. For example, if there were initially 1000 A groups and there are now only 10, each polymer must be 100 units long on average. Because we can express [A] in terms of p (eqn 7), the average number of monomers per polymer molecule, $\langle n \rangle$, is

$$\langle n \rangle = \frac{[A]_0}{[A]} = \frac{1}{1-p} \qquad (26.8a)$$

This result is illustrated in Fig. 26.7. When we express p in terms of the rate constant k (eqn 7), we find

$$\langle n \rangle = 1 + kt[A]_0 \qquad (26.8b)$$

The average length grows linearly with time. Therefore, the longer a stepwise polymerization proceeds, the higher the average molar mass of the product.

26.4 Chain polymerization

Chain polymerization occurs by addition of monomers to a growing polymer, often by a radical chain process. It results in the rapid growth of an individual polymer chain for each activated monomer. Examples include the addition polymerizations of ethene, methyl methacrylate, and styrene, as in

$$\text{—CH}_2\text{CHX}\cdot + \text{CH}_2\text{=CHX} \longrightarrow \text{—CH}_2\text{CHXCH}_2\text{CHX}\cdot$$

and subsequent reactions. The central feature of the kinetic analysis (which is summarized in the *Justification* below) is that the rate of polymerization is proportional to the square root of the initiator concentration:

$$v = k[I]^{1/2}[M] \qquad (26.9)$$

Justification 26.1

There are three basic types of reaction step in a chain polymerization process:

(a) Initiation:

$$I \longrightarrow R\cdot + R\cdot \qquad v_i = k_i[I]$$
$$M + R\cdot \longrightarrow \cdot M_1 \qquad \text{(fast)}$$

where I is the initiator, R· the radical I forms, and $\cdot M_1$ is a monomer radical. We have shown a reaction in which a radical is produced, but in some polymerizations the initiation step leads to the formation of an ionic chain carrier. The rate-determining step is the formation of the radicals R· by homolysis of the initiator, so the rate of initiation is equal to the v_i given above.

(b) Propagation:

$$M + \cdot M_1 \longrightarrow \cdot M_2$$
$$M + \cdot M_2 \longrightarrow \cdot M_3$$
$$\vdots$$
$$M + \cdot M_{n-1} \longrightarrow \cdot M_n \qquad v_p = k_p[M][\cdot M]$$

If we assume that the rate of propagation is independent of chain size for sufficiently large chains, then we may use only the equation given above to describe the propagation process. Consequently, for sufficiently large chains, the rate of propagation is equal to the overall rate of polymerization.

Because this chain of reactions propagates quickly, the rate at which the total concentration of radicals grows is equal to the rate of the rate-determining initiation step. It follows that

$$\left(\frac{d[\cdot M]}{dt}\right)_{\text{production}} = 2fk_i[I] \tag{26.10}$$

where f is the fraction of radicals R· that successfully initiate a chain.

(c) Termination:

$$\cdot M_n + \cdot M_m \longrightarrow M_{n+m} \quad \text{(mutual termination)}$$
$$\cdot M_n + \cdot M_m \longrightarrow M_n + M_m \quad \text{(disproportionation)}$$
$$M + \cdot M_n \longrightarrow \cdot M + M_n \quad \text{(chain transfer)}$$

In **mutual termination** two growing radical chains combine. In termination by **disproportionation** a hydrogen atom transfers from one chain to another, corresponding to the oxidation of the donor and the reduction of the acceptor. In **chain transfer**, a new chain initiates at the expense of the one currently growing.

Here we suppose that only mutual termination occurs. If we assume that the rate of termination is independent of the length of the chain, the rate law for termination is

$$v_t = k_t[\cdot M]^2$$

and the rate of change of radical concentration by this process is

$$\left(\frac{d[\cdot M]}{dt}\right)_{\text{depletion}} = -2k_t[\cdot M]^2$$

The steady-state approximation gives:

$$\frac{d[\cdot M]}{dt} = 2fk_i[I] - 2k_t[\cdot M]^2 = 0$$

The steady-state concentration of radical chains is therefore

$$[\cdot M] = \left(\frac{fk_i}{k_t}\right)^{1/2}[I]^{1/2} \tag{26.11}$$

Because the rate of propagation of the chains is the negative of the rate at which the monomer is consumed, we can write $v_p = -d[M]/dt$ and

$$v_p = k_p[\cdot M][M] = k_p\left(\frac{fk_i}{k_t}\right)^{1/2}[I]^{1/2}[M] \tag{26.12}$$

This rate is also the rate of polymerization, which has the form of eqn 9.

The **kinetic chain length**, ν, is the ratio of the number of monomer units consumed per activated centre produced in the initiation step:

$$\nu = \frac{\text{number of monomer units consumed}}{\text{number of activated centres produced}} \tag{26.13}$$

The kinetic chain length can be expressed in terms of the rate expressions in *Justification 26.1*. To do so, we recognize that monomers are consumed at the rate that chains propagate. Then,

$$\nu = \frac{\text{rate of propagation of chains}}{\text{rate of production of radicals}}$$

By making the steady-state approximation, we set the rate of production of radicals equal to the termination rate. Therefore, we can write the expression for the kinetic chain length as

$$\nu = \frac{k_p[\cdot M][M]}{2k_t[M\cdot]^2} = \frac{k_p[M]}{2k_t[\cdot M]}$$

When we substitute the steady-state expression, eqn 11, for the radical concentration, we obtain

$$\nu = k[M][I]^{-1/2} \qquad k = \tfrac{1}{2}k_p(fk_ik_t)^{-1/2} \qquad (26.14)$$

Consider a polymer produced by a chain mechanism with mutual termination. In this case, the average number of monomers in a polymer molecule, $\langle n \rangle$, produced by the reaction is the sum of the numbers in the two combining polymer chains. The average number of units in each chain is ν. Therefore,

$$\langle n \rangle = 2\nu = 2k[M][I]^{-1/2} \qquad (26.15)$$

with k given in eqn 14. We see that, the slower the initiation of the chain (the smaller the initiator concentration and the smaller the initiation rate constant), the greater the kinetic chain length, and therefore the higher the average molar mass of the polymer. Some of the consequences of molar mass for polymers were explored in Chapter 22: now we see how we can exercise kinetic control over them.

Homogeneous catalysis

A **catalyst** is a substance that accelerates a reaction but undergoes no net chemical change. The catalyst lowers the activation energy of the reaction by providing an alternative path that avoids the slow, rate-determining step of the uncatalysed reaction (Fig. 26.8). Catalysts can be very effective; for instance, the activation energy for the decomposition of hydrogen peroxide in solution is 76 kJ mol^{-1}, and the reaction is slow at room temperature. When a little iodide ion is added, the activation energy falls to 57 kJ mol^{-1} and the rate constant increases by a factor of 2000. **Enzymes**, which are biological catalysts, are very specific and can have a dramatic effect on the reactions they control. For example, the enzyme catalase reduces the activation energy for the decomposition of hydrogen peroxide to 8 kJ mol^{-1}, corresponding to an acceleration of the reaction by a factor of 10^{15} at 298 K.

A **homogeneous catalyst** is a catalyst in the same phase as the reaction mixture. For example, the decomposition of hydrogen peroxide in aqueous solution is catalysed by bromide ion or catalase (Sections 26.5 and 26.6). A **heterogeneous catalyst** is a catalyst in a different phase from the reaction mixture. For example, the hydrogenation of ethene to ethane, a gas-phase reaction, is accelerated in the presence of a solid catalyst such as palladium, platinum, or nickel. The metal provides a surface upon which the reactants bind; this binding facilitates encounters between reactants and increases the rate of the reaction. We examine heterogeneous catalysis in Chapter 28 and consider only homogeneous catalysis here.

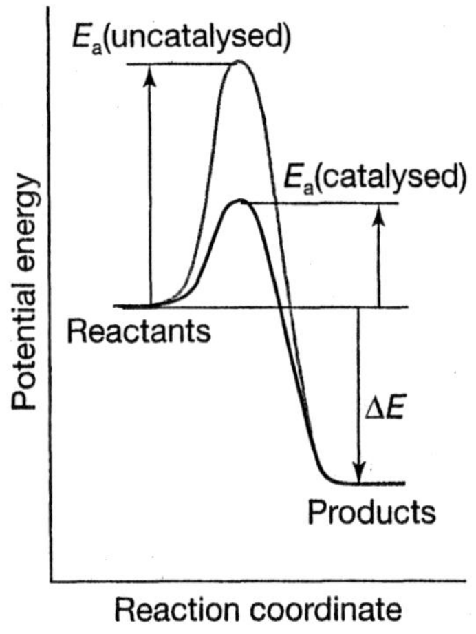

26.8 A catalyst provides a different path with a lower activation energy. The result is an increase in the rate of formation of products.

26.5 Features of homogeneous catalysis

We can obtain some idea of the mode of action of homogeneous catalysts by examining the kinetics of the bromide-catalysed decomposition of hydrogen peroxide:

$$2\,H_2O_2(aq) \longrightarrow 2\,H_2O(l) + O_2(g)$$

The reaction is believed to proceed through the following pre-equilibrium:[1]

$$H_3O^+ + H_2O_2 \rightleftharpoons H_3O_2^+ + H_2O \qquad K = \frac{[H_3O_2^+]}{[H_2O_2][H_3O^+]}$$

$$H_3O_2^+ + Br^- \longrightarrow HOBr + H_2O \qquad v = k[H_3O_2^+][Br^-]$$

$$HOBr + H_2O_2 \longrightarrow H_3O^+ + O_2 + Br^- \qquad \text{(fast)}$$

The second step is rate-determining. Therefore, we can obtain the rate law of the overall reaction by setting the overall rate equal to the rate of the second step and using the equilibrium constant to express the concentration of $H_3O_2^+$ in terms of the reactants. The result is

$$\frac{d[O_2]}{dt} = k_{eff}[H_2O_2][H_3O^+][Br^-]$$

with $k_{eff} = kK$, in agreement with the observed dependence of the rate on the Br^- concentration and the pH of the solution. The observed activation energy is that of the effective rate constant kK.

In **acid catalysis** the crucial step is the transfer of a proton to the substrate:

$$X + HA \longrightarrow HX^+ + A^- \qquad HX^+ \longrightarrow \text{products}$$

Acid catalysis is the primary process in the solvolysis of esters and keto–enol tautomerism. In **base catalysis**, a proton is transferred from the substrate to a base:

$$XH + B \longrightarrow X^- + BH^+ \qquad X^- \longrightarrow \text{products}$$

Base catalysis is the primary step in the isomerization and halogenation of organic compounds, and of the Claisen and aldol condensation reactions.

26.6 Enzymes

Enzymes are homogeneous biological catalysts. These ubiquitous compounds are special proteins or nucleic acids that contain an **active site**, which is responsible for binding the **substrate**, the reactant, and processing it into products. As is true of any catalyst, the active site returns to its original state after the products are released. The structure of the active site is specific to the reaction that it catalyses. For example, the active sites of the catalases have *d*-metal ions for binding of the substrate and for mediating the electron transfer reactions that lead to the decomposition of H_2O_2.

Enzyme-catalysed reactions are prone to inhibition by molecules that interfere with the formation of product. Many drugs for the treatment of disease function by inhibiting enzymes. For example, an important strategy in the treatment of acquired immune deficiency syndrome (AIDS) involves the steady administration of a specially designed protease inhibitor. The drug inhibits an enzyme that is key to the formation of the protein envelope surrounding the genetic material of the human immunodeficiency virus (HIV). Without a properly formed envelope, HIV cannot replicate in the host organism.

1. We have set the activity of H_2O in the equilibrium constant equal to 1 and assumed that the thermodynamic properties of the other substances are ideal.

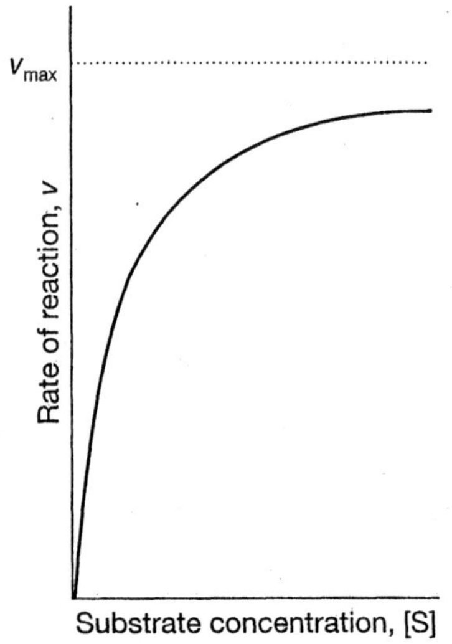

26.9 The variation of the rate of an enzyme-catalysed reaction with substrate concentration. The approach to a maximum rate, v_{max}, for large [S] is explained by the Michaelis–Menten mechanism.

(a) The Michaelis–Menten mechanism of enzyme catalysis

Experimental studies of enzyme kinetics are typically conducted by monitoring the initial rate of product formation in a solution in which the enzyme is present at very low concentration. Indeed, enzymes are such efficient catalysts that significant accelerations may be observed even when their concentration is more than three orders of magnitude smaller than that of the substrate.

The principal features of many enzyme-catalysed reactions are as follows:

1 For a given initial concentration of substrate, $[S]_0$, the initial rate of product formation is proportional to the total concentration of enzyme, $[E]_0$.

2 For a given $[E]_0$ and low values of $[S]_0$, the rate of product formation is proportional to $[S]_0$.

3 For a given $[E]_0$ and high values of $[S]_0$, the rate of product formation becomes independent of $[S]_0$, reaching a maximum value known as the **maximum velocity**, v_{max}.

The **Michaelis–Menten mechanism** accounts for these features (Fig. 26.9). According to this mechanism, an enzyme–substrate complex is formed in the first step and either the substrate is released unchanged or after modification to form products:

$$\begin{aligned} &\mathrm{E} + \mathrm{S} \rightleftharpoons \mathrm{ES} \qquad k_a, k_a' \\ &\mathrm{ES} \longrightarrow \mathrm{P} \qquad k_b \end{aligned} \tag{26.16}$$

The rate of product formation according to this mechanism is

$$v = k_b[\mathrm{ES}] \tag{26.17}$$

We can obtain the concentration of the enzyme–substrate complex by invoking the steady-state approximation and writing[2]

$$\frac{d[\mathrm{ES}]}{dt} = k_a[\mathrm{E}][\mathrm{S}] - k_a'[\mathrm{ES}] - k_b[\mathrm{ES}] = 0$$

It follows that

$$[\mathrm{ES}] = \left(\frac{k_a}{k_a' + k_b}\right)[\mathrm{E}][\mathrm{S}] \tag{26.18}$$

where [E] and [S] are the concentrations of *free* enzyme and substrate, respectively. To express the rate law in terms of the concentrations of enzyme and substrate added, we note that $[E]_0 = [E] + [ES]$. Moreover, because the substrate is typically in large excess relative to the enzyme, the free substrate concentration is approximately equal to the initial substrate concentration and we can write $[S] \approx [S]_0$. It then follows that:

$$[\mathrm{ES}] = \frac{[\mathrm{E}]_0}{1 + \left(\dfrac{k_a' + k_b}{k_a}\right)\dfrac{1}{[\mathrm{S}]_0}}$$

By substituting this result into eqn 17, we obtain the **Michaelis–Menten equation** for the rate of an enzyme-catalysed reaction:

$$v = \frac{k_b[\mathrm{E}]_0}{1 + \left(\dfrac{k_a' + k_b}{k_a}\right)\dfrac{1}{[\mathrm{S}]_0}} \tag{26.19}$$

2. In their original formulation, Michaelis and Menten supposed that the initial step was a pre-equilibrium. The treatment as a steady-state problem was suggested later by G.E. Briggs and J.B.S. Haldane.

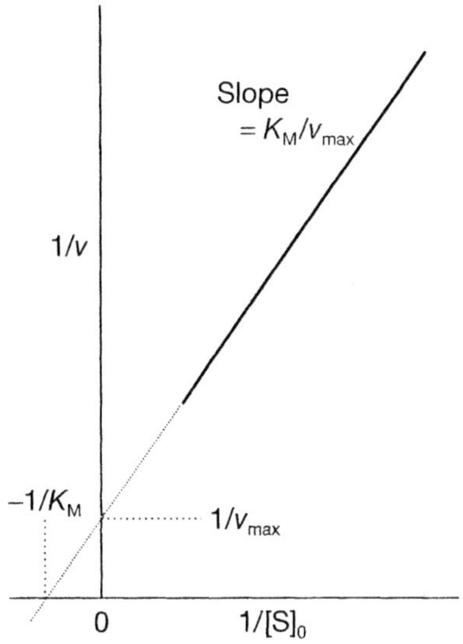

26.10 A Lineweaver–Burk plot for the analysis of an enzyme-catalysed reaction that proceeds by a Michaelis–Menten mechanism and the significance of the intercepts and the slope.

The constant $(k_a' + k_b)/k_a$ is characteristic of a given enzyme acting on a given substrate; it is called the **Michaelis constant** and written K_M. From eqn 18 it is also possible to express the Michaelis constant as $K_M = [E][S]/[ES]$. It follows that K_M has units of concentration.

Equation 19 shows that, in accord with experimental observations:

1 When $[S]_0 \ll K_M$, the rate is proportional to $[S]_0$:

$$v = \frac{k_a k_b}{k_a' + k_b}[S]_0[E]_0 \tag{26.20a}$$

2 When $[S]_0 \gg K_M$, the rate reaches its maximum value and is independent of $[S]_0$:

$$v = v_{max} = k_b[E]_0 \tag{26.20b}$$

Substitution of the definitions of K_M and v_{max} into eqn 19 gives:

$$v = \frac{v_{max}}{1 + K_M/[S]_0} \tag{26.21}$$

We can rearrange this expression into a form that is amenable to data analysis by linear regression:

$$\frac{1}{v} = \frac{1}{v_{max}} + \left(\frac{K_M}{v_{max}}\right)\frac{1}{[S]_0} \tag{26.22}$$

A **Lineweaver–Burk plot** is a plot of $1/v$ versus $1/[S]_0$, and according to eqn 22 it should yield a straight line with slope of K_M/v_{max}, a y-intercept at $1/v_{max}$, and an x-intercept at $-1/K_M$ (Fig. 26.10). The value of k_b is then calculated from the y-intercept and eqn 20b. However, the plot cannot give the individual rate constants k_a and k_a' that appear in the expression for K_M. The stopped-flow technique described in Section 25.1b can give the additional data needed, because we can find the rate of formation of the enzyme–substrate complex by monitoring the concentration after mixing the enzyme and substrate. This procedure gives a value for k_a, and k_a' is then found by combining this result with the values of k_b and K_M.

(b) The catalytic efficiency of enzymes

The **turnover number**, or **catalytic constant**, of an enzyme, k_{cat}, is the number of catalytic cycles (turnovers) performed by the active site in a given interval divided by the duration of the interval. This quantity has units of a first-order rate constant and, in terms of the Michaelis–Menten mechanism, is numerically equivalent to k_b, the rate constant for release of product from the enzyme–substrate complex. It follows from the identification of k_{cat} with k_b and from eqn 20b that

$$k_{cat} = k_b = \frac{v_{max}}{[E]_0} \tag{26.23}$$

The **catalytic efficiency**, ε (epsilon), of an enzyme is the ratio k_{cat}/K_M. The higher the value of ε, the more efficient is the enzyme. We can think of the catalytic activity as the effective rate constant of the enzymatic reaction. From $K_M = (k_a' + k_b)/k_a$ and eqn 23, it follows that

$$\varepsilon = \frac{k_{cat}}{K_M} = \frac{k_a k_b}{k_a' + k_b} \tag{26.24}$$

The efficiency reaches its maximum value of k_a when $k_b \gg k_a'$. Because k_a is the rate constant for the formation of a complex from two species that are diffusing freely in solution, the maximum efficiency is related to the maximum rate of diffusion of E

The web site contains links to databases of enzymes.

and S in solution. This limit (which is discussed further in Section 27.2) leads to rate constants of about 10^8–10^9 L mol^{-1} s^{-1} for molecules as large as enzymes at room temperature. The enzyme catalase has $\varepsilon = 4.0 \times 10^8$ L mol^{-1} s^{-1} and is said to have attained 'catalytic perfection'.

Example 26.3 Determining the catalytic efficiency of an enzyme

The enzyme carbonic anhydrase catalyses the hydration of CO_2 in red blood cells to give bicarbonate ion:

$$CO_2 + H_2O \longrightarrow HCO_3^- + H^+$$

The following data were obtained for the reaction at pH = 7.1, 273.5 K, and an enzyme concentration of 2.3 nmol L^{-1}:

$[CO_2]$/(mmol L^{-1})	1.25	2.5	5.0	20.0
rate/(mol L^{-1} s^{-1})	2.78×10^{-5}	5.00×10^{-5}	8.33×10^{-5}	1.67×10^{-4}

Determine the catalytic efficiency of carbonic anhydrase at 273.5 K.

Method Prepare a Lineweaver–Burk plot and determine the values of K_M and v_{max} graphically or from linear regression analysis. From eqn 23 and the enzyme concentration, calculate k_{cat} and the catalytic efficiency from eqn 24.

Answer Figure 26.11 shows the Lineweaver–Burk plot for the data. The slope is 40 s and the y-intercept is 4.0×10^3 L mol^{-1} s. Hence,

$$v_{max} = \frac{1}{4.0 \times 10^3 \text{ L mol}^{-1}\text{ s}} = 2.5 \times 10^{-4} \text{ mol L}^{-1}\text{ s}^{-1}$$

$$K_M = (2.5 \times 10^{-4} \text{ mol L}^{-1}\text{ s}^{-1}) \times (40 \text{ s}) = 1.0 \times 10^{-2} \text{ mol L}^{-1}$$

$$k_{cat} = \frac{2.5 \times 10^{-4} \text{ mol L}^{-1}\text{ s}^{-1}}{2.3 \times 10^{-9} \text{ mol L}^{-1}} = 1.1 \times 10^5 \text{ s}^{-1}$$

$$\varepsilon = \frac{k_{cat}}{K_M} = \frac{1.1 \times 10^5 \text{ s}^{-1}}{1.0 \times 10^{-2} \text{ mol L}^{-1}} = 1.1 \times 10^7 \text{ L mol}^{-1}\text{ s}^{-1}$$

Self-test 26.3 Calculate the substrate concentration required to decrease the rate to 50 per cent of v_{max}. Note that your answer is equal to K_M. This is a general result for the Michaelis–Menten mechanism. [1.0×10^{-2} mol L^{-1}]

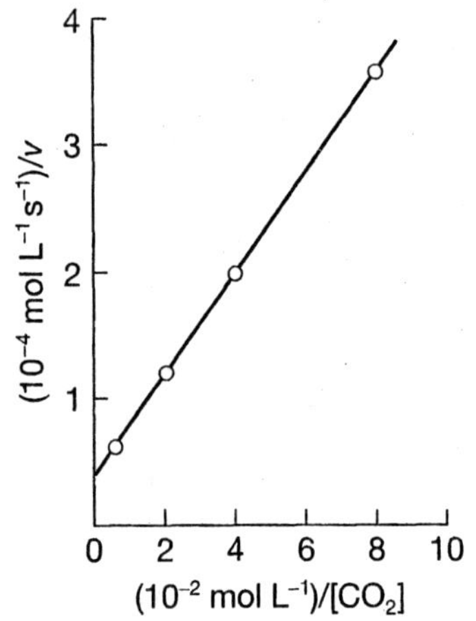

26.11 The Lineweaver–Burk plot of the data for Example 26.3.

(c) Mechanisms of enzyme inhibition

An inhibitor, I, decreases the rate of product formation from the substrate by binding to the enzyme, to the ES complex, or to the enzyme and ES complex simultaneously. The most general kinetic scheme for enzyme inhibition is then:

$$E + S \rightleftharpoons ES \qquad k_a, k_a'$$

$$ES \longrightarrow E + P \qquad k_b$$

$$EI \rightleftharpoons E + I \qquad K_I = \frac{[E][I]}{[EI]} \qquad (26.25a)$$

$$ESI \rightleftharpoons ES + I \qquad K_I' = \frac{[ES][I]}{[ESI]} \qquad (26.25b)$$

The lower the values of K_I and K_I' the more efficient are the inhibitors. The rate of product formation is then given by eqn 17 above, since only ES leads to product. As shown in the following *Justification*, the rate of reaction in the presence of an inhibitor is

$$v = \frac{v_{\max}}{\alpha' + \alpha K_M/[S]_0} \tag{26.26}$$

where $\alpha = 1 + [I]/K_I$ and $\alpha' = 1 + [I]/K_I'$. This equation is very similar to the Michaelis–Menten equation for the uninhibited enzyme (eqn 21) and is also amenable to analysis by a Lineweaver–Burk plot:

$$\frac{1}{v} = \frac{\alpha'}{v_{\max}} + \left(\frac{\alpha K_M}{v_{\max}}\right)\frac{1}{[S]_0} \tag{26.27}$$

Justification 26.2

By mass balance, the total concentration of enzyme is:

$$[E]_0 = [E] + [EI] + [ES] + [ESI]$$

By using eqns 25a and 25b and the definitions

$$\alpha = 1 + \frac{[I]}{K_I} \quad \text{and} \quad \alpha' = 1 + \frac{[I]}{K_I'}$$

it follows that

$$[E]_0 = [E]\alpha + [ES]\alpha'$$

By using $K_M = [E][S]/[ES]$ and assuming that $[S] \approx [S]_0$, we can write

$$[E]_0 = \frac{K_M[ES]}{[S]_0}\alpha + [ES]\alpha' = [ES]\left(\frac{\alpha K_M}{[S]_0} + \alpha'\right)$$

The expression for the rate of product formation is then:

$$v = k_b[ES] = \frac{k_b[E]_0}{\dfrac{\alpha K_M}{[S]_0} + \alpha'}$$

which, upon rearrangement, gives eqn 26.

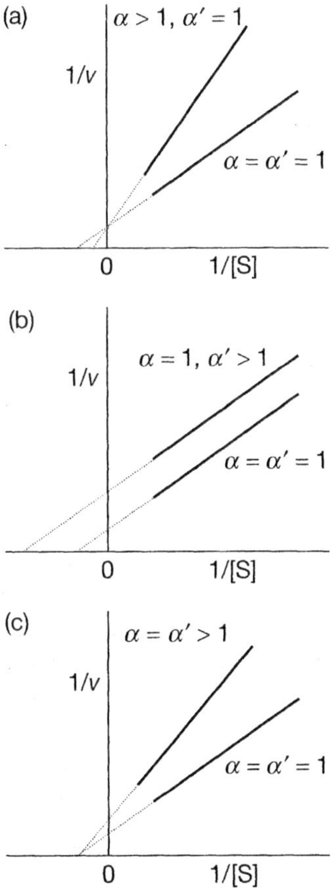

26.12 Lineweaver–Burk plots characteristic of the three major modes of enzyme inhibition: (a) competitive inhibition, (b) uncompetitive inhibition, and (c) non-competitive inhibition, showing the special case $\alpha = \alpha' > 1$.

There are three major modes of inhibition that give rise to distinctly different kinetic behaviour (Fig. 26.12). In **competitive inhibition** the inhibitor binds only to the active site of the enzyme and thereby inhibits the attachment of the substrate. This condition corresponds to $\alpha > 1$ and $\alpha' = 1$ (because ESI does not form). The slope of the Lineweaver–Burk plot increases by a factor of α relative to the slope for data on the uninhibited enzyme ($\alpha = \alpha' = 1$). The *y*-intercept does not change as a result of competitive inhibition. In **uncompetitive inhibition** the inhibitor binds to a site of the enzyme that is removed from the active site, but only if the substrate is already present. The inhibition occurs because ESI reduces the concentration of ES, the active type of complex. In this case $\alpha = 1$ (because EI does not form) and $\alpha' > 1$. The *y*-intercept of the Lineweaver–Burk plot increases by a factor of α' relative to the *y*-intercept for data on the uninhibited enzyme but the slope does not change. In **non-competitive inhibition**[3] the inhibitor binds to a site other than the active site,

3. Non-competitive inhibition is also known as *mixed inhibition*.

and its presence reduces the ability of the substrate to bind to the active site. Inhibition occurs at both the E and ES sites. This condition corresponds to $\alpha > 1$ and $\alpha' > 1$. Both the slope and y-intercept of the Lineweaver–Burk plot increase upon addition of the inhibitor. Figure 26.12c shows the special case of $K_I = K_I'$ and $\alpha = \alpha'$, which results in intersection of the lines at the x-axis.

In all cases, the efficiency of the inhibitor may be obtained by determining K_M and v_{max} from a control experiment with uninhibited enzyme and then repeating the experiment with a known concentration of inhibitor. From the slope and y-intercept of the Lineweaver–Burk plot for the inhibited enzyme (eqn 27), the mode of inhibition, the values of α or α', and the values of K_I or K_I' may be obtained.

Oscillating reactions

The concentrations of reactants, intermediates, and products of certain chemical reactions can vary periodically either in space or in time (Fig. 26.13). These oscillations occur in the presence of a feedback mechanism in which a product either increases or decreases the reaction rate. They can be sustained indefinitely only if the reaction is carried out under conditions that keep it far from equilibrium. This condition can be achieved by supplying reactants continuously and removing products continuously from a reaction vessel. Examples of such open systems are flow reactors (Sections 26.9 and 26.10) and biological cells. A biological cell, for instance, is an open system that can take in nutrients and excrete waste products of enzyme-catalysed reactions. Therefore, long-lasting chemical oscillations can occur only if a proper feedback mechanism is present. Such is the case in the glycolytic cycle, in which enzymes catalyse the breakdown of glucose with production of ATP and NADH. Feedback can occur if an enzyme of the glycolytic cycle, such as phosphofructokinase, is inhibited by high concentrations of ATP and activated by ADP. During the early stages of glycolysis, when ATP levels are low and ADP levels are high, the enzyme is active and the rate of production of NADH is high. However, as the concentration of ATP, a net product of glycolysis, rises in the cell, the enzyme becomes inhibited and the rate of production of NADH decreases. After some time, ATP levels drop again and a new cycle begins. Figure 26.14 shows how the concentration of NADH oscillates in yeast cells under anaerobic conditions.

26.13 Some reactions show oscillations in time; some show spatially periodic variations. This sequence of photographs shows the emergence of a spatial pattern.

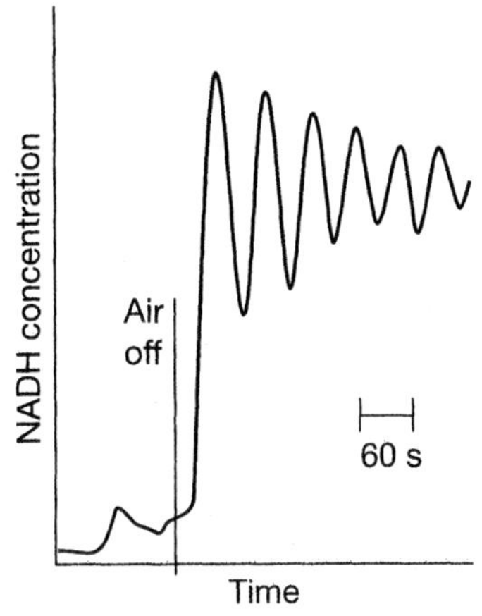

26.14 Under anaerobic conditions, oscillations in the concentration of NADH can occur in yeast cells because of inhibition of the enzyme phosphofructokinase by high levels of ATP and its activation by high levels of ADP.

26.7 Autocatalysis

Autocatalysis is the catalysis of a reaction by the products. For example, in a reaction A → P it may be found that the rate law is

$$v = k[A][P] \tag{26.28}$$

and the reaction rate is proportional to the concentration of P.[4]

An example of autocatalysis is provided by the **Belousov-Zhabotinskii reaction** (BZ reaction), one of the first oscillating reactions to be reported and studied systematically. The BZ reaction occurs in a mixture of potassium bromate, malonic acid, and a cerium(IV) salt in acidic solution. Two steps of the mechanism, which will be discussed more fully in Section 26.8c, show the role of $HBrO_2$ as an autocatalyst:

$$BrO_3^- + HBrO_2 + H_3O^+ \longrightarrow 2\,BrO_2\cdot + 2\,H_2O$$

$$2\,BrO_2\cdot + 2\,Ce(III) + 2\,H_3O^+ \longrightarrow 2\,HBrO_2 + 2\,Ce(IV) + 2\,H_2O$$

The product $HBrO_2$ is a reactant in the first step and provides a feedback mechanism that enhances the rate of formation of $HBrO_2$.

Example 26.4 Calculating concentrations in an autocatalytic reaction

Integrate the rate equation for the A → P autocatalytic reaction in eqn 28.

Method The rate law is given in eqn 28, with $v = -d[A]/dt$. It is convenient to write $[A] = [A]_0 - x$, $[P] = [P]_0 + x$ and then to write down the expression for the rate of change of either species in terms of x.

Answer With the substitution indicated, the rate law becomes

$$\frac{dx}{dt} = k([A]_0 - x)([P]_0 + x)$$

Integration by using the relation

$$\frac{1}{([A]_0 - x)([P]_0 + x)} = \frac{1}{[A]_0 + [P]_0}\left(\frac{1}{[A]_0 - x} + \frac{1}{[P]_0 + x}\right)$$

(see also Section 25.3c) gives

$$\frac{1}{[A]_0 + [P]_0}\ln\left(\frac{([P]_0 + x)[A]_0}{[P]_0([A]_0 - x)}\right) = kt$$

We can rearrange this expression into

$$\frac{x}{[P]_0} = \frac{e^{at} - 1}{1 + be^{at}}$$

where $a = ([A]_0 + [P]_0)k$ and $b = [P]_0/[A]_0$.

Comment The solution is plotted in Fig. 26.15. The rate of reaction is slow initially (little P present), then fast (when P and A are both present), and finally slow again (when A has disappeared).

Self-test 26.4 At what time is the reaction rate a maximum?

[$t_{max} = -(1/a)\ln b$]

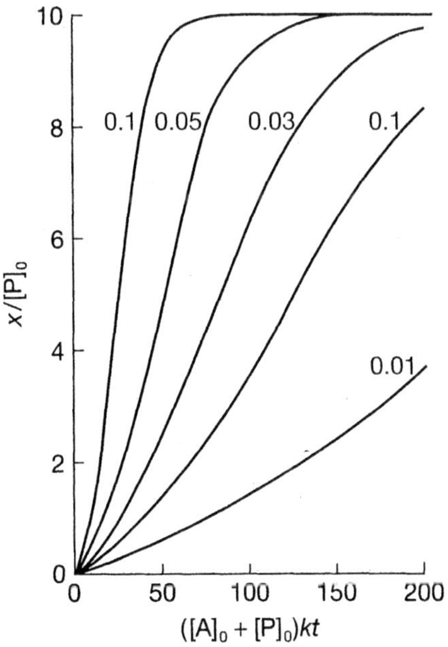

26.15 The concentration of product during the autocatalysed A → P reaction discussed in Example 26.4. Individual curves are labelled with the value of b.

4. The reaction gets started because there are usually other reaction routes for the formation of some P initially, which then takes part in the autocatalytic reaction proper.

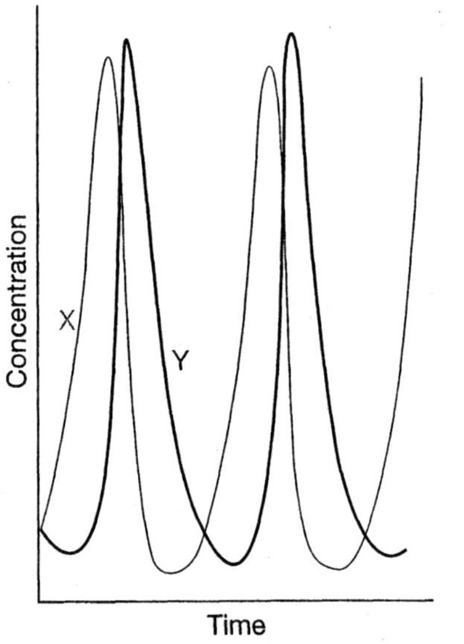

26.16 The periodic variation of the concentrations of the intermediates X and Y in a Lotka–Volterra reaction. The system is in a steady state, but not at equilibrium.

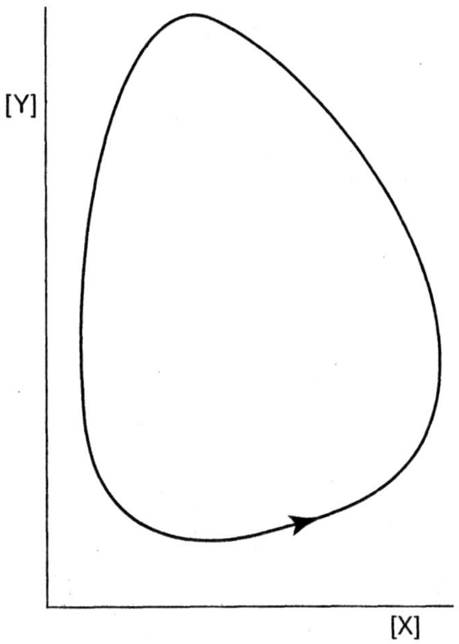

26.17 An alternative representation of the periodic variation of [X] and [Y] is to plot one against the other. Then the system describes closed orbits; different orbits are obtained with different starting conditions.

26.8 Autocatalytic mechanisms of oscillating reactions

We now consider a few specific examples of chemical oscillations that result from autocatalysis.

(a) The Lotka–Volterra mechanism

We consider first an autocatalytic reaction of a particularly simple form that illustrates how chemical oscillations may occur. The actual chemical examples that have been discovered so far have a different mechanism, as we shall see. The **Lotka–Volterra mechanism** is as follows:

$$\text{(a)}\quad A + X \longrightarrow X + X \qquad \frac{d[A]}{dt} = -k_a[A][X]$$

$$\text{(b)}\quad X + Y \longrightarrow Y + Y \qquad \frac{d[X]}{dt} = -k_b[X][Y]$$

$$\text{(c)}\quad Y \longrightarrow B \qquad \frac{d[B]}{dt} = k_c[Y]$$

Steps (a) and (b) are autocatalytic. The concentration of A is held constant by supplying it to the reaction vessel as needed. The product B plays no part in the reaction once it has been produced, so it is unnecessary to remove it; in practice, though it would normally be removed. These constraints leave [X] and [Y], the concentrations of the intermediates, as variables. Note that we are considering a **steady-state condition**, which is maintained by the flow of A into the reactor. This steady-state condition must not be confused with the steady-state approximation made earlier: in the present case we solve the rate equations exactly for the variable concentrations of X and Y, but hold [A] at an arbitrary but constant value.

The Lotka–Volterra equations can be solved numerically, and the results can be depicted in two ways. One way is to plot [X] and [Y] against time (Fig. 26.16). The same information is displayed more succinctly by plotting one concentration against the other (Fig. 26.17).

The periodic variation of the concentrations of the intermediates can be explained as follows. At some stage there may be only a little X present, but reaction (a) provides more, and the production of X autocatalyses the production of even more X. There is therefore a surge of X. However, as X is formed, reaction (b) can begin. It occurs slowly initially, because [Y] is small, but autocatalysis leads to a surge of Y. This surge, though, removes X, and so reaction (a) slows, and less X is produced. Because less X is now available, reaction (b) slows. As less Y becomes available to remove X, X has a chance to surge forward again, and so on.

(b) The brusselator

Another very interesting set of equations is called the **brusselator**:[5]

$$\text{(a)}\quad A \longrightarrow X \qquad \frac{d[X]}{dt} = k_a[A]$$

$$\text{(b)}\quad B + X \longrightarrow Y + C \qquad \frac{d[Y]}{dt} = k_b[B][X]$$

5. The mechanism is so-called because it is an oscillator that originated with Ilya Prigogine's group in Brussels.

(c) $X + X + Y \longrightarrow X + X + X \qquad \frac{d[Y]}{dt} = -k_c[X]^2[Y]$

(d) $X \longrightarrow D \qquad \frac{d[X]}{dt} = -k_d[X]$

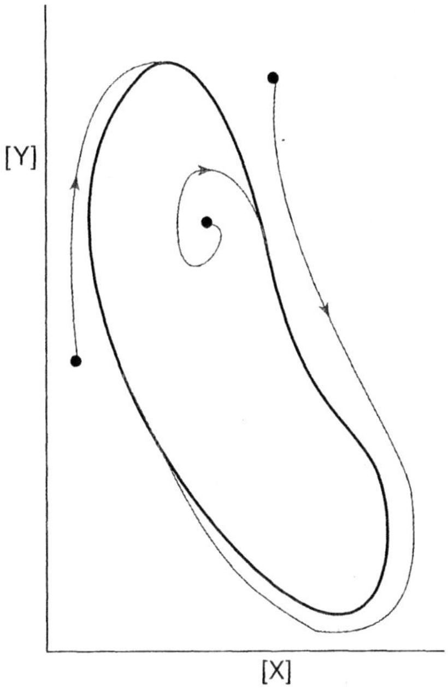

26.18 Some oscillating reactions approach a closed trajectory whatever their starting conditions. The closed trajectory (shown here in blue) is called a *limit cycle*.

Because the reactants (A and B) are maintained at constant concentration, the two variables are the concentrations of X and Y. These two concentrations may be calculated by solving the rate equations numerically, and the results are plotted in Fig. 26.18. The interesting feature is that, whatever the initial concentrations of X and Y, the system settles down into the same periodic variation of concentrations. The common trajectory to which the system migrates is called a **limit cycle**, and its period depends on the values of the rate constants. A limit cycle is an example of a structure called by mathematicians an **attractor** because it appears to attract to itself trajectories in its vicinity. In conventional chemistry, which deals with closed systems of reactants and products, the equilibrium state is an attractor determined by the minimum value of the Gibbs energy at a given temperature and pressure. In open systems maintained far from equilibrium, a limit cycle may act as an attractor; so, if the system is perturbed (by a surge of reactant, for instance), it relaxes back towards the time behaviour represented by the limit cycle.

(c) The oregonator

The mechanism of the BZ reaction was elucidated by Richard Noyes, and involves 18 elementary steps and 21 different chemical species. The main features of this awesomely complex mechanism can be reproduced by the following **oregonator**,[6] in which A stands for BrO_3^-, B for malonic acid, C for HOBr, P for additional oxidized and reduced products, X for $HBrO_2$, Y for Br^-, and Z for Ce^{4+}:

(a) $A + Y \longrightarrow X + C$

(b) $X + Y \longrightarrow C + C$

(c) $A + X \longrightarrow X + X + Z + Z$

(d) $X + X \longrightarrow A + C$

(e) $B + Z \longrightarrow Y + P$

The concentration of H_3O^+ is omitted because it is often held constant by buffering the reaction mixture. The concentrations of A, B, C, and D are also held constant by maintaining an influx of reactants and removal of products. The oscillations arise in a similar way to the brusselator, and can be traced to the autocatalysis in step (c) and the linkage between the reactions provided by the other steps.

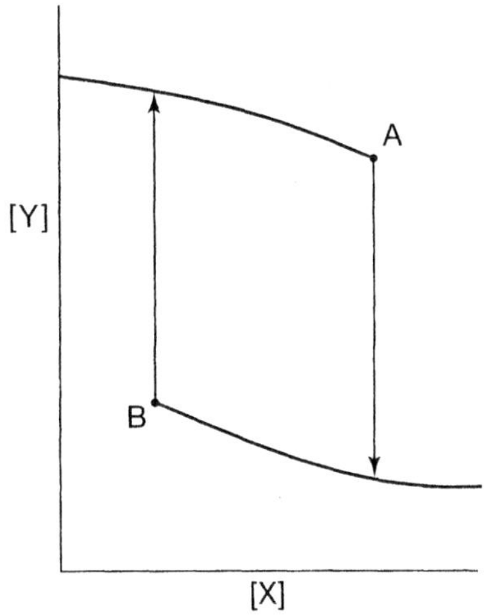

26.19 A system showing bistability. As the concentration of X is increased (by adding it to a reactor), the concentration of Y decreases along the upper curve but at A drops sharply to a low value. If X is then decreased, the concentration increases along the lower curve, but rises sharply to a high value at B.

26.9 Bistability

As shown in the *Justification* below, a number of chemical reactions carried out in flow reactors can have multiple steady states with very different chemical compositions. This possibility is depicted in Fig. 26.19 for a hypothetical reaction with two intermediates, X and Y. The compositions of the steady states are indicated by the lower and upper lines of a plot of [Y] versus [X]. The plot shows that, if the concentration of Y is at some high value in the reactor, and X is added, then the concentration of Y might decrease as shown by the upper line. If X is at some high value, then as Y is added the reaction might result in the slow increase of Y as shown by the lower line. However, a concentration may be reached at which the concentration of

6. It is so-called because Noyes and his group work in Oregon.

Y will jump from one curve to the other at points A and B (just as a supercooled liquid might suddenly solidify). This jump occurs because neither steady state is an equilibrium state in the thermodynamic sense: they occur in systems that are well removed from equilibrium, and the concentrations of X and Y represent the consequences of reactants continuously flowing into and of products flowing out of the reactor. The behaviour just described for a system with two stable steady states is known as **bistability**.

Justification 26.3

We demonstrate the possibility of multiple steady states by considering a hypothetical autocatalytic process in a flow reactor:

$$\mathrm{A} + 2\,\mathrm{B} \longrightarrow 3\,\mathrm{B} \qquad \frac{\mathrm{d}[\mathrm{A}]}{\mathrm{d}t} = -k[\mathrm{A}][\mathrm{B}]^2$$

The rate law for this reaction must take into account not only the chemical event (reaction between A and B) but also the inflow rate of A and the outflow rates of B and unreacted A. Let $[\mathrm{A}]_0$ be the initial concentration of A at the inlet of the reactor and k_0 a rate constant for the flow of reactant (and equal to the reciprocal of the time spent by A in the reactor). Then,

$$\frac{\mathrm{d}[\mathrm{A}]}{\mathrm{d}t} = -k[\mathrm{A}][\mathrm{B}]^2 + k_0[\mathrm{A}]_0 - k_0[\mathrm{A}] \qquad (26.29)$$

where, on the right-hand side of the equation, the first term corresponds to the chemical reaction, the second term is the rate of flow into the reactor, and the third term is the rate of removal of unreacted A from the reactor.

Upon application of the steady-state approximation to eqn 29, substitution of the mass balance expression $[\mathrm{B}] = [\mathrm{A}]_0 - [\mathrm{A}]$, and some rearrangement, we obtain:

$$([\mathrm{A}]_0 - [\mathrm{A}])\{k_0 - k[\mathrm{A}]([\mathrm{A}]_0 - [\mathrm{A}])\} = 0$$

The roots of this equation are the steady-state concentrations of A, which we denote $[\mathrm{A}]_{\mathrm{ss}}$. Because the equation is cubic in [A], it has multiple roots, which indicates a potential for coexistence of steady states in the system. However, we must ascertain that the values of $[\mathrm{A}]_{\mathrm{ss}}$ make physical sense.

The first root is $[\mathrm{A}]_{\mathrm{ss}} = [\mathrm{A}]_0$, the unreacted state. We determine the remaining roots by setting $k_0 - k[\mathrm{A}]([\mathrm{A}]_0 - [\mathrm{A}]) = 0$. It follows that:

$$k[\mathrm{A}]^2 - k[\mathrm{A}]_0[\mathrm{A}] + k_0 = 0$$

The roots of this equation are:

$$[\mathrm{A}]_{\mathrm{ss}} = \tfrac{1}{2}[\mathrm{A}]_0 \pm \frac{(k[\mathrm{A}]_0^2 - 4k_0)^{1/2}}{2k^{1/2}}$$

The roots will be physically significant if they are real and positive, which is guaranteed if $k[\mathrm{A}]_0^2 > 4k_0$. Therefore, the system will exhibit three steady states if the initial reaction rate far exceeds the flow rate of the reactor. Consequently, many autocatalytic reactions occurring in flow reactors can have multiple stable steady states.

Figure 26.19 does not describe an oscillation because it lacks a feedback mechanism that would cause, for example, the concentration of X to decrease to the value at point B upon a jump down from point A. Now consider the addition of a species Z that reacts with X and Y according to:

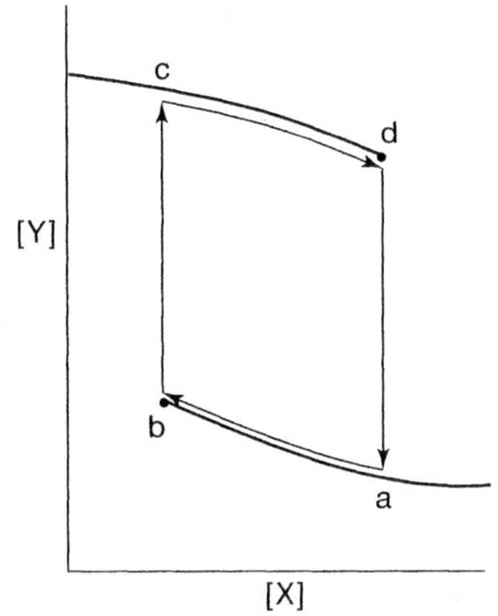

26.20 Chemical oscillation in a bistable system occurs as a result of the effect of a third substance Z which can react with X to produce Y and with Y to produce X. As a result, the system switches periodically between the upper and lower curves.

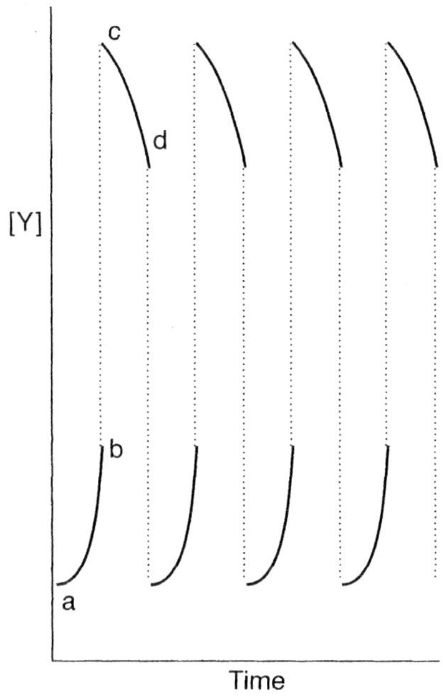

26.21 The leaping from one branch to another of a bistable system appears to the observer as a periodic surge and depletion of concentration.

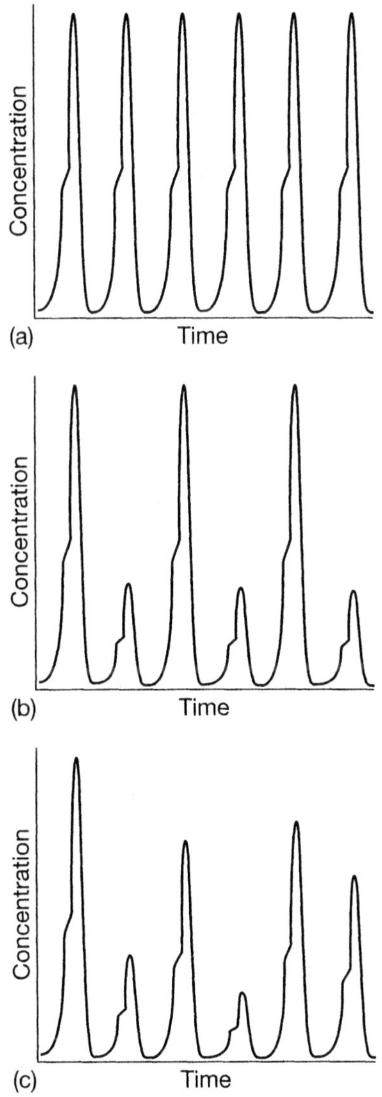

26.22 The onset of chaos as a result of period-doubling. (a) Steady-state oscillation, (b) the oscillation after one period-doubling step, (c) the chaotic regime after many period-doubling steps.

$$Y + Z \longrightarrow X$$
$$X + Z \longrightarrow Y$$

In the absence of Z the flow of material might correspond to the stable state on the upper curve of Fig. 26.20. However, as Z reacts with Y to produce X the state of the system moves along the curve (to the right, as Y decreases and X increases) until the sudden transition occurs to the lower curve. Then Z reacts with X and produces Y, which means that the composition moves to the left along the lower curve. There comes a point, however, when the concentration of X has been reduced so much, and that of Y has risen so much, that there is a sudden transition to the upper curve, when the process begins again. The leaping from one stable state to another is manifest as the sudden surge or depletion of the concentration of a species (Fig. 26.21). Therefore, with the addition of chemical feedback, a bistable system can become oscillatory.

26.10 Chemical chaos

For some reactions, the solutions of rate laws cannot be used to predict the composition of the system from its initial composition. Instead of showing periodic oscillatory behaviour, the concentrations burst into chaotic oscillation, characterized by unpredictable amplitudes or unpredictable frequencies (Fig. 26.22). Such behaviour can be literally a matter of life and death for, should the heartbeat become chaotic (the phenomenon of fibrillation), then death may result. The type of chaos that stems from the structure of well defined (and often quite simple) nonlinear differential equations is called **deterministic chaos**, for the behaviour of the solutions is predictable but infinitely sensitive to the initial conditions.

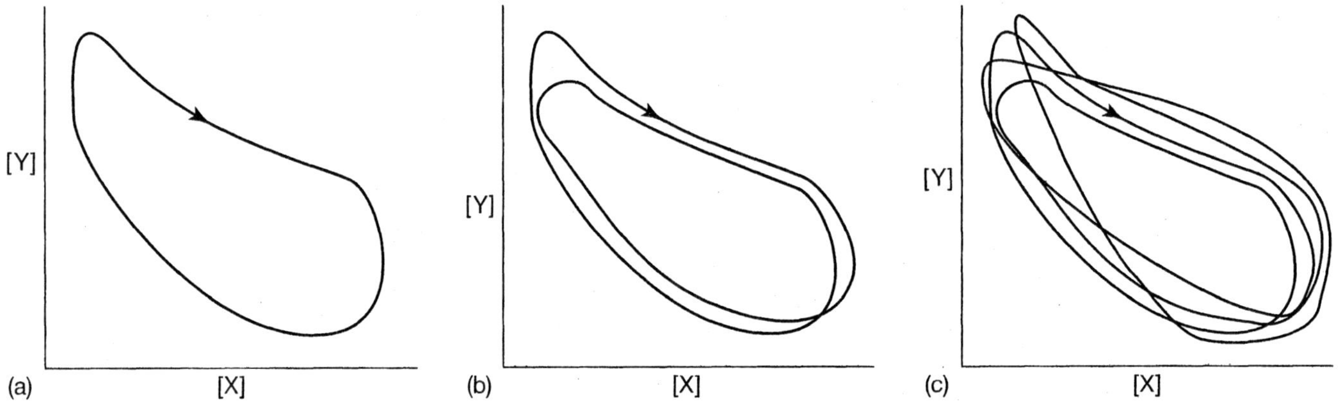

26.23 Successive trajectories in concentration space (a) before period-doubling, (b) after one period-doubling step, (c) after several period-doubling steps.

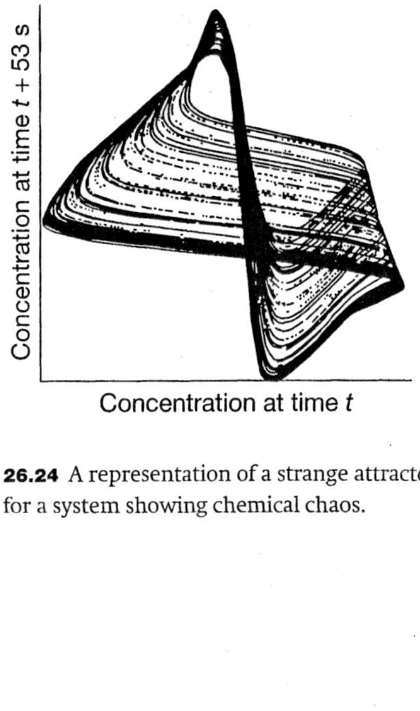

26.24 A representation of a strange attractor for a system showing chemical chaos.

There are several ways in which a reaction can be steered into a chaotic regime. For example, in certain systems the change in a parameter (such as a flow rate through the reaction vessel or the rate of stirring) may result in the successive doubling of the period of the limit cycle (Fig. 26.23). When this **period-doubling** occurs, the system must circulate twice round the cycle before an initial pair of concentrations is restored. Then, after a further modification of the parameter, the period doubles again, and the limit cycle circulates four times before repeating itself. As each period doubles, the periodicity of the motion becomes less apparent and finally appears like random fluctuations. The trajectory of the composition of the system is now highly complex, and may take a path that never retraces its path or crosses itself. Such a path is called a **strange attractor** (Fig. 26.24).

The term 'chaos' is in some respects misleading. First, the conditions under which certain systems of differential equations display period doubling can be specified exactly, and the progression through period-doubling cycles shows considerable regularity that is common to many classes of systems. In so far as the composition can be specified exactly, then a later composition can be predicted. Our inability to predict the composition of a system that is in a chaotic regime stems from our inability to know exactly the initial experimental conditions or measure the composition at an exact later instant. The unpredictability of a chaotic system lies not in the formulation or solution of the differential equations that describe the rates of processes, but in our ability to relate those solutions to the practical system of interest given the inherent imprecision of experimental observations.

Photochemistry

Many reactions can be initiated by the absorption of electromagnetic radiation by one of the mechanisms described in Section 17.2. The most important of all are the photochemical processes that capture the radiant energy of the Sun. Some of these reactions lead to the heating of the atmosphere during the daytime by absorption of ultraviolet radiation (Box 26.1). Others include the absorption of visible radiation during photosynthesis (Box 10.1). Without photochemical processes, the Earth would be simply a warm, sterile, rock. Table 26.1 summarizes common photochemical reactions.

Table 26.1 Examples of photochemical processes†

Process	General form	Example	
Ionization:	$A^* \longrightarrow A^+ + e^-$	$NO^* \longrightarrow NO^+ + e^-$	($\lambda = 134$ nm)
Electron transfer:	$A^* + B \longrightarrow A^+ + B^-$ (or $A^- + B^+$)	$[Ru(bpy)_3^{2+}]^* + Fe^{3+} \longrightarrow Ru(bpy)_3^{3+} + Fe^{2+}$	($\lambda = 452$ nm)
Dissociation:	$A^* \longrightarrow B + C$	$O_3^* \longrightarrow O_2 + O$	($\lambda < 1180$ nm)
	$A^* + B{-}C \longrightarrow A + B + C$	$Hg^* + CH_4 \longrightarrow Hg + CH_3 + H$	($\lambda = 254$ nm)
Addition:	$2\,A^* \longrightarrow B$ $A^* + B \longrightarrow AB$	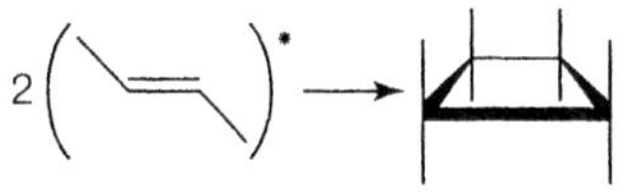	($\lambda = 230$ nm)
Abstraction:	$A^* + B{-}C \longrightarrow A{-}B + C$	$Hg^* + H_2 \longrightarrow HgH + H$	($\lambda = 254$ nm)
Isomerization or rearrangement:	$A^* \longrightarrow A'$	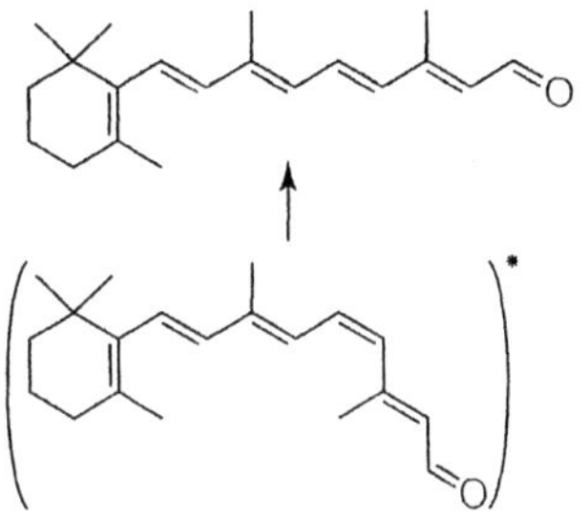	($\lambda = 380$ nm; see Box 17.1)

† The excited state is denoted by an asterisk. In each example, the photochemical process is initiated by irradiation at the wavelength in parentheses.

Box 26.1 The chemistry of stratospheric ozone

The Earth's atmosphere contains primarily N_2 and O_2 gas, with low concentrations of a large number of other species of both natural and anthropogenic origins. Indeed, many of the natural trace constituents of our atmosphere participate in complex chemical reactions that have contributed to the proliferation of life on the planet. The development of industrial societies has added new components to the Earth's atmosphere and has led to significant changes in the concentrations of some natural trace species. The negative consequences of these changes for the environment are either already being felt or, more disturbingly, are yet to be felt in the next few decades. Careful kinetic analysis allows us to understand the origins of our complex atmosphere and point to ways in which environmental problems can be solved or avoided.

The Earth's atmosphere consists of layers, as shown in the first illustration. The pressure decreases as altitude increases (see Problems 1.35 and 19.21), but the variation of temperature with altitude is complex, owing to processes that capture radiant energy from the Sun. We focus on the *stratosphere*, a region spanning from 15 km to 50 km above the surface of the Earth, and on the chemistry of the trace component ozone, O_3.

In the *troposphere*, the region between the Earth's surface and the stratosphere, temperature decreases with increasing altitude. This behaviour may be understood in terms of a model in which the boundary between the troposphere and the stratosphere, also called the *tropopause*, is considered adiabatic. Then we know from Section 2.6 that, as atmospheric gases are allowed to expand from layers close to the surface to higher layers, the temperature varies with pressure, and hence height, as

$$\frac{T_{\text{low altitude}}}{T_{\text{high altitude}}} = \left(\frac{p_{\text{low altitude}}}{p_{\text{high altitude}}}\right)^c \qquad c = \frac{C_{p,\text{m}}}{C_{V,\text{m}}} - 1$$

The model predicts a decrease in temperature with increasing altitude because $C_{p,\text{m}}/C_{V,\text{m}} \approx \frac{7}{5}$ for air. In the stratosphere,

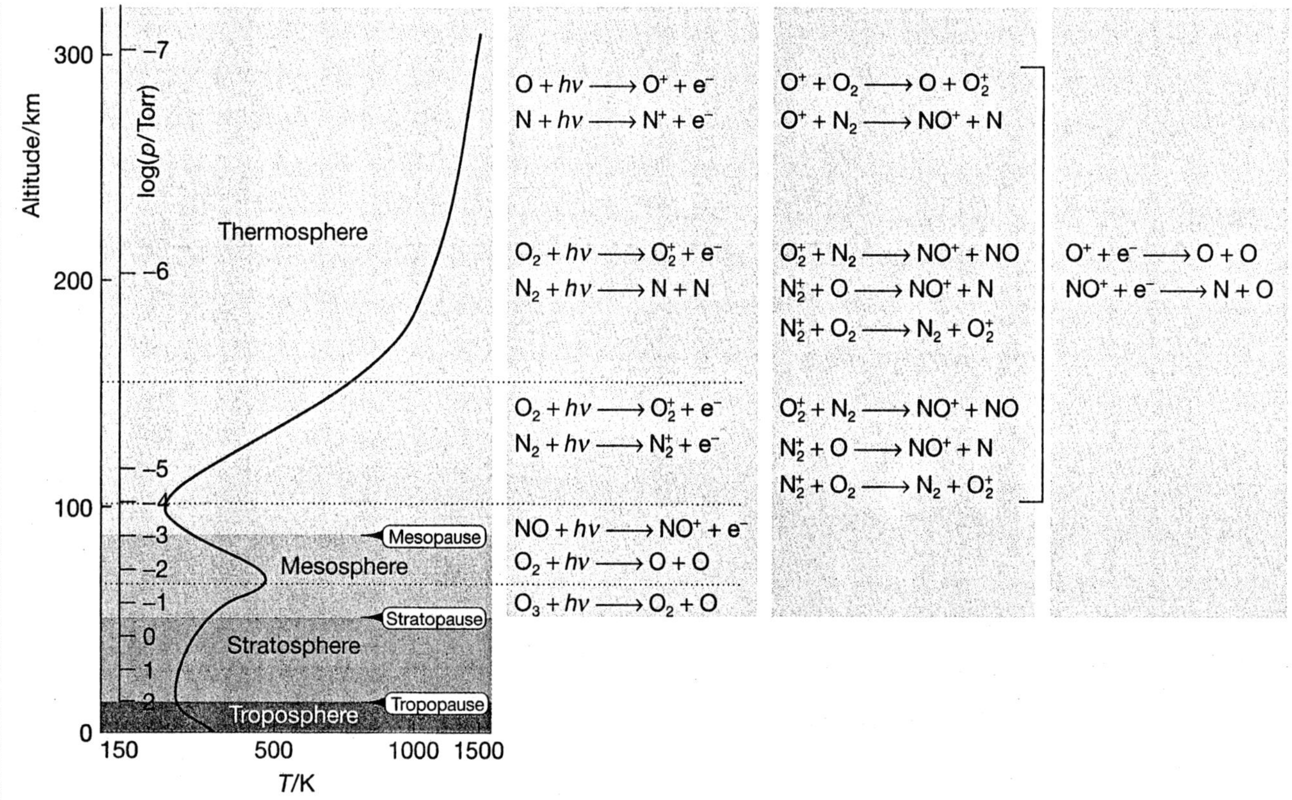

The temperature profile through the atmosphere and some of the reactions that occur in each region.

a *temperature inversion* is observed because of photochemical chain reactions that produce ozone from O_2. The *Chapman model* accounts for ozone formation and destruction in an atmosphere that contains only O_2:

Initiation: $O_2 + h\nu \longrightarrow O + O$
$185\ \text{nm} < \lambda < 220\ \text{nm}$ $\quad v = k_1[O_2]$

Propagation: $O + O_2 + M \longrightarrow O_3 + M$
$\Delta H_r = -106.6\ \text{kJ mol}^{-1}$ $\quad v = k_2[O][O_2][M]$
$O_3 + h\nu \longrightarrow O_2 + O$
$210\ \text{nm} < \lambda < 300\ \text{nm}$ $\quad v = k_3[O_3]$

Termination: $O + O_3 \longrightarrow O_2 + O_2$
$\Delta H_r = -391.9\ \text{kJ mol}^{-1}$ $\quad v = k_4[O][O_3]$
$O + O + M \longrightarrow O_2 + M$ $\quad v = k_5[O]^2[M]$

where M is an arbitrary third body, such as O_2 in an 'oxygen-only' atmosphere, which helps to conserve angular momentum. The mechanism shows that absorption of radiation by O_2 and O_3 during the daytime leads to the production of reactive oxygen atoms, which, in turn, participate in exothermic reactions that are responsible for the heating of the stratosphere.

Using values of the rate constants that are applicable to stratospheric conditions, the Chapman model predicts a net formation of trace amounts of ozone, as seen in the second illustration (see also *Discussion question 1*). However, the model overestimates the concentration of ozone in the stratosphere because other trace species X contribute to catalytic enhancement of the termination step $O_3 + O \rightarrow O_2 + O_2$ according to

$$X + O_3 \longrightarrow XO + O_2$$
$$XO + O \longrightarrow X + O_2$$

The catalyst X can be H, OH, NO, or Cl. Chlorine atoms are produced by photolysis of CH_3Cl which, in turn, is a by-product of reactions between Cl^- and decaying vegetation in oceans. Nitric oxide, NO, is produced in the stratosphere from reaction between excited oxygen atoms and N_2O, which is formed in the troposphere. The hydroxyl radical is a product, along with the methyl radical, of the reaction between excited oxygen atoms and methane gas, which is a by-product of a number of natural processes (such as digestion of cellulose in ruminant animals and anaerobic decomposition of organic waste matter) and industrial processes (such as food production and fossil fuel use). In spite of the

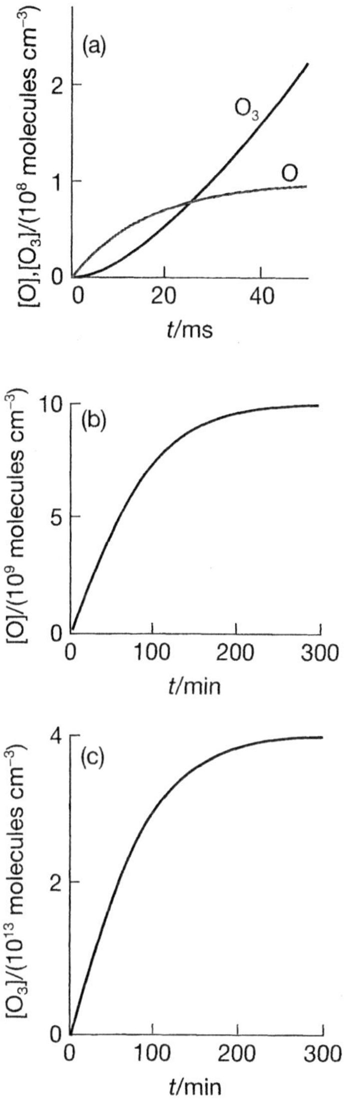

Net formation of ozone via the Chapman model in an stratospheric model containing only O_2, O, and O_3. The rate constants are consistent with reasonable stratospheric conditions. (a) Early reaction period after irradiation begins at $t = 0$. (b) Late reaction period, showing that the concentration of O atoms begins to level off after about four hours of continuous irradiation. (c) Late reaction period, showing that the ozone concentration also begins to level off similarly. For details of the calculation, see *Discussion question 1* and M.P. Cady and C.A. Trapp, *A Mathcad primer for physical chemistry*. Oxford University Press (1999).

presence of these catalysts, a natural stratosphere is still capable of maintaining a low concentration of ozone.

The chemistry outlined above shows that the photochemical reactions of the Chapman model account for absorption of a significant portion of solar ultraviolet radiation in the stratosphere. Hence, the surface of the Earth is bathed by

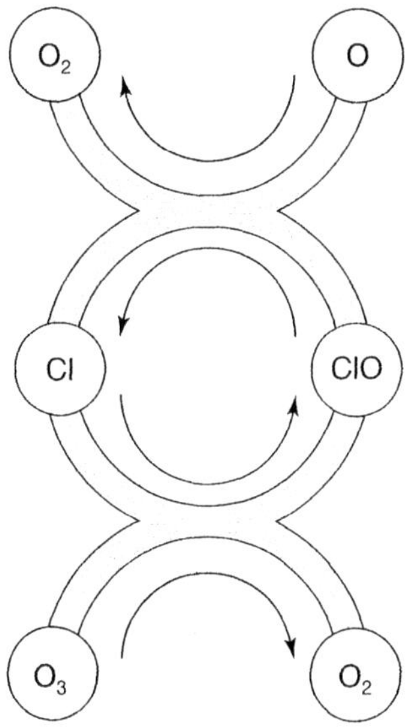

A catalytic cycle showing the propagation of ozone decomposition by chlorine atoms.

lower energy radiation, which does not damage biological tissue (see Box 26.2). However, some pollutants can lower the concentration of stratospheric ozone. For example, chlorofluorocarbons (CFCs) have been used as propellants and refrigerants over many years. As CFC molecules diffuse into the stratosphere, they are photolysed by ultraviolet radiation. For CF_2Cl_2, also known as CFC-12, the reaction is:

$$CF_2Cl_2 + h\nu \longrightarrow CF_2Cl + Cl$$

We already know that the resulting Cl atoms can participate in the decomposition of ozone according to the catalytic cycle shown in the third illustration. A number of experimental observations have linked this chemistry of CFCs to a dangerously rapid decline in the concentration of stratospheric ozone over the last three decades. Ozone depletion has increased the amount of ultraviolet radiation at the Earth's surface. The physiological consequences of prolonged exposure to ultraviolet radiation include DNA damage, genetic mutations, cell destruction, sunburning, and skin cancers. Consequently, there is concern that the depletion of stratospheric ozone may lead to an increase in mortality not only of animals but also the plants and lower organisms that form the base of the food chain.

Chlorofluorocarbons are being phased out according to international agreements and alternatives, such as the hydrofluorocarbon CH_2FCH_3, are already being used. However, the temperature inversion shown in the first illustration leads to trapping of gases in the troposphere, so CFCs are likely to continue to cause ozone depletion over many years as the molecules diffuse slowly into the stratosphere.

Discussion questions[7]

1. Use the Chapman model to explore the behaviour of a model atmosphere consisting of pure O_2 at 10 Torr and 298 K that is exposed to measurable frequencies and intensities of UV radiation. (a) Look up the values of k_2, k_4, and k_5 in a source such as the *CRC Handbook of chemistry and physics* or *Chemical kinetics and photochemical data for use in stratospheric modeling* (the URL is available in the web site of this book). The rate constants k_1 and k_3 depend upon the radiation conditions; assume values of 1.0×10^{-8} s^{-1} and 0.016 s^{-1}, respectively. If you cannot find a value for k_5, formulate chemically sound arguments for exclusion of the fifth step from the mechanism. (b) Write the rate expressions for the concentration of each chemical species. (c) Assume that the UV radiation is turned on at $t = 0$, and solve the rate expressions for the concentration of all species as a function of time over a period of 4 h. Examine relevant concentrations in the very early time period $t < 0.1$ s. State all assumptions. Is there any ozone present initially? Why must the pressure be low and the UV radiation intensities high for the production of ozone? Draw graphs of the time variations of both atomic oxygen and ozone on both the very short and the long timescales. What is the percentage of ozone after 4.0 h of irradiation? *Hint.* You will need a software package for solving a 'stiff' system of differential equations. Stiff differential equations have at least two rate constants with very different values and result in different behaviours on different timescales, so the solution usually requires that the total time period be broken into two or more periods; one may be very short and another very long. For help with using mathematical software to solve systems of differential equations, see M.P. Cady and C.A. Trapp, *A Mathcad primer for physical chemistry*. Oxford University Press (1999).

2.[7] Chlorine atoms react rapidly with ozone in the gas-phase bimolecular reaction $Cl + O_3 \rightarrow ClO + O_2$ with $k = (1.7 \times 10^{10}$ L mol^{-1} s$^{-1})e^{-260/(T/K)}$ (W.B. DeMore, S.P. Sander, D.M. Golden, R.F. Hampson, M.J. Kurylo, C.J. Howard, A.R. Ravishankara, C.E. Kolb, and M.J. Molina, Chemical kinetics and photochemical data for use in stratospheric modeling: Evaluation Number 11, JPL Publication 94-26 (1994)). Estimate the rate of this reaction at (a) 20 km, where $[Cl] = 5 \times 10^{-17}$ mol L^{-1}, $[O_3] = 8 \times 10^{-9}$ mol L^{-1}, and $T = 220$ K; (b) 45 km, where $[Cl] = 3 \times 10^{-15}$ mol L^{-1}, $[O_3] = 8 \times 10^{-11}$ mol L^{-1}, and $T = 270$ K.

Further reading

B.J. Finlayson-Pitts and J.N. Pitts, *Chemistry of the upper and lower atmosphere*. Academic Press, San Diego (2000).

R.P. Wayne, *Chemistry of atmospheres: an introduction to the chemistry of atmospheres of earth, the planets, and their satellites*. Oxford University Press (2000).

7. These problems were supplied by Charles Trapp, Carmen Giunta, and Marshall Cady.

26.11 Kinetics of photophysical and photochemical processes

Photochemical processes are initiated by the absorption of radiation by at least one component of a reaction mixture. In a **primary process**, products are formed directly from the excited state of a reactant. Examples include fluorescence (Section 17.3) and the *cis–trans* photoisomerization of retinal (Table 26.1, see also Box 17.1). Products of a **secondary process** originate from intermediates that are formed directly from the excited state of a reactant. Examples include photosynthesis and photochemical chain reactions (Section 26.12).

The web site contains links to databases on photochemical reactions.

Competing with the formation of photochemical products is a host of primary photophysical processes that can deactivate the excited state (Table 26.2). Therefore, it is important to consider the timescales of excited state formation and decay before describing the mechanisms of photochemical reactions.

(a) Timescales of photophysical processes

Electronic transitions caused by absorption of ultraviolet and visible radiation occur within 10^{-16}–10^{-15} s. We expect, then, that the upper limit for the rate constant of a first-order photochemical reaction is about 10^{16} s^{-1}. Fluorescence is slower than

Table 26.2 Common photophysical processes†

Primary absorption:	$S + h\nu \longrightarrow S^*$
Excited-state absorption:	$S^* + h\nu \longrightarrow S^{**}$
	$T^* + h\nu \longrightarrow T^{**}$
Fluorescence emission:	$S^* \longrightarrow S + h\nu$
Stimulated emission:	$S^* + h\nu \longrightarrow S + 2h\nu$
Intersystem crossing (ISC):	$S^* \longrightarrow T^*$
Phosphorescence emission:	$T^* \longrightarrow S + h\nu$
Internal conversion (IC):	$S^* \longrightarrow S$
Collision-induced emission:	$S^* + M \longrightarrow S + M + h\nu$
Collisional deactivation:	$S^* + M \longrightarrow S + M$
	$T^* + M \longrightarrow S + M$
Electronic energy transfer:	
singlet–singlet:	$S^* + S \longrightarrow S + S^*$
triplet–triplet:	$T^* + T \longrightarrow T + T^*$
Excimer formation:	$S^* + S \longrightarrow (SS)^*$
Energy pooling:	
singlet–singlet:	$S^* + S^* \longrightarrow S^{**} + S$
triplet–triplet:	$T^* + T^* \longrightarrow S^* + S$

† S denotes a singlet state and T a triplet state; M is a body that carries thermal energy. In photochemical studies, it is common to set 1 einstein = 1 mol photons.

absorption, with typical lifetimes of 10^{-12}–10^{-6} s. Therefore, the excited singlet state can initiate very fast photochemical reactions in the femtosecond (10^{-15} s) to picosecond (10^{-12} s) timescale. Examples of such ultrafast reactions are the initial events of vision (Box 17.1) and of photosynthesis. Typical intersystem crossing (ISC) and phosphorescence times for large organic molecules are 10^{-12}–10^{-4} s and 10^{-6}–10^{-1} s, respectively. As a consequence, excited triplet states are photochemically important. Indeed, because phosphorescence decay is several orders of magnitude slower than most typical reactions, species in excited triplet states can undergo a very large number of collisions with other reactants before deactivation. The interplay between reaction rates and excited state lifetimes is a very important factor in the determination of the kinetic feasibility of a photochemical process.

Illustration 26.1

To estimate whether the excited singlet or triplet state of the reactant is a suitable product precursor, we compare the emission lifetimes with the time constant, τ, of the reactant due to the chemical reaction. As an illustration, consider a unimolecular photochemical reaction with rate constant $k = 1.7 \times 10^4\ s^{-1}$ and time constant $\tau = 1/(1.7 \times 10^4\ s^{-1}) = 59\ \mu s$ that involves a reactant with an observed fluorescence lifetime of 1.0 ns and an observed phosphorescence lifetime of 1.0 ms. The excited singlet state is too short-lived and is not expected to be a major source of product in this reaction. On the other hand, the excited triplet state is a good candidate for a precursor.

(b) The primary quantum yield

We shall see that the rates of deactivation of the excited state by radiative, nonradiative, and chemical processes determine the yield of product in a photochemical reaction. The **primary quantum yield**, ϕ, is defined as the number of photophysical or photochemical events that lead to primary products divided by the number of photons absorbed by the molecule in the same interval. It follows that the primary quantum yield is also the rate of radiation-induced primary events divided by the rate of photon absorption. Because the rate of photon absorption is equal to the intensity of light absorbed by the molecule (Section 16.2), we write

$$\phi = \frac{\text{number of events}}{\text{number of photons absorbed}} = \frac{\text{rate of process}}{\text{intensity of light absorbed}} = \frac{v}{I_{abs}} \qquad [26.30]$$

A molecule in an excited state must either decay to the ground state or form a photochemical product. Therefore, the total number of molecules deactivated by radiative processes, nonradiative processes, and photochemical reactions must be equal to the number of excited species produced by absorption of light. We conclude that the sum of primary quantum yields ϕ_i for *all* photophysical and photochemical events i must be equal to 1, regardless of the number of reactions involving the excited state. It follows that

$$\sum_i \phi_i = \sum_i \frac{v_i}{I_{abs}} = 1 \qquad (26.31)$$

Illustration 26.2

For an excited singlet state that decays to the ground state only via the photophysical processes described in Section 26.11a, we write $\phi_f + \phi_{IC} + \phi_{ISC} + \phi_p = 1$, where ϕ_f, ϕ_{IC}, ϕ_{ISC}, and ϕ_p are the quantum yields of fluorescence, internal conversion, intersystem crossing, and phosphorescence, respectively. The quantum yield of photon emission by fluorescence and phosphorescence is $\phi_{emission} = \phi_f + \phi_p$, which is less than 1. If the excited singlet state also participates in a primary photochemical reaction with quantum yield ϕ_R, we write $\phi_f + \phi_{IC} + \phi_{ISC} + \phi_p + \phi_R = 1$.

We can now strengthen the link between reaction rates and primary quantum yield already established by eqns 30 and 31. By taking the constant I_{abs} out of the summation in eqn 31 and rearranging, we obtain $I_{abs} = \sum_i v_i$. Substituting this result into eqn 31 gives the general result

$$\phi_i = \frac{v_i}{\sum_i v_i} \qquad (26.32)$$

Therefore, the primary quantum yield may be determined directly from the experimental rates of *all* photophysical and photochemical processes that deactivate the excited state.

(c) Mechanism of decay of excited singlet states

Consider the formation and decay of an excited singlet state:

Absorption:	$S + h\nu_i \longrightarrow S^*$	$v_{abs} = I_{abs}$
Fluorescence:	$S^* \longrightarrow S + h\nu_f$	$v_f = k_f[S^*]$

Internal conversion:	$S^* \longrightarrow S$	$v_{IC} = k_{IC}[S^*]$
Intersystem crossing:	$S^* \longrightarrow T^*$	$v_{ISC} = k_{ISC}[S^*]$

in which S is an absorbing species, S^* an excited singlet state, T^* an excited triplet state, and $h\nu_i$ and $h\nu_f$ are the energies of the incident and fluorescent photons, respectively. If the absorbance of the sample is low and the incident light intensity is relatively high, we may assume that $[S^*]$ is small and constant. That is, we may invoke the steady-state approximation for $[S^*]$:

$$\frac{d[S^*]}{dt} = I_{abs} - k_f[S^*] - k_{IC}[S^*] - k_{ISC}[S^*] = I_{abs} - (k_f + k_{IC} + k_{ISC})[S^*] = 0$$

Consequently,

$$I_{abs} = (k_f + k_{IC} + k_{ISC})[S^*]$$

By using this expression and eqn 30, we obtain

$$\phi_f = \frac{v_f}{I_{abs}} = \frac{k_f[S^*]}{(k_f + k_{IC} + k_{ISC})[S^*]}$$

The final expression for the quantum yield of fluorescence is

$$\phi_f = \frac{k_f}{k_f + k_{IC} + k_{ISC}} \qquad (26.33)$$

We can summarize the results of these calculations as follows. Under experimental conditions that justify the steady-state approximation for $[S^*]$, the sample emits fluorescence photons with a constant intensity, I, which is proportional to $k_f[S^*]$ and hence also proportional to ϕ_f. There are many commercial instruments designed to make such steady-state fluorescence measurements of I. When the light is turned off, the excited state decays exponentially:

$$[S^*]_t = [S^*]_0\, e^{-t/\tau_0}$$

The observed fluorescence lifetime, τ_0, is given by:

$$\tau_0 = \frac{1}{k_f + k_{IC} + k_{ISC}} = \frac{\phi_f}{k_f}$$

The observed fluorescence lifetime can therefore be measured with a pulsed laser technique (Section 17.7). First, the sample is excited with a short light pulse from a laser at a wavelength where S absorbs strongly. Then, the decay of the fluorescence intensity after the pulse is monitored with a fast photodetector system.

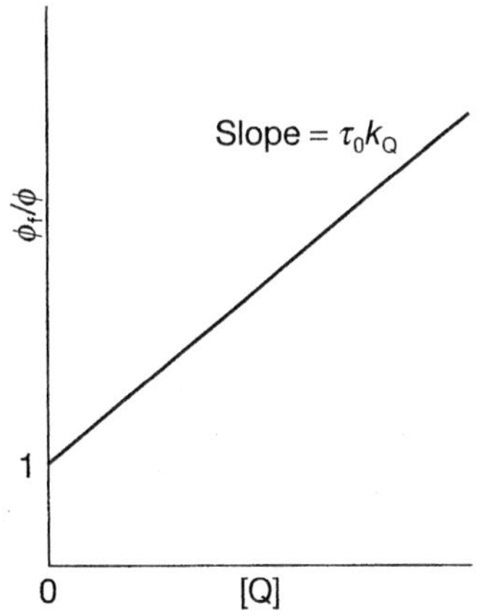

26.25 The format of a Stern–Volmer plot and the interpretation of the slope in terms of the rate constant for quenching and the observed fluorescence lifetime in the absence of quenching.

(d) Quenching

The shortening of the lifetime of the excited state is called **quenching**. Quenching may be either a desired process, such as in energy or electron transfer (Box 26.2), or an undesired side reaction that can decrease the quantum yield of a desired photochemical process. Quenching effects may be studied by monitoring the emission from the excited state that is involved in the photochemical reaction.

The addition of a quencher, Q, opens an additional channel for deactivation of S^*:

Quenching: $S^* + Q \longrightarrow S + Q \qquad v_Q = k_Q[Q][S^*]$

The steady-state approximation for $[S^*]$ now gives:

$$\frac{d[S^*]}{dt} = I_{abs} - (k_f + k_{IC} + k_{ISC} + k_Q[Q])[S^*] = 0$$

Box 26.2 Photobiology and photomedicine

Up to 1 kJ s^{-1} m^{-2} of solar radiation reaches the Earth's surface, with the exact intensity depending on latitude, time of day, and weather. A significant amount of this energy is harnessed during photosynthesis (see Box 10.1). Other photochemical processes also occur in both photosynthetic and non-photosynthetic organisms. Among the beneficial processes in humans are vision, which is initiated by the photoisomerization of 11-*cis*-retinal in the protein rhodopsin (Section 17.1), and the biosynthesis of vitamin D_3 from 7-dehydrocholesterol in skin. Other processes, such as DNA damage caused by prolonged exposure to ultraviolet (UV) radiation, are deleterious to both higher and lower organisms. When controlled carefully, potentially harmful photochemical processes may be turned into beneficial forms of therapy. We shall explore plant photosynthesis, UV-induced DNA damage, and one variety of laser-based tumour therapy in some detail.

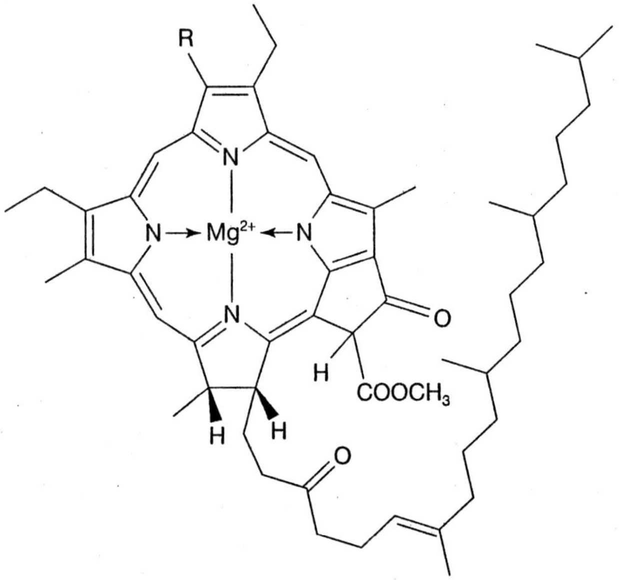

1 Chlorophylls *a* (R = CH_3) and *b* (R = CHO)

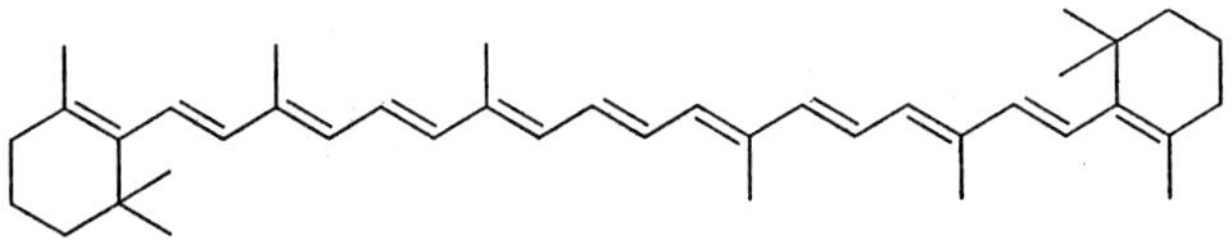

2 *β*-Carotene

Plant photosynthesis

A large proportion of solar radiation with wavelengths below 400 nm and above 1000 nm is absorbed by atmospheric gases such as ozone and O_2, which absorb ultraviolet radiation, and CO_2 and H_2O, which absorb infrared radiation. As a result, plants, algae, and some species of bacteria evolved photosynthetic apparatus that capture visible and near-infrared radiation. Plants use radiation in the wavelength range of 400–700 nm to drive the transfer of electrons from water to carbon dioxide. We have already examined the thermodynamics of plant photosynthesis in Box 10.1; here we shall describe the kinetics of the capture and utilization of solar energy.

In the chloroplast, chlorophylls *a* and *b* (**1**) and carotenoids (of which *β*-carotene (**2**) is an example) bind to integral proteins called *light-harvesting complexes*, which absorb solar energy and transfer it to protein complexes known as *reaction centres*, where light-induced electron transfer reactions occur. The combination of a light-harvesting complex and a reaction centre complex is called a *photosystem*. Plants have two photosystems that drive the reduction of $NADP^+$ by water (Box 10.1):

$$H_2O + NADP^+ \xrightarrow{\text{light, photosystems I and II}} \tfrac{1}{2}O_2 + NADPH + H^+$$

Light-harvesting complexes bind large numbers of pigments in order to provide a sufficiently large area for capture of radiation. In photosystems I and II, absorption of a photon raises a chlorophyll or carotenoid molecule to an excited singlet state and within 0.1–5 ps the energy hops to a nearby pigment via the Förster mechanism (Section 26.11e). About 100–200 ps later, which corresponds to thousands of hops within the light-harvesting complex, more than 90 per cent of the absorbed energy reaches the reaction centre. There, a chlorophyll *a* dimer becomes electronically excited and initiates ultrafast electron transfer reactions. For example, the transfer of an electron from the excited singlet state of P680, the chlorophyll dimer of the photosystem II reaction centre, to its immediate electron acceptor, a phaeophytin *a* molecule,[8] occurs within 3 ps. Once the excited state of P680 has been quenched efficiently by this first reaction, subsequent steps that lead to the oxidation of water and reduction of plastoquinone occur more slowly, with reaction times varying from 200 ps to 1 ms. The electrochemical reactions within the photosystem I reaction centre also occur in this time interval.

In summary, the initial energy and electron transfer events of photosynthesis are under tight kinetic control.

8. Phaeophytin *a* is a chlorophyll *a* molecule where the central Mg^{2+} ion is replaced by two protons, which are bound to two of the pyrrole nitrogens in the ring.

Photosynthesis captures solar energy efficiently because the excited singlet state of chlorophyll is quenched rapidly by processes that occur with relaxation times that are much shorter than the fluorescence lifetime, which is about 5 ns in diethyl ether at room temperature.

DNA damage by UV radiation

We saw in Box 26.1 that the depletion of ozone from the Earth's stratosphere increases the level of UVB radiation at the surface, with deleterious consequences for the biosphere. One example of a biological effect induced by UVB radiation is DNA damage. The principal mechanism of DNA damage involves the photodimerization of adjacent thymine bases (**3**) to yield either a cyclobutane thymine dimer (left) or a so-called 6–4 photoproduct (right). The former has been linked directly to cell death and the latter may lead to DNA mutations and, consequently, to the formation of tumours. There are several natural mechanisms for protection from and repair of photochemical damage. For example, the enzyme DNA photolyase catalyses the destruction of cyclobutane thymine dimers. Also, ultraviolet radiation can induce the production of the pigment melanin (in a process more commonly known as tanning), which shields the skin from damage. However, repair and protective mechanisms become increasingly less effective with persistent and prolonged exposure to solar radiation.

Photodynamic therapy of tumour

In photodynamic therapy (PDT), laser radiation, which is usually delivered to the tumour via a fibre optic cable, is absorbed by a drug which, in its first excited triplet state ^3P, photosensitizes the formation of an excited singlet state of O_2, $^1O_2$. The $^1O_2$ molecules are very reactive and destroy cellular components and it is thought that cell membranes are the primary cellular targets. Hence, the following photochemical cycle leads to the shrinkage (and sometimes total destruction) of tumours:

Absorption:	$P + h\nu \longrightarrow P^*$
Intersystem crossing:	$P^* \longrightarrow {}^3P$
Photosensitization:	${}^3P + {}^3O_2 \longrightarrow P + {}^1O_2$
Oxidation reactions:	${}^1O_2 + \text{reactants} \longrightarrow \text{products}$

The photosensitizer is hence a 'photocatalyst' for the production of $^1O_2$. It is common practice to use a porphyrin photosensitizer, such as compounds derived from haematoporphyrin (**4**). However, much effort is being expended to develop better drugs with enhanced photochemical properties. A potential PDT drug must meet many criteria. From the point of view of pharmacological effectiveness, the drug must be soluble in tissue fluids, so it can be transported to the tumour through blood and secreted from the body through urine. The therapy should also result in very few side-effects. The drug must also have unique photophysical and photochemical properties. It must be activated photochemically at wavelengths that are not absorbed by blood and skin. In practice, this means that the drug should have a strong absorption band at $\lambda > 650$ nm. Drugs based on haematoporphyrin do not meet this criterion very well, so much effort is being expended in the design of novel porphyrin and related macrocycles with more desirable electronic properties. At the same time, the quantum yield of triplet formation and of $^1O_2$ formation must be high, so many drug molecules can be activated and many tumour oxidation reactions can occur during a short period of laser irradiation. Photodynamic therapy has been used successfully in the treatment of a number of cancers, including those of the lung, bladder, skin, oesophagus, and retina.

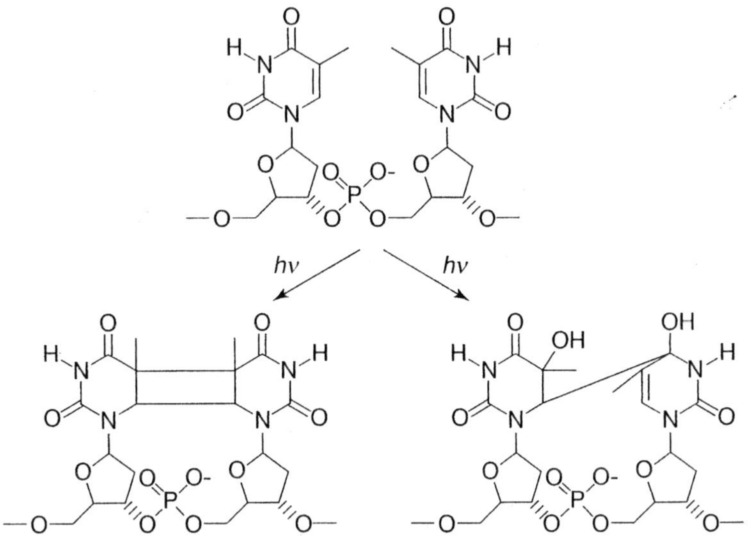

3

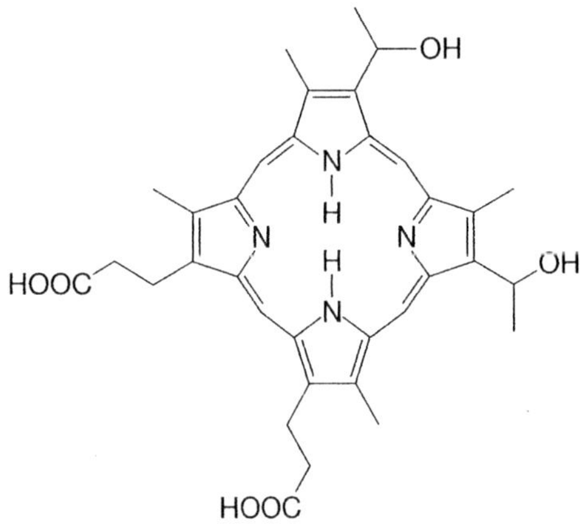

4

Discussion questions

1. The light-induced electron transfer reactions in photosynthesis occur because chlorophyll molecules (whether in monomeric or dimeric forms) are better reducing agents in their electronic excited states. Justify this observation with the help of molecular orbital theory.

2. The emission spectrum of a porphyrin dissolved in O_2-saturated water shows a strong band at 650 nm and a weak band at 1270 nm. In separate experiments, it was observed that the electronic absorption spectrum of the porphyrin sample showed bands at 420 nm and 550 nm, and the electronic absorption spectrum of O_2-saturated water showed no bands in the visible range of the spectrum (and therefore no emission spectrum when excited in the same range). Based on these data alone, make a preliminary assignment of the emission band at 1270 nm. Propose additional experiments that test your hypothesis. *Hint.* Review Section 17.3 and Problem 17.21.

Further reading

E. Kohen, R. Santus and J.G. Hirshberg, *Photobiology*. Academic Press, San Diego (1995).

J.B. Miller, Photodynamic therapy: the sensitization of cancer cells to light. *J. Chem. Educ.* **76**, 592 (1999).

and the fluorescence quantum yield in the presence of the quencher decreases to:

$$\phi = \frac{k_f}{k_f + k_{IC} + k_{ISC} + k_Q[Q]}$$

The ratio ϕ_f/ϕ is then given by:

$$\frac{\phi_f}{\phi} = 1 + \tau_0 k_Q[Q] \qquad (26.34)$$

Therefore, a plot of the left-hand side of this expression against [Q] should be a straight line with slope $\tau_0 k_Q$. Such a plot is called a **Stern–Volmer plot** (Fig. 26.25). The method is quite general and may also be applied to the quenching of phosphorescence emission.

Because the fluorescence intensity and lifetime are both proportional to the fluorescence quantum yield, plots of I_0/I and τ_0/τ (where the subscript 0 indicates a measurement in the absence of quencher) versus [Q] should also be linear with the same slope and intercept as those shown for eqn 34.

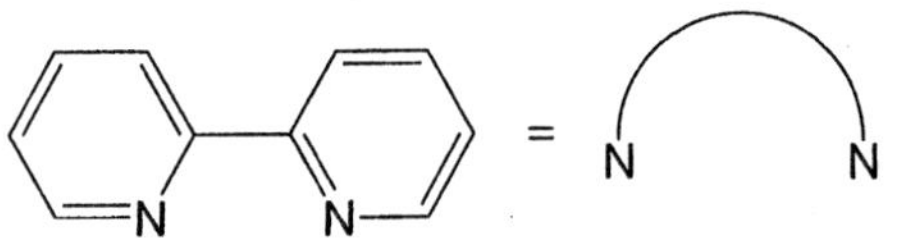

5

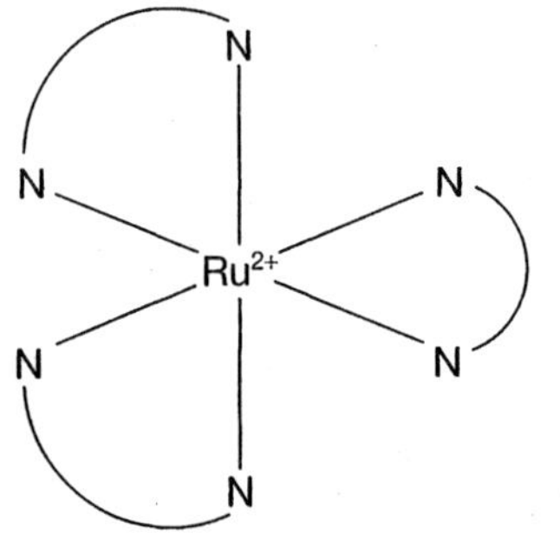

6

Example 26.5 Determining the quenching rate constant

The molecule 2,2′-bipyridine (**5**) forms a complex with the Ru^{2+} ion. Ruthenium(II) tris-(2,2′-bipyridyl), $Ru(bpy)_3^{2+}$ (**6**), has a strong metal-to-ligand charge transfer (MLCT) transition (Section 17.2) at 450 nm. The quenching of the $^*Ru(bpy)_3^{2+}$ excited state by $Fe(H_2O)_6^{3+}$ in acidic solution was monitored by measuring emission lifetimes at 600 nm. Determine the quenching rate constant for this reaction from the following data:

$[Fe(H_2O)_6^{3+}]/(10^{-4}\ mol\ L^{-1})$	0	1.6	4.7	7	9.4
$\tau/(10^{-7}\ s)$	6	4.05	3.37	2.96	2.17

Method Re-write the Stern–Volmer equation (eqn 34) for use with lifetime data; then fit the data to a straight-line, either graphically or by performing a linear-regression analysis with a computer.

Answer Upon substitution of τ_0/τ for ϕ_f/ϕ in eqn 34 and after rearrangement, we obtain:

$$\frac{1}{\tau} = \frac{1}{\tau_0} + k_Q[Q] \qquad (26.35)$$

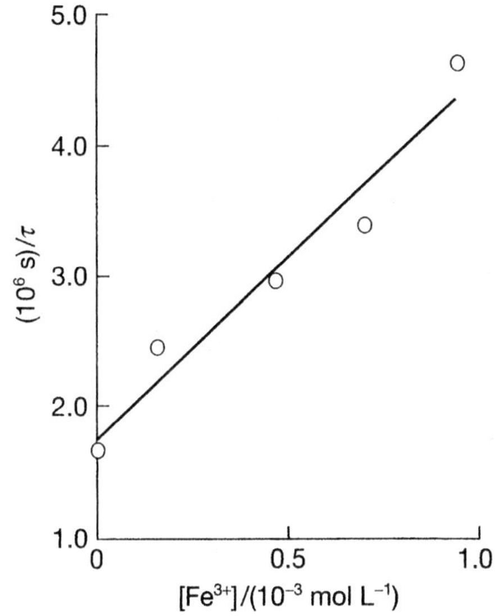

26.26 The Stern–Volmer plot of the data for Example 26.5.

Figure 26.26 shows a plot of $1/\tau$ versus $[Fe^{3+}]$ and the results of a fit to eqn 35. The slope of the line is $k_Q = 2.8 \times 10^9$ L mol^{-1} s^{-1}.

Comment This example shows that measurements of emission lifetimes are preferred because they yield the value of k_Q directly. To determine the value of k_Q from intensity or quantum yield measurements, we need to make an independent measurement of τ_0.

Self-test 26.5 From the data above, predict the value of $[Fe^{3+}]$ required to decrease the intensity of $Ru(bpy)_3^{2+}$ emission to 50 per cent of the unquenched value. [6.0×10^{-4} mol L^{-1}]

Three common mechanisms for bimolecular quenching of an excited singlet (or triplet) state are:

Collisional deactivation:	$S^* + Q \longrightarrow S + Q$
Energy transfer:	$S^* + Q \longrightarrow S + Q^*$
Electron transfer:	$S^* + Q \longrightarrow S^+ + Q^-$ or $S^- + Q^+$

Collisional quenching is particularly efficient when Q is a heavy species, such as iodide ion, which receives energy from S* and then decays primarily by internal conversion to the ground state. This fact may be used to determine the accessibility of amino acid residues of a folded protein to solvent. For example, fluorescence from a tryptophan residue ($\lambda_{abs} \approx 290$ nm, $\lambda_{fluor} \approx 350$ nm) is quenched by iodide ion when the residue is on the surface of the protein and hence accessible to the solvent. Conversely, residues in the hydrophobic interior of the protein are not quenched effectively by I^-.

The quenching rate constant itself does not give much insight into the mechanism of quenching. For the system of Example 26.5, it is known that the quenching of the excited state of $Ru(bpy)_3^{2+}$ is a result of light-induced electron transfer to Fe^{3+}, but the quenching data do not allow us to prove the mechanism. However, there are some criteria that govern the relative efficiencies of energy and electron transfer.

(e) Energy transfer processes

According to the **Förster theory** of resonance energy transfer, which was proposed by T. Förster in 1959, energy transfer is efficient when:

1 The energy donor and acceptor are separated by a short distance, in the nanometre scale.

2 Photons emitted by the excited state of the donor can be absorbed directly by the acceptor.

We show in the following *Justification* that, for donor–acceptor systems that are held rigidly either by covalent bonds or by a protein 'scaffold', the efficiency of energy transfer, E_T, increases with decreasing distance, R, according to

$$E_T = \frac{R_0^6}{R_0^6 + R^6} \tag{26.36}$$

where R_0 is a parameter (with units of distance) that is characteristic of each donor–acceptor pair. Equation 36 has been verified experimentally and values of R_0 are available for a number of donor–acceptor pairs (Table 26.3).

Table 26.3 Values of R_0 for some donor–acceptor pairs*

Donor†	Acceptor	R_0/nm
Naphthalene	Dansyl	2.2
Dansyl	ODR	4.3
Pyrene	Coumarin	3.9
IAEDANS	FITC	4.9
Tryptophan	IAEDANS	2.2
Tryptophan	Haem	2.9

* Additional values may be found in: J.R. Lacowicz in *Principles of fluorescence spectroscopy*, Kluwer Academic/Plenum Publishers, New York (1999), p. 388.
† Abbreviations: Dansyl, 5-dimethylamino-1-naphthalenesulfonic acid; FITC, fluorescein 5-isothiocyanate; IAEDANS, 5-(((2-iodoacetyl)amino)ethyl)amino)naphthalene-1-sulfonic acid; ODR, octadecyl-rhodamine.

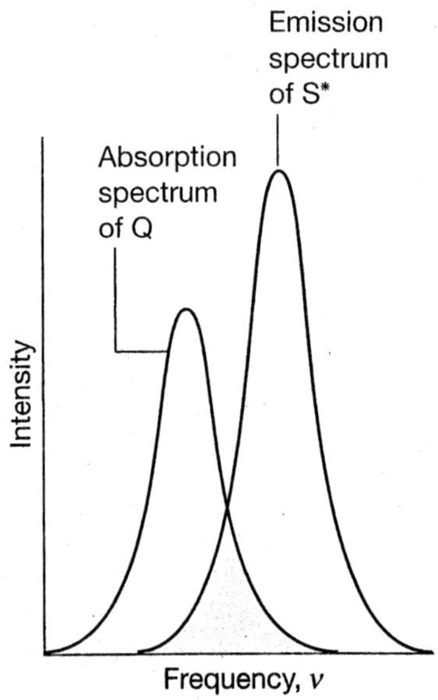

26.27 According to the Förster theory, the rate of energy transfer from a molecule S* in an excited state to a quencher molecule Q is optimized at radiation frequencies in which the emission spectrum of S* overlaps with the absorption spectrum of Q, as shown in the shaded region.

The emission and absorption spectra of molecules span a range of wavelengths (Section 16.3), so the second requirement of the Förster theory is met when the emission spectrum of the donor molecule overlaps significantly with the absorption spectrum of the acceptor. In the overlap region, photons emitted by the donor have the proper energy to be absorbed by the acceptor (Fig. 26.27).

In many cases, it is possible to prove that energy transfer is the predominant mechanism of quenching if the excited state of the acceptor fluoresces or phosphoresces at a characteristic wavelength. In a pulsed laser experiment, the rise in fluorescence intensity from Q* with a characteristic time which is the same as that for the decay of the fluorescence of S* is often taken as indication of energy transfer from S to Q.

Justification 26.4

We visualize the energy transfer process $S^* + Q \rightarrow S + Q^*$ as follows. The oscillating electric field of electromagnetic radiation induces an oscillating electric dipole moment, μ_S, in S. Energy will be absorbed by S if the frequency of the incident radiation, ν, is such that $\Delta E_S = h\nu$, where ΔE_S is the energy separation between the ground and excited electronic states of S and h is Planck's constant. This is the 'resonance condition' for absorption of radiation. The oscillating dipole on S now can affect electrons bound to a nearby Q molecule by inducing an oscillating dipole moment, μ_Q. If the frequency of oscillation of μ_S is such that $\Delta E_Q = h\nu$, then Q will absorb energy from S.

According to this picture, resonance energy transfer arises from the interaction between two oscillating dipoles with moments μ_S and μ_Q. From Section 21.5, the energy of the dipole–dipole interaction, $V_{\text{dipole–dipole}}$, is

$$V_{\text{dipole–dipole}} \propto \frac{\mu_S \mu_Q}{R^3}$$

where R is the distance between the dipoles. We saw in Section 12.10 that the rate of a transition from a state i to a state f at a radiation frequency ν is proportional to the square modulus of the matrix element of the perturbation between the two states:

$$w_{f \leftarrow i} \propto |H_{fi}^{(1)}|^2$$

For energy transfer, the wavefunctions of the initial and final states may be denoted as $\psi_S^* \psi_Q$ and $\psi_S \psi_Q^*$, respectively, and $H^{(1)}$ may be written from $V_{\text{dipole–dipole}}$. It follows that the rate of energy transfer, w_T, at a fixed distance R is given by

$$w_T \propto \frac{1}{R^6} |\langle \psi_S \psi_{Q^*} | \mu_S \mu_Q | \psi_{S^*} \psi_Q \rangle|^2 = \frac{1}{R^6} |\langle \psi_S | \mu_S | \psi_{S^*} \rangle|^2 |\langle \psi_{Q^*} | \mu_Q | \psi_Q \rangle|^2$$

where we have used the fact that the terms related to S are functions of coordinates that are independent of those for the functions related to Q. In the last expression, the integrals are squares of transition dipole moments at the radiation frequency ν, the first corresponding to emission of S* to S and the second to absorption of Q to Q*.

We interpret the expression for w_T as follows. The rate of energy transfer is proportional to R^{-6}, so it decreases sharply with increasing separation between the energy donor and acceptor. Furthermore, the energy transfer rate is optimized when both emission of radiation by S* and absorption of radiation by Q are efficient at the frequency ν. Because the absorption and emission spectra of large molecules in condensed phases are broad, it follows that the energy transfer rate is optimal at radiation frequencies in which the emission spectrum of the donor and the absorption spectrum of the acceptor overlap significantly.

In practice, it is more convenient to measure the efficiency of energy transfer and not the rate itself. In much the same way that we defined the quantum yield as a ratio of rates, we can also define the efficiency of energy transfer, E_T, as the ratio

$$E_T = \frac{w_T}{w_T + w_0} \qquad w_0 = (k_f + k_{IC} + k_{ISC})[S^*] \tag{26.37}$$

where w_0 is the rate of deactivation of S* in the absence of the quencher. It is useful to express w_0 in terms of the parameter R_0, the characteristic distance at which $w_T = w_0$ for a specified pair of S and Q (Table 26.3). Using $w_T \propto R^{-6}$ and $w_0 \propto R_0^{-6}$, the expression for E_T rearranges into eqn 36.

The dependence of E_T on R forms the basis of **fluorescence resonance energy transfer** (FRET), a technique that can be used to measure distances in biological systems. In a typical FRET experiment, a site on a biopolymer or membrane is covalently labelled with an energy donor and another site is labelled covalently with an energy acceptor. In certain cases, the donor or acceptor may be natural constituents of the system, such as amino acid groups, co-factors, or enzyme substrates (see Illustration 26.3).

The efficiency, E_T, may be expressed in terms of the experimental fluorescence quantum yields ϕ_f and ϕ of the donor in the absence and presence of the acceptor, respectively. To show this, we use eqn 32 to write:

$$\phi_f = \frac{v_f}{w_0} \quad \text{and} \quad \phi = \frac{v_f}{w_0 + w_T}$$

where v_f is the rate of fluorescence. Using eqn 37 and after a little algebra, we obtain

$$E_T = 1 - \frac{\phi}{\phi_f} \tag{26.38}$$

The distance between the labels is then calculated from the known value of R_0 and eqn 36. Several tests have shown that the FRET technique is useful for measuring distances ranging from 1 to 9 nm.

Illustration 26.3

The energy donor 1.5-I AEDANS (7) has a fluorescence quantum yield of 0.75 in aqueous solution. The visual pigment 11-*cis*-retinal (Box 17.1) is a quencher of fluorescence and R_0 = 5.4 nm for the 1.5-I AEDANS/11-*cis*-retinal pair. When an amino acid on the surface of the protein rhodopsin, which binds 11-*cis*-retinal in its interior, was labelled covalently with 1.5-AEDANS, the fluorescence quantum yield of the label decreased to 0.68. From eqn 38, we calculate $E_T = 1 - (0.68/0.75) = 0.093$ and from eqn 36 we calculate R = 7.9 nm, which is taken as the distance between the surface of the protein and 11-*cis*-retinal.

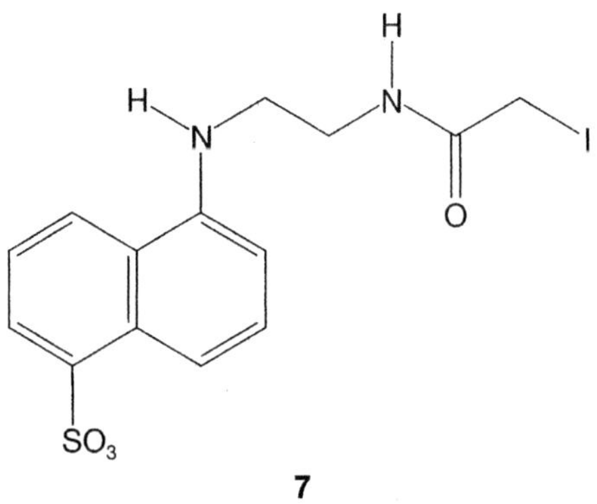

7

If donor and acceptor diffuse in solution or in the gas phase, Förster theory predicts that the efficiency of quenching by energy transfer increases as the average distance travelled between collisions of donor and acceptor decreases. That is, the quenching efficiency increases with concentration of quencher, as predicted by the Stern–Volmer equation.

(f) Electron transfer reactions

According to the **Marcus theory** of electron transfer, which was proposed by R.A. Marcus in 1965 and is discussed fully in Section 29.1, the rates of electron transfer (from ground or excited states) depend on:

1 The distance between the donor and acceptor, with electron transfer becoming more efficient as the distance between donor and acceptor decrease.

2 The reaction Gibbs energy, $\Delta_r G$, with electron transfer becoming more efficient as the reaction becomes more exergonic. For example, efficient photooxidation of S requires that the reduction potential of S* be lower than the reduction potential of Q.

3 The reorganization energy, the energy cost incurred by molecular rearrangements of donor, acceptor, and medium during electron transfer. The electron transfer rate is predicted to increase as this reorganization energy is matched closely by the reaction Gibbs energy.

Electron transfer can also be studied by time-resolved spectroscopy (Section 17.7e). The oxidized and reduced products often have electronic absorption spectra distinct from those of their neutral parent compounds. Therefore, the rapid appearance of such known features in the absorption spectrum after excitation by a laser pulse may be taken as indication of quenching by electron transfer.

26.12 Complex photochemical processes

Many photochemical processes have complex mechanisms that may be examined by considering the concepts developed above. In this section, we consider two examples: photochemical chain reactions and photosensitization.

(a) The overall quantum yield of a photochemical reaction

For complex reactions involving secondary processes, such as chain reactions initiated by photolysis, many reactant molecules might be consumed as a result of absorption of a single photon. The **overall quantum yield**, Φ, for such reactions, the number of reactant molecules consumed per photon absorbed, might exceed 1. In the photolysis of HI, for example, the processes are

$$\mathrm{HI} + h\nu \longrightarrow \mathrm{H\cdot} + \mathrm{I\cdot}$$

$$\mathrm{H\cdot} + \mathrm{HI} \longrightarrow \mathrm{H_2} + \mathrm{I\cdot}$$

$$\mathrm{I\cdot} + \mathrm{I\cdot} + \mathrm{M} \longrightarrow \mathrm{I_2} + \mathrm{M^*}$$

The overall quantum yield is 2 because the absorption of one photon leads to the destruction of two HI molecules. In a chain reaction, Φ may be very large, and values of about 10^4 are common. In such cases the chain acts as a chemical amplifier of the initial absorption step.

Example 26.6 Determining the quantum yield of a photochemical reaction

When a sample of 4-heptanone was irradiated for 100 s with 313 nm radiation with a power output of 50 W under conditions of total absorption, it was found that 2.8 mmol C_2H_4 was formed. What is the quantum yield for the formation of ethene?

Method First, calculate the amount of photons generated in an interval Δt: see Example 11.1. Then divide the amount of ethene molecules formed by the amount of photons absorbed.

Answer From Example 11.1, the amount (in moles) of photons absorbed is

$$n = \frac{P\Delta t}{(hc/\lambda)N_A}$$

If $n_{C_2H_4}$ is the amount of ethene formed, the quantum yield is

$$\Phi = \frac{n_{C_2H_4}}{n} = \frac{n_{C_2H_4}N_A hc}{\lambda P\Delta t}$$

$$= \frac{(2.8\times10^{-3}\ \text{mol})\times(6.022\times10^{23}\ \text{mol}^{-1})\times(6.626\times10^{-34}\ \text{J s})\times(2.997\times10^{8}\ \text{m s}^{-1})}{(3.13\times10^{-7}\ \text{m})\times(50\ \text{J s}^{-1})\times(100\ \text{s})}$$

$$= 0.21$$

Self-test 26.6 The overall quantum yield for another reaction at 290 nm is 0.30. For what length of time must irradiation with a 100 W source continue in order to destroy 1.0 mol of molecules? [3.8 h]

(b) Rate laws of complex photochemical reactions

As an example of how to write a rate law for a complex photochemical process, consider the photochemical activation of the reaction

$$H_2(g) + Br_2(g) \longrightarrow 2\,HBr(g)$$

In place of the first step in the thermal reaction we have

$$Br_2 + h\nu \longrightarrow Br\cdot + Br\cdot \qquad v = I_{abs}$$

where I_{abs} is the number of photons of the appropriate frequency absorbed divided by the volume in which absorption occurs and the time interval. We are assuming a primary quantum yield of unity for the photodissociation of Br_2. It follows that I_{abs} should take the place of $k_i[Br_2][M]$ in the thermal reaction scheme, so from Example 26.1 we can write

$$\frac{d[HBr]}{dt} = \frac{2k_p(1/k_t[M])^{1/2}[H_2][Br_2]I_{abs}^{1/2}}{[Br_2] + (k_r/k_p')[HBr]} \qquad (26.39)$$

We predict that the reaction rate should depend on the square root of the absorbed light intensity, which is confirmed experimentally.

(c) Photosensitization

The reactions of a molecule that does not absorb directly can be made to occur if another absorbing molecule is present, because the latter may be able to transfer its energy to the former during a collision. An example of this **photosensitization** is the reaction used to generate excited state O_2 in a type of tumour treatment known as *photodynamic therapy*. Another example is the reaction used to generate atomic hydrogen, the irradiation of hydrogen gas containing a trace of mercury vapour using radiation of wavelength 254 nm from a mercury discharge lamp. The Hg atoms are excited (to Hg*) by resonant absorption of the radiation, and then collide with H_2 molecules. Two reactions then take place:

$$Hg^* + H_2 \longrightarrow Hg + H\cdot + H\cdot$$

$$Hg^* + H_2 \longrightarrow HgH + H\cdot$$

The latter reaction is the initiation step for other mercury photosensitized reactions, such as the synthesis of formaldehyde from carbon monoxide and hydrogen:

$$H\cdot + CO \longrightarrow HCO\cdot$$
$$HCO\cdot + H_2 \longrightarrow HCHO + H\cdot$$
$$HCO\cdot + HCO\cdot \longrightarrow HCHO + CO$$

Note that the last step is termination by disproportionation rather than by combination.

Checklist of key ideas

Chain reactions

- ☐ chain reaction, a reaction in which an intermediate produced in one step generates an intermediate in a subsequent step, and so on
- ☐ chain carrier, the intermediates in a chain reaction
- ☐ radical chain reaction, a chain reaction in which the chain carriers are radicals

26.1 The rate laws of chain reactions

- ☐ pyrolysis, thermal decomposition in the absence of air
- ☐ Rice–Herzfeld mechanism, a chain reaction for the thermal decomposition of CH_3CHO
- ☐ initiation step, a step in which chain carriers are produced
- ☐ propagation step, a step in which a chain carrier reacts to produce another chain carrier
- ☐ termination step, a reaction in which radicals combine and end a chain or reactions
- ☐ hydrogen-bromine reaction, $d[HBr]/dt = k[H_2][Br_2]^{3/2}/([Br_2] + k'[HBr])$
- ☐ thermolysis, a reaction initiated by heat or by vigorous intermolecular collisions
- ☐ retardation step, a step in a chain reaction that reduces the net rate of formation of product

26.2 Explosions

- ☐ thermal explosion, a very rapid reaction arising from a rapid increase of reaction rate with increasing temperature
- ☐ chain-branching explosion, occurs when the number of chain centres grows exponentially
- ☐ branching step, an elementary reaction that produces more than one chain carrier
- ☐ explosion region, the conditions (of pressure and temperature) under which explosion occurs
- ☐ first explosion limit, the minimum pressure at a given temperature at which explosion occurs
- ☐ second explosion limit, the maximum pressure at a given temperature at which explosion occurs
- ☐ third explosion limit, the minimum pressure at a given temperature at which a thermal explosion occurs

Polymerization kinetics

- ☐ stepwise polymerization, a polymerization reaction in which any two monomers present in the reaction mixture can link together at any time and the growth of the polymer is not confined to chains that are already forming
- ☐ chain polymerization, a polymerization reaction in which an activated monomer attacks another monomer, links to it, then that unit attacks another monomer, and so on

26.3 Stepwise polymerization

- ☐ degree of polymerization, the average number of monomer residues per polymer molecule, $\langle n\rangle = 1/(1-p)$, where p is the average number of monomers per polymer molecule; $\langle n\rangle = 1 + kt[A]_0$

26.4 Chain polymerization

- ☐ rate of chain polymerization, $v = k[I]^{1/2}[M]$
- ☐ mutual termination, an elementary reaction in which two growing radical chains combine
- ☐ disproportionation (in polymerization), an elementary reaction in which a hydrogen atom transfers from one chain to another
- ☐ chain transfer, an elementary reaction in which a new chain initiates at the expense of the one currently growing
- ☐ kinetic chain length, ν, the ratio of the number of monomer units consumed per activated centre produced in the initiation step; $\nu = k[M][I]^{-1/2}$

Homogeneous catalysis

- ☐ catalyst, a substance that accelerates a reaction but undergoes no net chemical change
- ☐ enzyme, a biological catalyst
- ☐ homogeneous catalyst, a catalyst in the same phase as the reaction mixture
- ☐ heterogeneous catalyst, a catalyst in a different phase from the reaction mixture

26.5 Features of homogeneous catalysis

- ☐ acid catalysis, catalysis by the transfer of a proton from an acid to the substrate
- ☐ base catalysis, catalysis by the transfer of a proton from the substrate to a base

26.6 Enzymes

- ☐ active site, the region of an enzyme molecule at which reaction takes place
- ☐ substrate, the species that reacts in the presence of an enzyme
- ☐ maximum velocity, the greatest reaction rate for a given concentration of substrate; $v_{max} = k_b[E]_0$
- ☐ Michaelis–Menten mechanism, a mechanism for enzyme-catalysed reactions, $E + S \rightleftarrows ES$, $ES \rightarrow P$
- ☐ Michaelis–Menten equation, $v = v_{max}/(1 + K_M/[S]_0)$
- ☐ Michaelis constant, $K_M = (k_a' + k_b)/k_a$
- ☐ Lineweaver–Burk plot, $1/v = 1/v_{max} + (K_M/v_{max})(1/[S]_0)$

- ☐ turnover number (catalytic constant), the number of catalytic cycles (turnovers) performed by the active site in a given interval divided by the duration of the interval; $k_{cat} = v_{max}/[E]_0$
- ☐ catalytic efficiency, $\varepsilon = k_{cat}/K_M$
- ☐ enzyme inhibition, $v = v_{max}/(\alpha' + \alpha K_M/[S]_0)$, $\alpha = 1 + [I]/K_I$, $\alpha' = 1 + [I]/K_I'$
- ☐ competitive inhibition, inhibition in which the inhibitor binds only to the active site of the enzyme; $\alpha > 1$ and $\alpha' = 1$
- ☐ uncompetitive inhibition, inhibition in which the inhibitor binds to a site of the enzyme that is removed from the active site, but only if the substrate is already present; $\alpha = 1$ and $\alpha' > 1$
- ☐ non-competitive inhibition, inhibition in which the inhibitor binds to a site other than the active site; $\alpha > 1$ and $\alpha' > 1$

Oscillating reactions

26.7 Autocatalysis

- ☐ autocatalysis, the catalysis of a reaction by the products
- ☐ Belousov–Zhabotinskii reaction, an oscillating reaction involving potassium bromate, malonic acid, and a cerium(IV) salt in acidic solution

26.8 Autocatalytic mechanisms of oscillating reactions

- ☐ Lotka–Volterra mechanism, a model of a reaction showing oscillating behaviour
- ☐ steady-state condition, a non-equilibrium state in which the time behaviour of a reaction mixture is maintained (with a composition that may be oscillating steadily)
- ☐ brusselator, a model of a reaction showing oscillating behaviour
- ☐ limit cycle (attractor), a common trajectory to which a system migrates
- ☐ oregonator, a model of a reaction showing oscillating behaviour; a mechanism for the BZ reaction

26.9 Bistability

- ☐ bistability, the ability to exist in two steady states

26.10 Chemical chaos

- ☐ deterministic chaos, temporal and spatial solutions that are predictable but infinitely sensitive to the initial conditions
- ☐ period-doubling, a process in which the system must circulate twice round the cycle before an initial pair of concentrations is restored
- ☐ strange attractor, a path that never retraces its path or crosses itself

Photochemistry

26.11 Kinetics of photophysical and photochemical processes

- ☐ primary process, a process in which products are formed directly from the excited state of a reactant
- ☐ secondary process, a process in which products originate from intermediates formed directly from the excited state of a reactant
- ☐ primary quantum yield, ϕ, the number of photophysical or photochemical events that lead to primary products divided by the number of photons absorbed by the molecule in the same interval, $\phi = v/I_{abs}$
- ☐ observed fluorescence lifetime, $\tau_0 = \phi_f/k_f$
- ☐ quenching, shortening of the lifetime of an excited state
- ☐ Stern–Volmer plot, $\phi_f/\phi = 1 + \tau_0 k_Q[Q]$
- ☐ Förster theory, a theory of resonance energy transfer, efficiency $E_T = R_0^6/(R_0^6 + R^6)$
- ☐ fluorescence resonance energy transfer (FRET), a technique used to measure distances in biological systems
- ☐ Marcus theory, a theory of electron transfer

26.12 Complex photochemical processes

- ☐ overall quantum yield, the number of reactant molecules consumed per photon absorbed
- ☐ photosensitization, a process in which a photon is absorbed by a second species and then the excitation energy is transferred to the molecule of interest

Further reading

Articles of general interest

J. Andraos, How mathematics figures in chemistry: some examples. *J. Chem. Educ.* **76**, 258 (1999).

S. Bluestone and K.Y. Yan, A method to find the rate constants in chemical kinetics of a complex reaction. *J. Chem. Educ.* **72**, 884 (1995).

J.G. Eberhardt and E. Levin, A simplified integration technique for reaction rate laws of integral order in several substances. *J. Chem. Educ.* **72**, 193 (1995).

I.R. Epstein, The role of flow systems in far-from-equilibrium dynamics. *J. Chem. Educ.* **66**, 191 (1989).

R.J. Field, The language of dynamics. *J. Chem. Educ.* **66**, 188 (1989).

R.J. Field and F.W. Schneider, Oscillating chemical reactions and nonlinear dynamics. *J. Chem. Educ.* **66**, 195 (1989).

G.I. Gellene, Application of kinetic approximations to the A + B ⇄ C reaction system. *J. Chem. Educ.* **72**, 196 (1995).

J.M. Goodman, How do approximations affect the solutions to kinetic equations? *J. Chem. Educ.* **76**, 275 (1999).

E. Hughes, Jr., Solving differential equations in kinetics by using power series. *J. Chem. Educ.* **66**, 46 (1989).

W. Jahnke and A.T. Winfree, Recipes for Belousov–Zhabotinskii reagents. *J. Chem. Educ.* **68**, 320 (1991).

K.J. Laidler, Rate-controlling step: a necessary or useful concept? *J. Chem. Educ.* **65**, 250 (1988).

S.R. Logan, Does a photochemical reaction have a reaction order? *J. Chem. Educ.* **74**, 1303 (1997).

F. Mata-Perez and J.F. Perez-Benito, The kinetic rate law for autocatalytic reactions. *J. Chem. Educ.* **64**, 925 (1987).

D.B. Northrop, On the meaning of K_m and V/K in enzyme kinetics. *J. Chem. Educ.* **75**, 1153 (1998).

R.M. Noyes, Some models of chemical oscillators. *J. Chem. Educ.* **66**, 190 (1989).

R.S. Ochs, Understanding enzyme inhibition. *J. Chem. Educ.* **77**, 1453 (2000).

E. Peacock-López, Introduction to chemical oscillation using a modified Lotka model. *Chem. Educator* **5**, 216 (2000).

R.T. Raines and D.E. Hansen, An intuitive approach to steady-state kinetics. *J. Chem. Educ.* **65**, 757 (1988).

H.G. Schuster, Chaotic phenomena. In *Encyclopedia of applied physics* (ed. G.L. Trigg), **3**, 189. VCH, New York (1992).

A.F. Shaaban, The integrated rate equation of the nitric oxide–oxygen reaction. *J. Chem. Educ.* **67**, 869 (1990).

L.J. Solzberg, Self-organization in chemistry. *J. Chem. Educ.* **66**, 187 (1989).

J.N. Spencer, Competitive and coupled reactions. *J. Chem. Educ.* **69**, 281 (1992).

G.F. Swiegers, Applying the principles of chemical kinetics to population growth problems. *J. Chem. Educ.* **70**, 364 (1993).

X. Tan, S. Lindenbaum, and N. Meltzer, A unified equation for chemical kinetics. *J. Chem. Educ.* **71**, 566 (1994).

S. Toby and F.S. Toby, The simulation of dynamic systems. *J. Chem. Educ.* **76**, 1584 (1999).

Texts and sources of data and information

M. Boudart, *Kinetics of chemical processes.* Butterworth, London (1991).

C.H. Bamford, C.F. Tipper, and R.G. Compton (ed.), *Comprehensive chemical kinetics*, Vols 1–38. Elsevier, Amsterdam (1969–2001).

M.P. Cady and C.A. Trapp, *A Mathcad primer for physical chemistry*. Oxford University Press (1999).

I.M. Campbell, *Introduction to synthetic polymers*. Oxford University Press (1997).

C.E. Carraher, Jr., *Seymour/Carraher's polymer chemistry*. Marcel Dekker (2000).

I.R. Epstein and J.A. Pojman, *An introduction to nonlinear chemical dynamics*. Oxford University Press (1998).

A. Fersht, *Structure and mechanism in protein science: a guide to enzyme catalysis and protein folding*. W.H. Freeman, New York (1999).

W.M. Horspool and P.-S. Song (ed.), *CRC handbook of organic photochemistry and photobiology*. CRC Press, Boca Raton (1995).

D.R. Lide (ed.), *CRC handbook of chemistry and physics*, Section 5. CRC Press, Boca Raton (2000).

J.C. Lindon, G.E. Tranter, and J.L. Holmes (ed.), *Encyclopedia of spectroscopy and spectrometry*. Academic Press, San Diego (2000).

S.R. Logan, *Chemical kinetics*. Longman, Harlow (1996).

S.L. Murov, I. Carmichael, and G.L. Hug, *Handbook of photochemistry*. Marcel Dekker, New York (1993).

NDRL/NIST solution kinetics database, NIST standard reference database 40, National Institute of Standards and Technology, Gaithersburg (1994). The URL is available on the web site for this book.

NIST chemical kinetics database, NIST standard reference database 17, National Institute of Standards and Technology, Gaithersburg (1998). The URL is available on the web site for this book.

M.J. Pilling and P.W. Seakins, *Reaction kinetics*. Oxford University Press (1996).

S.K. Scott, *Oscillations, waves, and chaos in chemical kinetics*. Oxford Chemistry Primers, Oxford University Press (1994).

J.I. Steinfeld, J.S. Francisco, and W.L. Hase, *Chemical kinetics and dynamics*. Prentice Hall, Englewood Cliffs (1998).

N.J. Turro, *Modern molecular photochemistry*. University Science Books, Sausalito (1991).

G.R. Van Hecke and K.K. Karukstis, *A guide to lasers in chemistry*. Jones and Bartlett, Boston (1998).

K.E. van Holde, W.C. Johnson and P.H. Ho, *Principles of physical biochemistry*. Prentice Hall, Upper Saddle River (1998).

D. Voet and J.G. Voet, *Biochemistry*. Wiley, New York (1995).

R.P. Wayne, *Principles and applications of photochemistry*. Oxford University Press (1988).

C.E. Wayne and R.P. Wayne, *Photochemistry*. Oxford Chemistry Primers, Oxford University Press (1996).

Exercises

Discussion questions

26.1 (a) Identify any initiation, propagation, retardation, inhibition, and termination steps in the following chain mechanisms:

(a) (1) $AH \longrightarrow A\cdot + H\cdot$

(2) $A\cdot \longrightarrow B\cdot + C$

(3) $AH + B\cdot \longrightarrow A\cdot + D$

(4) $A\cdot + B\cdot \longrightarrow P$

(b) (1) $A_2 \longrightarrow A\cdot + A\cdot$

(2) $A\cdot \longrightarrow B\cdot + C$

(3) $A\cdot + P \longrightarrow B\cdot$

(4) $A\cdot + B\cdot \longrightarrow P$

26.1 (b) Bearing in mind distinctions between the mechanisms of stepwise and chain polymerization, describe ways in which it is possible to control the molar mass of a polymer by manipulating the kinetic parameters of polymerization.

26.2 (a) Discuss the features, advantages, and limitations of the Michaelis–Menten mechanism of enzyme action.

26.2 (b) Distinguish between competitive, non-competitive, and uncompetitive inhibition of enzymes. Discuss how these modes of inhibition may be detected experimentally.

26.3 (a) How can autocatalysis lead to chemical oscillations?

26.3 (b) Distinguish between a steady-state approximation and a steady-state condition. Discuss the role of bistability.

26.4 (a) Distinguish between the primary quantum yield and overall quantum yield of a chemical reaction. Describe an experimental procedure for the determination of the quantum yield.

26.4 (b) Discuss experimental procedures that make it possible to differentiate between quenching by energy transfer, collisions, or electron transfer.

Numerical exercises

In the following exercises and problems, it is recommended that rate constants are labelled with the number of the step in the proposed reaction mechanism, and any reverse steps are labelled similarly but with a prime.

26.5 (a) Derive the rate law for the decomposition of ozone in the reaction $2\,O_3(g) \rightarrow 3\,O_2(g)$ on the basis of the following proposed mechanism:

(1) $O_3 \rightleftharpoons O_2 + O$ $\quad k_1, k_1'$

(2) $O + O_3 \longrightarrow O_2 + O_2$ $\quad k_2$

26.5 (b) On the basis of the following proposed mechanism, account for the experimental fact that the rate law for the decomposition $2\,N_2O_5(g) \rightarrow 4\,NO_2(g) + O_2(g)$ is $v = k[N_2O_5]$.

(1) $N_2O_5 \rightleftharpoons NO_2 + NO_3$ $\quad k_1, k_1'$

(2) $NO_2 + NO_3 \longrightarrow NO_2 + O_2 + NO$ $\quad k_2$

(3) $NO + N_2O_5 \longrightarrow NO_2 + NO_2 + NO_2$ $\quad k_3$

26.6 (a) A slightly different mechanism for the decomposition of N_2O_5 from that in Exercise 26.5b has also been proposed. It differs only in the last step, which is replaced by

(3) $NO + NO_3 \longrightarrow NO_2 + NO_2$ $\quad k_3$

Show that this mechanism leads to the same overall rate law.

26.6 (b) Consider the following mechanism for the thermal decomposition of R_2:

(1) $R_2 \longrightarrow R + R$

(2) $R + R_2 \longrightarrow P_B + R'$

(3) $R' \longrightarrow P_A + R$

(4) $R + R \longrightarrow P_A + P_B$

where R_2, P_A, P_B are stable hydrocarbons and R and R′ are radicals. Find the dependence of the rate of decomposition of R_2 on the concentration of R_2.

26.7 (a) Refer to Fig. 26.3 and determine the pressure range for a chain-branching explosion in the hydrogen–oxygen reaction at 800 K.

26.7 (b) Refer to Fig. 26.3 and determine the pressure range for a chain-branching explosion in the hydrogen–oxygen reaction at (a) 700 K, (b) 900 K.

26.8 (a) The condensation reaction of propanone, $(CH_3)_2CO$, in aqueous solution is catalysed by bases, B, which react reversibly with propanone to form the carbanion $C_3H_5O^-$. The carbanion then reacts with a molecule of propanone to give the product. A simplified version of the mechanism is

(1) $AH + B \longrightarrow BH^+ + A^-$

(2) $A^- + BH^+ \longrightarrow AH + B$

(3) $A^- + AH \longrightarrow$ product

where AH stands for propanone and A^- denotes its carbanion. Use the steady-state approximation to find the concentration of the carbanion and derive the rate equation for the formation of the product.

26.8 (b) Consider the acid-catalysed reaction

(1) $HA + H^+ \rightleftharpoons HAH^+$ $\quad k_1, k_1'$, both fast

(2) $HAH^+ + B \longrightarrow BH^+ + AH$ $\quad k_2$, slow

Deduce the rate law and show that it can be made independent of the specific term $[H^+]$.

26.9 (a) Consider the following chain mechanism:

(1) $AH \longrightarrow A\cdot + H\cdot$

(2) $A\cdot \longrightarrow B\cdot + C$

(3) $AH + B\cdot \longrightarrow A\cdot + D$

(4) $A\cdot + B\cdot \longrightarrow P$

Use the steady-state approximation to deduce that the decomposition of AH is first-order in AH.

26.9 (b) Consider the following chain mechanism:

(1) $A_2 \longrightarrow A\cdot + A\cdot$

(2) $A\cdot \longrightarrow B\cdot + C$

(3) $A\cdot + P \longrightarrow B\cdot$

(4) $A\cdot + B\cdot \longrightarrow P$

Use the steady-state approximation to deduce that the rate law for the consumption of A_2.

26.10 (a) The enzyme-catalysed conversion of a substrate at 25°C has a Michaelis constant of 0.035 mol L^{-1}. The rate of the reaction is 1.15×10^{-3} mol L^{-1} s^{-1} when the substrate concentration is 0.110 mol L^{-1}. What is the maximum velocity of this enzymolysis?

26.10 (b) The enzyme-catalysed conversion of a substrate at 25°C has a Michaelis constant of 0.042 mol L^{-1}. The rate of the reaction is 2.45×10^{-4} mol L^{-1} s^{-1} when the substrate concentration is 0.890 mol L^{-1}. What is the maximum velocity of this enzymolysis?

26.11 (a) In a photochemical reaction $A \rightarrow 2\,B + C$, the quantum efficiency with 500 nm light is 2.1×10^2 mol einstein^{-1}. After exposure of 300 mmol of A to the light, 2.28 mmol of B is formed. How many photons were absorbed by A?

26.11 (b) In a photochemical reaction $A \rightarrow B + C$, the quantum efficiency with 500 nm light is 1.2×10^2 mol einstein^{-1}. After exposure of 200 mmol A to the light, 1.77 mmol B is formed. How many photons were absorbed by A?

26.12 (a) In an experiment to measure the quantum efficiency of a photochemical reaction, the absorbing substance was exposed to 490 nm light from a 100 W source for 45 min. The intensity of the transmitted light was 40 per cent of the intensity of the incident light. As a result of irradiation, 0.344 mol of the absorbing substance decomposed. Determine the quantum efficiency.

26.12 (b) In an experiment to measure the quantum efficiency of a photochemical reaction, the absorbing substance was exposed to 320 nm radiation from a 87.5 W source for 28.0 min. The intensity of the transmitted light was 0.257 that of the incident light. As a result of irradiation, 0.324 mol of the absorbing substance decomposed. Determine the quantum efficiency.

Problems†

Numerical problems

26.1 Studies of combustion reactions depend on knowing the concentrations of H atoms and HO radicals. Measurements on a flow system using ESR for the detection of radicals gave information on the reactions

(1) $H + NO_2 \longrightarrow OH + NO$ $\quad k_1 = 2.9 \times 10^{10}\ L\ mol^{-1}\ s^{-1}$

(2) $OH + OH \longrightarrow H_2O + O$ $\quad k_2 = 1.55 \times 10^{9}\ L\ mol^{-1}\ s^{-1}$

(3) $O + OH \longrightarrow O_2 + H$ $\quad k_3 = 1.1 \times 10^{10}\ L\ mol^{-1}\ s^{-1}$

(J.N. Bradley, W. Hack, K. Hoyermann, and H.G. Wagner, *J. Chem. Soc. Faraday Trans.* I, 1889 (1973)). Using initial H atom and NO_2 concentrations of 4.5×10^{-10} mol cm^{-3} and 5.6×10^{-10} mol cm^{-3}, respectively, compute and plot curves showing the O, O_2, and OH concentrations as a function of time in the range 0–10 ns.

26.2 In a flow study of the reaction between O atoms and Cl_2 (J.N. Bradley, D.A. Whytock, and T.A. Zaleski, *J. Chem. Soc. Faraday Trans.* I, 1251 (1973)) at high chlorine pressures, plots of $\ln[O]_0/[O]$ against distances l along the flow tube, where $[O]_0$ is the oxygen concentration at zero chlorine pressure, gave straight lines. Given the flow velocity as 6.66 m s^{-1} and the data below, find the rate coefficient for the reaction $O + Cl_2 \rightarrow ClO + Cl$.

l/cm	0	2	4	6	8
$\ln([O]_0/[O])$	0.27	0.31	0.34	0.38	0.45
l/cm	10	12	14	16	18
$\ln([O]_0/[O])$	0.46	0.50	0.55	0.56	0.60

with $[O]_0 = 3.3 \times 10^{-8}$ mol L^{-1}, $[Cl_2] = 2.54 \times 10^{-7}$ mol L^{-1}, $p = 1.70$ Torr.

26.3‡ J.D. Chapple-Sokol, C.J. Giunta, and R.G. Gordon (*J. Electrochem. Soc.* **136**, 2993 (1989)) proposed the following radical chain mechanism for the initial stages of the gas-phase oxidation of silane by nitrous oxide:

(1) $N_2O \longrightarrow N_2 + O$

(2) $O + SiH_4 \longrightarrow SiH_3 + OH$

(3) $OH + SiH_4 \longrightarrow SiH_3 + H_2O$

(4) $SiH_3 + N_2O \longrightarrow SiH_3O + N_2$

(5) $SiH_3O + SiH_4 \longrightarrow SiH_3OH + SiH_3$

(6) $SiH_3 + SiH_3O \longrightarrow (H_3Si)_2O$

Label each step with its role in the chain. Use the steady-state approximation to show that this mechanism predicts the following rate law for SiH_4 consumption (provided k_1 and k_6 are in some sense small):

$$\frac{d[SiH_4]}{dt} = -k[N_2O][SiH_4]^{1/2}$$

26.4‡ Arylmercury compounds have played an ever-increasing role in organic synthesis in recent years because of their ability to accommodate almost all important organic functional groups and their chemical and thermal stability. Y. Wang (*Int. J. Chem. Kinet.* **25**, 91 (1993)) has studied the kinetics and mechanism of the dimerization of arylmercurials (A), with formulas XC_6H_4HgCl (X = H, p-CH_3, m-CH_3, o-CH_3, p-Cl, and others) catalysed by $[ClRh(CO)_2]_2$ (C) in hexamethylphosphoramide (HMPA). The observed dimerization in each case is represented by the equation $2\ ArHgCl \rightarrow ArAr + HgCl_2 + Hg$, and plots of $x/([A]_0 - x)$, where x is the molar concentration of the arylmercury compound reacted and $[A]_0$ the initial concentration, were found to be linear functions of time. In each case, $[A]_0 = 0.100$ mol L^{-1} and $[C] = 5.00 \times 10^{-4}$ mol L^{-1}. Concentration data as a function of time at 60°C for the compounds with X = H and Cl are

X = H $x/([A]_0 - x)$	0.14	0.48	0.63	0.84	1.12
t/min	25	85	115	150	205
X = p-Cl $x/([A]_0 - x)$	0.04	0.27	0.60	1.05	
t/min	30	123	236	405	

Similar results were obtained for the other compounds in the series. (a) Plot these data and confirm that the reactions are second-order. Would the first-order integrated rate law fit the data too? (b) The proposed mechanism for these reactions is

(1) $[ClRh(CO)_2]_2 + ArHgCl \rightleftharpoons [ArRh(CO)_2]_2 + HgCl_2$ $\quad k_1, k_1'$, fast

(2) $[ArRh(CO)_2]_2 + ArHgCl \rightleftharpoons Ar_2Rh(CO)_2HgCl$ $\quad k_2, k_2'$, slow

(3) $Ar_2Rh(CO)_2HgCl \longrightarrow [ClRh(CO)_2]_2 + ArAr + Hg$ $\quad k_3$, fast

Derive the rate law for the rate of disappearance of ArHgCl and determine the conditions under which this mechanism supports the experimental results.

26.5‡ The water formation reaction has been studied many times and continues to be of interest. Despite the many studies there is not uniform agreement on the mechanism. But, as explosions are known to occur at certain critical values of the pressure, any proposed mechanism that is to be considered plausible must be consistent with the existence of these critical explosion limits. One such plausible mechanism is that of Example 26.2. Another is the following

(1) $H_2 \longrightarrow H + H$

(2) $H + O_2 \longrightarrow OH + O$

(3) $O + H_2 \longrightarrow OH + H$

(4) $H + O_2 \longrightarrow HO_2$

(5) $HO_2 + H_2 \longrightarrow H_2O + OH$

(6) $HO_2 + \text{wall} \longrightarrow \text{destruction}$

(7) $H + M \longrightarrow \text{destruction}$

In a manner similar to that in Example 26.2, determine whether or not this mechanism can lead to explosions under appropriate conditions.

26.6‡ For many years the reaction $H_2(g) + I_2(g) \rightarrow 2\ HI(g)$ and its reverse were assumed to be elementary bimolecular reactions. However, J.H. Sullivan (*J. Chem. Phys.* **46**, 73 (1967)) suggested that the following mechanism for the reaction, originally proposed by M. Bodenstein (*Z. Physik. Chem.* **29**, 56 (1898)), provides a better explanation of the experimental results:

(1) $I_2 \longrightarrow I + I$ $\quad k_1, k_1'$

(2) $I + I + H_2 \longrightarrow HI + HI$ $\quad k_2$

Obtain the expression for the rate of formation of HI based on this mechanism. Under what conditions does this rate law reduce to the one for the originally accepted mechanism?

† Problems denoted with the symbol ‡ were supplied by Charles Trapp, Carmen Giunta, and Marshall Cady.

26.7 The number of photons falling on a sample can be determined by a variety of methods, of which the classical one is chemical actinometry. The decomposition of oxalic acid $(COOH)_2$, in the presence of uranyl sulfate, $(UO_2)SO_4$, proceeds according to the sequence

(1) $UO^{2+} + h\nu \longrightarrow (UO^{2+})^*$

(2) $(UO^{2+})^* + (COOH)_2 \longrightarrow UO^{2+} + H_2O + CO_2 + CO$

with a quantum efficiency of 0.53 at the wavelength used. The amount of oxalic acid remaining after exposure can be determined by titration (with $KMnO_4$) and the extent of decomposition used to find the number of incident photons. In a particular experiment, the actinometry solution consisted of 5.232 g anhydrous oxalic acid, 25.0 cm^3 water (together with the uranyl salt). After exposure for 300 s the remaining solution was titrated with 0.212 M $KMnO_4$(aq), and 17.0 cm^3 was required for complete oxidation of the remaining oxalic acid. What is the rate of incidence of photons at the wavelength of the experiment? Express the answer in photons/second and einstein/second.

26.8 When benzophenone is illuminated with ultraviolet light it is excited into a singlet state. This singlet changes rapidly into a triplet, which phosphoresces. Triethylamine acts as a quencher for the triplet. In an experiment in methanol as solvent, the phosphorescence intensity varied with amine concentration as shown below. A time-resolved laser spectroscopy experiment had also shown that the half-life of the fluorescence in the absence of quencher is 29 μs. What is the value of k_q?

$[Q]/(\text{mol L}^{-1})$	0.0010	0.0050	0.0100
I_f/(arbitrary units)	0.41	0.25	0.16

26.9 An electronically excited state of Hg can be quenched by N_2 according to

$$Hg^*(g) + N_2(g, v = 0) \longrightarrow Hg(g) + N_2(g, v = 1)$$

in which energy transfer from Hg^* excites N_2 vibrationally. Fluorescence lifetime measurements of samples of Hg with and without N_2 present are summarized below (T = 300 K):

p_{N_2} = 0.0 atm

Relative fluorescence intensity	1.000	0.606	0.360	0.22	0.135
τ/μs	0.0	5.0	10.0	15.0	20.0

$p_{N_2} = 9.74 \times 10^{-4}$ atm

Relative fluorescence intensity	1.000	0.585	0.342	0.200	0.117
τ/μs	0.0	3.0	6.0	9.0	12.0

You may assume that all gases behave ideally. Determine the rate constant for the energy transfer process.

26.10 The Förster theory of resonance energy transfer and the basis for the FRET technique can be tested by performing fluorescence measurements on a series of compounds in which an energy donor and an energy acceptor are covalently linked by a rigid molecular linker of variable and known length. L. Stryer and R.P. Haugland (*Proc. Natl. Acad. Sci. USA* **58**, 719 (1967)) collected the following data on a family of compounds with the general composition dansyl-(L-prolyl)$_n$-naphthyl, in which the distance R between the naphthyl donor and the dansyl acceptor was varied from 1.2 nm to 4.6 nm by increasing the number of prolyl units in the linker:

R/nm	1.2	1.5	1.8	2.8	3.1	3.4	3.7	4.0	4.3	4.6
E_T	0.99	0.94	0.97	0.82	0.74	0.65	0.40	0.28	0.24	0.16

Are the data described adequately by eqn 36? If so, what is the value of R_0 for the naphthyl-dansyl pair?

Theoretical problems

26.11 The Rice–Herzfeld mechanism for the dehydrogenation of ethane is specified in Section 26.1, and it was noted there that it led to first-order kinetics. Confirm this remark, and find the approximations that lead to the rate law quoted there. How may the conditions be changed so that the reaction shows different orders?

26.12 The following mechanism has been proposed for the thermal decomposition of acetaldehyde (ethanal):

(1) $CH_3CHO \longrightarrow \cdot CH_3 + CHO$

(2) $\cdot CH_3 + CH_3CHO \longrightarrow CH_4 + \cdot CH_2CHO$

(3) $\cdot CH_2CHO \longrightarrow CO + \cdot CH_3$

(4) $\cdot CH_3 + \cdot CH_3 \longrightarrow CH_3CH_3$

Find an expression for the rate of formation of methane and the rate of disappearance of acetaldehyde.

26.13 Express the root mean square deviation $\{\langle M^2\rangle - \langle \bar{M}\rangle^2\}^{1/2}$ of the molar mass of a condensation polymer in terms of p, and deduce its time dependence.

26.14 Calculate the ratio of the mean cube molar mass to the mean square molar mass in terms of (a) the fraction p, (b) the chain length.

26.15 Calculate the average polymer length in a polymer produced by a chain mechanism in which termination occurs by a disproportionation reaction of the form $M\cdot + \cdot M \rightarrow M + {:}M$.

26.16 Derive an expression for the time dependence of the degree of polymerization for a stepwise polymerization in which the reaction is acid-catalysed by the —COOH acid functional group. The rate law is $d[A]/dt = -k[A]^2[OH]$.

26.17 An autocatalytic reaction $A \rightarrow P$ is observed to have the rate law $d[P]/dt = k[A]^2[P]$. Solve the rate law for initial concentrations $[A]_0$ and $[P]_0$. Calculate the time at which the rate reaches a maximum.

26.18 Another reaction with the stoichiometry $A \rightarrow P$ has the rate law $d[P]/dt = k[A][P]^2$; integrate the rate law for initial concentrations $[A]_0$ and $[P]_0$. Calculate the time at which the rate reaches a maximum.

26.19 Use mathematical software to integrate the Lotka–Volterra equations and plot the concentration of Y against that of X. Explore the consequences of varying the starting concentrations for the integration. Identify the conditions (the concentrations of X and Y) corresponding to the steady state of the Lotka–Volterra equations (the point at the centre of the orbits).

26.20 Set up the overall rate equations for the concentrations of X and Y in terms of the oregonator and explore the periodic properties of the solutions.

26.21 Conventional equilibrium considerations do not apply when a reaction is being driven by light absorption. Thus the steady-state concentration of products and reactants might differ significantly from equilibrium values. For instance, suppose the reaction $A \rightarrow B$ is driven by light absorption, and that its rate is I_a, but that the reverse reaction $B \rightarrow A$ is bimolecular and second-order with a rate $k[B]^2$. What is the stationary state concentration of B? Why does this 'photostationary state' differ from the equilibrium state?

26.22 Derive an expression for the rate of disappearance of a species A in a photochemical reaction for which the mechanism is:

(1) initiation with light of intensity I: $A \longrightarrow R\cdot + R\cdot$

(2) propagation: $A + R\cdot \longrightarrow R\cdot + B$

(3) termination: $R\cdot + R\cdot \longrightarrow R_2$

Hence, show that rate measurements will give only a combination of k_2 and k_3 if a steady state is reached, but that both may be obtained if a steady state is not reached.

26.23 The photochemical chlorination of chloroform in the gas has been found to follow the rate law $d[CCl_4]/dt = k[Cl_2]^{1/2} I_a^{1/2}$. Devise a mechanism that leads to this rate law when the chlorine pressure is high.

26.24 Photolysis of $Cr(CO)_6$ in the presence of certain molecules M, can give rise to the following reaction sequence:

(1) $Cr(CO)_6 + h\nu \longrightarrow Cr(CO)_5 + CO$

(2) $Cr(CO)_5 + CO \longrightarrow Cr(CO)_6$

(3) $Cr(CO)_5 + M \longrightarrow Cr(CO)_5M$

(4) $Cr(CO)_5M \longrightarrow Cr(CO)_5 + M$

Suppose that the absorbed light intensity is so weak that $I \ll k_4[Cr(CO)_5M]$. Find the factor f in the equation $d[Cr(CO)_5M]/dt = -f[Cr(CO)_5M]$. Show that a graph of $1/f$ against [M] should be a straight line.

Applications: to biology and environmental science

26.25 Models of population growth are analogous to chemical reaction rate equations. In the model due to Malthus (1798) the rate of change of the population N of the planet is assumed to be given by dN/dt = births − deaths. The numbers of births and deaths are proportional to the population, with proportionality constants b and d. Obtain the integrated rate law. How well does it fit the (very approximate) data below on the population of the planet as a function of time?

Year	1750	1825	1922	1960	1974	1987	2000
$N/10^9$	0.5	1	2	3	4	5	6

26.26 Many enzyme-catalysed reactions are consistent with a modified version of the Michaelis–Menten mechanism in which the second step is also reversible. For this mechanism obtain an expression for the rate of formation of product and find its limiting behaviour for large and small concentrations of substrate.

26.27 The initial rate of O_2 production by the action of an enzyme on a substrate was measured for a range of substrate concentrations; the data are below. Evaluate the Michaelis constant for the reaction.

$[S]/(\text{mol L}^{-1})$	0.050	0.017	0.010	0.0050	0.0020
$v/(\text{mm}^3\,\text{min}^{-1})$	16.6	12.4	10.1	6.6	3.3

26.28 Enzyme-catalysed reactions are sometimes analysed by use of the *Eadie–Hofstee plot*, in which $v/[S]_0$ is plotted against v. Using the simple Michaelis–Menten mechanism, derive a relation between $v/[S]_0$ and v. Discuss how the values of K_M and v_{max} are obtained from analysis of the Eadie–Hofstee plot.

26.29 In general, the catalytic efficiency of an enzyme depends on the pH of the medium in which it operates. One way to account for this behaviour is to propose that the enzyme and the enzyme–substrate complex are active only in specific protonation states. This situation can be summarized by the following mechanism:

$$EH + S \rightleftharpoons ESH \qquad k_a, k_a'$$

$$ESH \longrightarrow EH + P \qquad k_b$$

$$EH \rightleftharpoons E^- + H^+ \qquad K_{E,a} = \frac{[E^-][H^+]}{[EH]}$$

$$EH_2^+ \rightleftharpoons EH + H^+ \qquad K_{E,b} = \frac{[EH][H^+]}{[EH_2^+]}$$

$$ESH \rightleftharpoons ES^- + H^+ \qquad K_{ES,a} = \frac{[ES^-][H^+]}{[ESH]}$$

$$ESH_2 \rightleftharpoons ESH + H^+ \qquad K_{ES,b} = \frac{[ESH][H^+]}{[ESH_2]}$$

in which only the EH and ESH forms are active. (a) For the mechanism above, show that

$$v = \frac{v'_{max}}{1 + K'_M[S]_0}$$

with

$$v'_{max} = \frac{v_{max}}{1 + \dfrac{[H^+]}{K_{ES,b}} + \dfrac{K_{ES,a}}{[H^+]}} \qquad K'_M = K_M \frac{1 + \dfrac{[H^+]}{K_{E,b}} + \dfrac{K_{E,a}}{[H^+]}}{1 + \dfrac{[H^+]}{K_{ES,b}} + \dfrac{K_{ES,a}}{[H^+]}}$$

where v_{max} and K_M correspond to the form EH of the enzyme. (b) For pH values ranging from 0 to 14, plot v'_{max} against pH for a hypothetical reaction for which $v_{max} = 1.0 \times 10^{-6}$ mol L^{-1} s^{-1}, $K_{ES,b} = 1.0 \times 10^{-6}$ mol L^{-1}, and $K_{ES,a} = 1.0 \times 10^{-8}$ mol L^{-1}. Is there a pH at which v'_{max} reaches a maximum value? If so, determine the pH. (c) Redraw the plot in part (b) by using the same value of v_{max}, but $K_{ES,b} = 1.0 \times 10^{-4}$ mol L^{-1} and $K_{ES,a} = 1.0 \times 10^{-10}$ mol L^{-1}. Account for any differences between this plot and the plot from part (b).

26.30 The following results were obtained when the rate of an enzymolysis was monitored (a) without inhibitor, (b) with inhibitor at a concentration of 15 μmol L^{-1}:

$[S]_0/(10^{-4}\,\text{mol L}^{-1})$	1.0	3.0	7.0	12.0	18.0
(a) $v/(\mu\text{mol L}^{-1}\,\text{s}^{-1})$	0.49	0.95	1.3	1.5	1.6
(b) $v/(\mu\text{mol L}^{-1}\,\text{s}^{-1})$	0.27	0.52	0.71	0.81	0.86

Is the inhibition competitive, uncompetitive, or non-competitive?

26.31 Many biological and biochemical processes involve autocatalytic steps. In the SIR model of the spread and decline of infectious diseases the population is divided into three classes; the susceptibles, S, who can catch the disease, the infectives, I, who have the disease and can transmit it, and the removed class, R, who have either had the disease and recovered, are dead, are immune or isolated. The model mechanism for this process implies the following rate laws:

$$\frac{dS}{dt} = -rSI \qquad \frac{dI}{dt} = rSI - aI \qquad \frac{dR}{dt} = aI$$

What are the autocatalytic steps of this mechanism? Find the conditions on the ratio a/r that decide whether the disease will spread (an epidemic) or die out. Show that a constant population is built into this system, namely, that S + I + R = N, meaning that the timescales of births, deaths by other causes, and migration are assumed large compared to that of the spread of the disease.

26.32‡ Ultraviolet radiation photolyses O_3 to O_2 and O. Determine the rate at which ozone is consumed by 305 nm radiation in a layer of the stratosphere of thickness 1 km. The quantum efficiency is 0.94 at 220 K, the concentration about 8×10^{-9} mol L^{-1}, the molar absorption coefficient 260 L mol^{-1} cm^{-1}, and the flux of 305 nm radiation about 1×10^{14} photons cm^{-2} s^{-1}. Data from W.B. DeMore, S.P. Sander, D.M. Golden, R.F. Hampson, M.J. Kurylo, C.J. Howard, A.R. Ravishankara, C.E. Kolb, and M.J. Molina, *Chemical kinetics and* *photochemical data for use in stratospheric modeling: Evaluation Number 11*, JPL Publication 94–26 (1994).

26.33‡ Because of its importance in atmospheric chemistry, the thermal decomposition of nitric oxide, $2\,NO(g) \rightarrow N_2(g) + O_2(g)$, has been amongst the most thoroughly studied of gas-phase reactions. The commonly accepted mechanism has been that of H. Wise and M.F. Freech (*J. Chem. Phys.* **22**, 1724 (1952)):

(1) $NO + NO \longrightarrow N_2O + O$ $\quad k_1$

(2) $O + NO \longrightarrow O_2 + N$ $\quad k_2$

(3) $N + NO \longrightarrow N_2 + O$ $\quad k_3$

(4) $O + O + M \longrightarrow O_2 + M$ $\quad k_4$

(5) $O_2 + M \longrightarrow O + O + M$ $\quad k_4'$

(a) Label the steps of this mechanism as initiation, propagation, etc. (b) Write down the full expression for the rate of disappearance of NO. What does this expression for the rate become on the basis of the assumptions that $v_2 = v_3$ when [N] reaches its steady-state concentration, that the rate of the propagation step is more rapid than the rate of the initiation step, and that oxygen atoms are in equilibrium with oxygen molecules? (c) Find an expression for the effective activation energy, $E_{a,eff}$, for the overall reaction in terms of the activation energies of the individual steps of the reaction. (d) Estimate $E_{a,eff}$ from the bond energies of the species involved. (e) It has been pointed out by R.J. Wu and C.T. Yeh (*Int. J. Chem. Kinet.* **28**, 89 (1996)) that the reported experimental values of $E_{a,eff}$ obtained by different authors have varied from 253 to 357 kJ mol^{-1}. They suggest that the assumption of oxygen atoms and oxygen molecules being in equilibrium is unwarranted and that the steady-state approximation needs to be applied to the entire mechanism. Obtain the overall rate law based on the steady-state approximation and find the forms that it assumes for low NO conversion (low O_2 concentration). (f) When the reaction conversion becomes significant, Wu and Yeh suggest that two additional elementary steps,

(6) $O_2 + M \longrightarrow O + O + M$ $\quad k_6$

(7) $NO + O_2 \longrightarrow O + NO_2$ $\quad k_7$

start to compete with step (1) as the initiation step. Obtain the rate laws based on these alternative mechanisms and again estimate the apparent activation energies. Is the range of these different theoretically estimated values for $E_{a,eff}$ consistent with the range of values obtained experimentally?

Further information 1

Mathematical techniques

Basic procedures

1.1 Logarithms and exponentials

The **natural logarithm** of a number x is denoted $\ln x$, and is defined as the power to which $e = 2.718\ldots$ must be raised for the result to be equal to x. It follows from the definition of logarithms that

$$\ln x + \ln y + \cdots = \ln xy \cdots \quad \text{(FI1.1)}$$

$$\ln x - \ln y = \ln(x/y) \quad \text{(FI1.2)}$$

$$a \ln x = \ln x^a \quad \text{(FI1.3)}$$

We also encounter the **common logarithm** of a number, $\log x$, the logarithm compiled with 10 in place of e. Common logarithms follow the same rules of addition and subtraction as natural logarithms. Common and natural logarithms are related by

$$\ln x = \ln 10 \log x \approx 2.303 \log x \quad \text{(FI1.4)}$$

The **exponential function**, e^x, plays a very special role in the mathematics of chemistry. The following properties are important:

$$e^x e^y e^z \cdots = e^{x+y+z+\cdots} \quad \text{(FI1.5)}$$

$$e^x/e^y = e^{x-y} \quad \text{(FI1.6)}$$

$$(e^x)^a = e^{ax} \quad \text{(FI1.7)}$$

1.2 Combinatorial functions

Combinatorial functions allow us to express the number of ways in which a system of particles may be configured; they are especially useful in statistical thermodynamics (Chapters 19 and 20). Consider a simple coin-toss problem. If n coins are tossed, the number $N(n, i)$ of outcomes that have i heads and $(n - i)$ tails, regardless of the order of the results, is given by the coefficients of the **binomial expansion** of $(1 + x)^n$:

$$(1 + x)^n = 1 + \sum_{i=1}^{n} N(n, i)x^i, \qquad N(n, i) = \frac{n!}{(n - i)!\,i!} \quad \text{(FI1.8)}$$

where $n!$ denotes a **factorial**, given by

$$n! = n(n-1)(n-2)\ldots 1 \quad \text{(FI1.9)}$$

By definition $0! = 1$. The numbers $N(n, i)$, which are sometimes denoted $\binom{n}{i}$, are also called **binomial coefficients**.

Suppose that, unlike the coin-toss problem, there are more than two possible results for each event. For example, there are six possible results for the roll of a die. For n rolls of the die, the number of ways, W, that correspond to n_1 occurrences of the number 1, n_2 occurrences of the number 2, and so on, is given by

$$W = \frac{n!}{n_1!n_2!n_3!n_4!n_5!n_6!}, \qquad n = \sum_{i=1}^{6} n_i$$

This is an example of a **multinomial coefficient**, which has the form

$$W = \frac{n!}{n_1!n_2!\ldots n_m!}, \qquad n = \sum_{i=1}^{m} n_i \quad \text{(FI1.10)}$$

where W is the number of ways of achieving an outcome, n is the number of events, and m is the number of possible results. In Chapter 19 we use the multinomial coefficient to determine the number of ways to configure a system of identical particles given a specific distribution of particles into discrete energy levels.

In chemistry it is common to deal with a very large number of particles and outcomes and it is useful to express factorials in different ways. We can simplify factorials of large numbers by using **Stirling's approximation**:

$$n! \approx (2\pi)^{1/2} n^{n+\frac{1}{2}} e^{-n} \quad \text{(FI1.11)}$$

The approximation is in error by less than 1 per cent when n is greater than about 10. For very large values of n, it is possible to use another form of the approximation:

$$\ln n! \approx n \ln n - n \quad \text{(FI1.12)}$$

1.3 Complex numbers and complex functions

Complex numbers have the form

$$z = x + iy \quad \text{(FI1.13)}$$

where $i = (-1)^{1/2}$. The real numbers x and y are, respectively, the real and imaginary parts of z, denoted $\mathrm{Re}(z)$ and $\mathrm{Im}(z)$. We write the **complex conjugate** of z, z^*, by replacing i by −i:

$$z^* = x - iy \quad \text{(FI1.14)}$$

The **absolute value** or **modulus** of the complex number z is denoted $|z|$ and is given by:

$$|z| = (z^*z)^{1/2} = (x^2 + y^2)^{1/2} \quad \text{(FI1.15)}$$

The following rules apply for arithmetic operations involving complex numbers:

1 Addition. If $z = x + iy$ and $z' = x' + iy'$, then,

$$z \pm z' = (x \pm x') + i(y \pm y') \quad \text{(FI1.16)}$$

2 Multiplication. For z and z' defined above,

$$z \times z' = (x + iy)(x' + iy') = (xx' - yy') + i(xy' + yx') \quad \text{(FI1.17)}$$

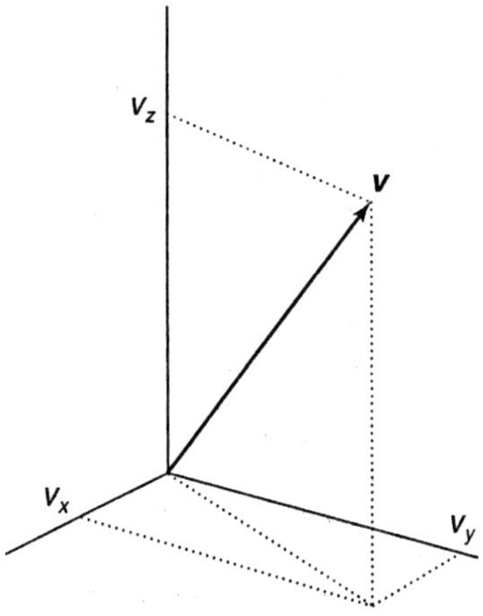

FI1.1 The vector v has components v_x, v_y, and v_z on the x-, y-, and z-axes with magnitudes v_x, v_y, and v_z, respectively.

3 Division. For z and z' defined above,

$$\frac{z}{z'} = \frac{z(z')^*}{|z'|^2} \quad \text{(FI1.18)}$$

Functions of complex arguments are useful in the discussion of wave equations (Chapter 11). We write the complex conjugate, f^*, of a complex function, f, by replacing i wherever it occurs by –i. For instance, the complex conjugate of e^{ix} is e^{-ix}.

Complex exponential functions may be written in terms of trigonometric functions. For example,

$$e^{\pm ix} = \cos x \pm i \sin x \quad \text{(FI1.19)}$$

which implies that

$$\cos x = \tfrac{1}{2}(e^{ix} + e^{-ix}) \quad \text{(FI1.20)}$$

$$\sin x = -i\tfrac{1}{2}(e^{ix} - e^{-ix}) \quad \text{(FI1.21)}$$

1.4 Vectors

A vector quantity has both magnitude and direction. The vector shown in Fig. FI1.1 has components on the x-, y-, and z-axes with magnitudes v_x, v_y, and v_z, respectively. The vector may be represented as

$$\boldsymbol{v} = v_x\boldsymbol{i} + v_y\boldsymbol{j} + v_z\boldsymbol{k} \quad \text{(FI1.22)}$$

where $\boldsymbol{i}$, $\boldsymbol{j}$, and $\boldsymbol{k}$ are **unit vectors**, vectors of magnitude 1, pointing along the positive directions on the x-, y-, and z-axes. The magnitude of the vector is denoted v or $|\boldsymbol{v}|$ and is given by

$$v = (v_x^2 + v_y^2 + v_z^2)^{1/2} \quad \text{(FI1.23)}$$

Using this representation, we can define the following vector operations:

1 Addition and subtraction.[1] If $\boldsymbol{v} = v_x\boldsymbol{i} + v_y\boldsymbol{j} + v_z\boldsymbol{k}$ and $\boldsymbol{u} = u_x\boldsymbol{i} + u_y\boldsymbol{j} + u_z\boldsymbol{k}$, then

$$\boldsymbol{v} \pm \boldsymbol{u} = (v_x \pm u_x)\boldsymbol{i} + (v_y \pm u_y)\boldsymbol{j} + (v_z \pm u_z)\boldsymbol{k} \quad \text{(FI1.24)}$$

2 Multiplication.[2] There are two ways to multiply vectors. In one procedure, the **cross-product** of two vectors $\boldsymbol{u}$ and $\boldsymbol{v}$ is a vector defined as

$$\boldsymbol{u} \times \boldsymbol{v} = (uv \sin \theta)\boldsymbol{l} \quad \text{(FI1.25a)}$$

where θ is the angle between the two vectors and $\boldsymbol{l}$ is a unit vector perpendicular to both $\boldsymbol{u}$ and $\boldsymbol{v}$, with a direction determined as in Fig. FI1.2. An equivalent definition is

$$\boldsymbol{u} \times \boldsymbol{v} = \begin{vmatrix} \boldsymbol{i} & \boldsymbol{j} & \boldsymbol{k} \\ u_x & u_y & u_z \\ v_x & v_y & v_z \end{vmatrix} = (u_y v_z - u_z v_y)\boldsymbol{i} - (u_x v_z - u_z v_x)\boldsymbol{j} - (u_x v_y - u_y v_x)\boldsymbol{k} \quad \text{(FI1.25b)}$$

where the structure in the middle is a determinant (see below). The second type of vector multiplication is the **scalar product** (or **dot product**) of two vectors $\boldsymbol{u}$ and $\boldsymbol{v}$:

$$\boldsymbol{u} \cdot \boldsymbol{v} = uv \cos \theta \quad \text{(FI1.26)}$$

As its name suggests, the scalar product of two vectors is a scalar, a quantity with magnitude, not direction.

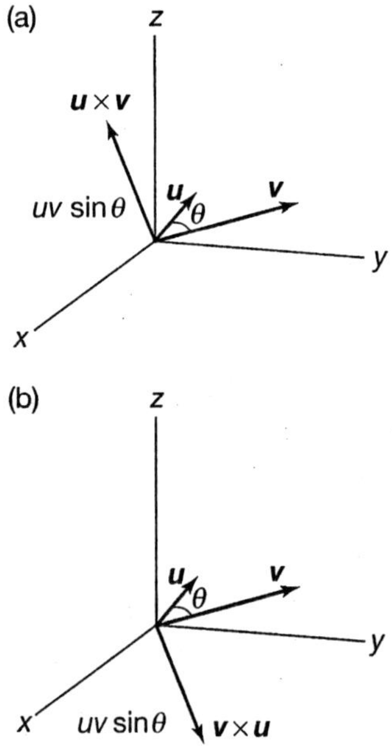

FI1.2 The direction of the cross-products of two vectors $\boldsymbol{u}$ and $\boldsymbol{v}$ with an angle θ between them: (a) $\boldsymbol{u} \times \boldsymbol{v}$ and (b) $\boldsymbol{v} \times \boldsymbol{u}$. Note that the cross-product, and the unit vector $\boldsymbol{l}$ of eqn 25, are perpendicular to both $\boldsymbol{u}$ and $\boldsymbol{v}$ but the direction depends on the order in which the product is taken.

1. Graphical methods for vector addition and subtraction are demonstrated in Section 13.8.
2. Vector division is not defined.

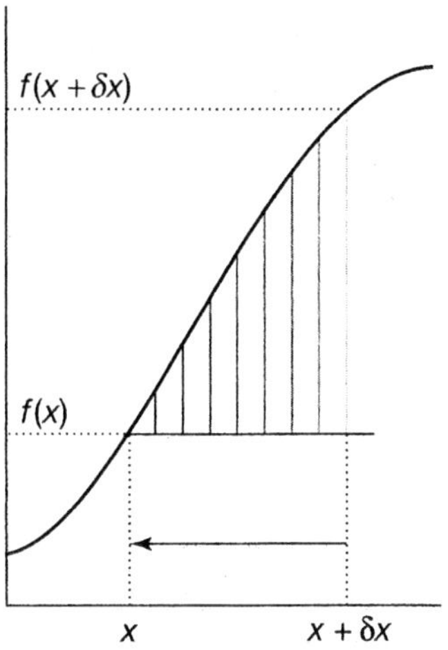

FI1.3 The slope of $f(x)$ at x, df/dx, is obtained by making a series of approximations to the value of $f(x + \delta x) - f(x)$ divided by the change in x, denoted δx, and allowing x to approach 0 (as indicated by the vertical lines getting closer to x).

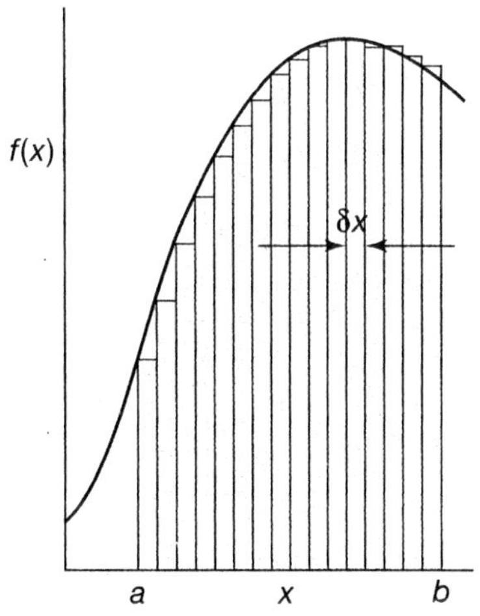

FI1.4 The shaded area is equal to the definite integral of $f(x)$ between the limits a and b.

Calculus

1.5 Differentiation and integration

Rates of change of functions—slopes of their graphs—are best discussed in terms of the infinitesimal calculus. The slope of a function, like the slope of a hill, is obtained by dividing the rise of the hill by the horizontal distance (Fig. FI1.3). However, because the slope may vary from point to point, we should make the horizontal distance between the points as small as possible. In fact, we let it become infinitesimally small—hence the name *infinitesimal* calculus. The values of a function f at two locations x and $x + \delta x$ are $f(x)$ and $f(x + \delta x)$, respectively. Therefore, the slope of the function f at x is the vertical distance, which we write δf, divided by the horizontal distance, which we write δx:

$$\text{Slope} = \frac{\text{rise in value}}{\text{horizontal distance}} = \frac{\delta f}{\delta x} = \frac{f(x + \delta x) - f(x)}{\delta x} \tag{FI1.27}$$

The slope at x itself is obtained by letting the horizontal distance become zero, which we write $\lim \delta x \to 0$. In this limit, the δ is replaced by a d, and we write

$$\text{Slope at } x = \frac{df}{dx} = \lim_{\delta x \to 0} \frac{f(x + \delta x) - f(x)}{\delta x} \tag{FI1.28}$$

To work out the slope of any function, we work out the expression on the right: this process is called **differentiation** and the expression for df/dx is the **derivative** of the function f with respect to the variable x. Some important derivatives are given inside the front cover of the text. Most of the functions encountered in chemistry can be differentiated by using the following rules:[3]

Rule 1 For two functions f and g:

$$d(f + g) = df + dg \tag{FI1.29}$$

Rule 2 (the product rule) For two functions f and g:

$$d(fg) = f\,dg + g\,df \tag{FI1.30}$$

Rule 3 (the quotient rule) For two functions f and g:

$$d\frac{f}{g} = \frac{1}{g}df - \frac{f}{g^2}dg \tag{FI1.31}$$

Rule no. 4 (the chain rule) For a function $f = f(g)$, where $g = g(t)$,

$$\frac{df}{dt} = \frac{df}{dg}\frac{dg}{dt} \tag{FI1.32}$$

The area under a graph of any function f is found by the techniques of **integration**. For instance, the area under the graph of the function f drawn in Fig. FI1.4 can be written as the value of f evaluated at a point multiplied by the width of the region, δx, and then all those products $f(x)\,\delta x$ summed over all the regions:

$$\text{Area between } a \text{ and } b = \sum f(x)\,\delta x$$

When we allow δx to become infinitesimally small, written dx, and sum an infinite number of strips, we write:

3. In these expressions, derivatives df/dx are written as df.

$$\text{Area between } a \text{ and } b = \int_a^b f(x)\,dx \qquad \text{(FI1.33)}$$

The elongated S symbol on the right is called the **integral** of the function f. When written as $\int$ alone, it is the **indefinite integral** of the function. When written with limits (as in eqn 33), it is the **definite integral** of the function. The definite integral is the indefinite integral evaluated at the upper limit (b) minus the indefinite integral evaluated at the lower limit (a). The **mean value theorem** states that the average value of a function $f(x)$ over the range $x = a$ to $x = b$ is

$$\text{Average value of } f(x) \text{ from } a \text{ to } b = \frac{1}{b-a}\int_a^b f(x)\,dx \qquad \text{(FI1.34)}$$

Integration is the inverse of differentiation. That is, if we integrate a function and then differentiate the result, we get back the original function. Some important integrals are given on the front back cover of the text. Many other standard forms are found in tables (see *Further reading*) and it is also possible to calculate definite and indefinite integrals with mathematical software. Two integration techniques are useful:

Technique 1 (integration by parts) For two functions f and g:

$$\int f\frac{dg}{dx}dx = fg - \int g\frac{df}{dx}dx \qquad \text{(FI1.35)}$$

Technique 2 (method of partial fractions) To solve an integral of the form $\int \frac{1}{(a-x)(b-x)}dx$, where a and b are constants, we write

$$\frac{1}{(a-x)(b-x)} = \frac{1}{b-a}\left(\frac{1}{a-x} - \frac{1}{b-x}\right)$$

and integrate the expression on the right. It follows that

$$\begin{aligned}\int \frac{dx}{(a-x)(b-x)} &= \frac{1}{b-a}\left[\int\frac{dx}{a-x} - \int\frac{dx}{b-x}\right] \\ &= \frac{1}{b-a}\left(\ln\frac{1}{a-x} - \ln\frac{1}{b-x}\right) + \text{constant}\end{aligned} \qquad \text{(FI1.36)}$$

1.6 Power series and Taylor expansions

A **power series** has the form

$$c_0 + c_1(x-a) + c_2(x-a)^2 + \cdots + c_n(x-a)^n + \cdots = \sum_{n=0}^{\infty} c_n(x-a)^n \qquad \text{(FI1.37)}$$

where c_n and a are constants. It is often useful to express a function $f(x)$ in the vicinity of $x = a$ as a special power series called the **Taylor series**, or **Taylor expansion**, which has the form:

$$\begin{aligned}f(x) &= f(a) + \left(\frac{df}{dx}\right)_a(x-a) + \frac{1}{2!}\left(\frac{d^2f}{dx^2}\right)_a(x-a)^2 + \cdots + \frac{1}{n!}\left(\frac{d^nf}{dx^n}\right)_a(x-a)^n \\ &= \sum_{n=0}^{\infty}\frac{1}{n!}\left(\frac{d^nf}{dx^n}\right)_a(x-a)^n\end{aligned} \qquad \text{(FI1.38)}$$

The following Taylor expansions are often useful:

$$\frac{1}{1+x} = 1 - x + x^2 \cdots$$

$$e^x = 1 + x + \tfrac{1}{2}x^2 + \cdots$$
$$\ln x = (x-1) - \tfrac{1}{2}(x-1)^2 + \tfrac{1}{3}(x-1)^3 - \tfrac{1}{4}(x-1)^4 + \cdots$$
$$\ln(1+x) = x - \tfrac{1}{2}x^2 + \tfrac{1}{3}x^3 \cdots$$

If $x \ll 1$, then $(1+x)^{-1} \approx 1 - x$, $e^x \approx 1 + x$, and $\ln(1+x) \approx x$.

1.7 Partial derivatives

A **partial derivative** of a function of more than one variable, such as $f(x, y)$, is the slope of the function with respect to one of the variables, all the other variables being held constant (see Fig. 3.2). Although a partial derivative shows how a function changes when one variable changes, it may be used to determine how the function changes when more than one variable changes by an infinitesimal amount. Thus, if f is a function of x and y, then when x and y change by dx and dy, respectively, f changes by

$$df = \left(\frac{\partial f}{\partial x}\right)_y dx + \left(\frac{\partial f}{\partial y}\right)_x dy \qquad \text{(FI1.39)}$$

where the symbol ∂ is used (instead of d) to denote a partial derivative. The quantity df is also called the **differential** of f. For example, if $f = ax^3y + by^2$, then

$$\left(\frac{\partial f}{\partial x}\right)_y = 3ax^2y \qquad \left(\frac{\partial f}{\partial y}\right)_x = ax^3 + 2by$$

Then, when x and y undergo infinitesimal changes, f changes by

$$df = 3ax^2y\, dx + (ax^3 + 2by)\, dy$$

Partial derivatives may be taken in any order:

$$\frac{\partial^2 f}{\partial x \partial y} = \frac{\partial^2 f}{\partial y \partial x} \qquad \text{(FI1.40)}$$

For the function f given above, it is easy to verify that

$$\left(\frac{\partial}{\partial y}\left(\frac{\partial f}{\partial x}\right)_y\right)_x = 3ax^2 \qquad \left(\frac{\partial}{\partial x}\left(\frac{\partial f}{\partial y}\right)_x\right)_y = 3ax^2$$

In the following, z is a variable on which x and y depend (for example, x, y, and z might correspond to p, V, and T).

Relation 1 When x is changed at constant z:

$$\left(\frac{\partial f}{\partial x}\right)_z = \left(\frac{\partial f}{\partial x}\right)_y + \left(\frac{\partial f}{\partial y}\right)_x \left(\frac{\partial y}{\partial x}\right)_z \qquad \text{(FI1.41)}$$

Relation 2

$$\left(\frac{\partial y}{\partial x}\right)_z = \frac{1}{(\partial x/\partial y)_z} \qquad \text{(FI1.42)}$$

Relation 3

$$\left(\frac{\partial x}{\partial y}\right)_z = -\left(\frac{\partial x}{\partial z}\right)_y \left(\frac{\partial z}{\partial y}\right)_x \qquad \text{(FI1.43)}$$

By combining this relation and Relation 2 we obtain the **Euler chain relation**:

$$\left(\frac{\partial y}{\partial x}\right)_z\left(\frac{\partial x}{\partial z}\right)_y\left(\frac{\partial z}{\partial y}\right)_x = -1 \qquad \text{(FI1.44)}$$

Relation 4 This relation establishes whether or not df is an **exact differential.**

$$df = g(x, y)\,dx + h(x, y)\,dy \qquad \text{is exact if} \qquad \left(\frac{\partial g}{\partial y}\right)_x = \left(\frac{\partial h}{\partial x}\right)_y \qquad \text{(FI1.45)}$$

If df is exact, its integral between specified limits is independent of the path.

1.8 Undetermined multipliers

Suppose we need to find the maximum (or minimum) value of some function f that depends on several variables $x_1, x_2, \ldots, x_n$. When the variables undergo a small change from x_i to $x_i + \delta x_i$ the function changes from f to $f + \delta f$, where

$$\delta f = \sum_i^n \left(\frac{\partial f}{\partial x_i}\right)\delta x_i \qquad \text{(FI1.46)}$$

At a minimum or maximum, $\delta f = 0$, so then

$$\sum_i^n \left(\frac{\partial f}{\partial x_i}\right)\delta x_i = 0 \qquad \text{(FI1.47)}$$

If the x_i were all independent, all the δx_i would be arbitrary, and this equation could be solved by setting each $(\partial f/\partial x_i) = 0$ individually. When the x_i are not all independent, the δx_i are not all independent, and the simple solution is no longer valid. We proceed as follows.

Let the constraint connecting the variables be an equation of the form $g = 0$. For example, in Chapter 19, one constraint was $n_0 + n_1 + \cdots = N$, which can be written

$$g = 0, \qquad \text{with} \quad g = (n_0 + n_1 + \cdots) - N$$

The constraint $g = 0$ is always valid, so g remains unchanged when the x_i are varied:

$$\delta g = \sum_i \left(\frac{\partial g}{\partial x_i}\right)\delta x_i = 0 \qquad \text{(FI1.48)}$$

Because δg is zero, we can multiply it by a parameter, λ, and add it to eqn 47:

$$\sum_i^n \left\{\left(\frac{\partial f}{\partial x_i}\right) + \lambda\left(\frac{\partial g}{\partial x_i}\right)\right\}\delta x_i = 0 \qquad \text{(FI1.49)}$$

This equation can be solved for one of the δx, δx_n for instance, in terms of all the other δx_i. All those other δx_i ($i = 1, 2, \ldots, n-1$) are independent, because there is only one constraint on the system. But here is the trick: λ is arbitrary; therefore we can choose it so that the coefficient of δx_n in eqn 49 is zero. That is, we choose λ so that

$$\left(\frac{\partial f}{\partial x_n}\right) + \lambda\left(\frac{\partial g}{\partial x_n}\right) = 0 \qquad \text{(FI1.50)}$$

Then eqn 49 becomes

$$\sum_i^{n-1} \left\{\left(\frac{\partial f}{\partial x_i}\right) + \lambda\left(\frac{\partial g}{\partial x_i}\right)\right\}\delta x_i = 0 \qquad \text{(FI1.51)}$$

Now the $n-1$ variations δx_i *are* independent, so the solution of this equation is

$$\left(\frac{\partial f}{\partial x_i}\right) + \lambda\left(\frac{\partial g}{\partial x_i}\right) = 0 \qquad i = 1, 2, \ldots, n-1 \tag{FI1.52}$$

However, eqn 50 has exactly the same form as this equation, so the maximum or minimum of f can be found by solving

$$\left(\frac{\partial f}{\partial x_i}\right) + \lambda\left(\frac{\partial g}{\partial x_i}\right) = 0 \qquad i = 1, 2, \ldots, n \tag{FI1.53}$$

The use of this approach was illustrated in the text for two constraints and therefore two undetermined multipliers λ_1 and λ_2 (α and $-\beta$).

The multipliers λ cannot always remain undetermined. One approach is to solve eqn 50 instead of incorporating it into the minimization scheme. In Chapter 19 we used the alternative procedure of keeping λ undetermined until a property was calculated for which the value was already known. Thus, we found that $\beta = 1/kT$ by calculating the internal energy of a perfect gas.

1.9 Differential equations

(a) Ordinary differential equations

An **ordinary differential equation** is a relation between derivatives of a function of one variable and the function itself, as in

$$a\frac{d^2y}{dx^2} + b\frac{dy}{dx} + cy = 0 \tag{FI1.54}$$

The coefficients a, b, etc. may be functions of x. The **order** of the equation is the order of the highest derivative that occurs in it, so eqn 54 is a second-order equation. Only rarely in science is a differential equation of order higher than 2 encountered. A solution of a differential equation is an expression for y as a function of x. The process of solving a differential equation is commonly termed 'integration', and in simple cases simple integration can be employed to find $y(x)$. A **general solution** of a differential equation is the most general solution of the equation and is expressed in terms of a number of constants. When the constants are chosen to accord with certain specified **initial conditions** (if one variable is the time) or certain **boundary conditions** (to fulfil certain spatial restrictions on the solutions), we obtain the **particular solution** of the equation. A first-order differential equation requires the specification of *one* boundary (or initial) condition; a second-order differential equation requires the specification of *two* such conditions, and so on.

First-order differential equations may often be solved by direct integration. For example, the equation

$$\frac{dy}{dx} = axy$$

with a constant may be rearranged into

$$\frac{dy}{y} = ax\,dx$$

and then integrated to

$$\ln y = \tfrac{1}{2}ax^2 + A$$

where A is a constant. If we know that $y = y_0$ when $x = 0$ (for instance), then it follows that $A = \ln y_0$, and hence the particular solution of the equation is

$$\ln y = \tfrac{1}{2}ax^2 + \ln y_0$$

This expression rearranges to

$$y = y_0 e^{ax^2/2}$$

First-order equations of a more complex form can often be solved by an appropriate substitution. For example, it is sensible to try the substitution $y = sx$, and to change the variables from x and y to x and s. An alternative useful transformation is to write $x = u + a$ and $y = v + b$, and then to select a and b to simplify the form of the resulting expression.

Solutions to complex differential equations may also be found by referring to tables (see *Further reading*). For example, first-order equations of the form

$$\frac{dy}{dx} + yf(x) = g(x) \qquad \text{(FI1.55)}$$

appear in the study of chemical kinetics. The solution is given by

$$ye^{\int f(x)dx} = \int e^{\int f(x)dx} g(x)\,dx + \text{constant} \qquad \text{(FI1.56)}$$

Mathematical software is now capable of finding analytical solutions of a wide variety of differential equations.

Second-order differential equations are in general much more difficult to solve than first-order equations. One powerful approach commonly used to lay siege to second-order differential equations is to express the solution as a power series:

$$y = \sum_{n=0}^{\infty} c_n x^n \qquad \text{(FI1.57)}$$

and then to use the differential equation to find a relation between the coefficients. This approach results, for instance, in the Hermite polynomials that form part of the solution of the Schrödinger equation for the harmonic oscillator (Section 12.5). All the second-order differential equations that occur in this text can be found tabulated in compilations of solutions or can be solved with mathematical software, and the specialized techniques that are needed to establish the form of the solutions may be found in mathematical texts.

(b) Numerical integration of differential equations

Many of the differential equations that describe physical phenomena are so complicated that their solutions cannot be cast as functions. In such cases, we resort to numerical methods, in which approximations are made in order to integrate the differential equation. Software packages are now readily available that can be used to solve almost any equation numerically. The general form of such programs to solve $df/dx = g(x)$, for instance, replaces the infinitesimal quantity $df = g(x)\,dx$ by the small quantity $\Delta f = g(x)\Delta x$, so that

$$f(x + \Delta x) \approx f(x) + g(x)\Delta x$$

and then proceeds numerically to step along the x-axis, generating $f(x)$ as it goes. The actual algorithms adopted are much more sophisticated than this primitive scheme, but stem from it. Among the simple numerical methods, the fourth-order **Runge–Kutta method** is one of the most accurate.

The *Further reading* section lists monographs that discuss the derivation of the fourth-order Runge–Kutta method. Here we illustrate the procedure with a first-order differential equation of the form:

$$\frac{dy}{dx} = f(x, y) \tag{FI1.58}$$

One example of this differential equation in chemical kinetics is eqn 25.32, which describes the time dependence of the concentration of an intermediate I in the reaction sequence A → I → P.

To obtain an approximate value of the integral of eqn 58, we proceed by rewriting it in terms of finite differences instead of differentials:

$$\frac{\Delta y}{\Delta x} = f(x, y)$$

where Δy may also be written as $y(x + \Delta x) - y(x)$. The fourth-order Runge–Kutta method is based on the following approximation:

$$y(x + \Delta x) = y(x) + \tfrac{1}{6}(k_1 + 2k_2 + 2k_3 + k_4) \tag{FI1.59}$$

where

$$k_1 = f(x, y)\Delta x \tag{FI1.60a}$$

$$k_2 = f(x + \tfrac{1}{2}\Delta x, y + \tfrac{1}{2}k_1)\Delta x \tag{FI1.60b}$$

$$k_3 = f(x + \tfrac{1}{2}\Delta x, y + \tfrac{1}{2}k_2)\Delta x \tag{FI1.60c}$$

$$k_4 = f(x + \Delta x, y + k_3)\Delta x \tag{FI1.60d}$$

Therefore, if we know the functional form of $f(x, y)$ and $y(0)$, we can use eqns 60(a–d) to calculate values of y for a range of x values. The process can be automated easily with an electronic spreadsheet or with mathematical software. The accuracy of the calculation increases with decreasing values of the increment Δx.

(c) Partial differential equations

A **partial differential equation** is a differential in more than one variable. An example is

$$\frac{\partial^2 y}{\partial t^2} = a\frac{\partial^2 y}{\partial x^2} \tag{FI1.61}$$

with y a function of the two variables x and t. In certain cases, partial differential equations may be separated into ordinary differential equations. Thus, the Schrödinger equation for a particle in a two-dimensional square well (Section 12.2) may be separated by writing the wavefunction, $\psi(x, y)$, as the product $X(x)Y(y)$, which results in the separation of the second-order partial differential equation into two second-order differential equations in the variables x and y. A good guide to the likely success of such a **separation of variables** procedure is the symmetry of the system.

Matrix algebra

A **matrix** is an array of numbers. Matrices may be combined together by addition or multiplication according to generalizations of the rules for ordinary numbers. Most numerical matrix manipulations are now carried out with mathematical software.

Consider a square matrix M of n^2 numbers arranged in n columns and n rows. These n^2 numbers are the **elements** of the matrix, and may be specified by stating the row, r, and column, c, at which they occur. Each element is therefore denoted M_{rc}. For example, in the matrix

$$M = \begin{pmatrix} 1 & 2 \\ 3 & 4 \end{pmatrix}$$

the elements are $M_{11} = 1$, $M_{12} = 2$, $M_{21} = 3$, and $M_{22} = 4$. This is an example of a 2×2 matrix. The **determinant**, $|M|$, of this matrix is

$$|M| = \begin{vmatrix} 1 & 2 \\ 3 & 4 \end{vmatrix} = 1 \times 4 - 2 \times 3 = -2$$

A **diagonal matrix** is a matrix in which the only nonzero elements lie on the major diagonal (the diagonal from M_{11} to M_{nn}). Thus, the matrix

$$D = \begin{pmatrix} 1 & 0 & 0 \\ 0 & 2 & 0 \\ 0 & 0 & 1 \end{pmatrix}$$

is diagonal. The condition may be written

$$M_{rc} = m_r \delta_{rc} \qquad \text{(FI1.62)}$$

where δ_{rc} is the **Kronecker delta**, which is equal to 1 for $r = c$ and to 0 for $r \neq c$. In the above example, $m_1 = 1$, $m_2 = 2$, and $m_3 = 1$. The **unit matrix**, **1** (and occasionally **I**), is a special case of a diagonal matrix in which all nonzero elements are 1.

The **transpose** of a matrix M is denoted M^T and is defined by

$$M^T_{mn} = M_{nm} \qquad \text{(FI1.63)}$$

Thus, for the matrix M we have been using,

$$M^T = \begin{pmatrix} 1 & 3 \\ 2 & 4 \end{pmatrix}$$

Matrices are very useful in chemistry. They simplify some mathematical tasks, such as solving systems of simultaneous equations, the treatment of molecular symmetry (Chapter 15), and quantum mechanical calculations (Chapter 14).

1.10 Matrix addition and multiplication

Two matrices M and N may be added to give the sum $S = M + N$, according to the rule

$$S_{rc} = M_{rc} + N_{rc} \qquad \text{(FI1.64)}$$

(that is, corresponding elements are added). Thus, with M given above and

$$N = \begin{pmatrix} 5 & 6 \\ 7 & 8 \end{pmatrix}$$

the sum is

$$S = \begin{pmatrix} 1 & 2 \\ 3 & 4 \end{pmatrix} + \begin{pmatrix} 5 & 6 \\ 7 & 8 \end{pmatrix} = \begin{pmatrix} 6 & 8 \\ 10 & 12 \end{pmatrix}$$

Two matrices may also be multiplied to give the product $P = MN$ according to the rule

$$P_{rc} = \sum_n M_{rn} N_{nc} \qquad \text{(FI1.65)}$$

For example, with the matrices given above,

$$P = \begin{pmatrix} 1 & 2 \\ 3 & 4 \end{pmatrix}\begin{pmatrix} 5 & 6 \\ 7 & 8 \end{pmatrix}$$

$$= \begin{pmatrix} 1\times5+2\times7 & 1\times6+2\times8 \\ 3\times5+4\times7 & 3\times6+4\times8 \end{pmatrix}$$

$$= \begin{pmatrix} 19 & 22 \\ 43 & 50 \end{pmatrix}$$

It should be noticed that in general $\boldsymbol{MN} \neq \boldsymbol{NM}$, and matrix multiplication is in general non-commutative.

The **inverse** of a matrix $\boldsymbol{M}$ is denoted $\boldsymbol{M}^{-1}$, and is defined so that

$$\boldsymbol{MM}^{-1} = \boldsymbol{M}^{-1}\boldsymbol{M} = \mathbf{1} \qquad \text{(FI1.66)}$$

The inverse of a matrix can be constructed by using mathematical software, but in simple cases the following procedure can be carried through without much effort:

1 Form the determinant of the matrix. For example, for our matrix $\boldsymbol{M}$, $|\boldsymbol{M}| = -2$.

2 Form the transpose of the matrix. For example, $\boldsymbol{M}^{\mathrm{T}} = \begin{pmatrix} 1 & 3 \\ 2 & 4 \end{pmatrix}$.

3 Form $\tilde{\boldsymbol{M}}'$, where $\tilde{M}'_{rc}$ is the **cofactor** of the element M_{rc}; that is, it is the determinant formed from $\boldsymbol{M}$ with the row r and column c deleted. For example, $\tilde{\boldsymbol{M}}' = \begin{pmatrix} 4 & -2 \\ -3 & 1 \end{pmatrix}$.

4 Construct the inverse as $\boldsymbol{M}^{-1} = \tilde{\boldsymbol{M}}'/|\boldsymbol{M}|$. For example,

$$\boldsymbol{M}^{-1} = \frac{1}{-2}\begin{pmatrix} 4 & -2 \\ -3 & 1 \end{pmatrix} = \begin{pmatrix} -2 & 1 \\ \frac{3}{2} & -\frac{1}{2} \end{pmatrix}$$

1.11 Simultaneous equations

A set of n simultaneous equations

$$\begin{aligned} a_{11}x_1 + a_{12}x_2 + \cdots + a_{1n}x_n &= b_1 \\ a_{21}x_1 + a_{22}x_2 + \cdots + a_{2n}x_n &= b_2 \\ &\vdots \\ a_{n1}x_1 + a_{n2}x_2 + \cdots + a_{nn}x_n &= b_n \end{aligned} \qquad \text{(FI1.67)}$$

can be written in matrix notation if we introduce the **column vectors** $\boldsymbol{x}$ and $\boldsymbol{b}$:

$$\boldsymbol{x} = \begin{pmatrix} x_1 \\ x_2 \\ \vdots \\ x_n \end{pmatrix} \qquad \boldsymbol{b} = \begin{pmatrix} b_1 \\ b_2 \\ \vdots \\ b_n \end{pmatrix}$$

Then, with the $\boldsymbol{a}$ matrix of coefficients a_{rc}, the n equations may be written as

$$\boldsymbol{ax} = \boldsymbol{b} \qquad \text{(FI1.68)}$$

The formal solution is obtained by multiplying both sides of this matrix equation by $\boldsymbol{a}^{-1}$, for then

$$\boldsymbol{x} = \boldsymbol{a}^{-1}\boldsymbol{b} \qquad \text{(FI1.69)}$$

1.12 Eigenvalue equations

An **eigenvalue equation** is a special case of eqn 68 in which

$$ax = \lambda x \tag{FI1.70}$$

where λ is a constant, the **eigenvalue**, and x is the **eigenvector**. In general, there are n eigenvalues $\lambda^{(i)}$, and they satisfy the n simultaneous equations

$$(a - \lambda 1)x = 0 \tag{FI1.71}$$

There are n corresponding eigenvectors $x^{(i)}$. Equation 71 has a solution only if the determinant of the coefficients is zero. However, this determinant is just $|a - \lambda 1|$, so the n eigenvalues may be found from the solution of the **secular equation**:

$$|a - \lambda 1| = 0 \tag{FI1.72}$$

The n eigenvalues the secular equation yields may be used to find the n eigenvectors. These eigenvectors (which are $n \times 1$ matrices), may be used to form an $n \times n$ matrix X. Thus, because

$$x^{(1)} = \begin{pmatrix} x_1^{(1)} \\ x_2^{(1)} \\ \vdots \\ x_n^{(1)} \end{pmatrix} \qquad x^{(2)} = \begin{pmatrix} x_1^{(2)} \\ x_2^{(2)} \\ \vdots \\ x_n^{(2)} \end{pmatrix} \qquad \text{etc.}$$

we may form the matrix

$$\mathrm{X} = (x^{(1)}, x^{(2)}, \ldots, x^{(n)}) = \begin{pmatrix} x_1^{(1)} & x_1^{(2)} & \cdots & x_1^{(n)} \\ x_2^{(1)} & x_2^{(2)} & \cdots & x_1^{(n)} \\ \vdots & \vdots & & \vdots \\ x_n^{(1)} & x_n^{(2)} & \cdots & x_n^{(n)} \end{pmatrix}$$

so that $X_{rc} = x_r^{(c)}$. If further we write $\Lambda_{rc} = \lambda_r \delta_{rc}$, so that Λ is a diagonal matrix with the elements $\lambda_1, \lambda_2, \ldots, \lambda_n$ along the diagonal, then all the eigenvalue equations $ax^{(i)} = \lambda_i x^{(i)}$ may be confined into the single equation

$$a\mathrm{X} = \mathrm{X}\Lambda \tag{FI1.73}$$

because this expression is equal to

$$\sum_n a_{rn} X_{nc} = \sum_n X_{rn} \Lambda_{nc}$$

or

$$\sum_n a_{rn} x_n^{(c)} = \sum_n x_r^{(n)} \lambda_n \delta_{nc} = \lambda_c x_r^{(c)}$$

as required. Therefore, if we form X^{-1} from X, we construct a **similarity transformation**

$$\Lambda = \mathrm{X}^{-1} a \mathrm{X} \tag{FI1.74}$$

that makes a diagonal (because Λ is diagonal). It follows that if the matrix X that causes $\mathrm{X}^{-1}a\mathrm{X}$ to be diagonal is known, then the problem is solved: the diagonal matrix so produced has the eigenvalues as its only nonzero elements, and the matrix X used to bring about the transformation has the corresponding eigenvectors as its columns. The solutions of eigenvalue equations are best found by using mathematical software.

Further reading

J.R. Barrante, *Applied mathematics for physical chemistry*. Prentice Hall, Upper Saddle River (1998).

M.P. Cady and C.A. Trapp, *A Mathcad primer for physical chemistry*. W.H. Freeman and Company, New York (2000).

W.H. Cropper, *Mathematical computer programs for physical chemistry*. Springer, New York (1998).

D.R. Lide (ed.), *CRC handbook of chemistry and physics*, Appendix A. CRC Press, Boca Raton (2000).

R.G. Mortimer, *Mathematics for physical chemistry*. Academic Press, San Diego (1999).

D.S. Sivia and S.G. Rawlings, *Foundations of science mathematics*. Oxford Chemistry Primers, Oxford University Press (1999).

D. Zwillinger (ed.), *CRC standard mathematical tables and formulae*. CRC Press, Boca Raton (1996).

Further information 2

Essential concepts of physics

Energy

The central concept of all explanations in physical chemistry, as in so many other branches of physical science, is that of **energy**, the capacity to do work. We make use of the apparently universal law of nature that *energy is conserved*; that is, energy can neither be created nor destroyed. Although energy can be transferred from one location to another, the total energy is constant.

2.1 Kinetic and potential energy

The **kinetic energy**, E_K, of a body is the energy the body possesses as a result of its motion. For a body of mass m travelling at a speed v,

$$E_K = \tfrac{1}{2}mv^2 \qquad \text{(FI2.1)}$$

The **potential energy**, E_P or V, of a body is the energy it possesses as a result of its position. The zero of potential energy is arbitrary. For example, the gravitational potential energy of a body is often set to zero at the surface of the Earth; the electrical potential energy of two charged particles is set to zero when their separation is infinite. No universal expression for the potential energy can be given because it depends on the type of interaction the body experiences. One example that gives rise to a simple expression is the potential energy of a body of mass m in the gravitational field close to the surface of the Earth (a gravitational field acts on the mass of a body). If the body is at a height h above the surface of the Earth, then its potential energy is mgh, where g is a constant called the **acceleration of free fall**, $g = 9.81\ \text{m s}^{-2}$, and $V = 0$ at $h = 0$ (the arbitrary zero mentioned previously).

The total energy is the sum of the kinetic and potential energies of a particle:

$$E = E_K + E_P \qquad \text{(FI2.2)}$$

2.2 Energy units

The SI unit of energy is the *joule* (J), which is defined as

$$1\ \text{J} = 1\ \text{kg m}^2\ \text{s}^{-2} \qquad \text{(FI2.3)}$$

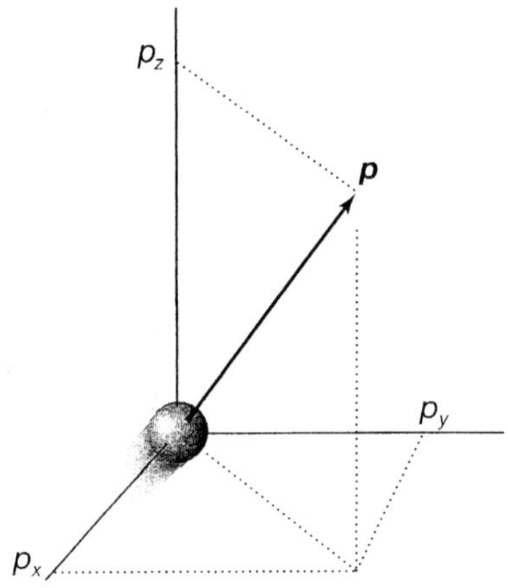

FI2.1 The linear momentum of a particle is a vector property and points in the direction of motion.

Calories (cal) and kilocalories (kcal) are still encountered in the chemical literature: by definition, 1 cal = 4.184 J. An energy of 1 cal is enough to raise the temperature of 1 g of water by 1°C.

The rate of change of energy is called the **power**, P, expressed as joules per second, or *watt*, W:

$$1\ \text{W} = 1\ \text{J s}^{-1} \tag{FI2.4}$$

Classical mechanics

Classical mechanics describes the behaviour of objects in terms of two equations. One expresses the fact that the total energy is constant in the absence of external forces; the other expresses the response of particles to the forces acting on them.

2.3 The trajectory in terms of the energy

The **velocity**, $\boldsymbol{v}$, of a particle is the rate of change of its position:

$$\boldsymbol{v} = \frac{\mathrm{d}\boldsymbol{r}}{\mathrm{d}t} \tag{FI2.5}$$

The velocity is a vector, with both direction and magnitude. The magnitude of the velocity is the **speed**, v. The **linear momentum**, $\boldsymbol{p}$, of a particle of mass m is related to its velocity, $\boldsymbol{v}$, by

$$\boldsymbol{p} = m\boldsymbol{v} \tag{FI2.6}$$

Like the velocity vector, the linear momentum vector points in the direction of travel of the particle (Fig. FI2.1). In terms of the linear momentum, the total energy of a particle is

$$E = \frac{p^2}{2m} + V(x) \tag{FI2.7}$$

This equation can be used to show that a particle will have a definite **trajectory**, or definite position and momentum at each instant. For example, consider a particle free to move in one direction (along the x-axis) in a region where $V = 0$ (so the energy is independent of position). Because $v = \mathrm{d}x/\mathrm{d}t$, it follows from eqns 6 and 7 that

$$\frac{\mathrm{d}x}{\mathrm{d}t} = \left(\frac{2E_\mathrm{K}}{m}\right)^{1/2} \tag{FI2.8}$$

A solution of this equation is

$$x(t) = x(0) + \left(\frac{2E_\mathrm{K}}{m}\right)^{1/2} t \tag{FI2.9}$$

The linear momentum is a constant:

$$p(t) = mv(t) = m\frac{\mathrm{d}x}{\mathrm{d}t} = (2mE_\mathrm{K})^{1/2} \tag{FI2.10}$$

Hence, if we know the initial position and momentum, we can predict all later positions and momenta exactly.

2.4 Newton's second law

The **force**, F, experienced by a particle free to move in one dimension is related to its potential energy, V, by

$$F = -\frac{dV}{dx} \qquad \text{(FI2.11a)}$$

This relation implies that the direction of the force is towards decreasing potential energy (Fig. FI2.2). In three dimensions,

$$\boldsymbol{F} = -\nabla V \qquad \nabla = \boldsymbol{i}\frac{\partial}{\partial x} + \boldsymbol{j}\frac{\partial}{\partial y} + \boldsymbol{k}\frac{\partial}{\partial z} \qquad \text{(FI2.11b)}$$

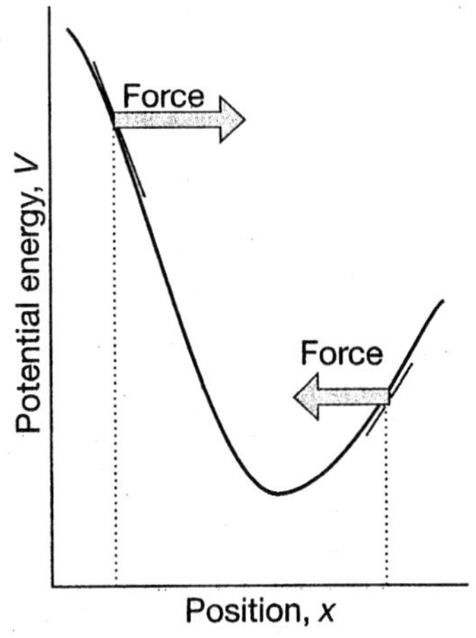

FI2.2 The force acting on a particle is determined by the slope of the potential energy at each point. The force points in the direction of lower potential energy.

Newton's second law of motion states that the rate of change of momentum is equal to the force acting on the particle. In one dimension:

$$\frac{dp}{dt} = F \qquad \text{(FI2.12a)}$$

Because $p = m(dx/dt)$ in one dimension, it is sometimes more convenient to write this equation as

$$m\frac{d^2x}{dt^2} = F \qquad \text{(FI2.12b)}$$

The second derivative, d^2x/dt^2, is the **acceleration** of the particle, its rate of change of velocity (in this instance, along the x-axis). It follows that, if we know the force acting everywhere and at all times, then solving eqn 12 will also give the trajectory. This calculation is equivalent to the one based on E, but is more suitable in some applications. For example, it can be used to show that if a particle of mass m is initially stationary and is subjected to a constant force F for a time τ, then its kinetic energy increases from zero to

$$E_K = \frac{F^2\tau^2}{2m} \qquad \text{(FI2.13)}$$

and then remains at that energy after the forces ceases to act. Because the applied force, F, and the time, τ, for which it acts may be varied at will, the solution implies that the energy of the particle may be increased to any value.

2.5 Rotational motion

The rotational motion of a particle about a central point is described by its **angular momentum, J**. The angular momentum is a vector: its magnitude gives the rate at which a particle circulates and its direction indicates the axis of rotation (Fig. FI2.3). The magnitude of the angular momentum, J, is given by the expression

$$J = I\omega \qquad \text{(FI2.14)}$$

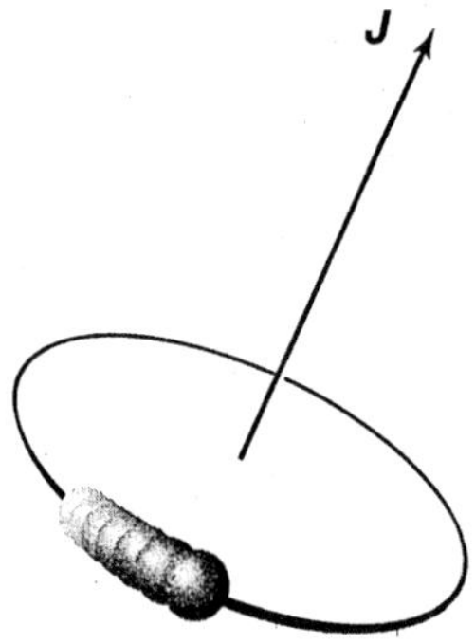

FI2.3 The angular momentum of a particle is represented by a vector along the axis of rotation and perpendicular to the plane of rotation. The length of the vector denotes the magnitude of the angular momentum. The direction of motion is clockwise to an observer looking in the direction of the vector.

where ω is the **angular velocity** of the body, its rate of change of angular position (in radians per second), and I is the **moment of inertia**. The analogous roles of m and I, of v and ω, and of p and J in the translational and rotational cases, respectively, should be remembered, because they provide a ready way of constructing and recalling equations. For a point particle of mass m moving in a circle of radius r, the moment of inertia about the axis of rotation is given by the expression

$$I = mr^2 \qquad \text{(FI2.15)}$$

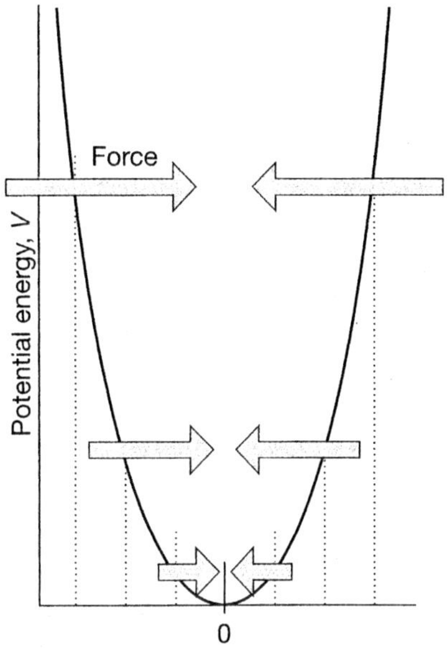

FI2.4 The force acting on a particle that undergoes harmonic motion. The force is directed towards zero displacement and is proportional to the displacement. The corresponding potential energy is parabolic (proportional to x^2).

To accelerate a rotation it is necessary to apply a **torque**, T, a twisting force. Newton's equation is then

$$\frac{\mathrm{d}J}{\mathrm{d}t} = T \tag{FI2.16}$$

If a constant torque is applied for a time τ, the rotational energy of an initially stationary body is increased to

$$E_K = \frac{T^2\tau^2}{2I} \tag{FI2.17}$$

The implication of this equation is that an appropriate torque and period for which it is applied can excite the rotation to an arbitrary energy.

2.6 The harmonic oscillator

A **harmonic oscillator** consists of a particle that experiences a restoring force proportional to its displacement from its equilibrium position:

$$F = -kx \tag{FI2.18}$$

An example is a particle joined to a rigid support by a spring. The constant of proportionality k is called the **force constant**, and the stiffer the spring the greater the force constant. The negative sign in F signifies that the direction of the force is opposite to that of the displacement (Fig. FI2.4).

The motion of a particle that undergoes harmonic motion is found by substituting the expression for the force, eqn 18, into Newton's equation, eqn 12b. The resulting equation is

$$m\frac{\mathrm{d}^2x}{\mathrm{d}t^2} = -kx$$

A solution is

$$x(t) = A \sin \omega t \qquad p(t) = m\omega A \cos \omega t \qquad \omega = (k/m)^{1/2} \tag{FI2.19}$$

These solutions show that the position of the particle varies **harmonically** (that is, as sin ωt) with a frequency $\nu = \omega/2\pi$. They also show that the particle is stationary ($p = 0$) when the displacement, x, has its maximum value, A, which is called the **amplitude** of the motion.

The total energy of a classical harmonic oscillator is proportional to the square of the amplitude of its motion. To confirm this remark we note that the kinetic energy is

$$E_K = \frac{p^2}{2m} = \frac{(m\omega A \cos \omega t)^2}{2m} = \tfrac{1}{2}m\omega^2 A^2 \cos 2\omega t \tag{FI2.20}$$

Then, because $\omega = (k/m)^{1/2}$, this expression may be written

$$E_K = \tfrac{1}{2}kA^2 \cos^2 \omega t \tag{FI2.21}$$

The force on the oscillator is $F = -kx$, so it follows from the relation $F = -\mathrm{d}V/\mathrm{d}x$ that the potential energy of a harmonic oscillator is

$$V = \tfrac{1}{2}kx^2 = \tfrac{1}{2}kA^2 \sin^2 \omega t \tag{FI2.22}$$

The total energy is therefore

$$E = \tfrac{1}{2}kA^2 \cos^2 \omega t + \tfrac{1}{2}kA^2 \sin^2 \omega t = kA^2 \tag{FI2.23}$$

(We have used $\cos^2\omega t + \sin^2\omega t = 1$.) That is, the energy of the oscillator is constant and, for a given force constant, is determined by its maximum displacement. It follows that the energy of an oscillating particle can be raised to any value by stretching the spring to any desired amplitude A. Note that the frequency of the motion depends only on the inherent properties of the oscillator (as represented by k and m) and is independent of the energy; the amplitude governs the energy, through $E = \frac{1}{2}kA^2$, and is independent of the frequency. In other words, the particle will oscillate at the same frequency regardless of the amplitude of its motion.

Waves

Waves are disturbances that travel through space with a finite velocity. Examples of disturbances include the collective motion of water molecules in ocean waves and of gas particles in sound waves. Waves can be characterized by a **wave equation**, a differential equation that describes the motion of the wave in space and time. **Harmonic waves** are waves with displacements that can be expressed as sine or cosine functions. These concepts are used in classical physics to describe the wave character of electromagnetic radiation, which is the focus of the following discussion.

2.7 The electromagnetic field

In classical physics, electromagnetic radiation is understood in terms of the **electromagnetic field**, an oscillating electric and magnetic disturbance that spreads as a harmonic wave through empty space, the vacuum. The wave travels at a constant speed called the *speed of light*, c, which is about 3×10^8 m s^{-1}. As its name suggests, an electromagnetic field has two components, an **electric field** that acts on charged particles (whether stationary of moving) and a **magnetic field** that acts only on moving charged particles. The electromagnetic field is characterized by a **wavelength**, λ (lambda), the distance between the neighbouring peaks of the wave, and its **frequency**, ν (nu), the number of times per second at which its displacement at a fixed point returns to its original value (Fig. FI2.5). The frequency is measured in *hertz*,

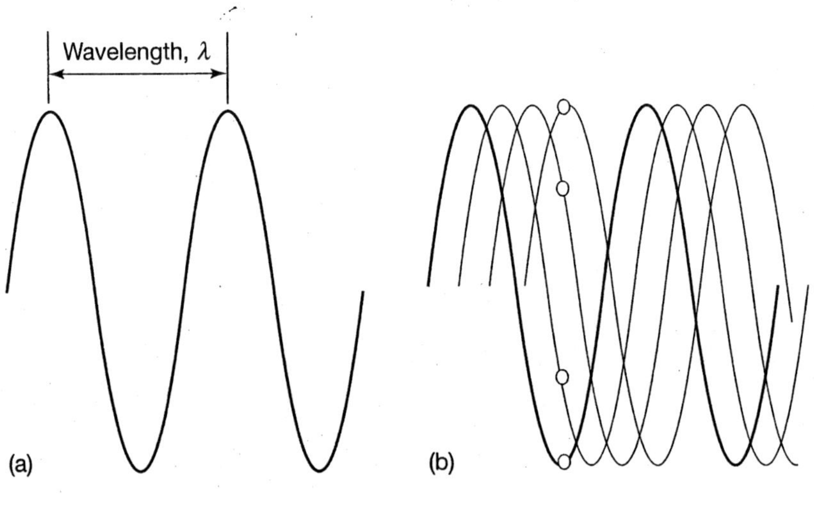

FI2.5 (a) The wavelength, λ, of a wave is the peak-to-peak distance. (b) The wave is shown travelling to the right at a speed c. At a given location, the instantaneous amplitude of the wave changes through a complete cycle (the four dots show half a cycle) as it passes a given point. The frequency, ν, is the number of cycles per second that occur at a given point. Wavelength and frequency are related by $\lambda\nu = c$.

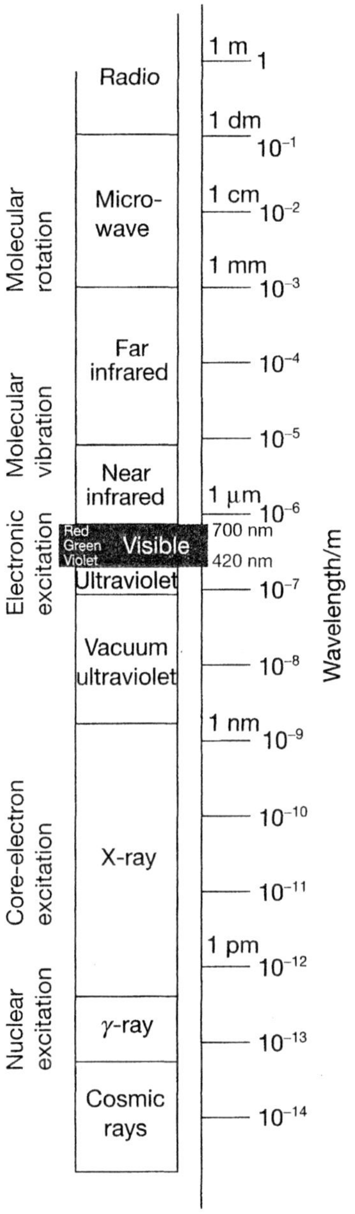

FI2.6 The regions of the electromagnetic spectrum and the types of excitation that give rise to each region.

where 1 Hz = 1 s^{-1}. The wavelength and frequency of an electromagnetic wave are related by

$$\lambda \nu = c \tag{FI2.24}$$

Therefore, the shorter the wavelength, the higher the frequency. The characteristics of a wave are also reported by giving the **wavenumber**, $\tilde{\nu}$ (nu tilde), of the radiation, where

$$\tilde{\nu} = \frac{\nu}{c} = \frac{1}{\lambda} \tag{FI2.25}$$

A wavenumber can be interpreted as the number of complete wavelengths in a given length. Wavenumbers are normally reported in reciprocal centimetres (cm^{-1}), so a wavenumber of 5 cm^{-1} indicates that there are 5 complete wavelengths in 1 cm. The classification of the electromagnetic field according to its frequency and wavelength is summarized in Fig. FI2.6.

2.8 Features of electromagnetic radiation

Consider an electromagnetic disturbance travelling along the x-direction with wavelength λ and frequency ν. The functions that describe the oscillating electric field, $\mathcal{E}(x, t)$, and magnetic field, $B(x, t)$, may be written as

$$\mathcal{E}(x, t) = \mathcal{E}_0 \cos\{2\pi\nu t - (2\pi/\lambda)x + \phi\} \tag{FI2.26a}$$

$$B(x, t) = B_0 \cos\{2\pi\nu t - (2\pi/\lambda)x + \phi\} \tag{FI2.26b}$$

where $\mathcal{E}_0$ and B_0 are the amplitudes of the electric and magnetic fields, respectively, and the parameter ϕ is the **phase** of the wave, which varies from $-\pi$ to π and gives the relative location of the peaks of two waves. If two waves, in the same region of space, with the same wavelength are shifted by $\phi = \pi$ or $-\pi$ (so the peaks of one wave coincide with the troughs of the other), then the resultant wave will have diminished amplitudes. The waves are said to interfere destructively. A value of $\phi = 0$ (coincident peaks) corresponds to constructive interference, or the enhancement of the amplitudes.

Equations 26a and 26b represent electromagnetic radiation that is **plane polarized**; it is so called because the electric and magnetic fields each oscillate in a single plane (in this case the xy- and xz-planes, respectively, Fig. FI2.7). The plane of polarization may be orientated in any direction around the direction of propagation (the x-direction in Fig. FI2.7), with the electric and magnetic fields perpendicular to that direction (and perpendicular to each other). An alternative mode of polarization is **circular polarization**, in which the electric and magnetic fields rotate around the direction of propagation in either a clockwise or a counter-clockwise sense but remain perpendicular to it and each other.

It is easy to show by differentiation that eqns 26a and 26b satisfy the following equations:

$$\frac{\partial^2}{\partial x^2}\psi(x, t) = -\frac{4\pi^2}{\lambda^2}\psi(x, t) \qquad \frac{\partial^2}{\partial t^2}\psi(x, t) = -4\pi^2\nu^2\psi(x, t) \tag{FI2.27}$$

where $\psi(x, t)$ is either $\mathcal{E}(x, t)$ or $B(x, t)$.

According to classical electromagnetic theory, the intensity of electromagnetic radiation is proportional to the square of the amplitude of the wave. For example, the light detectors discussed in Chapter 16 are based on the interaction between the electric field of the incident radiation and the detecting element, so light intensities are proportional to $\mathcal{E}_0^2$.

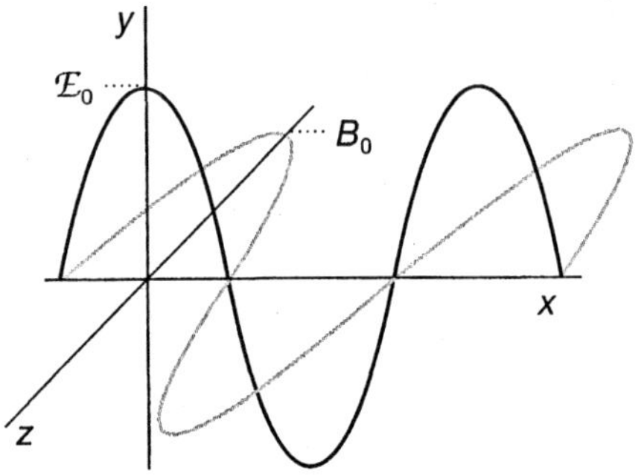

FI2.7 Electromagnetic radiation consists of a wave of electric and magnetic fields perpendicular to the direction of propagation (in this case the *x*-direction), and mutually perpendicular to each other. This illustration shows a plane-polarized wave, with the electric and magnetic fields oscillating in the *xy*- and *xz*-planes, respectively.

Electrostatics

Electrostatics is the study of the interactions of stationary electric charges. The elementary charge, the magnitude of charge carried by a single electron or proton, is $e \approx 1.60 \times 10^{-19}$ C. The magnitude of the charge per mole is the Faraday constant: $F = N_A e = 9.65 \times 10^4$ C mol^{-1}.

2.9 The Coulomb interaction

If a point charge q_1 is at a distance r in a vacuum from another point charge q_2, then their potential energy is

$$V = \frac{q_1 q_2}{4\pi\varepsilon_0 r} \qquad \text{(FI2.29)}$$

The constant ε_0 is the **vacuum permittivity**, a fundamental constant with the value 8.85×10^{-12} C^2 J^{-1} m^{-1}. (Note that with r in metres, m, and the charges in coulombs, C, the potential energy is in joules, J.) This very important relation is called the **Coulomb potential energy** and the interaction it describes is called the **Coulomb interaction** of two charges. The Coulomb potential energy is equal to the work that must be done to bring up a charge q_1 from infinity to a distance r from a charge q_2.

It follows from eqns 11 and 29 that the electrical force, F, exerted by a charge q_1 on a second charge q_2 has magnitude

$$F = \frac{q_1 q_2}{4\pi\varepsilon_0 r^2} \qquad \text{(FI2.30)}$$

The force itself is a vector directed along the line joining the two charges. With charge in coulombs and distance in metres, the force is obtained in newtons.

In a medium other than a vacuum, the potential energy of interaction between two charges is reduced, and the vacuum permittivity is replaced by the **permittivity**, ε, of the medium. It is common to express the permittivity as a multiple of the vacuum permittivity, and to write $\varepsilon = \varepsilon_r \varepsilon_0$, where ε_r is the **relative permittivity** (or dielectric constant) of the medium. For water at 25°C, $\varepsilon_r = 78.54$.

2.10 The Coulomb potential

The potential energy of a charge q_1 in the presence of another charge q_2 can be expressed in terms of the **Coulomb potential**, ϕ:

$$V = q_1\phi \qquad \phi = \frac{q_2}{4\pi\varepsilon_0 r} \qquad \text{(FI2.31)}$$

The units of potential are joules per coulomb, J C^{-1} so, when ϕ is multiplied by a charge in coulombs, the result is in joules. The combination joules per coulomb occurs widely in electrostatics, and is called a *volt*, V:

$$1\ \text{V} = 1\ \text{J C}^{-1} \qquad \text{(FI2.32)}$$

If there are several charges $q_2, q_3, \ldots$ present in the system, the total potential experienced by the charge q_1 is the sum of the potential generated by each charge:

$$\phi = \phi_2 + \phi_3 + \cdots \qquad \text{(FI2.33)}$$

When the charge distribution is more complex than a single point-like object, the Coulomb potential is described in terms of a charge density, ρ. With charge in coulombs and length in metres, the charge density is expressed in coulomb per metre-cubed (C m^{-3}). The electric potential arising from a charge distribution with density ρ is the solution to **Poisson's equation**:

$$\nabla^2\phi = -\rho/\varepsilon_0 \qquad \text{(FI2.34)}$$

where $\nabla^2 = (\partial^2/\partial x^2 + \partial^2/\partial y^2 + \partial^2/\partial z^2)$. If the distribution is spherically symmetrical, then so too is ϕ and eqn 34 reduces to the form used in *Justification 10.2.*

2.11 The strength of the electric field

Just as the potential energy of a charge q_1 can be written $V = q_1\phi$, so the magnitude of the force on q_1 can be written $F = q_1\mathcal{E}$, where $\mathcal{E}$ is the magnitude of the electric field strength arising from q_2 or from some more general charge distribution. The electric field strength (which, like the force, is actually a vector quantity) is the negative gradient of the electric potential:

$$\boldsymbol{\mathcal{E}} = -\nabla\phi \qquad \text{(FI2.35)}$$

2.12 The dipole–dipole interaction

An important problem in physical chemistry is the calculation of the potential energy of interaction between two point dipoles with moments $\boldsymbol{\mu}_1$ and $\boldsymbol{\mu}_2$, separated by a vector $\boldsymbol{r}$ (Sections 21.5 and 26.11). From classical electromagnetic theory, the potential energy of $\boldsymbol{\mu}_2$ in the electric field $\boldsymbol{\mathcal{E}}_1$ generated by $\boldsymbol{\mu}_1$ is given by the dot product

$$V = -\boldsymbol{\mathcal{E}}_1 \cdot \boldsymbol{\mu}_2 \qquad \text{(FI2.36)}$$

To calculate $\boldsymbol{\mathcal{E}}_1$, we consider a distribution of point charges q_i located at x_i, y_i, and z_i from the origin (that is, at r_i). It follows from eqns 31 and 33 that the Coulomb potential due to this distribution at a point with coordinates x, y, and z is:

$$\phi = \sum_i \frac{q_i}{4\pi\varepsilon_0} \frac{1}{\{(x-x_i)^2 + (y-y_i)^2 + (z-z_i)^2\}^{1/2}} \qquad \text{(FI2.37)}$$

where $\boldsymbol{r}$ is the location of the point of interest and the $\boldsymbol{r}_i$ are the locations of the charges q_i. If we suppose that all the charges are close to the origin (in the sense that $r_i \ll r$), we can use a Taylor expansion to write

$$\begin{aligned}\phi(\boldsymbol{r}) &= \sum_i \frac{q_i}{4\pi\varepsilon_0}\left\{\frac{1}{r} + \left(\frac{\partial\{(x-x_i)^2 + (y-y_i)^2 + (z-z_i)^2\}^{1/2}}{\partial x_i}\right)_{x_i=0} x_i + \cdots\right\} \\ &= \sum_i \frac{q_i}{4\pi\varepsilon_0}\left\{\frac{1}{r} + \frac{xx_i}{r^3} + \cdots\right\}\end{aligned} \qquad \text{(FI2.38)}$$

the ellipses (...) include the terms arising from derivatives with respect to y_i and z_i and higher derivatives. If the charge distribution is electrically neutral, the first term disappears because $\Sigma_i q_i = 0$. Next we note that $\Sigma_i q_i x_i = \mu_x$, and likewise for the y- and z-components. That is,

$$\begin{aligned}\phi &= \frac{1}{4\pi\varepsilon_0 r^3}(\mu_x x + \mu_y y + \mu_z z) \\ &= \frac{1}{4\pi\varepsilon_0 r^3}\boldsymbol{\mu}_1 \cdot \boldsymbol{r}\end{aligned} \qquad \text{(FI2.39)}$$

From eqn 35, the electric field strength is

$$\mathcal{E}_1 = \frac{1}{4\pi\varepsilon_0}\nabla\frac{\boldsymbol{\mu}_1\cdot\boldsymbol{r}}{r^3} = -\frac{\boldsymbol{\mu}_1}{4\pi\varepsilon_0 r^3} - \frac{\boldsymbol{\mu}_1\cdot\boldsymbol{r}}{4\pi\varepsilon_0}\nabla\frac{1}{r^3} \quad \text{(FI2.40)}$$

It follows from eqns 36 and 40 that

$$V = \frac{\boldsymbol{\mu}_1\cdot\boldsymbol{\mu}_2}{4\pi\varepsilon_0 r^3} - 3\frac{(\boldsymbol{\mu}_1\cdot\boldsymbol{r})(\boldsymbol{\mu}_2\cdot\boldsymbol{r})}{4\pi\varepsilon_0 r^5} \quad \text{(FI2.41)}$$

For the arrangement shown in Fig. 21.10, in which $\boldsymbol{\mu}_1\cdot\boldsymbol{r} = \mu_1 r\cos\theta$ and $\boldsymbol{\mu}_2\cdot\boldsymbol{r} = \mu_2 r\cos\theta$, eqn 41 becomes:

$$V = \frac{\mu_1\mu_2 f(\theta)}{4\pi\varepsilon_0 r^3} \qquad f(\theta) = 1 - 3\cos^2\theta \quad \text{(FI2.42)}$$

which is eqn 21.24.

2.13 Electric current and power

The motion of charge gives rise to an electric **current**, I. Electric current is measured in *ampere*, A, where

$$1\text{ A} = 1\text{ C s}^{-1} \quad \text{(FI2.43)}$$

If the current flows from a region of potential ϕ_i to ϕ_f, through a **potential difference** $\Delta\phi = \phi_f - \phi_i$, the rate of doing work is the current (the rate of transfer of charge) multiplied by the potential difference, $I\Delta\phi$. The rate of doing electrical work is the **electrical power**, P, so

$$P = I\Delta\phi \quad \text{(FI2.44)}$$

With current in amperes and the potential difference in volts, the power works out in watts. The total energy, E, supplied in an interval Δt is the power (the rate of energy supply) multiplied by the duration of the interval:

$$E = P\Delta t = I\Delta\phi\Delta t \quad \text{(FI2.45)}$$

The energy is obtained in joules with the current in amperes, the potential difference in volts, and the time in seconds.

Further reading

R.P. Feynman, R.B. Leighton, and M. Sands, *The Feynman lectures on physics*, Vols 1–3. Addison-Wesley, Reading (1966).

G.A.D. Ritchie and D.S. Sivia, *Foundations of physics for chemists*. Oxford Chemistry Primers, Oxford University Press (2000).

W.S. Warren, *The physical basis of chemistry*. Academic Press, San Diego (2000).

R. Wolfson and J.M. Pasachoff, *Physics for scientists and engineers with modern physics*. Addison-Wesley, Reading (1999).

Data section

The following tables reproduce and expand the data given in the short tables in the text, and follow their numbering. Standard states refer to a pressure of $p^{\ominus} = 1$ bar. The general references are as follows:

AIP: D.E. Gray (ed.), *American Institute of Physics handbook*. McGraw-Hill, New York (1972).

AS: M. Abramowitz and I.A. Stegun (ed.), *Handbook of mathematical functions*. Dover, New York (1963).

E: J. Emsley, *The elements*. Oxford University Press (1991).

HCP: D.R. Lide (ed.), *Handbook of chemistry and physics*. CRC Press, Boca Raton (2000).

JL: A.M. James and M.P. Lord, *Macmillan's chemical and physical data*. Macmillan, London (1992).

KL: G.W.C. Kaye and T.H. Laby (ed.), *Tables of physical and chemical constants*. Longman, London (1973).

LR: G.N. Lewis and M. Randall, revised by K.S. Pitzer and L. Brewer, *Thermodynamics*. McGraw-Hill, New York (1961).

NBS: *NBS tables of chemical thermodynamic properties*, published as *J. Phys. and Chem. Reference Data*, **11**, Supplement 2 (1982).

RS: R.A. Robinson and R.H. Stokes, *Electrolyte solutions*. Butterworth, London (1959).

TDOC: J.B. Pedley, J.D. Naylor, and S.P. Kirby, *Thermochemical data of organic compounds*. Chapman & Hall, London (1986).

Contents

The following is a directory of all tables in the text; those included in this *Data section* are marked with an asterisk. The remainder will be found on the pages indicted in parentheses.

Physical properties of selected materials

	ρ/(g cm^{-3}) at 293 K†	T_f/K	T_b/K
Elements			
Aluminium(s)	2.698	933.5	2740
Argon(g)	1.381	83.8	87.3
Boron(s)	2.340	2573	3931
Bromine(l)	3.123	265.9	331.9
Carbon(s, gr)	2.260	3700s	
Carbon(s, d)	3.513		
Chlorine(g)	1.507	172.2	239.2
Copper(s)	8.960	1357	2840
Fluorine(g)	1.108	53.5	85.0
Gold(s)	19.320	1338	3080
Helium(g)	0.125		4.22
Hydrogen(g)	0.071	14.0	20.3
Iodine(s)	4.930	386.7	457.5
Iron(s)	7.874	1808	3023
Krypton(g)	2.413	116.6	120.8
Lead(s)	11.350	600.6	2013
Lithium(s)	0.534	453.7	1620
Magnesium(s)	1.738	922.0	1363
Mercury(l)	13.546	234.3	629.7
Neon(g)	1.207	24.5	27.1
Nitrogen(g)	0.880	63.3	77.4
Oxygen(g)	1.140	54.8	90.2
Phosphorus(s, wh)	1.820	317.3	553
Potassium(s)	0.862	336.8	1047
Silver(s)	10.500	1235	2485
Sodium(s)	0.971	371.0	1156
Sulfur(s, α)	2.070	386.0	717.8
Uranium(s)	18.950	1406	4018
Xenon(g)	2.939	161.3	166.1
Zinc(s)	7.133	692.7	1180

	ρ/(g cm^{-3}) at 293 K†	T_f/K	T_b/K
Inorganic compounds			
$CaCO_3$(s, calcite)	2.71	1612	1171d
$CuSO_4\cdot 5H_2O$(s)	2.284	383($-H_2O$)	423($-5H_2O$)
HBr(g)	2.77	184.3	206.4
HCl(g)	1.187	159.0	191.1
HI(g)	2.85	222.4	237.8
H_2O(l)	0.997	273.2	373.2
D_2O(l)	1.104	277.0	374.6
NH_3(g)	0.817	195.4	238.8
KBr(s)	2.750	1003	1708
KCl(s)	1.984	1049	1773s
NaCl(s)	2.165	1074	1686
H_2SO_4(l)	1.841	283.5	611.2
Organic compounds			
Acetaldehyde, CH_3CHO(l, g)	0.788	152	293
Acetic acid, CH_3COOH(l)	1.049	289.8	391
Acetone, $(CH_3)_2CO$(l)	0.787	178	329
Aniline, $C_6H_5NH_2$(l)	1.026	267	457
Anthracene, $C_{14}H_{10}$(s)	1.243	490	615
Benzene, C_6H_6(l)	0.879	278.6	353.2
Carbon tetrachloride, CCl_4(l)	1.63	250	349.9
Chloroform, $CHCl_3$(l)	1.499	209.6	334
Ethanol, C_2H_5OH(l)	0.789	156	351.4
Formaldehyde, HCHO(g)		181	254.0
Glucose, $C_6H_{12}O_6$(s)	1.544	415	
Methane, CH_4(g)		90.6	111.6
Methanol, CH_3OH(l)	0.791	179.2	337.6
Naphthalene, $C_{10}H_8$(s)	1.145	353.4	491
Octane, C_8H_{18}(l)	0.703	216.4	398.8
Phenol, C_6H_5OH(s)	1.073	314.1	455.0
Sucrose, $C_{12}H_{22}O_{11}$(s)	1.588	457d	

d: decomposes; s: sublimes; Data: AIP, E, HCP, KL; † For gases, at their boiling points.

Masses and natural abundances of selected nuclides

Nuclide		m/u	Abundance/%
H	^{1}H	1.0078	99.985
	^{2}H	2.0140	0.015
He	^{3}He	3.0160	0.000 13
	^{4}He	4.0026	100
Li	^{6}Li	6.0151	7.42
	^{7}Li	7.0160	92.58
B	^{10}B	10.0129	19.78
	^{11}B	11.0093	80.22
C	^{12}C	12*	98.89
	^{13}C	13.0034	1.11
N	^{14}N	14.0031	99.63
	^{15}N	15.0001	0.37
O	^{16}O	15.9949	99.76
	^{17}O	16.9991	0.037
	^{18}O	17.9992	0.204
F	^{19}F	18.9984	100
P	^{31}P	30.9738	100
S	^{32}S	31.9721	95.0
	^{33}S	32.9715	0.76
	^{34}S	33.9679	4.22
Cl	^{35}Cl	34.9688	75.53
	^{37}Cl	36.9651	24.4
Br	^{79}Br	78.9183	50.54
	^{81}Br	80.9163	49.46
I	^{127}I	126.9045	100

* Exact value.

Table 1.3 Second virial coefficients, $B/(\text{cm}^3\ \text{mol}^{-1})$

	100 K	273 K	373 K	600 K
Air	−167.3	−13.5	3.4	19.0
Ar	−187.0	−21.7	−4.2	11.9
CH_4		−53.6	−21.2	8.1
CO_2		−142	−72.2	−12.4
H_2	−2.0	13.7	15.6	
He	11.4	12.0	11.3	10.4
Kr		−62.9	−28.7	1.7
N_2	−160.0	−10.5	6.2	21.7
Ne	−6.0	10.4	12.3	13.8
O_2	−197.5	−22.0	−3.7	12.9
Xe		−153.7	−81.7	−19.6

Data: AIP, JL. The values relate to the expansion in eqn 1.22 of Section 1.3b; convert to eqn 1.21 using $B' = B/RT$.
For Ar at 273 K, $C = 1200\ \text{cm}^6\ \text{mol}^{-1}$

Table 1.4 Critical constants of gases

	p_c/atm	$V_c/(\text{cm}^3\ \text{mol}^{-1})$	T_c/K	Z_c	T_B/K
Ar	48.00	75.25	150.72	0.292	411.5
Br_2	102	135	584	0.287	
C_2H_4	50.50	124	283.1	0.270	
C_2H_6	48.20	148	305.4	0.285	
C_6H_6	48.6	260	562.7	0.274	
CH_4	45.6	98.7	190.6	0.288	510.0
Cl_2	76.1	124	417.2	0.276	
CO_2	72.85	94.0	304.2	0.274	714.8
F_2	55	144			
H_2	12.8	64.99	33.23	0.305	110.0
H_2O	218.3	55.3	647.4	0.227	
HBr	84.0	363.0			
HCl	81.5	81.0	324.7	0.248	
He	2.26	57.76	5.21	0.305	22.64
HI	80.8	423.2			
Kr	54.27	92.24	209.39	0.291	575.0
N_2	33.54	90.10	126.3	0.292	327.2
Ne	26.86	41.74	44.44	0.307	122.1
NH_3	111.3	72.5	405.5	0.242	
O_2	50.14	78.0	154.8	0.308	405.9
Xe	58.0	118.8	289.75	0.290	768.0

Data: AIP, KL

Table 1.5 Van der Waals coefficients

	a/(atm L^2 mol^{-2})	b/(10^{-2} L mol^{-1})		a/(atm L^2 mol^{-2})	b/(10^{-2} L mol^{-1})
Ar	1.337	3.20	H_2S	4.484	4.34
C_2H_4	4.552	5.82	He	0.0341	2.38
C_2H_6	5.507	6.51	Kr	5.125	1.06
C_6H_6	18.57	11.93	N_2	1.352	3.87
CH_4	2.273	4.31	Ne	0.205	1.67
Cl_2	6.260	5.42	NH_3	4.169	3.71
CO	1.453	3.95	O_2	1.364	3.19
CO_2	3.610	4.29	SO_2	6.775	5.68
H_2	0.2420	2.65	Xe	4.137	5.16
H_2O	5.464	3.05			

Data: HCP

Table 2.2 Temperature variation of molar heat capacities†

	a	b/(10^{-3} K^{-1})	c/(10^5 K^2)
Monatomic gases			
	20.78	0	0
Other gases			
Br_2	37.32	0.50	−1.26
Cl_2	37.03	0.67	−2.85
CO_2	44.22	8.79	−8.62
F_2	34.56	2.51	−3.51
H_2	27.28	3.26	0.50
I_2	37.40	0.59	−0.71
N_2	28.58	3.77	−0.50
NH_3	29.75	25.1	−1.55
O_2	29.96	4.18	−1.67
Liquids (from melting to boiling)			
$C_{10}H_8$, naphthalene	79.5	0.4075	0
I_2	80.33	0	0
H_2O	75.29	0	0
Solids			
Al	20.67	12.38	0
C (graphite)	16.86	4.77	−8.54
$C_{10}H_8$, naphthalene	−115.9	3.920×10^3	0
Cu	22.64	6.28	0
I_2	40.12	49.79	0
NcCl	45.94	16.32	0
Pb	22.13	11.72	0.96

† For $C_{p,m}$/(J K^{-1} mol^{-1}) $= a + bT + c/T^2$
Source: LR.

Table 2.3 Standard enthalpies of fusion and vaporization at the transition temperature, $\Delta_{trs}H^{\ominus}/(kJ\ mol^{-1})$

	T_f/K	Fusion	T_b/K	Vaporization		T_f/K	Fusion	T_b/K	Vaporization
Elements					CO_2	217.0	8.33	194.6	25.23 s
Ag	1234	11.30	2436	250.6	CS_2	161.2	4.39	319.4	26.74
Ar	83.81	1.188	87.29	6.506	H_2O	273.15	6.008	373.15	40.656
Br_2	265.9	10.57	332.4	29.45					44.016 at 298 K
Cl_2	172.1	6.41	239.1	20.41	H_2S	187.6	2.377	212.8	18.67
F_2	53.6	0.26	85.0	3.16	H_2SO_4	283.5	2.56		
H_2	13.96	0.117	20.38	0.916	NH_3	195.4	5.652	239.7	23.35
He	3.5	0.021	4.22	0.084	**Organic compounds**				
Hg	234.3	2.292	629.7	59.30	CH_4	90.68	0.941	111.7	8.18
I_2	386.8	15.52	458.4	41.80	CCl_4	250.3	2.5	350	30.0
N_2	63.15	0.719	77.35	5.586	C_2H_6	89.85	2.86	184.6	14.7
Na	371.0	2.601	1156	98.01	C_6H_6	278.61	10.59	353.2	30.8
O_2	54.36	0.444	90.18	6.820	C_6H_{14}	178	13.08	342.1	28.85
Xe	161	2.30	165	12.6	$C_{10}H_8$	354	18.80	490.9	51.51
K	336.4	2.35	1031	80.23	CH_3OH	175.2	3.16	337.2	35.27
Inorganic compounds									37.99 at 298 K
CCl_4	250.3	2.47	349.9	30.00	C_2H_5OH	158.7	4.60	352	43.5

Data: AIP; s denotes sublimation

Table 2.5 Thermodynamic data for organic compounds (all values are for 298 K)

	$M/(g\ mol^{-1})$	$\Delta_fH^{\ominus}/(kJ\ mol^{-1})$	$\Delta_fG^{\ominus}/(kJ\ mol^{-1})$	$S^{\ominus}_m/(J\ K^{-1}\ mol^{-1})$†	$C^{\ominus}_{p,m}/(J\ K^{-1}\ mol^{-1})$	$\Delta_cH^{\ominus}/(kJ\ mol^{-1})$
C(s) (graphite)	12.011	0	0	5.740	8.527	−393.51
C(s) (diamond)	12.011	+1.895	+2.900	2.377	6.113	−395.40
CO_2(g)	44.040	−393.51	−394.36	213.74	37.11	
Hydrocarbons						
CH_4(g), methane	16.04	−74.81	−50.72	186.26	35.31	−890
CH_3(g), methyl	15.04	+145.69	+147.92	194.2	38.70	
C_2H_2(g), ethyne	26.04	+226.73	+209.20	200.94	43.93	−1300
C_2H_4(g), ethene	28.05	+52.26	+68.15	219.56	43.56	−1411
C_2H_6(g), ethane	30.07	−84.68	−32.82	229.60	52.63	−1560
C_3H_6(g), propene	42.08	+20.42	+62.78	267.05	63.89	−2058
C_3H_6(g), cyclopropane	42.08	+53.30	+104.45	237.55	55.94	−2091
C_3H_8(g), propane	44.10	−103.85	−23.49	269.91	73.5	−2220
C_4H_8(g), 1-butene	56.11	−0.13	+71.39	305.71	85.65	−2717
C_4H_8(g), *cis*-2-butene	56.11	−6.99	+65.95	300.94	78.91	−2710
C_4H_8(g), *trans*-2-butene	56.11	−11.17	+63.06	296.59	87.82	−2707
C_4H_{10}(g), butane	58.13	−126.15	−17.03	310.23	97.45	−2878
C_5H_{12}(g), pentane	72.15	−146.44	−8.20	348.40	120.2	−3537
C_5H_{12}(l)	72.15	−173.1				
C_6H_6(l), benzene	78.12	+49.0	+124.3	173.3	136.1	−3268

Table 2.5 (Continued)

	M/(g mol^{-1})	$\Delta_f H^\ominus$/(kJ mol^{-1})	$\Delta_f G^\ominus$/(kJ mol^{-1})	$S^\ominus_m$/(J K^{-1} mol^{-1})†	$C^\ominus_{p,m}$/(J K^{-1} mol^{-1})	$\Delta_c H^\ominus$/(kJ mol^{-1})
Hydrocarbons (Continued)						
C_6H_6(g)	78.12	+82.93	+129.72	269.31	81.67	−3302
C_6H_{12}(l), cyclohexane	84.16	−156	+26.8	204.4	156.5	−3920
C_6H_{14}(l), hexane	86.18	−198.7		204.3		−4163
$C_6H_5CH_3$(g), methylbenzene (toluene)	92.14	+50.0	+122.0	320.7	103.6	−3953
C_7H_{16}(l), heptane	100.21	−224.4	+1.0	328.6	224.3	
C_8H_{18}(l), octane	114.23	−249.9	+6.4	361.1		−5471
C_8H_{18}(l), iso-octane	114.23	−255.1				−5461
$C_{10}H_8$(s), naphthalene	128.18	+78.53				−5157
Alcohols and phenols						
CH_3OH(l), methanol	32.04	−238.66	−166.27	126.8	81.6	−726
CH_3OH(g)	32.04	−200.66	−161.96	239.81	43.89	−764
C_2H_5OH(l), ethanol	46.07	−277.69	−174.78	160.7	111.46	−1368
C_2H_5OH(g)	46.07	−235.10	−168.49	282.70	65.44	−1409
C_6H_5OH(s), phenol	94.12	−165.0	−50.9	146.0		−3054
Carboxylic acids, hydroxy acids, and esters						
HCOOH(l), formic	46.03	−424.72	−361.35	128.95	99.04	−255
CH_3COOH(l), acetic	60.05	−484.5	−389.9	159.8	124.3	−875
CH_3COOH(aq)	60.05	−485.76	−396.46	178.7		
$CH_3CO_2^-$(aq)	59.05	−486.01	−369.31	+86.6	−6.3	
$(COOH)_2$(s), oxalic	90.04	−827.2			117	−254
C_6H_5COOH(s), benzoic	122.13	−385.1	−245.3	167.6	146.8	−3227
$CH_3CH(OH)COOH$(s), lactic	90.08	−694.0				−1344
$CH_3COOC_2H_5$(l), ethyl acetate	88.11	−479.0	−332.7	259.4	170.1	−2231
Alkanals and alkanones						
HCHO(g), methanal	30.03	−108.57	−102.53	218.77	35.40	−571
CH_3CHO(l), ethanal	44.05	−192.30	−128.12	160.2		−1166
CH_3CHO(g)	44.05	−166.19	−128.86	250.3	57.3	−1192
CH_3COCH_3(l), propanone	58.08	−248.1	−155.4	200.4	124.7	−1790
Sugars						
$C_6H_{12}O_6$(s), α-D-glucose	180.16	−1274				−2808
$C_6H_{12}O_6$(s), β-D-glucose	180.16	−1268	−910	212		
$C_6H_{12}O_6$(s), β-D-fructose	180.16	−1266				−2810
$C_{12}H_{22}O_{11}$(s), sucrose	342.30	−2222	−1543	360.2		−5645
Nitrogen compounds						
$CO(NH_2)_2$(s), urea	60.06	−333.51	−197.33	104.60	93.14	−632
CH_3NH_2(g), methylamine	31.06	−22.97	+32.16	243.41	53.1	−1085
$C_6H_5NH_2$(l), aniline	93.13	+31.1				−3393
$CH_2(NH_2)COOH$(s), glycine	75.07	−532.9	−373.4	103.5	99.2	−969

Data: NBS, TDOC; † Standard entropies of ions may be either positive or negative because the values are relative to the entropy of the hydrogen ion.

Table 2.6 Thermodynamic data for elements and inorganic compounds (all values relate to 298 K)

	$M/(\text{g mol}^{-1})$	$\Delta_f H^{\ominus}/(\text{kJ mol}^{-1})$	$\Delta_f G^{\ominus}/(\text{kJ mol}^{-1})$	$S_m^{\ominus}/(\text{J K}^{-1}\text{ mol}^{-1})$†	$C_{p,m}^{\ominus}/(\text{J K}^{-1}\text{ mol}^{-1})$
Aluminium (aluminum)					
Al(s)	26.98	0	0	28.33	24.35
Al(l)	26.98	+10.56	+7.20	39.55	24.21
Al(g)	26.98	+326.4	+285.7	164.54	21.38
Al^{3+}(g)	26.98	+5483.17			
Al^{3+}(aq)	26.98	−531	−485	−321.7	
Al_2O_3(s, α)	101.96	−1675.7	−1582.3	50.92	79.04
$AlCl_3$(s)	133.24	−704.2	−628.8	110.67	91.84
Argon					
Ar(g)	39.95	0	0	154.84	20.786
Antimony					
Sb(s)	121.75	0	0	45.69	25.23
SbH_3(g)	124.77	+145.11	+147.75	232.78	41.05
Arsenic					
As(s, α)	74.92	0	0	35.1	24.64
As(g)	74.92	+302.5	+261.0	174.21	20.79
As_4(g)	299.69	+143.9	+92.4	314	
AsH_3(g)	77.95	+66.44	+68.93	222.78	38.07
Barium					
Ba(s)	137.34	0	0	62.8	28.07
Ba(g)	137.34	+180	+146	170.24	20.79
Ba^{2+}(aq)	137.34	−537.64	−560.77	+9.6	
BaO(s)	153.34	−553.5	−525.1	70.43	47.78
$BaCl_2$(s)	208.25	−858.6	−810.4	123.68	75.14
Beryllium					
Be(s)	9.01	0	0	9.50	16.44
Be(g)	9.01	+324.3	+286.6	136.27	20.79
Bismuth					
Bi(s)	208.98	0	0	56.74	25.52
Bi(g)	208.98	+207.1	+168.2	187.00	20.79
Bromine					
Br_2(l)	159.82	0	0	152.23	75.689
Br_2(g)	159.82	+30.907	+3.110	245.46	36.02
Br(g)	79.91	+111.88	+82.396	175.02	20.786
Br^-(g)	79.91	−219.07			
Br^-(aq)	79.91	−121.55	−103.96	+82.4	−141.8
HBr(g)	90.92	−36.40	−53.45	198.70	29.142
Cadmium					
Cd(s, γ)	112.40	0	0	51.76	25.98
Cd(g)	112.40	+112.01	+77.41	167.75	20.79
Cd^{2+}(aq)	112.40	−75.90	−77.612	−73.2	
CdO(s)	128.40	−258.2	−228.4	54.8	43.43
$CdCO_3$(s)	172.41	−750.6	−669.4	92.5	

Table 2.6 (Continued)

	M/(g mol^{-1})	$\Delta_f H^\ominus$/(kJ mol^{-1})	$\Delta_f G^\ominus$/(kJ mol^{-1})	$S_m^\ominus$/(J K^{-1} mol^{-1})†	$C_{p,m}^\ominus$/(J K^{-1} mol^{-1})
Caesium (cesium)					
Cs(s)	132.91	0	0	85.23	32.17
Cs(g)	132.91	+76.06	+49.12	175.60	20.79
Cs^+(aq)	132.91	−258.28	−292.02	+133.05	−10.5
Calcium					
Ca(s)	40.08	0	0	41.42	25.31
Ca(g)	40.08	+178.2	+144.3	154.88	20.786
Ca^{2+}(aq)	40.08	−542.83	−553.58	−53.1	
CaO(s)	56.08	−635.09	−604.03	39.75	42.80
$CaCO_3$(s) (calcite)	100.09	−1206.9	−1128.8	92.9	81.88
$CaCO_3$(s) (aragonite)	100.09	−1207.1	−1127.8	88.7	81.25
CaF_2(s)	78.08	−1219.6	−1167.3	68.87	67.03
$CaCl_2$(s)	110.99	−795.8	−748.1	104.6	72.59
$CaBr_2$(s)	199.90	−682.8	−663.6	130	
Carbon (for 'organic' compounds of carbon, see Table 2.5)					
C(s) (graphite)	12.011	0	0	5.740	8.527
C(s) (diamond)	12.011	+1.895	+2.900	2.377	6.113
C(g)	12.011	+716.68	+671.26	158.10	20.838
C_2(g)	24.022	+831.90	+775.89	199.42	43.21
CO(g)	28.011	−110.53	−137.17	197.67	29.14
CO_2(g)	44.010	−393.51	−394.36	213.74	37.11
CO_2(aq)	44.010	−413.80	−385.98	117.6	
H_2CO_3(aq)	62.03	−699.65	−623.08	187.4	
HCO_3^-(aq)	61.02	−691.99	−586.77	+91.2	
CO_3^{2-}(aq)	60.01	−677.14	−527.81	−56.9	
CCl_4(l)	153.82	−135.44	−65.21	216.40	131.75
CS_2(l)	76.14	+89.70	+65.27	151.34	75.7
HCN(g)	27.03	+135.1	+124.7	201.78	35.86
HCN(l)	27.03	+108.87	+124.97	112.84	70.63
CN^-(aq)	26.02	+150.6	+172.4	+94.1	
Chlorine					
Cl_2(g)	70.91	0	0	223.07	33.91
Cl(g)	35.45	+121.68	+105.68	165.20	21.840
Cl^-(g)	34.45	−233.13			
Cl^-(aq)	35.45	−167.16	−131.23	+56.5	−136.4
HCl(g)	36.46	−92.31	−95.30	186.91	29.12
HCl(aq)	36.46	−167.16	−131.23	56.5	−136.4
Chromium					
Cr(s)	52.00	0	0	23.77	23.35
Cr(g)	52.00	+396.6	+351.8	174.50	20.79
CrO_4^{2-}(aq)	115.99	−881.15	−727.75	+50.21	
$Cr_2O_7^{2-}$(aq)	215.99	−1490.3	−1301.1	+261.9	

Table 2.6 (Continued)

	$M/(g\ mol^{-1})$	$\Delta_f H^\ominus/(kJ\ mol^{-1})$	$\Delta_f G^\ominus/(kJ\ mol^{-1})$	$S_m^\ominus/(J\ K^{-1}\ mol^{-1})$†	$C_{p,m}^\ominus/(J\ K^{-1}\ mol^{-1})$
Copper					
Cu(s)	63.54	0	0	33.150	24.44
Cu(g)	63.54	+338.32	+298.58	166.38	20.79
Cu^+(aq)	63.54	+71.67	+49.98	+40.6	
Cu^{2+}(aq)	63.54	+64.77	+65.49	−99.6	
Cu_2O(s)	143.08	−168.6	−146.0	93.14	63.64
CuO(s)	79.54	−157.3	−129.7	42.63	42.30
$CuSO_4$(s)	159.60	−771.36	−661.8	109	100.0
$CuSO_4 \cdot H_2O$(s)	177.62	−1085.8	−918.11	146.0	134
$CuSO_4 \cdot 5H_2O$(s)	249.68	−2279.7	−1879.7	300.4	280
Deuterium					
D_2(g)	4.028	0	0	144.96	29.20
HD(g)	3.022	+0.318	−1.464	143.80	29.196
D_2O(g)	20.028	−249.20	−234.54	198.34	34.27
D_2O(l)	20.028	−294.60	−243.44	75.94	84.35
HDO(g)	19.022	−245.30	−233.11	199.51	33.81
HDO(l)	19.022	−289.89	−241.86	79.29	
Fluorine					
F_2(g)	38.00	0	0	202.78	31.30
F(g)	19.00	+78.99	+61.91	158.75	22.74
F^-(aq)	19.00	−332.63	−278.79	−13.8	−106.7
HF(g)	20.01	−271.1	−273.2	173.78	29.13
Gold					
Au(s)	196.97	0	0	47.40	25.42
Au(g)	196.97	+366.1	+326.3	180.50	20.79
Helium					
He(g)	4.003	0	0	126.15	20.786
Hydrogen (see also deuterium)					
H_2(g)	2.016	0	0	130.684	28.824
H(g)	1.008	+217.97	+203.25	114.71	20.784
H^+(aq)	1.008	0	0	0	0
H^+(g)	1.008	+1536.20			
H_2O(s)	18.015			37.99	
H_2O(l)	18.015	−285.83	−237.13	69.91	75.291
H_2O(g)	18.015	−241.82	−228.57	188.83	33.58
H_2O_2(l)	34.015	−187.78	−120.35	109.6	89.1
Iodine					
I_2(s)	253.81	0	0	116.135	54.44
I_2(g)	253.81	+62.44	+19.33	260.69	36.90
I(g)	126.90	+106.84	+70.25	180.79	20.786
I^-(aq)	126.90	−55.19	−51.57	+111.3	−142.3
HI(g)	127.91	+26.48	+1.70	206.59	29.158

Table 2.6 (Continued)

	M/(g mol^{-1})	$\Delta_f H^\ominus$/(kJ mol^{-1})	$\Delta_f G^\ominus$/(kJ mol^{-1})	$S_m^\ominus$/(J K^{-1} mol^{-1})†	$C_{p,m}^\ominus$/(J K^{-1} mol^{-1})
Iron					
Fe(s)	55.85	0	0	27.28	25.10
Fe(g)	55.85	+416.3	+370.7	180.49	25.68
Fe^{2+}(aq)	55.85	−89.1	−78.90	−137.7	
Fe^{3+}(aq)	55.85	−48.5	−4.7	−315.9	
Fe_3O_4(s) (magnetite)	231.54	−1118.4	−1015.4	146.4	143.43
Fe_2O_3(s) (haematite)	159.69	−824.2	−742.2	87.40	103.85
FeS(s, α)	87.91	−100.0	−100.4	60.29	50.54
FeS_2(s)	119.98	−178.2	−166.9	52.93	62.17
Krypton					
Kr(g)	83.80	0	0	164.08	20.786
Lead					
Pb(s)	207.19	0	0	64.81	26.44
Pb(g)	207.19	+195.0	+161.9	175.37	20.79
Pb^{2+}(aq)	207.19	−1.7	−24.43	+10.5	
PbO(s, yellow)	223.19	−217.32	−187.89	68.70	45.77
PbO(s, red)	223.19	−218.99	−188.93	66.5	45.81
PbO_2(s)	239.19	−277.4	−217.33	68.6	64.64
Lithium					
Li(s)	6.94	0	0	29.12	24.77
Li(g)	6.94	+159.37	+126.66	138.77	20.79
Li^+(aq)	6.94	−278.49	−293.31	+13.4	68.6
Magnesium					
Mg(s)	24.31	0	0	32.68	24.89
Mg(g)	24.31	+147.70	+113.10	148.65	20.786
Mg^{2+}(aq)	24.31	−466.85	−454.8	−138.1	
MgO(s)	40.31	−601.70	−569.43	26.94	37.15
$MgCO_3$(s)	84.32	−1095.8	−1012.1	65.7	75.52
$MgCl_2$(s)	95.22	−641.32	−591.79	89.62	71.38
Mercury					
Hg(l)	200.59	0	0	76.02	27.983
Hg(g)	200.59	+61.32	+31.82	174.96	20.786
Hg^{2+}(aq)	200.59	+171.1	+164.40	−32.2	
Hg_2^{2+}(aq)	401.18	+172.4	+153.52	+84.5	
HgO(s)	216.59	−90.83	−58.54	70.29	44.06
Hg_2Cl_2(s)	472.09	−265.22	−210.75	192.5	102
$HgCl_2$(s)	271.50	−224.3	−178.6	146.0	
HgS(s, black)	232.65	−53.6	−47.7	88.3	
Neon					
Ne(g)	20.18	0	0	146.33	20.786

Table 2.6 (Continued)

	M/(g mol^{-1})	$\Delta_f H^\ominus$/(kJ mol^{-1})	$\Delta_f G^\ominus$/(kJ mol^{-1})	$S_m^\ominus$/(J K^{-1} mol^{-1})†	$C_{p,m}^\ominus$/(J K^{-1} mol^{-1})
Nitrogen					
N_2(g)	28.013	0	0	191.61	29.125
N(g)	14.007	+472.70	+455.56	153.30	20.786
NO(g)	30.01	+90.25	+86.55	210.76	29.844
N_2O(g)	44.01	+82.05	+104.20	219.85	38.45
NO_2(g)	46.01	+33.18	+51.31	240.06	37.20
N_2O_4(g)	92.01	+9.16	+97.89	304.29	77.28
N_2O_5(s)	108.01	−43.1	+113.9	178.2	143.1
N_2O_5(g)	108.01	+11.3	+115.1	355.7	84.5
HNO_3(l)	63.01	−174.10	−80.71	155.60	109.87
HNO_3(aq)	63.01	−207.36	−111.25	146.4	−86.6
NO_3^-(aq)	62.01	−205.0	−108.74	+146.4	−86.6
NH_3(g)	17.03	−46.11	−16.45	192.45	35.06
NH_3(aq)	17.03	−80.29	−26.50	111.3	
NH_4^+(aq)	18.04	−132.51	−79.31	+113.4	79.9
NH_2OH(s)	33.03	−114.2			
HN_3(l)	43.03	+264.0	+327.3	140.6	43.68
HN_3(g)	43.03	+294.1	+328.1	238.97	98.87
N_2H_4(l)	32.05	+50.63	+149.43	121.21	139.3
NH_4NO_3(s)	80.04	−365.56	−183.87	151.08	84.1
NH_4Cl(s)	53.49	−314.43	−202.87	94.6	
Oxygen					
O_2(g)	31.999	0	0	205.138	29.355
O(g)	15.999	+249.17	+231.73	161.06	21.912
O_3(g)	47.998	+142.7	+163.2	238.93	39.20
OH^-(aq)	17.007	−229.99	−157.24	−10.75	−148.5
Phosphorus					
P(s, wh)	30.97	0	0	41.09	23.840
P(g)	30.97	+314.64	+278.25	163.19	20.786
P_2(g)	61.95	+144.3	+103.7	218.13	32.05
P_4(g)	123.90	+58.91	+24.44	279.98	67.15
PH_3(g)	34.00	+5.4	+13.4	210.23	37.11
PCl_3(g)	137.33	−287.0	−267.8	311.78	71.84
PCl_3(l)	137.33	−319.7	−272.3	217.1	
PCl_5(g)	208.24	−374.9	−305.0	364.6	112.8
PCl_5(s)	208.24	−443.5			
H_3PO_3(s)	82.00	−964.4			
H_3PO_3(aq)	82.00	−964.8			
H_3PO_4(s)	94.97	−1279.0	−1119.1	110.50	106.06
H_3PO_4(l)	94.97	−1266.9			
H_3PO_4(aq)	94.97	−1277.4	−1018.7	−222	
PO_4^{3-}(aq)	94.97	−1277.4	−1018.7	−221.8	
P_4O_{10}(s)	283.89	−2984.0	−2697.0	228.86	211.71
P_4O_6(s)	219.89	−1640.1			

Table 2.6 (Continued)

	M/(g mol^{-1})	$\Delta_f H^\ominus$/(kJ mol^{-1})	$\Delta_f G^\ominus$/(kJ mol^{-1})	$S_m^\ominus$/(J K^{-1} mol^{-1})†	$C_{p,m}^\ominus$/(J K^{-1} mol^{-1})
Potassium					
K(s)	39.10	0	0	64.18	29.58
K(g)	39.10	+89.24	+60.59	160.336	20.786
K^+(g)	39.10	+514.26			
K^+(aq)	39.10	−252.38	−283.27	+102.5	21.8
KOH(s)	56.11	−424.76	−379.08	78.9	64.9
KF(s)	58.10	−576.27	−537.75	66.57	49.04
KCl(s)	74.56	−436.75	−409.14	82.59	51.30
KBr(s)	119.01	−393.80	−380.66	95.90	52.30
KI(s)	166.01	−327.90	−324.89	106.32	52.93
Silicon					
Si(s)	28.09	0	0	18.83	20.00
Si(g)	28.09	+455.6	+411.3	167.97	22.25
SiO_2(s, α)	60.09	−910.94	−856.64	41.84	44.43
Silver					
Ag(s)	107.87	0	0	42.55	25.351
Ag(g)	107.87	+284.55	+245.65	173.00	20.79
Ag^+(aq)	107.87	+105.58	+77.11	+72.68	21.8
AgBr(s)	187.78	−100.37	−96.90	107.1	52.38
AgCl(s)	143.32	−127.07	−109.79	96.2	50.79
Ag_2O(s)	231.74	−31.05	−11.20	121.3	65.86
$AgNO_3$(s)	169.88	−129.39	−33.41	140.92	93.05
Sodium					
Na(s)	22.99	0	0	51.21	28.24
Na(g)	22.99	+107.32	+76.76	153.71	20.79
Na^+(aq)	22.99	−240.12	−261.91	59.0	46.4
NaOH(s)	40.00	−425.61	−379.49	64.46	59.54
NaCl(s)	58.44	−411.15	−384.14	72.13	50.50
NaBr(s)	102.90	−361.06	−348.98	86.82	51.38
NaI(s)	149.89	−287.78	−286.06	98.53	52.09
Sulfur					
S(s, α) (rhombic)	32.06	0	0	31.80	22.64
S(s, β) (monoclinic)	32.06	+0.33	+0.1	32.6	23.6
S(g)	32.06	+278.81	+238.25	167.82	23.673
S_2(g)	64.13	+128.37	+79.30	228.18	32.47
S^{2-}(aq)	32.06	+33.1	+85.8	−14.6	
SO_2(g)	64.06	−296.83	−300.19	248.22	39.87
SO_3(g)	80.06	−395.72	−371.06	256.76	50.67
H_2SO_4(l)	98.08	−813.99	−690.00	156.90	138.9
H_2SO_4(aq)	98.08	−909.27	−744.53	20.1	−293
SO_4^{2-}(aq)	96.06	−909.27	−744.53	+20.1	−293
HSO_4^-(aq)	97.07	−887.34	−755.91	+131.8	−84

Table 2.6 (Continued)

	M/(g mol^{-1})	$\Delta_f H^\ominus$/(kJ mol^{-1})	$\Delta_f G^\ominus$/(kJ mol^{-1})	$S^\ominus_m$/(J K^{-1} mol^{-1})†	$C^\ominus_{p,m}$/(J K^{-1} mol^{-1})
Sulfur (Continued)					
$H_2S(g)$	34.08	−20.63	−33.56	205.79	34.23
$H_2S(aq)$	34.08	−39.7	−27.83	121	
$HS^-(aq)$	33.072	−17.6	+12.08	+62.08	
$SF_6(g)$	146.05	−1209	−1105.3	291.82	97.28
Tin					
Sn(s, β)	118.69	0	0	51.55	26.99
Sn(g)	118.69	+302.1	+267.3	168.49	20.26
$Sn^{2+}(aq)$	118.69	−8.8	−27.2	−17	
SnO(s)	134.69	−285.8	−256.9	56.5	44.31
$SnO_2(s)$	150.69	−580.7	−519.6	52.3	52.59
Xenon					
Xe(g)	131.30	0	0	169.68	20.786
Zinc					
Zn(s)	65.37	0	0	41.63	25.40
Zn(g)	65.37	+130.73	+95.14	160.98	20.79
$Zn^{2+}(aq)$	65.37	−153.89	−147.06	−112.1	46
ZnO(s)	81.37	−348.28	−318.30	43.64	40.25

Source: NBS; † Standard entropies of ions may be either positive or negative because the values are relative to the entropy of the hydrogen ion.

Table 2.6a Standard enthalpies of hydration at infinite dilution, $\Delta_{hyd}H^\ominus$/(kJ mol^{-1})

	Li^+	Na^+	K^+	Rb^+	Cs^+
F^-	−1026	−911	−828	−806	−782
Cl^-	−884	−783	−685	−664	−640
Br^-	−856	−742	−658	−637	−613
I^-	−815	−701	−617	−596	−572

Entries refer to $X^+(g) + Y^-(g) \rightarrow X^+(aq) + Y^-(aq)$.
Data: Principally J.O'M. Bockris and A.K.N. Reddy, *Modern electrochemistry*, Vol. 1. Plenum Press, New York (1970).

Table 2.6b Standard ion hydration enthalpies, $\Delta_{hyd}H^\ominus$/(kJ mol^{-1}) at 298 K

Cations					
H^+	(−1090)	Ag^+	−464	Mg^{2+}	−1920
Li^+	−520	NH_4^+	−301	Ca^{2+}	−1650
Na^+	−405			Sr^{2+}	−1480
K^+	−321			Ba^{2+}	−1360
Rb^+	−300			Fe^{2+}	−1950
Cs^+	−277			Cu^{2+}	−2100
				Zn^{2+}	−2050
				Al^{3+}	−4690
				Fe^{3+}	−4430

Anions							
OH^-	−460						
F^-	−506	Cl^-	−364	Br^-	−337	I^-	−296

Entries refer to $X^{\pm}(g) \rightarrow X^{\pm}(aq)$ based on $H^+(g) \rightarrow H^+(aq)$; $\Delta H^\ominus = -1090$ kJ mol^{-1}.
Data: Principally J.O'M. Bockris and A.K.N. Reddy, *Modern electrochemistry*, Vol. 1. Plenum Press, New York (1970).

Table 2.7 Benson thermochemical groups

Group	$\Delta_f H^\ominus/(\text{kJ mol}^{-1})$	$S^\ominus_m/(\text{J K}^{-1}\text{ mol}^{-1})$	$C^\ominus_{p,m}/(\text{J K}^{-1}\text{ mol}^{-1})$
$C(H)_3(C)$	−42.17	127.2	25.9
$C(H)_2(C)_2$	−20.7	39.4	22.8
$C(H)(C)_3$	−6.19	−50.50	18.7
$C(C)_4$	+8.16	−146.9	18.2
$C(Cl)(H)_2(C)$	−65.7	158	37
$C(Br)(H)_2(C)$	−22	169	38
$C(I)(H)_2(C)$	+37		
$C(Cl)(H)(C)_2$	−60.2	74.1	36
$C(Br)(H)(C)_2$	−9.6	84.9	
$C(Cl)(H)(C)_3$	−53.1	−32	36
$C(Br)(H)(C)_3$		−18	39
$C(Cl)_3(C)$		210	66.1
$[O{-}(C)(H)] + [C{-}(O)(H)_3]$	−48.1	59.5	10.5
$[O{-}(C)(C)] + 2[C{-}(O)(H)_3]$	−45.3	69.4	15.7
$[C{-}(O)(C)(H)_2] + 2[C{-}(O)(H)_3]$	+2.0	−20.4	−1.1

Data: S.W. Benson, *Thermochemical kinetics*. McGraw-Hill, New York (1976).

Table 3.1 Expansion coefficients, α, and isothermal compressibilities, κ_T

	$\alpha/(10^{-4}\text{ K}^{-1})$	$\kappa_T/(10^{-6}\text{ atm}^{-1})$
Liquids		
Benzene	12.4	92.1
Carbon tetrachloride	12.4	90.5
Ethanol	11.2	76.8
Mercury	1.82	38.7
Water	2.1	49.6
Solids		
Copper	0.501	0.735
Diamond	0.030	0.187
Iron	0.354	0.589
Lead	0.861	2.21

The values refer to 20°C
Data: AIP(α), KL(κ_T).

Table 3.2 Inversion temperatures, normal freezing and boiling points, and Joule–Thomson coefficients at 1 atm and 298 K

	T_I/K	T_f/K	T_b/K	$\mu_{JT}/(\text{K atm}^{-1})$
Air	603			0.189 at 50°C
Argon	723	83.8	87.3	
Carbon dioxide	1500	194.7s		1.11 at 300 K
Helium	40		4.22	−0.062
Hydrogen	202	14.0	20.3	−0.03
Krypton	1090	116.6	120.8	
Methane	968	90.6	111.6	
Neon	231	24.5	27.1	
Nitrogen	621	63.3	77.4	0.27
Oxygen	764	54.8	90.2	0.31

s: sublimes
Data: AIP, JL, and M.W. Zemansky, *Heat and thermodynamics*. McGraw-Hill, New York (1957).

Table 4.1 Standard entropies (and temperatures) of phase transitions, $\Delta_{trs}S^{\ominus}/(\mathrm{J\,K^{-1}\,mol^{-1}})$

	Fusion (at T_f)	Vaporization (at T_b)
Ar	14.17 (at 83.8 K)	74.53 (at 87.3 K)
Br_2	39.76 (at 265.9 K)	88.61 (at 332.4 K)
C_6H_6	38.00 (at 278.6 K)	87.19 (at 353.2 K)
CH_3COOH	40.4 (at 289.8 K)	61.9 (at 391.4 K)
CH_3OH	18.03 (at 175.2 K)	104.6 (at 337.2 K)
Cl_2	37.22 (at 172.1 K)	85.38 (at 239.0 K)
H_2	8.38 (at 14.0 K)	44.96 (at 20.38 K)
H_2O	22.00 (at 273.2 K)	109.0 (at 373.2 K)
H_2S	12.67 (at 187.6 K)	87.75 (at 212.0 K)
He	4.8 (at 1.8 K and 30 bar)	19.9 (at 4.22 K)
N_2	11.39 (at 63.2 K)	75.22 (at 77.4 K)
NH_3	28.93 (at 195.4 K)	97.41 (at 239.73 K)
O_2	8.17 (at 54.4 K)	75.63 (at 90.2 K)

Data: AIP

Table 4.2 Standard entropies of vaporization of liquids at their normal boiling point

	$\Delta_{vap}H^{\ominus}/(\mathrm{kJ\,mol^{-1}})$	$\theta_b/°C$	$\Delta_{vap}S^{\ominus}/(\mathrm{J\,K^{-1}\,mol^{-1}})$
Benzene	30.8	80.1	+87.2
Carbon disulfide	26.74	46.25	+83.7
Carbon tetrachloride	30.00	76.7	+85.8
Cyclohexane	30.1	80.7	+85.1
Decane	38.75	174	+86.7
Dimethyl ether	21.51	−23	+86
Ethanol	38.6	78.3	+110.0
Hydrogen sulfide	18.7	−60.4	+87.9
Mercury	59.3	356.6	+94.2
Methane	8.18	−161.5	+73.2
Methanol	35.21	65.0	+104.1
Water	40.7	100.0	+109.1

Data: JL

Table 4.3 Standard Third-Law entropies at 298 K: see Tables 2.5 and 2.6

Table 4.4 Standard Gibbs energies of formation at 298 K: see Tables 2.5 and 2.6

Table 5.2 The fugacity coefficient of nitrogen at 273 K

p/atm	ϕ	p/atm	ϕ
1	0.999 55	300	1.0055
10	0.9956	400	1.062
50	0.9812	600	1.239
100	0.9703	800	1.495
150	0.9672	1000	1.839
200	0.9721		

Data: LR

Table 6.1 Surface tensions of liquids at 293 K

	γ/(mN m^{-1})
Benzene	28.88
Carbon tetrachloride	27.0
Ethanol	22.8
Hexane	18.4
Mercury	472
Methanol	22.6
Water	72.75
	72.0 at 25°C
	58.0 at 100°C

Data: KL

Table 7.1 Henry's law constants for gases at 298 K, K/(10 MPa)

	Water	Benzene
CH_4	4.19	0.0569
CO_2	0.167	0.0114
H_2	7.12	0.367
N_2	8.68	0.239
O_2	4.40	

Data: R.J. Silbey and R.A. Alberty, *Physical chemistry*, Wiley, New York (2001).

Table 7.2 Cryoscopic and ebullioscopic constants

	K_f/(K kg mol^{-1})	K_b/(K kg mol^{-1})
Acetic acid	3.90	3.07
Benzene	5.12	2.53
Camphor	40	
Carbon disulfide	3.8	2.37
Carbon tetrachloride	30	4.95
Naphthalene	6.94	5.8
Phenol	7.27	3.04
Water	1.86	0.51

Data: KL

Table 9.1 Acidity constants for aqueous solutions at 298 K. (a) In order of acid strength

Acid	HA	A^-	K_a	pK_a
Hydriodic	HI	I^-	10^{11}	−11
Hydrobromic	HBr	Br^-	10^9	−9
Hydrochloric	HCl	Cl^-	10^7	−7
Sulfuric	H_2SO_4	HSO_4^-	10^2	−2
Perchloric*	$HClO_4$	ClO_4^-	4.0×10^1	−1.6
Hydronium ion	H_3O^+	H_2O	1	0.0
Oxalic	$(COOH)_2$	$HOOCCO_2^-$	5.6×10^{-2}	1.25
Sulfurous	H_2SO_3	HSO_3^-	1.4×10^{-2}	1.85
Hydrogensulfate ion	HSO_4^-	SO_4^{2-}	1.0×10^{-2}	1.99
Phosphoric	H_3PO_4	$H_2PO_4^-$	6.9×10^{-3}	2.16
Glycinium ion	$^+NH_3CH_2COOH$	NH_2CH_2COOH	4.5×10^{-3}	2.35
Hydrofluoric	HF	F^-	6.3×10^{-4}	3.20
Formic	HCOOH	HCO_2^-	1.8×10^{-4}	3.75
Hydrogenoxalate ion	$HOOCCO_2^-$	$C_2O_4^{2-}$	1.5×10^{-5}	3.81
Lactic	$CH_3CH(OH)COOH$	$CH_3CH(OH)CO_2^-$	1.4×10^{-4}	3.86
Acetic (ethanoic)	CH_3COOH	$CH_3CO_2^-$	1.4×10^{-5}	4.76
Butanoic	$CH_3CH_2CH_2COOH$	$CH_3CH_2CH_2CO_2^-$	1.5×10^{-5}	4.83
Propanoic	CH_3CH_2COOH	$CH_3CH_2CO_2^-$	1.4×10^{-5}	4.87
Anilinium ion	$C_6H_5NH_3^+$	$C_6H_5NH_2$	1.3×10^{-5}	4.87
Pyridinium ion	$C_5H_5NH^+$	C_6H_5N	5.9×10^{-6}	5.23
Carbonic	H_2CO_3	HCO_3^-	4.5×10^{-7}	6.35
Hydrosulfuric	H_2S	HS^-	8.9×10^{-8}	7.05
Dihydrogenphosphate ion	$H_2PO_4^-$	HPO_4^{2-}	6.2×10^{-8}	7.21
Hypochlorous	HClO	ClO^-	4.0×10^{-8}	7.40
Hydrazinium ion	$NH_2NH_3^+$	NH_2NH_2	8×10^{-9}	8.1
Hypobromous	HBrO	BrO^-	2.8×10^{-9}	8.55
Hydrocyanic	HCN	CN^-	6.2×10^{-10}	9.21
Ammonium ion	NH_4^+	NH_3	5.6×10^{-10}	9.25
Boric*	$B(OH)_3$	$B(OH)_4^-$	5.4×10^{-10}	9.27
Trimethylammonium ion	$(CH_3)_3NH^+$	$(CH_3)_3N$	1.6×10^{-10}	9.80
Phenol	C_6H_5OH	$C_6H_5O^-$	1.0×10^{-10}	9.99
Hydrogencarbonate ion	HCO_3^-	CO_3^{2-}	4.8×10^{-11}	10.33
Hypoiodous	HIO	IO^-	3×10^{-11}	10.5
Ethylammonium ion	$CH_3CH_2NH_3^+$	$CH_3CH_2NH_2$	2.2×10^{-11}	10.65
Methylammonium ion	$CH_3NH_3^+$	CH_3NH_2	2.2×10^{-11}	10.66
Dimethylammonium ion	$(CH_3)_2NH_2^+$	$(CH_3)_2NH$	1.9×10^{-11}	10.73
Triethylammonium ion	$(CH_3CH_2)_3NH^+$	$(CH_3CH_2)_3N$	1.8×10^{-11}	10.75
Diethylammonium ion	$(CH_3CH_2)_2NH_2^+$	$(CH_3CH_2)_2NH$	1.4×10^{-11}	10.84
Hydrogenarsenate ion	$HAsO_4^{2-}$	AsO_4^{3-}	5.1×10^{-12}	11.29
Hydrogenphosphate ion	HPO_4^{2-}	PO_4^{3-}	4.8×10^{-13}	12.32
Hydrogensulfide ion	HS^-	S^{2-}	1.0×10^{-19}	19.00

* At 293 K.

Table 9.1 Acidity constants for aqueous solutions at 298 K. (b) In alphabetical order

Acid	HA	A^-	K_a	pK_a
Acetic (ethanoic)	CH_3COOH	$CH_3CO_2^-$	1.4×10^{-5}	4.76
Ammonium ion	NH_4^+	NH_3	5.6×10^{-10}	9.25
Anilinium ion	$C_6H_5NH_3^+$	$C_6H_5NH_2$	1.3×10^{-5}	4.87
Boric*	$B(OH)_3$	$B(OH)_4^-$	5.4×10^{-10}	9.27
Butanoic	$CH_3CH_2CH_2COOH$	$CH_3CH_2CH_2CO_2^-$	1.5×10^{-5}	4.83
Carbonic	H_2CO_3	HCO_3^-	4.5×10^{-7}	6.35
Diethylammonium ion	$(CH_3CH_2)_2NH_2^+$	$(CH_3CH_2)_2NH$	1.4×10^{-11}	10.84
Dihydrogenphosphate ion	$H_2PO_4^-$	HPO_4^{2-}	6.2×10^{-8}	7.21
Dimethylammonium ion	$(CH_3)_2NH_2^+$	$(CH_3)_2NH$	1.9×10^{-11}	10.73
Ethylammonium ion	$CH_3CH_2NH_3^+$	$CH_3CH_2NH_2$	2.2×10^{-11}	10.65
Formic	$HCOOH$	HCO_2^-	1.8×10^{-4}	3.75
Glycinium ion	$^+NH_3CH_2COOH$	NH_2CH_2COOH	4.5×10^{-3}	2.35
Hydrazinium ion	$NH_2NH_3^+$	NH_2NH_2	8×10^{-9}	8.1
Hydriodic	HI	I^-	10^{11}	−11
Hydrobromic	HBr	Br^-	10^9	−9
Hydrochloric	HCl	Cl^-	10^7	−7
Hydrocyanic	HCN	CN^-	6.2×10^{-10}	9.21
Hydrofluoric	HF	F^-	6.3×10^{-4}	3.20
Hydrogenarsenate ion	$HAsO_4^{2-}$	AsO_4^{3-}	5.1×10^{-12}	11.29
Hydrogencarbonate ion	HCO_3^-	CO_3^{2-}	4.8×10^{-11}	10.33
Hydrogenoxalate ion	$HOOCCO_2^-$	$C_2O_4^{2-}$	1.5×10^{-5}	3.81
Hydrogenphosphate ion	HPO_4^{2-}	PO_4^{3-}	4.8×10^{-13}	12.32
Hydrogensulfate ion	HSO_4^-	SO_4^{2-}	1.0×10^{-2}	1.99
Hydrogensulfide ion	HS^-	S^{2-}	1.0×10^{-19}	19.00
Hydronium ion	H_3O^+	H_2O	1	0.0
Hydrosulfuric	H_2S	HS^-	8.9×10^{-8}	7.05
Hypobromous	HBrO	BrO^-	2.8×10^{-9}	8.55
Hypochlorous	HClO	ClO^-	4.0×10^{-8}	7.40
Hypoiodous	HIO	IO^-	3×10^{-11}	10.5
Lactic	$CH_3CH(OH)COOH$	$CH_3CH(OH)CO_2^-$	1.4×10^{-4}	3.86
Methylammonium ion	$CH_3NH_3^+$	CH_3NH_2	2.2×10^{-11}	10.66
Oxalic	$(COOH)_2$	$HOOCCO_2^-$	5.6×10^{-2}	1.25
Perchloric*	$HClO_4$	ClO_4^-	4.0×10^{1}	−1.6
Phenol	C_6H_5OH	$C_6H_5O^-$	1.0×10^{-10}	9.99
Phosphoric	H_3PO_4	$H_2PO_4^-$	6.9×10^{-3}	2.16
Propanoic	CH_3CH_2COOH	$CH_3CH_2CO_2^-$	1.4×10^{-5}	4.87
Pyridinium ion	$C_5H_5NH^+$	C_6H_5N	5.9×10^{-6}	5.23
Sulfuric	H_2SO_4	HSO_4^-	10^2	−2
Sulfurous	H_2SO_3	HSO_3^-	1.4×10^{-2}	1.85
Triethylammonium ion	$(CH_3CH_2)_3NH^+$	$(CH_3CH_2)_3N$	1.8×10^{-11}	10.75
Trimethylammonium ion	$(CH_3)_3NH^+$	$(CH_3)_3N$	1.6×10^{-10}	9.80

* At 293 K.

Table 10.1 Standard thermodynamic functions of ions in aqueous solution at 298 K. See Table 2.6

Table 10.2 Relative permittivities (dielectric constants) at 293 K

Nonpolar molecules		Polar molecules	
Methane (at −173°C)	1.655	Water	78.54 (at 298 K)
			80.10
Carbon tetrachloride	2.238	Ammonia	16.9 (at 298 K)
			22.4 at −33°C
Cyclohexane	2.024	Hydrogen sulfide	9.26 at −85°C
			5.93 (at 283 K)
Benzene	2.283	Methanol	33.0
		Ethanol	25.3
		Nitrobenzene	35.6

Data: HCP

Table 10.3 Standard molar entropies of ions in aqueous solution at 298 K. See Table 2.6

Table 10.5 Mean activity coefficients in water at 298 K

$b/b^{\ominus}$	HCl	KCl	$CaCl_2$	H_2SO_4	$LaCl_3$	$In_2(SO_4)_3$
0.001	0.966	0.966	0.888	0.830	0.790	
0.005	0.929	0.927	0.789	0.639	0.636	0.16
0.01	0.905	0.902	0.732	0.544	0.560	0.11
0.05	0.830	0.816	0.584	0.340	0.388	0.035
0.10	0.798	0.770	0.524	0.266	0.356	0.025
0.50	0.769	0.652	0.510	0.155	0.303	0.014
1.00	0.811	0.607	0.725	0.131	0.387	
2.00	1.011	0.577	1.554	0.125	0.954	

Data: RS, HCP, and S. Glasstone, *Introduction to electrochemistry*. Van Nostrand (1942).

Table 10.7 Standard potentials at 298 K. (a) In electrochemical order

Reduction half-reaction	$E^{\ominus}/V$	Reduction half-reaction	$E^{\ominus}/V$
Strongly oxidizing		$Cu^{2+} + e^- \longrightarrow Cu^+$	+0.16
$H_4XeO_6 + 2H^+ + 2e^- \longrightarrow XeO_3 + 3H_2O$	+3.0	$Sn^{4+} + 2e^- \longrightarrow Sn^{2+}$	+0.15
$F_2 + 2e^- \longrightarrow 2F^-$	+2.87	$AgBr + e^- \longrightarrow Ag + Br^-$	+0.07
$O_3 + 2H^+ + 2e^- \longrightarrow O_2 + H_2O$	+2.07	$Ti^{4+} + e^- \longrightarrow Ti^{3+}$	0.00
$S_2O_8^{2-} + 2e^- \longrightarrow 2SO_4^{2-}$	+2.05	$2H^+ + 2e^- \longrightarrow H_2$	0, by definition
$Ag^{2+} + e^- \longrightarrow Ag^+$	+1.98	$Fe^{3+} + 3e^- \longrightarrow Fe$	−0.04
$Co^{3+} + e^- \longrightarrow Co^{2+}$	+1.81	$O_2 + H_2O + 2e^- \longrightarrow HO_2^- + OH^-$	−0.08
$H_2O_2 + 2H^+ + 2e^- \longrightarrow 2H_2O$	+1.78	$Pb^{2+} + 2e^- \longrightarrow Pb$	−0.13
$Au^+ + e^- \longrightarrow Au$	+1.69	$In^+ + e^- \longrightarrow In$	−0.14
$Pb^{4+} + 2e^- \longrightarrow Pb^{2+}$	+1.67	$Sn^{2+} + 2e^- \longrightarrow Sn$	−0.14
$2HClO + 2H^+ + 2e^- \longrightarrow Cl_2 + 2H_2O$	+1.63	$AgI + e^- \longrightarrow Ag + I^-$	−0.15
$Ce^{4+} + e^- \longrightarrow Ce^{3+}$	+1.61	$Ni^{2+} + 2e^- \longrightarrow Ni$	−0.23
$2HBrO + 2H^+ + 2e^- \longrightarrow Br_2 + 2H_2O$	+1.60	$Co^{2+} + 2e^- \longrightarrow Co$	−0.28
$MnO_4^- + 8H^+ + 5e^- \longrightarrow Mn^{2+} + 4H_2O$	+1.51	$In^{3+} + 3e^- \longrightarrow In$	−0.34
$Mn^{3+} + e^- \longrightarrow Mn^{2+}$	+1.51	$Tl^+ + e^- \longrightarrow Tl$	−0.34
$Au^{3+} + 3e^- \longrightarrow Au$	+1.40	$PbSO_4 + 2e^- \longrightarrow Pb + SO_4^{2-}$	−0.36
$Cl_2 + 2e^- \longrightarrow 2Cl^-$	+1.36	$Ti^{3+} + e^- \longrightarrow Ti^{2+}$	−0.37
$Cr_2O_7^{2-} + 14H^+ + 6e^- \longrightarrow 2Cr^{3+} + 7H_2O$	+1.33	$Cd^{2+} + 2e^- \longrightarrow Cd$	−0.40
$O_3 + H_2O + 2e^- \longrightarrow O_2 + 2OH^-$	+1.24	$In^{2+} + e^- \longrightarrow In^+$	−0.40
$O_2 + 4H^+ + 4e^- \longrightarrow 2H_2O$	+1.23	$Cr^{3+} + e^- \longrightarrow Cr^{2+}$	−0.41
$ClO_4^- + 2H^+ + 2e^- \longrightarrow ClO_3^- + H_2O$	+1.23	$Fe^{2+} + 2e^- \longrightarrow Fe$	−0.44
$MnO_2 + 4H^+ + 2e^- \longrightarrow Mn^{2+} + 2H_2O$	+1.23	$In^{3+} + 2e^- \longrightarrow In^+$	−0.44
$Br_2 + 2e^- \longrightarrow 2Br^-$	+1.09	$S + 2e^- \longrightarrow S^{2-}$	−0.48
$Pu^{4+} + e^- \longrightarrow Pu^{3+}$	+0.97	$In^{3+} + e^- \longrightarrow In^{2+}$	−0.49
$NO_3^- + 4H^+ + 3e^- \longrightarrow NO + 2H_2O$	+0.96	$U^{4+} + e^- \longrightarrow U^{3+}$	−0.61
$2Hg^{2+} + 2e^- \longrightarrow Hg_2^{2+}$	+0.92	$Cr^{3+} + 3e^- \longrightarrow Cr$	−0.74
$ClO^- + H_2O + 2e^- \longrightarrow Cl^- + 2OH^-$	+0.89	$Zn^{2+} + 2e^- \longrightarrow Zn$	−0.76
$Hg^{2+} + 2e^- \longrightarrow Hg$	+0.86	$Cd(OH)_2 + 2e^- \longrightarrow Cd + 2OH^-$	−0.81
$NO_3^- + 2H^+ + e^- \longrightarrow NO_2 + H_2O$	+0.80	$2H_2O + 2e^- \longrightarrow H_2 + 2OH^-$	−0.83
$Ag^+ + e^- \longrightarrow Ag$	+0.80	$Cr^{2+} + 2e^- \longrightarrow Cr$	−0.91
$Hg_2^{2+} + 2e^- \longrightarrow 2Hg$	+0.79	$Mn^{2+} + 2e^- \longrightarrow Mn$	−1.18
$Fe^{3+} + e^- \longrightarrow Fe^{2+}$	+0.77	$V^{2+} + 2e^- \longrightarrow V$	−1.19
$BrO^- + H_2O + 2e^- \longrightarrow Br^- + 2OH^-$	+0.76	$Ti^{2+} + 2e^- \longrightarrow Ti$	−1.63
$Hg_2SO_4 + 2e^- \longrightarrow 2Hg + SO_4^{2-}$	+0.62	$Al^{3+} + 3e^- \longrightarrow Al$	−1.66
$MnO_4^{2-} + 2H_2O + 2e^- \longrightarrow MnO_2 + 4OH^-$	+0.60	$U^{3+} + 3e^- \longrightarrow U$	−1.79
$MnO_4^- + e^- \longrightarrow MnO_4^{2-}$	+0.56	$Sc^{3+} + 3e^- \longrightarrow Sc$	−2.09
$I_2 + 2e^- \longrightarrow 2I^-$	+0.54	$Mg^{2+} + 2e^- \longrightarrow Mg$	−2.36
$Cu^+ + e^- \longrightarrow Cu$	+0.52	$Ce^{3+} + 3e^- \longrightarrow Ce$	−2.48
$I_3^- + 2e^- \longrightarrow 3I^-$	+0.53	$La^{3+} + 3e^- \longrightarrow La$	−2.52
$NiOOH + H_2O + e^- \longrightarrow Ni(OH)_2 + OH^-$	+0.49	$Na^+ + e^- \longrightarrow Na$	−2.71
$Ag_2CrO_4 + 2e^- \longrightarrow 2Ag + CrO_4^{2-}$	+0.45	$Ca^{2+} + 2e^- \longrightarrow Ca$	−2.87
$O_2 + 2H_2O + 4e^- \longrightarrow 4OH^-$	+0.40	$Sr^{2+} + 2e^- \longrightarrow Sr$	−2.89
$ClO_4^- + H_2O + 2e^- \longrightarrow ClO_3^- + 2OH^-$	+0.36	$Ba^{2+} + 2e^- \longrightarrow Ba$	−2.91
$[Fe(CN)_6]^{3-} + e^- \longrightarrow [Fe(CN)_6]^{4-}$	+0.36	$Ra^{2+} + 2e^- \longrightarrow Ra$	−2.92
$Cu^{2+} + 2e^- \longrightarrow Cu$	+0.34	$Cs^+ + e^- \longrightarrow Cs$	−2.92
$Hg_2Cl_2 + 2e^- \longrightarrow 2Hg + 2Cl^-$	+0.27	$Rb^+ + e^- \longrightarrow Rb$	−2.93
$AgCl + e^- \longrightarrow Ag + Cl^-$	+0.22	$K^+ + e^- \longrightarrow K$	−2.93
$Bi^{3+} + 3e^- \longrightarrow Bi$	+0.20	$Li^+ + e^- \longrightarrow Li$	−3.05

Table 10.7 Standard electrode potentials at 298 K. (b) In alphabetical order

Reduction half-reaction	$E^{\ominus}/V$	Reduction half-reaction	$E^{\ominus}/V$
$Ag^+ + e^- \longrightarrow Ag$	+0.80	$I_2 + 2e^- \longrightarrow 2I^-$	+0.54
$Ag^{2+} + e^- \longrightarrow Ag^+$	+1.98	$I_3^- + 2e^- \longrightarrow 3I^-$	+0.53
$AgBr + e^- \longrightarrow Ag + Br^-$	+0.0713	$In^+ + e^- \longrightarrow In$	−0.14
$AgCl + e^- \longrightarrow Ag + Cl^-$	+0.22	$In^{2+} + e^- \longrightarrow In^+$	−0.40
$Ag_2CrO_4 + 2e^- \longrightarrow 2Ag + CrO_4^{2-}$	+0.45	$In^{3+} + 2e^- \longrightarrow In^+$	−0.44
$AgF + e^- \longrightarrow Ag + F^-$	+0.78	$In^{3+} + 3e^- \longrightarrow In$	−0.34
$AgI + e^- \longrightarrow Ag + I^-$	−0.15	$In^{3+} + e^- \longrightarrow In^{2+}$	−0.49
$Al^{3+} + 3e^- \longrightarrow Al$	−1.66	$K^+ + e^- \longrightarrow K$	−2.93
$Au^+ + e^- \longrightarrow Au$	+1.69	$La^{3+} + 3e^- \longrightarrow La$	−2.52
$Au^{3+} + 3e^- \longrightarrow Au$	+1.40	$Li^+ + e^- \longrightarrow Li$	−3.05
$Ba^{2+} + 2e^- \longrightarrow Ba$	+2.91	$Mg^{2+} + 2e^- \longrightarrow Mg$	−2.36
$Be^{2+} + 2e^- \longrightarrow Be$	−1.85	$Mn^{2+} + 2e^- \longrightarrow Mn$	−1.18
$Bi^{3+} + 3e^- \longrightarrow Bi$	+0.20	$Mn^{3+} + e^- \longrightarrow Mn^{2+}$	+1.51
$Br_2 + 2e^- \longrightarrow 2Br^-$	+1.09	$MnO_2 + 4H^+ + 2e^- \longrightarrow Mn^{2+} + 2H_2O$	+1.23
$BrO^- + H_2O + 2e^- \longrightarrow Br^- + 2OH^-$	+0.76	$MnO_4^- + 8H^+ + 5e^- \longrightarrow Mn^{2+} + 4H_2O$	+1.51
$Ca^{2+} + 2e^- \longrightarrow Ca$	−2.87	$MnO_4^- + e^- \longrightarrow MnO_4^{2-}$	+0.56
$Cd(OH)_2 + 2e^- \longrightarrow Cd + 2OH^-$	−0.81	$MnO_4^{2-} + 2H_2O + 2e^- \longrightarrow MnO_2 + 4OH^-$	+0.60
$Cd^{2+} + 2e^- \longrightarrow Cd$	−0.40	$Na^+ + e^- \longrightarrow Na$	−2.71
$Ce^{3+} + 3e^- \longrightarrow Ce$	−2.48	$Ni^{2+} + 2e^- \longrightarrow Ni$	−0.23
$Ce^{4+} + e^- \longrightarrow Ce^{3+}$	+1.61	$NiOOH + H_2O + e^- \longrightarrow Ni(OH)_2 + OH^-$	+0.49
$Cl_2 + 2e^- \longrightarrow 2Cl^-$	+1.36	$NO_3^- + 2H^+ + e^- \longrightarrow NO_2 + H_2O$	−0.80
$ClO^- + H_2O + 2e^- \longrightarrow Cl^- + 2OH^-$	+0.89	$NO_3^- + 4H^+ + 3e^- \longrightarrow NO + 2H_2O$	+0.96
$ClO_4^- + 2H^+ + 2e^- \longrightarrow ClO_3^- + H_2O$	+1.23	$NO_3^- + H_2O + 2e^- \longrightarrow NO_2^- + 2OH^-$	+0.10
$ClO_4^- + H_2O + 2e^- \longrightarrow ClO_3^- + 2OH^-$	+0.36	$O_2 + 2H_2O + 4e^- \longrightarrow 4OH^-$	+0.40
$Co^{2+} + 2e^- \longrightarrow Co$	−0.28	$O_2 + 4H^+ + 4e^- \longrightarrow 2H_2O$	+1.23
$Co^{3+} + e^- \longrightarrow Co^{2+}$	+1.81	$O_2 + e^- \longrightarrow O_2^-$	−0.56
$Cr^{2+} + 2e^- \longrightarrow Cr$	−0.91	$O_2 + H_2O + 2e^- \longrightarrow HO_2^- + OH^-$	−0.08
$Cr_2O_7^{2-} + 14H^+ + 6e^- \longrightarrow 2Cr^{3+} + 7H_2O$	+1.33	$O_3 + 2H^+ + 2e^- \longrightarrow O_2 + H_2O$	+2.07
$Cr^{3+} + 3e^- \longrightarrow Cr$	−0.74	$O_3 + H_2O + 2e^- \longrightarrow O_2 + 2OH^-$	+1.24
$Cr^{3+} + e^- \longrightarrow Cr^{2+}$	−0.41	$Pb^{2+} + 2e^- \longrightarrow Pb$	−0.13
$Cs^+ + e^- \longrightarrow Cs$	−2.92	$Pb^{4+} + 2e^- \longrightarrow Pb^{2+}$	+1.67
$Cu^+ + e^- \longrightarrow Cu$	+0.52	$PbSO_4 + 2e^- \longrightarrow Pb + SO_4^{2-}$	−0.36
$Cu^{2+} + 2e^- \longrightarrow Cu$	+0.34	$Pt^{2+} + 2e^- \longrightarrow Pt$	+1.20
$Cu^{2+} + e^- \longrightarrow Cu^+$	+0.16	$Pu^{4+} + e^- \longrightarrow Pu^{3+}$	+0.97
$F_2 + 2e^- \longrightarrow 2F^-$	+2.87	$Ra^{2+} + 2e^- \longrightarrow Ra$	−2.92
$Fe^{2+} + 2e^- \longrightarrow Fe$	−0.44	$Rb^+ + e^- \longrightarrow Rb$	−2.93
$Fe^{3+} + 3e^- \longrightarrow Fe$	−0.04	$S + 2e^- \longrightarrow S^{2-}$	−0.48
$Fe^{3+} + e^- \longrightarrow Fe^{2+}$	+0.77	$S_2O_8^{2-} + 2e^- \longrightarrow 2SO_4^{2-}$	+2.05
$[Fe(CN)_6]^{3-} + e^- \longrightarrow [Fe(CN)_6]^{4-}$	+0.36	$Sc^{3+} + 3e^- \longrightarrow Sc$	−2.09
$2H^+ + 2e^- \longrightarrow H_2$	0, by definition	$Sn^{2+} + 2e^- \longrightarrow Sn$	−0.14
$2H_2O + 2e^- \longrightarrow H_2 + 2OH^-$	−0.83	$Sn^{4+} + 2e^- \longrightarrow Sn^{2+}$	+0.15
$2HBrO + 2H^+ + 2e^- \longrightarrow Br_2 + 2H_2O$	+1.60	$Sr^{2+} + 2e^- \longrightarrow Sr$	−2.89
$2HClO + 2H^+ + 2e^- \longrightarrow Cl_2 + 2H_2O$	+1.63	$Ti^{2+} + 2e^- \longrightarrow Ti$	−1.63
$H_2O_2 + 2H^+ + 2e^- \longrightarrow 2H_2O$	+1.78	$Ti^{3+} + e^- \longrightarrow Ti^{2+}$	−0.37
$H_4XeO_6 + 2H^+ + 2e^- \longrightarrow XeO_3 + 3H_2O$	+3.0	$Ti^{4+} + e^- \longrightarrow Ti^{3+}$	0.00
$Hg_2^{2+} + 2e^- \longrightarrow 2Hg$	+0.79	$Tl^+ + e^- \longrightarrow Tl$	−0.34
$Hg_2Cl_2 + 2e^- \longrightarrow 2Hg + 2Cl^-$	+0.27	$U^{3+} + 3e^- \longrightarrow U$	−1.79
$Hg^{2+} + 2e^- \longrightarrow Hg$	+0.86	$U^{4+} + e^- \longrightarrow U^{3+}$	−0.61
$2Hg^{2+} + 2e^- \longrightarrow Hg_2^{2+}$	+0.92	$V^{2+} + 2e^- \longrightarrow V$	−1.19
$Hg_2SO_4 + 2e^- \longrightarrow 2Hg + SO_4^{2-}$	+0.62	$V^{3+} + e^- \longrightarrow V^{2+}$	−0.26
		$Zn^{2+} + 2e^- \longrightarrow Zn$	−0.76

Table 12.2 The error function

z	erf z	z	erf z
0	0	0.45	0.475 48
0.01	0.011 28	0.50	0.520 50
0.02	0.022 56	0.55	0.563 32
0.03	0.033 84	0.60	0.603 86
0.04	0.045 11	0.65	0.642 03
0.05	0.056 37	0.70	0.677 80
0.06	0.067 62	0.75	0.711 16
0.07	0.078 86	0.80	0.742 10
0.08	0.090 08	0.85	0.770 67
0.09	0.101 28	0.90	0.796 91
0.10	0.112 46	0.95	0.820 89
0.15	0.168 00	1.00	0.842 70
0.20	0.222 70	1.20	0.910 31
0.25	0.276 32	1.40	0.952 28
0.30	0.328 63	1.60	0.976 35
0.35	0.379 38	1.80	0.989 09
0.40	0.428 39	2.00	0.995 32

Data: AS

Table 13.2 Screening constants for atoms; values of $Z_{eff} = Z - \sigma$ for neutral ground-state atoms

	H							He
1*s*	1							1.6875
	Li	Be	B	C	N	O	F	Ne
1*s*	2.6906	3.6848	4.6795	5.6727	6.6651	7.6579	8.6501	9.6421
2*s*	1.2792	1.9120	2.5762	3.2166	3.8474	4.4916	5.1276	5.7584
2*p*			2.4214	3.1358	3.8340	4.4532	5.1000	5.7584
	Na	Mg	Al	Si	P	S	Cl	Ar
1*s*	10.6259	11.6089	12.5910	13.5745	14.5578	15.5409	16.5239	17.5075
2*s*	6.5714	7.3920	8.3736	9.0200	9.8250	10.6288	11.4304	12.2304
2*p*	6.8018	7.8258	8.9634	9.9450	10.9612	11.9770	12.9932	14.0082
3*s*	2.5074	3.3075	4.1172	4.9032	5.6418	6.3669	7.0683	7.7568
3*p*			4.0656	4.2852	4.8864	5.4819	6.1161	6.7641

Data: E. Clementi and D.L. Raimondi, *Atomic screening constants from SCF functions*. IBM Res. Note NJ-27 (1963). *J. chem. Phys* **38**, 2686 (1963).

Table 13.3 Ionization energies, $I/(\text{kJ mol}^{-1})$

H							He
1312.0							2372.3
							5250.4
Li	Be	B	C	N	O	F	Ne
513.3	899.4	800.6	1086.2	1402.3	1313.9	1681	2080.6
7298.0	1757.1	2427	2352	2856.1	3388.2	3374	3952.2
Na	Mg	Al	Si	P	S	Cl	Ar
495.8	737.7	577.4	786.5	1011.7	999.6	1251.1	1520.4
4562.4	1450.7	1816.6	1577.1	1903.2	2251	2297	2665.2
		2744.6		2912			
K	Ca	Ga	Ge	As	Se	Br	Kr
418.8	589.7	578.8	762.1	947.0	940.9	1139.9	1350.7
3051.4	1145	1979	1537	1798	2044	2104	2350
		2963	2735				
Rb	Sr	In	Sn	Sb	Te	I	Xe
403.0	549.5	558.3	708.6	833.7	869.2	1008.4	1170.4
2632	1064.2	1820.6	1411.8	1794	1795	1845.9	2046
		2704	2943.0	2443			
Cs	Ba	Tl	Pb	Bi	Po	At	Rn
375.5	502.8	589.3	715.5	703.2	812	930	1037
2420	965.1	1971.0	1450.4	1610			
		2878	3081.5	2466			

Data: E

Table 13.4 Electron affinities, $E_{ea}/(\text{kJ mol}^{-1})$

H							He
72.8							−21
Li	Be	B	C	N	O	F	Ne
59.8	≤0	23	122.5	−7	141	322	−29
					−844		
Na	Mg	Al	Si	P	S	Cl	Ar
52.9	≤0	44	133.6	71.7	200.4	348.7	−35
					−532		
K	Ca	Ga	Ge	As	Se	Br	Kr
48.3	2.37	36	116	77	195.0	324.5	−39
Rb	Sr	In	Sn	Sb	Te	I	Xe
46.9	5.03	34	121	101	190.2	295.3	−41
Cs	Ba	Tl	Pb	Bi	Po	At	Rn
45.5	13.95	30	35.2	101	186	270	−41

Data: E

Table 14.2 Bond lengths, R_e/pm

(a) Bond lengths in specific molecules

Br_2	228.3
Cl_2	198.75
CO	112.81
F_2	141.78
H_2^+	106
H_2	74.138
HBr	141.44
HCl	127.45
HF	91.680
HI	160.92
N_2	109.76
O_2	120.75

(b) Mean bond lengths from covalent radii*

H	37						
C	77(1) 67(2) 60(3)	N	74(1) 65(2)	O	66(1) 57(2)	F	64
Si	118	P	110	S	104(1) 95(2)	Cl	99
Ge	122	As	121	Se	104	Br	114
		Sb	141	Te	137	I	133

* Values are for single bonds except where indicated otherwise (values in parentheses). The length of an A—B covalent bond (of given order) is the sum of the corresponding covalent radii.

Table 14.3a Bond dissociation enthalpies, $\Delta H^{\ominus}$(A—B)/(kJ mol^{-1}) at 298 K

Diatomic molecules

H—H	436	F—F	155	Cl—Cl	242	Br—Br	193	I—I	151
O=O	497	C=O	1076	N≡N	945				
H—O	428	H—F	565	H—Cl	431	H—Br	366	H—I	299

Polyatomic molecules

H—CH_3	435	H—NH_2	460	H—OH	492	H—C_6H_5	469
H_3C—CH_3	368	H_2C=CH_2	720	HC≡CH	962		
HO—CH_3	377	Cl—CH_3	352	Br—CH_3	293	I—CH_3	237
O=CO	531	HO—OH	213	O_2N—NO_2	54		

Data: HCP, KL

Table 14.3b Mean bond enthalpies, $\Delta H^{\ominus}$(A—B)/(kJ mol^{-1})

	H	C	N	O	F	Cl	Br	I	S	P	Si
H	436										
C	412	348(i) 612(ii) 838(iii) 518(a)									
N	388	305(i) 613(ii) 890(iii)	163(i) 409(ii) 946(iii)								
O	463	360(i) 743(ii)	157	146(i) 497(ii)							
F	565	484	270	185	155						
Cl	431	338	200	203	254	242					
Br	366	276				219	193				
I	299	238				210	178	151			
S	338	259			496	250	212		264		
P	322									201	
Si	318		374	466							226

(i) Single bond, (ii) double bond, (iii) triple bond, (a) aromatic.
Data: HCP and L. Pauling, *The nature of the chemical bond*. Cornell University Press (1960).

Table 14.4 Pauling (*italics*) and Mulliken electronegativities

H							**He**
2.20							
3.06							
Li	**Be**	**B**	**C**	**N**	**O**	**F**	**Ne**
0.98	*1.57*	*2.04*	*2.55*	*3.04*	*3.44*	*3.98*	
1.28	1.99	1.83	2.67	3.08	3.22	4.43	4.60
Na	**Mg**	**Al**	**Si**	**P**	**S**	**Cl**	**Ar**
0.93	*1.31*	*1.61*	*1.90*	*2.19*	*2.58*	*3.16*	
1.21	1.63	1.37	2.03	2.39	2.65	3.54	3.36
K	**Ca**	**Ga**	**Ge**	**As**	**Se**	**Br**	**Kr**
0.82	*1.00*	*1.81*	*2.01*	*2.18*	*2.55*	*2.96*	*3.0*
1.03	1.30	1.34	1.95	2.26	2.51	3.24	2.98
Rb	**Sr**	**In**	**Sn**	**Sb**	**Te**	**I**	**Xe**
0.82	*0.95*	*1.78*	*1.96*	*2.05*	*2.10*	*2.66*	*2.6*
0.99	1.21	1.30	1.83	2.06	2.34	2.88	2.59
Cs	**Ba**	**Tl**	**Pb**	**Bi**			
0.79	*0.89*	*2.04*	*2.33*	*2.02*			

Data: Pauling values: A.L. Allred, *J. Inorg. Nucl. Chem.* **17**, 215 (1961); L.C. Allen and J.E. Huheey, *ibid.*, **42**, 1523 (1980). Mulliken values: L.C. Allen, *J. Am. Chem. Soc.* **111**, 9003 (1989). The Mulliken values have been scaled to the range of the Pauling values.

Table 16.2 Properties of diatomic molecules

	$\tilde{\nu}_0$/cm^{-1}	θ_V/K	B/cm^{-1}	θ_R/K	r/pm	k/(N m^{-1})	D/(kJ mol^{-1})	σ
$^1H_2^+$	2321.8	3341	29.8	42.9	106	160	255.8	2
$^1H_2$	4400.39	6332	60.864	87.6	74.138	574.9	432.1	2
$^2H_2$	3118.46	4487	30.442	43.8	74.154	577.0	439.6	2
$^1H^{19}F$	4138.32	5955	20.956	30.2	91.680	965.7	564.4	1
$^1H^{35}Cl$	2990.95	4304	10.593	15.2	127.45	516.3	427.7	1
$^1H^{81}Br$	2648.98	3812	8.465	12.2	141.44	411.5	362.7	1
$^1H^{127}I$	2308.09	3321	6.511	9.37	160.92	313.8	294.9	1
$^{14}N_2$	2358.07	3393	1.9987	2.88	109.76	2293.8	941.7	2
$^{16}O_2$	1580.36	2274	1.4457	2.08	120.75	1176.8	493.5	2
$^{19}F_2$	891.8	1283	0.8828	1.27	141.78	445.1	154.4	2
$^{35}Cl_2$	559.71	805	0.2441	0.351	198.75	322.7	239.3	2
$^{12}C^{16}O$	2170.21	3122	1.9313	2.78	112.81	1903.17	1071.8	1
$^{79}Br^{81}Br$	323.2	465	0.0809	10.116	283.3	245.9	190.2	1

Data: AIP

Table 16.3 Typical vibrational wavenumbers, $\tilde{\nu}$/cm^{-1}

C—H stretch	2850–2960
C—H bend	1340–1465
C—C stretch, bend	700–1250
C═C stretch	1620–1680
C≡C stretch	2100–2260
O—H stretch	3590–3650
H-bonds	3200–3570
C═O stretch	1640–1780
C≡N stretch	2215–2275
N—H stretch	3200–3500
C—F stretch	1000–1400
C—Cl stretch	600–800
C—Br stretch	500–600
C—I stretch	500
CO_3^{2-}	1410–1450
NO_3^-	1350–1420
NO_2^-	1230–1250
SO_4^{2-}	1080–1130
Silicates	900–1100

Data: L.J. Bellamy, *The infrared spectra of complex molecules* and *Advances in infrared group frequencies*. Chapman and Hall.

Table 17.1 Colour, frequency, and energy of light

Colour	λ/nm	ν/(10^{14} Hz)	$\tilde{\nu}$/(10^4 cm^{-1})	E/eV	E/(kJ mol^{-1})
Infrared	>1000	<3.00	<1.00	<1.24	<120
Red	700	4.28	1.43	1.77	171
Orange	620	4.84	1.61	2.00	193
Yellow	580	5.17	1.72	2.14	206
Green	530	5.66	1.89	2.34	226
Blue	470	6.38	2.13	2.64	254
Violet	420	7.14	2.38	2.95	285
Near ultraviolet	300	10.0	3.33	4.15	400
Far ultraviolet	<200	>15.0	>5.00	>6.20	>598

Data: J.G. Calvert and J.N. Pitts, *Photochemistry*. Wiley, New York (1966).

Table 17.3 Absorption characteristics of some groups and molecules

Group	$\tilde{\nu}_{max}/(10^4\ cm^{-1})$	λ_{max}/nm	$\varepsilon_{max}/(L\ mol^{-1}\ cm^{-1})$
C=C ($\pi^* \leftarrow \pi$)	6.10	163	1.5×10^4
	5.73	174	5.5×10^3
C=O ($\pi^* \leftarrow n$)	3.7–3.5	270–290	10–20
—N=N—	2.9	350	15
	>3.9	<260	Strong
$—NO_2$	3.6	280	10
	4.8	210	1.0×10^4
C_6H_5—	3.9	255	200
	5.0	200	6.3×10^3
	5.5	180	1.0×10^5
$[Cu(OH_2)_6]^{2+}(aq)$	1.2	810	10
$[Cu(NH_3)_4]^{2+}(aq)$	1.7	600	50
H_2O ($\pi^* \leftarrow n$)	6.0	167	7.0×10^3

Table 18.2 Nuclear spin properties

Nuclide	Natural abundance %	Spin I	Magnetic moment μ/μ_N	g-value	$\gamma/(10^7\ T^{-1}\ s^{-1})$	NMR frequency at 1 T, ν/MHz
^1n*		$\frac{1}{2}$	−1.9130	−3.8260	−18.324	29.164
^1H	99.9844	$\frac{1}{2}$	2.792 85	5.5857	26.752	42.576
^2H	0.0156	1	0.857 44	0.857 45	4.1067	6.536
^3H*		$\frac{1}{2}$	2.978 96	−4.2553	−20.380	45.414
^{10}B	19.6	3	1.8006	0.6002	2.875	4.575
^{11}B	80.4	$\frac{3}{2}$	2.6886	1.7923	8.5841	13.663
^{13}C	1.108	$\frac{1}{2}$	0.7024	1.4046	6.7272	10.708
^{14}N	99.635	1	0.403 76	0.403 56	1.9328	3.078
^{17}O	0.037	$\frac{5}{2}$	−1.893 79	−0.7572	−3.627	5.774
^{19}F	100	$\frac{1}{2}$	2.628 87	5.2567	25.177	40.077
^{31}P	100	$\frac{1}{2}$	1.1316	2.2634	10.840	17.251
^{33}S	0.74	$\frac{3}{2}$	0.6438	0.4289	2.054	3.272
^{35}Cl	75.4	$\frac{3}{2}$	0.8219	0.5479	2.624	4.176
^{37}Cl	24.6	$\frac{3}{2}$	0.6841	0.4561	2.184	3.476

* Radioactive.

μ is the magnetic moment of the spin state with the largest value of m_I: $\mu = g_I \mu_N I$ and μ_N is the nuclear magneton (see inside front cover).

Data: KL and HCP

Table 18.3 Hyperfine coupling constants for atoms, a/mT

Nuclide	Spin	Isotropic coupling	Anisotropic coupling
^{1}H	$\frac{1}{2}$	50.8(1s)	
^{2}H	1	7.8(1s)	
^{13}C	$\frac{1}{2}$	113.0(2s)	6.6(2p)
^{14}N	1	55.2(2s)	4.8(2p)
^{19}F	$\frac{1}{2}$	1720(2s)	108.4(2p)
^{31}P	$\frac{1}{2}$	364(3s)	20.6(3p)
^{35}Cl	$\frac{3}{2}$	168(3s)	10.0(3p)
^{37}Cl	$\frac{3}{2}$	140(3s)	8.4(3p)

Data: P.W. Atkins and M.C.R. Symons, *The structure of inorganic radicals*. Elsevier, Amsterdam (1967).

Table 21.1 Dipole moments, polarizabilities, and polarizability volumes

	$\mu/(10^{-30}$ C m)	μ/D	$\alpha'/(10^{-30}$ m^3)	$\alpha/(10^{-40}$ J^{-1} C^2 m^2)
Ar	0	0	1.66	1.85
C_2H_5OH	5.64	1.69		
$C_6H_5CH_3$	1.20	0.36		
C_6H_6	0	0	10.4	11.6
CCl_4	0	0	10.5	11.7
CH_2Cl_2	5.24	1.57	6.80	7.57
CH_3Cl	6.24	1.87	4.53	5.04
CH_3OH	5.70	1.71	3.23	3.59
CH_4	0	0	2.60	2.89
$CHCl_3$	3.37	1.01	8.50	9.46
CO	0.390	0.117	1.98	2.20
CO_2	0	0	2.63	2.93
H_2	0	0	0.819	0.911
H_2O	6.17	1.85	1.48	1.65
HBr	2.67	0.80	3.61	4.01
HCl	3.60	1.08	2.63	2.93
He	0	0	0.20	0.22
HF	6.37	1.91	0.51	0.57
HI	1.40	0.42	5.45	6.06
N_2	0	0	1.77	1.97
NH_3	4.90	1.47	2.22	2.47
1,2-$C_6H_4(CH_3)_2$	2.07	0.62		

Data: HCP and C.J.F. Böttcher and P. Bordewijk, *Theory of electric polarization*. Elsevier, Amsterdam (1978).

Table 21.2 Refractive indices relative to air at 20°C

	434 nm	589 nm	656 nm
Benzene	1.5236	1.5012	1.4965
Carbon tetrachloride	1.4729	1.4676	1.4579
Carbon disulfide	1.6748	1.6276	1.6182
Ethanol	1.3700	1.3618	1.3605
KCl(s)	1.5050	1.4904	1.4973
KI(s)	1.7035	1.6664	1.6581
Methanol	1.3362	1.3290	1.3277
Methylbenzene	1.5170	1.4955	1.4911
Water	1.3404	1.3330	1.3312

Data: AIP

Table 21.4 Lennard-Jones (12,6)-potential parameters

	(ε/k)/K	r_0/pm
Ar	111.84	362.3
C_2H_2	209.11	463.5
C_2H_4	200.78	458.9
C_2H_6	216.12	478.2
C_6H_6	377.46	617.4
CCl_4	378.86	624.1
Cl_2	296.27	448.5
CO_2	201.71	444.4
F_2	104.29	357.1
Kr	154.87	389.5
N_2	91.85	391.9
O_2	113.27	365.4
Xe	213.96	426.0

Source: F. Cuadros, I. Cachadiña, and W. Ahamuda, *Molec. Engineering*, **6**, 319 (1996).

Table 22.1 Radius of gyration of some macromolecules

	M/(kg mol^{-1})	R_g/nm
Serum albumin	66	2.98
Myosin	493	46.8
Polystyrene	3.2×10^3	50 (in poor solvent)
DNA	4×10^3	117.0
Tobacco mosaic virus	3.9×10^4	92.4

Data: C. Tanford, *Physical chemistry of macromolecules*. Wiley, New York (1961).

Table 22.2 Diffusion coefficients of macromolecules in water at 20°C

	M/(kg mol^{-1})	D/(10^{-10} m^2 s^{-1})
Sucrose	0.342	4.586
Ribonuclease	13.7	1.19
Lysozyme	14.1	1.04
Serum albumin	65	0.594
Haemoglobin	68	0.69
Urease	480	0.346
Collagen	345	0.069
Myosin	493	0.116

Data: C. Tanford, *Physical chemistry of macromolecules*. Wiley, New York (1961).

Table 22.3 Frictional coefficients and molecular geometry

Major axis/Minor axis	Prolate	Oblate
2	1.04	1.04
3	1.11	1.10
4	1.18	1.17
5	1.25	1.22
6	1.31	1.28
7	1.38	1.33
8	1.43	1.37
9	1.49	1.42
10	1.54	1.46
50	2.95	2.38
100	4.07	2.97

Data: K.E. Van Holde, *Physical biochemistry*. Prentice-Hall, Englewood Cliffs (1971).

Sphere; radius a, $c = a f_0$

Prolate ellipsoid; major axis $2a$, minor axis $2b$, $c = (ab^2)^{1/3}$

$$f = \left\{ \frac{(1 - b^2/a^2)^{1/2}}{(b/a)^{2/3} \ln\{[1 + (1 - b^2/a^2)^{1/2}]/(b/a)\}} \right\} f_0$$

Oblate ellipsoid; major axis $2a$, minor axis $2b$, $c = (a^2 b)^{1/3}$

$$f = \left\{ \frac{(a^2/b^2 - 1)^{1/2}}{(a/b)^{2/3} \arctan[(a^2/b^2 - 1)^{1/2}]} \right\} f_0$$

Long rod; length l, radius a, $c = (3a^2/4)^{1/3}$

$$f = \left\{ \frac{(1/2a)^{2/3}}{(3/2)^{1/3}\{2 \ln(l/a) - 0.11\}} \right\} f_0$$

In each case $f_0 = 6\pi\eta c$ with the appropriate value of c.

Table 22.4 Intrinsic viscosity

Macromolecule	Solvent	θ/°C	$K/(10^{-3}\ cm^3\ g^{-1})$	a
Polystyrene	Benzene	25	9.5	0.74
	Cyclohexane	34†	81	0.50
Polyisobutylene	Benzene	23†	83	0.50
	Cyclohexane	30	26	0.70
Amylose	0.33 M KCl(aq)	25†	113	0.50
Various proteins‡	Guanidine hydrochloride + $HSCH_2CH_2OH$		7.16	0.66

† The θ temperature.
‡ Use $[\eta] = KN^a$; N is the number of amino acid residues.
Data: K.E. Van Holde, *Physical biochemistry*. Prentice-Hall, Englewood Cliffs (1971).

Table 23.3 Ionic radii (r/pm)†

Li^+(4)	Be^{2+}(4)	B^{3+}(4)	N^{3-}	O^{2-}(6)	F^-(6)		
59	27	12	171	140	133		
Na^+(6)	Mg^{2+}(6)	Al^{3+}(6)	P^{3-}	S^{2-}(6)	Cl^-(6)		
102	72	53	212	184	181		
K^+(6)	Ca^{2+}(6)	Ga^{3+}(6)	As^{3-}(6)	Se^{2-}(6)	Br^-(6)		
138	100	62	222	198	196		
Rb^+(6)	Sr^{2+}(6)	In^{3+}(6)		Te^{2-}(6)	I^-(6)		
149	116	79		221	220		
Cs^+(6)	Ba^{2+}(6)	Tl^{3+}(6)					
167	136	88					
***d*-block elements (high-spin ions)**							
Sc^{3+}(6)	Ti^{4+}(6)	Cr^{3+}(6)	Mn^{3+} (6)	Fe^{2+}(6)	Co^{3+}(6)	Cu^{2+}(6)	Zn^{2+}(6)
73	60	61	65	63	61	73	75

† Numbers in parentheses are the coordination numbers of the ions. Values for ions without a coordination number stated are estimates.
Data: R.D. Shannon and C.T. Prewitt, *Acta Cryst.* **B25**, 925 (1969).

Table 23.5 Lattice enthalpies, $\Delta H_L^\ominus/(\text{kJ mol}^{-1})$

	F	Cl	Br	I
Halides				
Li	1037	852	815	761
Na	926	787	752	705
K	821	717	689	649
Rb	789	695	668	632
Cs	750	676	654	620
Ag	969	912	900	886
Be		3017		
Mg		2524		
Ca		2255		
Sr		2153		

Oxides							
MgO	3850	CaO	3461	SrO	3283	BaO	3114
Sulfides							
MgS	3406	CaS	3119	SrS	2974	BaS	2832

Entries refer to $MX(s) \rightarrow M^+(g) + X^-(g)$.
Data: Principally D. Cubicciotti, *J. Chem. Phys.* **31**, 1646 (1959).

Table 23.6 Magnetic susceptibilities at 298 K

	$\chi/10^{-6}$	$\chi_m/(10^{-4}\ \text{cm}^3\ \text{mol}^{-1})$
Water	−90	−16.0
Benzene	−7.2	−6.4
Cyclohexane	−7.9	−8.5
Carbon tetrachloride	−8.9	−8.4
NaCl(s)	−13.9	−3.75
Cu(s)	−96	−6.8
S(s)	−12.9	−2.0
Hg(l)	−28.5	−4.2
$CuSO_4 \cdot 5H_2O(s)$	+176	+192
$MnSO_4 \cdot 4H_2O(s)$	+2640	$+2.79 \times 10^3$
$NiSO_4 \cdot 7H_2O(s)$	+416	+600
$FeSO_4(NH_4)_2SO_4 \cdot 6H_2O(s)$	+755	$+1.51 \times 10^3$
Al(s)	+22	+2.2
Pt(s)	+262	+22.8
Na(s)	+7.3	+1.7
K(s)	+5.6	+2.5

Data: KL and $\chi_m = \chi M/\rho$.

Table 24.1 Collision cross-sections, σ/nm^2

Ar	0.36
C_2H_4	0.64
C_6H_6	0.88
CH_4	0.46
Cl_2	0.93
CO_2	0.52
H_2	0.27
He	0.21
N_2	0.43
Ne	0.24
O_2	0.40
SO_2	0.58

Data: KL

Table 24.2 Transport properties of gases at 1 atm

	$\kappa/(\text{J K}^{-1}\ \text{m}^{-1}\ \text{s}^{-1})$	$\eta/\mu\text{P}$	
	273 K	273 K	293 K
Air	0.0241	173	182
Ar	0.0163	210	223
C_2H_4	0.0164	97	103
CH_4	0.0302	103	110
Cl_2	0.079	123	132
CO_2	0.0145	136	147
H_2	0.1682	84	88
He	0.1442	187	196
Kr	0.0087	234	250
N_2	0.0240	166	176
Ne	0.0465	298	313
O_2	0.0245	195	204
Xe	0.0052	212	228

Data: KL

Table 24.4 Viscosities of liquids at 298 K, $\eta/(10^{-3}\ kg\ m^{-1}\ s^{-1})$

Benzene	0.601
Carbon tetrachloride	0.880
Ethanol	1.06
Mercury	1.55
Methanol	0.553
Pentane	0.224
Sulfuric acid	27
Water†	0.891

† The viscosity of water over its entire liquid range is represented with less than 1 per cent error by the expression

$$\log(\eta_{20}/\eta) = A/B,$$

$$A = 1.370\,23(t-20) + 8.36 \times 10^{-4}(t-20)^2$$

$$B = 109 + t \qquad t = \theta/°C$$

Convert $kg\ m^{-1}\ s^{-1}$ to centipoise (cP) by multiplying by 10^3 (so $\eta \approx 1$ cP for water).
Data: AIP, KL

Table 24.5 Limiting ionic conductivities in water at 298 K, $\lambda/(mS\ m^2\ mol^{-1})$

Cations		Anions	
Ba^{2+}	12.72	Br^-	7.81
Ca^{2+}	11.90	$CH_3CO_2^-$	4.09
Cs^+	7.72	Cl^-	7.635
Cu^{2+}	10.72	ClO_4^-	6.73
H^+	34.96	CO_3^{2-}	13.86
K^+	7.350	$(CO_2)_2^{2-}$	14.82
Li^+	3.87	F^-	5.54
Mg^{2+}	10.60	$[Fe(CN)_6]^{3-}$	30.27
Na^+	5.010	$[Fe(CN)_6]^{4-}$	44.20
$[N(C_2H_5)_4]^+$	3.26	HCO_2^-	5.46
$[N(CH_3)_4]^+$	4.49	I^-	7.68
NH_4^+	7.35	NO_3^-	7.146
Rb^+	7.78	OH^-	19.91
Sr^{2+}	11.89	SO_4^{2-}	16.00
Zn^{2+}	10.56		

Data: KL, RS

Table 24.6 Ionic mobilities in water at 298 K, $u/(10^{-8}\ m^2\ s^{-1}\ V^{-1})$

Cations		Anions	
Ag^+	6.24	Br^-	8.09
Ca^{2+}	6.17	$CH_3CO_2^-$	4.24
Cu^{2+}	5.56	Cl^-	7.91
H^+	36.23	CO_3^{2-}	7.46
K^+	7.62	F^-	5.70
Li^+	4.01	$[Fe(CN)_6]^{3-}$	10.5
Na^+	5.19	$[Fe(CN)_6]^{4-}$	11.4
NH_4^+	7.63	I^-	7.96
$[N(CH_3)_4]^+$	4.65	NO_3^-	7.40
Rb^+	7.92	OH^-	20.64
Zn^{2+}	5.47	SO_4^{2-}	8.29

Data: Principally Table 24.4 and $u = \lambda/zF$.

Table 24.7 Debye–Hückel–Onsager coefficients for (1,1)-electrolytes at 25°C

Solvent	$A/(mS\ m^2\ mol^{-1}/(mol\ L^{-1})^{1/2})$	$B/(mol\ L^{-1})^{-1/2}$
Acetone (propanone)	3.28	1.63
Acetonitrile	2.29	0.716
Ethanol	8.97	1.83
Methanol	15.61	0.923
Nitrobenzene	4.42	0.776
Nitromethane	111	0.708
Water	6.020	0.229

Data: J.O'M. Bockris and A.K.N. Reddy, *Modern electrochemistry*. Plenum, New York (1970).

Table 24.8 Diffusion coefficients at 25°C, $D/(10^{-9}\ m^2\ s^{-1})$

Molecules in liquids				Ions in water			
I_2 in hexane	4.05	H_2 in CCl_4(l)	9.75	K^+	1.96	Br^-	2.08
in benzene	2.13	N_2 in CCl_4(l)	3.42	H^+	9.31	Cl^-	2.03
CCl_4 in heptane	3.17	O_2 in CCl_4(l)	3.82	Li^+	1.03	F^-	1.46
Glycine in water	1.055	Ar in CCl_4(l)	3.63	Na^+	1.33	I^-	2.05
Dextrose in water	0.673	CH_4 in CCl_4(l)	2.89			OH^-	5.03
Sucrose in water	0.5216	H_2O in water	2.26				
		CH_3OH in water	1.58				
		C_2H_5OH in water	1.24				

Data: AIP and (for the ions) $\lambda = zuF$ in conjunction with Table 24.5.

Table 25.1 Kinetic data for first-order reactions

	Phase	θ/°C	k/s^{-1}	$t_{1/2}$
$2\ N_2O_5 \longrightarrow 4\ NO_2 + O_2$	g	25	3.38×10^{-5}	5.70 h
	HNO_3(l)	25	1.47×10^{-6}	131 h
	Br_2(l)	25	4.27×10^{-5}	4.51 h
$C_2H_6 \longrightarrow 2\ CH_3$	g	700	5.36×10^{-4}	21.6 min
Cyclopropane $\longrightarrow$ propene	g	500	6.71×10^{-4}	17.2 min
$CH_3N_2CH_3 \longrightarrow C_2H_6 + N_2$	g	327	3.4×10^{-4}	34 min
Sucrose $\longrightarrow$ glucose + fructose	aq(H^+)	25	6.0×10^{-5}	3.2 h

g: High pressure gas-phase limit.
Data: Principally K.J. Laidler, *Chemical kinetics*. Harper & Row, New York (1987); M.J. Pilling and P.W. Seakins, *Reaction kinetics*. Oxford University Press (1995); J. Nicholas, *Chemical kinetics*. Harper & Row, New York (1976). See also JL.

Table 25.2 Kinetic data for second-order reactions

	Phase	θ/°C	$k/(L\ mol^{-1}\ s^{-1})$
$2\ NOBr \longrightarrow 2\ NO + Br_2$	g	10	0.80
$2\ NO_2 \longrightarrow 2\ NO + O_2$	g	300	0.54
$H_2 + I_2 \longrightarrow 2\ HI$	g	400	2.42×10^{-2}
$D_2 + HCl \longrightarrow DH + DCl$	g	600	0.141
$2\ I \longrightarrow I_2$	g	23	7×10^{9}
	hexane	50	1.8×10^{10}
$CH_3Cl + CH_3O^-$	methanol	20	2.29×10^{-6}
$CH_3Br + CH_3O^-$	methanol	20	9.23×10^{-6}
$H^+ + OH^- \longrightarrow H_2O$	water	25	1.35×10^{11}
	ice	−10	8.6×10^{12}

Data: Principally K.J. Laidler, *Chemical kinetics*. Harper & Row, New York (1987); M.J. Pilling and P.W. Seakins, *Reaction kinetics*. Oxford University Press (1995); J. Nicholas, *Chemical kinetics*. Harper & Row, New York (1976).

Table 25.4 Arrhenius parameters

First-order reactions	A/s^{-1}	$E_a/(kJ\ mol^{-1})$
Cyclopropane $\longrightarrow$ propene	1.58×10^{15}	272
$CH_3NC \longrightarrow CH_3CN$	3.98×10^{13}	160
cis-CHD=CHD $\longrightarrow$ *trans*-CHD=CHD	3.16×10^{12}	256
Cyclobutane $\longrightarrow 2\ C_2H_4$	3.98×10^{13}	261
$C_2H_5I \longrightarrow C_2H_4 + HI$	2.51×10^{17}	209
$C_2H_6 \longrightarrow 2\ CH_3$	2.51×10^{7}	384
$2\ N_2O_5 \longrightarrow 4\ NO_2 + O_2$	4.94×10^{13}	103
$N_2O \longrightarrow N_2 + O$	7.94×10^{11}	250
$C_2H_5 \longrightarrow C_2H_4 + H$	1.0×10^{13}	167

Second-order, gas-phase	$A/(L\ mol^{-1}\ s^{-1})$	$E_a/(kJ\ mol^{-1})$
$O + N_2 \longrightarrow NO + N$	1×10^{11}	315
$OH + H_2 \longrightarrow H_2O + H$	8×10^{10}	42
$Cl + H_2 \longrightarrow HCl + H$	8×10^{10}	23
$2\ CH_3 \longrightarrow C_2H_6$	2×10^{10}	ca. 0
$NO + Cl_2 \longrightarrow NOCl + Cl$	4.0×10^{9}	85
$SO + O_2 \longrightarrow SO_2 + O$	3×10^{8}	27
$CH_3 + C_2H_6 \longrightarrow CH_4 + C_2H_5$	2×10^{8}	44
$C_6H_5 + H_2 \longrightarrow C_6H_6 + H$	1×10^{8}	ca. 25

Second-order, solution	$A/(L\ mol^{-1}\ s^{-1})$	$E_a/(kJ\ mol^{-1})$
$C_2H_5ONa + CH_3I$ in ethanol	2.42×10^{11}	81.6
$C_2H_5Br + OH^-$ in water	4.30×10^{11}	89.5
$C_2H_5I + C_2H_5O^-$ in ethanol	1.49×10^{11}	86.6
$CH_3I + C_2H_5O^-$ in ethanol	2.42×10^{11}	81.6
$C_2H_5Br + OH^-$ in ethanol	4.30×10^{11}	89.5
$CO_2 + OH^-$ in water	1.5×10^{10}	38
$CH_3I + S_2O_3^{2-}$ in water	2.19×10^{12}	78.7
Sucrose + H_2O in acidic water	1.50×10^{15}	107.9
$(CH_3)_3CCl$ solvolysis		
in water	7.1×10^{16}	100
in methanol	2.3×10^{13}	107
in ethanol	3.0×10^{13}	112
in acetic acid	4.3×10^{13}	111
in chloroform	1.4×10^{4}	45
$C_6H_5NH_2 + C_6H_5COCH_2Br$		
in benzene	91	34

Data: Principally J. Nicholas, *Chemical kinetics*. Harper & Row, New York (1976) and A.A. Frost and R.G. Pearson, *Kinetics and mechanism*. Wiley, New York (1961).

Table 27.1 Arrhenius parameters for gas-phase reactions

	A/(L mol^{-1} s^{-1})		E_a/(kJ mol^{-1})	P
	Experiment	Theory		
$2\,NOCl \longrightarrow 2\,NO + Cl_2$	9.4×10^{9}	5.9×10^{10}	102.0	0.16
$2\,NO_2 \longrightarrow 2\,NO + O_2$	2.0×10^{9}	4.0×10^{10}	111.0	5.0×10^{-2}
$2\,ClO \longrightarrow Cl_2 + O_2$	6.3×10^{7}	2.5×10^{10}	0.0	2.5×10^{-3}
$H_2 + C_2H_4 \longrightarrow C_2H_6$	1.24×10^{6}	7.4×10^{11}	180	1.7×10^{-6}
$K + Br_2 \longrightarrow KBr + Br$	1.0×10^{12}	2.1×10^{11}	0.0	4.8

Data: Principally M.J. Pilling and P.W. Seakins, *Reaction kinetics*. Oxford University Press (1995).

Table 27.2 Arrhenius parameters for reactions in solution. See Table 25.4

Table 28.1 Maximum observed enthalpies of physisorption, $\Delta_{ad}H^{\ominus}$/(kJ mol^{-1})

C_2H_2	−38	H_2	−84
C_2H_4	−34	H_2O	−59
CH_4	−21	N_2	−21
Cl_2	−36	NH_3	−38
CO	−25	O_2	−21
CO_2	−25		

Data: D.O. Haywood and B.M.W. Trapnell, *Chemisorption*. Butterworth (1964).

Table 28.2 Enthalpies of chemisorption, $\Delta_{ad}H^{\ominus}$/(kJ mol^{-1})

Adsorbate	Adsorbent (substrate)											
	Ti	Ta	Nb	W	Cr	Mo	Mn	Fe	Co	Ni	Rh	Pt
H_2		−188			−188	−167	−71	−134			−117	
N_2		−586						−293				
O_2						−720					−494	−293
CO	−640							−192	−176			
CO_2	−682	−703	−552	−456	−339	−372	−222	−225	−146	−184		
NH_3				−301				−188		−155		
C_2H_4		−577		−427	−427			−285		−243	−209	

Data: D.O. Haywood and B.M.W. Trapnell, *Chemisorption*. Butterworth (1964).

Table 28.3 Activation energies of catalysed reactions

	Catalyst	E_a/(kJ mol^{-1})
$2\,HI \longrightarrow H_2 + I_2$	None	184
	Au(s)	105
	Pt(s)	59
$2\,NH_3 \longrightarrow N_2 + 3\,H_2$	None	350
	W(s)	162
$2\,N_2O \longrightarrow 2\,N_2 + O_2$	None	245
	Au(s)	121
	Pt(s)	134
$(C_2H_5)_2O$ pyrolysis	None	224
	I_2(g)	144

Data: G.C. Bond, *Heterogeneous catalysis*. Clarendon Press, Oxford (1986).

Table 29.1 Exchang current densities and transfer coefficients at 298 K

Reaction	Electrode	j_0/(A cm^{-2})	α
$2\,H^+ + 2\,e^- \longrightarrow H_2$	Pt	7.9×10^{-4}	
	Cu	1×10^{-6}	
	Ni	6.3×10^{-6}	0.58
	Hg	7.9×10^{-13}	0.50
	Pb	5.0×10^{-12}	
$Fe^{3+} + e^- \longrightarrow Fe^{2+}$	Pt	2.5×10^{-3}	0.58
$Ce^{4+} + e^- \longrightarrow Ce^{3+}$	Pt	4.0×10^{-5}	0.75

Data: Principally J.O'M. Bockris and A.K.N. Reddy, *Modern electrochemistry*. Plenum, New York (1970).

Character tables

The groups C_1, C_s, C_i

C_1 (1)	E	$h = 1$
A	1	

$C_s = C_h$ (*m*)	E	σ_h	$h = 2$	
A′	1	1	x, y, R_z	x^2, y^2, z^2, xy
A″	1	−1	z, R_x, R_y	yz, xz

$C_i = S_2$ ($\bar{1}$)	E	i	$h = 2$	
A_g	1	1	R_x, R_y, R_z	$x^2, y^2, z^2, xy, xz, yz$
A_u	1	−1	x, y, z	

The groups C_{nv}

C_{2v}, 2*mm*	E	C_2	σ_v	σ'_v	$h = 4$	
A_1	1	1	1	1	z, z^2, x^2, y^2	
A_2	1	1	−1	−1	xy	R_z
B_1	1	−1	1	−1	x, xz	R_y
B_2	1	−1	−1	1	y, yz	R_x

C_{3v}, $3m$	E	$2C_3$	$3\sigma_v$	$h = 6$	
A_1	1	1	1	$z, z^2, x^2 + y^2$	
A_2	1	1	−1		R_z
E	2	−1	0	$(x, y), (xy, x^2 - y^2)$ (xz, yz)	(R_x, R_y)

C_{4v}, $4mm$	E	C_2	$2C_4$	$2\sigma_v$	$2\sigma_d$	$h = 8$	
A_1	1	1	1	1	1	$z, z^2, x^2 + y^2$	
A_2	1	1	1	−1	1		R_z
B_1	1	1	−1	1	−1	$x^2 - y^2$	
B_2	1	1	−1	−1	1	xy	
E	2	−2	0	0	0	$(x, y), (xz, yz)$	(R_x, R_y)

C_{5v}	E	$2C_5$	$2C_5^2$	$5\sigma_v$	$h = 10$, $\alpha = 72°$	
A_1	1	1	1	1	$z, z^2, x^2 + y^2$	
A_2	1	1	1	−1		R_z
E_1	2	$2\cos\alpha$	$2\cos 2\alpha$	0	$(x, y), (xz, yz)$	(R_x, R_y)
E_2	2	$2\cos 2\alpha$	$2\cos\alpha$	0	$(xy, x^2 - y^2)$	

C_{6v}, $6mm$	E	C_2	$2C_3$	$2C_6$	$3\sigma_d$	$3\sigma_v$	$h = 12$	
A_1	1	1	1	1	1	1	$z, z^2, x^2 + y^2$	
A_2	1	1	1	1	−1	1		R_z
B_1	1	−1	1	−1	−1	1		
B_2	1	−1	1	−1	1	−1		
E_1	2	−2	−1	1	0	0	$(x, y), (xz, yz)$	(R_x, R_y)
E_2	2	2	−1	−1	0	0	$(xy, x^2 - y^2)$	

$C_{\infty v}$	E	$2C_\phi$†	$\infty\sigma_v$	$h = \infty$	
A_1 (Σ^+)	1	1	1	$z, z^2, x^2 + y^2$	
A_2 (Σ^-)	1	1	−1		R_z
E_1 (Π)	2	$2\cos\phi$	0	$(x, y), (xz, yz)$	(R_x, R_y)
E_2 (Δ)	2	$2\cos 2\phi$	0	$(xy, x^2 - y^2)$	

† There is only one member of this class if $\phi = \pi$.

The groups D_n

D_2, 222	E	C_2^z	C_2^y	C_2^x	$h = 4$	
A_1	1	1	1	1	x^2, y^2, z^2	
B_1	1	1	−1	−1	z, xy	R_z
B_2	1	−1	1	−1	y, xz	R_y
B_3	1	−1	−1	1	x, yz	R_x

D_3, 32	E	$2C_3$	$3C_2'$	$h = 6$	
A_1	1	1	1	$z^2, x^2 + y^2$	
A_2	1	1	−1	z	R_z
E	2	−1	0	$(x, y), (xz, yz), (xy, x^2 - y^2)$	(R_x, R_y)

D_4, 422	E	C_2	$2C_4$	$2C_2'$	$2C_2''$	$h = 8$	
A_1	1	1	1	1	1	$z^2, x^2 + y^2$	
A_2	1	1	1	−1	−1	z	R_z
B_1	1	1	−1	1	−1	$x^2 - y^2$	
B_2	1	1	−1	−1	1	xy	
E	2	−2	0	0	0	$(x, y), (xz, yz)$	(R_x, R_y)

The groups D_{nh}

D_{3h}, $\bar{6}2m$	E	σ_h	$2C_3$	$2S_3$	$3C_2'$	$3\sigma_v$	$h = 12$	
A_1'	1	1	1	1	1	1	$z^2, x^2 + y^2$	
A_2'	1	1	1	1	−1	−1		R_z
A_1''	1	−1	1	−1	1	−1		
A_2''	1	−1	1	−1	−1	1	z	
E′	2	2	−1	−1	0	0	$(x, y), (xy, x^2 - y^2)$	
E″	2	−2	−1	1	0	0	(xz, yz)	(R_x, R_y)

D_{4h}, $4/mmm$	E	$2C_4$	C_2	$2C_2'$	$2C_2''$	i	$2S_4$	σ_h	$2\sigma_v$	$2\sigma_d$	$h = 16$	
A_{1g}	1	1	1	1	1	1	1	1	1	1	$x^2 + y^2, z^2$	
A_{2g}	1	1	1	−1	−1	1	1	1	−1	−1		R_z
B_{1g}	1	−1	1	1	−1	1	−1	1	1	−1	$x^2 - y^2$	
B_{2g}	1	−1	1	−1	1	1	−1	1	−1	1	xy	
E_g	2	0	−2	0	0	2	0	−2	0	0	(xz, yz)	(R_x, R_y)
A_{1u}	1	1	1	1	1	−1	−1	−1	−1	−1		
A_{2u}	1	1	1	−1	−1	−1	−1	−1	1	1	z	
B_{1u}	1	−1	1	1	−1	−1	1	−1	−1	1		
B_{2u}	1	−1	1	−1	1	−1	1	−1	1	−1		
E_u	2	0	−2	0	0	−2	0	2	0	0	(x, y)	

D_{5h}	E	$2C_5$	$2C_5^2$	$5C_2$	σ_h	$2S_5$	$2S_5^3$	$5\sigma_v$	$h = 20$	$\alpha = 72°$
A_1'	1	1	1	1	1	1	1	1	$x^2 + y^2, z^2$	
A_2'	1	1	1	−1	1	1	1	−1		R_z
E_1'	2	2 cos α	2 cos 2α	0	2	2 cos α	2 cos 2α	0	(x, y)	
E_2'	2	2 cos 2α	2 cos α	0	2	2 cos 2α	2 cos α	0	$(x^2–y^2, xy)$	
A_1''	1	1	1	1	−1	−1	−1	−1		
A_2''	1	1	1	−1	−1	−1	−1	1	z	
E_1''	2	2 cos α	2 cos 2α	0	−2	−2 cos α	−2 cos 2α	0	(xz, yz)	(R_x, R_y)
E_2''	2	2 cos 2α	2 cos α	0	−2	−2 cos 2α	−2 cos α	0		

$D_{\infty h}$	E	$2C_\phi$	$\infty C_2'$	i	$2iC_\infty$	iC_2'	$h = \infty$	
A_{1g} (Σ_g^+)	1	1	1	1	1	1	$z^2, x^2 + y^2$	
A_{1u} (Σ_u^+)	1	1	1	−1	−1	−1	z	
A_{2g} (Σ_g^-)	1	1	−1	1	1	−1		R_z
A_{2u} (Σ_u^-)	1	1	−1	−1	1	1		
E_{1g} (Π_g)	2	2 cos ϕ	0	2	−2 cos ϕ	0	(xz, yz)	(R_x, R_y)
E_{1u} (Π_u)	2	2 cos ϕ	0	−2	2 cos ϕ	0	(x, y)	
E_{2g} (Δ_g)	2	2 cos 2ϕ	0	2	2 cos 2ϕ	0	$(xy, x^2 - y^2)$	
E_{2u} (Δ_u)	2	2 cos 2ϕ	0	−2	−2 cos 2ϕ	0		
⋮								

The cubic groups

T_d, $\bar{4}3m$	E	$8C_3$	$3C_2$	$6\sigma_d$	$6S_4$	$h = 24$	
A_1	1	1	1	1	1	$x^2 + y^2 + z^2$	
A_2	1	1	1	−1	−1		
E	2	−1	2	0	0	$(3z^2 - r^2, x^2 - y^2)$	
T_1	3	0	−1	−1	1		(R_x, R_y, R_z)
T_2	3	0	−1	1	−1	$(x, y, z), (xy, xz, yz)$	

O_h $(m3m)$	E	$8C_3$	$6C_2$	$6C_2$	$3C_2$ $(= C_4^2)$	i	$6S_4$	$8S_6$	$3\sigma_h$	$6\sigma_d$	$h = 48$	
A_{1g}	1	1	1	1	1	1	1	1	1	1		$x^2 + y^2 + z^2$
A_{2g}	1	1	−1	−1	1	1	−1	1	1	−1		
E_g	2	−1	0	0	2	2	0	−1	2	0		$(2z^2 - x^2 - y^2, x^2 - y^2)$
T_{1g}	3	0	−1	1	−1	3	1	0	−1	−1	(R_x, R_y, R_z)	
T_{2g}	3	0	1	−1	−1	3	−1	0	−1	1		(xy, yz, xy)
A_{1u}	1	1	1	1	1	−1	−1	−1	−1	−1		
A_{2u}	1	1	−1	−1	1	−1	1	−1	−1	1		
E_u	2	−1	0	0	2	−2	0	1	−2	0		
T_{1u}	3	0	−1	1	−1	−3	−1	0	1	1	(x, y, z)	
T_{2u}	3	0	1	−1	−1	−3	1	0	1	−1		

The icosahedral group

I	E	$12C_5$	$12C_5^2$	$20C_3$	$15C_2$	$h = 60$	
A	1	1	1	1	1		$z^2 + y^2 + z^2$
T_1	3	$\frac{1}{2}(1+\sqrt{5})$	$\frac{1}{2}(1-\sqrt{5})$	0	−1	(x, y, z) (R_x, R_y, R_z)	
T_2	3	$\frac{1}{2}(1-\sqrt{5})$	$\frac{1}{2}(1+\sqrt{5})$	0	−1		
G	4	−1	−1	1	0		
G	5	0	0	−1	1		$(2z^2 - x^2 - y^2, x^2 - y^2, xy, yz, zx)$

Further information: P.W. Atkins, M.S. Child, and C.S.G. Phillips, *Tables for group theory*. Oxford University Press (1970).

Answers to exercises

Detailed solutions to the (a) exercises can be found in the *Student's Solutions Manual for Physical Chemistry*, seventh edition, by P.W. Atkins, C.A. Trapp., M. Cady, and C. Giunta

Answers to 'a' exercises

1.4 10 atm.
1.5 (a) 24 atm; (b) 22 atm.
1.6 (a) 2.57 kTorr; (b) 3.38 atm.
1.7 30 K.
1.8 30 lb in^{-2}.
1.9 4.20×10^{-2} atm.
1.10 8.3147 J K^{-1} mol^{-1}.
1.11 S_8.
1.12 6.2 kg.
1.13 (a) 0.758, 0.242, 561 Torr, 179 Torr;
(b) 0.780, 0.210, 0.010, 577 Torr, 155 Torr, 7 Torr.
1.14 169 g mol^{-1}.
1.15 −272°C.
1.16 (a) 1.0 atm, 8.2×10^2 atm; (b) 1.0 atm, 1.7×10^3 atm.
1.17 67.8 mL mol^{-1}, 54.5 atm, 120 K.
1.18 (a) 0.88; (b) 1.2 L.
1.19 140 atm.
1.20 (a) 0.1353 L mol^{-1}, 0.6957; (b) 0.658.
1.21 (a) 50.7 atm; (b) 35.2 atm, 0.695.
1.22 (a) 0.67, 0.33; (b) 2.0 atm, 1.0 atm; (c) 3.0 atm.
1.23 32.9 cm^3 mol^{-1}, 1.33 L^2 atm mol^{-2}, 0.24 nm.
1.24 (a) 1.41×10^3 K; (b) 0.59 nm.
1.25 (a) 3.64 kK, 8.7 atm; (b) 2.60 kK, 4.5 atm; (c) 46.7 K, 0.18 atm.
1.26 4.6×10^{-5} m^3 mol^{-1}, 0.66.

2.4 (a) 98 J; (b) 16 J.
2.5 2.6 kJ.
2.6 -1.0×10^2 J.
2.7 (a) $\Delta U = 0$, $\Delta H = 0$, $q = +1.57$ kJ, $w = -1.57$ kJ;
(b) $\Delta U = 0$, $\Delta H = 0$, $q = +1.13$ kJ, $w = -1.13$ kJ; (c) all 0.
2.8 1.33 atm, $\Delta U = +1.25$ kJ, $w = 0$, $q = +1.25$ kJ.
2.9 (a) −88 J; (b) −167 J.
2.10 +123 J.
2.11 $\Delta U = -37.55$ kJ, $\Delta H = -40.656$ kJ, $q = -40.656$ kJ, $w = +3.10$ kJ.
2.12 −1.5 kJ.

2.13 84.8 MJ.
2.14 (a) $\Delta U = +26.8$ kJ, $\Delta H = +28.3$ kJ, $q = +28.3$ kJ, $w = -1.45$ kJ;
(b) $\Delta U = +26.8$ kJ, $\Delta H = +28.3$ kJ, $q = +26.8$ kJ, $w = 0$.
2.15 131 K.
2.16 194 J.
2.17 22 kPa.
2.18 0.45 atm.
2.19 -125 kJ mol^{-1}.
2.20 $C_{p,m} = 30$ J K^{-1} mol^{-1}, $C_{V,m} = 22$ J K^{-1} mol^{-1}.
2.21 $q_p = -1.2$ kJ, $\Delta H = -1.2$ kJ, $C_p = 80$ J K^{-1}.
2.22 $\Delta U = +1.6$ kJ, $\Delta H = +2.2$ kJ, $q = +2.2$ kJ.
2.23 $w = -3.2$ kJ, $q = 0$, $\Delta T = -38$ K, $\Delta U = -3.2$ kJ, $\Delta H = -4.5$ kJ.
2.24 $w = +4.1$ kJ, $q = 0$, $\Delta U = +4.1$ kJ, $\Delta H = +5.4$ kJ,
$p_f = 5.2$ atm, $V_f = 11.8$ L.
2.25 9.4 L, 288 K, -0.46 kJ.
2.26 $+0.9$ mm^3.
2.27 $q = 0$, $w = \Delta U = -20$ J, $\Delta T = -0.35$ K, $\Delta H = -26$ J.
2.28 (a) 226 K; (b) 238 K.
2.29 $\Delta U = +12$ kJ, $\Delta H = +13$ kJ, $q = +13$ kJ, $w = -1.0$ kJ.
2.30 -4564.7 kJ mol^{-1}.
2.31 -126 kJ mol^{-1}.
2.32 $+53$ kJ mol^{-1}, -33 kJ mol^{-1}.
2.33 -432 kJ mol^{-1}.
2.34 641 J K^{-1}.
2.35 1.58 kJ K^{-1}, +2.05 K.
2.36 $+65.49$ kJ mol^{-1}.
2.37 -383 kJ mol^{-1}.
2.38 (a) -2205 kJ mol^{-1}; (b) -2200 kJ mol^{-1}.
2.39 (a) -57.20 kJ mol^{-1}; (b) -176.01 kJ mol^{-1}.
2.40 (a) -114.40 kJ mol^{-1}, -109.44 kJ mol^{-1};
(b) -92.31 kJ mol^{-1}, -241.82 kJ mol^{-1}.
2.41 -1368 kJ mol^{-1}.
2.42 (a) -392.1 kJ mol^{-1}; (b) -946.6 kJ mol^{-1}; (c) $+52.5$ kJ mol^{-1}.
2.43 -56.98 kJ mol^{-1}.
2.44 (a) $+131.29$ kJ mol^{-1}, $+128.81$ kJ mol^{-1};
(b) $+132.56$ kJ mol^{-1}, $+129.42$ kJ mol^{-1}.
2.45 -1892.2 kJ mol^{-1}.
2.46 (a) -124.2 kJ mol^{-1}; (b) -222.46 kJ mol^{-1}.

3.4 (a) $\partial^2 f/\partial y\partial x = \partial/\partial y(2xy) = 2x$, $\partial^2 f/\partial x\partial y = \partial/\partial x(x^2 + 6y) = 2x$;
(b) $\partial^2 f/\partial y\partial x = \partial/\partial y(\cos xy - xy \sin xy)$
$= -x \sin xy - x \sin xy - x^2 y \cos xy$
$= -2x \sin xy - x^2 y \cos xy$,
$\partial^2 f/\partial x\partial y = \partial/\partial x(-x^2 \sin xy) = -2x \sin xy - x^2 y \cos xy$.
3.5 $dz = 2axy^3\, dx + 3ax^2y^2\, dy$.
3.6 (a) $dz = (2x - 2y + 2)\, dx + (4y - 2x - 4)\, dy$.
3.7 $dz = (y + 1/x)\, dx + (x - 1)\, dy$.
3.8 $(\partial C_V/\partial V)_T = (\partial(\partial U/\partial V)_T \partial T)_V$.
3.9 $(\partial H/\partial U)_p = 1 + p(\partial V/\partial U)_p$.
3.10 $dV = (\partial V/\partial p)_T\, dp + (\partial V/\partial T)_p\, dT$; $d \ln V = -\kappa_T\, dp + \alpha\, dT$.
3.11 0, 0.
3.13 0.71 K atm^{-1}.
3.14 $\Delta U_m = +131$ J mol^{-1}, $q = +8.05 \times 10^3$ J mol^{-1},
$w = -7.92 \times 10^3$ J mol^{-1}.
3.15 1.31×10^{-3} K^{-1}.
3.16 1×10^3 atm.
3.17 -7.2 J atm^{-1} mol^{-1}, $+8.1$ kJ.
3.18 -4.2 atm.

4.4 (a) $+92$ J K^{-1}; (b) $+67$ J K^{-1}.
4.5 152.67 J K^{-1} mol^{-1}.
4.6 $+8.92$ J K^{-1}.
4.7 -22.1 J K^{-1}.
4.8 $w = +4.1$ kJ, $q = 0$, $\Delta U = +4.1$ kJ, $\Delta H = +5.4$ kJ, $\Delta S = 0$.
4.9 $+12.9$ J K^{-1}.
4.10 Not reversible.
4.11 (a) -54.9 kJ; (b) -195 J K^{-1}.
4.12 $+26$ J K^{-1}.
4.13 6.6 L.
4.14 $+2.8$ J K^{-1}.
4.15 ΔH(overall) = 0, ΔH(individual) = $\pm 1.9 \times 10^2$ kJ,
ΔS (overall) = $+93.4$ J K^{-1}.
4.16 (a) $q = 0$; (b) $w = -20$ J; (c) $\Delta U = -20$ J;
(d) $\Delta T = -0.35$ K; (e) $\Delta S = +0.60$ J K^{-1}.
4.17 (a) $+87.8$ J K^{-1} mol^{-1}; (b) -87.8 J K^{-1} mol^{-1}.
4.18 (a) -386.1 J K^{-1} mol^{-1}; (b) $+92.6$ J K^{-1} mol^{-1};
(c) -153.1 J K^{-1} mol^{-1}.
4.19 (a) -521.5 kJ mol^{-1}; (b) $+25.8$ kJ mol^{-1}; (c) -178.7 kJ mol^{-1}.
4.20 (a) -522.1 kJ mol^{-1}; (b) $+25.78$ kJ mol^{-1};
(c) -178.6 kJ mol^{-1}.
4.21 -93.05 kJ mol^{-1}.
4.22 -50 kJ mol^{-1}.
4.23 (a) $+2.9$ J K^{-1}; -2.9 J K^{-1}, 0;
(b) $+2.9$ J K^{-1}, 0, $+2.9$ J K^{-1}; (c) 0, 0, 0.
4.24 $\Delta S = n(C_{V,m} - R) \ln 2$.
4.25 817.90 kJ mol^{-1}.
4.26 0.11, 0.38.

5.4 $(\partial S/\partial V)_T = \alpha/\kappa_T$.
5.5 -3.8 J.
5.6 -36.5 J K^{-1}.
5.7 $+10$ kJ.
5.8 (a) 15.7 atm; (b) $+8.25$ kJ.
5.9 $+7.3$ kJ mol^{-1}.
5.10 -0.55 kJ mol^{-1}.
5.11 -2.63×10^{-8} Pa^{-1}, 0.88.
5.12 $+10$ kJ.
5.13 $+11$ kJ mol^{-1}.
5.14 $p - RT/(V_m \ \ b) - a/V_m^2$.
5.15 $(\partial S/\partial V)_T = nR/(V - nb)$, ΔS greater for van der Waals gas.

6.2 At 373 K, water vapour condenses to liquid. At 273 K, liquid water freezes. Ice remains at 260 K. There is a pause in the rate of cooling at 373 K and at 273 K.
6.4 303 K (30°C).
6.5 $+16$ kJ mol^{-1}, $+45.2$ J K^{-1} mol^{-1}.
6.6 $+20.80$ K J mol^{-1}.
6.7 (a) $+34.08$ kJ mol^{-1}; (b) 350.5 K.

6.8 281.8 K (8.7°C).
6.9 25 g s^{-1}.
6.10 (a) 1.7 kg; (b) 31 kg; (c) 1.4 g.
6.11 (a) +49 kJ mol^{-1}; (b) 216°C; (c) +99 J K^{-1} mol^{-1}.
6.12 272.80 K.
6.13 0.076 30 (7.6 per cent).
6.14 2.6 kPa.
6.15 72.8 mN m^{-1}.
6.16 728 kPa.

7.4 886.8 cm^3.
7.5 56.3 cm^3 mol^{-1}.
7.6 6.4 MPa.
7.7 0.13 MPa.
7.8 $K_f = 32$ K kg mol^{-1}, $K_b = 5.22$ K kg mol^{-1}.
7.9 82 g mol^{-1}.
7.10 381 g mol^{-1}.
7.11 −0.09°C.
7.12 +1.2 J K^{-1}, −0.35 kJ.
7.13 +4.7 J K^{-1} mol^{-1}.
7.14 (a) 1:1; (b) 0.8600.
7.15 (a) 3.4 mmol kg^{-1}; (b) 34 mmol kg^{-1}.
7.16 0.17 mol L^{-1}.
7.17 −0.16°C.
7.18 24 g kg^{-1}.
7.19 87 kg mol^{-1}.
7.20 Raoult's law basis: $a_A = 0.833$, $\gamma_A = 0.93$; Henry's law basis (concentration in mole fractions), $a_B = 0.125$, $\gamma_B = 1.25$; (concentration as molality) $a_B = 2.8$, $\gamma_B = 1.25$.
7.21 $p(CCl_4) = 32.2$ Torr, $p(Br_2) = 6.1$ Torr, p(Total) = 38.3 Torr, $y(CCl_4) = 0.841$, $y(Br_2) = 0.16$.
7.22 $a_A = 0.499$, $a_M = 0.668$, $\gamma_A = 1.25$, $\gamma_M = 1.11$.

8.4 $x_A = 0.920$, $y_A = 0.968$.
8.5 440 Torr, $x_A = 0.268$.
8.6 (a) yes; (b) $y_A = 0.830$.
8.7 (a) 154 Torr; (b) $y_{DE} = 0.67$.
8.8 (a) $y_M = 0.36$; (b) $y_M = 0.82$.
8.9 (a) 2; (b) 2.
8.10 2, 2.
8.11 (a) 3, 2; (b) 1.
8.13 At b_3, $C = 2$, $P = 2$, $F = 2$. The compositions of the phases are $x_A = 0.18$ and $x_A = 0.70$. At b_2, $C = 2$, $P = 1$, $F = 3$. Between the liquid line and b_1, $C = 2$, $P = 2$, $F = 2$. Above b_1, $C = 1$, $P = 1$, $F = 3$.
8.14 Incongruent melting occurs at 460°C. The composition of the eutectic is 4 per cent by mass of silver; it melts at $\theta_e = 215$°C.
8.16 (a) ≈ 80 per cent silver by mass; (b) compound decomposes; (c) ≈ 80 per cent silver by mass.
8.17 (b) 620 Torr; (c) 490 Torr; (d) $x_{Hexane} = 0.50$, $y_{Hexane} = 0.72$; (e) $y_{Hexane} = 0.50$, $x_{Hexane} = 0.30$; (f) $n_{vap} \approx 1.7$ mol, $n_{liq} \approx 0.3$ mol.
8.20 (a) The mixture has a single liquid phase at all compositions.
(b) At $x(C_6F_{14}) = 0.24$, two liquid phases separate with compositions $x = 0.24$ and $x = 0.48$. At $x > 0.48$, a single phase forms.

9.5 −2.42 kJ mol^{-1}.
9.6 3.01.
9.7 (a) 2.85×10^{-6}; (b) +240 kJ mol^{-1}; (c) 0.
9.8 (a) 0.1411; (b) +4.855 kJ mol^{-1}; (c) 14.556.
9.9 (a) −68.26 kJ mol^{-1}, 9.2×10^{11}; (b) −69.7 kJ mol^{-1}, 1.3×10^9.
9.10 (a) 0.087 (A), 0.370 (B), 0.196 (C), 0.438 (D); (b) 0.33; (c) 0.33; (d) +2.8 kJ mol^{-1}.
9.11 1.5 kK.
9.12 +2.77 kJ mol^{-1}, −16.5 J K^{-1} mol^{-1}.
9.13 +12.3 kJ mol^{-1}.
9.14 50 per cent.
9.15 0.904, 0.096.
9.16 (a, c).
9.17 (b).
9.18 (a) +53 kJ mol^{-1}; (b) −53 kJ mol^{-1}.
9.19 −14.38 kJ mol^{-1}, product formation.
9.20 1110 K.
9.21 (a) 5.40; (b) 3.61.
9.22 (a) 5.13; (b) 8.88; (c) 2.88.
9.23 8.3.
9.25 (a) Na_2HPO_4/H_3PO_4; (b) $H_2PO_4^-/HPO_4^{2-}$.

10.4 −218.66 kJ mol^{-1}.
10.5 1.25×10^{-5} mol L^{-1}.
10.6 −291 kJ mol^{-1}.
10.7 (a) $I(KCl) = b/b^\ominus$; (b) $I(FeCl_3) = 6b/b^\ominus$; (c) $I(CuSO_4) = 4b/b^\ominus$.
10.8 0.90.
10.9 (a) 2.73 g; (b) 2.92 g.
10.10 0.25 mol kg^{-1}.
10.11 $\gamma_\pm = (\gamma_+\gamma_-^2)^{1/3}$.
10.12 0.56.
10.13 1×10^4 per cent.
10.14 2.01.
10.15 −1108 kJ mol^{-1}.
10.16 +34.2 mV.
10.17 −1.18 V.
10.18 (a) $Ag^+(aq) + e^- \rightarrow Ag(s)$, $Zn^{2+}(aq) + 2e^- \rightarrow Zn(s)$, $2Ag^+(aq) + Zn(s) \rightarrow 2Ag(s) + Zn^{2+}(aq)$, +1.56 V;
(b) $H^+(aq) + e^- \rightarrow \frac{1}{2}H_2(g)$, $Cd^{2+}(aq) + 2e^- \rightarrow Cd(s)$, $Cd(s) + 2H^+(aq) \rightarrow Cd^{2+}(aq) + H_2(g)$, +0.40 V;
(c) $Cr^{3+}(aq) + 3e^- \rightarrow Cr(s)$, $[Fe(CN)_6]^{3-}(aq) + e^- \rightarrow [Fe(CN)_6]^{4-}(aq)$, $Cr^{3+}(aq) + 3[Fe(CN)_6]^{4-}(aq) \rightarrow Cr(s) + 3[Fe(CN)_6]^{3-}(aq)$, −1.10 V.
10.19 (a) $Zn(s)|ZnSO_4(aq)||CuSO_4(aq)|Cu(s)$, +1.10 V;
(b) $Pt|H_2(g)|HCl(aq)|AgCl(s)|Ag(s)$, +0.22 V;
(c) $Pt|H_2(g)|H^+(aq), H_2O(l)|O_2(g)|Pt$, +1.23 V.
10.20 (a) +1.56 V; (b) +0.40 V; (c) −1.10 V.
10.21 (a) +1.10 V; (b) +0.22 V; (c) +1.23 V.
10.22 (a) −363 kJ mol^{-1}; (b) −405 kJ mol^{-1}.
10.23 (a) +0.324 V; (b) +0.45 V.
10.24 +1.92 V.

10.25 −0.62 V.
10.26 (a) 6.5×10^9; (b) 1.5×10^{12}.
10.27 +0.49 V; 4×10^{16}.
10.28 $1.80 \times 10^{-10} \rightarrow 1.78 \times 10^{-10}$, $9.04 \times 10^{-7} \rightarrow 5.1 \times 10^{-7}$.
10.29 $E = E^{\ominus} - (RT/6F)\ln[(a(Cr^{3+})^2)/(a(Cr_2O_7^{2-})a(H^+)^{14})]$.
10.30 0.86.
10.31 0.
10.32 (a) 1×10^{-8} mol kg^{-1}; (b) 1×10^{-16}.

11.4 1.7 MW.
11.5 262 nm.
11.6 2.42 cm s^{-1}.
11.7 332 pm.
11.8 8.83×10^{-28} kg m s^{-1}, 0.969 km s^{-1}.
11.9 50.6 nm.
11.10 0.70 nm.
11.11 $E/(10^{-19}$ J), $E/$(kg mol^{-1}): (a) 3.31, 199; (b) 3.61, 218; (c) 4.97, 299.
11.12 (a) 0.66 m s^{-1}; (b) 0.72 m s^{-1}; (c) 0.99 m s^{-1}.
11.13 21 m s^{-1}.
11.14 (a) 2.8×10^{18}; (b) 2.8×10^{20}.
11.15 6 kK.
11.16 (a) no ejection; (b) 3.19×10^{-19} J, 837 km s^{-1}.
11.17 (a) 7×10^{-19} J, 400 kJ mol^{-1}; (b) 7×10^{-20} J, 40 kJ mol^{-1}; (c) 7×10^{-34} J, 4×10^{-13} kJ mol^{-1}.
11.18 (a) 6.6×10^{-29} m; (b) 6.6×10^{-36} m; (c) 99.7 pm.
11.19 1.1×10^{-28} m s^{-1}, 1×10^{-27} m.
11.20 1.12×10^{-15} J.

12.4 (a) 1.81×10^{-19} J, 110 kJ mol^{-1}, 1.1 eV, 9.1×10^3 cm^{-1}; (b) 6.6×10^{-19} J, 400 kJ mol^{-1}, 4.1 eV, 3.3×10^4 cm^{-1}.
12.5 (a) 0.04; (b) 0.
12.6 0, $h^2/4L^2$.
12.7 $L/6$, $L/2$, $5L/6$.
12.8 3.
12.9 23 per cent.
12.10 4.30×10^{-21} J.
12.11 278 N m^{-1}.
12.12 2.63 μm.
12.13 3.72 μm.
12.14 (a) 3.3×10^{-34} J; (b) 3.3×10^{-33} J.
12.16 5.61×10^{-21} J.
12.17 $N = 1/(2\pi)^{1/2}$.
12.18 1.49×10^{-34} J s; 0, $\pm 1.05 \times 10^{-34}$ J s.

13.5 14.0 eV.
13.6 $r = 4a_0$, 0.
13.7 101 pm, 376 pm.
13.8 $N = 2/a_0^{3/2}$.
13.9 $\langle V \rangle = 2E(1s)$, $\langle T \rangle = -E(1s)$.
13.10 $r^* = 5.24a_0/Z$.
13.11 (Angular momentum/$\hbar$, angular nodes, radial nodes) = (a) 0, 0, 0; (b) 0, 0, 2; (c) $6^{1/2}$, 2, 0.
13.12 (a) $\frac{5}{2}$, $\frac{3}{2}$; (b) $\frac{7}{2}$, $\frac{5}{2}$.
13.13 1, 1
13.14 (a) 1; (b) 9; (c) 25.
13.15 $L = 2$, $S = 0$, $J = 2$.
13.16 $r = 0.35a_0$.
13.17 (b, c).
13.18 (a) 2; (b) 6; (c) 10; (d) 18.
13.19 (a) [Ar]$3d^8$; (b) $S = 1$, $M_S = 0, \pm 1$, $S = 0$, $M_S = 0$.
13.20 (a) 1 (3), 0 (1); (b) $\frac{3}{2}$ (4), $\frac{1}{2}$ (2), $\frac{1}{2}$ (2).
13.21 $^3D_3$, $^3D_2$, $^3D_1$, $^1D_2$ with $^3D < {}^1D$.
13.22 (a) 0 (1); (b) $\frac{3}{2}$ (4), $\frac{1}{2}$ (2); (c) 2 (5), 1 (3), 2 (1).
13.23 (a) $^2S_{1/2}$; (b) $^2P_{3/2}$, $^2P_{1/2}$.

14.5 (a) $1\sigma^2$ (1); (b) $1\sigma^2 2\sigma^{*2}$ (0); (c) $1\sigma^2 2\sigma^{*2} 1\pi^4$ (2).
14.6 (a) $1\sigma^2 2\sigma^{*2} 1\pi^4 3\sigma^2$; (b) $1\sigma^2 2\sigma^{*2} 1\pi^4 3\sigma^2 2\pi^{*1}$; (c) $1\sigma^2 2\sigma^{*2} 1\pi^4 3\sigma^2$.
14.7 C_2.
14.8 XeF^+ is shorter.
14.9 N_2 is shorter.
14.12 $E_{trial} = 5\hbar^2/mL^2 = 1.013\,E_{true}$.
14.13 $E_{trial} = 3a\hbar^2/2\mu - (e^2/a_0)(a/2\pi^3)^{1/2}$.

15.5 E, C_3, $3\sigma_v$.
15.6 (a, b).
15.7 Yes.
15.10 i, σ_h.
15.12 (a) R_3; (b) C_{2v}; (c) D_{3h}; (d) $D_{\infty h}$.
15.13 (a) C_{2v}; (b) $C_{\infty v}$; (c) C_{3v}; (d) D_{2h}.
15.14 (a) C_{2v}; (b) C_{2h}.
15.15 Polar: NO_2, N_2O, $CHCl_3$, and *cis*-CHBr═CHBr. Chiral: none.
15.16 No, d_{xy}.
15.17 $B_1(x)$, $B_2(y)$, $A_1(z)$.
15.18 (a) E_{1u}, A_{2u}; (b) B_{3u}, B_{2u}, B_{1u}.

16.4 (a) 0.0469 J m^{-3} s; (b) 1.33×10^{-13} J m^{-3} s; (c) 4.50×10^{-16} J m^{-3} s.
16.5 0.999 999 925 × 660 nm.
16.6 (a) 5 × 10 ps; (b) 5 ps.
16.7 (a) 53 cm^{-1}; (b) 0.53 cm^{-1}.
16.8 0.409 THz.
16.9 (a) 2.642×10^{-47} kg m^2; (b) 127.4 pm.
16.10 4.442×10^{-47} kg m^2, 165.9 pm.
16.11 232.1 pm.
16.12 106.5 pm, 115.6 pm.
16.13 20 475 cm^{-1}.
16.14 198.9 pm.
16.15 b, d, and *cis*-CH_2Cl_2.
16.16 a, b and d.
16.17 0.16 kN m^{-1}.
16.18 1.089 per cent.
16.19 328.7 N m^{-1}.
16.20 (a) 0.067; (b) 0.20.
16.21 HF (967.0 N m^{-1}), HCI (515.6 N m^{-1}), HBr (411.8 N m^{-1}), HI (314.2 N m^{-1}).

16.22 1580.38 cm^{-1}, 7.644×10^{-3}.
16.23 5.15 eV.
16.24 2699.77 cm^{-1}.
16.25 b, c, and d.
16.26 (a) 3; (b) 6; (c) 12.
16.27 (a) All; (b) symmetric stretch: Raman, antisymmetric stretch and bends: IR.
16.28 Raman active.
16.29 $4A_1 + A_2 + 2B_1 + 2B_2$.

17.5 (a) g; (c) g; (d) u.
17.6 $\frac{1}{2}$, 0.
17.7 80 per cent.
17.8 6.28×10^3 L mol^{-1} cm^{-1}.
17.9 1.5 mmol L^{-1}.
17.10 5.44×10^7 L mol^{-1} cm^{-2}.
17.11 Diene: 243 nm; butene: 192 nm.
17.12 450 L mol^{-1} cm^{-1}.
17.13 159 L mol^{-1} cm^{-1}, 23 per cent.
17.14 (a) 0.9 m; (b) 3 m.
17.15 (a) 5×10^7 L mol^{-1} cm^{-2}; (b) 2.5×10^6 L mol^{-1} cm^{-2}.

18.4 600 MHz.
18.5 $(-1.625 \times 10^{-26}$ J$) \times m_I$.
18.6 154 MHz.
18.7 (a) proton.
18.8 6.116×10^{-26} J.
18.9 (i) (a) 5.87 T; (b) 38.3 T; (c) 23.4 T.
(ii) (a) 11.7 T; (b) 76.6 T; (c) 46.8 T.
18.10 (a) 1×10^{-6}; (b) 5.1×10^{-6}; (c) 3.4×10^{-5}.
18.11 13.
18.12 (a) 11 μT; (b) 110 μT.
18.14 6.7×10^2 s^{-1}.
18.17 (b).
18.18 0.59 mT, 20 μs.
18.19 0.2 kT, 10 mT.
18.20 2.0022.
18.21 2.3 mT, 2.003.
18.22 330.2 mT, 332.2 mT, 332.8 mT, 334.8 mT, 1:1:1:1.
18.23 (a) 1:3:3:1; (b) 1:3:6:7:6:3:1.
18.24 (a) 331.9 mT; (b) 1.201 T.
18.25 $\frac{3}{2}$.

19.5 1.
19.6 (a) 2.57×10^{27}; (b) 7.26×10^{27}.
19.7 2.83.
19.8 3.156.
19.9 2.45 kJ mol^{-1}.
19.10 354 K.
19.11 (a) 0.71; (b) 0.996.
19.12 (a) 5×10^{-5}, 0.4, 0.905; (b) 1.4; (c) 22 J mol^{-1};
(d) 1.6 J K^{-1} mol^{-1}; (e) 4.8 J K^{-1} mol^{-1}.
19.13 4303 K.
19.14 (a) 138 J K^{-1} mol^{-1}; (b) 146 J K^{-1} mol^{-1}.
19.15 5.18 J K^{-1} mol^{-1}.
19.16 a, b, and d.

20.4 (a) $5R/2$; (b) $3R$; (c) $3R$.
20.5 NH_3: 1.33 and 1.11 (1.31); CH_4: 1.33 and 1.08 (1.31).
20.6 (a) 19.6; (b) 34.3.
20.7 (a) 1; (b) 2; (c) 2; (d) 12; (e) 3.
20.8 43.1, 23.36 K.
20.9 43.76 J K^{-1} mol^{-1}.
20.10 (a) 36.95, 80.08; (b) 36.7, 79.7.
20.11 72.5.
20.12 (a) 14.93 J K^{-1} mol^{-1}; (b) 25.65 J K^{-1} mol^{-1}.
20.13 −13.8 kJ mol^{-1}, −0.20 kJ mol^{-1}.
20.14 (a) $0.236R$; (b) $0.193R$.
20.15 11.5 J K^{-1} mol^{-1}.
20.16 (a) 9 J K^{-1} mol^{-1}; (b) 13 J K^{-1} mol^{-1};
(c) 15 J K^{-1} mol^{-1}.
20.17 3.70×10^{-3}.

21.5 O_3 and H_2O_2, but rotation about the O—O bond in H_2O_2 averages out the polarity.
21.6 1.01×10^{-39} J^{-1} C^2 m^2 (9.1×10^{-24} cm^3), 1.7 D.
21.7 4.8.
21.8 1.42×10^{-39} J^{-1} C^2 m^2 (1.28×10^{-23} cm^3).
21.9 17 per cent, 23 per cent; no correlation.
21.10 (a) 0 (by symmetry); (b) 0.7 D; (c) 0.4 D.
21.11 37 D at 11.7° to x-axis.
21.12 4.9 μD.
21.13 1.34.
21.14 18.
21.15 6.9×10^{-6}.

22.5 24 nm.
22.6 3.08 μm, 3.08 nm.
22.7 70 kg mol^{-1}, 71 kg mol^{-1}.
22.8 1.4×10^4.
22.9 (a) 18 kg mol^{-1}; (b) 20 kg mol^{-1}.
22.10 24 ns, 14 pm.
22.11 100.
22.12 63 kg mol^{-1}.
22.13 0.73 mm s^{-1}.
22.14 31 kg mol^{-1}.
22.15 3.4 Mg mol^{-1}.
22.16 4.3×10^5 g.

23.6 $(1, \frac{1}{2}, 0)$, $(1, 0, \frac{1}{2})$, $(\frac{1}{2}, \frac{1}{2}, \frac{1}{2})$.
23.7 (323), (110).
23.8 249 pm, 176 pm, 432 pm.
23.9 70.7 pm.
23.10 16°, 23°, 28°.
23.11 0.215 cm.
23.12 3.96×10^{-28} m^3.

23.13 4, 4.01 g cm^{-3}.
23.14 190 pm.
23.15 (111), (200), (311).
23.16 8.17°, 4.82°, 11.75°.
23.17 fcc.
23.18 $F_{hkl} = f$.
23.19 0.9069.
23.20 0.414.
23.21 (a) 58.0 pm; (b) 102 pm.
23.22 0.340.
23.23 +1.6 per cent.
23.25 7.9 km s^{-1}.
23.26 4.6 kV.
23.29 0.010.
23.30 9.3×10^{-4} cm^3.
23.31 3.
23.32 -6.4×10^{-5} cm^3 mol^{-1}.
23.33 2.
23.34 4.326.
23.35 +0.016 cm^3 mol^{-1}.
23.36 222 T.

24.4 (a) 9.975; (b) 1.
24.5 (a) 72 K; (b) 0.95 km s^{-1}; (c) 72 K.
24.6 81 mPa.
24.7 9.7×10^{-7} m.
24.8 (a) 5×10^{10} s^{-1}; (b) 5×10^{9} s^{-1}; (c) 5×10^{3} s^{-1}.
24.9 (a) 6.7 nm; (b) 67 nm; (c) 6.7 cm.
24.10 9.06×10^{-3}.
24.11 1.9×10^{20}.
24.12 104 mg.
24.13 4.1×10^{-2} J m^{-2} s^{-1}.
24.14 0.056 nm^2.
24.15 17 W, 17 W.
24.16 43 g mol^{-1}.
24.17 30 h.
24.18 0.142 nm^2.
24.19 205 kPa.
24.20 (a) 130 μP; (b) 130 μP; (c) 240 μP.
24.21 (a) 5.4 mJ K^{-1} m^{-1} s^{-1}, 8.1 mW;
(b) 29 mJ K^{-1} m^{-1} s^{-1}, 44 mW.
24.22 138 μP, 390 pm.
24.23 5.4×10^{-3} J K^{-1} m^{-1} s^{-1}.
24.24 (a) 1.1 m^2 s^{-1}, 4.4×10^{2} mol m^{-2} s^{-1};
(b) 1.1×10^{-5} m^2 s^{-1}, 4.4×10^{-3} mol m^{-2} s^{-1};
(c) 1.1×10^{-7} m^2 s^{-1}, 4.4×10^{-5} mol m^{-2} s^{-1}.
24.25 7.63×10^{-3} S m^2 mol^{-1}.
24.26 347 μm s^{-1}.
24.27 0.331.
24.28 13.83 mS m^2 mol^{-1}.
24.29 4.01×10^{-8} m^2 V^{-1} s^{-1}, 5.19×10^{-8} m^2 V^{-1} s^{-1},
7.62×10^{-8} m^2 V^{-1} s^{-1}.
24.30 1.90×10^{-9} m^2 s^{-1}.
24.31 1.3 ks.
24.32 420 pm.
24.33 27 ps.
24.34 113 μm; 5.594×10^{-5} m.

25.5 C: 3.0 mol L^{-1} s^{-1}, D: 1.0 mol L^{-1} s^{-1}, A: 1.0 mol L^{-1} s^{-1},
B: 2.0 mol L^{-1} s^{-1}.
25.6 v: 0.50 mol L^{-1} s^{-1}, D: 1.5 mol L^{-1} s^{-1}, A: 1.0 mol L^{-1} s^{-1},
B: 0.50 mol L^{-1} s^{-1}.
25.7 L mol^{-1} s^{-1}; (a) $k[A][B]$; (b) $3k[A][B]$.
25.8 $v = \frac{1}{2}k[A][B][C]$, L^2 mol^{-2} s^{-1}.
25.9 2.
25.10 2.
25.11 10.3 ks; (a) 499.7 Torr; (b) 480 Torr.
25.12 (a) 4.1×10^{-3} L mol^{-1} s^{-1}; (b) A: 2.6 ks, B: 7.4 ks.
25.13 (a) L mol^{-1} s^{-1}, L^2 mol^{-2} s^{-1}; (b) kPa^{-1} s^{-1}, kPa^{-2} s^{-1}.
25.14 (a) 45, 95 mmol L^{-1}; (b) 1, 51 mmol L^{-1}.
25.15 124 ks.
25.16 64.9 kJ mol^{-1}, 4.32×10^{8} mol L^{-1} s^{-1}.
25.17 $k_D/k_H \approx 0.15$.
25.19 1.9 MPa^{-1} s^{-1}.
25.20 7.1×10^{5} s^{-1}, 7.61 ns.

26.1 (b) initiation, (3) retardation, (4) termination
26.5 $v = k_1k_2[O_3]^2/(k_1'[O_2] + k_2[O_3])$.
26.7 0.16 to 4.0 kPa.
26.8 $d[P]/dt = (k_1/k_3[AH]^2[B])/(k_2[BH]^+ + k_3[AH])$.
26.10 1.52 mmol L^{-1} s^{-1}.
26.11 3.3×10^{18}.
26.12 0.518.

27.4 9.5×10^{9} s^{-1}, 1.2×10^{35} m^{-3} s^{-1}, 1.7%.
27.5 (a) 0.018, 0.30; (b) 3.9×10^{-18}, 6.0×10^{-6}.
27.6 (a) 13 per cent, 1.2 per cent; (b) 130 per cent, 12 per cent.
27.7 1.7×10^{-2} L mol^{-1} s^{-1}.
27.8 3×10^{10} L mol^{-1} s^{-1}.
27.9 (a) 6.61×10^{6} m^3 mol^{-1} s^{-1}; (b) 3.0×10^{7} m^3 mol^{-1} s^{-1}.
27.10 7.4×10^{9} L mol^{-1} s^{-1}; 0.14 μs.
27.11 1.2×10^{-3}.
27.12 1.9×10^{8} mol L^{-1} s^{-1}.
27.13 69.7 kJ mol^{-1}, −25 J K^{-1} mol^{-1}.
27.14 +71.9 kJ mol^{-1}.
27.15 −96.6 J K^{-1} mol^{-1}.
27.16 −76 J K^{-1} mol^{-1}.
27.17 (a) −45.8 J K^{-1} mol^{-1}; (b) +5.0 kJ mol^{-1}; (c) +18.7 kJ mol^{-1}.
27.18 20.9 L^2 mol^{-2} min^{-1}.

28.5 (a) 1.07×10^{25} m^{-2} s^{-1}, 1.4×10^{18} m^{-2} s^{-1};
(b) 2.35×10^{24} m^{-2} s^{-1}, 3.1×10^{17} m^{-2} s^{-1}.
28.6 1.3×10^{4} Pa.
28.7 3.4×10^{5} s^{-1}.
28.8 12.7 m^2.
28.9 20.5 cm^3.

28.10 Chemisorption, 50 s.
28.11 $E_d = 610$ kJ mol^{-1}, $\tau_0 = 1.13 \times 10^{-13}$ s, $A = 6.15 \times 10^{12}$ s^{-1}.
28.12 (a) 0.21 kPa; (b) 22 kPa.
28.13 0.83, 0.36.
28.14 (a) 40 ps, 0.6 ps; (b) 20 Ts, 7 μs.
28.15 15 kPa.
28.16 0 on gold, 1 on platinum.
28.17 −13 kJ mol^{-1}.
28.18 650 kJ mol^{-1}; (a) 1.1×10^{97} min; (b) 2.9×10^{-6} min.

29.5 1.2 eV.
29.6 12 nm^{-1}.
29.7 0.24 GV m^{-1}.
29.8 138 mV.
29.9 2.8 mA cm^{-2}.
29.10 Increase × 50.
29.11 (a) 0.17 mA cm^{-2}; (b) 0.17 mA cm^{-2}.
29.12 0.99 A m^{-2}.
29.13 0.2 μmol L^{-1}.
29.14 (a) 0.31 mA cm^{-2}; (b) 5.41 mA cm^{-2}; (c) −2.19 mA cm^{-2}.
29.15 $a(Fe^{3+})/a(Fe^{2+}) = 0.1$: 684 mA cm^{-2};
$a(Fe^{3+})/a(Fe^{2+}) = 1$: 215 mA cm^{-2};
$a(Fe^{3+})/a(Fe^{2+}) = 10$: 68 mA cm^{-2}.
29.16 108 mV.
29.17 (a) 4.9×10^{15} cm^{-2} s^{-1}, 3.8 s^{-1}; (b) 1.6×10^{16} cm^{-2} s^{-1}, 12 s^{-1}; (c) 3.1×10^{7} cm^{-2} s^{-1}, 2.4×10^{-8} s^{-1}.
29.18 (a) 33 Ω; (b) 33 GΩ.
29.21 Yes.
29.22 No.
29.23 +1.30 V, 0.13 W.
29.24 (a) +1.23 V; (b) +1.06 V.
29.25 1.2 mm year^{-1}.

Answers to problems

Detailed solutions for these selected problems can be found in the *Student's Solutions Manual for Physical Chemistry*, seventh edition, by P.W. Atkins, C.A. Trapp, M. Cady, and C. Giunta.

1.1 $0.50\ m^3$.

1.3 −233°N.

1.5 $p = \rho RT/M$, 45.9 g mol^{-1}.

1.8 (a) 0.184 Torr; (b) 68.6 Torr; (c) 0.184 Torr.

1.11 0.927, 0.208 L.

1.12 (a) 0.944 L mol^{-1}, $Z = 0.911$; (b) 2.67 L mol^{-1}, $Z = 0.992$;
(c) 5.11 L mol^{-1}, $Z = 1.002$. N_2 is closest to perfection at T_I.

1.14 (a) 0.1353 L mol^{-1}; (b) 0.6957;
(c) 0.59 (from expansion in $1/V_m$), 0.72 (from expansion in p).

1.17 $Z < 1$ when $\frac{a}{RT} > b$ and attractive forces predominate; $Z > 1$ when $\frac{a}{RT} < b$ and repulsive forces predominate

1.19 $V_c = 2b$, $T_c = a/4Rb$, $P_c = (a/4b^2)e^{-2}$, $Z_c = 0.271$, $Z_c\,(\text{vdw}) = 0.375$.

1.20 $V_c = 3C/B$, $T_c = B^2/3RC$, $p_c = B^3/27C^2$, $\frac{1}{3}$.

1.21 $B' = B/RT$, $C' = (C - B^2)/R^2T^2$.

1.24 no.

1.26 8.54, 15.1.

1.28 (a) $B = -1.32 \times 10^{-2}$ L mol^{-1};
(b) $B = -1.51 \times 10^{-2}$ L mol^{-1}, $C = 1.07 \times 10^{-3}$ $L^2\ mol^{-2}$.

1.30 (a) 4.6×10^3 mol; (b) 1.3×10^2 kg; (c) 1.2×10^2 kg.

1.32 (a) 1.1×10^{-11} mol L^{-1}, 2.2×10^{-11} mol L^{-1};
(b) 8.0×10^{-13} mol L^{-1}, 1.6×10^{-12} mol L^{-1}.

1.33 51.5 km, 3.0×10^{-3} bar.

2.1 $q = \Delta H = \Delta U = 2.6$ MJ.

2.2 (a) −3.46 kJ; (b) 0; (c) −3.46 kJ; (d) +24.0 kJ; (e) +27.5 kJ.

2.3 $T_1 = 273$ K, $T_2 = 546$ K, $T_3 = 273$ K;
Step 1→2: $w = -2.27$ kJ, $q = +5.67$ kJ, $\Delta U = +3.40$ kJ, $\Delta H = +5.67$ kJ;
Step 2→3: $w = 0$, $q = -3.40$ kJ, $\Delta U = -3.40$ kJ, $\Delta H = -5.67$ kJ;
Step 3→1: $w = +1.57$ kJ, $q = -1.57$ kJ, $\Delta U = 0$, $\Delta H = 0$;
Cycle: $w = -0.70$ kJ, $q = +0.70$ kJ, $\Delta U = 0$, $\Delta H = 0$.

2.6 $w = 0$, $\Delta U = +2.35$ kJ, $\Delta H = +3.03$ kJ.

2.9 +98.7 kJ mol^{-1}, +95.8 kJ mol^{-1}.

2.10 (a) $w = 0$, $q = +6.19$ kJ, $\Delta U = +6.19$ kJ, $\Delta H = +8.67$ kJ;
(b) $w = -6.19$ kJ, $q = 0$, $\Delta U = -6.19$ kJ, $\Delta H = -8.67$ kJ;
(c) $w = +4.29$ kJ, $q = -4.29$ kJ, $\Delta U = 0$, $\Delta H = 0$;
cycle: $w = -1.90$ kJ, $q = +1.90$ kJ, $\Delta U = 0$, $\Delta H = 0$.

2.12 $-87.33\ \text{kJ mol}^{-1}$.

2.13 $-2.13\ \text{MJ mol}^{-1}$, $-1.267\ \text{MJ mol}^{-1}$.

2.15 $\Delta_f H^{\ominus}(\text{Na}^+, \text{aq}) = -241\ \text{kJ mol}^{-1}$.

2.16 $-25\,968\ \text{kJ mol}^{-1}$, $+2357\ \text{kJ mol}^{-1}$.

2.18 (a) $+240\ \text{kJ mol}^{-1}$; (b) $+228\ \text{kJ mol}^{-1}$.

2.20 (a) (1) isochoric step: $n = \infty$; isobaric step: $n = 0$; (2) adiabatic step: $n = \gamma$; isochoric step: $n = \infty$; (b) (1) isochoric step: $w = 0$, $\Delta U = q = +55.8$ kJ, $\Delta H = +78.1$ kJ; isobaric step: $w = +22.3$ kJ, $\Delta U = -55.8$ kJ, $\Delta H = q = -78.1$ kJ; overall: $w = -q = 22.3$ kJ, $\Delta U = \Delta H = 0$; (2) adiabatic step: $q = 0$, $\Delta U = w = +9.37$ kJ, $\Delta H = +13.1$ kJ; isochoric step: $w = 0$, $\Delta U = q = -9.37$ kJ, $\Delta H = -13.1$ kJ; overall: $w = -q = 9.37$ kJ, $\Delta U = \Delta H = 0$.

2.22 (a) no, as a plot of $\ln p$ against $\ln V$ does not yield a straight line; (b) numerical integration yields $w = +685$ J; (c) fitting the data to the van der Waals equation yields $T = 350$ K.

2.24 $w = -nRT\ln(V_2/V_1) + n^2RTB(1/V_2 - 1/V_1) + \tfrac{1}{2}n^3RTC(1/V_2^2 - 1/V_1^2)$; (a) -1.5 kJ; (b) -1.6 kJ.

2.26 $w = -nRT\ln\{(V_2 - nb)/(V_1 - nb)\} - n^2a(1/V_2 - 1/V_1)$. (a) -1.7 kJ; (b) -1.8 kJ; (c) -1.5 kJ.

2.28 $\Delta H = nC_{p,\text{m}}(T_f - T_i)$.

2.31 (a) $-2.80\ \text{MJ mol}^{-1}$; (b) more exothermic by $5376\ \text{kJ mol}^{-1}$.

2.32 $\Delta_{\text{trs}}H - \Delta_{\text{trs}}U = -93.9\ \text{kJ mol}^{-1}$.

2.33 (a) 2878, 3537, 5471 kJ mol^{-1}; (b) 49.51, 49.02, 47.89 kJ g^{-1}.

2.34 $k = -67.44\ \text{kJ mol}^{-1}$, $n = 0.9253$, $R = 1.000$ (an excellent fit). Predicted $\Delta_c H^{\ominus}(\text{decane}) = -6625.5\ \text{kJ mol}^{-1}$; experimental value, $-6772.5\ \text{kJ mol}^{-1}$; error, 2.17 per cent.

2.36 (a) $16.2\ \text{kJ mol}^{-1}$; (b) $114.6\ \text{kJ mol}^{-1}$; (c) $122.0\ \text{kJ mol}^{-1}$.

3.2 (a) $+0.75\ \text{kJ mol}^{-1}$; (b) $+0.75\ \text{kJ mol}^{-1}$.

3.3 $41.40\ \text{J K}^{-1}\ \text{mol}^{-1}$.

3.4 $-30.5\ \text{J mol}^{-1}$.

3.6 $T_1 = 842$ K, $T_2 = T_3 = 348$ K; $p_1 = p_2 = p_3 = 1.72$ bar; $V_1 = 40.7$ L, $V_2 = V_3 = 16.8$ L; $\Delta U_1 = 11.3$ kJ, $\Delta U_2 = \Delta U_3 = 1.04$ kJ; ΔU (total) $= 13.4$ kJ.

3.8 increase.

3.9 not exact.

3.10 not exact; $\mathrm{d}q/T$ is exact.

3.14 $C_{p,\text{m}} - C_{V,\text{m}} = R$.

3.19 $+3.60$ kJ.

3.22 $\alpha = RV^2(V - nb)/\{RTV^3 - 2na(V - nb)^2\}$, $\kappa_T = V^2(V - nb)^2/\{RTV^3 - 2na(V - nb)^2\}$.

3.24 $-T\left(\frac{\partial V}{\partial T}\right)_p + V$.

3.26 $322\ \text{m s}^{-1}$.

3.28 0.80 m, 1.6 m, 2.8 m.

3.30 (a) $23.5\ \text{K MPa}^{-1}$; (b) $14.0\ \text{K MPa}^{-1}$.

4.3 (a) 57.0°C, -43.9 kJ, $-118\ \text{J K}^{-1}$, $+146\ \text{J K}^{-1}$, $+28\ \text{J K}^{-1}$; (b) 49.9°C.

4.4 (a) $+50.74\ \text{J K}^{-1}$, $-11.5\ \text{J K}^{-1}$; (b) ΔA_A is indeterminate, $\Delta A_B = +3.46$ kJ; (c) ΔG_A is indeterminate, $\Delta G_B = +3.46$ kJ; (d) $+39.2\ \text{J K}^{-1}$, $-39.2\ \text{J K}^{-1}$.

4.5 Step 1: $w = -11.5$ kJ, $q = +11.5$ kJ, $\Delta U = 0$, $\Delta H = 0$, $\Delta S = +19.1\ \text{J K}^{-1}$, $\Delta S_{\text{tot}} = 0$; Step 2: $w = -3.74$ kJ, $q = 0$, $\Delta U = -3.74$ kJ, $\Delta H = -6.23$ kJ, $\Delta S = 0$, $\Delta S_{\text{sur}} = 0$, Step 3: $w = +5.74$ kJ, $q = -5.74$ kJ, $\Delta U = 0$, $\Delta H = 0$, $\Delta S = -19.1\ \text{J K}^{-1}$, $\Delta S_{\text{tot}} = 0$; Step 4: $w = +3.74$ kJ, $q = 0$, $\Delta U = +3.74$ kJ, $\Delta H = +6.23$ kJ, $\Delta S = 0$, $\Delta S_{\text{tot}} = 0$; Cycle: $w = -5.8$ kJ, $q = +5.8$ kJ, $\Delta U = 0$, $\Delta H = 0$, $\Delta S = 0$, $\Delta S_{\text{tot}} = 0$.

4.6 Path (a): $w = -2.74$ kJ, $q = +2.74$ kJ, $\Delta U = 0$, $\Delta H = 0$, $\Delta S = +9.13\ \text{J K}^{-1}$, $\Delta S_{\text{sur}} = -9.13\ \text{J K}^{-1}$, $\Delta S_{\text{tot}} = 0$; Path (b): $w = -1.66$ kJ, $q = +1.66$ kJ, $\Delta U = 0$, $\Delta H = 0$, $\Delta S = +9.13\ \text{J K}^{-1}$, $\Delta S_{\text{sur}} = -5.53\ \text{J K}^{-1}$, $\Delta S_{\text{tot}} = +3.60\ \text{J K}^{-1}$.

4.7 Path (a): $q = 0$, $w = -9.1 \times 10^2$ J, $\Delta H = -1.5$ kJ, $\Delta S = 0$, $\Delta S_{\text{sur}} = 0$, $\Delta S_{\text{tot}} = 0$; Path (b): $q = 0$, $w = -7.5$ kJ, $\Delta H = -1.2$ kJ, $\Delta S = +1.12\ \text{J K}^{-1}$, $\Delta S_{\text{sur}} = 0$, $\Delta S_{\text{tot}} = +1.12\ \text{J K}^{-1}$.

4.11 $-160.07\ \text{kJ mol}^{-1}$.

4.13 (a) $0.11\ \text{kJ mol}^{-1}$; (b) $0.11\ \text{kJ mol}^{-1}$.

4.14 (a) $63.88\ \text{J K}^{-1}\ \text{mol}^{-1}$; (b) $66.08\ \text{J K}^{-1}\ \text{mol}^{-1}$.

4.16 (a) $+76.9\ \text{J K}^{-1}\ \text{mol}^{-1}$; (b) $+96.864\ \text{J K}^{-1}\ \text{mol}^{-1}$.

4.18 $+34.4\ \text{kJ mol}^{-1}$, $243\ \text{J K}^{-1}\ \text{mol}^{-1}$ at 298 K.

4.19 $46.60\ \text{J K}^{-1}\ \text{mol}^{-1}$, $46.73\ \text{J K}^{-1}\ \text{mol}^{-1}$.

4.24 -21 K, $+35.9\ \text{J K}^{-1}\ \text{mol}^{-1}$.

4.29 $+247.8\ \text{J K}^{-1}\ \text{mol}^{-1}$, $+336.6\ \text{J K}^{-1}\ \text{mol}^{-1}$, $+314.7\ \text{J K}^{-1}\ \text{mol}^{-1}$.

4.32 Step 1: $\Delta S = 0$, $\Delta S_{\text{sur}} = 0$; Step 2: $\Delta S = +33\ \text{J K}^{-1}$, $\Delta S_{\text{sur}} = -33\ \text{J K}^{-1}$; Step 3: $\Delta S = 0$, $\Delta S_{\text{sur}} = 0$; Step 4: $\Delta S = -33\ \text{J K}^{-1}$, $\Delta S_{\text{sur}} = +33\ \text{J K}^{-1}$.

5.1 $-501\ \text{kJ mol}^{-1}$.

5.3 73 atm.

5.4 (a) 0.750; (b) 0.372.

5.7 $(\partial S/\partial V)_T = \alpha/\kappa_T$ and $(\partial V/\partial S)_p = \alpha TV/C_p$.

5.16 $\pi_T = 4.85\ \text{J L}^{-1}$, $\Delta S_\text{m} = -41.1\ \text{J K}^{-1}\ \text{mol}^{-1}$, $\Delta H_\text{m} = -1.23\ \text{kJ mol}^{-1}$, $\Delta G_\text{m} = 11.0\ \text{kJ mol}^{-1}$.

5.18 (a) $\Delta_r G' = \tau\Delta_r G + (1 - \tau)\Delta_r H$; (b) $\Delta_r G' = \tau\Delta_r G + (1 - \tau)\Delta_r H - T\Delta_r C_p) - T'\Delta_r C_p \ln \tau$, $\tau = T'/T$.

5.20 $q_{\text{rev}} = nRT\ln\{(V_f - nb)/(V_i - nb)\}$.

5.22 As pressure increases, entropy increases, remains constant, and decreases, respectively.

5.23 The second option leads to $p = nRT/(V - nb) + \text{constant}$.

5.24 $f = 10.2$ atm.

5.26 $\ln \phi = Bp/RT + (C - B^2)p^2/2R^2T^2 + \cdots$, 0.999 atm.

5.29 13 per cent.

5.33 $G' = G + p^*V_0(1 - \mathrm{e}^{-p/p^*})$, expansion.

6.1 196.0 K, 11.1 Torr.
6.2 9 atm.
6.4 (a) $-22.0\ \mathrm{J\,K^{-1}\,mol^{-1}}$; (b) $-109.0\ \mathrm{J\,K^{-1}\,mol^{-1}}$; (c) $+110\ \mathrm{J\,mol^{-1}}$.
6.6 234.4 K.
6.8 (a) 357 K (84°C); (b) $+38.0\ \mathrm{kJ\,mol^{-1}}$.
6.13 $\Delta_{\mathrm{sub}}H(1\ \mathrm{Torr}) = +80.84\ \mathrm{kJ\,mol^{-1}}$, $\Delta_{\mathrm{sub}}S(1\ \mathrm{Torr}) = +201.1\ \mathrm{J\,K^{-1}\,mol^{-1}}$.
6.18 $(\partial^2\mu/\partial T^2)_p = -C_{p,\mathrm{m}}/T$.
6.19 Both ΔV and ΔS are zero.
6.21 (b) 112 K; (c) $+8.07\ \mathrm{kJ\,mol^{-1}}$.

7.1 $K_\mathrm{A} = 15.58$ kPa, $K_\mathrm{B} = 47.03$ kPa.
7.2 $17.5\ \mathrm{cm^3\,mol^{-1}}$ (NaCl), $18.07\ \mathrm{cm^3\,mol^{-1}}$ (H_2O).
7.4 $12.0\ \mathrm{cm^3\,mol^{-1}}$.
7.6 36 to 62 $\mathrm{g\,mol^{-1}}$.
7.8 (b) $K_\mathrm{A} = 450$ Torr, $K_\mathrm{I} = 465$ Torr.
7.10 $V_{\mathrm{cyclo}} = 109.0\ \mathrm{cm^3\,mol^{-1}}$, $V_{\mathrm{polymer}} = 279.3\ \mathrm{cm^3\,mol^{-1}}$.
7.12 For $x_\mathrm{T} = 0.228, 0.511, 0.810$, $\gamma_\mathrm{T} = 0.490, 0.723, 0.966$, and $\gamma_\mathrm{E} = 1.031, 0.920, 0.497$.
7.14 $S_0 = 19.89\ \mathrm{mol\,L^{-1}}$, $\tau = 165$ K; fits well, $R = 0.999\,78$.
7.17 $\mu_\mathrm{A} = \mu_\mathrm{A}^* - RT\ln x_\mathrm{A} + gRTx_\mathrm{B}^2$.
7.19 $V_\mathrm{B}(x_\mathrm{A}, x_\mathrm{B}) = V_\mathrm{B}(0,1) - \int_{V_\mathrm{A}(0)}^{V_\mathrm{A}(x_\mathrm{A})}\{x_\mathrm{A}\,\mathrm{d}V_\mathrm{A}/(1-x_\mathrm{A})\}$, $80.4\ \mathrm{cm^3\,mol^{-1}}$.
7.23 $[\mathrm{Cl^-}]_\mathrm{R} = [\mathrm{Na^+}]_\mathrm{R} = 0.201\ \mathrm{mol\,L^{-1}}$; $[\mathrm{Cl^-}]_\mathrm{L} = [\mathrm{Na^+}]_\mathrm{L} - 6[\mathrm{P}] = 0.199\ \mathrm{mol\,L^{-1}}$.
7.25 $1.26\times10^5\ \mathrm{g\,mol^{-1}}$, $1.23\times10^4\ \mathrm{L\,mol^{-1}}$.

8.1 (b) 391.0 K; (c) $n_{\mathrm{liq}}/n_{\mathrm{vap}} = 0.532$.
8.3 Temperature (γ): 78 K (0.9); 80 K (1.08); 82 K (1.04); 84 K (1.00); 86 K (0.99); 88 K (0.99); 90.2 K (0.99). To within experimental uncertainties the solution appears ideal.
8.4 (a) $\chi_\mathrm{S} = 0.43(P_4S_3), 0.64(P_4S_7), 0.71(P_4S_{10})$.
8.5 A compound with probable formula A_3B exists. It melts incongruently at 700°C. The proportions of A and B in the product are dependent upon the overall composition and the temperature. A eutectic exists at 400°C and $x_\mathrm{B} \approx 0.83$.
8.7 $T_{\mathrm{uc}} = 122$°C, $T_{\mathrm{lc}} = 8$°C.
8.9 (a) $F(1500\ \mathrm{K}) = 3$, $F(1100\ \mathrm{K}) = 2$; (c) 0.67 g Pb per g Cu; 0.050 g Cu per g Pb.
8.11 No phases are in equilibrium as the system cools. The phases through which the system passes upon cooling are: liquid, liquid + solid K_2FeCl_4, solid K_2FeCl_4 + solid $KFeCl_3$.
8.15 (a) 2150°C; (b) $x(\mathrm{MgO}) = 0.35$, $y(\mathrm{MgO}) = 0.18$, ratio 0.4; (c) 2640°C.
8.18 (a) $\mathrm{mp}(SiO_2) = 1934.1$ K, $\mathrm{mp}(\mathrm{Si}) = 1715.5$ K.

9.1 (a) $+4.48\ \mathrm{kJ\,mol^{-1}}$; (b) 0.101 atm.
9.4 $\Delta_\mathrm{r}H^\ominus = +3.00\times10^5\ \mathrm{J\,mol^{-1}}$ and $\Delta_\mathrm{r}S^\ominus = +102\ \mathrm{J\,K^{-1}\,mol^{-1}}$ throughout this temperature range; at 1395 K: $K = 1.22\times10^{-6}$, $\Delta_\mathrm{r}G^\ominus = +158\ \mathrm{kJ\,mol^{-1}}$; at 1443 K: $K = 2.80\times10^{-6}$, $\Delta_\mathrm{r}G^\ominus = +153\ \mathrm{kJ\,mol^{-1}}$; at 1498 K: $K = 7.23\times10^{-6}$, $\Delta_\mathrm{r}G^\ominus = +147\ \mathrm{kJ\,mol^{-1}}$.
9.6 1.69×10^{-5}.
9.8 7 mmol H_2, 107 mmol I_2, 786 mmol HI.
9.11 0.23, 0.46, 0.30.
9.12 $+76.8\ \mathrm{kJ\,mol^{-1}}$.
9.19 $+14.7\ \mathrm{kJ\,mol^{-1}}$, $+18.8\ \mathrm{kJ\,mol^{-1}}$.
9.23 now: 5.64; then: 5.70.
9.25 (b) maximum carbon mass in ocean $\approx 2.67\times10^{12}$ tonne.
9.27 (a) $p(450°\mathrm{C}) = 156.5$ bar, $p(400°\mathrm{C}) = 81.8$ bar; (b) $p(450°\mathrm{C}) = 132.5$ bar, $p(400°\mathrm{C}) = 210$ bar.

10.2 (a) 4.0×10^{-3}, 1.2×10^{-2}; (b) 0.74, 0.60; (c) 5.9; (d) +1.102 V; (e) +1.079 V.
10.3 2.0.
10.4 (a) $E = E^\ominus - (38.54\ \mathrm{mV})\times\{\ln(4^{1/3}b) + \ln\gamma_\pm\}$; (b) 1.0304 V; (c) 6.84×10^{34}; (d) 0.763; (e) 0.75; (f) $-87.2\ \mathrm{J\,K^{-1}\,mol^{-1}}$, $-262.4\ \mathrm{kJ\,mol^{-1}}$.
10.5 +0.268 38 V.
10.8 (a) +0.2223 V, +0.2223 V; (b) 1.10, 0.796.
10.13 (a) $\Delta E = -(RT/\nu F)\ln p/(\mathrm{atm})$; fits well below 100 atm, but deviates above that pressure;
(b) $\Delta E = (RT/\nu F)\{\ln p/(\mathrm{atm}) + Cp\}$, $(\partial E/\partial p)_T = (RT/\nu F)\{1/p + C\}$, $C = 6.665\times10^{-4}\ \mathrm{atm^{-1}}$, $R = 0.999\,40$;
(c) From empirical virial equation: $\Delta E = (RT/\nu F)\ln p/(\mathrm{atm}) + 0.000\,537(p-1)/(\mathrm{atm}) + 1.75\times10^{-8}(p^2-1)/(\mathrm{atm}^2)$;
(d) p, $\phi/\phi_{1\,\mathrm{atm}}$: 1.00 atm, 1.00; 10 atm, 0.994; 38 atm, 1.02; 51 atm, 1.04; 108 atm, 1.07; 210 atm, 1.03; 380 atm, 1.23; 430 atm, 1.38; 560 atm, 1.44; 720 atm, 1.64; 900 atm, 1.81; 1020 atm, 2.02.
10.16 $\mathrm{p}K_\mathrm{a}(25°\mathrm{C}) = 6.736$.
10.17 $a_0 = -14.102$, $a_1 = 2461$ K, $a_2 = 2.209$.
10.26 $5 < \mathrm{pH} < 8$.
10.28 (a) +1.23 V; (b) +1.09 V.

11.1 (a) $1.6\times10^{-33}\ \mathrm{J\,m^{-3}}$; (b) $2.5\times10^{-4}\ \mathrm{J\,m^{-3}}$.
11.2 6.29×10^{-34} J s.
11.4 (a) 0.020; (b) 0.007; (c) 7×10^{-6}; (d) 0.5; (e) 0.61.
11.6 π.
11.8 $i\hbar$.
11.11 (a) $a = 8\pi hc$, $b = hc$; (b) $c_2 = hc/k$.
11.13 (a) $N = (2/L)^{1/2}$; (b) $1/c(2L)^{1/2}$; (c) $1/(\pi a^3)^{1/2}$; (d) $1/(32\pi a^5)^{1/2}$.
11.15 (a) ik; (c) 0.
11.17 (a) $-k^2$; (b) $-k^2$; (c) 0; (d) 0.
11.18 (a) $\cos^2\chi$; (b) $\sin^2\chi$; (c) $0.95e^{ikx} \pm 0.32e^{-ikx}$.
11.21 (a) $6a_0$, $42a_0^2$; (b) $5a_0$, $30a_0^2$.
11.22 (a) $-e^2/4\pi\varepsilon_0 a_0$; (b) $\hbar^2/2m_\mathrm{e}a_0^2$.
11.24 (a) 1; (b) $2x$; (c) $\hbar$.
11.25 500 nm, blue green.
11.26 255 K, 11.3 μm.
11.28 (a) Assuming $\Delta_\mathrm{r}C_p \approx$ constant, methane becomes unstable above ≈ 825 K; the authors' statement is confirmed;

(b) λ_{max} (1000 K) = 2880 nm;
(c) ρ(brown dwarf)/ρ(Sum) = 8.8×10^{-3} at λ_{max} (brown dwarf), M(brown dwarf)/M(Sun) = 7.7×10^{-4};
(d) 2.31×10^{-7}, it hardly shines.

12.1 1.24×10^{-39} J, 2.2×10^{9}, 1.8×10^{-30} J.
12.2 CO (1900 N m^{-1}) > NO (1600 N m^{-1}) > HCl (516 N m^{-1}) > HBr (412 N m^{-1}) > HI (314 N m^{-1}).
12.3 1.30×10^{-22} J, $\pm\hbar$.
12.5 (a) $\dfrac{\varepsilon a}{L} + \dfrac{\varepsilon}{\pi}\sin\left(\dfrac{\pi a}{L}\right)$; (b) 0.1984ε.
12.7 With $a = mgL/(h^2/8mL^2)$, we have $E^{(2)} = -0.010\,83\, amgL$, $\Psi^{(1)} = a\{0.0600\Psi_2 + 0.000\,96\Psi_4 + 0.000\,13\Psi_6 + \cdots\}$.
12.11 (a) $N^2/2\kappa$; (b) $N^2/4\kappa^2$.
12.13 $\langle T\rangle = \frac{1}{2}(v+\frac{1}{2})\hbar\omega$.
12.15 (a) $L((1/12) - (1/2\pi^2 n^2))^{1/2}$, $nh/2L$;
(b) $\{(v\frac{1}{2})\hbar/\omega m\}^{1/2}$, $\{(v+\frac{1}{2})\hbar\omega m\}^{1/2}$.
12.16 $\mu_{v+1,v} = \alpha\{(v+1)/2\}^{1/2}$, $\mu_{v-1,v} = \alpha(v/2)^{1/2}$.
12.19 (a) 0, 0; (b) $3\hbar^2/I$, $6^{1/2}\hbar$; (c) $6\hbar^2/I$, $2(3^{1/2})\hbar$.
12.21 $\cos\theta = m_l/\{l(l+1)\}^{1/2}$, 54°44′.
12.23 $l_x = (\hbar/i)(y\partial/\partial z - z\partial/\partial y)$ and cyclic permutations, $[l_x, l_y] = i\hbar l_z$.
12.26 (a) 1.60×10^{-19} J; (b) 2.42×10^{14} Hz;
(c) total number of electrons = 1.123.
12.28 1.2×10^{6}.

13.1 $n_2 \to 6$, 7503, 5908, 5129, … 3908 nm.
13.3 987 663 cm^{-1}, 137 175 cm^{-1}, 185 187 cm^{-1}, 122.5 eV.
13.6 $A = 38.50$ cm^{-1}.
13.8 7621 cm^{-1}, 10 228 cm^{-1}, 11 552 cm^{-1}, 6.80 eV.
13.9 (a) 0.9 cm^{-1}; (b) small.
13.11 2s.
13.16 $p_x \pm ip_y$.
13.17 Exact knowledge of energy, total orbital angular momentum, and its projection along an arbitrary axis are possible.
13.18 $2.66a_0$.
13.19 $d_H \approx 0.314$ nm, $d_U \approx 0.301$ nm.
13.22 $v = 2.19 \times 10^{6}$ m s^{-1}, $\varepsilon = 5.14 \times 10^{11}$ V m^{-1}, $\mathcal{H} = 9.98 \times 10^{6}$ A m^{-1}.
13.23 $a_{Ps} = 2a_0$, $E_{1,Ps} = \frac{1}{2}E_{1,H}$.
13.24 (b) 1663 (from wavelength data), 1840 (from Rydberg constants).
13.27 $\Delta E = 4.29 \times 10^{-24}$ J = 0.216 cm^{-1}, $\langle r\rangle_{100} = 529$ nm, $I_{100} = 10.9677$ cm^{-1}; yes, thermal energy is 207 cm^{-1} and is sufficient, $v_{min} = 511$ m s^{-1}.

14.3 $R_{max} = 2.1a_0$.
14.8 1.9 eV, 130 pm.
14.14 $\Delta E = 2.7$ eV, $\lambda = 460$ nm, orange.
14.15 (b) $\beta \approx -21\,135$ cm^{-1}; (c) $E_{deloc} = 1.5176\beta$.
14.16 (b) $\Delta E/eV \approx 3.3534 + 1.3791 \times 10^{-4}\,\tilde{\nu}/\text{cm}^{-1}$;
(d) 30 937 cm^{-1}.
14.18 (b) at the PM$_3$ level: C_2H_4, 32.6%; C_4H_6, 19.3%; C_4H_8, 12.2%; C_8H_{10}, 16.6%.
14.22 (a) $E = -hc\mathcal{R}_H$; (b) $-(8/3\pi)hc\mathcal{R}_H$.
14.24 (a) $E = (\alpha - \beta)$ (twice), $\alpha + 2\beta$; binding energies: $\alpha + 2\beta$, $2\alpha + 4\beta$, $3\alpha + 3\beta$, $4\alpha + 2\beta$, respectively;
(b) $\Delta_r H = -413$ kJ mol^{-1}, slightly less than binding energy of H_2 (435.94 kJ mol^{-1});
(c) $\beta = (-849\text{ kJ mol}^{-1} - 2\alpha)/4$; H_3^{2+}: -425 kJ mol^{-1}; H_3^{+}: -849 kJ mol^{-1}; H_3: $3(\alpha/2 - 212\text{ kJ mol}^{-1})$; H_3^{-}: $3\alpha - 425$ kJ mol^{-1}.
14.26 (a) linear relation exists; (b) -0.122 V; (c) -0.174, better.

15.1 (a) D_{3d}; (b) D_{3d}, C_{2v}; (c) D_{2h}; (d) D_3; (e) D_{4d}.
15.2 *trans*-CHCl=CHCl.
15.7 the matrices do not form a group.
15.13 (a) $2A_1 + A_2 + 2B_1 + 2B_2$; (b) $A_1 + 3E$; (c) $A_1 + T_1 + T_2$; (d) $A_{2u} + T_{1u} + T_{2u}$.
15.15 irreducible representations: $4A_1 + 2B_1 + 3B_2 + A_2$.
15.18 neither transition allowed; with vibration transition to T_{1g} becomes allowed, to G_g remains foridden.
15.19 (a) symmetry elements; E, $2C_3$, $3C_2$, σ_h, $2S_3$, $3\sigma_v$; point group: D_{3h}; (d) reduces to: $A_1' + E'$.
15.20 The basis spans $A_{1g} + B_{1g} + E_u$.

16.4 596 GHz, 19.9 cm^{-1}, 0.503 mm, $B = 9.941$ cm^{-1}.
16.6 R(CC) = 139.6 pm, R(CH) = R(CD) = 108.5 pm.
16.7 2.728×10^{-47} kg m^2, 129.5 pm; DCl lines at 10.56, 21.11, 31.67, … cm^{-1}.
16.9 R(CuBr) = 218 pm.
16.11 14.35 m^{-1}, 26, 15.
16.13 linear, $\tilde{\nu}_1$: 1400 cm^{-1}, $\tilde{\nu}_2$ (bend): 540 cm^{-1}, $\tilde{\nu}_3$: 2360 cm^{-1} combination band, $\tilde{\nu}_1 + \tilde{\nu}_3$: 3735 cm^{-1}.
16.17 (a) 152 m^{-1}, 2.72×10^{-4} kg s^{-2}, 2.93×10^{-46} kg m^2, 95.5 m^{-1}; (b) 293 m^{-1}, 0.96.
16.18 (b) $\tilde{\nu} = 2169.3$ cm^{-1}, $\chi_e\tilde{\nu} = 13.14$ cm^{-1}, $B = 1.930$ cm^{-1}, $a = 0.017\,47$ cm^{-1}, $D_J = 4.178 \times 10^{-6}$ cm^{-1}.
(c) $I_e = 1.450 \times 10^{-46}$ kg m^2;
$R_e = 112.8$ pm;
$D_e = 89\,533$ cm^{-1}, $D_0 = 88\,452$ cm^{-1};
$I_{v=0} = 1.457 \times 10^{-46}$ kg m^2, $R_{v=0} = 113.1$ pm;
$I_{v=2} = 1.484 \times 10^{-46}$ kg m^2, $R_{v=2} = 114.1$ pm;
$I_{v=1} = 1.470 \times 10^{-46}$ kg m^2, $R_{v=1} = 113.6$ pm;
$I_{v=3} = 1.498 \times 10^{-46}$ kg m^2, $R_{v=3} = 114.7$ pm.
16.20 (a) C_{3v}; (b) 9; (c) $2A_1 + A_2 + 3E$; (d) all but A_2; (e) all but A_2.
16.24 (a) spherical rotor, (b) symmetric rotor, (c) linear rotor, (d) asymmetric rotor, (e) symmetric rotor, (f) asymmetric rotor.
16.25 $J_{max} = (kT/2hcB)^{1/2} - \frac{1}{2}$, 30, $J_{max} = (kT/hcB)^{1/2} - \frac{1}{2}$, 6.
16.27 (b) $A/hc = 46\,119$ cm^{-1}, $B = 1.4275$, $C = 2.3896$, $D = 88.99$ pm, $\varepsilon/hc = 1750$ cm^{-1}.
16.30 (c) 281.9 K, 81.1 per cent.
16.32 2.30×10^{7} m s^{-1}, 8.34×10^{5} K.
16.33 2.35 K.

17.1 49 364 cm^{-1}.
17.2 5.1147 eV.
17.4 4.8×10^4 L mol^{-1} cm^{-2}.
17.6 $^2\Sigma_g^+$, $^2\Pi_g$.
17.7 5.06 eV.
17.10 65 MW.
17.18 $f = (RS/a_0)^2 f_0$.
17.26 (a) 2.42×10^5 L mol^{-1} cm^{-2}; (b) 0.185; (c) 6.97, 135 L mol^{-1} cm^{-1}.
17.28 $V_1 - V_0 = 3.1938$ eV, $\tilde{\nu}_1 - \tilde{\nu}_0 = 79.538$ cm^{-1}, $\tilde{\nu}_1 = 2113.8$ cm^{-1}, $\tilde{\nu}_0 = 2034.3$ cm^{-1}, relative populations = 10, $T_{eff} = 1.3 \times 10^3$ K.

18.1 10.3 T, 2.42×10^{-5}, β.
18.3 300×10^6 Hz ± 10 Hz, 0.29 s.
18.4 4×10^2 s^{-1}, 3.7 kJ mol^{-1}, 16 kJ mol^{-1}.
18.6 (a) yes; (b) $^3J_{SnSn}/\text{Hz} = 580 - 79\cos\phi + 395\cos 2\phi$; (c) a staggered conformation with the two $SnMe_3$ groups at 180° to each other relative to the C—C bond.
18.7 1.992, 2.002.
18.8 −0.89 mT.
18.9 1:2:3:2:1 quintet of 1:4:6:4:1 quintets.
18.12 $\sigma_d = Ze^2\mu_0/12\pi m_e a_0$.
18.13 (b) There is a roughly linear correlation if phenol is omitted from the data set.
18.14 158 pm.
18.16 $I = \frac{1}{2}A\tau/\{1 + (\omega_0 - \omega)^2\tau^2\}$.
18.18 (a) R_ν(nuclide)/R_ν(proton): 0.409, 0.251, 0.193, 0.941, 0.405; $R_{\mathcal{B}}$(nuclide)/$R_{\mathcal{B}}$(proton) = 0.009 65, 0.015 90, 0.001 01, 0.833 50, 0.066 54.
18.19 average dipolar field = 0.58 mT.
18.22 (a) Using density functional methods $P_2 = P_6 = 0.27$; $P_3 = P_5 = 0.10$.
(b) $a_2 = a_6 = 0.61$ mT; $a_3 = a_5 = 0.23$ mT; $a_4 = 0.46$ mT.

19.1 $W = 2 \times 10^{40}$, $S = 1.282 \times 10^{-21}$ J K^{-1}, $S_1 = 0.637 \times 10^{-21}$ J K^{-1}, $S_2 = 0.645 \times 10^{-21}$ J K^{-1}. Entropy is an extensive property.
19.2 ^{4}He: 6.30×10^5, 7, 2; ^{3}He: 9.63×10^5, 11, 4.
19.3 $\Delta W/W \approx 2.4 \times 10^{25}$.
19.5 3.5 fK, 7.41.
19.7 (a) 5.00, 6.26; (b) 1.00 at 298 K, 0.80 at 5000 K; 6.5×10^{-11} at 298 K and 0.12 at 5000 K; (c) 13.38 J K^{-1} mol^{-1}, 18.07 J K^{-1} mol^{-1}.
19.9 (a) 0.64, 0.36; (b) 0.52 kJ mol^{-1}.
19.12 (a) $10^{5.44\times10^{23}}$; (b) $10^{5.69\times10^{23}}$, the heat capacity is temperature-dependent; (c) this is a spontaneous process, $\Delta S_{U,V} > 0$.
19.14 (a) 1; (b) {2,2,0,1,0,0} and {2,1,2,0,0,0}.
19.17 (a) $q = 1 + 3e^{-\varepsilon/kT}$, 2.104; (b) $0.5245RT$, 2.074 J K^{-1} mol^{-1}, 10.55 J K^{-1} mol^{-1}.
19.18 (a) 104 K; (b) $q = 1 + a$; (c) $2Nk \ln 2$.
19.20 (b) $\ln N(0)/N(h) = \nu(\rho - \rho_0) gh/kT$; (c) $k = 1.36 \times 10^{-23}$ J K^{-1}; $N_A = 6.11 \times 10^{23}$ mol^{-1}.

20.1 (a) 0.351R; (b) 0.079R; (c) 0.029R.
20.3 4.2, 15 J K^{-1} mol^{-1}.
20.5 19.89.
20.6 191.4 J K^{-1} mol^{-1}, residual entropy is negligible.
20.7 199.4 J K^{-1} mol^{-1}.
20.8 28, 258 J K^{-1} mol^{-1}.
20.11 (a) 3.89; (b) 2.41.
20.12 $S_m = R\ln(2\pi e^2 m\sigma_m/h^2 N_A\beta)$, $\sigma_m = \sigma/n$; $\Delta S_m = R\ln\{(\sigma_m/V_m)(h^2\beta/2\pi me)^{1/2}\}$.
20.16 (b) 5.41 J K^{-1} mol^{-1}.
20.18 350 m s^{-1}.
20.19 9.57×10^{-15} J K^{-1}.
20.21 45.76 kJ mol^{-1}.

21.1 (a) 0.11 GV m^{-1}; (b) 4 GV m^{-1}; (c) 4 kV m^{-1}.
21.3 1.2×10^{-23} cm^3, 0.86 D.
21.5 2.24×10^{-24} cm^3, 1.58 D.
21.8 $R = 360.71$ pm, $\tilde{\nu} = 21.84$ cm^{-1}, $k = 0.3746$ Nm^{-1}.
21.10 (a) 1.51×10^{-23} J, 265 pm.
21.13 (a) $6Q_1Q_2/\pi\varepsilon_0 r^5$; (b) $9Q_1Q_2/4\pi\varepsilon_0 r^5$.
21.14 $n_r = 1 + p(2\pi\alpha'/kT)$.
21.16 the relative permittivity should decrease.
21.22 1.0 J mol^{-1}.
21.23 (a) −39 J mol^{-1}; (b) ∞.

22.2 5.0 Sv.
22.5 0.0716 L g^{-1}.
22.6 0.21 Mg mol^{-1}.
22.7 158 kg mol^{-1}.
22.8 $K = 2.73$ cm^3 g^{-1} $kg^{-1/2}$ $mol^{1/2}$, $a = 0.500$, $M = 1.34 \times 10^3$ kg mol^{-1}.
22.9 (a) $K = 0.0117$ cm^3 g^{-1}, $a = 0.717$, (b) THF is polar. The constants depend upon both solute and solvent and their interactions.
22.10 23.1 kg mol^{-1}, 1.02 m^3 mol^{-1}.
22.13 (a) $lN^{1/2}$, 9.74 nm; (b) $(8N/3\pi)^{1/2}l$, 8.97 nm; (c) $(2N/3)^{1/2}l$, 7.95 nm.
22.16 (a) $B = 4.22 \times 10^{23}$ $mol^{-1} \times (l\sqrt{N})^3$; 0.39 m^3 mol^{-1}; (b) $B = 1.19 \times 10^{24}$ $mol^{-1} \times (l\sqrt{N})^3$; 1.1 m^3 mol^{-1}.
22.18 (a) Two of three converged; different conformations represent local energy minima. (b) None converged with the MMFF force field.
22.19 $v_p = 8\, v_{mol}$; BSV: 28 m^3 mol^{-1}, 2.6%. Hb: 0.33 m^3 mol^{-1}, 5%.
22.20 serum albumin and bushy stunt virus resemble spheres; DNA does not.

23.2 P, 342 pm.
23.3 yes.
23.5 10.51 g cm^{-3}.
23.7 628 pm, yes.
23.9 neutron: 0°, 14.0°; electron: 0°, 0.72°.
23.11 1.385 g cm^{-3}, 1.578 g cm^{-3}.
23.13 $s = 2$, 0.127 cm^3 mol^{-1};
$s = 3$, 0.254 cm^3 mol^{-1};
$s = 4$, 0.423 cm^3 mol^{-1};
Overall molar susceptibility corresponds to $s = 3$ state value.
23.19 $dG = -SdT - ldt$; $dA = -SdT - tdl$; $\left(\frac{\partial U}{\partial l}\right)_t = -T\left(\frac{\partial t}{\partial T}\right)_l + t$
23.20 $t = -T\left(\frac{\partial S}{\partial l}\right)_T$
23.24 the susceptibility varies as $\alpha = 1 - d$, where d = degree of dimerization; $d = 1 - \left(\frac{1}{4(p/k)+1}\right)^{1/2}$;
susceptibility decreases as pressure increases;
susceptibility increases as temperature increases.
23.26 3.61×10^5 g mol^{-1}.
23.27 1.01 g cm^{-3}.
23.30 7.654 g cm^{-3}.

24.2 (a) 2.8 km h^{-1}; (b) 86 km h^{-1}; (c) 86 km h^{-1}.
24.5 7.3 mPa.
24.6 (a) 100 Pa; (b) 24 Pa.
24.8 5.3 S cm^2 mol^{-1}.
24.9 $\Lambda_m^\circ = 12.6$ mS m^2 mol^{-1}; $\kappa = 7.30$ mS m^2 mol^{-1} M$^{1/2}$;
(a) 11.96 mS m^2 mol^{-1}; (b) 119.6 mS m^{-1}; (c) 172.5 Ω.
24.11 40 μm s^{-1}, 52 μm s^{-1}, 76 μm s^{-1}; 250 s, 190 s, 130 s;
(a) 13 nm, 17 nm, 24 nm. (b) 43, 55, 81.
24.13 0.48, 7.5×10^{-8} m^2 s^{-1} V^{-1}, 72 S cm^2 mol^{-1}.
24.15 (a) 12 kN mol^{-1}, 2.1×10^{-20} N molecule^{-1};
(b) 17 kN mol^{-1}, 2.8×10^{-20} N molecule^{-1};
(c) 25 kN mol^{-1}, 4.1×10^{-20} N molecule^{-1}.
24.16 Li^+: 4 water molecules; Na^+: 1 to 2 water molecules.
24.18 (a) ≈ 0, (b) 0.063 M.
24.20 Λ_m°(NaI) = 60.7 S cm^2 mol^{-1},
Λ_m°(KI) = 58.9 S cm^2 mol^{-1}, $\lambda^\circ(Na^+) - \lambda^\circ(K^+) =$
1.8 S cm^2 mol^{-1}; the analogous quantities in water are, respectively, 126.9, 150.3, and −23.4 S cm^2 mol^{-1}.
24.23 $c^* = (2kT/m)^{1/2}$.
24.24 $c_{mean} = (\pi kT/2m)^{1/2}$.
24.26 (a) 0.39; (b) 0.61; (c) 0.47, 0.53.
24.30 (a) 0; (b) 0.016; (c) 0.054.
24.34 1.6×10^{16} m^2 s^{-1}, 0.34 J K^{-1} m^{-1} s^{-1}.
24.35 100 Ms.
24.36 1.7×10^{-2} s.

25.1 2, 59 mL mol^{-1} min^{-1}, 2.94 g.
25.3 1, 1.2×10^{-4} s^{-1}.
25.5 97.0 kJ mol^{-1}.
25.6 55.4 per cent.
25.9 1, 7.2×10^{-4} s^{-1}.
25.10 propene: 1; HCl: 3.
25.12 rate = $kK_2K_1[HCl]^3[CH_3CH{=}CH_2]$.
25.14 1.14×10^{10} L mol^{-1} s^{-1}, 16.7 kJ mol^{-1}.
25.16 121.2 kJ mol^{-1}, 247.0 J K^{-1} mol^{-1}, 148.3 kJ mol^{-1}.
25.17 (a) $a = 1$, $b = \frac{1}{2}$;
(b) in units of cm$^{3/2}$ mol$^{-1/2}$ s^{-1}
k(433.15 K) = 0.077
k(453.15 K) = 0.504
k(473.15 K) = 2.26
k(493.15 K) = 12.0.
E_a = 148 kJ mol^{-1}.
$A = 5.00 \times 10^{16}$ cm$^{3/2}$ mol$^{-3/2}$ s^{-1}.
25.19 $[B]_\infty/[A]_\infty = k/k'$.
25.21 (a) $kt = \frac{2x(A_0 - x)}{A_0^2(A_0 - 2x)^2}$;
(b) $kt = \frac{2x}{A_0^2(A_0 - 2x)} + \frac{1}{A_0^2}\ln\frac{A_0 - 2x}{A_0 - x}$.
25.22 when B is a reactive intermediate.
25.27 2720 y.
25.29 3.65×10^{-3} min^{-1}, 274 min.
25.31 E_a = 105 kJ mol^{-1}, ΔG = −26.6 kJ mol^{-1},
ΔH = −34.3 kJ mol^{-1}; still favourable under prebiotic conditions.
25.32 1.03×10^9 L mol^{-1} s^{-1}, 13.9 kJ mol^{-1}.
25.34 $k_1 = 3.82 \times 10^6$ L mol^{-1} s^{-1}; $k_2 = 5.1 \times 10^5$ L mol^{-1} s^{-1};
$k_3 = 4.17 \times 10^6$ L mol^{-1} s^{-1}.
25.35 9×10^{-10} mol L^{-1} s^{-1}, 3×10^5 s.

26.4 (a) For both X = H and X = Cl, the fit of the data to the second-order integrated rate expression is superior to the fit to the first-order rate expressions as evidenced by comparing correlation coefficients and standard deviations of the slopes and intercepts.
(b) rate = $-k_2k_1[(ClRh(CO)_2)_2][ArHgCl]^2/[HgCl_2]$.
26.6 When step (b) is rate-determining and step (a) is a rapid equilibrium so that [I] is in a steady state.
26.7 1.9×10^{20} s^{-1}, 3.1×10^{-4} einstein s^{-1}.
26.9 $k_q = 1.98 \times 10^9$ L mol^{-1} s^{-1}.
26.10 3.52 nm.
26.13 $\delta M = M\{kt[A]_0(1 + kt[A]_0)\}^{1/2}$.
26.15 $\langle n \rangle = \nu = k[\cdot M][I]^{1/2}$
26.17 $([A]_0 + [P]_0)^2 kt_{max} = \frac{1}{2} - p - \ln 2p$.
26.19 $[X] = k_c/k_b$, $[Y] = k_a[A]/k_b$.
26.22 $\frac{d[A]}{dt} = -I - k_2\left(\frac{I}{k_3}\right)^{1/2}[A]$.
26.24 $f = k_2k_4[CO]/\{k_2[CO] + k_3[M]\}$.
26.27 0.010 mol L^{-1}.
26.29 (b) pH = 7.0.
26.30 Non-competitive.
26.31 Step 1 is autocatalytic. $a/r < S_0$: infection spreads;
$a/r > S_0$: infection dies out.

26.33 (a) Initiation, propagation, propagation, termination, initiation; (b) $d[NO]/dt = -2k_b(k_{-d}/k_d)^{1/2}[O_2]^{1/2}[NO]$; (c) $E_{a,eff} = E_b + \frac{1}{2}E_{-d} - \frac{1}{2}E_d$; (d) $E_{a,eff} = 381$ kJ mol^{-1}, consistent with high end of range; (e) $d[NO]/dt = -2k_b(k_a/2k_d[M])^{1/2}[NO]^2$; (f) $d[NO]/dt = -2k_b(k_e/2k_d[M])^{1/2}[O]^{1/2}[NO]^{3/2}$, where k_e is the rate constant for $NO + O_2 \rightarrow O + NO_2$; $E_{a,eff} =$ 253 kJ mol^{-1}, consistent with low end of range.

27.2 0.007, 0.0040 nm^2.
27.4 +83.8 kJ mol^{-1}, +19.1 J K^{-1} mol^{-1}, 85.9 kJ mol^{-1}, +79.0 kJ mol^{-1}.
27.5 2–.
27.8 A complex of two univalent ions of the same sign.
27.9 $k = k_1$, $k' = k_2\left(\frac{k_1}{k_2}\right)^{1/2}$; (b) $D(\text{FO—F}) \approx E_{a1} = 160.9$ kJ mol^{-1}, $D(\text{O—F}) = 224.4$ kJ mol^{-1}, $E_{a2} = 60$ kJ mol^{-1}.
27.11 ln k fits the Arrhenius equation with $A = 3.12 \times 10^{14}$ L mol^{-1} s^{-1}, $E_a = 193$ kJ mol^{-1}, $R = 0.999\,76$; ln k' fits with $A' = 7.29 \times 10^{11}$ L mol^{-1} s^{-1}, $E'_a = 175$ kJ mol^{-1}, $R = 0.998\,48$.
27.15 $P = 5.2 \times 10^{-6}$.
27.17 (a) 2.7×10^{-15} m^2 s^{-1}; (b) 1.1×10^{-14} m^2 s^{-1}.
27.21 k_2 (atoms)/k_2(molecules) $\approx 3 \times 10^7$.
27.23 6.23×10^9 L mol^{-1} s^{-1}, 0.37 nm.

28.1 $Z_w = 2.69 \times 10^{23}$ cm^{-2} s^{-1} at 100 kPa, $Z_w = 2.69 \times 10^{18}$ cm^{-2} s^{-1} at 1.000 Pa, $Z_{atom} = 2.0 \times 10^8$ s^{-1} at 100 kPa, $Z_{atom} = 2.0 \times 10^3$ s^{-1} at 1.000 Pa.
28.3 0.37 Torr^{-1}.
28.5 1.4 mL, 5.9 m^2.
28.8 (a) $v = gkT/p$; (b) $R = 0.959$ for the linear equation which yields $g = 1.0 \times 10^{21}$ m^{-3} s^{-1}. $z(t)$ is seen to be nonlinear; $z(t) = a(e^{bt} - 1)$, with $a = 5.71$ cm, $b = 0.35$ s^{-1}, $R = 0.994$.
28.10 −20.1 kJ mol^{-1}, −63.6 kJ mol^{-1}.
28.13 $X_{max} = 0.380$, specific surface area of carbon = 922 m^2 mol^{-1}.
28.14 $U \propto C_6/R^3$, 294 pm.
28.19 40.4.
28.20 (a) R values range from 0.975 to 0.998; the fit is good at all temperatures. (b) $k_a = 3.68 \times 10^{-3}$, $k_b = 2.62 \times 10^{-5}$ ppm^{-1}, $\Delta_{ad}H = -8.67$ kJ mol^{-1}, $\Delta_b H = -15.7$ kJ mol^{-1}.
28.21 (a) $k = 0.2289$ and $n = 0.6180$; $R = 0.9995$. (c) $k = 0.5227$, $n = 0.7273$; $R = 0.996$.

29.1 5.68×10^8 L mol^{-1} s^{-1}.
29.2 0.38, 0.78 mA cm^{-2}.
29.4 0.25 mm.
29.5 87 mA.
29.7 (a) $E_0 = -0.618$ V, η/mV = −84, −109, −134, −194; (b) j_c/(μA cm^{-2}) = 0.0324, 0.0469, 0.0663, 0.154; (c) $j_0 = 0.009\,97$ μA cm^{-2}, $\alpha = 0.363$; excellent fit, $R = 0.999\,94$.
29.10 (a) No linear region exists; the Tafel equation cannot be used to calculate j_0 and α.
29.15 7.2 μA.
29.18 2.3 eV.
29.23 0.28 mg cm^{-2} d^{-1}.

Answers to Box questions

1.1.1 (a) 0.00; (b) −0.05.

2.2.1 +37 K, 4.09 kg.

4.1.1 (a) $nC_{p,m}|T_f - T_i - T_i \ln(T_f/T_i)|$; (b) 6.86 KJ; 68.6 s, but this time is meaningless.
4.2.1 (a) 106; (b) 318.

6.1.1 (a) $U_r(T_r, p_r) = -\int_{V_r}^{\infty} \frac{2p_r(T_r, V_r)}{T_r V_r} dV_r$; (c) $0.85 < p_r < 0.40$ when $T_r = 1$.

9.1.1 1300°C.

10.1.1 Yes.

11.1.1 For neutrons: 5.8^B, 1.7^B. For electrons: 0.3^B, 0.9^B.

12.1.1 0.235.

13.1.1 $K = 60$, therefore there are few H atoms present and the hydrogen spectrum will be weak.

16.1.1 B_0(CN, calculated) = 1.892 cm^{-1}; thus it has transitions that span the background radiation.

17.1.1 Values depend on method, but all show lower frequency for **5** than for the 11-cis form.
17.2.1 (a) 1.7 nmol^{-1}; (b) 6.0 H 10^2 molecules.

18.1.1 29 μT m^{-1}.
18.2.1 The deeper the crevice, the broader the ESR line.

23.2.1 6.0 K.

24.1.1 0.25 J cm^{-3}.

26.1.1 (c) 0.0123% O_3.

27.2.1 Rb atoms must hit the I side of CH_3I.

28.1.1 (a) −0.11; (b) −0.51.

Index

The Periodic Table

Period	1	2	3	4	5	6	7	8	9	10	11	12	13/III	14/IV	15/V	16/VI	17/VII	18/VIII
								1 H 1.0079										2 He 4.003
2	3 Li 6.941	4 Be 9.012											5 B 10.81	6 C 12.01	7 N 14.01	8 O 16.00	9 F 19.00	10 Ne 20.18
3	11 Na 22.99	12 Mg 24.31											13 Al 26.98	14 Si 28.09	15 P 30.97	16 S 32.07	17 Cl 35.45	18 Ar 39.95
4	19 K 39.10	20 Ca 40.08	21 Sc 44.96	22 Ti 47.87	23 V 50.94	24 Cr 52.00	25 Mn 54.94	26 Fe 55.85	27 Co 58.93	28 Ni 58.68	29 Cu 63.55	30 Zn 65.39	31 Ga 69.72	32 Ge 72.61	33 As 74.92	34 Se 78.96	35 Br 79.90	36 Kr 83.80
5	37 Rb 85.47	38 Sr 87.62	39 Y 88.91	40 Zr 91.22	41 Nb 92.91	42 Mo 95.94	43 Tc (98)	44 Ru 101.07	45 Rh 102.91	46 Pd 106.42	47 Ag 107.87	48 Cd 112.41	49 In 114.82	50 Sn 118.71	51 Sb 121.76	52 Te 127.60	53 I 126.90	54 Xe 131.29
6	55 Cs 132.91	56 Ba 137.33	La-Lu	72 Hf 178.49	73 Ta 180.95	74 W 183.84	75 Re 186.21	76 Os 190.23	77 Ir 192.22	78 Pt 195.08	79 Au 196.97	80 Hg 200.59	81 Tl 204.38	82 Pb 207.2	83 Bi 208.98	84 Po (209)	85 At (210)	86 Rn (222)
7	87 Fr (223)	88 Ra (226)	Ac-Lr	104 Rf (261)	105 Db (262)	106 Sg (263)	107 Bh (262)	108 Hs (265)	109 Mt (266)	110 Uun	111 Uuu	112 Uub	113 Uut	...				
	S block		d block										p block					

Lanthanides	57 La 138.91	58 Ce 140.12	59 Pr 140.91	60 Nd 144.24	61 Pm (145)	62 Sm 150.36	63 Eu 151.96	64 Gd 157.25	65 Tb 158.93	66 Dy 162.50	67 Ho 164.93	68 Er 167.26	69 Tm 168.93	70 Yb 173.04	71 Lu 174.97
Actinides	89 Ac (227)	90 Th 232.04	91 Pa 231.04	92 U 238.03	93 Np (237)	94 Pu (244)	95 Am (243)	96 Cm (247)	97 Bk (247)	98 Cf (251)	99 Es (252)	100 Fm (257)	101 Md (258)	102 No (259)	103 Lr (262)
	f block														